P9-AOE-155

The New York Botanical Garden Illustrated Encyclopedia of Horticulture

ERRATA

Volume 4

Page 1061, column 1, line 26:
"HYDRANGACEAE" should read "HYDRANGEACEAE"

Page 1109, upper left picture caption:
"*Dodecathem*" should read "*Dodecatheon*"

Page 1159, upper right picture caption:
"*Echballium*" should read "*Ecballium*"

Page 1193, column 2, line 17:
Insert "or hardy-ageratum" after "Mist flower"

Page 1215, column 3, lines 20, 21:
"temperate parts" should read "arctic, subarctic, and alpine"

Page 1272, column 1, picture caption:
"a young grafted specimen" should read "a crest variety grafted"

Page 1311, column 3, line 45:
"SAPOTACEAE" should read "SAPINDACEAE"

Color plate, lower left, facing page 1346:
"*sagitata*" should read "*sagittata*"

Page 1364, upper right picture caption:
"*elastic*" should read "*elastica*"

Page 1388, column 1, line 23:
"Virginia" should read "Philadelphia, Pennsylvania"

Page 1404, lower right picture caption:
"*americana*" should read "*pennsylvanica*"

The New York Botanical Garden Illustrated Encyclopedia of Horticulture

Thomas H. Everett

Volume 4
Di-Fu

Garland Publishing, Inc.
New York & London

15 14 13 12 11 10 9 8 7 6 5 4 3 2 1

Library of Congress Cataloging in Publication Data

Everett, Thomas H
The New York Botanical Garden illustrated encyclopedia of horticulture.

1. Horticulture—Dictionaries. 2. Gardening—Dictionaries. 3. Plants, Ornamental—Dictionaries. 4. Plants, Cultivated—Dictionaries. I. New York (City). Botanical Garden. II. Title.
SB317.58.E94 635.9'03'21 80-65941
ISBN 0-8240-7234-0

PHOTO CREDITS

Black and White
The British Travel & Holidays Association, London, England: A well-located border of perennials, p. 1380. Malak, Ottawa, Canada: *Erythronium tuolumnense,* p. 1261. J. F. Michajluk, West Hatfield, Massachusetts: *Ficus benghalensis,* p. 1363. Netherlands Flower Institute: *Fritillaria imperialis,* p. 1411. The New York Botanical Garden: *Dichorisandra thyrsiflora,* p. 1060; A thicket of unidentified *Dicranopteris,* p. 1063; *Dictamnus albus* (flowers), p. 1064; Dieffenbachias with other tropical plants, p. 1067; *Dieffenbachia maculata,* p. 1068; *Dieffenbachia maculata* 'Janet Weidner', p. 1068; *Dieffenbachia maculata jenmannii,* p. 1068; *Dieffenbachia maculata* 'Rudolph Roehrs', p. 1068; *Dimorphotheca pluvialis ringens,* p. 1077; *Dimorphotheca cuneata,* p. 1077; *Dipcadi serotinum,* p. 1086; *Dipcadi serotinum* (flowers), p. 1086; *Dirca palustris,* p. 1091; *Dodecatheon pulchellum* variety, p. 1109; *Dodecatheon hendersonii,* p. 1109; *Dorstenia contrajerva,* p. 1117; *Dracaena masseffana* in bloom, p. 1129; *Drosera filiformis,* p. 1141; *Drosera capensis,* p. 1141; *Dryopteris marginalis,* p. 1145; *Dryopteris austriaca spinulosa,* p. 1146; *Dryopteris austriaca intermedia,* p. 1146; *Dryopteris filix-mas,* p. 1146; *Echinocereus enneacanthus,* p. 1169; *Echinocereus pectinatus rigidissimus,* p. 1170; *Eichhornia crassipes* (rooted in mud), p. 1184; *Epidendrum prismatocarpum,* p. 1211; *Epimedium grandiflorum,* p. 1216; *Erigeron compositus,* p. 1242; *Erigeron simplex,* p. 1242; *Erythronium californicum,* p. 1261; *Erythronium albidum,* p. 1262; *Euphorbia bupleurifolia,* p. 1308; *Faucaria tuberculosa,* p. 1336; Collecting spores (parts a & b), p. 1351; Raising ferns from spores, pp. 1352–1353; *Forsythia suspensa,* p. 1392; *Fraxinus americana* (leaf), p. 1404; *Fraxinus americana* (young leaves and flowers), p. 1404. Southport Visiter: An exhibit of vegetables at Southport Flower Show, p. 1385. United Nations: Formal espaliered fruit trees in the gardens of the United Nations (parts a & b), p. 1271. Other photographs by Thomas H. Everett.

Color
Harold Frisch: *Exochorda racemosa, Fothergilla* (flowers). The New York Botanical Garden: *Diphylleia cymosa, Dodecatheon alpinum, Dracaena sanderana, Drosera intermedia* (leaf), *Elmera racemosa, Epilobium latifolium, Eriophorum* species, *Erysimum capitatum, Eschscholzia californica, Fatsia japonica, Fouquieria splendens, Franklinia alatamaha, Fritillaria recurva.* Other photographs by Thomas H. Everett.

Published by Garland Publishing, Inc.
136 Madison Avenue, New York, New York 10016

Printed in the United States of America

This work is dedicated to the honored memory of the distinguished horticulturists and botanists who most profoundly influenced my professional career: Allan Falconer of Cheadle Royal Gardens, Cheshire, England; William Jackson Bean, William Dallimore, and John Coutts of the Royal Botanic Gardens, Kew, England; and Dr. Elmer D. Merrill and Dr. Henry A. Gleason of The New York Botanical Garden.

Foreword

According to Webster, an encyclopedia is a book or set of books giving information on all or many branches of knowledge generally in articles alphabetically arranged. To the horticulturist or grower of plants, such a work is indispensable and one to be kept close at hand for frequent reference.

The appearance of *The New York Botanical Garden Illustrated Encyclopedia of Horticulture* by Thomas H. Everett is therefore welcomed as an important addition to the library of horticultural literature. Since horticulture is a living, growing subject, these volumes contain an immense amount of information not heretofore readily available. In addition to detailed descriptions of many thousands of plants given under their generic names and brief description of the characteristics of the more important plant families, together with lists of their genera known to be in cultivation, this Encyclopedia is replete with well-founded advice on how to use plants effectively in gardens and, where appropriate, indoors. Thoroughly practical directions and suggestions for growing plants are given in considerable detail and in easily understood language. Recommendations about what to do in the garden for all months of the year and in different geographical regions will be helpful to beginners and will serve as reminders to others.

The useful category of special subject entries (as distinct from the taxonomic presentations) consists of a wide variety of topics. It is safe to predict that one of the most popular will be Rock and Alpine Gardens. In this entry the author deals helpfully and adequately with a phase of horticulture that appeals to a growing group of devotees, and in doing so presents a distinctly fresh point of view. Many other examples could be cited.

The author's many years as a horticulturist and teacher well qualify him for the task of preparing this Encyclopedia. Because he has, over a period of more than a dozen years, written the entire text (submitting certain critical sections to specialists for review and suggestions) instead of farming out sections to a score or more specialists to write, the result is remarkably homogeneous and cohesive. The Encyclopedia is fully cross referenced so that one may locate a plant by either its scientific or common name.

If, as has been said, an encyclopedia should be all things to all people, then the present volumes richly deserve that accolade. Among the many who call it "friend" will be not only horticulturists ("gardeners," as our author likes to refer to them), but growers, breeders, writers, lecturers, arborists, ecologists, and professional botanists who are frequently called upon to answer questions to which only such a work can provide answers. It seems safe to predict that it will be many years before lovers and growers of plants will have at their command another reference work as authoritative and comprehensive as T. H. Everett's Encyclopedia.

John M. Fogg, Jr.
Director Emeritus, Arboretum of the Barnes Foundation
Emeritus Professor of Botany, University of Pennsylvania

Preface

The primary objective of *The New York Botanical Garden Illustrated Encyclopedia of Horticulture* is a comprehensive description and evaluation of horticulture as it is known and practiced in the United States and Canada by amateurs and by professionals, including those responsible for botanical gardens, public parks, and industrial landscapes. Although large-scale commercial methods of cultivating plants are not stressed, much of the content of the Encyclopedia is as basic to such operations as it is to other horticultural enterprises. Similarly, although landscape design is not treated on a professional level, landscape architects will find in the Encyclopedia a great deal of importance and interest to them. Emphasis throughout is placed on the appropriate employment of plants both outdoors and indoors, and particular attention is given to explaining in considerable detail the how- and when-to-do-it aspects of plant growing.

It may be useful to assess the meanings of two words I have used. Horticulture is simply gardening. It derives from the Latin *hortus,* garden, and *cultura,* culture, and alludes to the intensive cultivation in gardens and nurseries of flowers, fruits, vegetables, shrubs, trees, and other plants. The term is not applicable to the extensive field practices that characterize agriculture and forestry. Amateur, as employed by me, retains its classic meaning of a lover from the Latin *amator;* it refers to those who garden for pleasure rather than for financial gain or professional status. It carries no implication of lack of knowledge or skills and is not to be equated with novice, tyro, or dabbler. In truth, amateurs provide the solid basis upon which American horticulture rests; without them the importance of professionals would diminish. Numbered in millions, amateur gardeners are devotees of the most widespread avocation in the United States. This avocation is serviced by a great complex of nurseries, garden centers, and other suppliers; by landscape architects and landscape contractors; and by garden writers, garden lecturers, Cooperative Extension Agents, librarians, and others who dispense horticultural information. Numerous horticultural societies, garden clubs, and botanical gardens inspire and promote interest in America's greatest hobby and stand ready to help its enthusiasts.

Horticulture as a vocation presents a wide range of opportunities which appeal equally to women and men. It is a field in which excellent prospects still exist for capable entrepreneurs. Opportunities at professional levels occur too in nurseries and greenhouses, in the management of landscaped grounds of many types, and in teaching horticulture.

Some people confuse horticulture with botany. They are not the same. The distinction becomes more apparent if the word gardening is substituted for horticulture. Botany is the science that encompasses all systematized factual knowledge about plants, both wild and cultivated. It is only one of the several disciplines upon which horticulture is based. To become a capable gardener or a knowledgeable plantsman or plantswoman (I like these designations for gardeners who have a wide, intimate, and discerning knowledge of plants in addition to skill in growing them) it is not necessary to study botany formally, although such study is likely to add greatly to one's pleasure. In the practice of gardening many botanical truths are learned from experience. I have known highly competent gardeners without formal training in botany and able and indeed distinguished botanists possessed of minimal horticultural knowledge and skills.

Horticulture is primarily an art and a craft, based upon science, and at some levels perhaps justly regarded as a science in its own right. As an art it calls for an appreciation of beauty and form as expressed in three-dimensional spatial relationships and an ability to translate aesthetic concepts into reality. The chief materials used to create gardens are living plants, most of which change in size and form with the passing of time and often show differences in color and texture and in other ways from season to season. Thus it is important that designers of gardens have a wide familiarity with the sorts of plants that lend themselves to their purposes and with plants' adaptability to the regions and to the sites where it is proposed to plant them.

As a craft, horticulture involves special skills often derived from ancient practices passed from generation to generation by word of mouth and apprenticeship-like contacts. As a technology it relies on this backlog of empirical knowledge supplemented by that acquired by scientific experiment and investigation, the results of which often serve to explain rather than supplant old beliefs and practices but sometimes point the way to more expeditious methods of attaining similar results. And from time to time new techniques are developed that add dimensions to horticultural practice; among such of fairly recent years that come to mind are the manipulation of blooming season by artificial day-length, the propagation of orchids and some other plants by meristem tissue culture, and the development of soilless growing mixes as substitutes for soil.

One of the most significant developments in American horticulture in recent decades is the tremendous increase in the number of different kinds of plants that are cultivated by many more people than formerly. This is particularly true of indoor plants or houseplants, the sorts grown in homes, offices, and other interiors, but is by no means confined to that group. The relative affluence of our society and the freedom and frequency of travel both at home and abroad has contributed to this expansion, a phenomenon that will surely continue as avid collectors of the unusual bring into cultivation new plants from the wild and promote wider interest in sorts presently rare. Our garden flora is also constantly and beneficially expanded as a result of the work of both amateur and professional plant breeders.

It is impracticable in even the most comprehensive encyclopedia to describe or even list all plants that somewhere within a territory as large as the United States and Canada are grown in gardens. In this Encyclopedia the majority of genera known to be in cultivation are described, and descriptions and often other pertinent information about a complete or substantial number of their species and lesser categories are given. Sorts likely to be found only in collections of botanical gardens or in those of specialists may be omitted.

The vexing matter of plant nomenclature inevitably presents itself when an encyclopedia of horticulture is contemplated. Conflicts arise chiefly between the very understandable desire of gardeners and others who deal with cultivated plants to retain long-familiar names and the need to reflect up-to-date botanical interpretations. These points of view are basically irreconcilable and so accommodations must be reached.

As has been well demonstrated in the past, it is unrealistic to attempt to standardize the horticultural usage of plant names by decree or edict. To do so would negate scientific progress. But it is just as impracticable to expect gardeners, nurserymen, arborists, seedsmen, dealers in bulbs, and other amateur and professional horticulturists to keep current with the interpretations and recommendations of plant taxonomists; particularly as these sometimes fail to gain the acceptance even of other botanists and it is not unusual for scientists of equal stature and competence to prefer different names for the same plant.

In practice time is the great leveler. Newly proposed plant names accepted in botanical literature are likely to filter gradually into horticultural usage and eventually gain currency value, but this sometimes takes several years. The complete up-to-dateness and niceties of botanical naming are less likely to bedevil horticulturists than uncertainties concerned with correct plant identification. This is of prime importance. Whether a tree is labeled *Pseudotsuga douglasii, P. taxifolia,* or *P. menziesii* is of less concern than that the specimen so identified is indeed a Douglas-fir and not some other conifer.

After reflection I decided that the most sensible course to follow in *The New York Botanical Garden Illustrated Encyclopedia of Horticulture* was to accept almost in its entirety the nomenclature adopted in *Hortus Third* published in 1976. By doing so, much of the confusion that would result from two major comprehensive horticultural works of the late twentieth century using different names for the same plant is avoided, and it is hoped that for a period of years a degree of stability will be attained. Always those deeply concerned with critical groups of plants can adopt the recommendations of the latest monographers. Exceptions to the parallelism in nomenclature in this Encyclopedia and *Hortus Third* are to be found in the CACTACEAE for which, with certain reservations but for practical purposes, as explained in the Encyclopedia entry Cactuses, the nomenclature of Curt Backeburg's *Die Cactaceae,* published in 1958–62, is followed; and the ferns, where I mostly accepted the guidance of Dr. John T. Mickel of The New York Botanical Garden. The common or colloquial names employed are those deemed to have general acceptance. Cross references and synonomy are freely provided.

The convention of indicating typographically whether or not plants of status lesser than species represent entities that propagate and persist in the wild or are sorts that persist only in cultivation is not followed. Instead, as explained in the Encyclopedia entry Plant Names, the word variety is employed for all entities below specific rank and if in Latin form the name is written in italic, if in English or other modern language, in Roman type, with initial capital letter, and enclosed in single quotation marks.

Thomas H. Everett
Senior Horticulture Specialist
The New York Botanical Garden

Acknowledgments

I am indebted to many people for help and support generously given over the period of more than twelve years it has taken to bring this Encyclopedia to fruition. Chief credit belongs to four ladies. They are Lillian M. Weber and Nancy Callaghan, who besides accepting responsibility for the formidable task of filing and retrieving information, typing manuscript, proofreading, and the management of a vast collection of photographs, provided much wise council; Elizabeth C. Hall, librarian extraordinary, whose superb knowledge of horticultural and botanical literature was freely at my disposal; and Ellen, my wife, who displayed a deep understanding of the demands on time called for by an undertaking of this magnitude, and with rare patience accepted inevitable inconvenience. I am also obliged to my sister, Hette Everett, for the valuable help she freely gave on many occasions.

Of the botanists I repeatedly called upon for opinions and advice and from whom I sought elucidation of many details of their science abstruse to me, the most heavily burdened have been my friends and colleagues at The New York Botanical Garden, Dr. Rupert C. Barneby, Dr. Arthur Cronquist, and Dr. John T. Mickel. Other botanists and horticulturists with whom I held discussions or corresponded about matters pertinent to my text include Dr. Theodore M. Barkley, Dr. Lyman Benson, Dr. Ben Blackburn, Professor Harold Davidson, Dr. Otto Degener, Harold Epstein, Dr. John M. Fogg, Jr., Dr. Alwyn H. Gentry, Dr. Alfred B. Graf, Brian Halliwell, Dr. David R. Hunt. Dr. John P. Jessop, Dr. Tetsuo Koyama, Dr. Bassett Maguire, Dr. Roy A. Mecklenberg, Everitt L. Miller, Dr. Harold N. Moldenke, Dr. Dan H. Nicolson, Dr. Pascal P. Pirone, Dr. Ghillean Prance, Don Richardson, Stanley J. Smith, Ralph L. Snodsmith, Marco Polo Stufano, Dr. Bernard Verdcourt, Dr. Edgar T. Wherry, Dr. Trevor Whiffin, Dr. Richard P. Wunderlin, Dr. John J. Wurdack, Yuji Yoshimura, and Rudolf Ziesenhenne.

Without either exception or stint these conferees and correspondents shared with me their knowledge, thoughts, and judgments. Much of the bounty so gleaned is reflected in the text of the Encyclopedia but none other than I am responsible for interpretations and opinions that appear there. To all who have helped, my special thanks are due and are gratefully proferred.

I acknowledge with much pleasure the excellent cooperation I have received from the Garland Publishing Company and most particularly from its President, Gavin Borden. To Ruth Adams, Nancy Isaac, Carol Miller, and Melinda Wirkus, I say thank you for working so understandingly and effectively with me and for shepherding my raw typescript through the necessary stages.

How to Use This Encyclopedia

A vast amount of information about how to use, propagate, and care for plants both indoors and outdoors is contained in the thousands of entries that compose the *New York Botanical Garden Illustrated Encyclopedia of Horticulture.* Some understanding of the Encyclopedia's organization is necessary in order to find what you want to know.

Arrangement of the Entries

Genera

The entries are arranged in alphabetical order. Most numerous are those that deal with taxonomic groups of plants. Here belong approximately 3,500 items entered under the genus name, such as ABIES, DIEFFENBACHIA, and JUGLANS. If instead of referring to these names you consult their common name equivalents of FIR, DUMB CANE, and WALNUT, you will find cross references to the genus names.

Bigeneric Hybrids & Chimeras

Hybrids between genera that have names equivalent to genus names—most of these belonging in the orchid family—are accorded separate entries. The same is true for the few chimeras or graft hybrids with names of similar status. Because bigeneric hybrids frequently have characteristics similar to those of their parents and require similar care, the entries for them are often briefer than the regular genus entries.

Families

Plant families are described under their botanical names, with their common name equivalents also given. Each description is followed by a list of the genera accorded separate entries in this Encyclopedia.

Vegetables, Fruits, Herbs, & Ornamentals

Vegetables and fruits that are commonly cultivated, such as broccoli, cabbage, potato, tomato, apple, peach, and raspberry; most culinary herbs, including basil, chives, parsley, sage, and tarragon; and a few popular ornamentals, such as azaleas, carnations, pansies, and poinsettias, are treated under their familiar names, with cross references to their genera. Discussions of a few herbs and some lesser known vegetables and fruits are given under their Latin scientific names with cross references to the common names.

Other Entries

The remaining entries in the Encyclopedia are cross references, definitions, and more substantial discussions of many subjects of interest to gardeners and others concerned with plants. For example, a calendar of gardening activity, by geographical area, is given under the names of the months and a glossary of frequently applied species names (technically, specific epithets) is provided in the entry Plant Names. A list of these general topics, which may provide additional information about a particular plant, is provided at the beginning of each volume of the Encyclopedia (see pp. xvii–xx).

Cross References & Definitions

The cross references are of two chief types: those that give specific information, which may be all you wish to know at the moment:
Boojam Tree is *Idria columnaris.*
Cobra plant is *Darlingtonia californica.*
and those that refer to entries where fuller explanations are to be found:
Adhatoda. See Justicia.
Clubmoss. See Lycopodium and Selaginella.

Additional information about entries of the former type can, of course, be found by looking up the genus to which the plant belongs—*Idria* in the case of the boojam tree and *Darlingtonia* for the cobra plant.

ORGANIZATION OF THE GENUS ENTRIES

Pronunciation

Each genus name is followed by its pronunciation in parentheses. The stressed syllable is indicated by the diacritical mark ′ if the vowel sound is short as in man, pet, pink, hot, and up; or by ` if the vowel sound is long as in mane, pete, pine, home, and fluke.

Genus Common Names
Family Common Names
General Characteristics

Following the pronunciation, there may be one or more common names applicable to the genus as a whole or to certain of its kinds. Other names may be introduced later with the descriptions of the species or kinds. Early in the entry you will find the common and botanical names of the plant family to which the genus belongs, the number of species the genus contains, its natural geographical distribution, and the derivation of its name. A description that stresses the general characteristics of the genus follows, and this may be supplemented by historical data, uses of some or all of its members, and other pertinent information.

Identification of Plants

Descriptions of species, hybrids, and varieties appear next. The identification of unrecognized plants is a fairly common objective of gardeners; accordingly, in this Encyclopedia various species have been grouped within entries in ways that make their identification easier. The groupings may bring into proximity sorts that can be adapted for similar landscape uses or that require the same cultural care, or they may emphasize geographical origins of species or such categories as evergreen and deciduous or tall and low members of the same genus. Where the description of a species occurs, its name is designated in ***bold italic.*** Under this plan, the description of a particular species can be found by referring to the group to which it belongs, scanning the entry for the species name in bold italic, or referring to the opening sentences of paragraphs which have been designed to serve as lead-ins to descriptive groupings.

Gardening & Landscape Uses
Cultivation
Pests & Diseases

At the end of genus entries, subentries giving information on garden and landscape uses, cultivation, and pests or diseases or both are included, or else reference is made to other genera or groupings for which these are similar.

General Subject Listings

The lists below organize some of the encyclopedia entries into topics which may be of particular interest to the reader. They are also an aid in finding information other than Latin or common names of plants.

PLANT ANATOMY AND TERMS USED IN PLANT DESCRIPTIONS

All-America Selections
Alternate
Annual Rings
Anther
Apex
Ascending
Awl-shaped
Axil, Axillary
Berry
Bloom
Bracts
Bud
Bulb
Bulbils
Bulblet
Bur
Burl
Calyx
Cambium Layer
Capsule
Carpel
Catkin
Centrals
Ciliate
Climber
Corm
Cormel
Cotyledon
Crown
Deciduous
Disk or Disc
Double Flowers
Drupe
Florets
Flower
Follicle
Frond
Fruit
Glaucous
Gymnosperms
Head
Hips
Hose-in-Hose
Inflorescence
Lanceolate
Leader
Leaf
Leggy
Linear
Lobe
Midrib
Mycelium
Node
Nut and Nutlet
Oblanceolate
Oblong
Obovate
Offset
Ovate
Palmate
Panicle
Pedate
Peltate
Perianth
Petal
Pinnate
Pip
Pistil
Pit
Pod
Pollen
Pompon
Pseudobulb
Radials
Ray Floret
Rhizome
Runners
Samara
Scion or Cion
Seeds
Sepal
Set
Shoot
Spore
Sprigs
Spur
Stamen
Stigma
Stipule
Stolon
Stool
Style
Subshrub
Taproot
Tepal
Terminal
Whorl

GARDENING TERMS AND INFORMATION

Acid and Alkaline Soils
Adobe
Aeration of the Soil
Air and Air Pollution
Air Drainage
Air Layering
Alpine Greenhouse or Alpine House
Amateur Gardener
April, Gardening Reminders For
Aquarium
Arbor
Arboretum
Arch
Asexual or Vegetative Propagation
Atmosphere
August, Gardening Reminders For
Balled and Burlapped
Banks and Steep Slopes
Bare-Root
Bark Ringing
Baskets, Hanging
Bed
Bedding and Bedding Plants
Bell Jar
Bench, Greenhouse
Blanching
Bleeding
Bog
Bolting
Border
Bottom Heat
Break, Breaking
Broadcast
Budding
Bulbs or Bulb Plants

Gardening Terms and Information (Continued)

Gardening Terms and Information (Continued)

State Agricultural Experimental Stations
Stock or Understock
Straightedge
Strawberry Jars
Strike
Stunt
Succession Cropping
Sundials
Syringing
Thinning or Thinning Out
Tillage
Tilth
Tools
Top-Dressing
Topiary Work
Training Plants
Tree Surgery
Tree Wrapping
Trenching
Trowels
Tubs
Watering
Weeds and Their Control
Window Boxes

FERTILIZERS AND OTHER SUBSTANCES RELATED TO GARDENING

Algicide
Aluminum Sulfate
Ammonium Nitrate
Ammonium Sulfate
Antibiotics
Ashes
Auxins
Basic Slag
Blood Meal
Bonemeal
Bordeaux Mixture
Calcium Carbonate
Calcium Chloride
Calcium Metaphosphate
Calcium Nitrate
Calcium Sulfate
Carbon Disulfide
Chalk
Charcoal
Coal Cinders
Cork Bark
Complete Fertilizer
Compost and Composting
Cottonseed Meal
Creosote
DDT
Dormant Sprays
Dried Blood
Fermate or Ferbam
Fertilizers
Fishmeal
Formaldehyde
Fungicides
Gibberellic Acid
Green Manuring
Growth Retardants
Guano
Herbicides or Weed-Killers
Hoof and Horn Meal
Hormones
Humus
Insecticide
John Innes Composts
Lime and Liming
Liquid Fertilizer
Liquid Manure
Manures
Mulching and Mulches
Muriate of Potash
Nitrate of Ammonia
Nitrate of Lime
Nitrate of Potash
Nitrate of Soda
Nitrogen
Orchid Peat
Organic Matter
Osmunda Fiber or Osmundine
Oyster Shells
Peat
Peat Moss
Permagnate of Potash
Potassium
Potassium Chloride
Potassium-Magnesium Sulfate
Potassium Nitrate
Potassium Permagnate
Potassium Sulfate
Pyrethrum
Rock Phosphate
Rotenone
Salt Hay or Salt Marsh Hay
Sand
Sawdust
Sodium Chloride
Sprays and Spraying
Sulfate
Superphosphate
Trace Elements
Urea
Urea-Form Fertilizers
Vermiculite
Wood Ashes

TECHNICAL TERMS

Acre
Alternate Host
Annuals
Antidessicant or Antitranspirant
Biennals
Binomial
Botany
Chromosome
Climate
Clone
Composite
Conservation
Cross or Crossbred
Cross Fertilization
Cross Pollination
Cultivar
Decumbent
Dicotyledon
Division
Dormant
Endemic
Environment
Family
Fasciation
Fertility
Fertilization
Flocculate
Floriculture
Genus
Germinate
Habitat
Half-Hardy
Half-Ripe
Hardy Annual
Hardy Perennial
Heredity
Hybrid
Indigenous
Juvenile Forms
Juvenility
Legume
Monocotyledon
Monoecious
Mutant or Sport
Mycorrhiza or Mycorhiza
Nitrification
Perennials
pH
Plant Families
Photoperiodism
Photosynthesis
Pollination
Pubescent
Saprophyte
Self-Fertile
Self-Sterile
Species
Standard
Sterile
Strain
Terrestrial
Tetraploid
Transpiration
Variety

TYPES OF GARDENS AND GARDENING

Alpine Garden
Artificial Light Gardening
Backyard Gardens
Biodynamic Gardening
Bog Gardens
Botanic Gardens and Arboretums
Bottle Garden
City Gardening
Colonial Gardens
Conservatory
Container Gardening
Cutting Garden
Desert Gardens
Dish Gardens
Flower Garden
Fluorescent Light Gardening
Formal and Semiformal Gardens
Greenhouses and Conservatories
Heath or Heather Garden
Herb Gardens
Hydroponics or Nutriculture
Indoor Lighting Gardening
Japanese Gardens
Kitchen Garden
Knot Gardens

Types of Gardens and Gardening (Continued)

Miniature Gardens
Native Plant Gardens
Naturalistic Gardens
Nutriculture
Organic Gardening
Rock and Alpine Gardens
Roof and Terrace Gardening
Salads or Salad Plants
Seaside Gardens
Shady Gardens
Sink Gardening
Terrariums
Vegetable Gardens
Water and Waterside Gardens
Wild Gardens

PESTS, DISEASES, AND OTHER TROUBLES

Ants
Aphids
Armyworms
Bagworms
Bees
Beetles
Billbugs
Biological Control of Pests
Birds
Blight
Blindness
Blotch
Borers
Budworms and Bud Moths
Bugs
Butterflies
Canker
Cankerworms or Inchworms
Casebearers
Caterpillars
Cats
Centipede, Garden
Chinch Bugs
Chipmunks
Club Root
Corn Earworm
Crickets
Cutworms
Damping Off
Deer
Die Back
Diseases of Plants
Downy Mildew
Earthworms
Earwigs
Edema
Fairy Rings
Fire Blight
Flies
Fungi or Funguses
Galls
Gas Injury
Gophers
Grasshoppers
Grubs
Gummosis
Hornworms
Inchworms
Insects
Iron Chelates
Iron Deficiency
Lace Bugs
Lantana Bug
Lantern-Flies
Larva
Leaf Blight
Leaf Blister
Leaf Blotch
Leaf Curl
Leaf Cutters
Leaf Hoppers
Leaf Miners
Leaf Mold
Leaf Rollers
Leaf Scorch
Leaf Skeletonizer
Leaf Spot Disease
Leaf Tiers
Lightening Injury
Maggots
Mantis or Mantid
Mealybugs
Mice
Midges
Milky Disease
Millipedes
Mites
Mold
Moles
Mosaic Diseases
Moths
Muskrats
Needle Cast
Nematodes or Nemas
Parasite
Pests of Plants
Plant Hoppers
Plant Lice
Praying Mantis
Psyllids
Rabbits
Red Spider Mite
Rootworms
Rots
Rust
Sawflies
Scab Diseases
Scale Insects
Scorch or Sunscorch
Scurf
Slugs and Snails
Smut and White Smut Diseases
Sowbugs or Pillbugs
Spanworms
Spittlebugs
Springtails
Squirrels
Stunt
Suckers
Sun Scald
Thrips
Tree Hoppers
Virus
Walking-Stick Insects
Wasps
Webworms
Weevils
Wilts
Witches' Brooms
Woodchucks

GROUPINGS OF PLANTS

Accent Plants
Aquatics
Aromatic Plants
Bedding and Bedding Plants
Berried Trees and Shrubs
Bible Plants
Broad-leaved and Narrow-leaved Trees and Shrubs
Bulbs or Bulb Plants
Bush Fruits
Carnivorous or Insectivorous Plants
Dried Flowers, Foliage, and Fruits
Edging Plants
Epiphyte or Air Plant
Evergreens
Everlastings
Fern Allies
Filmy Ferns
Florists' Flowers
Foliage Plants
Fragrant Plants and Flowers
Gift Plants
Graft Hybrids
Grasses, Ornamental
Hard-Wooded Plants
Houseplants or Indoor Plants
Japanese Dwarfed Trees
Medicinal or Drug Plants
Night-Blooming Plants
Ornamental-Fruited Plants
Pitcher Plants
Poisonous Plants
Shrubs
State Flowers
State Trees
Stone Fruits
Stone or Pebble Plants
Stove Plants
Succulents
Tender Plants
Trees
Windowed Plants

The New York Botanical Garden Illustrated Encyclopedia of Horticulture

DICHELOSTEMMA (Dicheló-stemma)—Firecracker Plant, Snake-Lily, Blue Dicks. By some botanists this genus of the lily family LILIACEAE is included in *Brodiaea*. It consists of six species, natives of western North America. The name is derived from the Greek *dicha*, in two, and *stemma*, a crown, in allusion to the staminodes. From *Brodiaea* the genus *Dichelostemma* differs in its leaves having longitudinal keels along their undersides, in the flowers having three-lobed rather than three-winged stigmas, and in more recondite technical details. Also, dichelostemmas favor semishady locations, brodiaeas sunny ones.

Dichelostemmas have underground, fibrous-coated, bulblike organs called corms that, like those of crocuses and gladioluses, last for one year and are replaced by others that develop during the growing season, from early fall to late spring. All foliage is basal. The two to five leaves are long-linear and have flat upper surfaces. The flowers are in umbels, each with a collar of papery bracts at its base, that top slender, erect stalks. Each bloom has an individual jointed stalk and a perianth of six segments, their bases united for at least one-half their lengths to form a tube. Their upper parts are separate and are commonly called petals. There are three fertile stamens and three aborted ones (staminodes), the latter appendages to the outer petals. The stigma is three-lobed. The fruits are capsules.

Firecracker plant (***D. ida-maia*** syns. *Brodiaea ida-maia, B. coccinea, Brevoortia ida-maia*) is well named for its brightly colored blooms, which indeed suggest firecrackers. Native from Oregon to California, this sort has leaves shorter than the flowering stalks and up to ⅓ inch wide. The slender flowering stalks, 1 foot to 3 feet in height, are topped with umbels of five to twenty or sometimes more mostly nodding blooms with individual stalks ½ to 1½ inches in length. The perianth tubes ¾ to 1 inch long are scarlet to crimson or rarely yellow. The petals, about ¼ inch long, are light green. The whitish staminodes, broader than long, are minutely dented at their apexes.

Sweetly-scented, violet to lilac flowers are borne by ***D. multiflorum*** (syn. *Brodiaea multiflora*). Native from Oregon to California, this comparatively late bloomer has flowers suggestive of miniature hyacinths. Its flowering stalks are from under 1 foot to 2½ feet tall. Compact and nearly spherical, the umbels are of short-stalked blooms with constricted throats. The swollen perianth tubes are ⅓ inch or a little more in length, the petals about as long. The white staminodes are more or less tinged with purple or violet. From the last, ***D. congestum*** (syn. *Brodiaea congesta*) differs in the stalks of the individual flowers being joined at their bases so that the blooms appear to be in short, crowded racemes instead of umbels, and in the perianth tube being scarcely constricted at its throat. This, native from Washington to California, has flowering stalks 1 foot to 3 feet tall.

Snake-lily (***D. volubile*** syn. *Brodiaea volubilis*), a curious Californian, differs from other sorts in having flexuous, contorted or twining flowering stalks 1½ to 5 feet in length. The leaves, 1 foot to 2 feet long, are up to ½ inch wide. The pink flowers have six-angled perianth tubes and spreading petals, each approximately ¼ inch long. The staminodes are narrow, white, and notched at their apexes.

Dichelostemma volubile

Blue dicks or wild-hyacinth (***D. pulchellum*** syns. *Brodiaea pulchella, B. capitata*) has a greater range in the wild than any of the other species. It occurs from Oregon to Utah and central Baja California. The leaves, 6 inches to nearly 1½ feet long, are ¼ to ½ inch wide. The violet or rarely white flowers in tight, headlike umbels, each with a collar of metallic-purple bracts, top stalks 1 foot to 2 feet tall or sometimes taller. They have a narrowly-bell-shaped perianth tube up to ⅓ inch long and not constricted at its mouth, and ascending petals up to nearly ½ inch long. Blue dicks is distinct in its corms bearing cormlets on short stolons so that they are never clustered around the base of the mother corms as all or many are in other species.

Garden and Landscape Uses and Cultivation. These are as for *Brodiaea*.

DICHONDRA (Dichón-dra)—Lawn-Leaf. The morning glory family CONVOLVULACEAE includes as members the nine species of low, mat-forming herbaceous plants of the tropics and subtropics that constitute *Dichondra*. One is popular in California and other dry-climate regions as a substitute for lawn grass. The name, from the Greek *dis*, double, and *chondros*, a grain, alludes to the fruits. Dichondras are annuals or perennials with slender stems and small, stalked, silky-hairy or hairless leaves with round to kidney-shaped blades. The tiny blooms, with stalks usually shorter than the leaves, are solitary in the leaf axils. They have five persistent, spreading, oblong or spatula-shaped sepals, deeply-five-lobed corollas, five slender-stalked stamens, and two-parted ovaries. The fruits are one- or two-seeded capsules.

The sort commonly cultivated is ***D. micrantha***, a native of Texas, Mexico, the West Indies, Japan, the Ryukyu Islands, and China. In gardens it is frequently misnamed *D. repens* and *D. carolinensis* both valid names of sorts probably not cultivated. A perennial, this forms a close mat of creeping stems and attractive foliage.

Dichondra micrantha

The leaves, usually not more than ½ inch across and sparingly hairy on their undersides, and roundish to hoof-shaped, have curved stalks usually under ½ inch long. The whitish flowers are tiny and of no ornamental merit.

Garden and Landscape Uses. In the Southwest and elsewhere where climates are suitable, *D. micrantha* is extensively used for lawns. It prospers in sun, adapts well to reasonable shade, and needs slightly acid soil. Unlike grass, it will not stand much walking upon, but on sandy soils withstands some foot traffic. A great advantage is that mowing needs to be done less frequently than with grass lawns. Some gardeners leave it unmown, but then it presents a more irregular and less neat surface, and weeds may be more troublesome. This plant stands several degrees of frost, but is killed if the ground freezes to a depth of 3 to 4 inches.

Cultivation. Dichondra lawns are started from seeds or from 1-inch-wide plugs of sod. The latter, spaced on six-inch centers, will usually form a good cover the first season. The most favorable time for planting plugs is spring. Seeds are sown at the rate of one pound to 1,000 square feet in summer. Before sowing it is well to treat the ground with a weedkiller containing diphenamid. This inhibits the growth of most weed seeds, but not those of dichondra. In preparing the ground liberal amounts of compost or other decayed organic matter, as well as a dressing of a slow-acting complete fertilizer, should be incorporated with the upper few inches. It is important to keep newly seeded areas evenly moist; if they dry for only a few hours at critical stages the germinating seeds or tiny plants may be lost. Mulching is sometimes resorted to to minimize drying, but is unnecessary if the area is irrigated regularly. Although somewhat less critical, it is also important to keep newly plugged lawns moist.

Care of established dichondra lawns also calls for regular watering especially away from natural humid areas. Frequency of application will depend upon a number of variables including the character of the soil and the prevailing weather. On porous soils daily watering in hot, dry weather may be advisable; on more retentive earths and in cooler or more humid weather intervals between soakings may be three days or longer. Watering by flooding or with a permanently installed sprinkler system is best. Movable sprinklers, the use of which involves walking over the lawn, are unsatisfactory. Regular stimulation with soluble fertilizer applied in liquid form is highly beneficial. Chemical fertilizers applied dry are apt to damage the foliage, but slow-acting organic fertilizers can be used dry. Unless the lawn is adequately cared for with respect to watering, fertilizing, and making sure the soil remains on the acid side, weeds may become a problem. Crab grass can be controlled by the use of dimethyl arsonate without harming the dichondra, and certain other weeds by using diphenamid. Such weeds as dandelions and plantains may be killed by spot applications of weedkillers or by placing a small teaspoonful of nitrate of soda, urea, or sulfate of ammonia in the center of each. Mowing will ordinarily be needed whenever the surface shows signs of becoming uneven, once a month perhaps, or oftener if it so pleases.

Dichorisandra thyrsiflora

DICHORISANDRA (Dichoris-ándra). The approximately thirty-five species of the tropical American genus *Dichorisandra* belong in the spiderwort family COMMELINACEAE. Few are cultivated. The name is from the Greek *dis,* twice, *chorizo,* to part, and *aner,* male, or by analogy anther; it alludes to the anthers opening with two valves.

Dichorisandras are evergreen and deciduous herbaceous perennials with erect or lax stems, in some cases clambering up trees to a height of 15 feet, and leaves with sheathing bases. The flowers are in raceme-like or spikelike panicles, the individuals arising from the axils of bracts. Each bloom has three persistent sepals of unequal size, three separate petals, and five or six short-stalked stamens without hairs. The fruits are capsules.

A magnificent flowering plant, sometimes called blue-ginger, ***D. thyrsiflora,*** of

Dichorisandra thyrsiflora (flowers)

Brazil, is 3 to 4 feet tall. It has erect, scarcely branched, canelike stems, and, mostly toward their apexes, stalked, lanceolate, dark lustrous green, hairless leaves 6 inches to 1 foot long, approximately 2 inches wide, and purplish on their undersides. In showy, crowded, erect, terminal panicles up to 7 inches long are borne the medium-sized, light blue to rich purple blooms. Their bright yellow anthers contrast pleasingly with the petals. This species is deciduous. Variety *D. t. variegata* has leaves with two longitudinal silvery bands and a reddish midrib.

A very handsome foliage plant, looking much like a giant-leaved *Zebrina,* is ***D. reginae*** (syn. *Tradescantia reginae*), a native of Peru. Evergreen, this species very rarely blooms. Because of this, although it had been in cultivation since 1890, it was not certainly identified until 1957. From fleshy roots it sends a cluster of erect or suberect, canelike stems to a height of up to 1 foot. Its short-stalked leaves, in two ranks, are pointed-elliptic or pointed-lanceolate and up to 7 inches long by 2½ inches wide. Except for hairs on the undersides, chiefly on the veins (veins so minute that a magnifying lens is needed to detect them readily), the foliage is hairless. In color the leaves are dark green, often, especially when young, suffused with reddish-purple on both sides. Short silver streaks angle outward from the mid-veins on the upper surface in fishbone pattern, and generally there are two broadish, longitudinal silvery bands. All variegation is much brighter and clearly defined on younger than on older foliage. The flowers are in compact, terminal, conical panicles. Individually short-stalked, they have fleshy sepals, and blue-violet petals with white lower halves, a little under ½ inch long. There are six white-stalked stamens with blue anthers, a blue style, and a three-angled stigma.

Dichorisandra reginae

A low plant with purplish-black, red-veined leaves known in cultivation as *D.* 'Blackie' may be a *Geogenanthus.* The plant sometimes misnamed *D. musaica undata* is a *Geogenanthus.*

Garden and Landscape Uses and Cultivation. These very beautiful plants love warmth. They are admirably adapted for humid greenhouses where minimum temperatures of 60 to 70°F are maintained and for outdoor cultivation in the tropics. They respond to deep, fertile soil and part-shade. In greenhouses *D. thyrsiflora* is kept dry during its winter dormancy, at other times it is watered freely. The soil of *D. reginae* is always kept moist. Both are helped by liberal applications of dilute liquid fertilizer when they are in active growth. Repotting is done annually in late winter or spring. Propagation is simple by division and by cuttings.

DICHROA (Di-chròa). A close relative of *Hydrangea,* the genus *Dichroa* belongs in the saxifrage family SAXIFRAGACEAE, or if one accepts splitting of that group, in the hydrangea family HYDRANGACEAE. It consists of what for long was considered one variable species, but is now accounted as about a dozen. It inhabits southeastern Asia and Indonesia. The name, alluding to the bicolored blooms, comes from the Greek *dis,* twice, and *chroa,* color.

Dichroas are nonhardy shrubs with opposite, undivided leaves and terminal, pyramidal panicles of many small white to blue flowers. The latter have five or six each sepals and petals, about ten stamens, and three to five styles. The fruits are small berries.

As known in cultivation ***D. febrifuga*** is an evergreen, hydrangea-like shrub 3 to 7 feet tall. It has pointed-elliptic, toothed leaves 4 to 8 inches long and approximately one-third as wide. The panicled heads of bloom may be 6 inches, or more, long and wide. The flower buds and the outsides of the petals are white. The insides of the petals are bright blue. The decorative fruits also are bright blue.

Dichroa febrifuga

Garden and Landscape Uses and Cultivation. Although uncommon, this should make a beautiful addition to outdoor shrub plantings in warm-temperate and subtropical regions where little frost is experienced. It is also attractive in pots and tubs in greenhouses. It prospers in sun or light shade in ordinary fertile soil kept moderately moist throughout the year. It is easily propagated by seeds and by cuttings. Indoors a winter night temperature of 50 to 55°F with an increase of five to ten degrees by day suits.

DICKSONIA (Dick-sònia)—Tree Fern. This group of thirty species of ferns enjoys wide distribution. Representatives occur in the native floras of the mountains of Malaysia and in Australia, New Zealand, New Caledonia, the Americas, and St. Helena. It belongs in the dicksonia family DICKSONIACEAE and was named in honor of the English botanist James Dickson, who died in 1822.

Most, but not all species have trunks, and thus qualify as tree ferns. The trunks are clothed with a thick layer of coarse rootlets. The large, firm fronds (leaves), paler and somewhat more glaucous on their undersides than above, have short stalks and two- or three-times-pinnately-divided blades. Those of the species described below are very beautiful, and of lacy appearance. The clusters of spore capsules are at the tips of the veins near the margins of the leaflets. Their coverings (indusia) have two clamlike valves, the outer formed of the incurved apex of the leaf segment.

Hardiest of tree ferns, New Zealand ***Dicksonia fibrosa*** not uncommonly masquerades in cultivation as closely similar *D. antarctica.* From that species it differs in being smaller, in its leaflets being more crowded, and in having smaller spore clusters. Its maximum height not more than 20 feet, its trunk, which may attain 2 feet in diameter, is densely clothed with matted rootlets. The twice- or thrice-divided leaves are 4 to 6 feet long or sometimes longer by up to about 2 feet wide.

Dicksonia fibrosa

Much like the last, ***D. antarctica*** attains heights of 25 to 50 feet and develops a magnificent crown of thrice-pinnate leaves up to 8 feet long by 2 to 3 feet wide. Their stalks are clothed with tan hairs. This species is native to New Zealand and Tasmania.

Dicksonia antarctica in Ireland

Dicksonia antarctica, mature specimens in California

New Zealand ***D. squarrosa*** is slender-trunked, and up to 20 feet tall. It sends runners to distances of 3 feet or more from the main trunk, and these give rise to secondary trunks that eventually form groves. The trunks are covered with the persistent, long bases of fallen leaves, at the bottoms of which are dormant buds that often give rise to small plants or branches. The fronds, twice- or thrice-divided, are 4 to 8 feet long and 2 to 3½ feet wide. Their stalks at first have abundant dark hairs. These fall, but leave the stalks roughened with their persistent swollen bases.

Garden and Landscape Uses and Cultivation. Dicksonias are strikingly handsome. The cultivated kinds are hardier than most tree ferns, withstanding occasional light frosts. Outdoors, in mild, humid climates, they are at their best where sheltered from wind, and in light shade.

Dicksonia squarrosa, young specimen

They need deep soil that contains abundant organic matter and that, although well drained, is never dry. These stately ferns are seen to particular advantage when planted at irregular intervals in groups or groves, with adequate space between individuals so that the tops can spread without crowding. The most pleasing effects are obtained when the specimens in each group or grove are of varied heights.

In greenhouses and conservatories dicksonias are best accommodated in ground beds. They are less suited for pots or tubs, but if great care is taken to never allow the soil to become dry they can be grown in them. It is very important to keep the trunks moist. This may be done by wetting them two or three times a day. It is good practice to wrap them in sphagnum moss, tied into place with thin wire or nylon thread or held in a corset of chicken wire. The moss must be kept moist. Dicksonias thrive best in well-drained containers rather small for the size of the plant. They must be watered generously, especially from spring to fall. Specimens that have filled their pots or tubs with roots benefit from regular applications of a dilute liquid fertilizer. Soil for dicksonias should be coarse, turfy, and contain abundant organic matter, as well as a generous lacing of coarse sand and some crushed charcoal. Propagation is by spores and offsets. For more information see Ferns.

A grove of *Dicksonia antarctica* in the Strybing Arboretum, San Francisco, California

DICKSONIACEAE—Dicksonia Family. This family of five genera of often tall tree ferns and a few lower ferns with rhizomes comprises about thirty species, natives of Central America, Juan Fernandez, Hawaii, eastern Asia, and Indonesia. From the cyathea family CYATHACEAE, in which some authorities have included it, the dicksonia family differs in the bases of its usually very big, pinnate fronds (leaves) being surrounded by a dense mass of woolly hairs without admixture of scales and by the clusters of spore capsules being at the ends of veinlets generally at the frond margins. Genera cultivated are *Cibotium* and *Dicksonia.* By some authorities the DICKSONIACEAE is included in the pteris family PTERIDACEAE.

DICLIPTERA (Di-clíptera). Only one of the more than 150 species of *Dicliptera,* of the acanthus family ACANTHACEAE, seems to be cultivated. The generic name, alluding to the two-winged fruits, derives from the Greek *diklis,* double-doored, and *pteron,* a wing. The genus is a widely distributed native of tropical, subtropical, and temperate regions.

Diclipteras are subshrubs, herbaceous perennials, and annuals with more or less angled stems and opposite, undivided, sometimes wavy-edged leaves. The flowers, usually in branched, terminal or axillary, bracted clusters, more rarely are solitary. They have a five-parted calyx, a narrowly-tubular, two-lipped corolla, two stamens, and a slender style. The fruits are capsules containing two or four seeds.

Native to Uruguay, ***D. suberecta*** (syn. *Jacobinia suberecta*) is a nonhardy herbaceous perennial or subshrub with gray-velvety-hairy foliage. It has long, arching stems and short-stalked, broad-ovate leaves 1½ to 2½ inches in length. The flowers, with brick-red corollas about 1¾ inches long and protruding stamens, are in fairly compact clusters of several to many.

Dicliptera suberecta

Garden Uses and Cultivation. The species described is a useful plant for growing in pots and hanging baskets in tropical greenhouses and for planting outdoors in warm, frost-free climates. It appreciates just a little shade from strong sun, and porous, fertile soil containing an abundance of humus and moderately moist, but not wet. Cuttings and seeds afford easy means of propagation.

DICOTYLEDON. Botanists recognize two major groups or classes within the subdivision of the flowering or seed plants called angiosperms. They are the dicotyledons and monocotyledons, often referred to as dicots and monocots. These differ in the total of a few critical characteristics rather than indisputably in any one. Most constant is the feature suggested by the name, derived from the

Greek *dis*, two, and cotyledon. Dicotyledons have seeds with two, rarely one, three, or four cotyledons. Their leaves are generally net-veined. The parts of their flowers, when definite in number, are mostly in fours, fives, or multiples of those numbers. Trees and shrubs of the group have a cambium layer between the bark and wood tissues, and their trunks and branches thicken by the addition of annual rings. More recondite characteristics are associated with the roots and pollen. By far the larger number of familiar flowering plants are dicotyledons. For examples of some that are not see Monocotyledons.

Typical seedlings of a dicotyledon. The first two leaves represent the cotyledons

Dicranopteris linearis

DICRANOPTERIS (Dicran-ópteris). Less well known horticulturally than many ferns, *Dicranopteris* of the gleichenia family GLEICHENIACEAE comprises fewer than ten species, natives of sunny locations in warm-temperate and subtropical regions. The name, alluding to the manner in which the fronds branch, is from the Greek *dikranos*, two-branched, and *pteris*, a fern.

These form thickets with creeping rhizomes that spread widely over the surface or sometimes just below ground level. They have repeatedly-forking evergreen fronds, usually glaucous beneath, that lengthen throughout their lives, with deeply-pinnately-divided leaflets. Their spore clusters, usually in groups of two to four, on the undersides of the leaflets, are without coverings (indusia).

A thicket of an unidentified *Dicranopteris*

Native of the Old World tropics, ***D. linearis*** (syns. *D. dichotoma, Gleichenia linearis*) is an attractive species with zigzagged, forking, lustrous fronds the ultimate branches terminating in a pair of oblong-lanceolate leaflets 6 inches to 1½ feet long by about 2 inches wide. The segments have recurved margins.

Garden and Landscape Uses. The species described, like most of its fellows, needs excellent light and tolerates and even appreciates considerable sun. As with others of its genus it does not take kindly to cultivation.

Cultivation. Resentful of root disturbance, dicranopterises should not be potted or repotted oftener than necessary. Plant them with their rhizomes on the surface of coarse, well-drained, sandy soil, on the lean side rather than rich. Grow in a humid atmosphere with the soil kept dampish, but not for long periods wet. Successful cultivators avoid applications of fertilizers, but very dilute solutions of liquid manure may be used. A winter night temperature of 50°F with increases of a few degrees by day suits. Higher temperatures at other seasons are appropriate. Propagation is by spores. Attempts to divide these ferns almost surely result in failure and usually loss of the specimens. For additional information see Ferns.

DICRANOPYGIUM (Dicrano-pýgium). Formerly included in *Carludovica*, the genus *Dicranopygium* consists of forty-four species of the cyclanthus family CYCLANTHACEAE. In the wild it is confined to Central America and northern South America. The name, from the Greek *dikranon*, a fork, and *pyge*, a rump or tail, alludes to the cleft leaves.

Dicranopygiums, mostly small to medium in size, are stemless or have short, fleshy stems. Their leaves have stalks not exceeding twice the lengths of the blades. The latter, of palmlike texture and pleated, are divided primarily into two lobes, which in older leaves may irregularly split. The flowers are similar in aspect and arrangement to those of *Carludovica*, but the spadixes on which they are borne are smaller. There are two to four spathes (bracts) at the base of the spadix. The fruits are berries.

Perhaps the only species cultivated, ***D. atrovirens*** (syn. *Carludovica atrovirens*) is a handsome, stemless evergreen. Its leaves have stalks up to 1 foot or rarely 1½ feet long, and dark green blades divided for two-thirds of their lengths into narrow- or more broadly-lanceolate lobes. The flower stalks are thickened considerably in their upper parts. The cylindrical spadixes are about 1½ inches long and have at their bases two pale greenish-white spathes. The native home of this species is not known.

Garden and Landscape Uses and Cultivation. These are as for *Carludovica*.

DICRANOSTIGMA (Dicrano-stígma). Closely related to the horned-poppy (*Glaucium*) and to the greater celandine (*Chelidonium*), this genus of the poppy family PAPAVERACEAE, indigenous to western China and the Himalayas, contains two or three species. It is distinguished from *Glaucium* by its seed capsules being always under four inches long, by its seeds not being embedded in a spongy matrix, and by its petals spreading widely or curving only slightly to form shallow, saucer-shaped flowers. The blooms are considerably larger than those of *Chelidonium.* Dicranostigmas are hardy herbaceous perennials with light green, pinnately-divided basal and stem leaves and erect stems bearing at the ends of their branches upturned poppy-like yellow blooms. The flowers have two early deciduous sepals, four petals, numerous stamens, and more or less three-lobed stigmas. The name is from the Greek *dikranos,* two-branched, and *stigma.*

The most commonly grown species, usually misidentified in gardens as *Dicranostigma franchetianum,* is *D. leptopodum.* Interestingly enough, it is not unusual to find the true *D. franchetianum* masquerading under the name *Glaucium vitellianum,* which is a synonym for *G. oxylobum.* Easily observable differences between *D. leptopodum* and *D. franchetianum* are the smaller blooms and linear seed pods not tapering at the ends, of the former, and the almost or quite 2-inch-wide flowers and tapered seed pods of the latter. The flowers of ***D. leptopodum,*** ¾ inch to 1¼ inches in diameter, are on plants ordinarily not more than 1 foot in height; the other species, ***D. franchetianum,*** is up to 3 feet tall. The blooms of both are yellow.

Dicranostigma franchetianum

Garden Uses and Cultivation. These plants are most appropriate for informal areas, low-growing *D. leptopodum* for rock gardens. They give no trouble in any ordinary garden soil that is not excessively dry and are easy to raise from seed or by division in spring. A little shade from very strong sun is beneficial, but not essential, provided the soil is fairly moist. Unlike many members of the poppy family they can be transplanted easily and successfully. Spacing of about 1 foot between individuals is satisfactory.

DICTAMNUS (Dictám-nus)—Gas Plant, Burning Bush, Fraxinella, Dittany. A peculiarity of this plant that never fails to interest children and curious adults is responsible for two of its common names. If on a calm, sultry evening a flame is brought close to the flower stem just beneath the cluster of blooms a volatile "gas" given off from the plant ignites in an easily seen flash of flame. The gas plant, burning bush, fraxinella, or dittany, the only species of its genus, belongs in the rue family RUTACEAE. As a native it extends from southern Europe to northern China. Its name, an ancient Greek one, is believed to allude to its foliage resembling that of the ash tree.

Dictamnus albus

The gas plant (***Dictamnus albus*** syn. *D. fraxinella*), is a sturdy, hardy, deciduous, herbaceous perennial; it is in fact one of the most perennial of all perennials. Specimens have been known to flourish for more than fifty years without transplanting. Longevity and tenacity to life of that order will satisfy most of us so far as our garden plants are concerned. The gas plant from a somewhat woody rootstock sends up erect stems to a height of 2 or 3 feet. Its dark green leaves, alternate and pinnate, are composed of nine or eleven toothed, ovate leaflets 1 inch to 3 inches long. As with many members of the rue family, the foliage is aromatic and is marked with tiny translucent dots that are most easily seen against the light. It is lemon-scented. The asymmetrical flowers, about 1 inch long and showy, are white, pink, old rose, or purplish, and in erect terminal racemes. They have five inconspicuous, narrow sepals and five pointed, narrow petals, the lowest one bent downward as are the ten stamens. The latter usually curve upward at their ends and are as long or longer than the petals. The fruits are capsules. This is a variable species. Especially distinct forms include one with white flowers, *D. a. rubra,* with purplish rose-red blooms, *D. a. purpureus,* with dark red flowers, and *D. a. caucasicus,* which is a bigger plant with longer racemes of flowers.

Garden and Landscape Uses. Provided it is accorded an acceptable location, the gas plant is virtually indestructable. Yet it is not aggressive. It covets not its neighbors' territory and increases in size slowly. It is a thoroughly dependable, backbone-of-the-flower-border type of plant that should find a place in most gardens. Not the least of its attractive features is its clean-looking, quite beautiful, dark green glossy foliage that remains in good condi-

Dictamnus albus (flowers)

Dictamnus albus caucasicus

tion throughout the season and forms a nicely patterned, substantial clump, topped in June or July by spikelike racemes of bloom. If these are not removed they are succeeded by attractive seed pods. In olden times the latter were employed as small candle holders. They are useful in dried flower arrangements. The blooms are of minor importance as cut flowers. One thing the gas plant is not, it is not a plant for the maker of instant gardens who wants immediate results from his (or her) planting efforts, or for those insecure and uncertain gardeners who are impelled to dig up plants and move them around frequently. It takes time, three or four years perhaps, for newly planted young gas plants to amount to much, but once settled and established they are rewarding for unpredictable decades. They need full sun and well-drained, fertile soil and tolerate occasional periods of dryness better than constant wetness.

Cultivation. Because this plant should remain put for long years it is important to prepare the soil thoroughly before planting. Spade it deeply and mix in sufficient compost, old rotted manure, humus, or peat moss, as well as a generous lacing of bonemeal, to assure a good start and to provide for as long a future as possible. Later amendments will be restricted to surface applications. The plants set out will normally be two or three years old and should have a ball of soil retained about their roots; older specimens do not usually recover well from transplanting. They should be allowed ample room, 3 to 4 feet between individuals, and even further from other strong-growing perennials. To show their full beauty gas plants need space. Once established the plants can be left pretty much on their own.

DICTYOGRAMME JAPONICA is *Coniogramme japonica*.

DICTYOSPERMA (Dictyo-spérma)—Princess Palm. Two or three species of nonspiny, feather-leaved palms of the palm family PALMAE constitute *Dictyosperma*. They are natives of the Mascarene Islands. The name is from the Greek *diktyon*, a net, and *sperma*, seed, and refers to the netted seed coats.

Princess palms have a solitary trunk 40 to 50 feet in height and 6 to 8 inches in diameter, and a beautiful crown of spreading, curved leaves that are 10 to 12 feet long and have on each side of their center axes fifty to seventy long, narrow, drooping leaflets. The sheathing lower parts of the leafstalks surround each other to form a gray-green crownshaft that looks like a continuation of the trunk and is 2½ to 3 feet long. The flower clusters develop from below the crownshaft and soon shed the two spathes that at first enclose them. The reddish, orange, or yellowish flowers are in groups of three on the yellowish branches of the clusters (the branches become green when in fruit). Each group of flowers consists of a female flanked by two males, the latter with six stamens. The cartridge-shaped, blue-black fruits, about ½ inch long, are seated in shallow, green cups.

Quite common in cultivation in the tropics and in places in the subtropics not subject to frost, ***D. album*** has whitish leafstalks and whitish-green leaf veins. It is a rather variable plant. A sort with young leaves dark green with reddish margins is sometimes called *D. a. rubrum*, a shorter, more slender kind with yellow or orange-yellow leaf stalks and veins is known as *D. a. aureum*.

Dictyosperma album rubrum in Hope Botanic Gardens, Jamaica, West Indies

Garden and Landscape Uses. These are beautiful palms for planting in groups or singly in full sun in any ordinary garden soil not excessively dry. They are useful as decorative pot plants and for growing in greenhouses and conservatories.

Dictyosperma album aureum

Cultivation. No special difficulties attend the cultivation of princess palms. They are propagated by fresh seed sown in sandy, peaty soil in a temperature of 80°F. Under greenhouse cultivation they need a minimum winter night temperature of 60 to 65°F, a humid atmosphere, shade from strong summer sun, and a rich, porous, soil kept evenly moist, but never waterlogged. Well-rooted specimens benefit from weekly or biweekly applications of dilute liquid fertilizer. For more information see Palms.

DIDIEREA (Didièr-ea). The two species of this curious genus of the remarkable didierea family DIDIEREACEAE are endemic to Madagascar. Their name commemorates Alfred Grandidier, editor of a comprehensive natural history of Madagascar, who discovered the genus about 1880.

In aspect didiereas resemble certain cactuses and cactus-like euphorbias. The relationship of their family to other plant families has long been a matter of speculation and sometimes dispute among botanists. That both species of *Didierea* have been successfully grafted onto understocks of the cactus genera *Pereskia*, *Pereskiopsis*, and *Trichocereus* suggests a near relationship between the didierea and the cactus families. This is one of the very few authenticated instances of grafting proving successful between members of different plant families.

Didiereas have succulent stems with clusters of sometimes solitary, prominent spines sprouting from spirally disposed leaf cushions much like the areoles of cactuses. The tufts of alternate leaves and the flowers also develop from the younger cushions. The tiny blooms are unisexual. They have two sepals and four petals. The males have eight or ten stamens, the females a style with an expanded, lobed stigma. The fruits are small, dry, one-seeded nutlets not enclosed by the calyx.

The first species discovered, ***D. madagascariensis*** is a variable tree 12 to 25 feet in height, branched or not. Its spines are

Didierea madagascariensis

½ inch to 1½ inches long, solitary or in groups of up to twenty. The leaves are obovate, ¾ to 1 inch long and approximately ¼ inch wide. The species previously identified as *D. mirabilis* is thought to be a juvenile form of *D. madagascariensis.* Distinctive ***D. trollii,*** when young, clothes the ground with a cushion of horizonal, thorny branches. Much later there grow from the center of this maze erect stems several feet tall. When they approach full height these develop numerous horizontal branches. The long-elliptic leaves are ⅓ inch to 1⅓ inches long. Other plants previously named *Didierea* belong in *Alluaudia.*

Didierea trollii

Garden Uses and Cultivation. These plants, so remindful of fouquierias, have much the same uses in landscapes and respond to the same care. They belong in collections of cactuses and other succulents, outdoors in warm desert regions and in greenhouses. They need well-drained sandy soil with a fair organic content and exposure to full sun. If there is any danger of frost winter protection is advisable. Moderate watering is required while the plants are in leaf, none when they are resting. Propagation is by seeds, cuttings, and grafting.

DIDIEREACEAE—Didierea Family. This family of dicotyledons, peculiar in the wild to Madagascar, comprises eleven species distributed in four genera. Its members are succulent, often spiny trees and shrubs with opposite, undivided leaves. In clusters, the flowers are small and unisexual, with the sexes on different individuals. They have two petal-like sepals, four petals, and the males eight to ten stamens. Female flowers are similar except that the stamens are represented by staminodes (nonfunctional stamens), and there is one style tipped with a usually enlarged three- or four-lobed stigma. The dry, three-angled fruits contain a single seed. Genera cultivated are *Alluaudia, Decaryia,* and *Didierea.* For comments on the relationship of this family to the cactus family CACTACEAE see Didierea.

DIDYMAOTUS (Didy-maòtus). One stemless, very succulent, nonhardy perennial previously named *Mesembryanthemum lapidiforme* is the only representative of this South African genus of the carpetweed family AIZOACEAE. Closely related to *Gibbaeum,* it differs in technical details of its flowers and in that they arise from between pairs of little bracts. Usually there are two pairs of bracts and two blooms. This explains the name, from the Greek *didymos,* twin, and *aotus,* a flower.

In appearance ***Didymaotus lapidiformis*** looks much like the more common *Pleiospilos bolusii,* but its flowers and manner of producing them are quite different. So also are its very small, smooth seeds, which, as in other members of the family, are borne in capsules. This stonelike plant is up to about 1½ inches high and has a pair of reddish-brown, purplish, or grayish gray-green, very fleshy, flat-topped leaves, keeled on their backs and joined at their bases, and up to 2 inches long. In the wild this species grows under desert conditions of scanty rainfall and blooms and makes its new growth in the South African spring. As new leaves grow the old ones shrivel and die, and at the same time new pairs of bracts develop with, concealed within them, the flower buds that will produce the next year's bloom. The flowers are shining white, usually tinged with purplish-rose at their centers, and have purple-red stamens. They are about 1½ inches across. They resemble those of mesembryanthemums, have many narrow petals, and six stigmas, and originate from each side of the pair of leaves.

Garden Uses and Cultivation. This species is considered a treasure by collectors of choice succulents. Unfortunately it has proved resistant to cultivation and considerably more failures than permanent successes with it are reported. About all that can be suggested is that it be afforded the conditions and care that suit *Pleiospilos, Lithops,* and similar plants of the *Mesembryanthemum* relationship. Its resting season is early summer. For further information see Succulents.

DIDYMOCARPUS (Didymo-càrpus). The Asian genus *Didymocarpus* belongs in the gesneria family GESNERIACEAE and consists of about 120 species. The name, alluding to the paired fruit capsules, is derived from the Greek *didymos,* twin, and *karpos,* fruit. From *Chirita,* in which by some authorities it is included, it differs chiefly in its flowers having a lobeless or scarcely-lobed stigma.

The sorts of this genus are nonhardy, stemmed or stemless herbaceous perennials with hairy, often wrinkled, round-toothed, generally heart-shaped leaves. The flowers, on fork-branched stalks, have a usually five-lobed calyx, a funnel-shaped, asymmetrical corolla with a swollen throat and five spreading lobes (petals), two stamens, two or three staminodes (nonfunctional stamens), and one style ending in an oblique, heart-shaped stigma scarcely or not at all lobed. The fruits are long, slender capsules.

Kinds cultivated include these: ***D. mortonii,*** of the Himalayan region, has hairy stems 2 to 6 inches long with up to three pairs of opposite, heart-shaped, sharply-toothed leaves up to 6 inches long and two-thirds as broad, and hairy on their under surfaces. On branched stalks, the flowers have corollas with slender, whitish tubes almost ½ inch long and a face ½ inch across of spreading, purple petals. ***D. podocarpus,*** of the Himalayan region, has hairy stems up to 9 inches high, each with two pairs of minutely-hairy, coarsely-round-toothed, ovate to elliptic leaves up to 6 inches long by 4 inches wide. On stalks 1½ to 2½ inches long from the leaf axils, the 1-inch-long flowers are purple. ***D. pulcher,*** of the Himalayan region, is up to 1 foot tall. It has oblong to elliptic, toothed or toothless leaves up to 4½ inches long by approximately one-half as wide, and with minutely-hairy upper surfaces. From the stem ends and from the leaf axils, the flowering stalks have colored bracts. The blooms, dark purple and hairy on their outsides, are 1 inch to 1¼ inches long.

Less well known are *D. biserratus* and *D. kerri,* both natives of Thailand. A somewhat succulent inhabitant of rocks in forests, ***D. biserratus*** has softly-hairy, pendulous stems up to 6 inches long and similarly hairy, blunt broad-ovate leaves.

Didymocarpus biserratus

Its dark purple flowers, 1¼ to 1½ inches long, are few together at the tops of slender stalks 1½ to 3½ inches long. From 3 to 8 inches tall and stemless or stemmed, ***D. kerri*** has broad-ovate or broad-elliptic to nearly circular, crenately-toothed, hairy leaves up to 2 inches long. Its ¾ inch to 1½ inch-long dark purple flowers are in loose panicle-like clusters displayed on slender stalks above the foliage.

Garden Uses and Cultivation. These are as for *Chirita.*

Didymocarpus kerri

DIDYMOCHLAENA (Didymo-chlaèna). Consisting of one species, *Didymochlaena* is a fern of the aspidium family ASPIDIACEAE. Its name, derived from the Greek *didymos,* twin, and *chlaina,* a cloak, alludes to the covers (indusia) of the clusters of spore capsules.

Native of tropical Africa and Central and South America, ***D. truncatula*** has an upright stem and crowded tufts of twice-pinnate leaves 4 to 6 feet long and up to 2 feet wide. They have narrow-linear leaflets with ultimate segments blunt, oblongish, ¾ to 1 inch long with slightly wavy margins. The few more or less elliptic clusters of spore capsules are in one row at right angles to the upper edge of the segments, except for usually one cluster near the tip of the lower edge.

Garden and Landscape Uses and Cultivation. This is attractive in collections of tropical ferns. As long as its roots are not allowed to dry it prospers with little care in well-drained soil that contains an abundance of organic matter, such as compost, leaf mold, or peat moss. Propagation is by spores and division. For more information see Ferns.

DIDYMOSPERMA. See Arenga.

DIE-BACK. The death of the tips or sometimes longer terminal portions of shoots that normally should be alive and healthy is commonly known as die-back. The term describes a condition or symptom rather than any specific cause. Die-back may come from infection by bacteria, fungi, or viruses, from infestations of such sucking insects as aphids and mealybugs, from exposure to excessively low or high temperatures, especially if the relative humidity is low, as a result of drought, and from exposure to drying winds. Mending measures consist of cutting out affected parts and correcting underlying causes, or in the case of virus-caused die-back, discarding all affected plants. Trees and shrubs, deciduous and evergreen, are most subject to die-back.

DIEFFENBACHIA (Dieffen-báchia)—Dumb Cane, Mother-In-Law Plant. Tropical South American, Central American, and West Indian relatives of the jack-in-the-pulpit (*Arisaema*) and calla-lily (*Zantedeschia*), the about thirty species of *Dieffenbachia* belong to the arum family ARACEAE. They include several familiar as indoor ornamentals in the north as well as in warmer climates, and much used for outdoor landscaping in the tropics and warm subtropics.

Dieffenbachias with other tropical plants

The vernacular names dumb cane and mother-in-law plant, the latter of rather peevish implication, allude to the stems and foliage if chewed being painfully irritating to the tongue and mouth, possibly to the extent that speech may become difficult or impossible. The cause is largely mechanical, the result of sharp, almost glasslike slivers of calcium oxalate contained in the plant becoming embedded in the flesh, but toxic proteins are also involved. The botanical name commemorates the German botanist and physician Johann Friedrich Dieffenbach, who died in 1847.

Dieffenbachias are evergreen perennials of noble aspect with thick, somewhat woody, erect, often clustered stems, sometimes horizontal at their bases, and leafy chiefly in their upper parts. They are ringed in canelike fashion and prominently marked with scars indicating the places from which leaves have fallen. The alternate, commonly paddle-shaped, oblong to ovate, lobeless and toothless leaves have stout stalks usually not longer than the blades, the latter with prominent midribs. The bases of the leafstalks clasp or sheathe the stems, and like them contain milky or yellowish juice, which can be highly irritating. The inflorescences, commonly called

An inflorescence of a *Dieffenbachia*

"flowers," are of the familiar calla-lily pattern and come from the leaf axils on stalks shorter than the leafstalks. They consist of an erect, clublike spadix (spike) of very crowded, small true flowers, from the base of which arises a spathe (petal-like or leaflike bract) that encloses the lower part of the spadix which has only female flowers, and provides a background foil for the upper part of the spadix which has only male flowers. The males are separated from the females by a brief section of naked spadix. Male flowers have four united stamens, females four or five staminodes (nonfunctional stamens) and a two- or three-lobed, stalkless stigma. The fruits are berries, bright red or orange-red at maturity. Most cultivated dieffenbachias are notable for the handsome variegations of their foliage, and even in the wild, variants from the species having leaves blotched with white or creamy-yellow frequently occur. The most commonly cultivated kinds are variants of *D. maculata* and *D. seguine.*

Native to South America, highly variable ***D. maculata*** (syn. *D. picta*) has stems often more or less prostrate and rooting at their bases, then erect and 3 to 4 feet tall. They are up to 1 inch in diameter. The oblong, oblong-elliptic, or oblong-lanceolate leaves have broadly-grooved stalks and blades often irregularly-blotched and spotted with white or yellowish-white and with fifteen to twenty chief lateral veins from each side of the midrib. Most cultivated dieffenbachias with white- or cream-spotted foliage are referable to this species. Notable among them are *D. m. baraquiniana,* oblong to lanceolate leaves with white stalks, conspicuous white midribs, and a few white spots; *D. m.* 'Janet Weidner', tall and with leaves freely marbled and spotted with white; *D. m. jenmannii,* tall and slender, its narrowly-oblong leaves clearly marked with white bands angling outward from the mid-vein; *D. m. lancifolium* (syn. *D. eburnea*), with red-stalked, slender-oblong leaves narrowed at both ends and thickly spotted with white; *D. m.* 'Rudolph Roehrs', which has pale yellowish-green to chartreuse leaves variously blotched with ivory-white and narrowly and irregularly bordered with deep green and with dark green midribs, or occasionally with a whole leaf or part of one reverting to the typical *D. maculata* pattern of dark green spotted with white; *D. m. shuttleworthii,* with oblong-lanceolate leaves with pea-green blotches and a narrow irregular grayish band along the midrib; *D. m. superba,* compact, with dark green leaves very freely blotched and spotted creamy-white; and *D. m. viridis,* with rather thin, plain green, pointed-oblong leaves.

Others that may belong here include *D.* 'Exotica' (syn. *D.* 'Arvida') and *D. leonii.* Slender and compact, the first has smallish

Dieffenbachia maculata

Dieffenbachia maculata 'Janet Weidner'

Dieffenbachia maculata jenmannii

Dieffenbachia maculata 'Rudolph Roehrs'

Dieffenbachia 'Exotica'

leaves irregularly and very freely blotched with ivory-white. The leaves of the other are elliptic-oblong, dark green, satiny and have faintly cream midribs generously spotted and blotched with yellowish-white.

Hybrids of *D. maculata* include ***D. bausei,*** its parents *D. maculata* and *D. weiri.* This has pale yellowish-green leaves nar-

Dieffenbachia bausei

rowly edged with dark green and spotted with dark green and white. A hybrid between *D. maculata* and *D. wallisii,* named ***D. memoria-corsii,*** has grayish-green leaves with dark veins, green blotches, and a few white spots. Leaves with ivory-white midribs and spots are characteristic of ***D. splendens,*** a hybrid between *D. maculata* and *D. leopoldii.*

Differing from *D. maculata* in its leafstalks being not or scarcely channeled and in having leaf blades generally with nine to fifteen side veins extending from the midrib, very variable ***D. seguine*** has erect stems 3 to 6 feet tall. Its ovate-oblong to oblong leaves have stalks streaked and dotted with white, and blades paler beneath than above, sometimes spotted, and with markedly depressed veins. Varieties of this are *D. s. liturata,* with large, broad, dark green leaves irregularly narrowly banded along their centers with yellowish-green, and *D. s. nobilis,* with white-striped stalked leaves with broad blades and irregular patches of yellowish-green.

Additional sorts include these: ***D. amoena,*** as known in gardens, but different from the species to which that name

Dieffenbachia amoena

rightfully belongs is up to 6 feet tall or taller with leaves with elliptic-oblong blades 1½ feet or so long that are dark green irregularly zoned with creamy-white along the side veins. ***D. bowmannii,*** of Brazil, has narrowly-ovate-elliptic leaves up to 2 feet long with light and dark green blotches paralleled with the side veins and with deeply-channeled stalks. ***D. chelsonii,*** of Colombia, has broad-oblong satiny green leaves clearly relieved by a broadish, irregular central stripe of gray-green and many irregular greenish-yellow blotches. ***D. daguensis,*** up to 4 feet tall or taller, has firm, lustrous, leathery, short-stalked plain green leaves with raised midribs and depressed side veins. It is native to Colombia. ***D. delecta,*** also from Colombia, as known in cultivation is distinct because of its narrowly-elliptic, pendulous, grass-green leaves enhanced with an irregular lengthwise central vein of pale yellow. ***D. fosteri*** is a horticultural name applied to a

Dieffenbachia fosteri

botanically unidentified sort, believed to be native to Costa Rica, that is similar to *D. oerstedii* but is smaller and has leaves with a satiny sheen. ***D. fournieri,*** of Colombia, is robust, has broad-oblong-lanceolate, dark green leaves with blades up to 1¼ feet long, with pale green and ivory-white mid-veins and spots and blotches of the same hue paralleling the side veins. ***D. hoffmannii*** is a horticultural name for a sort with satiny green leaves the blades

Dieffenbachia hoffmannii

and stalks of which are abundantly marbled with white and have white midribs. ***D. imperialis*** is a tall native of Peru with ovate to broad-elliptic, dark green leaves striped irregularly lengthwise down their centers with silvery-gray and more or less blotched with pale yellow. ***D. leopoldii,*** of Costa Rica, has broad-elliptic to ovate, dark green, satiny leaves with ivory-white midribs. ***D. longispatha,*** of Panama and Colombia, has plain green leaves with oblong-lanceolate blades up to nearly 2 feet in length. ***D. oerstedii,*** of Mexico and Central America, has plain dark green, oblong-ovate, slender-stalked leaves. *D. o. variegata* is distinguished by its leaves hav-

Dieffenbachia oerstedii variegata

ing very distinct ivory-white midribs. ***D. parlatorei,*** of Colombia, is up to about 3 feet tall, has leaves with broadly-oblanceolate blades up to 2 feet long, which are sometimes spotted yellowish-green. *D. p. marmorea* has leaves with clearly defined white midribs and silvery blotches especially noticeable on young leaves. ***D. pittieri,*** endemic to Panama, has oblong leaves with pale-based stalks, sometimes irregularly blotched with pale green or white.

Garden and Landscape Uses. In the humid tropics and warm subtropics dieffenbachias are among the most esteemed decorative foliage plants for massing in shaded areas and for underplantings. They are also highly regarded for pots and tubs for patio, porch, and room decoration. In temperate regions they are much admired pot and tub plants widely grown as indoor ornamentals and in greenhouses.

Cultivation. Where conditions favor their growth outdoors dieffenbachias luxuriate with little attention. They prefer moderately moist soil that contains an abundance of compost, leaf mold, or other decayed organic matter, and need shade. In greenhouses they succeed with little difficulty if the atmosphere is humid and the minimum night temperature in winter is 60 to 65°F and that by day, and at other seasons by night and day, higher. A cozy, warm, humid atmosphere produces the best dieffenbachias, yet if out of drafts and where there is reasonable light, but not exposure to strong sun, they tolerate drier room conditions remarkably well and are durable.

Hearty, fertile, loamy soil, well supplied with organic matter in the form of leaf mold, peat moss, or rotted manure, and sufficiently porous to assure the free passage of water is to the liking of dieffenbachias in containers. It should be packed fairly firmly, and watered each time it approaches dryness, but not before, by saturating it thoroughly. Well-rooted specimens benefit from moderate fertilization, but more than this is likely to stimulate vigorous growth, with accompanying loss of lower leaves, that may not be desirable with container specimens. Repotting is needed by sizable plants at quite infrequent intervals, more often by small ones. It may be done at any time, spring being the most favorable season. It is usual to set three plants or more in large containers, thus assuring from the beginning a full, well-foliaged specimen. Increase is by division, terminal leafy cuttings, and by cuttings 2 or 3 inches long of sections of the stems. Air layering is also a certain and easy method of increase and of securing shorter plants from specimens that have become too tall and leggy.

Three plants potted in one container soon form an attractive specimen

Dieffenbachias from stem cuttings: (a) Cut stout stems into short sections

(b) Plant the sections horizontally in sand, perlite, or vermiculite

(c) When leaves and roots have developed, pot the cuttings in sandy soil

Air layering is a sure method of propagating dieffenbachias

DIERAMA (Dier-àma)—Wand Flower. This African genus of the iris family IRIDACEAE is cultivated outdoors in California and other mild climates, and in greenhouses. The total number of species differs, according to opinions of individual botanists, from two to twenty-five. A realistic appraisal is four to six. The name is from the Greek *dierama,* a funnel, and alludes to the shape of the flowers.

Dieramas are cormous plants, that is they have solid, underground, bulblike organs called corms similar to those of gladioluses and crocuses. From closely related *Ixia* they differ in the branches of their styles being wedge- or club-shaped, rather than threadlike, and from allied *Streptanthera* in their blooms being more numerous and not having spirally twisted anthers. Dieramas bloom in summer or in mild climates in late spring. They have deciduous to nearly evergreen foliage. Their more or less spherical corms are 1 inch or more in diameter. The stems are slender. The leaves are rigid, narrow, grasslike, and two-ranked. The attractive flowers are in pendulous, branching spikes. They have short perianth tubes and six perianth segments (petals, or more correctly, tepals) of nearly equal size and shape. There are three short stamens. The style has three very short, spreading branches. The fruits are capsules.

The most commonly cultivated dieramas are ***D. pulcherrimum*** and its selected horticultural varieties and hybrids. These have stems 4 to 6 feet tall, rigid leaves 1½ to 2 feet long by about ½ wide, and flowers with perianth tubes about ½ inch long. The petals, 1½ inches long, in the species are bright wine-red, but in varieties and hybrids vary from white to deep rose-pink, and reddish-violet. Rather less vigorous, ***D. pendulum*** has bluish-white, pink, or rosy-purple blooms with perianth tubes under ½ inch and petals about ¾ inch long. The leaves are up to 2 feet, and the wiry flower stalks up to 3 feet long. Hybrids between *D. pulcherrimum* and *D. pendulum* that combine the characteristics of both have been grouped as *D. hybridum*. These have occurred spontaneously at the Strybing Arboretum, San Francisco, and probably elsewhere.

Dierama pendulum

Garden and Landscape Uses. Dieramas are about as hardy as gladioluses. If their corms (bulbs) do not freeze, they survive outdoors as perennials. Where winters are too severe for this they may be cultivated in pots in greenhouses. In climates as mild or milder than that of Washington, D.C., they are useful in flower borders and are especially effective when planted near ponds or pools. By the waterside their graceful arching stalks of attractive flowers show to special advantage. They are useful for cutting.

Cultivation. A freely drained, fairly moist, sandy, fertile soil best suits dieramas. They need full sun. The bulbs (corms) should be planted in fall, about three inches deep and approximately the same distance apart. They may be lifted, separated, and replanted about every third year, but because they are somewhat resentful of transplanting some gardeners prefer to sow seeds directly where the plants are to be established. In some gardens self-sown seedlings develop freely.

In greenhouses the bulbs are planted in September about 3 inches apart in well-drained 6- or 7-inch pots of porous, fertile soil. They are grown in full sun where the night temperature is 45 to 50°F. Day temperatures may be five to ten degrees higher. Water sparsely at first, but with increasing frequency as the roots permeate the soil. After flowering is through and the foliage begins to wither naturally, water less frequently and finally stop completely. During their dormant period the bulbs are kept quite dry. In greenhouses repotting in new soil is done annually.

Pests. Mice and gophers are given to destroying the bulbs of dieramas. To circumvent this it may be necessary to plant the corms completely enclosed in wire baskets.

DIERVILLA (Diervíl-la) — Bush-Honeysuckle. The genus *Diervilla*, of the honeysuckle family CAPRIFOLIACEAE, differs from nearly related *Weigela* (which at one time was included in it) in its flowers being always yellow, decidedly two-lipped, and borne on shoots of the current season's growth. Those of *Weigela*, white, pink, crimson, or in *W. middendorfiana* yellow, are much more symmetrical and are carried on short shoots that sprout from the branches of the previous year's growth. There are three species of *Diervilla*, natives of eastern North America. Weigelas are all from eastern Asia. The name commemorates a French surgeon, M. Dierville, who traveled in Canada in 1699 and 1700 and introduced *D. lonicera* to Europe.

Diervillas are deciduous shrubs that spread by underground runners or stolons. Their leaves are opposite and toothed. Their flowers, in small axillary clusters, grouped in terminal panicles, come in summer. They are small, funnel-shaped, five-lobed and have five protruding stamens and a protruding style. The fruits are slender capsules containing numerous small seeds.

Most attractive in bloom, ***D. sessilifolia***, of North Carolina, Georgia, Tennessee, and Alabama, is hardy in southern New England. It grows to a height of up to 4 feet, has four-angled branches, nearly glabrous, nearly stalkless, ovate-lanceolate leaves 2 to 6 inches long, and sulfur-yellow flowers in groups of three to seven, often crowded in terminal panicles. The petals are of nearly equal size. One is somewhat hairy. Another species with the same natural geographical range, ***D. rivularis*** attains a height of 6 feet and has round, densely-pubescent branches, almost stalkless, ovate to oblong-lanceolate leaves hairy on both surfaces and up to 3 inches long, and crowded terminal panicles of lemon-yellow flowers. This is about as hardy as *D. sessilifolia*.

Diervilla sessilifolia

The hardiest kind is ***D. lonicera***, which occurs as a native from Newfoundland to Saskatchewan and North Carolina. This is a spreading shrub 2 to 4 feet tall, with nearly round, hairless branches and, except for a marginal fringe of hairs, nearly glabrous, ovate or ovate-oblong, distinctly-stalked leaves up to 4 inches long. The flowers, in groups of three to five, are yellow at first, but darken as they age.

Garden and Landscape Uses. Because their blooms are small and not very conspicuous diervillas have but small attraction as flowering ornamentals. They cannot compete favorably with other more showy shrubs that bloom at the same season. They are, however, useful for stabilizing soil on banks and slopes and for planting in areas where shrubby plants that renew and maintain themselves by suckering are needed. They do well in dryish soil.

Cultivation. Diervillas are readily increased by suckers dug from around the parent plants and transplanted, by leafy cuttings taken in summer, and by seeds. They respond well to ordinary garden soil in sun or light shade. Planting is best done in early spring or early fall. Pruning, done as soon as the flowers are past, consists of removing old, weak, and crowded branches and shortening others to just above the vigorous new shoots.

DIETES (Diè-tes). So closely related to *Moraea* is *Dietes* that some authorities include it as a subgroup of that genus. It belongs in the iris family IRIDACEAE and comprises three species native to South Africa, and one to Lord Howe Island and Australia. Its name derives from the Greek *dis*, double.

The chief differences between *Dietes* and *Moraea* are that the former has short, creeping rhizomes and evergreen foliage, whereas moraeas have bulblike organs called corms and deciduous foliage. The leaves of *Dietes* are narrowly-sword-shaped or iris-like. The blooms, very like those of irises, have tubeless perianths with six spreading petals (more correctly tepals), the outer three obovate and somewhat reflexed, the others smaller, but similar. The three stamens have anthers resembling those of *Iris,* and stalks joined to each other, at least toward their bases. The styles are divided into three petal-like branches with stigmatic surfaces in their undersides. The fruits are angled capsules. Together, *Dietes* and *Moraea* are the Southern Hemisphere counterpart of the Northern Hemisphere genus *Iris.* From the latter they differ in no single feature. Most commonly irises have a definite perianth tube above the ovary and their stamens are separate.

Best known are *D. vegeta* (syns. *D. iridoides, D. catenulata, Moraea iridoides*) and its variety *D. v. johnsonii.* The latter is taller and has larger blooms than the species. Native of South Africa, ***D. vegeta*** is about 2 feet tall. Its numerous stiff, narrow, erect

Dietes vegeta

leaves are in crowded fans. The flower stalks, furnished with sheathing bracts, somewhat overtop the foliage. From their spathes (floral bracts) emerge and open in succession three or four blooms. About 3 inches wide these have white petals with yellow marks at the bases of the outer ones. The crests of the styles are lilac-blue. In *D. v. prolongata* the blooms are pure white. A curious habit of *D. vegeta* is that all the plants over large areas open their blooms on the same day. This is repeated several times at intervals of one to three weeks. Some believe that the cause of this uniform periodicity is changes in night temperatures.

Up to 4 feet tall and with erect, not two-ranked leaves as long, South African ***D. grandiflora*** has white flowers 3 to 4½ inches wide. They have orange-yellow to brownish spots at the bases of the petals and yellow-bearded keels on the three outer petals. The style is decorated with bands of violet.

Bright yellow blooms with maroon to purplish-brown blotches at the bases of the outer petals and often with spots of the same color on the inner ones, are a feature

Dietes vegeta prolongata

of South African ***D. bicolor*** (syn. *Moraea bicolor*). The flowers, about 2 inches in diameter and several to many on branching stalks somewhat longer than the foliage, generally resemble those of *D. vegeta.*

The giant of the group, ***D. robinsoniana*** (syn. *Moraea robinsoniana*), from Lord Howe Island and Australia, when not in bloom looks much like New Zealand-flax (*Phormium tenax*), but has paler green foliage. Its leaves are up to 5 or 6 feet long by about 2 inches wide. This noble species has fragrant flowers up to 4 inches in diameter, but in contrast with the massive foliage they tend to look smaller. They are on branched stalks and are creamy-white with reddish-yellow basal blotches on the outer petals.

Dietes grandiflora

Oakhurst hybrids raised in California between *D. bicolor* and *D. vegeta* are larger than either parent. Compact, with grassy, green foliage overtopped by tall stalks, and blooms resembling those of *D. bicolor* in form, but larger and of firmer substance, they bloom freely through most of the year. Their flowers are cream to yellowish with yellow to orange basal blotches. Named varieties include 'Lemon Drops', 'Orange Drops', and 'Contrast'.

Garden and Landscape Uses. Not hardy in the north, members of this genus are sometimes grown in greenhouses and conservatories. They are highly satisfactory outdoor plants in California and the south and in other regions subject to little frost, especially where summers are dry or dryish. They lend themselves for use in beds, groups, and informal landscapes. They associate pleasingly with architectural features such as steps, walls, and paved areas. Because of its lax flower stalks *D. vegeta* is attractive in hanging baskets. Most tender to cold, and even where hardy usually taking several years before it blooms, *D. robinsoniana* is well worth growing for its imposing foliage alone.

Cultivation. With the possible exception of *D. robinsoniana,* members of this genus are among the easiest of plants to grow. They ask for little in the way of care and provide delightful displays of bloom. Although not fussy about soil, they appreciate one moderately fertile and well-drained, but not arid. Wet, stagnant soils are abhorrent to them. Those of a sandy peaty nature are to their liking. They flourish in sun or part-day shade. Once planted, these plants may remain undisturbed for many years. They benefit from having a complete fertilizer applied each year when new growth begins and from watering regularly during long dry periods. Except with *D. bicolor* and *D. robinsoniana*, the flowering stems are perennial and continue to produce blooms for several years. They must not be cut off after a flush of flowers has faded. If for any reason it is necessary to cut blooms, if the flowering stems are left with one, two or more nodes intact, they will grow new branches that in time bear flowers.

In greenhouses *Dietes* grow well in large pots, tubs, and ground beds. Routine care is minimal and consists chiefly of watering to keep the soil moderately moist, but slightly drier in winter than in summer, and of applying a dilute liquid fertilizer to well-rooted specimens in containers at about biweekly intervals from spring to fall. A sunny location in an airy greenhouse where the winter night temperature is 45 to 50°F and that by day a few degrees higher is ideal. In summer, conditions as nearly as possible like normal outdoor ones are needed. Propagation is easy by division and by seed.

DIGGING. In North America spading is usually the preferred name for the garden operation commonly known as digging in the British Isles. The American gardener spades his garden, the Briton digs his. For more information see Spading. Americans, however, refer to removing trees, shrubs, and sometimes other plants from the ground in readiness for transplanting as digging, for which operation the British gardener is more likely to employ the term lifting.

DIGITALIS (Digit-àlis) — Foxglove. Foxgloves are old-fashioned plants of easy cultivation, natives of Europe and western and central Asia. They number twenty to thirty species and constitute *Digitalis* of the figwort family, SCROPHULARIACEAE. The name, alluding to the shape of the flowers, is derived from the Latin, *digitus*, a finger. In addition to its use as an ornamental the common foxglove is cultivated as the source of the important drug digitalis used in cardiac medicine. Less important sources of this drug are *D. ferruginea*, *D. lanata*, and *D. lutea*.

Foxgloves are rarely small shrubs, generally biennials or herbaceous perennials. Most are hardy. They have alternate, mostly undivided, toothed or toothless leaves, the lower ones usually in rather crowded rosettes. The purple, white, or yellowish, more or less pendent blooms are in long, terminal, often one-sided, branchless or sometimes branched racemes. They have a five-parted calyx, a tubular to bell-shaped somewhat obliquely two-lipped, four-lobed corolla, often constricted near its base, and with the lower lip longer than the upper. There are four usually nonprotruding stamens and one style. The fruits are capsules containing numerous tiny seeds.

Common foxglove (***D. purpurea***) is a biennial that occasionally blooms in its first year or sometimes lives longer than its normal life-span of two seasons. A variable native of nonlimy soils in western Europe, it is divided into three subspecies. Typically *D. purpurea* is 2 to 7 feet tall. It has greenish to whitish, ovate to ovate-lanceolate, wrinkled, more or less hairy leaves, the lower ones long-stalked, those on the stems short-stalked to stalkless and diminishing in size upward. Characteristically in one-sided, branchless or sparingly branched racemes up to 2 feet long or sometimes longer, the 2- to 3-inch-long blooms are typically rosy-purple freely spotted inside with darker purple. The lobes of their calyxes are ovate-lanceolate to elliptic. The other subspecies are *D. p. mariana* (syn. *D. mariana*), native of Spain and Portugal, which has white-hairy, broad-ovate leaves abruptly narrowed at their bases and purple blooms with individual stalks longer than the calyx, and *D. p. heywoodii*, of Portugal, which is densely white-hairy and has white, pink-tinged or yellowish blooms, their individual stalks longer than the calyx. Well-marked horticultural varieties are *D. p. alba*, its flowers white, usually with some yellowish spots; *D. p. campanulata*, in which the uppermost blooms are joined into a very large, many-lobed bell-shaped flower; *D. p. gloxiniaeflora*, a robust variety with long racemes of conspicuously-spotted, wider blooms than those of the typical kind; *D. p. monstrosa*, with more or less double blooms; and *D. p. isabellina*, with buff-yellow flowers. The

Digitalis purpurea

name ***D. lutzii*** applies to a hybrid swarm the flowers of which are crimson to pink.

Closely related to the common foxglove are *D. dubia* and *D. thapsi,* the first endemic to nonlimy earths in the Balearic Islands, the other to limestone soils in Spain and Portugal. From the common foxglove ***D. dubia*** differs in having short-stalked, scarcely toothed or toothless leaves and being under 1 foot high. The other differs from the common foxglove in being wholly covered with yellowish, glandular hairs.

Yellow-flowered foxgloves, muted rather than bright of hue, include *D. grandiflora* (syn. *D. ambigua*), native of Europe and western Asia, *D. lutea* and *D. laevigata,* both of Europe. A biennial or perennial mostly 2 to 3 feet tall, ***D. grandiflora*** has firm, finely toothed, ovate-lanceolate leaves,

Digitalis grandiflora

usually glossy and hairless above, and sparsely hairy on their undersides. They are up to 10 inches long by somewhat over 2 inches wide. The flowers, 1½ to 2 inches long, have pointed-lanceolate calyx lobes.

Digitalis grandiflora, flowers

Its flowers less bell-shaped than those of the last, ***D. lutea*** is a hairless or slightly hairy perennial 2 to 3 feet tall, with oblong-lanceolate, toothed to nearly toothless leaves. Its racemes are of many pale yellow to whitish blooms ¾ to 1 inch long. Their calyxes have pointed-linear-lanceolate lobes fringed with glandular hairs.

Digitalis lutea

The racemes of *D. l. australis* of the Balkan Peninsula are cylindrical rather than one-sided. A natural hybrid between *D. purpurea* and *D. lutea* is named *D. purpurascens.* The yellowish flowers of ***D. laevigata,*** more globose than those of kinds previously discussed and ½ to 1¼ inches long, are veined or otherwise marked with purple-brown. The calyx lobes are pointed-ovate.

White to yellowish-white flowers marked with fine brownish to violet veins are borne by perennial or biennial *D. lanata,* of the Balkan Peninsula and adjacent regions, *D. leucophaea,* native of Greece, and *D. cariensis,* which hails from the Levant. Those of ***D. lanata,*** ¾ inch to 1¼ inches long, are in densely-crowded racemes carried on hairy, often reddish-purple stems to a height of 1 foot to 3 feet. They have glandular-hairy, pointed-lanceolate calyx lobes. From the last, ***D. leucophaea*** differs in its flowers not exceeding and usually under ¾ inch in length. About 3 feet tall, ***D. cariensis*** has stems that in their upper parts are clothed with white, glandular hairs. Its leaves are stalkless, hairless, and lanceolate. The 1-inch-long blooms, in loose, interrupted racemes or spikes, are white except for their finely red-striped, long lower lips.

The rusty foxglove (***D. ferruginea***) is a 1- to 4-feet-tall perennial or biennial native of southern central Europe. It has hairless stems and oblong-lanceolate to lanceolate leaves without hairs or with their undersides slightly hairy. The flowers, about ⅔ inch long and netted with darker veins, are yellowish- to reddish-brown and have the center lobe of the lower lip twice or more as long as the side lobes. The calyx lobes are blunt, ovate to oblong-elliptic. Native of the Balkan Peninsula, ***D. viridiflora*** is a perennial 1½ to 2½ feet tall. It has toothed, oblanceolate to elliptic leaves and crowded racemes of conspicuously-veined, dull greenish-yellow blooms ½ to ¾ inch long.

A shrub or subshrub, woody at least in its lower parts, ***D. obscura*** sprawls or is erect. From 1 foot to 4 feet tall, its stems, densely-leafy above, are without foliage below. The shiny, leathery, toothed or toothless leaves are oblong-lanceolate to lanceolate. The somewhat one-sided racemes are of flowers ¾ to 1 inch long, orange-yellow to brown, spotted or netted with a darker hue. Variety *D. o. laciniata* has narrower sepals than the typical species and usually deeply-toothed leaves. Probably not hardy in the north, this species and its variety are endemic to Spain.

Garden and Landscape Uses. Foxgloves, especially the common foxglove and its varieties, are stately and satisfying plants for grouping in flower borders and planting in naturalistic drifts in light

A border of foxgloves and perennial phloxes

woodlands and at the shaded fringes of tree and shrub plantings. They serve usefully as cut flowers and in cool greenhouses can be gently forced to bloom some weeks ahead of their natural summer seasons. The lower kinds are adaptable for rock gardens. Foxgloves grow most vigorously in deep, mellow soil that contains an abundance of organic matter and that is pleasantly moist without being wet. In the wild many kinds favor limy soils and for these the addition of lime to soils that are on the acid side may be beneficial, but does not seem to be essential.

Cultivation. No special difficulties attend the growing of foxgloves. Perennial kinds can be increased by careful division in early fall or spring and by seed. With the biennials and kinds treated as such, which for practical purposes means most commonly cultivated foxgloves, it is important to sow the seeds fairly early so that strong plants with large rosettes of foliage are had by fall. Too early sowing,

Foxgloves in a cutting garden

however, is likely to result in a fairly large proportion of the plants blooming in the fall of their first season. Mid-May is about the time to sow. Sow in a cold frame or shaded place outdoors, scarcely covering the fine seeds with soil. Keep the seed bed damp. As soon as the seedlings are big enough to handle and before they crowd unduly transplant them to nursery rows allowing 1 foot to 1½ feet between the rows and 9 inches to 1 foot between individuals in the rows. Summer care consists of keeping down weeds and watering generously in dry weather. In fall or early the following spring transplant to the places where the plants are to bloom, taking care to take with each a good ball of soil about the roots. Plants to be forced in greenhouses are lifted in early fall with large balls of roots and are planted in pots or tubs just large enough to contain them. They are then well watered and put in a shaded cold frame where they remain until January or February. In severe climates it is well to bury the pots to their rims in sand, peat moss, or similar material. Forcing is done in a sunny greenhouse where the night temperature is about 50°F and that by day is five to fifteen degrees higher depending upon the brightness of the weather. Ample supplies of water are needed and, when the flower stalks begin to push up, weekly applications of dilute liquid fertilizer.

Transplanting foxgloves to their flowering stations

Foxgloves in nursery rows at the end of their first summer

DILL. See Anethum.

DILLENIA (Dil-lènia)—Elephant-Apple. Of the sixty species of trees that constitute *Dillenia,* one, the elephant-apple, is cultivated in the warmest parts of the continental United States and in Hawaii and other tropical countries. The group, which belongs in the dillenia family DILLENIACEAE, is indigenous from tropical Asia to Australia. Its name commemorates John James Dillenius, an eighteenth-century professor of botany at Oxford, England.

Dillenias are tall and have large, strongly-pinnately-veined leaves and solitary or clustered white or yellow magnolia-like blooms. The blooms are in the leaf axils and have five sepals and the same number of spreading petals. There are many stamens. The fleshy fruits are enclosed by the persistent enlarging calyxes.

The elephant-apple (***D. indica***) is native to tropical Asia and Indonesia. Considered to be the most ornamental of the group, it attains a height of 25 to 40 feet and forms a rounded, broad head with handsome leaves located toward the ends of the branches and branchlets. In always moist climates this tree is evergreen, but where dry seasons occur it will lose its foliage for a period each year. The thick, stiff, leaves, with short, stem-sheathing stalks, are pointed-oblong to oblong-lanceolate and

Dillenia indica

Dillenia indica, in fruit

Dillenia indica (leaves and young fruit)

up to 1 foot in length. They are toothed. The beautiful, solitary, down-facing blooms are 6 to 8 inches or more across. They have thick sepals and obovate white petals. The center of the flower is occupied by a bunch of yellow stamens and an ovary with about twenty radiating styles. The edible fruits are 3 to 5 inches in diameter. They contain many kidney-shaped seeds and in Asia are sometimes used as a substitute for soap. A less well-known species, ***D. philippenensis,*** a native of the Philippine Islands, is generally similar, but has white flowers 5 or 6 inches in diameter with purple inner stamens. Its edible fruits are 2 to 3 inches in diameter. From the bark of this tree a red dye is obtained.

Garden and Landscape Uses and Cultivation. At its best as a single specimen where its handsome foliage and lovely blooms can be admired without being confused with other vegetation, *D. indica* is well adapted as a lawn and shade tree. It grows slowly and succeeds best in fertile, moist ground. It is tender to frost. Propagation is by seed.

DILLENIACEAE—Dillenia Family. A group of ten genera of dicotyledons comprising some 400 species of tropical and subtropical trees, shrubs, woody vines and rarely herbaceous plants compose the dillenia family. They have usually alternate, rarely opposite leaves, and flowers in clusters, often raceme-like, or solitary. Generally symmetrical, the yellow, white, or less commonly red blooms mostly have five each sepals and petals, the former persistent, many stamens, and one or more pistils with usually separate styles. The fruits are follicles or are berry-like.

Some members of the family are useful sources of lumber and tannin, some have medicinal qualities. The most familiar genera are *Dillenia, Hibbertia,* and *Saurauia.*

DILLONARA. This is the name of hybrid orchids the parents of which include *Epidendrum, Laelia,* and *Schomburgkia.*

DILLWYNIA (Dillwýn-ia). Australian and Tasmanian *Dillwynia* of the pea family LEGUMINOSAE consists of fifteen species of heathlike shrubs, none hardy. The name commemorates the English botanist Lewis Weston Dillwyn, who died in 1855.

Dillwynias are well furnished with linear to needle-like or nearly threadlike leaves usually not over 1 inch long and channeled along their upper surfaces rather than with revolute margins like those of related *Aotus.* Their pea-like blooms, usually few together in terminal or axillary racemes, rarely are solitary. They have five-lobed calyxes, the two upper lobes partly joined. Broader than long, and often kidney-shaped, the standard or banner petal is usually yellow and red. The other petals are similarly colored or red. The ten stamens are separate from each other. Short and thick, the style is hooked near its apex. The fruits are pods.

The most commonly cultivated sorts are *D. cinerascens* and *D. ericifolia.* From 2 to 4 feet tall, ***D. cinerascens*** has hairy younger parts. Its mature foliage is gray-hairy to hairless or almost so. The blunt, cylindrical leaves, ¼ to ½ inch long, spread outward from the stems and are often recurved at their tips. The flowers, in terminal clusters of three to eight, are in loose, terminal clusters. Nearly hairless and up to 6 feet tall, ***D. ericifolia*** has spreading, pointed-cylindrical leaves up to ¾ inch long and an abundance of yellow flowers, few together in usually terminal clusters or very short racemes. Each has a red spot at the bottom of the standard petal. The leaves of *D. e. glaberrima* are usually recurved near their tips, and its flowers are rather larger than those of the species.

About 5 feet tall, ***D. preissii*** has long, upright branches, and ¾-inch-long, rigid leaves. Up to three together, its yellow and red blooms come from the leaf axils. A shrub of similar size, also with erect branches, ***D. floribunda*** has slender leaves up to ¾ inch long. Its nearly stalkless red and yellow blooms, one to three together from the leaf axils, are crowded in cylindrical, leafy spikes 1 inch to 4 inches long. The standard petal is up to ½ inch long.

Garden and Landscape Uses and Cultivation. These are attractive shrubs for California and other places with warm, dry climates. They are suitable for planting in beds alone or with other shrubs and for informal grouping. Well-drained soil and sunny places are to their liking. They are propagated by seeds and by cuttings. As cool greenhouse plants they may be grown in pots or tubs under conditions appropriate for chorizemas.

DIMORPHORCHIS. See Arachnis.

DIMORPHOTHECA (Dimorpho-thèca)—Cape-Marigold. Some species previously included in this genus belong in *Castalis* and *Osteospermum.* The seven that remain in *Dimorphotheca* are annuals, herbaceous perennials, and subshrubs of the daisy family COMPOSITAE. All are natives of South Africa. The name, from the Greek *dis,* two, *morphe,* shape, and *theke,* a case, alludes to the two forms of fruits (achenes). From closely allied *Calendula* dimorphothecas differ in having straight instead of curved achenes.

Dimorphothecas at home in South Africa

Beautiful ornamentals are included in *Dimorphotheca,* some of which have been cultivated for a long time. As early as 1687 *D. pluvialis* was grown in Holland, and gardeners who object to modern scientific names may reflect how much greater would be their dismay if faced with *Calendula humilis africana, flore intus albo, foris violaceo simplici* which was the name of *D. pluvialis* before Linnaeus brought order to plant nomenclature. From *Castalis* and *Osteospermum* the genus *Dimorphotheca* differs in producing fertile fruits from both disk and ray florets.

Dimorphothecas have alternate leaves variously pinnately- or sinuously-lobed or toothed and often showing considerable variation. The flower heads are solitary at the ends of leafless stalks and have involucres (collars at the backs of the flower heads) of separate green bracts. The tubular, five-lobed disk florets (that form the centers of the daisy-like flower heads) are bisexual, the petal-like ray florets surrounding the disk are female. The seedlike fruits are flattened two-winged achenes.

A branching annual with ray florets white or creamy-white above and lilac or purple on their undersides, ***D. pluvialis*** (syn. *D. annua*) is erect or somewhat sprawling and up to about 1 foot tall. Its hairy, narrowly-oblong or ovate-oblong leaves are 1 inch to 3 inches long, and bluntly toothed. The flower heads, 2 to 2½ inches wide and with ray florets minutely-three-toothed at their apexes, have centers of golden-brown disk florets tipped with blue. In dull and cool weather and at night the flower heads close. Variety *D. p. rin-*

Dimorphotheca pluvialis ringens

gens has rays with blue-violet bases that form a prominent zone around the central eye.

Brilliant yellow or orange ray florets, sometimes violet at their bases, chiefly distinguish the plant commonly, and quite wrongly, known in gardens as *D. aurantiaca* from *D. pluvialis.* The correct identification of this beautiful annual, which has coarsely-toothed, oblong to oblanceolate leaves about 3 inches long and flower heads about 1½ inches in diameter, is ***D. sinuata*** (syn. *D. calendulacea*). Hybrids between it and *D. pluvialis,* named *D. hybrida* are intermediate between their parents and show a wide range of beautiful flower colors.

Dimorphotheca chrysanthemifolia

Dimorphotheca sinuata

Two somewhat shrubby perennials up to 3 feet in height are *D. chrysanthemifolia* and *D. cuneata.* Quite distinct from other dimorphothecas, ***D. chrysanthemifolia*** most closely resembles *D. sinuata,* but differs in not being an annual and in having the lower parts of its stems woody. Its obovate-oblong, coarsely-toothed to deeply-pinnately-lobed leaves are up to 3 inches long. This species has densely glandular-hairy younger parts and flower heads up to 3 inches in diameter with yellow centers and yellow ray florets.

The most widely distributed species in the wild, ***D. cuneata*** is much-branched and has sticky-hairy younger stems and foliage. Varying from broadly-ovate to nearly linear, and with wedge-shaped bases, its leaves are about 1 inch long and more or less coarsely-toothed. The flower heads are 2 inches wide and have white ray florets with their undersides blue-violet suffused with copper. Hybrids between *Dimorphotheca* and *Osteospermum* have been named *Dimorphospermum.*

Dimorphotheca cuneata

Dimorphospermum hybrid

Garden and Landscape Uses. Annual dimorphothecas are splendid for garden beds, and in pots for greenhouse decoration. In California and places with like climates, they bloom outdoors in winter and spring, where winters are more severe, in summer. In bloom they make quite dazzling displays. They are not hardy to frost, nor do they withstand torrid, humid weather well. Because the flower heads of most kinds close in dull weather and at night, the blooms are of little use as cut flowers. This is true of both annual and perennial kinds. The latter can be grown permanently outdoors only where there is little or no frost, and are at their best in Mediterranean-type climates such as that of California. They are not well suited to regions of high summer humidity. Very attractive as furnishings for flower beds, where they are not hardy they can be treated as annuals by raising them from seeds or from cuttings taken from plants wintered in greenhouses.

Cultivation. Dimorphothecas are sun-lovers. They appreciate porous, well-drained, not excessively fertile soil. For winter and spring flowering outdoors in mild climates and in greenhouses seeds are sown in September or October, or in January. As summer-blooming annuals they are raised from seeds sown outdoors in spring, or early indoors in a temperature of 60 to 65°F, and the young plants transplanted outside from pots or flats as soon after the last frost as weather permits. Until a week or ten days before planting out time the seedlings are grown in a greenhouse in a night temperature of 55°F, rising by five to fifteen degrees by day. Then they are hardened off by standing them in their pots or flats in a cold frame or sheltered place outdoors. An early start indoors is especially advisable where summers are hot and humid. It permits the plants to attain blooming size before the weather becomes too torrid. Spacings in the garden of about 6 inches apart for *D. pluvialis,* 9 inches for most others, and about 1 foot for *D. chrysanthemifolia* and *D. cuneata* when grown as annuals and more when they are grown as perennials, are satisfactory.

Dimorphothecas grown to bloom in pots in greenhouses are well satisfied during winter with a night temperature of 50°F, increased five to fifteen degrees during the day. An airy, not excessively humid atmosphere is needed, and watering must be done with some caution. Excessive wetness is soon fatal. Sizable specimens can be had in 5- or 6-inch pots, and from later sowings, attractive plants in 4-inch containers. As spring approaches and the final pots become filled with roots, occasional applications of dilute liquid fertilizer are beneficial.

Perennial dimorphothecas located permanently outdoors need little routine attention beyond the removal of spent flower heads and pruning to shape when flowering is through or at the beginning of a new growing season. Moderate watering may be needed, and once or twice a year light fertilizing. The care needed by perennial kinds grown to flower in greenhouses is similar. Conditions indoors from fall to spring should be those suggested above for annual kinds raised to bloom in greenhouses. In summer pot-grown perennial dimorphothecas are best kept in a sunny place outdoors with their pots buried almost to their rims in sand, ashes, peat moss, soil, or similar material.

DINTERANTHUS (Dinter-ánthus). None of the five or six species of this *Mesembryanthemum* relative is easy to grow. Comprising small, nonhardy succulents of South Africa, it belongs in the carpetweed family AIZOACEAE. Its name honors the South African Professor K. Dinter. Dinteranthuses are more or less like *Lithops* and, like members of that genus, much resemble pebbles or small stones. Differences between the two genera lie chiefly in the fruits and in the seeds of *Dinteranthus* being dust-fine.

Stemless, or when aged short-branched and forming clusters of plant bodies each with one to three pairs of very fleshy, rounded or more or less angled leaves with flattish upper sides and united at their bases, dinteranthuses have smooth, firm, whitish skins, sometimes marked with numerous tiny dots or brief lines. The solitary, large blooms, yellow and opening late in the day, are on stalks without bracts. They resemble those of *Lithops* and appear in spring or fall.

Up to 1½ inches in height and with leaves united to form a plant body similar to those of *Lithops,* charming ***D. vanzylii*** (syn. *Lithops vanzylii*) is hairless and gray-green with reddish or purplish markings. It has orange-yellow flowers slightly over ½ inch across. Also hairless, ***D. pole-evansii*** (syn. *Rimaria pole-evansii*) rarely branches. It has a roughened, dove-gray skin often tinged with yellow or red, without dots. Almost 1¾ inches in diameter, the flowers are lustrous yellow. Forming solitary plant bodies of two leaves, ***D. wilmotianus*** is hairless, gray tinged with pink and with dark violet dots. Its golden-yellow blooms are slightly over 1 inch across.

Some species are to a greater or less extent pubescent. Notable among these is ***D. puberulus,*** in which the one or two pairs of brownish-gray leaves, as well as other parts, are finely-hairy, and are conspicuously dotted with green. The golden-yellow blooms are about 1 inch in diameter. The flower stalks and sepals of *D. inexpectatus* and *D. microspermus* are softly hairy, but their leaves are not. In the former the upper leaf surfaces are rounded at the tips, those of the latter are pointed. Very compact, ***D. inexpectatus*** is smaller than *D. microspermus* and is gray with greenish, translucent dots. Its golden-yellow flowers are about 1 inch wide. With usually solitary plant bodies 1 inch tall or somewhat taller, ***D. microspermus*** is prevailingly reddish- or purplish-gray with darker dots; young leaves are chalky-white to olive-gray. Golden-yellow with reddish-tipped petals, the flowers are 1¼ to 1¾ inches wide.

Garden Uses and Cultivation. More challenging to grow than the generally easier to satisfy *Lithops,* dinteranthuses need essentially the same conditions and are propagated in the same ways. Plenty of light is required. They are very adversely sensitive to excessive moisture, so that even during their summer period of growth water them sparingly. In winter keep the soil dry. See Succulents.

DIOCLEA. See Camptosema.

DIOECIOUS. Most flowering plants have both functioning male and female elements in the same bloom. Such flowers are termed perfect. Imperfect flowers are functionally unisexual. Kinds of plants in which male and female blooms are on separate individuals are dioecious. This term, which is of Greek origin and means two households, contrasts with monoecious, used to describe plants with separate male and female flowers on the same plant. Where fruits are an important objective, as with hollies, it is important with dioecious plants to have mostly females with just enough males nearby or interplanted to ensure pollination and thus crops of fruit by the females. Males, of course, never bear fruits. Contrary to a rather common misunderstanding, the presence of a male plant near a female has no influence upon the blooming of the latter. Only if the female blooms, which under favorable circumstances it will whether a male is proximate or not, can a male affect her through pollination.

DIONAEA (Dio-naèa)—Venus' Fly Trap. One of the most astonishing plants in the world, the charming and intriguing Venus' fly trap as a native inhabits only a meager strip of coastal land approximately ten miles wide and 100 miles long in North Carolina and adjacent South Carolina. Its genus name is a modification of Dione, the Greek name for Venus. A relative of the sundews (*Drosera*), this species belongs in the sundew family DROSERACEAE and is a herbaceous perennial insectivorous plant. From the sundew Venus' fly trap differs in having more stamens and united styles and, most obviously, in its remarkable "rat-trap" leaves that snap shut to secure their prey.

There is only one species, ***Dionaea muscipula,*** a rosette plant with mostly eight or fewer spreading leaves 1 inch to 5 inches long. Each has a broad, flat stalk and a rounded, two-lobed blade fringed with long, stiff bristles. On the upper surface of each lobe are three highly sensitive trigger hairs arranged in triangles. When one of these is touched twice or when two are touched in succession the leaf folds quickly

Dionaea muscipula

Dionaea muscipula (flowers)

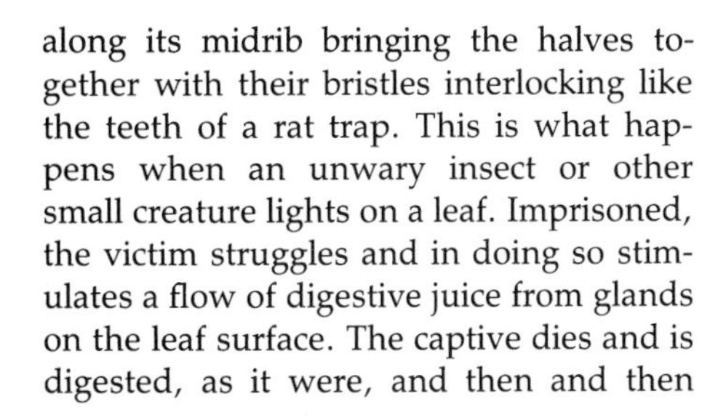

along its midrib bringing the halves together with their bristles interlocking like the teeth of a rat trap. This is what happens when an unwary insect or other small creature lights on a leaf. Imprisoned, the victim struggles and in doing so stimulates a flow of digestive juice from glands on the leaf surface. The captive dies and is digested, as it were, and then and then only, the lobes of the leaves open to await the next visitor. The flowers of Venus' fly trap are on stalks that develop in summer from the centers of the plants and attain heights of 6 inches to 1 foot. They terminate in umbels of usually four to ten white blooms, each about 1 inch across. The flowers have five sepals, five petals, from ten to twenty, but usually fifteen stamens, and five united styles with five fringed stigmas. The fruits are egg-shaped capsules under ½ inch long.

Dionaea muscipula, lower leaf with trap open, upper with trap closed

Garden Uses and Cultivation. Venus' fly trap is cultivated as an oddity and educational aid. It is easy to propagate and grow once its simple, but exacting needs are known. An inhabitant of sandy, acid bogs in open pine forests, to succeed it must be provided with conditions approximating those it knows in the wild. It is not hardy in the north, but may be grown in cool greenhouses, terrariums, and even in window gardens provided it is kept under a glass jar or polyethylene cover so that the air about it is constantly humid. There must be openings in the cover for ventilation, to prevent too high temperatures building up. A winter night temperature of 40 to 50°F is satisfactory. Day temperatures may be five to ten degrees higher. The soil may consist of leaf mold or peat moss mixed with coarse sand and topped with live sphagnum moss. The pots or pans (shallow pots) are kept standing in an inch or two of water so that moisture constantly seeps upward. Nonacid water, preferably rain water, must be used. Full sun, with perhaps a little light shade in high summer, is needed. Repot about every second year. Seeds afford an easy and entirely satisfactory means of raising new plants. They are sown on the top of sphagnum moss packed rather tightly, or on a mixture of peat moss and sand, in pans. The pans are stood in shallow water and kept covered with a bell or Mason jar. To secure fertile seeds it is necessary to transfer, with a small brush or tuft of cotton, pollen from the stamens of one plant to the pistils of another.

DIONYSIA (Dioný-sia). These primrose relatives of temperate central Asia are scarcely known in America. A few kinds are grown by connoisseurs of alpine plants in Great Britain. The genus *Dionysia* of the primrose family PRIMULACEAE numbers thirty-four species; it has a name derived from that of the Greek god, Dionysus. From *Primula* it is distinguished by a combination of characteristics rather than any particular one.

Dionysias, small and more or less woody, have fragrant foliage usually with a dense woolly farina, at least on the undersides of the leaves, and flowers with five sepals, a long narrow corolla with five spreading lobes, five stamens, and one style. The fruits are capsules.

Kinds most likely to be cultivated are ***D. aretioides,*** of Iran, which forms low, loose tufts of grayish-green foliage above which the solitary or paired, up-facing, clear yellow flowers about 1½ inches across stand on short stalks, and ***D. curviflora,*** also of Iran, which has pink to white, solitary flowers arising from compact cushions of foliage.

Garden Uses and Cultivation. These are plants for the experienced cultivator of alpines. British experience indicates that they are finicky and difficult to grow. They seem to prefer gritty, freely-drained soil and respond best to cultivation in pans (shallow pots) in alpine greenhouses. The plants are set in the pans wedged between two pieces of rock. Care must be exercised not to keep the soil too wet in winter. Propagation is by seeds or by cuttings.

DIOON (Dì-oon). This name, sometimes spelled *Dion,* belongs to an exclusively Mexican genus of three to five species of the cycad family CYCADACEAE. Derived from the Greek *dis,* two, and *oon,* an egg, it alludes to the seeds being paired.

Dioons are palmlike evergreens of impressive appearance, with short or long trunks clothed with the persistent remains of the leafstalks, and surmounted by crowns of stiff, leathery, pinnate leaves, the leaflets of which are broadest at their bases. As is characteristic of the family, the reproductive bodies are in cones, those of *Dioon* having flat, woolly scales. The seeds are about the size of chestnuts and contain much starch. When raw they are poisonous, but those of some kinds are ground into meal that is cooked and used locally for food.

Dioon edule

Its thick trunk up to 6 feet in height, ***D. edule,*** native to hot, sunny locations, has leaves 3 to 6 feet long and 6 to 7 inches wide, with numerous leaflets. The leaflets, except the youngest, which have teeth at their tips, are toothless, and have up to a dozen side veins. The male cones are cylindrical, the females ovoid.

Dioon edule with female cone

Dioon spinulosum at Garfield Park Conservatory, Chicago

Dionysia aretioides

More slender-trunked, and in its native rain forests becoming 50 feet tall, ***D. spinulosum*** has both margins of its 5- to 6-foot-long leaves with numerous spiny teeth. The leaflets have eighteen or more lateral veins on each side of the midrib. Female cones, sometimes over 1½ feet long, produce large seeds, which Mexicans grind into meal from which they make tortillas. Male cones are much smaller. This species in the wild grows in limestone regions. The leaflets of closely related, but short-trunked ***D. purpusii,*** have a few teeth on their upper margins, none on their lower.

Garden and Landscape Uses and Cultivation. Dioons are striking embellishments for creating palmlike effects in tropical and subtropical landscapes. They are also handsome in tubs and large pots. For the last purpose *D. spinulosum* is especially popular. For more about them and information on their cultivation see Cycads.

DIOSCOREA (Dios-corèa)—Yam, Yampi, Air-Potato, Cinnamon Vine, Elephant's Foot. The name yam, often in North America applied to varieties of sweet potato (*Ipomoea batatus*), is better reserved for members of the genus *Dioscorea* of the yam family DIOSCOREACEAE. This group, widely distributed in the wild, mostly in the tropics and subtropics, numbers 600 species of mostly twining, herbaceous or somewhat woody vines. Its name honors Dioscorides, the ancient Greek physician and naturalist.

Dioscoreas almost always have tuberous roots, which are often very large. Some kinds also produce aerial tubers. Their broad, alternate or opposite leaves are undivided or have separate leaflets. Unlike those of the majority of monocotyledons, they are net-veined. The small, unisexual flowers, in spikes or racemes, have six perianth segments, six or sometimes three stamens, and three styles. The fruits are capsules containing winged seeds.

The greater yam, winged yam, or water yam (***D. alata***), now cultivated wherever yams are important as food, is unknown in the wild. It apparently developed under cultivation centuries ago from an Asian species. Distinguishing characteristics are its nearly always solitary, cylindrical to nearly globose, more rarely serpentine, flattened, or knobby, subterranean tubers that may be 6 feet long and weigh up to 150 pounds, but commonly are much smaller. In some varieties the tubers have pink to purple flesh. The stems are square and have winged angles. The leaves, basically ovate or rounded, vary considerably in proportions and size. From the leaf axils come panicles, up to 1 foot long, of male flowers, and shorter spikes of female ones.

The yampi, yampee, or cush-cush (***D. trifida***), much cultivated in the Caribbean region, is the only American species used to any extent for food. It has many well-flavored tubers 6 to 8 inches long, with white, yellow, pink, or purple flesh. Its stems are square, sometimes winged, and without spines. Opposite or alternate, the leaves are large and divided into three or five lobes.

The air-potato, potato yam, or aerial yam (***D. bulbifera***) enjoys a wide natural distribution in tropical Africa and Asia, and is cultivated elsewhere. It has very small or no subterranean tubers, but instead develops large, flat, angled, more or less liver-shaped aerial tubers 1 pound to 4 pounds or more in weight. These are cooked as food or sometimes are eaten raw. The cylindrical, ordinarily spineless stems are up to 20 feet long. Alternate or opposite, the leaves are ovate-heart-shaped and up to 10 inches long by 7 inches wide. The flowers are bigger than those of other species discussed here, and have spreading perianths.

Cinnamon vine or Chinese yam (***D. batatas*** syn. *D. opposita*), a native of China, is about 10 feet tall. It has long, slender, spindle-shaped tubers up to 3 feet long that strike deeply into the ground. Its stems are slightly angled, often twisted, and nearly spineless. In the axils of its opposite, short-pointed, broad-based, ovate to triangular, small leaves, raceme-like spikes of cinnamon-scented flowers and small tubers develop. Although its above-ground parts are frost-tender, the roots of the cinnamon vine persist outdoors in the north. Native of Japan, and widely cultivated there and in China for food, ***D. japonica*** is very similar to *D. batatas* and by some authorities is accepted as a variety of it. Native to the Balkan Peninsula as its name indicates, ***D. balcanica*** is a slender-stemmed vine 2 to 3 feet tall having pointed-heart-shaped, nearly hairless leaves with blades 3 to 4 inches long and usually nine conspicuous, longitudinal veins. The little flowers are in pendulous spikes.

Dioscorea japonica

Dioscorea balcanica

A source of the drug cortisone and native to Mexico and Guatemala, ***D. composita*** is a vigorous vine, in the wild often occurring in limestone soils. It has rather glossy leaves with pointed-ovate to triangular-ovate blades 3 to 8 inches long and five to nine prominent, longitudinal veins. The tiny flowers are in long slender spikes from the leaf axils.

Dioscorea composita

Variegated-leaved dioscoreas include several splendid ornamentals. Here belongs ***D. discolor,*** of Surinam. This has broadly heart-shaped, dark olive-green leaves, purple on their undersides, and above beautifully marbled with pale green and silvery-gray, and with silvery to carmine-pink veins. Proportionately narrower and more arrow-shaped, the rich satiny green leaves of ***D. dodecaneura,*** of Brazil, have a brownish sheen. They are banded with gray along their midribs and marbled with the same color at their bases. Their undersides are purple. The plant known in horticulture as ***D. amarantoides*** 'Eldorado' (syn. *D. multicolor* 'Eldorado') is similar to if not identical with the last. Two varieties of variegated-leaved ***D. multicolor*** are cultivated. The emerald-green, long-pointed, narrowly-heart-shaped leaves of *D. m. argyraea* (syn. *D. argyraea*), of Colombia, are strikingly marked with broad longitudinal bands of silvery-gray, those of *D. m. melanoleuca* are wider, when young copper-tinted, and later become green marked with an irregular central band and spots of silvery-white. Their undersides are purple. North American species known by the names of wild yam and colic root

Dioscorea discolor

(***D. villosa*** and ***D. quaternata*** syn. *D. glauca*), were formerly used medicinally. Very similar, they have stems up to about 15 feet long, and pointed-heart-shaped leaves 2 to 4 inches long. The first inhabits moist open woodlands from southern New England to Minnesota, Florida, and Texas, the other similar woodlands from Pennsylvania to Missouri, Florida, and Louisiana.

Dioscorea quaternata

The elephant's foot or hottentot's bread (***D. elephantipes*** syn. *Testudinaria elephantipes*) is a decided curiosity sometimes included in collections of desert plants as an object of interest. Native to rocky, desert regions in South Africa, this kind develops an immense tuber, chiefly above the ground surface, covered with hard, woody bark deeply grooved in a turtle-back pattern. The tubers may be 3 feet or more in diameter. Specimens weighing more than 700 pounds have been reported. They are sometimes eaten by African natives. In contrast to the immense tubers, the stems of this species are slender and delicate. Up to 10 feet tall, they bear attractive, small, heart-shaped leaves and, in axillary racemes, little greenish-yellow or whitish flowers. The stems and foliage are deciduous and are in evidence for a few months only of each year. Another South African, this of very local distribution in the wild, ***D. hemicrypta*** (syn. *Testudinaria glauca*) is somewhat similar to *D. elephantipes*. From that species *D. hemicrypta* differs most obviously in having an erect tuber with its above-ground part elongated, instead of being flat or hemispherical, and less deeply grooved. With much of its very large, woody tuber above ground, *D. hemicrypta* has slender stems that spread rampantly among and over low shrubs. Broadly-heart-shaped, with a tiny sharp point at the apex, its short-stalked leaves are 1 to 1½ inches long. The tiny greenish flowers are in short racemes from the leaf axils. Also South African, ***D. cotinifolia*** is a vine with a woody tuber, slender stems, and alternate or opposite, broadly-ovate to pointed-triangular-heart-shaped, hairless leaves 2½ to 3½ inches long and with seven or nine conspicuous longitudinal veins. Its tiny flowers are in slender spikes.

Dioscorea elephantipes

A tuber of *Dioscorea elephantipes*

Dioscorea elephantipes (flowers and foliage)

Dioscorea hemicrypta

Dioscorea cotinifolia

Garden and Landscape Uses and Cultivation. In North America yams are cultivated chiefly as ornamentals and items of interest rather than for food. They may be used to decorate pergolas and other supports acceptable to their twining stems. So accommodated they form effective screens. In general, yams grow best in deep, rather loose, well-drained, fertile soil of average moistness, in sun or part-day shade. Most need tropical temperatures, but the cinnamon vine and the North American native ones described above are hardy outdoors. Greenhouse cultivation is largely restricted to tropical kinds with variegated foliage. Propagation is chiefly by offset root tubers, aerial tubers, and by cut pieces of root tubers. Seed may also be used. Variegated-leaved kinds are grown from cuttings.

The elephant's foot presents a special case. It prospers under conditions that suit cactuses and other succulents, outdoors in warm desert and semidesert regions, and in greenhouses. Its soil must be extremely porous and kept quite dry throughout the dormant season. When new growth begins, regular but not excessive watering is resumed, to be gradually reduced and finally stopped when, at the end of the growth cycle, the foliage begins to turn yellow naturally. Indoors this species does well where the winter night temperature is 45 to 50°F, and by day and at other seasons is higher. The elephant's foot is easy to raise from seed.

DIOSCOREACEAE — Yam Family. Five genera totaling approximately 750 species widely distributed in temperate, tropical, and subtropical regions constitute this family. They are vines with tubers (sometimes very large) or rhizomes and usually twining, herbaceous to somewhat woody, generally annual stems. The leaves, alternate or opposite, stalked and prevailingly heart- to arrow-shaped and sometimes lobed in finger-fashion, are net-veined, which is unusual for monocotyledons. The unisexual or bisexual flowers have six perianth segments, their bases joined as a tube, six stamens or three stamens and three staminodes, and three styles. The fruits are capsules or berries. The yam family includes several sorts the subterranean or aerial tubers of which are staple foods in warm countries. Some kinds are grown for ornament. The most familiar genera are *Dioscorea, Rajania,* and *Tamus.*

DIOSMA (Diós-ma). To the rue family RUTACEAE belong the fifteen species of low South African evergreen shrubs of the genus *Diosma.* Other plants once included here belong in *Adenandra, Agathosma, Audouinia, Barosma,* and *Coleonema.* The name *Diosma* is from the Greek *dios,* divine, and *osme,* smell. It alludes to the fragrance of the bruised leaves. The plant grown as *D. reevesii* is *Coleonema album.*

Diosmas are bushy, heathlike plants with slender branches and small, alternate or opposite leaves, with two longitudinal grooves on their undersides. Their white to reddish flowers are at the ends of the shoots, solitary or in small clusters. They have five petals, not constricted into long claws at their bases as are those of *Agathosma,* and five or ten stamens. Unlike those of nearly related *Agathosma* and *Barosma,* the stamens are all fertile. The style is short. The fruits are five-sectioned capsules.

A vigorous grower, 1 foot to 2 feet tall, ***D. ericoides*** has hairless shoots and leaves. The latter are oblong, up to ½ inch long, and are keeled and bespeckled with glandular dots on their undersides. Their lower parts hug the stems, their tips spread. The short-stalked blooms, in twos or threes, terminate the numerous short twigs and are borne profusely. They are white to reddish.

Diosma ericoides

Garden and Landscape Uses. Diosmas are attractive for beds and rock gardens in mild climates, such as that of California, and in pots in cool greenhouses. Flowering branches of *D. ericoides* are delightful for cutting and mixing with other flowers in arrangements. They add lightness and grace.

Cultivation. These plants need very much the same care as tender heaths (*Erica*), but are less finicky. They need full sun. Fertile, sandy peaty soil is most to their liking, but they are fairly tolerant and will grow in most average garden soils that are not alkaline. Routine care consists of cutting them back fairly severely after blooming. They are helped if a mulch of peat moss or other organic material is maintained about them. Propagation is easily accomplished by summer cuttings and by seeds sown in sandy peaty soil in a temperature of 55 to 60°F. The tips of the shoots of young specimens are pinched out occasionally to induce branching.

As pot plants in greenhouses diosmas succeed in a winter night temperature of 45 to 50°F with a daytime rise of five to ten degrees or thereabouts permitted. Whenever weather allows the greenhouse should be ventilated freely. Never should the atmosphere be dankly humid. In summer best results are had by putting the plants in a sunny place outdoors with their pots buried to their rims in sand, peat moss, or other material that will conserve moisture and keep the roots at a nearly even temperature. They are brought indoors before the advent of fall frosts. In spring, when blooming is over, they are cut back fairly severely, and two or three weeks later, when new shoots begin to sprout, are repotted. At no time must the soil be completely dry, but long periods of excessive wetness must be avoided.

DIOSPYROS (Dios-pỳros) — Persimmon, Date-Plum, Kaki, Black Sapote, Mabolo, Ebony. The ebony family EBENACEAE consists of a few genera of trees and shrubs remarkable for their hard woods. One of these is *Diospyros,* the source of ebony wood and edible fruits. Notable among the species esteemed for their fruits are the common persimmon, the Japanese persimmon, the date-plum, and the black sapote. There are 500 species of *Diospyros,* trees and shrubs mostly of tropical, subtropical, and warm-temperate regions around the globe. There is one notable hardy kind, the common persimmon of eastern North America. The name, derived from the Greek *dios,* divine, and *pyros,* a grain of wheat, refers to the edibility of the fruits. Other plants once named *Diospyros* are included in *Royena.*

Diospyroses have undivided, mostly leathery, alternate leaves and small, bell-shaped, urn-shaped, or tubular, unisexual blooms in few- to many-flowered clusters or, the females, solitary. There are usually four or five, but sometimes three or seven, calyx and corolla lobes. The males most commonly have sixteen stamens, but there may be more or as few as four. In the female blooms there are ordinarily four to eight nonfunctional stamens (staminodes) in addition to two to six styles. The fruits, often large, and usually with a large persistent calyx attached, are technically berries. They contain one to ten seeds and, until they are ripe, are often strongly astringent, causing the mouth to pucker unpleasantly if they are bitten into. The sap of some kinds is irritating to the skin and poisonous to fish.

The common persimmon (***D. virginiana***), native from Connecticut to Florida and Texas, is a deciduous, round-crowned tree with horizontal or drooping branches. Uncommonly it is 100 feet tall, more usually not over 50 feet. Its pointed-ovate leaves are 3 to 6 inches in length and may be pubescent at first, but later are hairless. Their top sides are lustrous dark green, beneath they are paler. In fall they become yellow.

Diospyros virginiana

Individual trees are usually unisexual. The greenish-yellow blooms make no conspicuous display nor do the fruits. The male flowers are about ½ inch, the females ¾ inch, long. The former have about sixteen stamens, the latter eight staminodes. Considerable variation in shape and eating quality of the plumlike fruit occurs. They are spherical to egg-shaped, 1 inch to 2 inches long, and orange, often tinged red on one side. A few superior varieties have been selected and given horticultural names, but are not widely cultivated.

Diospyros virginiana (bark)

Diospyros virginiana (fruits and foliage)

The Japanese persimmon or kaki (***D. kaki***) has been cultivated in the Orient for centuries and several hundred varieties are recognized. Fewer are grown in North America. The best are agreeable fruits that come into the markets regularly. Despite the implication of its common name and its popularity in Japan, this species is not a native of Japan, but was brought there originally from China. It may be of hybrid origin. This persimmon is not hardy in the north. It survives only in mild climates, such as those of California and parts of the south. It is not adapted to the humid tropics.

Deciduous, *D. kaki* is a round-headed tree up to 40 feet in height. Its younger branches are clothed with rusty hairs. Its leaves, 3 to 7 inches long, are ovate-elliptic to obovate. Their glossy upper surfaces are without hairs, their lower ones are pubescent. About ¾ inch long and yellowish, the flowers are usually unisexual, occasionally bisexual. The males have four-lobed calyxes and corollas, and fifteen to twenty-four stamens in clusters of three. The females have large leaflike calyxes, four corolla lobes, eight staminodes, and a four-parted style. Bisexual flowers are intermediate in their characteristics between males and females. Usually the sexes are on separate trees, but single trees sometimes have bisexual and unisexual blooms. The flowers are on shoots of the current season. The fruits, in some varieties 3 inches or more in diameter, in the wild species often are not over 1 inch across. They are tomato-like in appearance, orange-yellow to reddish-orange, thin-skinned, with soft, sweet, orange flesh, and up to eight flattened, elliptic seeds (there are some seedless varieties). The fruits are eaten out of hand or, in the Orient, dried. In China a varnish is made from the unripe fruits.

The date-plum or wild Oriental persimmon (***D. lotus***) ranges as a native from Japan to western Asia. Deciduous, round-headed, and up to 45 feet in height, it has shoots at first downy, but later hairless, and pointed, elliptic to oblong leaves 2 to 5 inches long, with hairs at least on the veins on the undersides. The upper surfaces are smooth. The blooms are greenish or reddish. The males have sixteen stamens and are under ¼ inch long, the females are slightly larger. The spherical to egg-shaped, sweet, edible fruits, at first yellow, but when mature blackish, are about ½ inch in diameter. In China this species is used as an understock upon which to graft cultivated varieties of the Japanese persimmon. Its wood is used for furniture and other articles of quality.

Diospyros lotus

The black sapote (***D. digyna***) has been misnamed *D. ebenaster*. Distinct from fruits called sapote that belong in the sapodilla family SAPOTACEAE, the black sapote is a tree of good appearance up to 60 feet tall, but often much lower, believed to be a native of Mexico and Central America. It is common in the Philippine Islands, but was probably introduced there at an early date. This species has leathery, elliptic to oblong, glossy leaves, 4 to 8 inches long, with well-marked mid-veins. The flowers, small and white, are much like those of the Japanese date-plum. Some are male, others bisexual. The fruits give reason for the colloquial name. They are tomato-shaped, 2 to 5 inches in diameter, olive-green, and have chocolate-colored flesh. Sometimes they are seedless, but more often contain one to ten flattened, oval seeds about ¾ inch long. The ripe fruits are eaten fresh or after being beaten with lemon or orange juice and chilled. They are reported to be poisonous when unripe and are said to be used for stupefying fish.

Less well known than the kinds discussed above is the mabolo (***D. discolor***), a native of Malaya and the Philippine Islands. A tree up to about 40 feet tall, it has pendulous twigs and pointed-oblong leaves, glossy above and pubescent beneath, and 4 to 9 inches long, and solitary, axillary, creamy-white flowers ¾ inch across. More or less globular, the purple-red or yellowish to brownish edible fruits 2 to 3 inches in diameter, have four persistent sepals at their stem ends. This species is a desirable tropical shade tree. Some individuals bear seedless fruits, but more commonly each has four to eight seeds. The aromatic flesh is whitish and dryish.

Ebony wood, highly prized for its black color, hardiness, and other desirable qualities, is the heartwood of several species. The chief commercial source is ***D. ebenum,*** a heavy-foliaged evergreen tree native of southern India and Ceylon. It has an irregular crown up to 60 feet in height, and firm, glossy, blunt-elliptic, short-stalked leaves, 2 to 3 inches long and with prominent mid-veins. The fruits, ¾ inch long, have a four-pointed calyx attached. They contain three to eight seeds.

A species sometimes misnamed *D. ebenum* is ***D. malabarica.*** It is distinguished by its proportionately longer, narrow-elliptic leaves and its fruits, which are 2 inches in diameter, and covered on their outsides with a rusty scurf, containing four to eight seeds. The large persistent calyx is four- or sometimes five-lobed. The flowers of this tree, which attains a height

Diospyros malabarica

of 50 feet and is indigenous to Ceylon, India, and Malaya, are attractive to bees. In its native lands fish lines and nets are tanned by treating them with the fruits.

Garden and Landscape Uses. The garden and landscape uses of the *Diospyros* species discussed, apart from their desirability for inclusion in arboretum collections and, in the case of the common persimmon, in areas devoted to native plants, fall into two categories. They deserve to be appraised as fruit trees, and as ornamentals and shade trees. Outstanding for the first purpose is the Japanese persimmon, a kind suitable for cultivation in the cotton belt of the United States and in California. It has merit also for shade and ornament.

The other tender bearers of edible fruits, the date-plum, black sapote, and mabolo are not likely to be much grown for their crops in the continental United States, but may be planted for interest, for ornament and shade. The date-plum is about as hardy as the Japanese persimmon, but the others will live only in southern Florida and other essentially tropical climates.

The common persimmon, the only hardy species, succeeds as far north as southern New England. As a fruit tree it is of minor importance. As an ornamental and a shade tree it is not without merit, but has no virtues that outshine those of many other available species.

Diospyros ebenum in Jamaica, West Indies

Diospyros ebenum (foliage)

Cultivation. The cultivation of the Japanese persimmon as a fruit tree is discussed under Persimmon. It and the other kinds will grow in a wide variety of soils. They need sunny locations. The common persimmon is tap-rooted and rather difficult to transplant except when small. Propagation is by seeds, cuttings of half-ripened or mature shoots, and by layering. Improved varieties are mostly increased by grafting or budding onto seedling understocks of *D. lotus, D. kaki,* or *D. virginiana.*

DIOSTEA (Di-óstea). Three species constitute this genus of the vervain family VERBENACEAE. Probably the only one in cultivation is ***Diostea juncea*** (syn. *Baillonia juncea*). The name comes from the Greek

Diostea juncea

di, double, and *ostea,* bone, and alludes to the fruits having two stones. Native of Chile and Peru, this shrub or tree, up to 18 feet tall, has slender rushlike branchlets, and rather distantly spaced pairs or whorls (circles of more than two) of stalkless, ovate-oblong, toothed, thickish leaves ½ inch to 1½ inches or more long. In 1-inch-long racemes terminating short lateral shoots, the lilac flowers are borne quite freely. They are about ⅓ inch long and have bell-shaped calyxes with four or five teeth and slightly two-lipped, tubular corollas with five small, spreading lobes (petals). There are four scarcely protruding stamens and a staminode (rudimentary stamen). The berry-like fruits are drupes.

Garden and Landscape Uses and Cultivation. Rare in cultivation and not hardy in the north, this curious plant is suitable for growing outdoors in mild, dryish climates. Its appeal is chiefly to collectors of the unusual. It favors well-drained soil and sunny locations. Because of its slender habit it is best planted among other shrubs to conceal the skimpiness of its lower parts. Propagation is by seeds and summer cuttings.

DIOTIS. See Otanthus.

DIPCADI (Dipcà-di). The genus *Dipcadi* is native to southern Europe, Africa, Malagasy (Madagascar), and India. It comprises fifty-five species of bulb plants of the lily family LILIACEAE. Its name derives from an oriental one for a kind of grape-hyacinth (*Muscari*) to which *Dipcardi* is akin.

Dipcadis have bulbs from which come linear leaves and leafless racemes of scilla-like flowers with six perianth segments (petals, or more correctly tepals), six stamens, and one style. The fruits are capsules.

Native of sandy and rocky places in southern Europe, ***D. serotinum*** (syn. *Uropetalum serotinum*) has channeled, linear leaves 6 to 8 inches long, and leafless flowering stalks with loose racemes of blooms slightly exceeding the leaves and suggestive of those of the English bluebell (*Endymion nonscriptus*). The flowers, approximately ½ inch long, are yellowish to brownish, nodding, and tubular-bell-shaped. Their petals are joined in their lower fourths. The outer three are recurved, the inner are erect or slightly incurved.

Garden Uses and Cultivation. Of modest appearance, the species described is not likely to have wide appeal, but is of interest to collectors who appreciate the unusual and uncommon. It responds to treatment favorable to scillas and endymions, but is not reliably hardy in climates as severe as that of New York City. In greenhouses it responds to conditions that suit lachenalias.

Dipcadi serotinum

Dipcadi serotinum (flowers)

DIPELTA (Di-pélta). Just why these highly ornamental deciduous shrubs are not better known and more widely grown is difficult to say. Others less worthy are more popular. Perhaps if an attractive common name for them was devised and made known, as was done with the beauty bush (*Kolkwitzia*), it would generate interest. Possibly what is needed is an enterprising nurseryman to propagate dipeltas in large numbers and introduce them through appropriate advertising to American gardeners. Most likely a combination of these promotional aids would bring quickest success.

Dipeltas are similar to weigelas and resemble the beauty bush and abelias. They are readily distinguished by the several conspicuous, unequal-sized bracts at the bases of the flowers. Two of these, much larger than the others, cover the ovary and enlarge to form dry disklike or shieldlike papery wings to the fruits. It is this feature that is responsible for the name *Dipelta,* derived from the Greek *dis,* two, and *pelte,* a shield. The genus belongs in the honeysuckle family CAPRIFOLIACEAE and comprises four species, all Chinese.

Dipeltas have opposite, short-stalked, toothed or smooth-edged leaves. Their beautiful flowers, borne on shoots of the previous year's growth, are solitary or in leafy, few-flowered racemes. They are funnel-shaped with five linear to lanceolate calyx lobes and five spreading corolla lobes disposed asymmetrically to produce a slightly two-lipped bloom. There are four stamens and a slender style. The fruits are capsules.

The hardiest species *D. floribunda* thrives in southern New England. The other cultivated kind, *D. ventricosa,* cannot be relied upon to survive winters north of New Jersey. Attaining a height of 10 to 15 feet, ***D. floribunda*** has peeling bark. Its young shoots are downy. Its long-pointed leaves, 2 to 4 inches long, ovate to ovate-lanceolate, and sometimes distantly toothed, are slightly hairy when young, but soon become smooth. The fragrant, nodding blooms are on slender stalks and are solitary or in racemes of up to six. Their corolla tubes narrow at their bases. The flowers are 1 inch to 1¼ inches long and about 1 inch in width. They have rounded corolla lobes and are pleasing light pink with yellow in their throats. The two largest of

Dipelta floribunda

the four floral bracts are ¾ inch long by more than ½ inch wide, and are attached at their centers.

From the above, ***D. ventricosa*** differs in having earlike floral bracts, the largest of which become ¾ inch long by one-half as wide. They are attached at their heart-shaped bases. Also, the corolla tube is distinctly pouched or bellied at its lower end to form a cup-shaped base. This species, 8 to 15 feet tall, has pubescent young shoots and ovate to ovate-lanceolate leaves, smooth-edged or sometimes with a few gland-tipped teeth. They are 2 to 6 inches long and slightly hairy on both sides. The flowers are usually solitary in the leaf axils and in threes at the terminations of short branchlets. Hanging on slender stalks, they are 1 inch to 1¼ inches long and about ¾ inch across the throat. On the outside they are deep rose-pink; inside they are paler pink marked with orange. The corolla lobes are rounded.

Garden and Landscape Uses. These attractive spring-blooming shrubs can be massed to provide screening and can also be used in mixed borders and beds. They thrive in any ordinary, well-drained, not excessively dry soil, in sun, and need little attention beyond judicious thinning out of old and crowded branches as soon as flowering is through. Propagation is easy by seeds, and by leafy cuttings in summer and hardwood cuttings in fall.

DIPHYLLEIA (Diphyl-lèia)—Umbrella Leaf. Three species of this eastern Asian and North American genus are recognized; one is American. By some authorities it is included in the barberry family BERBERIDACEAE, by others it is segregated with *Podophyllum*, *Jeffersonia*, and three other genera, in the may-apple family PODOPHYLLACEAE, which is considered to be somewhat intermediate between the barberry family and the buttercup family (RANUNCULACEAE). The name *Diphylleia* is from the Greek *dis*, two, and *phyllon*, a leaf and alludes to the two leaves borne on each flowering stem.

Members of this genus are deciduous, herbaceous perennials with thick, horizontal rhizomes. They differ from *Podophyllum* in having their flowers in clusters and from *Jeffersonia* in their leaves being lobed instead of divided into separate leaflets. Specimens of flowering size produce a single basal leaf with the stalk attached centrally to its blade. Flowering stalks have two similar, but smaller and more deeply-lobed leaves with their stalks attached closer to the margins. The stem leaves are alternate, the lower one the larger. The flowers have six deciduous sepals, six petals, and six stamens. The fruits are few-seeded berries.

The umbrella leaf (***D. cymosa***) is indigenous in moist places in cool mountain woods from Virginia to Georgia. Its solitary, long-stalked, basal leaves have blades 8 inches to 1½ feet across and deeply-cleft into two major lobes, each of which is again shallowly-lobed and sharply-toothed. They are peltate, that is, there is no sinus or opening in the blade from its edge to the top of the leafstalk. The stem leaves are up to 1¼ feet broad and are lobed and toothed similarly, but more deeply. The white flowers, each ½ to ¾ inch wide, are many together in long-stalked clusters. They are succeeded by red-stemmed, blue fruits about ⅓ inch long. The Asian species, ***D. grayi***, from Japan, and ***D. sinensis***, from central and western China, are not known to be cultivated in North America. They differ from the American species in having less-deeply-lobed leaves and in the stalks of their flower clusters having short hairs.

Diphylleia cymosa

Garden Uses and Cultivation. These plants are well worth cultivating in shady wild gardens and other informal areas where their large leaves can be displayed boldly and to advantage. A place where they are not subject to strong winds, where the air is humid, the soil deep, moist, and fat with an abundance of leaf mold, peat moss, or other decayed organic matter, is ideal. For the best effects the plants should be in fairly substantial clumps. Planting may be done in early spring or early fall. Established plants benefit from an annual mulching with well-decayed compost, leaf mold, or peat moss. Propagation is by division, by cuttings made from the rhizomes, or by seeds sown as soon as they are ripe in a cold frame or protected bed outdoors in sandy peaty soil kept always evenly moist.

DIPIDAX (Dipí-dax). Endemic to South Africa, this genus of two species of the lily family LILIACEAE is little known horticulturally. Its members are bulb plants with usually three leaves, and few- to many-flowered, terminal or apparently lateral, often one-sided, loose spikes of starry blooms. The flowers have six separate, spreading petals (more correctly, tepals), each with two nectaries at the base, and with a stamen arising from the claw (narrow basal part) of each petal. There are three slender styles. The fruits are top-shaped capsules. The name is from the Greek *dis*, two, and *pidax*, a spring. It alludes to the watery habitats of one species.

Delightful for planting permanently outdoors in mild, California-type climates, and for cool greenhouses, the larger species ***Dipidax triquetra*** is about 1 foot tall. It has semicylindrical leaves, the upper one of which appears to be a continuation of the stem, and a flower spike up to 4 inches in length that angles outward and upward and bears several ¾-inch-wide blooms with pink-flushed, white petals each with two purple spots at its base. This kind inhabits marshes and shallow ponds. The other species, ***D. punctata***, differs in having flat leaves and an obviously terminal spike of much smaller blooms. It is not a marsh or an aquatic plant.

Dipidax triquetra in South Africa

Dipidax triquetra (flowers)

Garden and Landscape Uses and Cultivation. As an embellishment for outdoor and indoor bog and water gardens *D. triquetra* is highly desirable, and is exquisite for growing in pots or pans (shallow pots) in greenhouses. It responds to the same general conditions as freesias, sparaxis, babianas, and many other bulb plants native to South Africa, but unlike those mentioned must have soil that is very moist or wet throughout its entire period of growth and until well after blooming. To ensure this, the pots or other containers may be submerged a few inches below the surface of a pool or in greenhouses may be kept standing in saucers of water. After blooming, when the foliage has ripened, drier conditions are needed. Then, plants in containers should be lifted from the water and kept dry until the beginning of the next growing season. Propagation is by offsets and seeds.

DIPLACUS. See Mimulus.

DIPLADENIA (Dipla-dènia). Plants commonly grown under this name belong in allied *Mandevilla*. The genus *Dipladenia* consists of thirty species of mostly vines that climb by hooks. They belong in the dogbane family APOCYNACEAE and are natives of tropical South America. It is improbable that any are cultivated.

DIPLARRHENA (Diplar-rhèna). One, or according to some botanists two, species of southern Australia and Tasmania represent this genus of the iris family IRIDACEAE. The name is from the Greek *diploos*, double, and *arrhen*, male; it alludes to the flowers having only two fertile stamens.

With beautiful, fragrant blooms that suggest those of *Libertia*, *Moraea*, or *Iris*, and having much the growth habit of those more familiar plants, ***Diplarrhena moraea*** has short, creeping rhizomes and pointed-linear, stiff leaves 1 foot to 1½ feet long, mostly in basal tufts, but with a few sheathing the lower parts of the slender, erect, sometimes branched flowering stems that are 1½ to 2 feet tall. The white and blue blooms, usually two or three together in loose terminal spikes with 2-inch-long bracts, last for only a short time, but are produced in succession over a long period. They have three big spreading perianth lobes (commonly called petals), the upper one the largest, and three very much smaller, inturned ones. There are two fertile stamens and one nonfunctional one. The fruits are three-angled, oblongish capsules containing flat seeds. Variety *D. m. latifolia* (syn. *D. latifolia*) is more robust and has usually five or six blooms in each flower spike.

Garden and Landscape Uses and Cultivation. These are quite excellent evergreen herbaceous perennials for flower borders and rock gardens. They need full sun and well-drained, sandy, but not too dry soil that contains a goodly amount of compost, peat moss, or other organic ingredient. They are not hardy in the north, but succeed in such favorable climates as that of California. Once well-established they are virtually indestructable and develop into quite large clumps. Propagation is very easy by seeds, division, and by plantlets that form on old stems.

DIPLAZIUM (Di-plàzium). The nearly 400 species of *Diplazium* of the aspidium family ASPIDIACEAE are by some authorities included in *Athyrium*. They are distinguished by the clusters of spore capsules and their coverings (indusia), or some clusters of spore capsules being on both sides of the free veins instead of along one side only. The name, from the Greek *diplasios*, double, alludes to this characteristic. One species, the paco, is the most esteemed of ferns used as human food. In its homelands its young fronds are eaten both cooked and raw. Diplaziums are natives of tropical and north temperate regions. Mostly of medium to rather large size, they commonly have much the appearance of aspleniums.

Tufted ***D. plantaginifolium*** has undivided fronds sometimes lobed toward their bases. Leathery, pointed-lanceolate, 1 foot to 1½ feet long by 2 to 3 inches wide, they have slightly wavy, toothed margins. This kind is native from Mexico to Brazil and the West Indies. Very different ***D. proliferum*** is a native of tropical Asia, Africa, and Polynesia. In tufts, its pinnate fronds are 3 to 5 feet long or longer by 1 foot to 2 feet wide. Each of the numerous leaflets or segments are 1 inch to 2 inches wide. Plantlets are frequently produced along the midribs. The clusters of spore capsules extend nearly to the margins of the leaflets. The paco (***D. esculentum*** syn. *Athyrium esculentum*) is a beautiful, low tree fern, native from India to Polynesia. It has a trunk up to 2 feet tall crowned with once-, twice-, or sometimes thrice-pinnate fronds 4 to 6 feet long. They are fresh green in color and have sharp-tipped, toothed segments.

Its slender rhizomes creeping, and its undivided, erect fronds spaced along them, ***D. lanceum*** (syns. *Athyrium dubium*, *A. lanceum*) is native from Japan and China to India and Ceylon. In gardens *Pyrrhosa lingua* has been misidentified as this species. The fronds of *D. lanceum* are lanceolate to pointed-linear-lanceolate and 4 to 8 inches or sometimes 1 foot long by up to 1 inch wide or slightly wider. They have slightly recurved, sometimes somewhat wavy margins, are dark green above, and paler on their undersides. In *D. l. crenatum* the fronds are pinnately-lobed or divided into round-ended segments. Differing from *D. lanceum* in its fronds being pinnate, variable ***D. japonicum*** (syn. *Athyrium japonicum*) is native from Japan and Taiwan to Korea, China, and India. It has slender, creeping rhizomes and fronds 1½ to 3 feet long by 3½ to 6 inches wide with broadly-lanceolate to narrowly-oblong-ovate or triangular mostly 4- to 9-inch-wide blades sometimes sparingly-hairy. The leaflets are deeply-pinnately-lobed.

Garden and Landscape Uses and Cultivation. Diplaziums are suitable for outdoor cultivation in the humid tropics and warm subtropics and in greenhouses. If their simple needs are met they are easy to grow. They luxuriate in well-drained, moderately moist soil that is rich with such organic matter as leaf mold, good compost, or peat moss. They need shade

Diplazium proliferum showing plantlets on fronds

from strong sun. Indoors a minimum winter night temperature of 60°F is appropriate with an increase of ten to fifteen degrees allowed on sunny days and less on dull ones. A humid atmosphere must be maintained. For additional information see Ferns.

DIPLOCAULOBIUM (Diplo-caulobìum). The genus *Diplocaulobium* of the orchid family ORCHIDACEAE consists of possibly seventy species and has a natural range extending from Malaya to Fiji. The name, from the Greek *diploos,* double, and *kaulos,* a stalk or stem, alludes to the two sorts of pseudobulbs.

Rare in cultivation, diplocaulobiums in the wild perch on trees as epiphytes or on rocks. Most sorts are rather small. Practically unique among orchids, most have two types of pseudobulbs, the ones that bear flowers markedly different from the sterile ones. A solitary leaf tops each pseudobulb. The solitary, usually white, yellow, or reddish, small to medium-large, somewhat spidery flowers last for only a few hours but are produced in succession over a fairly long period.

Native to New Guinea, ***D. chrysotropis*** is 2 to 3 inches tall and has clusters of tiny pseudobulbs each with a strap-shaped leaf about 1 inch long and slightly notched at its apex. The attractive, slender-stalked flowers, 1 inch wide or less, have cream sepals, petals deepening to yellow at their tips, and a cream lip with a yellow blotch and violet markings toward its base.

Diplocaulobium chrysotropis

Garden Uses and Cultivation. These are excellent orchids for the connoisseur. Somewhat challenging to grow, they can be expected to respond best to environments and care that suit tropical bulbophyllums and dendrobiums.

DIPLOCYATHA (Diplo-cỳatha). The one species of South African succulent that constitutes this genus resembles and is closely related to *Stapelia* and *Huernia.* It belongs in the milkweed family ASCLEPIADACEAE. Its name is from the Greek *diploos,* double, and *kyathos,* a cup. It refers to the flower structure.

Forming clumps of leafless, four-angled, coarsely conical-toothed, erect, fleshy stems up to about 2 inches tall, ***Diplocyatha ciliata*** (syn. *Stapelia ciliata*) has 3-inch-wide blooms on stalks approximately ½ inch long. They have bell-shaped corolla tubes and five pointed corolla lobes (petals), fringed with long white hairs. The lobes spread like the arms of a starfish. Their upper surfaces are nearly white to pale yellow stippled with red dots. A curious feature is that the center of the flower is occupied by a circular, deep-cup-shaped, inner corona that has much the relationship to the remainder of the bloom that the trumpet of a daffodil does to the other parts of its flower. In *Diplocyatha,* however, the inner corona does not protrude. Its exposed circular face is flush with the corolla lobes, and its rim is not frilled or crimped, but is rolled under. The stamens are united as a column around the ovary and at their tops are joined to the style. The flowers have the slightest suggestion of a disagreeable odor. The fruits are podlike follicles.

As this is a leafless plant, the food-elaborating functions of leaves are performed by the stems, which contain chlorophyll and are green, sometimes with a dull reddish tinge.

Garden Uses and Cultivation. This is an easy-to-grow plant. Its uses and cultural needs are those of *Stapelia.*

DIPLOCYCLOS (Diplo-cỳclos). Three of the four species of *Diplocyclos* are natives of tropical Africa, the other of that continent, tropical and subtropical Asia, and Indonesia. They belong in the gourd family CUCURBITACEAE. The name, from the Greek *diploos,* double, and *kyklos,* a circle, refers to the double borders around the seeds.

These are nonhardy, tendril-climbing, herbaceous perennial vines. They have palmately (in hand-fashion) five-lobed leaves, and usually once-forked tendrils. The flowers, clustered in the leaf axils, are unisexual. They have bell-shaped, five-lobed calyxes and broadly-bell-shaped, five-parted corollas. Male blooms have three separate stamens. Female blooms have three staminodes (nonfunctional stamens) and a style with two-lobed stigmas. The spherical or ovoid fruits are red berries, striped lengthwise or otherwise variegated with white or silver.

Native to Africa, Asia, and Indonesia, ***D. palmatus*** (syn. *Bryonopsis laciniosa*) is a hairless, slender-stemmed, tall vine with deeply-lobed, long-stalked, rough-surfaced leaves. Their blades are 2½ to 6 inches in diameter, and have tiny teeth where the veins reach their margins. The nearly spherical fruits are about ¾ inch in diameter.

Garden and Landscape Uses and Cultivation. The species described here is perennial only in the tropics and warm subtropics. It is useful for covering trellises and other surfaces that offer support its tendrils can grasp. The fruits are ornamental. This is easily grown as an annual under conditions that suit melons, cucumbers, gourds, and other warmth-loving members of the gourd family. Propagation is easy by seeds and by cuttings.

DIPLOGLOTTIS (Diplo-glóttis). One, or according to some authorities two species of trees endemic to Australia constitute

Diplocyclos palmatus

Diploglottis of the soapberry family SAPINDACEAE. The name, from the Greek *diploos*, double, and *glottis*, a tongue, alludes to the petals having two inner scales, not united. In its homeland the pulpy coverings of the seeds, which have an agreeable acid flavor, are esteemed for making jam.

Native in scrub lands, often in coastal regions and near rivers, ***D. cunninghamii*** is 30 to 40 feet tall or taller. It has shoots covered with rust-colored hairs. Its alternate, pinnate leaves, sometimes exceeding 2 feet in length, have eight to twelve oblong-elliptic to oblong-lanceolate leaflets 6 to 8 inches or, exceptionally, up to more than 1 foot long. They are pubescent only on their undersides. The little greenish-yellow blooms, in many-flowered panicles 1 foot long, or longer, from the leaf axils, have hairy, deeply five-lobed calyxes, four rounded petals, and eight unequal stamens that sometimes are longer than, sometimes shorter than, the petals. The short, curved style is tipped with a sometimes obscurely three-lobed stigma. The fruits are slightly fleshy, spherical capsules about ½ inch in diameter. Their seeds are covered with a fleshy coating called an aril. In Australia this tree is called the native tamarind.

Garden and Landscape Uses and Cultivation. Not much is recorded regarding the uses and cultivation of this quite handsome species in North America. It is suitable for climates such as that of southern California. Increase is by seeds and by cuttings.

DIPLOSOMA (Di-plòsoma). Two rare species of nonhardy, South African, perennial succulents of the *Mesembryanthemum* group of the carpetweed family AIZOACEAE are the sole constituents of this genus. The name, from the Greek *diploos*, double, and *soma*, a body, was given because each plant has only two leaves. These are opposite and rest on the ground. From their joined bases they spread like an open book, and together extend for only 1 inch or a little more. Very fleshy, and with slightly grooved or flat upper surfaces with tiny translucent markings, the leaves are green and hairless. At the conclusion of each growing season they shrivel and dry and only small, cone-shaped resting bodies remain. The stalkless, terminal blooms, of characteristic *Mesembryanthemum* form, are purplish and ½ to 1 inch across. They come from the centers of the pairs of leaves. The fruits are capsules.

Up to 1 inch tall, ***Diplosoma retroversum*** has oblong, blunt leaves joined along one side for about one-half of their lengths and for less than one-half of their opposite sides. They have translucent dots. Similar, but with the longest marginal joining of the leaves extending for only about one-quarter of their lengths, and with the leaves marked with translucent longitudinal lines, is ***D. leipoldtii.***

Garden Uses and Cultivation. These rare and choice miniatures are reported to be difficult to handle. Not a great deal is known about their requirements, but undoubtedly these are in general those that satisfy *Lithops, Conophytum,* and other dwarf, very succulent *Mesembryanthemum* relatives. In the wild diplosomas live in limestone regions and the addition of crushed limestone to the potting soil surely would be helpful and perhaps essential to success. Full sun and excellent drainage are obvious requirements, and care must be taken to keep the soil dry through the summer season of dormancy. It is recommended that the plants never be watered overhead, partial immersion of the pots so that moisture seeps from below being the preferred method of supplying moisture. Propagation is by seed. For further information see Succulents.

DIPLOTAXIS (Diplo-táxis)—Wall-Rocket. This group of the mustard family CRUCIFERAE accounts for more than twenty-five species of annuals, biennials, and perennials. It is indigenous to Europe and the Mediterranean region and is naturalized in North America. Closely related to *Moricandia,* its leaves are pinnately-lobed, its flowers white, pink, lilac, or yellow. The flowers have four sepals, four petals that spread to form a cross, six stamens, two of which are shorter than the others, and one style. The fruits are slender, flattened, beaked pods containing two rows of seeds in each of the two compartments. The name is from the Greek *diploos,* double, and *taxis,* arrangement, alluding to the seeds being in two rows.

The white wall-rocket (***D. erucoides***) is an annual 6 inches to 1½ feet tall. It has branching stems and a basal rosette of lobeless, pinnately-lobed, or toothed leaves, with the terminal lobe decidedly the largest. The upper leaves, shallowly-lobed and stem-clasping, are without stalks. About ¾ inch across, the flowers are white with purple veins; sometimes later they become entirely violet. The linear seed pods, with short, conical beaks, are up to 1¾ inches long.

Garden Uses and Cultivation. Of secondary importance as a garden plant, this species may be used in flower borders to give variety and in other places where the soil is well drained and the site sunny. Seeds are sown where the plants are to remain, in spring, or in regions of mild winters, in fall. The seedlings are thinned sufficiently that the plants are not unduly crowded. No special care is required.

DIPPING. This is a very effective way of applying insecticides, miticides, and fungicides to a few pot plants of sizes easy to handle, as is usually the case with houseplants. For such, dipping is often far less troublesome and messy than spraying.

Mix the solution to be used, at the dilution recommended for spraying, in a pail or other container. Let the amount be sufficient to allow the entire tops of the plants to be immersed. Then, one at a time, invert each plant and gently swish its aboveground parts for several seconds in the liquid. Shake it free of surplus solution and set it out of the sun to dry. With small plants one hand held over the top of the pot will prevent soil from dropping into the liquid. Cover the soil of larger plants with metal foil before upending them.

DIPSACEAE—Teasel Family. This group of dicotyledons includes about 150 species of annuals, biennials, and herbaceous perennials distributed among eight genera. Chiefly natives of north temperate Europe and Asia, tropical Africa, and South Africa, its members have opposite or rarely whorled leaves and interrupted spikes or compact heads of flowers, each backed by an involucre (collar) of bracts, that has a general resemblance to those of the daisy family COMPOSITAE.

Individual flowers (florets) of the heads have five- or sometimes four-cleft calyxes and corollas, four stamens, and a slender

Dipping: (a) Swishing the plant in an insecticide

(b) Gently withdrawing the plant and allowing it to drain

style. The fruits are achenes. The genera include *Cephalaria, Dipsacus, Knautia, Pterocephalus, Scabiosa,* and *Succisa.*

DIPSACUS (Díp-sacus)—Teasel. This, the type genus of the teasel family DIPSACACEAE, comprises fifteen species. It inhabits Europe, North Africa, and temperate Asia. The name *Dipsacus,* alluding to the water-holding leaf bases of some kinds, comes from the Greek, *dipsa,* thirst. Dried flower heads of the fuller's teasel are used for fulling (raising nap on) woolen cloth.

Teasels are coarse, thistle-like, hardy biennials with prickly or rough-hairy stems and foliage. Their leaves are opposite, pinnately-lobed or lobeless. The bases of each pair are usually joined to form an upturned cup. The small flowers are crowded in heads or short spikes with, at their bases, an involucre (collar) of several spiny bracts. Numerous slender, needle-like bracts sprout in pincushion fashion all over the heads. The blooms, those encircling the center of the flower head opening first, have calyxes with four teeth or lobes, corollas with four nearly equal lobes, four stamens, and a slender style. The fruits are achenes.

Fullers' teasel, ***D. sativus*** (syn. *D. fullonum*), is 4 to 6 feet tall and prickly-stemmed and has stalkless, lanceolate to oblong, toothed or toothless leaves, prickly along their midribs on the undersides, and up to 1 foot long. Held aloft on long, prickly stalks, the 2- to 4-inch-long heads of light lavender flowers have involucral (basal) bracts that at maturity point downward and are usually shorter than the heads. The numerous needle-like bracts that clothe the heads have strongly hooked apexes. This last characteristic, together with the fact that the bracts of the involucre curve upward and are as long or longer than the head, distinguishes the fullers' teasel from the common teasel (***D. sylvestris***) in which the slender spines of the head end in straight, barbed tips. These species, natives of Europe, are naturalized in parts of the United States.

Dipsacus sylvestris

Dipsacus sylvestris (heads of flowers)

Garden and Landscape Uses and Cultivation. Teasels are suitable for colonizing in semiwild landscapes, and the fullers' teasel for including in displays of plants useful to man. They succeed with little attention in sunny locations in any reasonably fertile, well-drained soil. Seeds sown about midsummer or a little earlier give plants that, set about 1 foot apart in nursery beds, are of a size suitable for transplanting to their flowering quarters in fall or early the following spring.

DIPTERONIA (Dipter-ònia). Except for *Acer* this is the only genus of the maple family ACERACEAE. It consists of two species of Chinese trees that differ from maples in that the wings of their fruits extend, like those of *Ptelea,* all around the seeds instead of from only one side of them as in *Acer.* Also, their leaves commonly have more leaflets than those of pinnate-leaved maples. The name *Dipteronia* comes from the Greek *dis,* twice, and *pteron,* a wing, and alludes to the fruits.

Dipteronias are deciduous trees or, in cultivation, sometimes tall shrubs. They have opposite leaves with an odd number, seven to fifteen, of toothed leaflets. Borne in large panicles at the branch ends, bisexual and male flowers are on the same plant. The tiny blooms have five each sepals and petals. The males have six to eight stamens and the bisexual flowers a compressed ovary, and a style with two stigmas as well as stamens.

The only cultivated kind, ***D. sinensis*** is up to 30 feet tall. Its leaves, up to 1 foot long, have mostly nine to thirteen short-stalked, coarsely-toothed, ovate to lanceolate leaflets 1½ to 4 inches long and one-third as wide. When young, like the shoots, the leaves have scattered hairs, and there are tufts of hair in the axils of the veins. The lowest pair of leaflets is sometimes divided into three smaller leaflets or one three-lobed. The greenish-white flowers are succeeded by fruits, ¾ to 1 inch long, in large clusters.

Garden and Landscape Uses and Cultivation. The species described is handsome, and hardy about as far north as Philadelphia. Its chief ornamental features are its foliage and fruits. Its blooms are insignificant. This species is effective as a small specimen tree. It is satisfied with ordinary soil of reasonable moisture content and is propagated by seeds, summer cuttings in a greenhouse propagating bench or under mist, and by layers.

DIRCA (Dír-ca)—Leatherwood. There are two species of this North American genus of the mezereum family THYMELAEACEAE. They are related to *Daphne* from which they differ in their flowers having extremely minute sepals (the showy parts of the blooms in daphnes) and the stamens and style much protruding. The name is from the Greek *Dirke,* the name of a fountain at Thebes and also one of mythological significance. Dircas are freely-branching shrubs with tough pliable shoots covered with strong fibrous bark, and with old leaf scars that give them a jointed appearance. They have undivided, alternate, short-stalked leaves. Their tiny, inconspicuous, yellow blooms, in clusters of two to four and short-stalked or stalkless, are without petals. There are usually four calyx lobes (sepals) and as many stamens, exceeded in length by the style. The flowers come in early spring before the foliage. The fruits are berry-like; each contains one seed. The Indians bent shoots of these shrubs to make hoops and used them as ties and thongs.

An inhabitant of rich, moist to wet, fertile woodlands from Quebec to Minnesota, Florida, Alabama, Arkansas, and Oklahoma, ***Dirca palustris*** is 3 to 6 feet tall, of neat habit, and has elliptic to obovate

Dirca palustris

leaves 2 to 3 inches long and one-half to two-thirds as wide as long. Their undersides are slightly glaucous and, when young, are pubescent. The flowers have very short stalks. The fruits, pale green or reddish, are egg-shaped and about ⅓ inch long. The Western species ***D. occidentalis*** is similar except that its flowers are stalkless and have more deeply-lobed calyxes.

Endemic to California, it is not hardy in the north.

Garden and Landscape Uses and Cultivation. Although by no means showy, leatherwoods are pleasing both in and out of bloom and are worth planting in a modest way for variety. They are especially well adapted to damp and even wet soils, but get along in any ordinary earth not too dry. They grow slowly, and, well-spaced in the open, are shapely. If crowded or in shade they become straggly. No regular pruning is needed, but any deemed desirable to shape them may be done in spring. Propagation is easy by seed and by layering.

DIRT GARDENER. This, an American expression scarcely known in Great Britain, is a colorful vernacular designation for gardeners who actually work with soil and plants in contrast to those who merely know about such things or only direct or supervise the labors of others. It is commonly applied in admiration and approbation of successful practical gardeners.

DISA (Dì-sa). One of the most challenging groups of orchids to grow, *Disa* inhabits Africa, Madagascar, and the Mascarene Islands. It includes about 130 species of the orchid family ORCHIDACEAE. The name, alluding to the somewhat netted pattern of the upper sepal of the flowers of the typical species, is that of mythical Queen Disa of Sweden who, it was believed, appeared before the King of the Sveas clad in a fishnet. Most familiar to gardeners, by reputation at least, is *D. uniflora,* a splendid endemic of South Africa that following many failures has occasionally been brought to flower in the United States.

Disas are tuberous-rooted ground orchids. They do not perch on trees. Their leaves may be all basal, or along the stems. The flowers, in clusters, racemes, dense spikes, or rarely solitary, have three separate sepals. The upper one is concave or hood- or helmet-shaped and has a basal pouch or spur. Usually much smaller, the two petals are generally more or less joined to the base of the column. The lip is commonly small and narrow.

The red disa (***D. uniflora*** syn. *D. grandiflora*), renowned inhabitant of Table Mountain, Cape Town, South Africa, is stringently protected by law from collectors and pickers of wild flowers. The largest-flowered and showiest of the clan, it is variable and occurs in acid soils along mountain streamsides that in winter are often inundated. Attaining heights of 1 foot to 3 feet and with leafy stems and leaves up to 8 inches long by one-half as wide, this species belies its name by frequently having two to six or sometimes more blooms on a stem. They are 3 to 4 inches in diameter and typically have vivid scarlet lower sepals and an upper paler

Disa uniflora

one tinged with yellow and strongly lined with crimson. The petals are inconspicuous, the lip small, the column white. There is a variant with rich rose-pink blooms. Much resembling a miniature *D. uniflora* is South African ***D. racemosa.*** This has rose-pink blooms with darker veins. Hybrids between *D. uniflora* and *D. racemosa,* some more robust, with more flowers, and more disease-resistant than either parent, have been raised, as have others between *D. uniflora* and other species.

Garden Uses and Cultivation. Disas are only for the most expert orchid fanciers, equipped to ensure the exacting environmental conditions these plants require. These include cool nights throughout the year. In the native haunts of *D. uniflora* temperatures on summer nights are often 45 to 50°F, and although days may be considerably warmer, the plants are then likely to be bathed in cool mist. A porous, acid rooting medium is essential. Watering should be from below by capillarity rather than by overhead sprinkling, but very light misting of the foliage, not so heavy that water collects in the centers of the plants, is in order on bright days. In most parts of the United States success is likely only in air-cooled greenhouses, such as have been employed with fine results at Longwood Gardens, Kennett Square, Pennsylvania, but disas have been grown and bloomed outdoors in favored locations in the Pacific Northwest. Particularly in their young stages, these orchids are subject to fungus infections that can be disastrous. These must be prevented or checked. The provision of a porous rooting medium is of help in this. For this purpose sphagnum moss, and sphagnum moss mixed with a little leaf mold and sand, have been used successfully. In addition, regular preventative spraying with a fungicide may be necessary. For further information see Orchids.

DISANTHUS (Dis-ánthus). As its specific epithet *cercidifolius* indicates, the foliage of the only species of this genus looks much like that of unrelated redbud (*Cercis*). It resembles its near kin, the witch-hazels (*Hamamelis*), less closely. Belonging in the witch-hazel family HAMAMELIDACEAE, and restricted in the wild to the mountains of Japan, *Disanthus cercidifolius* differs manifestly from witch-hazels in the way the veins of its leaves are arranged. In *Hamamelis* there is a strong mid-vein with parallel lateral ones branching from it; in *Disanthus* five to seven main veins diverge from the base of the leaf like the fingers of a widespread hand. The name derives from the Greek *dis,* twice, and *anthos,* a flower. It alludes to the arrangement of the blooms.

A deciduous shrub 8 to 10 feet tall or sometimes taller, ***D. cercidifolius*** has slender branches, and alternate, blunt or rounded, broadly-ovate to nearly circular, toothless, stalked leaves with more or less heart-shaped bases. They are 2 to 4½ inches long and nearly as wide, hairless, and dull glaucous blue-green. In fall they turn beautiful claret-red and orange. The deep purple, star-shaped flowers, ½ to ¾ inch wide, are in pairs, the individuals set back to back at the ends of short stalks from the leaf axils. They open in late fall and are faintly ill-scented. They have hairy, five-parted calyxes, five linear petals, the same number of stamens, and two short styles. The fruits are two-lobed capsules ½ inch across containing several shining black seeds.

Garden and Landscape Uses. Habit and foliage are the chief merits of *Disanthus.* Its floral display is meager, and its fruits are without decorative appeal. It is especially lovely in fall when its leaves make a grand show of color. This general-purpose shrub is well adapted for fertile, not excessively dry, slightly acid soils, and sunny locations. It is hardy about as far north as Philadelphia.

Cultivation. The cultivation of this pleasing shrub makes no great demands of the gardener. It needs no special care. Regular pruning is not needed, but any necessary to improve shape or limit size may be done in spring. Propagation is by seed.

DISBUDDING. The removal of young buds by pinching is called disbudding. Its purpose is to concentrate growth energy in the buds that remain in the expectation that they will be encouraged to produce superior blooms. The practice is commonly followed with large-flowered chrysanthemums and dahlias, and with calendulas, roses, peonies, snapdragons, stocks, and many other plants. Disbudding may involve, as with chrysanthemums, leaf buds that if left would develop into unwanted shoots, but more often is concerned only with reducing the number of flower buds. The pinching out of leaf buds changes the branching and form of the plant and so is akin to pruning. Taking out flower buds

Disbudding a carnation

(b) A disbudded rose

Disbudding chrysanthemums

(c) A rose not disbudded

does not change the shape of the plant appreciably. Its effects show in the placement, size, form, and occasionally improved color of the blooms.

Beyond the injunction to disbud early, as soon as the little buds can be rolled out with the ball of the thumb or nipped out with finger and thumb, rules generally applicable to disbudding can not be given. With a little experience and observation the gardener inexperienced in this technique soon learns how many buds can be safely left on a particular plant or branch with the expectation that they will mature to the quality desired. When disbudding take special care not to injure the remaining buds.

Disbudding a rose: (a) Removing tiny side buds

DISCARIA (Dis-cária). Most of the ten to fifteen species of *Discaria* of the buckthorn family RHAMNACEAE are natives of South America, but one is endemic to New Zealand, and a very similar one to Australia. The genus is related to, and resembles *Colletia*. The most clear distinction is that the stipules (appendages) at the bases of the leafstalks in *Discaria* are connected by two distinct lines or ridges, those of *Colletia* are not. The name comes from the Greek *diskos*, a disk, and alludes to the broad fleshy disks of the flowers.

Discarias are rigid-branched, spiny shrubs or small trees with leaves opposite, in opposite clusters, or sometimes absent. Morphologically the spines are modified branches. The flowers are in axillary clusters or are solitary. Each has a bell-shaped, four- or five-lobed calyx below which is a disk, frequently with a wavy margin. Often there are no petals, sometimes there are four or five. The stigma is three-lobed. The fruits are dry, three-lobed capsules.

In New Zealand called Irishman, wild Irishman, and tumatu-kuri, ***D. toumatou*** is a very spiny, much-branched shrub or small tree 15 to 20 feet tall. It has forking, intricately interlaced branches with green younger parts. The smaller lateral branches are represented by pairs of vicious, stout spines 1 inch to 2 inches long that spread at right angles to the branches. Sometimes there are no leaves, but more commonly this species has narrow, obovate or obovate-oblong, more or less deciduous, leathery leaves up to ¾ inch long. They are opposite on shoots of the current year, or in clusters from just beneath the spines on older shoots. The profuse, clustered, or rarely solitary blooms appear in spring with the leaves. They have four or five greenish-white sepals, but no petals. The deeply-three-lobed, approximately spherical fruits are about ⅕ inch across. The Australian ***D. australis*** differs chiefly in its flowers having petals.

Discaria toumatou

Native of Chile and Patagonia, ***D. crenata*** (syn. *D. serratifolia*) a shrub about 12 feet tall, has slender, pendulous branches armed with pairs of stiff, sharp, up-pointing spines ¾ to 1 inch long. Its opposite, shallowly-toothed, ovate-oblong, highly glossy, dark green leaves are ½ to 1 inch long. Crowded on short twigs that develop from the previous year's shoots, the fragrant greenish-white blooms, without petals, appear in late spring or early summer. A more recent introduction into cultivation, ***D. discolor,*** of Argentina and Patagonia, has succeeded outdoors in southern England. Up to 6 feet tall and 8 feet across, it has sometimes prostrate main branches, opposite, often spine-tipped lateral branchlets, and opposite, blunt, elliptic or elliptic-obovate, toothed or toothless leaves up to ¾ inch long. In twos or threes, the petal-less, white, sweetly scented flowers have four-lobed calyxes.

Garden and Landscape Uses and Cultivation. Unusual shrubs that add considerable interest to garden plantings because of their distinct appearance, discarias are suitable only for mild, nearly frost-free climates such as that of California. They may be included in shrub groupings, foundation plantings, and suchlike displays. They favor sunny locations and well-drained, dry or at least dryish soils. No pruning is needed. Propagation is by seeds and by summer cuttings.

DISCHIDIA (Dis-chídia). Species of this fascinating genus of the milkweed family ASCLEPIADACEAE are easy to grow in warm, humid greenhouses under conditions that suit many orchids and other tropicals. Because of their curious appearance and extraordinary way of life they are well worth cultivating by gardeners interested in the unusual. Dischidias are slender, evergreen vines many of which live naturally in rain forests where from time to time spells of dry weather reduce or eliminate their regular supplies of moisture. This difficulty is overcome by some species by storing water against times of scarcity. Their leaves are developed as large pitchers or pouches that vary in shape and size according to species. Some kinds have double pouches, one inside the other like a lined purse. It is believed that, in addition to functioning as reservoirs, the pouches absorb carbon dioxide from the air through their interior walls at times when it is too dry for ordinary leaves to function satisfactorily in this way.

The pouches collect water transpired through their inner walls, supplemented, in some cases, by moisture that finds its way in through the small orifice in each pouch that opens to the outside. The saving of transpired water for future use is a remarkable example of conservation. In addition to the water, the pouches often contain soil, brought in, apparently, grain by grain by ants. An extraordinary development of some species, for instance *Dischidia rafflesiana*, is that, from the leafstalk or stem, close to the orifice of the pouch, a root develops, enters the opening, and inside the pouch branches freely. There it absorbs moisture, as well as nutrients from any soil that is there. In other species similar roots originate from the neck of the inner bladder. This is true of *D. pectenoides.* The pouches are, in effect, living flower pots. In the wild, aggressive, stinging ants inhabit *Dischidia* pouches, but most botanists are of the opinion that their presence has no symbiotic significance other than the insects' need for shelter and places to raise their young.

Dischidias are related to the more familiar wax plant (*Hoya*). They are epiphytes, clinging to trees and shrubs and perching upon them without being connected to the ground. They do not take nourishment from their hosts. The genus contains about 10 species and is native from India to Australia. Its leaves are opposite, the pouches opposite or solitary. The small, urn-shaped, white, yellowish, or red flowers are in axillary clusters. They have five each sepals and petals, and five stamens united with the two carpels, which have their styles joined. The name comes from the Greek *dischides,* twice cleft, and alludes to the segments of the corona or crown of the minute flowers.

Kinds sometimes grown include ***D. pectenoides,*** a native of the Philippine Islands. This has solitary, double pouches alternating with pairs of ordinary leaves. The pouches are flattish and up to 2¾ inches long by 2 inches wide. Another kind met with in cultivation is ***D. rafflesiana,*** of Malaya and New Guinea. It has single-walled pouches, 2 to 5 inches long, and yellowish flowers. Another interesting species, ***D. imbricata,*** has no pouches, but does have shallowly-inflated, nearly round leaves that have their edges firmly pressed against the support upon which they grow and so protect and shelter the roots behind them.

Dischidia pectenoides

Dischidia rafflesiana

Dischidia imbricata

Garden Uses and Cultivation. These are choice, rare plants for keen gardeners who can provide a highly humid, tropical environment. They need a minimum winter temperature of 60 to 65°F and shade. They may be grown attached to sections of trunks of tree ferns, to posts wrapped around with a mixture of osmunda fiber and sphagnum moss, or in small hanging baskets containing a mixture of osmunda fiber, sphagnum moss, and some broken charcoal. The rooting medium should be kept always moist. Increase is by division and by cuttings.

DISCOCACTUS (Disco-cáctus). Generally considered by specialists to be among the more difficult cactuses to grow, the nine species of *Discocactus* of the cactus family CACTACEAE are natives of Brazil and Paraguay. They much resemble small melocactuses, but differ in having flowers that open only at night and in having more numerous perianth segments (petals). The name, which must not be confused with that of the entirely different cactus genus *Disocactus,* comes from the Greek *diskos,* a disk, and cactus. It alludes to the form of the plants.

Discocactuses have obese, flattened-spherical to shortly-cylindrical plant bodies with low, usually straight, not spiraled, slightly tubercled (lumpy) ribs. Like *Melocactus,* at maturity they are crowned with a topknot-like cephalium of hairs and spines around a woody core. From the cephaliums, which are smaller than those of melocactuses, the flowers develop. Quite large in comparison to the sizes of the plants, the funnel- to flat-saucer-shaped, tubular blooms are white to pinkish. The fruits are small and slightly fleshy.

Broader than tall, bluish-green ***D. placentiformis*** has ten to fourteen low ribs. The six or seven areoles (spine-producing cushions) on each have about the same number of more or less recurved, flattened, spreading, thick spines and sometimes an erect, central spine. The large blooms are rose-pink with the inner petals white. The cephalium from which they develop is woolly, but has few or no bristles. The spherical, juicy fruits are white. This is a native of Brazil. From the last, ***D. alteolens*** differs in the cephalium being of long woolly hairs with many erect bristles interspersed. Its gray-green plant body is solitary and up to about 4 inches in diameter. It has nine or ten low ribs. The areoles have five or six radial spines and occasionally an erect central one. The slender-tubed flowers have petals, the inner ones white, that spread widely. This is also Brazilian.

Garden Uses and Cultivation. Not likely to be found in other than choice collections of succulents, discocactuses are challenging to grow. The suggested care is that favorable to *Melocactus*. They require considerable warmth, with a minimum temperature of 55 or 60°F recommended. For more information see Cactuses.

DISEASES OF PLANTS. In the broad sense used by some plant pathologists diseases are defined as any interference with the normal development of a plant caused by continued irritation. Obviously infestations of many kinds of insects, mites, and other small creatures come within the scope of this definition. One might even include the constriction of a stem or branch caused by a too-tight tie, and the misshapen, dwarfed, bonsai-like aspect trees may assume at seashores and in mountain habitats as a result of exposure to wind. More often gardeners exclude from their concept of diseases the effects of insects and other creatures of the animal world, which they group as pests. They also omit the results of mechanical irritations from their definition of disease, and even certain virus-produced effects considered desirable or acceptable such as some leaf variegations, are excluded.

Restricted in these ways, as it is in this discussion, the term disease as applied to plants still covers conditions referable to a wide range of causes. Diseases may result from the presence and activities of living organisms of the plant world, such as bacteria and fungi, from infection with viruses, by the absence or lack of availability of a nutritional element, such as boron, iron, or zinc, by the presence of harmful substances in the soil or water or of pollutants in the air, or by other unsatisfactory environmental conditions, such as poorly drained soil or continued exposure to too-low temperatures.

Prevention and control of diseases are highly important phases of horticultural practice. Gardeners must ever be cognizant of the harm they can do and alert to prevent their onset and limit their destructiveness and spread. Approached intelligently, the problem of containing diseases is not nearly as insufferable as many amateurs fear, nor as casually achieved as others hope. The keys to success are a knowledge of kinds of diseases most likely to be encountered, frequent keen observation of the plants and evaluation of their health, and the timely and prompt use of preventative and control measures.

Prevention, where practicable better than cure, begins with the selection of kinds of plants best adapted to local environments, including climate, type of soil, and particular features, including exposure to sun, shade, and wind, condition of the subsurface drainage, and provision for satisfactory air circulation and air drainage. In some cases, as for instance with aster yellows and with rust disease of snapdragons, restricting plantings to immune or resistant varieties is of prime importance in areas where these diseases are troublesome. Diseases that depend for survival on alternate hosts, those with causal organisms that, like cedar-apple rust and white pine blister rust live on different kinds of plants at different seasons of the year, can be controlled by eliminating one of the hosts from the vicinity.

Aster yellows disease

Good cultural practices are of immense importance in limiting disease damage. These include draining soils too wet for the kinds of plants to be grown, intelligent use of fertilizers, selection of suitable sowing and planting dates, adequate spacing between plants, protection against excessive cold, heat, dryness, and wetness, and sanitation, including weed control and certain pruning.

Positive control measures are often necessary to supplement preventative ones. Despite extravagant claims sometimes made by way-out devotees of the organic school of gardening, disease-free or even reasonably disease-free gardens are rarely possible without the use of fungicides and other chemical controls. The unnecessary employment of these, however, is foolishly wasteful of time and materials and may in some circumstances even be harmful to the plants. As with human medication one must always weigh probable benefits against possible disadvantages. The chief of these last are cost, expenditure of time, and sometimes unsightly deposits on the plants. Carefully used according to directions, sprays, dusts, and other chemical disease controls recommended by State Agricultural Experiment Stations and other recognized authorities will not endanger the health or lives of humans or pets.

Many diseases can be controlled with timely spraying

The first need is for the gardener to become acquainted with the diseases com-

Among common plant diseases are: (a) Anthracnose of plane trees

(b) Black spot of roses

(c) Club root of cabbage

(d) Mildew of roses

mon and recurrent in his neighborhood. Depending upon where he (or she) gardens these may include, among many others, anthracnose of beans or plane trees, asparagus rust, black spot of roses, botrytis or gray mold, club root of cabbage and allied crops, damping-off of seedlings, hollyhock rust, mildews, and soft rot of irises. Fortunately, if suitable care is taken in selecting and caring for plants comparatively few diseases are likely to be bothersome in any one locality. They are the ones upon which to concentrate. Information about them is readily available from Cooperative Extension Agents and State Agricultural Experiment Stations.

Correct diagnosis of an ailing plant is basic to appropriate treatment. It is useless to apply a fungicide if the trouble is caused by bacteria or mites or to spray or dust if root knot nematodes or poor soil drainage have destroyed roots.

Early detection is of great importance in keeping plants healthy. Preventative measures are of course taken in advance of trouble, before visible symptoms come and, one hopes, precluding their appearance. Soil sterilization is an excellent example of such technique. Others are the application of fungicides in accordance with carefully timed schedules calculated to prevent rather than cure infection. Such spray schedules are commonly followed by fruit growers. With some diseases, black spot of roses, for example, regular, but less precisely timed applications of sprays or dusts give good results in preventing infection.

When preventive measures have not been taken or for some reason have failed, accurate diagnosis of what is wrong is imperative. There is little use fooling around on a trial basis, grounded in ignorance. Except with common diseases recognizable beyond reasonable doubt, the gardener should seek expert advice in determining the cause of trouble. This may come from experienced local gardeners or nurserymen, from accurate interpretation of information available in books, bulletins, and pamphlets dealing with particular groups of plants, from consultation with authorities at botanical gardens and similar institutions and State Agricultural Experiment Stations, or with Cooperative Extension Agents. It is important to remember that besides the few easily recognizable diseases that all gardeners should learn to identify there are many others that require microscopical examination and other sophisticated techniques, as well as expert knowledge beyond that the ordinary gardener can be expected to possess. If at all in doubt seek expert advice. But do that early. Hence the importance of early detection of symptoms. That is your business.

Some common types of diseases grouped according to easily observable symptoms are discussed in this Encyclopedia under the entries Anthracnose, Bacteria, Bleeding, Blight, Blotch, Canker, Chlorosis, Club Root, Crown Gall, Damping-Off, Die-Back, Downy Mildew, Edema, Fairy Rings, Fire Blight, Galls, Gas Injury, Gummosis, Leaf Blister, Leaf Blotch, Leaf Curl, Leaf Scorch, Leaf Spot Diseases, Mold, Mosaic Diseases, Needle Cast, Powdery Mildew, Rot, Rust, Scab, Scurf, Smut, Stunt, Virus, Wilt, and Witches' Broom.

DISH GARDENS. Bowls, saucers, and other dishes and dishlike containers in which grow small plants arranged as miniature landscapes are called dish gardens. From terrariums they differ in that the plants stand well above the rims of the receptacles instead of being enclosed or covered. Their environment is that of the place where they stand, not a special, locally controlled one.

Dish gardens are used chiefly as indoor decorations, and less often on porches, patios, and terraces. They may be employed as accents on window sills, mantles, coffee tables, end tables, and suchlike places and as dining table centerpieces. Usually the containers are porcelain, earthenware, glass, metal, or plastic. The plants used in each container are generally of more than one kind.

It is often stressed, quite soundly that dish garden containers be furnished with drainage holes to permit the ready escape of surplus water. Yet it is perfectly practicable to have plants, even cactuses and other desert succulents, prosper for long periods in containers without such accommodation. The explanation is that, especially in warm, arid rooms, loss of moisture from the soil by evaporation and transpiration from the foliage equals the amount of water supplied with the result that there is no buildup of water in the dish and the roots therefore do not remain sodden for long periods of time. Nevertheless drainage holes act as safety valves. Arrange for them if you can. Dish gardens with drainage holes should of course be kept standing in saucers if there is danger that leakage will damage furniture or other surfaces.

Containers may be of any shape or size, but not so big that they cannot readily be lifted. They should be deep enough to hold at least 3 inches of soil. Let this be appropriate for the type of plants used. For most a mix consisting of equal parts loamy top soil, peat moss, and perlite or coarse grit will suit, a little more perlite or grit for succulents, a somewhat higher proportion of peat moss and a sprinkling of crushed charcoal added for ferns and other woodlanders are appropriate modifications. Before filling in the soil put a 1-inch layer of perlite, grit, small crocks, or small pebbles in the container after first covering the drainage holes with larger crocks or fine wire mesh. If there are no drainage holes increase the depth of the under layer.

Select plants for individual containers that can be expected to prosper under similar conditions of light, temperature, humidity, and watering. Dish gardens planted in this way look right. Incongruous mixes of desert plants and shade-loving forest plants, of kinds needing much warmth and those that do well only under cool conditions do not. Of necessity plants to be set in dish gardens are small. Generally they will come from pots 2 or 2½ inches in diameter. Make sure that their balls are soaked with water an hour or two before planting and that they are free of scale insects, red spider mites, or other pests.

When planting aim to achieve interesting combinations of masses and forms and lines as well as types, textures, and foliage colors. It often helps to contour the soil surface by sloping it or establishing mounds and valleys rather than to have it flat, and perhaps to introduce an interesting rock or two or a piece of tree branch or driftwood. Avoid like the plague the dreadful figurines, plastic birds, trellises, pieces of mirror intended to simulate lakes, baubles, colored gravels, and other dressings beloved by the designers of some commer-

A selection of dish gardens planted with cactuses and succulents

cially produced dish gardens. Begin by putting a little soil over the drainage layer, then arrange and rearrange the plants until you have a satisfactory pleasing effect. Plant fairly thickly. A little crowding, but not to the extent of that often practiced by many commercial florists, is in order. As the plants are positioned, work soil around their root balls and press it firmly.

When planting is completed water gently with a fine spray. Cover the garden with a polyethylene plastic bag of the type used in freezers and set it in a light place out of direct sun. Keep the covering on, with holes or a tear or two in it to prevent humidity from building up too high, for three or four weeks and then remove it.

Now the garden is ready to be put where it is to serve its chief decorative purpose. This should not be a location exposed unduly to drafts or, if it be composed of shade-loving plants, near a sunny window. Often the reverse will be true. You may want to display your dish garden where there is insufficient light for the plants' needs. This can be overcome in part at least by keeping it near a lamp or some other source of light or by moving it for a few hours each day to a lighter location than the one it is to grace on occasion. Thus a dish garden used as a dinner table centerpiece may stand in a light window for part of each day. Or if that is impracticable or too much trouble you may rotate more than one garden on a weekly or monthly basis between favorable and less favorable locations. Yet another solution is to accept as expendable dish gardens in difficult locations and simply replant them with the most tolerant plants available when they begin to show signs of shabbiness. Even so they will please for a long time. In fact, all dish gardens are expendable in that they will not live and remain attractive forever. Eventually they exhaust their limited soil, in time they become overgrown. When this happens they cannot, as can pot plants, be repotted into bigger containers. The answer is to take them apart and replant, using new soil and, as needed, new plants.

Dish garden with peperomia and English ivies

Dish garden with *Philodendron, Acorus, Cordyline,* and *Aglaonema*

Dish garden with *Syngonium, Sansevieria,* and dracaenas

Dish garden with *Aglaonema, Dracaena,* and *Philodendron*

Dish garden, a ceramic log with bromeliads

Routine care consists of careful watering, promptly removing faded and dead plant parts, possibly occasional pruning to check exhuberant growth, and periodically sponging the leaves of smooth-leaved plants with soap and water to remove dust and to freshen them. Occasionally a plant may die or may outgrow its space and need replacing. If this happens remove as much of the soil about its roots as possible when you take it out, and pack new soil around the roots of the replacement. Fertilizing is usually not desirable. Certainly this is true of gardens in containers without drainage holes. Others may perhaps benefit from a mild application given two or three times from spring to fall. If plants are clean of them from the beginning, pests are unlikely to be troublesome, but those common to houseplants, such as aphids, mealybugs, scale insects, and red spider mites, may appear. See also Houseplants or Indoor Plants.

Plants suitable for dish gardens are numerous. Give your imagination more or less free rein and you will have no difficulty in coming up with suitable ones. A visit to a local greenhouse, garden center, or plant counter in a department store is likely to be profitable. Small cactuses and other succulents of all kinds are splendid. Tiny-leaved varieties of English ivy, peperomias, baby palms, creeping *Ficus,* baby's tears, dwarf bromeliads, small-leaved begonias, *Acorus gramineus,* African-violets, and little plants of boxwood and pittosporums all are suitable. There are many others.

DISK or DISC. This is an expanded end of a flower stalk or floral axis (receptacle or torus) or consists of more or less united staminodes or nectar glands at or around the base of a pistil. The centers of flower heads of plants of the daisy family COMPOSITAE, consist of such a development covered with tubular florets are called disk florets.

DISOCACTUS (Diso-cáctus). The name of this genus of the cactus family CACTACEAE, which must not be confused with that of quite different *Discocactus,* refers to the blooms having perianth parts (petals) in two rows with usually equal numbers in each. It is from the Greek *dis,* twice, *isos,* equal and *cactus.* The genera *Chiapasia* and *Pseudorhipsalis,* included by some authorities in *Disocactus* are treated separately in this Encyclopedia. When so limited, *Disocactus* consists of two species.

Disocactuses are epiphytes. Like many orchids and bromeliads and some few other cactuses they live on trees, but without taking sustenance from them. In this they differ from parasites. Disocactuses have much the appearance of epiphyllums. They have slender, wandlike stems with flattened, notched or toothed branches, and with areoles (specialized areas on cactus stems from which spines, when present, arise) along their edges. Comparatively small and red, their few-petaled flowers are open by day. The fruits, spherical or egg-shaped, are without angles.

Native of Honduras, ***D. biformis*** has stems 9 inches to 2 feet long, with lanceolate branches with conspicuous midribs. The short-tubed blooms, up to 2 inches long, in bud are slender and pointed. They are borne near the ends of the branches and open to reveal eight or nine pointed, light red to magenta petals, and slightly longer stamens and styles, the latter tipped with four- or five-lobed stigmas. The outer petals are recurved or spreading, the inner more erect. The pear-shaped fruits, a little more than ½ inch long, are wine-red.

Native to Guatemala, ***D. eichlamii*** is much branched from its base. Cylindrical below and flattened above, the stems are 9 inches to over 1 foot long by up to 2 inches wide. The carmine flowers, borne in great profusion along the upper margins of the branches, never spread their petals widely. They have slender, trumpet-shaped perianth tubes. The fruits are red, spherical, and a little over ⅓ inch across.

Garden Uses and Cultivation. These are as for *Rhipsalis.* See also Cactuses.

Disocactus biformis

DISPHYMA (Dis-phŷma). The rather unusual natural distribution of *Disphyma* of the *Mesembryanthemum* relationship of the carpetweed family AIZOACEAE includes Australia, Tasmania, New Zealand, and the Chatham Islands, as well as South Africa. The group consists of three species of low, creeping, tuft-forming or matting, succulent herbaceous perennials with flowers, daisy-like in aspect but not in structure, that open about noon. The name, from the Greek *dis,* double, and *phyma,* a tubercle, alludes to a feature of the fruits.

Disphymas have opposite leaves, the bases of each pair slightly united. In section they are semicylindrical or three-angled and may be smooth and have little translucent dots or slightly roughened with tiny projections. The rosy-violet, pink, or white flowers are solitary or in twos or threes. The fruits are capsules with a two-lobed tubercle at the opening of each of their compartments, a condition unique in the family.

South African ***D. crassifolium*** has fairly widely spaced pairs of bluntly three-angled, short-pointed leaves 1 inch to 1½ inches, or sometimes up to 3 inches, long. They are dark green, with slightly translucent dots, and have short, densely-leaved branchlets in their axils. The solitary rose-pink, mauve, or sometimes white blooms are about 1¼ inches across. Native to Australia, Tasmania, New Zealand, and the Chatham Islands, ***D. australe*** has prostrate, freely-branching, rooting stems about 1 foot long, with the pairs of leaves well separated on the long shoots, crowded on the short branches. The leaves are ¾ inch long, triangular to half-round in section, with smooth, slightly horny edges, somewhat bulging sides, and the keel broadened toward the apex. The solitary, short-stalked reddish blooms are approximately 1 inch across. They have five stigmas. This species in the wild favors brackish, often moist soils.

Garden and Landscape Uses and Cultivation. In warm, arid climates these are good rock garden plants. They are also appropriate for greenhouses conditioned to suit succulents, and for sunny windows. Indoors, a winter night temperature of 40 to 50°F is adequate, with up to fifteen degrees increase by day permitted. Disphymas may be grown in pots, pans (shallow pots), and hanging baskets. They need very well-drained soil, kept on the dryish side, and full sun. They are readily multiplied by division, cuttings, and seeds. For further information see Succulents.

DISPORUM (Dí-sporum)—Fairy Bells. Closely related to and having the general appearance of the bellworts (*Uvularia*), but generally inferior to them as ornamentals, disporums differ in having stamens with stalks longer than the anthers, a single style with three short-spreading stigmas, and fleshy, few-seeded berries. These characteristics also distinguish it from related *Tricyrtis,* of Asia. The genus *Disporum* is represented in the native floras of North America and eastern Asia and consists of twenty species. Belonging in the lily family LILIACEAE, it has a name derived from the Greek *dis,* double, and *spora,* a seed, alluding to the ovules of some kinds being in pairs.

Disporums are deciduous herbaceous perennials with rhizomes and erect stems branched above and having several stalkless leaves. The drooping or spreading flowers, bell-shaped or rotate (with wide-spreading segments), solitary or in terminal clusters, have six narrow, petal-like segments. There are six stamens.

North American species cultivated include *D. lanuginosum* and *D. maculatum,* of the eastern part of the continent and some from the West. The flowers of ***D. maculatum*** are white spotted with purple, have stamens nearly as long as the corolla, and are about twice as big as those of ***D. lanuginosum,*** which has unspotted, yellowish-green blooms, $\frac{1}{10}$ to ⅕ inch across, with stamens one-half to two-thirds as long as the corolla. Both have forked stems up to about 2 feet in height. The leaves of the former are ovate or ovate-lanceolate, those of the latter ovate-oblong to oblong. They are about 3 to 4 inches long. The densely-hairy berries of *D. maculatum* are yellow, those of *D. lanuginosum* are red and without hairs.

Of the Pacific Coast species ***D. smithii*** has the biggest flowers, ¾ to 1 inch long, cylindrical, and whitish. They are in clusters of up to five or occasionally solitary.

Disporum smithii

The fruits are pale orange to red. This kind, which has lanceolate-ovate to ovate leaves up to 4½ inches long, inhabits moist woods and is 1 foot to 3 feet in height. About as tall, but differing in having whitish blooms shaped like a top, narrowed at their bases and not more than ⅝ inch in length, is ***D. hookeri.*** Its flowers are solitary, or in twos or threes, and its leaves are ovate to oblong-ovate and up to 4 inches in length. The berries are scarlet. A probable natural hybrid between these two species is ***D. parvifolium.*** Rarely exceeding 1¼ feet in height, it has ovate to ovate-lanceolate leaves slightly over 1 inch in length, and creamy-white, cylindrical flowers under ½ inch long, in twos or threes. It does not produce seeds.

Native to woodlands and foothills in Japan and China, ***D. sessile*** has rather short rhizomes and erect, branchless or sparingly branched stems 1 foot to 2 feet tall. Its pointed, oblong to broad-lanceolate or elliptic leaves are 2 to 6 inches long and up to 1¾ inches wide. Solitary or in twos or threes, the tubular, more or less pendulous flowers 1 inch long or a little longer, are white or greenish. The leaves of *D. s. variegatum* are variegated with white.

Garden Uses and Cultivation. Disporums are appropriate in woodland and native plant gardens in regions with climates that approximate those in which they live in the wild, and they are easy to grow in shade in moist soil that contains an abundance of organic matter. They are increased by division in early spring or early fall and by fresh seeds sown in a cold frame, cool greenhouse, or bed outdoors protected from disturbance. Care is routine. They must be watered in extended periods of dry weather and they benefit from being kept mulched with peat moss, leaf mold, or sifted compost. A light application of a complete fertilizer each early spring increases vigor.

Disporum sessile variegatum

Disporum sessile variegatum, in flower

DISSOTIS (Diss-òtis). Native to Africa, including South Africa, *Dissotis,* of the melastoma family MELASTOMATACEAE, comprises 140 species. Its name, alluding to the anthers, is derived from the Greek *dissoi,* of two kinds.

Members of this genus are herbaceous plants or small shrubs, several handsome. They are usually hairy, have ovate to oblong or elliptic leaves with three or five longitudinal, pronounced veins. Solitary or in racemes or panicles, the mostly large, rose-pink, purple, or violet flowers have calyxes with four or five lobes, the same number of obovate, spreading petals, twice as many stamens of two different lengths, and one style. The fruits are capsules enclosed in the persistent calyx tubes.

Tropical African ***D. canescens*** (syn. *D. incana*), a subshrub up to 3 feet tall, has coarsely-hairy, four-angled stems and ovate to oblong-lanceolate, gray-green, hairy leaves 1 inch to 3 inches long. The ¾-inch-wide, crimson flowers are in panicles from the axils of the upper leaves. Tropical West African ***D. grandiflora*** is a semi-woody, tuberous-based perennial with a few four-angled, bristly-hairy, twiggy stems 1 foot to 2 feet tall. Its short-stalked, elliptic-oblong to elliptic-lanceolate, three-veined, toothed, sparsely-hairy leaves are 2 to 3½ inches long. The purplish blooms, 2 to 3 inches across, are few together in terminal racemes. Quite different ***D. rotundifolia*** (syn. *D. plumosa*), of tropical West Africa, has slender, creeping, rooting, four-angled stems with bristly nodes. Its broad-ovate, three-veined, hairy leaves are ½ inch to 1½ inches long. The blooms, phlox-purple and solitary, are about 2 inches wide. The variety *D. g. lambii* has pale pink-lilac flowers about 3 inches across.

Dissotis grandiflora lambii

Garden Uses and Cultivation. In Hawaii and other tropical places these are suitable for outdoor planting in partially shaded places. They are also of interest for warm and intermediate temperature greenhouses. They succeed in ordinary soil, preferably well supplied with humus, and moderately moist. Propagation is easy by seeds and by cuttings.

DISTICTIS (Dís-tictis). Natives of Mexico, Central America, and the West Indies, the nine species of *Distictis* of the bignonia family BIGNONIACEAE are mostly hairy, woody vines. The name comes from the Greek *dis,* twice, and *stiktos,* spotted. It alludes to the much-flattened seeds that appear as two rows of spots when the capsules are opened.

The species of this genus have opposite leaves of two or three leaflets. If two, the third is usually represented by a three-branched tendril. The purple, red, or pink, funnel-shaped blooms are in terminal racemes or panicles. The calyx and corolla are five-lobed. There are four stamens and a style with a slender, two-lobed stigma. From closely related *Pithecoctenium* these plants differ in having smooth instead of prickly, podlike seed capsules. The seeds are broadly-winged.

Sometimes called evergreen trumpet creeper, ***D. buccinatoria*** (syn. *Bignonia cherere, Phaedranthus buccinatorius*) is endemic to Mexico. A vigorous vine that attaches itself to supports by small flat disks at the ends of three-parted tendrils, this grows rapidly and extends itself for 100 feet or more. It blooms over a long period

Distictis buccinatoria

Distictis buccinatoria (flowers)

in summer and fall. Two forms seen to be cultivated, one with larger leaves relatively close together and with larger flowers, and an inferior sort with smaller blooms and smaller, more widely spaced leaves. The flowers pendulous and 1½ to 4 inches long, are in terminal clusters of from four to eight. They have a leathery, very hairy, unequally-five-toothed, bell-shaped calyx. The corolla is blood-red to purple-red with a yellow base. It has a curved tube and five spreading, notched lobes (petals). The four stamens protrude slightly. The stems of this vine are hairy and distinctly four-angled. Its leaves, hairy on their undersides at least when young, are of two ovate or ovate-oblong leaflets up to 3 inches long, often with a tendril between them, and with hairy stalks. The seed capsules are usually curved.

Beautiful ***D. laxiflora***, of Mexico, has heart-shaped to oblongish leaflets 1½ to 2½ inches long and purple to milky-white flowers varying greatly in size, often even on the same plant. They are 1½ to 3½ inches long by up to 2½ inches across the face of the bloom. On their outsides conspicuously hairy, the flowers, in loose, few-flowered racemes or panicles, are borne in summer. The seed capsules are straight.

A hybrid between *D. buccinatoria* and *D. laxiflora* is named *D.* 'Rivers' (syn. *D. riversii*). This has blooms similar to those of *D. laxiflora*, but nearly twice as big.

Garden and Landscape Uses. These plants are adapted for sunny locations outdoors in frost-free and nearly frost-free climates and for large greenhouses and conservatories. They are admirable for adorning walls and pillars and for screening, but because of their method of attachment are not suitable for growing against wooden buildings that must be painted or otherwise treated from time to time. They are well adapted for covering masonry, old tree stumps, bare trunks of trees, and similar supports and succeed in ordinary well-drained garden soil.

Cultivation. At planting time prune the young vines to within a few inches of the ground to induce the development of new shoots that will cling readily to their supports. Pruning in subsequent years consists of severely shortening in late winter or early spring all shoots not needed to cover additional space and of thinning out any obviously too crowded. Propagation is usually by cuttings or layering, but seed may also be used.

DISTYLIUM (Dis-tŷlium). Only one of the six or more species of this relative of the witch-hazels (*Hamamelis*) is cultivated, and that more as a curiosity than as an ornamental. The genus, native of Japan, China, and the Himalayas, consists of evergreen trees and shrubs of the witch-hazel family HAMAMELIDACEAE, with short-stalked, leathery, alternate, undivided, toothed or toothless leaves. Individual plants may bear flowers of one sex only, or unisexual and bisexual blooms. The flowers are in axillary racemes, with small bracts. They have no petals and one to five or no sepals. There are two to eight nearly spoon-shaped stamens, and two slender styles. The fruits are nearly woody, two-seeded capsules with the persistent styles attached. The name is from the Greek *dis*, two, and *stylos*, a style, in allusion to those conspicuous organs.

Of tree dimensions in its native Japan and Ryukyu Islands, and reported to attain 80 feet in height, as known in cultivation ***D. racemosum*** is a shrub only a few feet tall. It has leaves oblong to narrowly-obovate, 1½ to 3 inches long and about one-half as wide as long. Sometimes they are slightly toothed toward their pointed or blunt ends. When young they have stellate (star-shaped) hairs. The downy flower spikes, 2 to 3 inches long, come from the leaf axils. The blooms, displayed in spring, have reddish-pubescent calyxes, and purple-red anthers. The egg-shaped, shallowly-cleft capsules, ⅓ inch long, are covered with yellow-brown pubescence. Variety *D. r. variegatum* has leaves irregularly margined or blotched with yellowish-white and often deformed.

Garden Uses and Cultivation. Not hardy in the north, the species described and its variety are cultivated in botanical and other special collections in mild areas. The variegated kind is occasionally accommodated in greenhouses. These plants prosper in well-drained, reasonably fertile soil and need no special attention. Propagation is by cuttings taken in late summer and, the green-leaved kind, by seeds. In greenhouses a winter night temperature of 50°F is satisfactory, and light shade in summer is recommended. A fertile porous soil should be supplied and kept always moderately moist.

DITTANY. See Dictamnus. Dittany-of-Crete is *Origanum dictamnus*. Maryland-dittany is *Cunila origanoides*.

DIVI-DIVI is *Caesalpinia coriaria*.

DIVISION. One of the commonest and often easiest means of increasing many herbaceous perennials and a few clump-forming shrubs, such as edging boxwood and spireas, is by splitting or dividing one plant into two or more parts and replanting these as individuals. In its usual sense division is understood to mean that each fragment consists of roots and above-ground parts or of roots and a portion of a crown of buds that will develop into above-ground parts. It further implies that fairly considerable force must be exerted to effect the parting of the pieces. If the segments are naturally detachable or at least break apart very easily, as do those of clumps of daffodil bulbs and of houseleeks (sempervivums), the word separation is likely to be used instead. But this is by no means always true, and the terms are more or less interchangeable. Thus one gardener will speak of lifting and separating his narcissus bulbs, another of lifting and dividing them. In like fashion the words dividing and separating are used somewhat indiscriminatingly with reference to such tightly clumped plants as chrysanthemums and phloxes. The term division as used here applies to plants that show no obvious lines of cleavage, the parts of which are not naturally separable.

In its crudest form division consists of digging up a chrysanthemum, peony, phlox, rhubarb, shasta-daisy, or other clump-forming plant and with a spade, axe, or machete splitting it into two or more parts. Similar treatment with a sharp knife, trowel, or other implement may be accorded such pot plants as aspidistras, Boston ferns, and sansevierias.

A more refined technique, applicable in many cases, but not with such stout-rooted clumps as those of peonies and rhubarb, is to use a fork or forks instead of a spade or cutting implement to effect division. This permits pulling apart the clump in such a way that the maximum amount of

(a) Usıng a spade to divide a clump of *Stachys*

(b) The divisions ready for planting

roots is retained with each piece. With large clumps of herbaceous perennials, after removing them from the ground insert a pair of spading forks vertically and closely together. Then force the handles slowly away from each other so that the tines tear the clumps apart. Smaller clumps of herbaceous perennials and pot plants to be divided can be conveniently handled by the careful use of a knife or pruning shears.

(c) Prying the forks apart

(d) Dividing still further with a knife

Dividing daylilies: (a) Digging up an old clump

(b) Inserting two spading forks back to back

(e) Ultimate divisions ready for planting

Dividing perennial asters: (a) Pulling a clump apart in early spring

(b) Strong divisions

(c) Planting the divisions

(d) Taking single shoot divisions in early spring

(e) Potting them individually for planting in the garden after they are well rooted

(f) In fall, removing vigorous outer portions of an old clump for dividing

(g) Divisions ready for planting

Dividing bearded irises: (a) After blooming, digging up an old clump

(b) Cutting the rhizomes into sections, each with one fan of leaves

(c) Cutting back the leaves

(d) Shortening the roots

(e) Divisions ready for planting

(f) Planting a division

Dividing dahlias: (a) Removing single tubers, each with a small piece of stem attached

(b) Old clump of tubers with two divisions detached

Dividing a pot of *Acorus gramineus:* (a) Pulling the clump apart

(b) The first divisions

(c) Subdividing the first divisions with a knife

(d) The final divisions

(e) Three final divisions, with leaves cut back, potted in new soil

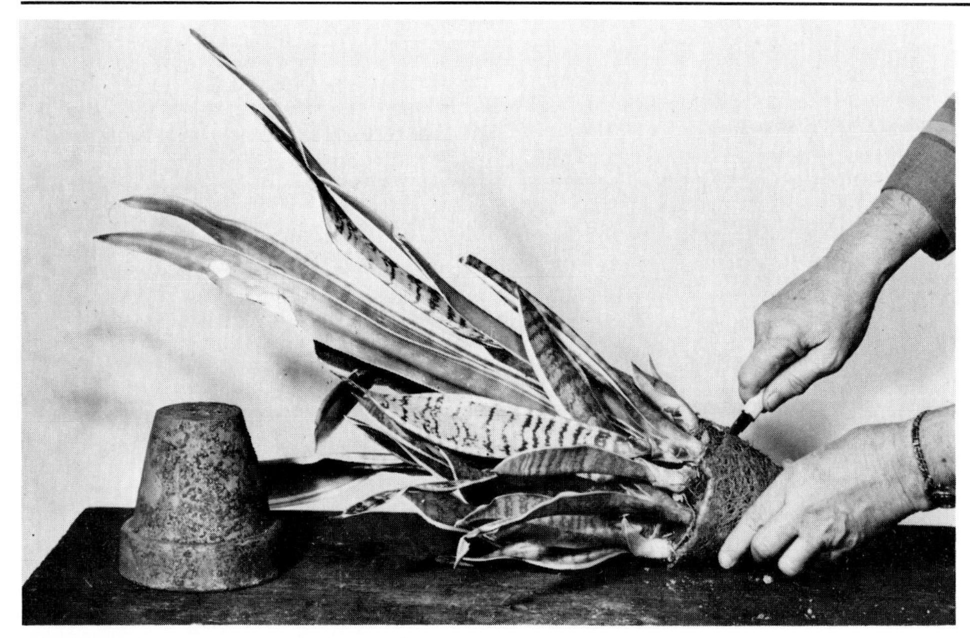

Dividing a sansevieria with extremely matted roots: (a) Cutting into the root ball with a knife

(b) Breaking the ball apart

Dividing a *Clivia:* (a) Loosening the root ball with a hand fork

(b) Pulling the clump into single-shoot divisions

(c) Potting a single-shoot division

Dividing Boston fern (*Nephrolepis*): (a) Trimming single-shoot divisions

(b) Divisions potted in small pots

A common mistake is to make the divisions too big. In the vast majority of cases, for the best results, they should not consist of more than three stems or growth buds and often one is enough. Small, vigorous divisions are nearly always to be preferred to large ones. With tightly packed clumps, such as those of Boston ferns, chrysanthemums, heleniums, perennial asters, and phloxes, make the divisions from the outer portions of the clumps and, unless as many as possible new plants are needed, discard the older inner parts.

The best time to divide most plants susceptible to this mode of propagation is early spring or early fall. Cut back the tops of most shrubby kinds fairly severely to compensate for the comparatively small amount of roots they have. For the same reason it may be necessary or desirable to reduce the amount of foliage of divisions of leafy evergreen sorts, such as Boston ferns.

Care after planting is that normal for the kind of plant with the added precaution with leafy divisions of shading them from bright sun and wind and misting them occasionally with water to reduce transpiration until adequate new roots develop.

DIZYGOTHECA (Dizygo-thèca)—False-Aralia. A dozen species native of New Caledonia, not very well understood botanically, and in gardens often cultivated under the name *Aralia,* comprise *Dizygotheca* of the aralia family ARALIACEAE. They are small evergreen trees and shrubs that, at least when young, are usually single-stemmed. The name comes from the Greek *di,* two, *zygos,* a yoke, and *theke,* a case. It alludes to the cells of the anthers being double the usual number.

Dizygothecas are without spines or thorns. They have alternate leaves with long stalks swollen at their bases, and seven to eleven short-stalked, smooth- or wavy-edged, or lobed leaflets radiating in palmate (handlike or finger-like) fashion from their apexes. The leaves of young specimens are often unlike those of adult ones. The bisexual flowers are in large terminal umbels of smaller umbels (they are in compound umbels). Each little bloom has a five-toothed to nearly toothless calyx, five each petals and stamens, and ten separate styles. The subspherical to ellipsoid fruits are berries containing flattened seeds.

Kinds in cultivation include well-named ***D. elegantissima*** (syn. *Aralia elegantissima*) which has leaves with cream-mottled stalks, and coarsely-toothed or lobed leaflets. In juvenile specimens, such as are commonly seen in cultivation as indoor plants, the leaflets are about 6 inches long by ½ inch wide. Those of adult specimens, which may be trees up to 25 feet tall, have leaves with leaflets approximately 10 inches in length by 3 inches wide. Quite as attrac-

Dizygotheca elegantissima

tive, ***D. veitchii*** (syn. *Aralia veitchii*) is a slender-stemmed shrub. Its leaves have narrowly long-elliptic, wavy-edged, coppery-green leaflets that are dark red on their undersides and have pinkish mid-veins. In *D. v. gracillima* the leaflets are even narrower and have white mid-veins.

Garden and Landscape Uses and Cultivation. In the humid tropics dizygothecas are graceful elements in shrub plantings, and as lawn specimens. As young, single-stemmed plants they are choice furnishings for tropical greenhouses and for use as houseplants. They succeed most surely indoors where minimum temperatures are 60 to 70°F at night and five to fifteen degrees higher by day. A humid atmosphere is favorable. Too-dry air and exposure to drafts results in premature loss of lower leaves. Fertile, well-drained earth maintained in a uniformly moist, but not constantly wet state is needed, as is shade from strong sun. Specimens that have filled their containers with healthy roots benefit from regular applications of dilute liquid fertilizer. When they become too tall the tops can be air layered, cut off, and potted as new plants. Sectional stem cuttings made of the less woody upper portions of the stems afford an alternative means of propagation, grafting a third. For understocks to graft on, young plants raised from cuttings or seeds grown to a size where their stems are pencil-thick are used. Their tops are cut off a little above ground level and the scion inserted, usually as a cleft or wedge graft. All propagation should be done in high temperatures and high humidities.

DOCK. Species of *Rumex,* chiefly the larger-leaved kinds, are called docks. A few are cultivated, more are deep-rooted weeds. They are most surely eradicated by hand digging. The plant called prairie-dock is *Silphium terebinthaceum.* Butterfly-dock is *Petasites hybridus.*

DOCKMACKIE is *Viburnum acerifolium.*

DOCTRINE OF SIGNATURES. This is the name of a belief that persisted for hundreds of years and was especially popular with the herbalists of the sixteenth and seventeenth centuries. It postulated that plants possessed of healing qualities bore indications of these in their forms, markings, or other characteristics. As stated by Robert Turner in *The British Physician* in 1664 "God hath imprinted upon the Plants, Herbs and Flowers, as it were in Hieroglyphics, the very signature of their Ver-

Plants significant in the doctrine of signatures: (a) *Pulmonaria*

(b) *Scutellaria*

(c) *Hepatica*

tues." Examples of the doctrine of signatures include lungwort (*Pulmonaria*) the spotted leaves of which were thought to resemble diseased lungs and hence useful for treating pulmonary troubles. Eyebright (*Euphrasia*), its flowers reminiscent of the human eye, were thought useful for relieving eye strain. Skullcap (*Scutellaria*) which has blooms shaped somewhat like a skull, was accepted as a cure for insomnia. The yellow juice of the celandine (*Chelidonium majus*) suggested its use as a cure for jaundice. Liverleaf (*Hepatica triloba*) by the shape of its leaves clearly indicated to believers in the doctrine its usefulness for alleviating afflictions of the liver. Numerous other examples could be cited to illustrate this once profoundly held belief.

DOCYNIA (Do-cỳnia). From closely related *Cydonia*, the name of which its name is an anagram, *Docynia*, of the rose family ROSACEAE, differs in the styles of its flowers being united at their bases. It comprises six species of evergreen, semievergreen, or nearly deciduous Asian trees, hardy in mild climates only.

Docynias have alternate, undivided, sometimes slightly lobed, toothed or toothless leaves. The short-stalked blooms, in umbels of up to five, come before or with the new foliage. They have a densely-hairy, five-lobed, persistent calyx, five petals, thirty to fifty stamens, and five hairy styles. The stigmas are two-lobed. From somewhat under to somewhat over 1 inch in diameter, the fruits are nearly spherical, ovate, or pear-shaped.

Half evergreen to nearly deciduous ***D. rufifolia*** (syn. *D. docynioides*) has elliptic to oblong-lanceolate, lustrous leaves 2 to 3¼ inches long and up to 1¾ inches wide. They are smooth-edged, or toothed toward their apexes and more or less hairy on their undersides. The nearly stalkless white blooms, about 1 inch in diameter, have approximately thirty stamens. The quince-type fruits are subspherical to ellipsoid, and edible. More decidedly evergreen ***D. delavayi*** differs in having leaves without teeth usually somewhat larger than those of *D. rufifolia*. Their undersides are white-tomentose. Both species are natives of southwestern China.

Native of the Himalayas and Burma, ***D. indica,*** up to 15 feet tall, has ovate to oblong-lanceolate leaves up to 4 inches long, those of young plants lobed or toothed, those of more mature specimens with or without teeth. There undersides may be woolly-hairy to nearly hairless. The densely-woolly flowers are 1½ inches wide.

Garden and Landscape Uses and Cultivation. Not much cultivated, docynias have good-looking foliage and pleasing displays of bloom. They thrive under conditions appropriate to apples and pears, but need a warm temperate or subtropical climate. They are not hardy in the north. Propagation is by seed, and perhaps by grafting onto understocks of apples (*Malus*) or other related genera.

DODDER. This is the common name of a genus (*Cuscuta*) of parasitic plants, many of which are pestiferous, hard-to-get-rid of weeds. They are degenerate relatives of morning glories and belong to the morning glory family CONVOLVULACEAE. There are numerous species, many capable of infesting a wide variety of hosts.

Dodders begin life in the manner of more respectable plants. Their seeds germinate in the soil, but the slender stems of young dodders soon attach themselves by numerous minute, rootlike absorbing organs to those of nearby plants after which they lose contact with the earth and tap their host for the water and nutrients they need, to the great detriment of the host. Growth is rapid, and soon they festoon the plants they parasitize with slender, usually yellow strands that are their stems. In the subtropics and tropics the hosts are often large shrubs and even trees, in temperate climates annuals, herbaceous perennials, and such plants as English ivy are commonly affected. Dodders have no leaves, only minute scales representing them. The tiny yellow or whitish flowers are in clusters along the stems.

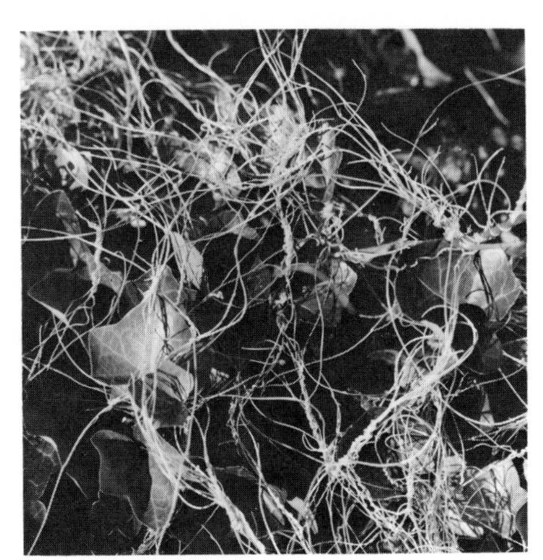

Dodder on English ivy

The gardener's only concern with dodder is to eliminate it. This is not easy. The only practical way is to destroy it before it flowers and sets and disperses a crop of seeds to give rise to future generations. Because of the tight hold dodders have on their hosts this is likely to involve cutting the latter back severely. If you elect this, be sure to act early. Even so, success does not come easily for it has been established that of dodder seeds distributed in any one season only a proportion of them germinate the next year; others germinate in later years. Therefore, to eliminate the pest, it is essential to repeat the destruction of dodders before they seed over a period of two or three or perhaps more years. With English ivy spraying the infested plants with a strong solution of sulfate of ammonia (a fertilizer), made by dissolving one pound in one gallon of water, kills the dodder. It also kills the foliage of the ivy, but not the roots or stems and the latter soon leaf out again.

DODECATHEON (Do-decàtheon)—Shooting Star, American-Cowslip. Except for one species that maintains a toehold in arctic northeastern Asia, although it chiefly inhabits western arctic North America, *Dodecatheon* of the primrose family PRIMULACEAE is endemic to the last-named continent. There, most of its fifteen species are confined to the West. The name, perhaps alluding to the supposed number of flowers in the umbels of the first species described, in an ancient Greek one for some other plant, perhaps a *Primula*. It derives from the Greek *dodeka*, twelve, and *theos*, god.

Dodecatheons are hardy herbaceous perennials. Their linear, elliptic, or spatula-shaped leaves, in basal rosettes, narrow to more or less winged stalks. Some little time after flowering they wither and the plants remain dormant until the following spring. Their nodding flowers, in umbels of few to many atop erect, leafless stalks are quite un-primrose-like. They much more resemble blooms of cyclamens, this because their petals are strongly reflexed (backward-spreading). Unlike cyclamens dodecatheons are without corms (bulblike organs) and their blooms are in umbels, not solitary. Each flower has a five-lobed calyx, a short-tubed corolla with fine unequal lobes (petals), five stamens the long, slender anthers of which hug each other to form a projecting cone, and one style. The fruits are cylindrical capsules. Species of this genus commonly exhibit considerable variation and are not always easy to identify as to kind. Often a helpful clue is given by the presence or absence of rice-grain-sized bulblets among the roots.

Eastern North American ***D. meadia***, the most commonly cultivated shooting star, inhabits dry woodlands and prairies from Pennsylvania to Wisconsin, Georgia, Alabama, Arkansas, and Texas. Variable and showy, this is without rice-grain bulblets. It has oblong to lanceolate or occasionally obovate leaves generally tinged with red toward their bases. They are 2½ to 8 inches long. The umbels of few to many white to lavender or lilac blooms top stalks 9 inches to 2 feet long. The flowers, erect in bud, but later hanging, are from less than ½ to 1 inch long. A white-flowered variety is known horticulturally as *D. m. album*, a red-flowered one *D. m. splendidum*. Very similar to *D. meadia*, differing technically in the walls of its seed capsules being thin and flexible instead of thick and rigid, ***D. pulchellum*** (syn. *D. amethy-*

Dodecathem pulchellum variety

Variety *D. h. cruciatum* has only four sepals and petals. Very similar to *D. hendersonii*, but entirely green and having not more than seven flowers to each umbel, ***D. hansenii***, of California, is 3 to 10 inches tall.

Dodecatheon hendersonii

stinum) is a variable kind to which many names have been applied. Except for *D. meadia* the only species native to eastern North America, this occurs on moist cliffs, hillsides, and river banks from Pennsylvania to Wisconsin, West Virginia, and Missouri. It is even more widespread and plentiful in the West, where it is native from Alaska to Mexico. Generally smaller than *D. meadia* and blooming about two weeks later, this is without the red staining at the bases of the leaves so often characteristic of that species. It is also without rice-grain bulblets. The smallest of the genus, *D. p. watsonii* (syn. *D. uniflorum*) has leaves ¾ inch to 1½ inches long and flowering stalks bearing an umbel of one to three blooms. This is a high mountain variety with a natural distribution for Montana to British Columbia, Wyoming, and Oregon.

Three western species are distinguished by their stigmas being twice as thick, instead of not appreciably thicker than, the styles (stalks) that support them. The alpine shooting star (*D. alpinum*), native of wet or very moist locations at high elevations in California, Nevada, and Oregon, belongs here, as also do *D. redolens* and *D. jeffreyi*. Purple-blossomed ***D. alpinum*** has narrowly-oblanceolate to linear, toothless leaves 2½ inches or less long and blooms solitary or two or three together, each with a four-parted calyx and corolla and four stamens. The corollas have whitish bases and are barred with a yellow zone and a darker purple band. The flower umbels are on stalks up to 5 inches tall. Variety *D. a. majus* is taller and has more flowers to the clusters. Distinguishing features of ***D. redolens*** are the yellow corolla tube that conceals the bases of the anthers and its unpleasant odor. Its calyx lobes, corolla lobes, and stamens number five. This robust species, native from California to Nevada and Utah, is densely-glandular-downy and 10 inches to 2 feet tall. The third kind with a large stigma, ***D. jeffreyi***, is reported to be the most variable of all shooting stars. This tendency manifests itself in its flowers having either four or five calyx lobes, corolla lobes, and stamens. From 6 inches to 2 feet in height, this native of wet soils from Montana to California and Alaska has purplish to pinkish-lavender, yellowish or whitish blooms. Its leaves may or may not be toothed or scalloped. Sometimes they are glandular-downy.

Problems of identification are likely to be presented by some western species. Three, *D. hendersonii*, *D. subalpinum*, and *D. hansenii*, are distinctive in that they produce rice-grain bulblets around the rootstock at flowering time. A reddish-stemmed plant with spoon-shaped to elliptic, glandular or hairless leaves 2 to 6 inches long, ***D. hendersonii*** has flowering stalks 6 inches to 1¼ feet tall. They carry umbels of three to seventeen magenta to deep lavender or white blooms with maroon and yellow corolla tubes. This is indigenous from California to British Columbia. Lower stature and white roots distinguish ***D. subalpinum*** from *D. hendersonii*. Native of high altitudes in the Sierra Nevada, this has toothless leaves, does not exceed 6 inches in height, and has one- to five-flowered umbels. The blooms, their petals sometimes nearly 1 inch long, are magenta to white with dark maroon and yellow corolla tubes.

Differing from the last three species in never having rice-grain bulblets among its roots, ***D. clevelandii*** is a variable native of California, Baja California, and islands off the coast. Glandular-pubescent, it has white roots and flowering stalks 8 inches to 1½ feet in height with umbels of five to sixteen yellow-stamened, magenta to white blooms with maroon and yellow corolla tubes. The petals are from almost ½ to ¾ inch long. From 1 inch to 7 inches long, the leaves are widest toward their apexes. Their margins are crisped and toothed. Botanists distinguish several varieties based on differences in the anthers.

Other species may be cultivated. Ranging from California to Washington, Wyoming, and Montana and in its glandular-leaved variety *D. c. viscidum* into adjacent Canada, ***D. conjugens*** has toothless leaves and umbels of usually one to seven white to magenta flowers atop stalks 3 to 10 inches tall. Related fairly closely to it, ***D. frigidum*** has lavender or magenta flowers. Indigenous from Canada and Alaska to Siberia, this has a specific name that appears to be well chosen. The only shooting star with flowers invariably white, although white-flowered varieties of other kinds occur, is ***D. dentatum***, native to Oregon, Washington, Idaho, and British Columbia. It has toothed or wavy-edged, long-stalked leaves. Its variety *D. d. ellisiae*, which differs from the typical kind in having yellow-stalked stamens, extends into New Mexico and Arizona. Always densely-pubescent, ***D. cusickii*** is distinguished from similar *D. pulchellum* by that and by its thick-walled fruits. It ranges in the wild from Oregon to British Columbia, Idaho, and Montana. Having a very limited natural range on both sides of the Columbia River in Oregon and Washington, ***D. poeticum*** somewhat resembles *D. hendersonii,* but is without the rice-grain bulblets of that species.

Garden Uses and Cultivation. Shooting stars are charming plants best adapted for rock gardens and native plant gardens in lightly shaded places where the soil is moderately rich and of a woodland character, not alkaline, and never too dry. Those from high altitudes are not easy to grow in lowland gardens, often being short-lived. Propagation is by seeds, sown as soon as possible after they ripen, and by division or separation in early fall or early spring. Established plants benefit from being kept mulched with peat moss or other organic material, such as leaf mold, and from watering copiously during dry weather until after their foliage dies down.

DODONAEA (Dodon-aèa)—Hop Bush. The great majority of the approximately sixty species of this genus of the soapberry family SAPINDACEAE are natives of Australia, but a few are indigenous in warm regions elsewhere and one, *D. viscosa,* is widespread, chiefly in coastal thickets, in tropical and subtropical regions. The name commemorates Rembert Dodoens, a physician and author, who died in 1585.

Dodonaeas are shrubs or trees with alternate, usually undivided, but sometimes pinnate leaves, and small, ornamentally unimportant, generally unisexual flowers, solitary or in racemes or panicles. The blooms have three to five sepals and five to ten, commonly eight, stamens. They are without petals. The fruits are capsules.

Quite frequently cultivated, ***D. viscosa,*** an evergreen shrub up to 15 feet in height, has usually sticky shoots and undivided, oblongish leaves 3 or 4 inches long and about 1 inch wide and greenish flowers in usually terminal clusters. The fruits are red or purple. More commonly cultivated than the species is its variety *D. v. purpurea,* which has dark purple foliage. This species has been used in native medicines and in Australia the fruits were formerly employed as a substitute for hops. The seeds are reported to be edible.

Another kind, ***D. cuneata,*** of Australia, is a freely-branched, usually sticky shrub with obovate or wedge-shaped, dark green leaves up to 1 inch long. Its inconspicuous blooms are in terminal or axillary racemes or clusters and are succeeded by red or purplish fruits.

Garden and Landscape Uses and Cultivation. The cultivated dodonaeas are attractive for general landscaping. The purple-leaved variety of *D. viscosa* is popular in California where it is used for planting at roadsides and other places. It with-

Dodonaea viscosa purpurea

stands drought and smog well and needs little or no care. Propagation is easy by cuttings, but seeds are slow to germinate. These shrubs may be pruned as needed to keep them shapely. They make good hedges.

DOG ROSE. The rose of this name is *Rosa canina.* In Australia *Bauera rubioides* is called dog-rose. African-dog-rose is *Oncoba kraussiana.*

DOGBANE. See Apocynum. Climbing-dogbane is *Trachelospermum difforme.*

DOGS. Friends of man though they may be, dogs are out of place in gardens. There, they can be decidedly destructive, harming plants by stepping on them, scampering among them, and with supreme disdain and, as it seems to gardeners, with a keen sense for selecting the choicest, urinating on them.

Unfortunately for those who garden in suburbs and cities their dog population is likely to be high. The problem is chiefly with the animals of neighbors. One's own dog is likely to be kept under fair control, at least in its home territory. The only adequate protection is to fence the intruders out, preferably by enclosing the entire garden, or if this is not possible, by encircling the most prized plants with wire guards. Commercial repellent sprays and dusts have some value, but must be applied fairly frequently to be long effective.

DOG'S MERCURY. See Mercurialis.

DOG'S-TOOTH-VIOLET. See Erythronium.

DOGWOOD. See Cornus.

DOLICHANDRA (Doli-chándra). One species of the bignonia family BIGNONIACEAE is the only representative of this genus. Native of Brazil, Argentina, and Patagonia, it is closely related to *Macfadyena* from which it differs in its flowers having long-protruding stamens and in technical details of their anthers. The name, derived from the Greek *dolichos,* long, and *andros,* a man, alludes to this characteristic of the stamens. Its "ch" is pronounced as "k."

A vigorous, hairless, evergreen vine, ***Dolichandra cynanchoides*** (syn. *Macfadyena cynanchoides*) has opposite, stalked, leathery leaves each with two oblong, ovate, or ovate-lanceolate, toothless leaflets 1 inch to 2 inches long and often a terminal, three-branched tendril. The showy red, tubular blooms, 2 to 2½ inches long and markedly obliquely asymmetrical, are solitary or in clusters of few on stalks from near the branch ends. They have spathelike calyxes split down one side to their middles, and long-tubular corollas with five rounded, more or less reflexed lobes (petals). There are four fertile and one rudimentary stamens. The fruits are broad, angled capsules 3 to 4 inches long and ½ to ¾ inch wide.

Garden and Landscape Use and Cultivation. In tropical and subtropical regions where frost is little known this handsome vine is admirable for planting outdoors. It can also be grown in greenhouses in ground beds or large pots or tubs. It responds to conditions that suit *Allamanda.*

DOLICHANDRONE (Dolichandrò-ne). The "ch" in *Dolichandrone* is pronounced as "k." The genus, native in Africa, Madagascar, tropical Asia, islands of the Pacific, and Australia, is of the bignonia family BIGNONIACEAE. Derived from the Greek *dolichos,* long, and *andros,* a man, the name alludes to the stamens.

Dolichandrones number nine species of trees. They have toothed or toothless, pinnate leaves with an uneven number of leaflets, or undivided leaves. The white, more or less trumpet-shaped flowers have tubular, spathelike calyxes, five-lobed corollas, and four fertile and one abortive stamen, and a slender style with a two-lobed stigma. The fruits are pods, often spirally twisted.

The mangrove trumpet tree (***D. spathacea***) is a narrow-headed, evergreen native from India to Malaya and New Caledonia. From 20 to 60 feet tall, it has leaves 6 inches to over 1 foot long, slightly pinkish when young, and with five to nine broad-elliptic, short-stalked, sometimes slightly toothed leaflets. In small clusters, the highly fragrant, slender-tubed, trumpet-shaped blooms are at the branch ends. One in each cluster opens at nightfall and drops by the following morning. Each is 5 to 7 inches long by 3 to 5 inches across its face. The seed pods are about 1 inch wide and 1½ feet long.

Other species sometimes cultivated include these: ***D. alba,*** of Mozambique, is a small tree, with opposite, hairless leaves with eight elliptic, ovate, or obovate leaflets up to 4 inches long. The 3-inch-long flowers are succeeded by pods up to 2 feet long. ***D. arcuata*** (syn. *D. crispa*), of India, is up to 60 feet in height, has opposite leaves up to 1 foot long with nine or eleven elliptic, 2- to 3-inch-long leaflets. The flowers are 3 inches, the pods up to 1½ feet, long. ***D. heterophylla*** is a hairless shrub or tree up to 15 feet in height, a native of Australia. Up to 6 inches long, its leaves have a single linear blade or three to seven ovate-lanceolate to linear leaflets up to 3 inches long. The 1½-inch-long flowers, in racemes of few, are succeeded by pods up to 2 feet long.

Garden and Landscape Uses and Cultivation. Suitable for cultivation in warm, humid climates, such as those of Hawaii and southern Florida, the mangrove trumpet tree is well adapted to moist and wet soils. The other species also need tropical environments, but are likely to be more insistent on fairly well drained soil. Propagation is usually by seed.

DOLICHOS (Dólich-os)—Hyacinth-Bean or Lablab or Bonavist, Australian-Pea. To the pea family LEGUMINOSAE belongs *Dolichos,* an assemblage of about 150 species of herbaceous perennials, subshrubs, and twining vines, some with large, carrot-like rootstocks. They are natives of the tropics and subtropics. The name is an ancient Greek one for a kind of bean. Varieties of *D. lablab* are widely cultivated in warm regions for their edible pods and seeds, and for forage.

The leaves of *Dolichos* have three leaflets, the center one stalked. The reddish, purple, or white, pea-like flowers differ from those of related *Phaseolus* and *Vigna* in having their narrow keels bent inward and upward at approximately right angles instead of being coiled or curved. Solitary or clustered, the blooms come from the leaf axils or are terminal. They have two-lipped calyxes, the upper lip with or without two lobes, the lower one three-lobed. The broad and rounded, standard or banner petal has at its base a pair of earlike projections. The wing petals are more or less united with the keel. There is one separate stamen and nine united. The swollen, bearded style is tipped with a head-like stigma. Often large, the fruits are flat, curved pods each ending in a sharp beak.

The hyacinth-bean, lablab, or bonavist (***D. lablab***), almost surely a native of the Old World tropics, is a very variable perennial, generally grown as an annual. Typically it is a vine, exceptionally 30 feet, more commonly 6 to 10 feet tall. There are lower, bushy varieties. The leaves have broad leaflets 3 to 6 inches long, the lateral ones lopsided. The purple or white blooms, ½ to nearly 1 inch long, are in groups of up to four arranged in elongating, axillary racemes. They are succeeded by hairy or hairless pods 1 inch to 2½ inches in length, which contain black or white seeds. A variety with white flowers and white seeds

Dolichos lablab

is called 'Daylight', one with dark purple-violet blooms and black seeds 'Darkness'.

The Australian-pea (***D. lignosus***), almost surely native of Asia rather than of the continent suggested by its common name, is a perennial, evergreen, slender-stemmed, somewhat hairy, woody vine. Its leaves have pointed, triangular-ovate leaflets up to 1½ inches long. Under ½ inch in length, the white or rosy-purple blooms are crowded in clusters at the ends of long stalks. The hairless, 1-inch-long pods contain black seeds.

Garden and Landscape Uses and Cultivation. The Australian-pea is suitable for outdoor cultivation in warm regions only. It serves well as a general purpose vine and is propagated by seeds. It needs no special care. In North America the hyacinth-bean is cultivated as an ornamental. It is excellent for growing on trellises, strings, or wires and for covering teepees of brushwood stakes the bottoms of which are pushed well into the ground. It grows quickly and makes an effective screen. For best results it needs a sunny location and moderately fertile, reasonably moist soil. Seeds are sown outdoors in spring where the plants are to remain or are started indoors in a temperature of 65 to 70°F to give plants in small pots ready for setting in the garden five or six weeks later. This is done when it is safe to plant tomatoes and other frost-tender plants. The plants are set about 1 foot apart.

DOLICHOTHELE (Dolicho-thè-le). Pronounced with a "k" sound for the "ch," the name of this genus derives from the Greek *dolichos,* long, and *thele,* a nipple. It alludes to the elongated tubercles of these members of the cactus family CACTACEAE. Native of Texas and Mexico, *Dolichothele* consists of approximately a dozen species very closely related to and by some authorities included in *Mammillaria.*

Dolichotheles are small, low plants without milky juice. They have, often in clusters, soft, rounded stems or plant bodies with long, conspicuous protrusions or tubercles without longitudinal grooves. The flowers, mostly bigger than those characteristic of *Mammillaria* and with long, funnel-shaped perianth tubes, are yellow. They spring from the axils of older tubercles. The fruits are spherical, smooth, green, or red.

Kinds in cultivation include these: ***D. albescens*** (syn. *Mammillaria albescens*) has clustered spherical to ovoid plant bodies up to 3½ inches high by 2 inches wide with slender, conical, soft tubercles. There are a few bristles in the axils of the tubercles. The white to cream spines, up to ½ inch long, are in clusters of mostly five, sometimes four to six radials and rarely one central. About ¾ inch wide, the white flowers have greenish-white inner petals. ***D. balsasoides*** (syn. *Mammillaria balsasoides*) has clustered, globular to ovoid plant bodies about 3½ inches in diameter. The spines are in clusters of ten or eleven white radials about ¼ inch long and four longer, hooked, dark-colored centrals. The flowers, about 1½ inches across, have bright green outer petals and orange inner ones. ***D. baumii*** (syn. *Mammillaria baumii*) forms flat clumps of plant bodies up to 3½ inches tall and 2½ inches wide with white, woolly hairs in the axils of the young tubercles. The spine clusters are of up to thirty-five white, hairlike radials of various lengths, the largest up to more than ½ inch long, and five or six stiffer, slender centrals, yellowish with paler apexes, one or more hooked. The flowers, under ½ inch wide, are greenish-yellow with yellow centers. ***D. beneckei*** (syn. *Mammillaria beneckei*) forms small clumps of small spherical plant bodies slightly hollowed at their tops when young. In the axils of their tubercles are sparse, white, woolly hairs. The spine clusters are of twelve to fifteen brown-tipped, smooth, white radials up to ⅓ inch long and two to six, but mostly four dark brown to black centrals, the lower ones hooked. About 1¼ inches wide, the flowers have yellow to orange-yellow inner petals, rose-pink to violet outer ones. ***D. campotricha*** (syn. *Mammillaria campotricha*) has clustered stems with slim, curved tubercles up to ¾ inch long and spine clusters without centrals, of up to eight slender, interlacing, curved or twisted radials. The ½-inch-long flowers have greenish-white outer petals with a darker mid-stripe and green-lined, white inner ones. ***D. decipiens*** (syn. *Mammillaria decipiens*) has usually clustered, spherical plant bodies with soft, cylindrical tubercles with two or three bristles in their axils. The slender, white or yellowish spines are in clusters of seven to nine radials much shorter than the one ½- to ¾-inch-long straight central. About ½ inch long, the flowers are white to delicate pink. ***D. longimamma*** (syn. *Mammillaria longimamma*) attains a height of about 4 inches. Its 2-inch-long, blunt, cylindrical tubercles, hairy

Dolichothele longimamma

or hairless in their axils, have spine clusters of five to twelve pale, spreading radials about 1 inch long and one to three dark-tipped, but otherwise similar centrals. The blooms are approximately 2 inches in length. Native of Texas and Mexico, ***D. sphaerica*** (syn. *Mammillaria sphaerica*) has a big, thickened rootstock and forms low clusters up to 8 or 9 inches wide of soft stems with tubercles approximately ½ inch long. The spine clusters are of twelve to fifteen radials and one central. The flowers, when fully open, are about 2½ inches wide. ***D. surculosa*** (syn. *Mammillaria surculosa*) has clusters of spherical plant bodies about 1½ inches in diameter, the axils of their tubercles practically naked of hairs. The spine clusters consist of about fifteen white to light yellow radials up to ⅜ inch long and generally only one strongly-hooked, amber-yellow central up to ¾ inch long tipped with reddish-brown. The flowers up to ¾ inch wide are yellow with the petals tipped with dark yellow, or have orange stripes. Mexican ***D. uberiformis*** (syn. *Mammillaria uberiformis*), considered by some to be merely a variety of *D. longimamma,* differs from that species in its tubercles being conical and its spine clusters of four or five yellowish or reddish radials and no centrals. ***D. zephyranthoides*** (syn. *Mammillaria zephyranthoides*) has flattened, spherical to short-cylindrical plant bodies up to about 3½ inches tall and wide and tubercles almost 1 inch long with few or no hairs in their axils. The spine clusters are of fifteen to eighteen white, stiff radials about ⅜ inch long and four slightly shorter centrals, the lower ones hooked. The flowers, approximately ¾ inch in diameter, are purple to reddish-violet with paler throats and purple-brown outer petals.

Garden Uses and Cultivation. These are as for *Mammillaria.* For more information see Cactuses.

DOMBEYA (Dom-bèya). An estimated 350 species belong in *Dombeya,* a genus of African, Madagascan, and tropical Asian trees and shrubs, by far the greater number natives of the last-named region. They belong in the sterculia family STERCULIACEAE. Their name commemorates the French medical man and botanist Joseph Dombey, who accompanied Spanish expeditions to Peru and Chile, made extensive collections of plants there, and died in Spain in prison in 1794.

Dombeyas have alternate, undivided leaves, mostly heart-shaped, lobed, or angled, with the main veins spreading in hand-fashion (palmately) from their bases. Often they are large. The flowers, usually showy, and sometimes fragrant, are mostly in dense heads or loose clusters from the leaf axils or ends of the branches. They have deeply-five-cleft, persistent calyxes, five petals, ten to twenty-five stamens, five staminodes (nonfunctional stamens) combined into a tube, and five styles. The fruits are capsules.

Handsome ***D. wallichii,*** native of Madagascar, is a broad-headed tree up to 30 feet tall. Its heart-shaped, toothed leaves, with seven or nine chief veins, are 6 inches to 1 foot in diameter. Their undersides are furnished with stellate (star-shaped) hairs. The scarlet or pink blooms, in spherical or nearly spherical heads,

Dombeya wallichii

hang on long stalks from the leaf axils. Fragrant, white, sometimes pink-tinged flowers in much smaller, looser clusters than those of *D. wallichii* are borne by ***D. burgessiae*** (syn. *D. mastersii*) of East and South Africa. This is a shrub 6 to 12 feet high. It has velvety-hairy, heart-shaped to broad-ovate or nearly round, toothed leaves, and from the leaf axils pendulous flower clusters. The blooms are nearly 1½ inches wide.

A splendid hybrid, more handsome than either parent, ***D. cayeuxii*** was raised toward the end of the nineteenth century by

Dombeya cayeuxii in the Huntington Botanical Garden, California

mating *D. wallichii* and *D. burgessiae*. Of robust growth, the offspring is a broad shrub up to 15 feet tall. Its stalked, three-lobed, broadly-heart-shaped, shallowly-toothed, hairy leaves have blades up to 1 foot long or longer and wide. Its dense, dome-shaped or spherical clusters of flowers, 4 to 5 inches in diameter suggest those of hortensia hydrangeas, except that they are pendulous on long stalks. They are pleasing rose-pink with white bases to the petals. Their anthers are yellow.

A plant cultivated as *D.* 'Rosemound' is very similar to *D. burgessiae*. A broad shrub 6 to 10 feet tall, it has densely-softly-hairy, rounded-heart-shaped, five-lobed, toothed leaves 4 to 9 inches in diameter. Well displayed in crowded clusters, its dusty rose-pink flowers are ¾ to 1 inch across. Quite different ***D. spectabilis***, of Africa and Madagascar, is a deciduous, round-topped tree 20 to 45 feet in height. This has roundish, sometimes slightly three-lobed, round-toothed leaves 6 inches or a little more long, covered with stellate (star-shaped) hairs. When young they are pinkish. Coming before the leaves the white, creamy-white, or greenish-white flowers are in dense, erect clusters, often aggregated into larger clusters. Because of its general aspect in bloom, in South Africa this is called wild-pear and wild-plum. A pink-flowered variant occurs in the wild.

Other dombeyas are cultivated, among them ***D. calantha***. This native of tropical Africa is 3 feet tall or taller. It has longish-pointed, ovate to narrow-ovate, three- to five-lobed or nearly lobeless, more or less hairy leaves up to 1 foot in diameter. They have wavy, toothed margins. The rose-pink flowers are 1¼ to 1½ inches across. Madagascan *D. mollis* has soft-velvety, shallow-lobed, toothed leaves 6 to 8 inches in diameter, with finely hairy stalks. In loose to dense, more or less hairy-stalked clusters, the pink blooms are 1 inch or more wide. Free-flowering, its 1-inch-wide, pink flowers solitary or few together in loose, hairy-stalked clusters approximately 3 inches across, ***D. tiliacea*** has slender-stalked, pointed-ovate leaves 2 to 3 inches in diameter. This is a native of Africa. From East Africa comes ***D. nyasica***, a shrub 4 to 6 feet tall. This has shallowly-

Dombeya nyasica in the Huntington Botanical Garden, California

lobed, toothed leaves up to 5 inches or more across, and from toward its branch ends long-stalked clusters approximately 3 inches wide of pink blooms 1¼ inches wide or wider. Similar ***D. tanganyikensis***, also of East Africa, has larger flowers in clusters at the ends of more densely-hairy stalks. Angular-lobed, toothed, softly-hairy leaves and similarly hairy stems are characteristic of ***D. coria***, native of Madagascar. From toward the ends of its branches this produces long-stalked, few-bloomed clusters of deep pink flowers over 1½ inches in diameter.

Garden and Landscape Uses and Cultivation. Dombeyas thrive outdoors only in warm regions where little or no frost occurs. They are splendid for shrub borders and as lawn specimens, and need minimum care. They prosper in sun or part-day shade, and respond to any ordinary, reasonably fertile soil that does not lack moisture. Pruning to thin the growth and admit light to the shoots that are left may be done as soon as flowering is through. Then, too, remove all faded flower clusters, otherwise they persist and look untidy. Propagation is easy by cuttings and seeds. Although not suitable for small greenhouses they are sometimes planted in ground beds in large conservatories.

DOODIA (Doòd-ia). Fewer than a dozen species are included in *Doodia*, a genus of ferns of the blechnum family BLECHNACEAE, that occurs in the wild from New Zealand to Australia, New Caledonia, Juan Fernandez, and Ceylon. Its name commemorates Samuel Doody, a botanist, apothecary, and Keeper of Chelsea Physic Garden in London, who died in 1706.

Doodias are small to medium in size and have black roots and ascending rhizomes. Their dark-stemmed, rigid fronds (leaves) are densely clustered, and pinnate, with the segments sharply-toothed. The oblong clusters of spore capsules are in one to several rows parallel to the mid-veins of the leaflets.

A variable native of new Zealand and Australia, ***D. caudata*** has fronds with slender stalks 2 to 6 inches long and lanceolate to linear-oblong blades divided into up to thirty rather distantly spaced, light green leaflets. Spore-bearing fronds are erect and have blades from 8 inches to over 1 foot long and up to 2¼ inches wide. They have slender, toothed leaflets that, except for the terminal one, which may be 6 inches long, are about 1 inch long. Sterile fronds spread horizontally and have blades about 6 inches long and rarely more than 1½ inches wide. The clusters of spore capsules are in single rows.

From the last ***D. media*** differs in its harsher, more leathery fronds, those that bear spores and the barren ones similar. They are 1 foot to 2 feet long, 2 to 4 inches wide, and linear-lanceolate to oblanceolate. Their leaflets are sharply toothed and up to 2 inches long or a little longer. This species inhabits New Zealand and islands of the Pacific, including Hawaii. The clusters of spore capsules are in single rows.

Doodia media

Garden and Landscape Uses and Cultivation. In frost-free or almost frost-free humid climates doodias are splendid as low groundcovers in shaded places where the soil has a high organic content and is well drained, but not dry. They are charming for open woodlands, rock gardens, and the shaded sides of buildings. They are also excellent greenhouse plants and are serviceable as houseplants.

In greenhouses doodias are useful for planting in ground beds and under benches as well as in pots. They succeed in a mixture of peat moss and coarse sand or vermiculite with a lacing, not too generous, of turfy topsoil and some crushed charcoal. It is important that the pots be well drained and of moderate, not excessive, size in comparison to the sizes of the plants. Stagnant water is abhorrent; watering should be done to keep the soil evenly moist, but not for long periods saturated. Shade from sun is necessary, as is a humid atmosphere. Minimum winter night temperatures of 50 to 55°F are satisfactory; by day they may be five to ten degrees higher. Well-rooted specimens benefit from moderate fertilization. For additional information see Ferns.

DORICENTRUM. This is the name of orchid hybrids the parents of which are *Ascocentrum* and *Doritis.*

DORIELLA. This is the name of orchid hybrids the parents of which are *Doritis* and *Kingiella.*

DORIELLAOPSIS. This is the name of hybrid orchids the parents of which include *Doritis, Kingiella,* and *Phalaenopsis.*

Doritis pulcherrima

DORITAENOPSIS (Doritaen-ópsis). Hybrids between *Doritis* and *Phalaenopsis* of the orchid family ORCHIDACEAE bear the name, composed of parts of those of the parent genera, *Doritaenopsis.* In appearance these orchids more resemble *Doritis* than their other parent. Their blooms range from blush-pink to rose-pink to almost fuchsia-pink, or are lavender, amethyst-purple or white. They are two-ranked in many-flowered, graceful sprays. Orchid growers find doritaenopsises easy to manage. They respond to conditions and care that suit *Phalaenopsis.* For more information see Orchids.

DORITIS (Dór-itis). Closely allied to *Phalaenopsis* and by some authorities included in that genus, *Doritis* of the orchid family ORCHIDACEAE consists of two evergreen, tree-perching species, natives of Indomalaysia. From *Phalaenopsis* it differs in having erect, instead of horizontal or pendulous stems and in its flowers having a column-foot about as long as the column and the basal parts of their lips slender and with two flat, narrow appendages. The name, from the Greek *doris,* a knife, refers to the stiff leaves. Hybrids between *Doritis* and *Phalaenopsis* are named *Doritaenopsis.*

The showiest species and the only one cultivated is ***D. pulcherrima*** (syns. *Phalaenopsis pulcherrima, P. esmeralda, P. antennifera, P. buyssoniana*). This has stems 6 inches or so tall. Its up to about twelve rigid, usually spreading leaves up to 8 inches in length, are often suffused with purple, particularly on their undersides. In fall and winter it develops erect, slender-stalked, elegant racemes, or rarely panicles that lengthen, as the blooms open in succession, to an eventual 2 to 3 feet. Each has ten to twenty blooms that may vary much in color and are ½ inch to 2 inches in diameter. Their oblong-elliptic sepals and petals, the latter much the smaller, are spreading or recurved.

In its typical form the flowers have dorsal sepals and petals up to ½ inch long, the sepals about one-half as wide, the petals slightly wider. In the form once recognized as *Phalaenopsis esmeralda* petals and sepals are brilliant amethyst and all lobes of the lip are deep purple. In the kind previously called *P. antennifera* petals and sepals are light rose-pink, the middle lobe of the lip amethyst, the side lobes striped orange-red. In the phase that was known as *P. buyssoniana* sepals and petals are bright purple, all lobes of the lip are bright scarlet inside, with the outsides of the side lobes ochre-yellow with scarlet lines.

Horticultural tetraploid varieties of *D. pulcherrima* have flowers conspicuously larger, their dorsal sepals and petals ¾ inch long and broader than those of the typical species. In gardens it has become common, but incorrect to identify large-flowered tetraploid plants as *P. buyssoniana* and small-flowered diploid ones as *P. esmeralda.* In addition to tetraploids there are triploid hybrids between them and diploid plants.

Garden Uses and Cultivation. These are as for *Phalaenopsis.* For more information see Orchids.

DORMANT. In garden parlance dormant commonly alludes to deciduous plants when they are without leaves, less often it is said of evergreens not in active growth. Semidormant is also used to describe evergreens in this stage and is perhaps preferable. Transplanting deciduous trees, shrubs, herbaceous, and bulb plants is frequently best done at the end of the dormant or resting season. Indoor plants and others in containers need less frequent watering when dormant than at other times. Many bulb plants need to be kept quite dry then.

DORMANT SPRAYS. Many deciduous trees and shrubs in their dormant stages withstand certain sprays used to control pests and diseases in stronger concentrations than they will without injury when in leaf. Applications at such concentrations are called dormant sprays. They are used especially to combat insects, mites, fungi, and other damaging organisms resistant to the weaker concentrations it is practicable to employ when the plants are in foliage. Among the most commonly used dormant sprays are miscible oils and lime-sulfur. Delayed-dormant sprays, especially effective in some instances, are those applied, chiefly to fruit trees, between the time when the buds become plump and silvery

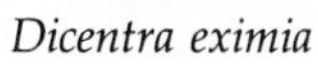

Dicentra eximia

Dieffenbachia maculata 'Rudolph Roehrs'

Dietes vegeta prolongata

Dictamnus albus

Digitalis purpurea

Dimorphotheca sinuata

Dimorphotheca sinuata, hybrid

Diphylleia cymosa

Distictis buccinatoria

Dodecatheon alpinum

and their green tips attain "the size of squirrels' ears." Up-to-date advice on the uses of dormant sprays is available from Cooperative Extension Agents and State Agricultural Experiment Stations.

DORONICUM (Dorόni-cum)—Leopard's Bane. Hardy herbaceous perennials with mostly large, showy, daisy-type flower heads constitute the genus *Doronicum* of the daisy family COMPOSITAE. There are thirty-five species, natives chiefly of mountains, in Europe, North Africa, and temperate Asia. The name is a Latinized form of the Arabic one of some kinds. The roots are said to be poisonous.

Doronicums have few-branched or branchless stems and undivided, generally coarsely-toothed or sometimes lobed leaves, the basal ones long-stalked, the usually rather distantly spaced stem ones alternate and shorter-stalked or stem-clasping. The yellow flower heads are solitary or few to several together at the ends of long stalks. They have a center or disk of bisexual florets surrounded by a single row of petal-like, female ray florets that, according to species, have or are without a pappus (tuft of hairs associated with the ovary and fruits). The seedlike fruits are ten-ribbed achenes.

Popular, early-flowering ***D. cordatum*** (syn. *D. caucasicum*), of from Sicily to Asia Minor, spreads, though not aggressively, by underground stolons. It is somewhat hairy, 1 foot to 2 feet tall, and has coarsely-toothed leaves. The basal ones have nearly round to kidney-shaped blades with heart-shaped bases. The lower stem leaves are stalked and ovate, those above are stalkless and stem-clasping, and change in shape upward from ovate to linear-lanceolate. From 1½ to 2 inches wide, the solitary flower heads have ray florets without pappuses. Variety *D. c. magnificum* is more robust and has larger flower heads. In variety 'Spring Beauty' (syn. 'Fruhlingspracht') the deep yellow heads of bloom are fully double. The often scarcely toothed or sometimes toothless lower leaves of ***D. clusii*** (syn. *Arnica clusii*), a 1- to 2-foot-tall species of the Pyrenees, European Alps, and Carpathian Mountains, have short, winged stalks and blunt oblong to broad-elliptic blades not heart-shaped at their bases. The higher lightly stem-clasping leaves, toothed only in their lower halves, are lanceolate. Solitary and 1½ to 2 inches wide, the flower heads have ray florets with pappuses.

More robust and later blooming than the kinds considered above, ***D. plantagineum*** is 2 to 3 feet tall. From *D. cordatum* it differs in its basal leaves not having heart-shaped bases, and from *D. clusii* in the ray florets of its flower heads being without pappuses. Native of southwestern Europe, *D. plantagineum* has somewhat tuberous roots, and toothed, ovate to ob-

Doronicum cordatum 'Spring Beauty'

long, hairy, wavy-edged leaves, the basal and lower stem ones stalked, those above stalkless. The flower heads, 2 to 3 inches wide, usually solitary are rarely two or three together. An especially tall, vigorous variant, its stems sometimes with one or two branches, and its flower heads 3 to 4 inches wide, is named *D. p. giganteum*. Variety 'Harpur Crewe', is about 3 feet tall and has golden-yellow flower heads. The name *D. excelsum* has been used both for this and the species. Variety 'Miss Mason', about 2 feet in height, is an exceptionally free-flowering variety of *D. cordatum*.

Doronicum cordatum 'Miss Mason'

Up to about 4 feet tall, their flower heads several on each stem, and their stems with leaves to their tops, *D. austriacum* and *D. pardalianches* are Europeans. Without underground stolons ***D. austriacum*** has stems that are much more leafy than those of the other. The leaves are longer than the spaces on the stems (internodes) between them. At flowering time the basal foliage, consisting of stalked, ovate-lozenge-shaped, toothed leaves with heart-shaped bases, has mostly died away. The branched stems have ovate, stem-clasping lower leaves, with winged stalks, and higher ones stalkless and clasping the stems with heart-shaped bases. The approximately 2-inch-wide flower heads have ray florets without pappuses. From the last, ***D. pardalianches*** differs in spreading by underground more or less tuberous stolons and in its basal foliage being present and green at flowering time. A selected variety with canary-yellow flowers is named 'Bunch of Gold'.

Doronicum pardalianches

Garden and Landscape Uses. Leopard's banes are appreciated as cheerful spring- and early-summer-flowering members of the vast daisy family, useful for flower gardens and less formal areas, and as cut blooms. Low-growing species can be used with good effect in rock gardens. In general, lower-growing kinds such as *D. clusii* are superior to taller ones. The latter, in their appearance, have a suspicion of coarseness. All grow readily in fertile, well-drained soil, which is preferably slightly acid. They appreciate just a little shade from the fiercest sun, although they will flourish without this, and sufficient moisture to keep their foliage from wilting. Because they lose all or most of their foliage in summer it is unwise to plant large groups of doronicums in flower beds or borders. If this is done, unsightly, difficult-to-conceal gaps in the plantings become evident. It is better to employ these plants in groups of not more than three, the individuals set 1 foot to 2 feet apart according to the vigor of their kind, so that neighboring plants can at least partially fill the gaps.

Cultivation. Care of doronicums makes no great demands of the gardener. It is well to dig up, divide, and replant them every third year and to apply a dressing of a complete garden fetilizer each early spring. The prompt removal of spent blooms sometimes encourages a sparser second flowering in fall. Seeds of leopard's banes usually germinate poorly. By far the commonest method of multiplication is by division.

In greenhouses low, early flowering doronicums such as *D. cordatum* and its varieties and *D. clusii* are very satisfactory for forcing in pots for late winter bloom. For this set strong young specimens in early fall into containers just big enough to hold their roots without undue crowding. Water well and bury the pots to their rims in peat moss, sand, or similar material in a cold frame. After freezing weather comes cover them with a good layer of dry leaves, salt hay, branches of evergreens or other such protective material that admits of free air circulation. In January or February bring the pots into a sunny greenhouse where the night temperature is about 50°F and that by day is five to fifteen degrees higher. Keep well watered, and once a week apply dilute liquid fertilizer. After blooming is through, if the plants are kept growing under cool condition, they may, when all danger of frost is passed, be planted outdoors. The same plants should not be forced two years in succession.

DOROTHEANTHUS (Doroth-eánthus)—Livingstone-Daisy. Formerly included in *Mesembryanthemum* and still in seed catalogs and gardens often referred to as such, this group of ten species includes three commonly cultivated. The genus belongs in the carpetweed family AIZOACEAE and consists of low annuals, natives of South Africa. Their name, from the feminine one Dorothea, and the Greek *anthus,* a flower, honors a German lady, Dorothea Schwantes. Dorotheanthuses have succulent, linear or spoon-shaped leaves, which, like those of the ice plant (*Cryophytum crystallinum*) to which *Dorotheanthus* is cousin, are covered with tiny glistening pustules. Because of this, these plants are sometimes called ice plant, although that designation is more properly reserved for the *Cryophytum.*

A difference between *C. crystallinum* and *Dorotheanthus* is that the leaves of the latter are not stalked. They narrow little or not at all to their bases, whereas those of *Cryophytum* have definite stalks. The flowers of *Dorotheanthus* are solitary, stalked, and in aspect daisy-like. Actually they are very different from daisies, which have heads composed of many florets that are truly miniature flowers. The *Dorotheanthus* bloom is one bisexual flower with numerous petals and stamens and five stigmas. The fruits are many-seeded capsules.

Among the showiest of garden annuals ***D. tricolor*** (syn. *Mesembryanthemum tricolor*) has many spreading branches 3 to 4 inches long that arise from a short stem. It forms mats or mounds 3 inches or so high. Linear, fleshy, and somewhat channeled, the leaves, 2 to 3 inches long, often broaden toward their apexes. The 1½-inch-wide flowers are borne freely over a long season. In color they vary from white through pink to red, often with their centers of a hue different from that of the petals. The stamens and anthers are black-purple.

The Livingstone-daisy (***D. bellidiformis*** syns. *D. criniflorus, Mesembryanthemum bellidiforme*) differs from the last chiefly in having flatter, obovate leaves up to 3

Dorotheanthus bellidiformis

inches long by ⅜ inch wide and with pimply surfaces. The long-stalked flowers are 1½ to 1¾ inches wide. They have dark centers and petals that are bright pink with dark margins or are white, pink, or orange, tipped with reddish-orange or red. From 3 to 4 inches tall, ***D. gramineus*** has pimply-surfaced, linear leaves up to 2 inches long. Its 1-inch-wide flowers on stalks up to 2½ inches long have a dark center eye and petals the basal halves of which are white and the upper halves rose-pink to cerise-pink.

Dorotheanthus gramineus

Garden and Lanscape Uses and Cultivation. The best uses for these plants are for edgings and the fronts of borders, and in rock gardens. They succeed in poor, dryish soils, as well as fertile, moister ones that are well drained. They require full sun. Seeds may be sown where the plants are to stay and the seedlings thinned to about six inches apart, or they may be started indoors, in a temperature of 60 to 65ωF earlier and be grown spaced 2 inches apart in flats, and planted in the garden later. When the latter plan is followed sowing is done eight to ten weeks before planting out time, which is about the same as for other frost-tender annuals and tomatoes. A planting distance of 6 inches is satisfactory.

DORSTENIA (Dors-tènia). As teaching aids dorstenias are frequently cultivated in botanical collections to illustrate plant morphology, otherwise they are little known horticulturally. Although of modest mien, some are quite attractive. They are milky-juiced, perennial, evergreen, herbaceous plants and subshrubs belonging to the mulberry family MORACEAE. Their name commemorates Theodore Dorsten, a German student of plants, who died in 1552. A modern tally lists 170 species, natives chiefly of tropical America and Africa, and one from India.

Botanical interest centers around the floral structure. This is intermediate between that of figs (*Ficus*) and that of most other plants. In figs the minute flowers are crowded on the inside of a hollow, spherical to pear-shaped receptacle with an opening at its apex. In *Dorstenia* the receptacle is flat or slightly saucered rather than globular, and the tiny flowers are disposed thickly over its upper surface. It is in effect an opened and flattened-out fig. The receptacles are circular or angled, plain-edged or variously lobed, and are held aloft on long stalks. The inconspicuous flowers are unisexual, usually with both sexes in the one receptacle. Behind and projecting outward from the receptacle are a number of leafy bracts. When ripe the seeds are ejected to a considerable distance as a result of the receptacle squeezing them as an orange pip might be pressed between finger and thumb and caused to shoot out. Dorstenias are reported to be poisonous. One, *D. contrajerva,* has been employed as an antidote for snake bite, and others have been used medicinally.

Most commonly cultivated is ***D. contrajerva,*** native from Mexico to northern South America, and in the West Indies. This has creeping rhizomes and broad-ly-ovate, often deeply-palmately-lobed, toothed, dark green leaves with heart-shaped bases. The floral structure is green, irregularly quadrangular, and about 1 inch in diameter.

Other species occasionally found in special collections include the following. ***D.***

Dorstenia contrajerva

argentata, from Brazil, has an erect, purplish, downy stem and lanceolate leaves up to 5 inches long, silvery-white at their centers and green at their margins. The flower head is concave and ringed with purple tubercles. Native of Egypt, ***D. barnimiana*** is a low, creeping plant with small, short-stalked, nearly circular leaves. Its narrow-elliptic flower heads, each with eight to ten spreading, slender rays, terminate erect stalks 1 to 3 inches long that lift them well above the foliage. ***D. barteri,*** of West Africa, and 2 feet tall, is hairy and has green, elliptic leaves up to 7 inches long. Its flat, grayish flower heads, 1 inch to 2 inches wide, are encircled with a green band and many variously-sized rays. ***D. gigas,*** a curious endemic of the island of Socotra, has a thick, gouty stem up to 4½ feet tall and 1½ feet thick that when wounded exudes a viscous yellow juice. The short-stalked leaves, clustered at the branch ends, and on young plants narrower, rougher, and with more decidedly recurved margins than on older specimens, are oblanceolate and 4 to 6 inches long. About ¾ inch in diameter, the short-stalked flower heads have about seven short rays. ***D. mannii,*** a West African species, has stems swollen at the nodes and elliptic or obovate, densely-hairy leaves paler beneath than on their upper sides. Its pale purplish, convex flower heads, 1 inch across, are fringed with many long, slender, green rays. ***D. multiradiata,*** from tropical Africa, has essentially hairless, obovate leaves and flower heads encircled by numerous, irregular-sized rays joined into a broadish band at their bases. ***D. nervosa,*** a Brazilian kind, has long stems, and green leaves with centers irregularly banded with leaden-gray. The concave flower heads are dark purple and without rays. ***D. phillipsiae*** of Somaliland has a stout trunk a few inches tall, shortly-branched above and furnished with short spines. Its oblong, sharply-toothed, leaden-gray, hair-

Dorstenia argentata

Dorstenia barnimiana

Dorstenia barteri

Dorstenia gigas

Dorstenia mannii

Dorstenia multiradiata

Dorstenia nervosa

Dorstenia phillipsiae

Dorstenia zanzibarica

less leaves 1½ to 2 inches long are in terminal clusters. The flower heads, on stalks up to 2 inches long, have cream receptacles ½ to ¾ inch in diameter, fringed by six to eight inch-long tan rays. ***D. turneraefolia,*** of tropical America, is a small species with obovate, toothed leaves and concave, irregularly rounded, green flower heads without rays. ***D. urceolata variegata,*** native to Brazil, has firm, lanceolate leaves mottled in two tones of green and banded with silver-gray along the midrib. The concave flower head is yellowish-green and without rays. ***D. yambuyaensis*** is indigenous in the Congo. Erect and bristly-hairy, it has glossy oblanceolate leaves, paler and duller beneath, irregularly toothed and up to 6 inches long. The flower heads have receptacles up to 1 inch across, roughly rounded, and with long rays. ***D. zanzibarica*** has upright stems and toothed, elliptic leaves. From the small gray-green flower heads project pale green rays. This species is native of Zanzibar.

Garden Uses. In addition to their botanical interest dorstenias are sufficiently decorative to warrant inclusion in collections of tropical ornamentals. Their leaf patterns are pleasing, their flower heads unusual. The lower kinds can be used as ground covers. They succeed under greenhouse benches. As outdoor plants they are of use only in the humid tropics and warm subtropics.

Cultivation. No difficulty attends the growing of these plants. All that the kinds described above ask is warmth, ordinary soil, and reasonable moisture. Shade from strong sun is needed. In greenhouses a minimum winter night temperature of 55 to 60°F is desirable. Higher temperatures by day, and at other seasons at night, are in order. Well-rooted specimens are stimulated by occasional applications of dilute liquid fertilizer. Pans (shallow pots) are suitable containers. The best effects are had when three plants or more are accommodated in each, at least that is true of the smaller growing kinds. Fertile soil that contains an abundance of humus is most to these plants' liking. Propagation is by division in late winter or spring just before new growth begins, and by seeds sown in spring in a temperature of 60 to 70°F. Self-sown seedlings often appear.

Dorstenia turneraefolia

Dorstenia yambuyaensis

DORYANTHES (Dory-ánthes)—Spear-Lily. This genus of three species of the amaryllis family AMARYLLIDACEAE is endemic to eastern Australia. Its name derives from the Greek *dory,* a spear, and *anthos,* a flower, and alludes to the tall blooming stems. In habit of growth resembling *Beschorneria* and *Furcraea,* these desert or semidesert plants form giant clumps of numerous large, fleshy, ribbed, sword-shaped leaves and send up stout stalks bearing dense terminal clusters of red flowers. The leaves of the flowering stalks are very small. The flowers differ from those of *Beschorneria* in having almost no perianth tube and spreading segments (petals). Often the clusters develop bulblets instead of blooms. Each flower has six petals and the same number of stamens, the latter attached to the bases of the petals and with stalks thickened, but not as markedly so as those of *Furcraea.* The fruits are top-shaped capsules containing slightly-winged seeds. The strong fibers of the leaves have been used for making matting and brushes.

Of impressive appearance, ***D. palmeri*** has up to 100 lanceolate leaves about 6 feet in length and 6 inches in width. Crowded at their bases, they arch gracefully and terminate in long, slender, cylindrical points. The flowers, interspersed with leafy bracts, are in pyramidal spikes 1 foot to 1½ feet long atop stalks 6 to 15 feet long. They are large, funnel-shaped, and red with paler centers. From this species ***D. excelsa*** differs chiefly in having narrower, less fleshy leaves, with shorter points, and flowers in spherical rather than elongated clusters about 1 foot in diameter. The blooms are bright red. The leaves of ***D. guilfoylei*** are up to 9 feet long by 9 inches wide. In other respects this kind closely resembles *D. palmeri.*

Doryanthes palmeri in southern California

Doryanthes palmeri (flowers)

Garden and Landscape Uses and Cultivation. As outdoor plants spear-lilies are satisfactory only where mild winters prevail. They succeed in southern California and elsewhere where little or no frost occurs. They withstand drought well and are impressive elements in the landscape, forming bold clumps and strong accents. They need deep, fertile, well-drained earth and full sun. They are usually increased by suckers. Seeds germinate readily, but the young plants grow slowly and several years elapse before they first bloom. Spear-lilies are also occasionally grown in large conservatories, usually among collections of succulent plants. They succeed under conditions that suit *Agave* and *Furcraea.*

DORYCNIUM (Dorýc-nium). Herbaceous perennials and subshrubs of the Mediterranean region and Canary Islands compose this genus of a few species of the pea family LEGUMINOSAE. The name is an ancient Greek one for some plant unknown.

Dorycniums have stalkless or short-stalked leaves of five leaflets, the lowest pair of which resemble leaf appendages called stipules. The true stipules are minute. The pea-like blooms are in stalked, headlike clusters from the leaf axils. They are white or pink and have dark red to nearly black keels. The stamens are ten, one free and the others joined. Containing one to several seeds, the fruits are thick, short pods.

A graceful, deciduous, much-branched subshrub with the lower, woody parts of its stems providing a permanent base from which grow annual leafy flowering shoots, ***Dorycnium pentaphyllum*** (syn. *D. suffruticosum*) is up to 1½ feet tall. Its slightly ribbed stems, when young, are gray-downy. The stalkless, grayish, hairy leaves have linear-oblanceolate leaflets ¼ to ½ inch long. In rounded heads about ½ inch in diameter, on stalks 1 inch to 2½ inches in length, the ¼-inch-long, white or pinkish flowers are borne in clusters of five to fifteen. They come in summer and are abundant. The ovoid, one-seeded pods are under ¼ inch across. This native of southwestern Europe is represented by minor variations in other parts of that continent. One variant that differs fairly consistently is *D. p. herbaceum* (syn. *D. herbaceum*). This is less woody than *D. pentaphyllum,* and has rather larger leaves. Its somewhat smaller blooms are in heads of from a dozen to twenty-five. This variety is endemic to central and southeastern Europe. Variety *D. p. germanicum* (syn. *D. suffruticosum sericeum*), of the Balkans and other parts of Europe, has upper leaves not smaller than the lower ones.

Reported to grow in the wild, where it favors moist soils, sometimes 5 feet in height, as known in cultivation ***D. rectum,*** of the Mediterranean region, is more usually 1 foot to 2 feet tall. It is subshrubby and has short-stalked leaves with obovate to obovate-oblong leaflets up to 1 inch long. Its white or pink flowers, about ¼ inch long, are in heads of twenty to forty, and are succeeded by linear-oblong pods ⅓ to ⅔ inch long.

Larger blooms are typical of ***D. hirsutum,*** which has been grown in gardens as *Cytisus lotus.* This has stems woody toward the base of the plant from which each year arise softer, leafy, flowering shoots to a height of 9 inches to 1½ feet. The leaflets are ⅓ to 1 inch long and oblong-obovate. From ⅓ to ¾ inch long, the white or pink flowers are in clusters of four to ten. The seed pods are oblong-ovoid and ¼ to nearly ½ inch in length.

Dorycnium hirsutum

Garden and Landscape Uses and Cultivation. The kinds described above are hardy in southern New England. They are suitable for rock gardens, and similar areas, and at the fronts of flower borders. Sun-lovers, they are best content with porous, loamy, not excessively fertile earth, that except for *D. rectum,* is dryish. Propagation is by seed.

DORYOPTERIS (Dory-ópteris). Most numerous as to species in Brazil, *Doryopteris* is represented in the floras of many parts of the Old World and the New World tropics. There are about thirty-five species, ferns of the pteris family PTERIDACEAE. The name derives from the Greek *dory,* a spear, and *pteris,* a fern. It alludes to the shape of the fronds (leaves) of some kinds.

Doryopterises are small to medium in size, compact and clustered or creeping. Their hairless, leathery fronds have glossy, generally dark purplish or black stalks.

The leaf blades are characteristically ovate, arrow-shaped, or more or less pentagonal in outline and palmately (in the manner of the fingers of a widespread hand) lobed, with the lobes often again variously lobed or cleft. The spore capsules are grouped in continuous or broken lines paralleling the leaf margins.

In the wild inhabiting the West Indies, ***Doryopteris pedata*** has short rhizomes and leaves with usually hairy stalks, winged at their angles. Those of the largest barren fronds are up to 6 inches long and support three- to seven-lobed blades up to 4 inches long. The lobes are often pinnately-cleft. Fertile fronds 3½ inches to 1½ feet long overall, have blades 2 to 8 inches long, commonly pentagonal with the central segment much larger than the others.

Doryopteris pedata

Rarely they are deeply-five-lobed. Variety *D. p. palmata,* native from Mexico to northern South America, differs most obviously in having buds that develop into plantlets at the bases of its leaf blades.

Doryopteris pedata palmata

A handsome native of Brazil and Colombia, ***D. nobilis*** has long-stalked fronds up to 1½ feet tall. The earliest produced by young plants have heart- to arrow-shaped blades. The later fertile ones are deeply-palmately-lobed with the lobes deeply-pinnately-cleft. Mature sterile fronds, less dissected than fertile ones, sometimes bear a plantlet where the blade joins the stalk. Distinctive ***D. sagittifolia,*** of Brazil and Venezuela, has fronds up to 1 foot long or a little longer, their stalks and blades of approximately equal lengths. The blades are arrow-shaped to long-triangular without further lobing or incising. Their two basal lobes spread widely outward or point backward. One of the loveliest and widely disseminated sorts, ***D. concolor*** inhabits the tropics of the Old World and the New World. Its long-stalked fronds, up to 1 foot tall or sometimes taller, have triangular to pentagonal blades rather small compared to the lengths of their stalks. They are deeply-palmately-divided into three main lobes, which are pinnately-lobed and with each of the lobes deeply-pinnately-cleft into narrow ultimate segements.

An extensively creeping species with slender rhizomes, ***D. ludens*** inhabits India, southeast Asia, China, Java, and the Philippine Islands. Its sterile fronds, up to 1½ feet long, have blades up to 6 inches across. They vary from ovate or arrow-shaped to deeply-palmately-lobed, with the lobes again lobed. Fertile fronds, which may be 2 feet long, have blades ordinarily deeply-palmately-lobed.

Garden and Landscape Uses and Cultivation. Only in the humid tropics and warm subtropics can these ferns be expected to succeed outdoors. They need shade and humus-rich, well-drained, but moist soil. They are appropriate for underplantings, rock gardens, and similar uses. They are attractive greenhouse plants, succeeding in a minimum winter night temperature of 60°F with daytime increases of five to fifteen degrees. In summer considerably higher temperatures are in order. Shade and high humidity are needed. Watering is done to keep the soil moderately moist, but not soggy. Judicious fertilization is helpful to well-rooted specimens. Propagation is by spores and by division. For further information see Ferns.

Plantlets on leaves of *Doryopteris pedata palmata*

DOSSINIA (Dos-sínia)—Jewel Orchid. Because of its beautiful foliage, for which rather than its flowers it is greatly admired, the only species of this genus of the orchid family ORCHIDACEAE shares with *Anoectochilus, Haemaria,* and *Macodes* the lovely colloquial name jewel orchid. Its botanical designation commemorates the Belgian botanist Pierre Etienne Dossin, who died in 1852.

Nearly related to *Anoectochilus,* and native to Borneo, ***Dossinia marmorata*** (syn. *Anoectochilus marmoratus*) is one of the most beautiful of the jewel orchids. An evergreen with a mostly basal, loose rosette of five or so broadly-elliptic to almost round leaves up to about 3½ inches in length, it produces in summer and fall stems with many-flowered racemes up to 1½ feet tall. The ½-inch-wide, light brown flowers have the ends of their sepals and petals pinkish-white or white, and a pink or white lip. The splendidly colored leaves are rich velvety green, beautifully netted with clear golden-yellow veins.

Garden Uses and Cultivation. These are as for *Anoectochilus.* For further information see Orchids.

DOSSINIMARIA. This is the name of bigeneric orchid hybrids the parents of which are *Dossinia* and *Haemaria.*

DOT PLANT. British gardeners employ this term to denote tall plants set among lower ones in formal flower beds. A bed of pink geraniums may be interplanted with standard heliotropes, or staked plumbagos may be dotted among deep purple petunias. Dot plants are sometimes standards or half standards, sometimes plants tied to stakes. Their purpose is to relieve the flatness of beds planted wholly to low plants. Carefully used as an integral part of good design they can be decidedly effective, but as with many gardening innovations (dot plants first became popular in Britain after World War I) the practice of using them has been much overdone and in public parks in Great Britain it is not uncommon to see what otherwise would be attractive beds spoiled by the tasteless inclusion of too many dot plants.

Standard lantanas (tree-trained) as dot plants in a bed of heliotropes edged with low lantanas, at The New York Botanical Garden

DOUBLE-COCONUT is *Lodoicea maldivica.*

DOUBLE FLOWERS. Flowers are said to be double when they have conspicuously more petals, or in the case of daisy-type flowers, more ray florets, than is normal for the species from which they are derived. Such multiplication of parts is commonly accompanied by a reduction in number or complete absence of stamens and sometimes of pistils. In a sense these organs are changed into additional petals or often sterile ray florets. If the transformation is complete the double flower cannot produce seeds. It has neither pollen nor stigmas. Not all double flowers are so sterile. Some retain some functioning sexual parts and thus are capable of seed production. Semidouble flowers are those in which there is some increase in number of petals or ray florets, but a sensible number of untransformed stamens or disk florets remain. Double flowers are extremely rare in the wild and the cause or causes of the phenomenon are not known. Possibly high soil fertility has something to do with it. Certainly in poor soils some double varieties tend to produce less fully double or fewer double blooms. Doubling can often be stabilized and increased by selecting and reselecting for several generations seeds from individual plants that show the greatest tendency in this direction. Double flowers are very common in such plants of the daisy family COMPOSITAE as asters, chrysanthemums, dahlias, shasta-daisies, sunflowers, and zinnias, but are also found in many other families and genera. There are, for instance, double-flowered African-violets, carnations, crab apples, day-lilies, geraniums, horse-chestnuts, lilacs, peonies, portulacas, and roses, to mention only a very small proportion of numerous familiar examples.

Double flowers: (a) Camellia

(b) Gloxinia (*Sinningia*)

(c) *Kerria japonica pleniflora*

(d) Sunflower (*Helianthus*)

(e) Tulip

DOUGLAS-FIR is *Pseudotsuga menziesii.*

DOUGLASIA (Doug-lásia). The great botanical explorer of the Pacific Northwest, David Douglas, who in 1834 met a tragic death in the Sandwich Islands, now the Hawaiian Islands, is commemorated not only by the common name of the splendid Douglas-fir, but also by the botanical one *Douglasia,* which identifies a genus first discovered by Douglas. It consists of six pretty alpine and arctic species of northwestern and arctic North America, with one perhaps also into Siberia, and one in Europe. It belongs in the primrose family PRIMULACEAE.

Douglasias are low herbaceous perennials after the fashion of *Androsace.* Tufted or cushion-forming, they have ascending or prostrate, branched stems ending in rosettes of small, undivided, toothed or toothless, usually narrow leaves. The flowers are solitary or in umbels at the tops of stalks leafless except for sometimes a few bracts at the bottoms of the umbels. They commonly have five-lobed calyxes and corollas, the latter with quite long funnel-shaped tubes and spreading lobes (petals). In the throats of the blooms are five scales, there are the same number of stamens, and one style ending in a head-like stigma. The fruits are capsules, few-seeded in contrast to the many-seeded ones of *Primula.*

The only yellow-flowered species, also the only European, is ***D. vitaliana.*** It occurs in the mountains of Spain and central Europe and makes cushions of freely-branching little stems and narrowly-linear, toothless, overlapping leaves about ½ inch long, edged with hairs. Its short-stalked, brilliant flowers, about ½ inch across, are solitary or sometimes in pairs at the branch ends.

American douglasias all have pink, rose-pink, or rosy-violet flowers, and in general, are less amenable to cultivation than their European relative. The most tractable, ***D. laevigata,*** of Washington and Oregon, is a matting kind with more or less oblanceolate, often few-toothed, firm, lustrous leaves ¼ to ¾ inch long, and hairless or nearly so. Its rather loose two- to ten flowered, short-stalked umbels are of individually nearly stalkless flowers ⅓ inch wide. The bracts at the bases of the umbels are generally not more than twice as long as wide. Variety *D. l. ciliolata* has more compact umbels and more obviously hair-fringed, thinner leaves.

Foliage gray with tiny star-shaped hairs, flowers with longer individual stalks, and in two- to eight-flowered umbels with bracts at their bases usually several times as long as broad, distinguish ***D. nivalis*** from *D. laevigata.* The typical species has toothless leaves. Those of *D. n. dentata* (syn. *D. dentata*) are toothed. This species is known from Washington to British Columbia, the Rocky Mountains, and Alberta. Flowers solitary or in pairs without bracts at their bases, are characteristic of ***D. montana,*** a tight-growing gem with foliage gray with branchless or forked, but not stellate hairs. It has clear pink blooms. Its linear-lanceolate leaves are up to ⅓ inch long, its flowers about ⅓ inch across. This kind is at home in limestone and non-limestone soils from the Rocky Mountains to British Columbia, but is difficult to manage in cultivation.

Garden Uses and Cultivation. These lovely alpines are strictly for skilled rock gardeners. They are impatient of hot, torrid summers and of poor soil drainage. Near New York City *D. vitaliana* and *D. laevigata* grow satisfactorily. They prosper in sun, but prefer a north-facing slope where summer heat is tempered, to more extreme conditions. Gritty soil suits them. Other kinds are most likely to succeed where summers are fairly cool, and under scree or moraine conditions. For more about this see Rock and Alpine Gardens. Propagation is by seed and by careful division in early spring.

DOVE ORCHID. This is a vernacular name for *Peristeria elata* and *Oncidium ornithorhynchum.*

DOVE TREE. See Davidia.

DOVYALIS (Dovy-àlis)—Kei-Apple or Umkokola, Ceylon-Gooseberry or Kitambilla. Shrubs and small trees of Africa and Ceylon constitute this genus of about eleven species of the flacourtia family FLACOURTIACEAE, the name of which is sometimes spelled *Doryalis.* They are hardy only in regions of very mild winters. From the closely related *Flacourtia* this group differs in having seeds not enclosed in stonelike shells. Its name is of unexplained origin. Cultivated kinds have edible fruits.

The species of *Dovyalis* (syn. *Aberia*) may or may not be spiny. They have short-stalked, pinnately-veined leaves and male and female petal-less flowers, commonly on separate plants. The males, many together in the leaf axils, have four or sometimes more hairy sepals and many stamens that alternate with large glands and grow from a fleshy disk. The female blooms are solitary or in small groups. Each has five to nine hairy sepals and sometimes staminodes (non-functional stamens). The fruits are berries.

The kei-apple or umkokola (***D. caffra***) is a rather dense, thorny native of South Africa that is reported to have lived through temperatures as low as 20°F in California without injury. A vigorous shrub or small tree attaining a height of 15 to 20 feet, it has glossy oblong-obovate, smooth-edged leaves 1½ to 2½ inches long, often clustered at the bases of the sharp spines. The inconspicuous greenish or yellowish flowers are in the leaf axils of the younger branches. The strongly acid fruits are aromatic and have the fragrance of ripe apricots. They have juicy pulp and five or more small flattened seeds. They are approximately spherical, 1 inch to 1½ inches in diameter, and are used chiefly for preserves, jelly, and sauces. They are not first-class fruits for eating out of hand.

The Ceylon-gooseberry or kitambilla (***D. hebecarpa***) is less hardy than the kei-apple, but can be grown satisfactorily in south-central Florida. Of upright growth, it attains a height of 15 to 20 feet and has slender branches, likely to droop when heavily fruited, and furnished with long, sharp spines. The pale green, pointed leaves, 2 to 4 inches long, sometimes slightly toothed, and when young more or less velvety-hairy, are lanceolate to ovate. The flowers are inconspicuous. Closely resembling gooseberries in size, appearance, and flavor, the maroon-purple fruits contain luscious, sweet, purplish pulp. They are of better quality than those of the kei-apple.

Less well known than the kei-apple and the Ceylon-gooseberry is ***D. abyssinica,*** which has fruits of better eating quality than either. This native of Ethiopia is a bushy, spiny or nonspiny shrub up to 10 feet tall and with ovate, hairless, glossy light green leaves 1 inch to 3 inches long. Male and female flowers are in the leaf axils and on separate plants. They are small, inconspicuous, and greenish, the females solitary, the males in clusters. The fruits are globose-ovoid, apricot-colored and flavored, and about 1 inch in diameter. They contain a few small seeds. A natural hybrid between this species and *D. hebecarpa* has yellowish-brown fruits, less acid than those of the Ceylon-gooseberry. Individual plants may produce all male or both male and female blooms.

Garden and Landscape Uses. In addition to their usefulness as fruit producers these species of *Dovyalis* have value as garden and landscape ornamentals. The kei-apple especially is recommended as a hedge plant and can scarcely be bettered for that purpose in regions where it thrives. It is a strong grower, makes a splendid barrier, and is decidedly drought-resistant. At its best in the subtropics rather than in intensely tropical climates, it appreciates soil that has a high organic content. The Ceylon-gooseberry is more tropical in its needs and responds well to humid conditions. It is less resistant to drought than the kei-apple.

Cultivation. Propagation of these plants is by seeds or, to multiply particularly desirable individuals, by grafting or shield budding onto seedling understocks, or by layering. When grown for their fruits the trees should be spaced 12 to 15 feet apart, and one male should be included among every twenty to thirty females. Plants set to form hedges should be 3 to 5 feet apart. Trees raised from seed fruit when four or five years old.

DOWNINGIA (Downíng-ia). One of the most illustrious American landscape architects and horticulturists, Andrew Jackson Downing who died in 1852, is honored by the name *Downingia,* that of a genus of eleven species of the bellflower family CAMPANULACEAE. It inhabits western North America, and one species inhabits southern South America. The cultivated kinds are North American.

The genus consists of hairless, soft and rather succulent annuals with alternate, lobeless, awl-shaped to lanceolate, or the upper ones somewhat broader, leaves. Those above gradually diminish in size until they are represented by vestigial leaves called bracts. The distinctly two-lipped, five-parted flowers, although actually stalkless appear to be stalked because of their long, twisted ovaries, which serve like stalks, to hold the blooms away from the stems. The flowers have narrow sepals and usually blue corollas with the three-lobed lower lip blotched with white or a color. The two upper lobes are usually smaller than those of the lower lip. There are five stamens and one style. The fruits are many-seeded, slender capsules.

Blue calico flower (***D. elegans***) grows in ditches and at pond sides where the soil is moist or wet early in the year even though it dries later. It is endemic from Washington to California. From 4 inches to 1 foot in height and with thick stems, it has slender leaves up to 1 inch long and spikes of bloom from 2 to 10 inches long. The flowers fancifully resemble the torsos of a bevy of girls attired in pale blue cotton prints with the white blotches on the lower lips of the blooms representing aprons. But sometimes the maids are attired in other colors. Rarely the blooms are a delicate chocolate-pink, occasionally they are white.

Native in Oregon and California, ***D. pulchella*** differs from the last species in that the lower lips of its flowers bend downward at a sharp angle to the corolla tube. It grows in similar habitats and has stems 2 to 10 inches long with slender leaves not over ½ inch long. The flowers, blotched on their lower lips with white and with two yellow spots and three alternating dark purple ones, are prevailing blue with purple corolla tubes.

Garden and Landscape Uses. In aspect cultivated downingias somewhat suggest large editions of the annual blue-flowered lobelias that are their relatives. They are delightful for edgings and the fronts of borders and are appropriate for window and porch boxes, rock gardens, and in regions where they are wild, native plant gardens. Because it is a looser-growing plant than the other, *D. pulchella* is better suited for hanging baskets than *D. elegans.* Both may be grown as pot plants for spring bloom in cool greenhouses, but *D. elegans* is of tidier habit and is generally preferred for this purpose. Downingias prosper in any well-drained garden soil.

Cultivation. Outdoors, seeds are sown where the plants are to bloom, in sun or where there is a little part-day shade. The seedlings are thinned to about 6 inches apart. Alternatively, seeds may be started indoors eight to ten weeks before planting out time and the seedlings transplanted 2 inches apart in flats and grown in a sunny, airy greenhouse where the night temperature is about 50°F. Planting outdoors is done as soon as there is no danger of frost.

Greenhouse plants for spring blooming are had by sowing seeds in September and growing the plants throughout in full sun

Downingia pulchella

in a night temperature of 50°F and day temperatures five to ten degrees higher. The plants are repotted into larger containers as growth requires. Receptacles 5 inches in diameter will usually be large enough for the final shift. After the last pots are filled with roots additional nourishment is provided by regular applications of dilute liquid fertilizer. Soil for downingias should be fertile and porous and the pots well drained. Watering must at all times be carefully done, since excessive wet easily causes damage or death.

DOWNY MILDEW. A number of diseases of plants are caused by fungi classed as downy or false mildews. These are distinct from powdery or true mildews from which they can generally be distinguished by the downy appearance of the white, gray, or lavender-violet patches they cause. Similar patches resulting from infections of powdery mildews have a white, felty, or powdery aspect.

Unlike powdery mildews, which live on the surfaces of plant parts and ordinarily extend only tiny, rootlike sucking organs into the cells, downy mildews live within the plant tissues, invading them deeply, and extend, through the pores or stomata, only their spore-bearing organs.

Among plants susceptible to downy mildews are beets, Boston-ivy, lettuce, poppies, spinach, stocks, sweet alyssum, Swiss chard, and violets.

Control is based on avoiding excessively humid conditions and crowding, spraying or dusting with Bordeaux mixture or other copper insecticide, not practicable with food crops soon to be eaten, and eradicating badly affected plants.

DOWNY-MYRTLE is *Rhodomyrtus tomentosa.*

DOXANTHA. See Macfadyena.

DRABA (Drà-ba). Chiefly denizens of mountains and other cool-climate, including arctic, parts of the Northern Hemisphere, the approximately 300 species of *Draba,* are poorly represented in gardens. True, many are too weedy in appearance to warrant inclusion, but a fair number are decidedly pretty and a few are choice. In the main they are easy to propagate and grow. They are hardy. The kinds described below are perennials.

The genus belongs in the mustard family CRUCIFERAE and like the vast majority of that mighty race has typically cross-shaped flowers with four sepals, four white, yellow, or purplish petals, six stamens two of which are shorter than the others, and a single style. In *Draba* the short, linear to almost spherical seed pods contain two rows of seeds in each of two compartments. Drabas are tufted, cushion- or mat-forming, hairy or hairless annuals, biennials, and perennials with undivided, toothed or toothless leaves, generally mostly in basal rosettes. Usually fairly numerous and small, the flowers are in racemes, which may or may not have leafy stalks. The name *Draba* is the ancient Greek one for cress (*Lepidium*).

One of the most obliging, ***D. sibirica*** (syn. *D. repens*) is native from the Caucasus to Siberia, and in Greenland. It is very distinct. It has long, slender, creeping stems from which arise branches to a height of 2 to 10 inches, bearing in spring eight to twenty bright yellow flowers about ⅓ inch across. The leaves are pointed oblong-lanceolate, not toothed, and somewhat hairy. They are scarcely in rosettes. The seed pods are oblong-elliptic, often somewhat curved, and 1½ to 3 inches long. A vigorous, easily contented plant, this is excellent for rock gardeners.

Another easy-to-grow kind, very different in appearance from the last, is ***D. lasiocarpa*** (syn. *D. aizoon*). This forms firm, dome-shaped cushions of usually tightly packed stems and foliage, 3 to 4 inches tall, and has racemes of sulfur-yellow flowers that attain a height of up to 8 inches. The stamens are shorter than the petals. The styles are about 1/20 inch long. The linear-lanceolate leaves, up to ¾ inch long by ⅛ inch wide, have bristle-haired margins. The oblong-elliptic seed pods are flat and up to ⅓ inch long. This is a variable native of central and southern Europe. Very similar to *D. lasiocarpa* and sometimes treated as a variety of it is ***D. compacta,*** of high mountains in the Balkans and the Carpathians. Extremely dwarf, it is technically distinguished from *D. lasiocarpa* by its minute style.

Draba lasiocarpa

A favorite species, ***D. aizoides*** inhabits the mountains of central and southern Europe. Similar to, but smaller than *D. lasiocarpa,* it forms dense hillocks of stiff, bristly rosettes above which rise leafless racemes of up to eighteen yellow blooms to heights of 2 to 4 inches. The stamens are approximately as long as the petals. Except for marginal bristles the linear, ½-inch-long, rigid, bristle-tipped leaves are hairless. This is a very variable species. Similar, but looser, softer, less definitely forming

Draba aizoides

humped cushions, with linear leaves only about ¼-inch-long, and stamens one-half as long as the petals, ***D. bruniifolia,*** is a yellow-flowered native of Asia Minor and the Caucasus. Its small seed pods are hairy. From it mat-forming *D. b. olympica* is distinguished by all of its leaves, instead of only those of the short branches, being fringed in comblike fashion with bristles.

From Spain comes ***D. dedeana.*** This mountain gem forms low tussocks of greenery from which arise hairy, 2-inch flower stems with three to ten white or cream-yellow blooms under ½ inch across, with stamens much shorter than the petals. The linear-lanceolate, hair-fringed leaves, about ¼ inch long, are in rosettes. Minor variants of this species are sometimes distinguished as *D. d. cantabrica,* with pale yellow blooms, and *D. d. zapateri,* in which the leaves are longer and narrower and the racemes looser than those of the typical species.

Forming tufts about 2 inches across, ***D. fladnizensis*** is a native of arctic and alpine regions of North America as well as of Europe and Asia. It has oblong-oblanceolate, rarely toothed, usually sparsely-hairy leaves about ¼ inch long, margined with stiffish hairs, in rather loose rosettes. Whitish or yellowish, its flowers, on stalks that scarcely lengthen as seed pods are formed, are two to twelve together. The seed pods are flattened, hairless, and oblong-elliptic. Sometimes perennial but often biennial, variety ***D. incana*** of northern North America, northern Europe, and northern Asia is 3 to 10 inches tall. It has blunt-lanceolate, sometimes toothed leaves up to 1 inch long and ten to forty small white flowers in hairy racemes.

Garden Uses. Drabas are best adapted for rock gardens, wall gardens, and similar intimate plantings and for growing in containers in alpine greenhouses and cold frames. They like gritty soil, reasonably nourishing, and neither excessively dry or wet. Those containing lime are generally highly acceptable. Dependable and ingratiating *D. sibirica* is a first-class carpeter where a flat mat of greenery, blanketed in season with small yellow blooms, is satisfactory. It is a good rock garden plant that spreads so that one plant soon fills an area 1 foot to 2 feet across, yet it is not so rampant that it is a threat to any but the tiniest neighbors. It is an excellent cover for small bulbs, either those that bloom in spring or kinds that flower at other seasons. The other drabas described need less room. They are neat plants suitable for miniature sloping plains in rock gardens, and crevices, vertical or nearly so, or other niches where they receive just a little shade from the strongest sun and where soil drainage is very thorough. Excessive wet in fall and winter is very harmful and for most kinds torrid, humid summer weather can be disastrous. One of the easiest of the cushion type to grow, more tolerant of sun, heat, and humidity than most, is *D. lasiocarpa.*

Cultivation. Drabas are increased readily by seed and division. The seeds are sown in fall or spring, for convenience usually in pots or pans (shallow pots) of porous soil. The receptacles may be buried to their rims in a cold frame or outdoors and covered with a sheet of shaded glass until germination takes place, or in spring they may be transferred to a cool greenhouse or similar place. The young seedlings are transplanted to flats, pans, or individually to small pots. In pots or pans in greenhouses or cold frames they may be grown with little trouble, always provided the bogey of excessive winter wetness is guarded against.

Draba incana in a rock crevice

Draba rigida

DRACAENA (Drac-aèna)—Dragon Tree, Corn-Plant. Many handsome foliage plants are found here. Some are familiar as room and greenhouse ornamentals, and they and others as first-rate furnishings for outdoor gardens in the tropics and near tropics. Not all plants commonly called dracaenas are correctly assigned. Some, including the popular ti plant, belong to the genus *Cordyline* and are treated under that heading in this Encyclopedia. Plants sometimes differentiated as *Pleomele* are here included in *Dracaena.* A distinction between *Dracaena* and *Cordyline* is that the ovules in each cell of the ovary in *Dracaena* are solitary, in *Cordyline* numerous. Both genera belong to the lily family LILIACEAE.

There are about 150 species of *Dracaena,* most of them natives of the Eastern Hemisphere. The name, from the Greek *drakaina,* a female dragon, alludes to the trunks and branches of the dragon tree (*D. draco*), of the Canary Islands, containing and exuding a red, resinous sap, called dragon's blood. This is dried and used, like the dragon's blood obtained from *Daemonorops draco,* for coloring varnishes, and pharmaceutically. A notable *Dracaena draco,* one that grew on Tenerife and was blown down in 1868, was famous for centuries. It was believed to be the oldest tree in the world. It was 70 feet high and had a trunk girth of nearly 45 feet.

Dracaenas are evergreen shrubs or trees with solitary or clustered trunks or stems that may branch or not. The leaves, of various shapes and colors, according to kind, may be in tufts at the tops of the stems or arranged along all or much of the lengths of the stems. In many species and varieties they are beautifully colored or variegated. In most kinds the flowers are of little decorative importance. Small, often fragrant, they are greenish, yellowish, reddish, or white and are in dense heads or in tight clusters arranged in panicles. Each bloom has a tubular or funnel-shaped perianth of six tepals, commonly called petals, that spread or are reflexed and may or may not be shortly united at their bases, six stamens, a slender style, and a rounded or somewhat lobed stigma. The fruits are spherical berries containing one to three seeds.

The dragon tree (**D. draco**), is one of the most massive species. It eventually be-

Dracaena draco at Huntington Botanical Gardens, San Marino, California

Dracaena draco at the Royal Botanic Gardens, Kew, England

comes 40 to 60 feet tall or even taller, and develops a huge trunk, but in cultivation it is likely to be considerably smaller. It branches freely and is well foliaged. The crowded, sword-shaped, long-pointed, glaucous-green leaves are soft and rather fleshy. They are erect or reflexed and 1½ to 2 feet long by 1¼ to 1¾ inches wide. The greenish flowers are succeeded by orange berries. Another large tree with branching trunks, ***D. arborea,*** of New Guinea, is popular for outdoor landscaping in southern Florida and other warm regions. Up to 40 feet in height, but often lower, this kind has broad, sword-shaped, fresh green leaves up to 3 feet long. The midribs are raised, the side veins depressed.

Endemic to Hawaii, **D. aurea** is a stout-trunked, slender-branched tree up to 20 to 35 feet tall. It has arching, leathery, stalkless, stem-clasping, glossy leaves, 1½ feet long by somewhat under 1½ inches wide, in terminal clusters. The greenish or yellowish flowers, in panicles, are followed by red berries. From the last, Hawaiian ***D. forbesii*** differs in being only about 15 feet tall and having leaves ⅓ inch wide, and Hawaiian ***D. fernaldii,*** which is 18 to 24 feet high, in its leaves being about 8 inches long by ½ inch wide.

The only New World dracaena is ***D. americana.*** This handsome native from Mexico to Costa Rica has much the aspect of a yucca or of *D. marginata.* From 10 to 30 feet tall, it has a trunk considerably swollen at its base, and with a few branches. The recurving, glossy, thin leaves, 8 inches to 1 foot long, are narrowly-strap-shaped, and stem-clasping. They are in terminal tufts. The white flowers form large terminal panicles. Yellowish-green to orange, the fruits are about ¾ inch in diameter.

Dracaena americana at Longwood Gardens, Kennett Square, Pennsylvania

The familiar corn-plant (**D. fragrans**) and its varieties, long popular in the north as durable indoor foliage plants remarkably tolerant of adverse environments, are useful for outdoor landscaping in more genial, frostless and essentially frostless climes. They have thick, sometimes branched, canelike stems up to 20 feet tall and gracefully recurved, smooth, glossy, stalkless leaves, in old specimens mostly along the upper reaches of the stems, but in younger ones clothing their entire lengths. The leaves have the texture of pliable leather. They are strap-shaped and up to 3 feet long by 4 inches broad. The specific *fragrans* alludes to the yellowish flowers, in panicles of globular clusters, being strongly and sweetly scented at night. The berries are orange-red. This is native to tropical Africa. Popular varieties are *D. f. lindenii,* with drooping leaves, green at their middles and with usually indistinct greenish-yellow bands paralleling the margins; *D. f. massangeana,* in which the leaves are striped down their centers with broad

Dracaena fragrans massangeana

Dracaena fragrans massangeana, a young plant

bands of yellow and light green; and most beautiful of all, but unfortunately much more exacting in its environmental needs than the others and much slower growing, *D. f.* 'Victoria', which has leaves with green centers streaked with silvery-gray, bordered by broad bands of rich cream-yellow to golden-yellow. Leathery leaves with white margins are typical of *D. f. rothiana.*

Dracaena fragrans 'Victoria'

Low to moderately tall dracaenas include many popular as house and greenhouse ornamentals, and for outdoor gardens in the tropics and subtropics. Those now to be described are of this category. A sort that passes as *D. marginata* although the Madagascan species to which that

Dracaena marginata of gardens

name properly belongs is perhaps not in cultivation, is probably ***D. concinna,*** of Mauritius, or ***D. cincta*** (syn. *D. gracilis*), of uncertain nativity. Possibly up to 12 feet tall, but often not more than one-half that height, it has erect, rarely branched, slender stems. Each branch ends in a rosette of red-edged, narrowly-linear, stalkless leaves 1 foot to 2 feet long and ½ inch wide or somewhat wider. They clasp the stems and spread horizontally or the lower ones droop. About as tall, ***D. angustifolia,*** of the Solomon Islands, has slender, willowy stems, and pliable, lustrous, leathery, stem-clasping, stalkless, narrowly-lanceolate leaves 6 to 10 inches in length that recurve gracefully. In Florida this species produces its yellow fruits in abundance, and self-sown seedlings often volunteer near fruiting specimens. Variety *D. a. honoriae* has leaves beautifully and distinctly banded along their margins with ivory-yellow.

Dracaena angustifolia honoriae

Second only to the corn-plant in popularity as an indoor decorative, ***D. deremensis*** is a native of tropical Africa. It occasionally becomes 15 feet tall. It has dark green, stalkless, sword-shaped leaves faintly longitudinally-striped with lighter green. They are about 1 foot to 2 feet long by 2 inches wide. The flowers are red. Much more commonly grown than the species are its varieties. These are *D. d. warneckei,* the green-margined leaves of which are streaked down their centers with pale gray-green bordered with white; *D. d. bausei,* in which the milky-white bands are much clearer, and separated by only a very narrow streak of pale gray-green; *D. d. longii,* which has lengthwise-pleated leaves that droop somewhat forlornly and have a conspicuous broad band of white down their centers; *D. d.* 'J. A.

Dracaena deremensis warneckei, a young plant

Dracaena deremensis warneckei in bloom

Dracaena deremensis bausei

Dracaena deremensis longii

Truffaut', somewhat similar, but with spreading rather than drooping leaves, with very narrow marginal bands of green to their white centers; *D. d.* 'Roehrs' Gold', which is similar to *D. d. warneckei,* but has creamy-yellow bands bordered with white down the middles of its leaves; and *D. d.* 'Janet Craig', a robust variety with lustrous, nearly black-green, slightly rippled, wavy-margined leaves.

Nearly as popular as *D. deremensis* for house and greenhouse cultivation and for embellishing gardens in the tropics and subtropics, are varieties of ***D. sanderana.*** This native of tropical Africa has slender,

Dracaena sanderana

erect, canelike stems 2 feet tall or a little taller furnished with regularly spaced, slightly twisted, spreading leaves up to 8 or 9 inches long by 1½ inches wide, but smaller in young plants, broadly edged with creamy-white, and with stalks 2 to 4 inches long. Varieties of this are *D. s. borinquensis,* robust, and with broadly green-margined leaves with milky-green centers with two white stripes; *D. s.* 'Margaret Berkery', the leaves of which are about 5 inches long and green with a broad, center stripe of white; and *D. s. virescens,* which has olive-green foliage unmarked by variegation except for a few paler green streaks.

Scarcely fitting the general idea of what a dracaena should look like, tropical African ***D. surculosa*** (syn. *D. godseffiana*) suggests an unusual, white-spotted *Aucuba.* About 2 feet tall, it and its varieties are popular for outdoor cultivation where climates permit, and as greenhouse and room plants. They are much branched and have wiry, spreading, slender stems with short-stalked, oblongish leaves 3 to 5 inches long by nearly one-half as wide. They are in rather widely spaced twos or threes and are irregularly and conspicuously spotted and blotched with white. The flowers, in short-stalked racemes, are succeeded by yellow to red berries. Thicker, somewhat more liberally spotted leaves are characteristic of *D. s. kelleri.* Outstanding *D. s.* 'Florida Beauty' has the variegation of the foliage much more strongly pronounced. An intermediate, slender-stemmed hybrid between *D. surculosa* and *D. fragrans massangeana,* ***D. masseffana,*** has small, soft-leathery, pointed-elliptic leaves in rosettes at the stem ends. They are dark green spotted with yellow.

Somewhat resembling *D. sanderana* but becoming much taller, sometimes 12 feet

Dracaena surculosa

Dracaena surculosa (fruits)

Dracaena surculosa 'Florida Beauty'

Dracaena masseffana

Dracaena masseffana in bloom

in height, popular ***D. reflexa*** is a native of India, Mauritius, and Madagascar. It has branching, canelike stems, and clasping them along almost their entire lengths, whorls (circles of more than two) of sharp-pointed, shortish, wavy leaves. With narrower leaves, bushy *D. r. gracilis* is native to Malaysia. Very beautiful, but slow-growing *D. r. variegata* (syn. *D. r.* 'Song of India'), about 10 feet high, branches and has leaves broadly and distinctly margined with yellow bands.

Dracaena reflexa variegata

South African ***D. hookerana*** has heavy trunks or stems, erect to a height of about 6 feet or snaking along the ground. At their ends are very large rosettes of thick, pointed, sword-shaped, glossy, dark green leaves up to 1½ to 2 inches wide or wider and narrowed to their bases. A variety with leaves about 3½ inches wide is *D. h. latifolia*. White-striped leaves are a distinguishing feature of *D. h. variegata*. A hybrid between *D. hookerana* and *D. grandis,* the latter not known to be in cultivation, is *D.* 'Gertrude Manda'. Handsome and durable, this has much the appearance of *D. hookerana*. Its thick leaves are more than 4 inches wide. Up to 6 feet in height, ***D. umbraculifera,*** of Java and Mauritius, has a usually branchless narrow trunk with a generous tuft of narrowly-strap-shaped leaves that spray upward and outward from its top.

Dracaena umbraculifera

Dracaena umbraculifera (flowers)

Several dracaenas have broad-bladed leaves with long slender stalks. One of the best, ***D. thalioides,*** native of tropical Africa and Ceylon, is robust and durable both outdoors and in. Its rich green, leathery-broadly-lanceolate or narrowly-oblong leaves, 1 inch to 3 inches wide at the bases of their blades, are strongly ribbed lengthwise. Their gray-spotted, channeled stalks clasp the stems with their bases. The clusters of slender, light beige flowers are in erect racemes. With longitudinally furrowed blades elliptic rather than lanceolate, and long, channeled stalks, the leaves of ***D. kindtiana*** have much the appearance of those of *D. thalioides*. Native to tropical Africa, this is a good houseplant and greenhouse plant, but tends to become straggly outdoors. Somewhat similar ***D. phrynioides,*** native of Fernando Po off the west coast of tropical Africa, has dark green leaves with blades up to 8 inches long liberally sprinkled with yellow spots margined with pale green.

Shorter-stalked, spreading leaves, broad-ovate, corrugated lengthwise, and with indistinct crossbands of paler green, are characteristic of slow-growing, tropical African ***D. glomerata.*** Much more intensely variegated, and one of the most beautiful dracaenas, ***D. goldieana,*** also is a native of tropical Africa. This choice kind grows slowly and under ideal conditions may become 10 feet tall, but is usually lower. Its stems are clothed with short-stalked, broad-ovate, lustrous, dark green leaves generously and distinctly spotted, the spots forming crossbands, and pale green changing to almost white. In dense, globular heads, the strongly fragrant white flowers open only at night. Other kinds sometimes cultivated are ***D. cantleyi,*** of Singapore, with a rosette of long, oblanceolate, leathery leaves, green relieved with oblong chartreuse blotches; ***D. dracaenoides,*** native of tropical Africa, which has clustered stems, and finely-toothed, narrow, arching leaves that taper to both ends. On their undersides the midribs are yellow.

Dracaena goldieana

Dracaena goldieana in bloom

Garden and Landscape Uses. Except for a few, notably *D. goldieana, D. fragrans* 'Victoria', and *D. deremensis bausei,* dracaenas are remarkably tolerant of unfavorable environments. Most survive and even prosper for long periods under difficult conditions. The longevity of the corn-plant in poorly lit locations in houses, hotel lobbies, halls, and elsewhere testifies to this, but others including *D. thalioides, D. deremensis warneckei, D. d.* 'Janet Craig', and *D. sanderana* are nearly or quite as tough. All of course respond better to more genial surroundings. Ideally, they need tropical warmth with moderate to high humidity and, most, some shade from strong sun. Indoors, dracaenas are grown in pots, tubs, and planters, and in display greenhouses in ground beds. In addition, *D. sanderana* and its varieties and *D. thalioides* are popular decoratives for standing without soil in containers of water. So accommodated they are not permanent, but often live for a year or more.

Outdoors in the tropics and warm subtropics these are among the most useful foliage plants. From their great variety kinds can be selected to suit many purposes and locations. Those of bold and noble aspect are magnificent and dramatic as single specimens and accents. Some, such as the corn-plant, are effective in clumps and miniature groves. Kinds, such as *D. sanderana* and *D. reflexa,* are useful for grouping. Wider-spreading dracaenas, such as *D. surculosa,* are admirable shrubs for beds, borders, foundation plantings, and similar uses. Most kinds, planted in ground beds or containers, are good terrace and patio plants.

Cultivation. There is not a great deal to be said about the cultivation of dracaenas, most grow so willingly. They appreciate loamy, well-drained, nourishing soil, moderately moist, but not for long periods wet. Outdoors, if the earth tends to be poorish, regular applications of fertilizer encourage desirable luxuriance, but indoors fertilizing should not be overdone otherwise too exuberant growth, often accompanied by loss of lower leaves, is likely to result.

Specimens that become too tall or leggy can be reduced to acceptable proportions by pruning to any desired extent, preferably in spring or when new growth is about to begin. New shoots that develop following pruning reach upward, a fact to keep in mind when determining where cuts are to be made. Watering container-grown dracaenas should be done to keep the soil moist, but it is advisable to let the earth become dryish and then to give enough to thoroughly saturate the entire body of soil. Specimens in large pots, tubs, or planters will usually go for several years without repotting or replanting, but it is advisable to remove some of the surface soil and replace it with fresh each spring. In greenhouses the foliage is kept clean by forcibly spraying it with water every day that the weather is favorable, but this is obviously impracticable in rooms. Remove dust and grime from house specimens by sponging the leaves occasionally with soapy water, to which if there is any evidence of the presence of red spider mites, mealybugs, or scale insects, a little insecticide has been added. Propagation is easy by sectional stem cuttings, terminal cuttings, air layering, and seeds. The young plants are first planted individually in small pots, and when well rooted in those are transferred to larger ones, either singly or, especially with slender kinds such as *D. sanderana,* three or more together.

DRACOCEPHALUM (Draco-céphalum)—Dragonhead. The common name of this genus of the mint family LABIATAE is a literal translation of its botanical one, derived from the Greek *drakon,* a dragon, and *kephale,* a head. The allusion is to the shape of the flowers. In the wild the forty-five species of *Dracocephalum* occur in Europe, temperate Asia, North Africa, and North America. The related American genus *Physostegia* is by some authorities included here. Although not among the most common cultivated plants, a few sorts are grown as ornamentals.

Dracocephalums include annuals, herbaceous perennials, and subshrubs, mostly with erect stems. Their leaves are opposite, deeply-incised, toothed, or smooth-edged. The asymmetrical flowers are in spikes at the ends of the stems or in whorls (tiers) from the axils of leaves or of leafy bracts. The blooms have a tubular, five-lobed, two-lipped calyx and a two-lipped, tubular, purple, blue, lavender, or less commonly white corolla. The upper lip of the latter is arched and at its apex notched. The lower lip is three-lobed with the center lobe notched or cleft. There are four stamens, one style, and two stigmas. The fruits consist of four seedlike nutlets. Unless otherwise indicated the sorts described here are perennials.

European ***D. ruyschiana*** is 1 foot to 2 feet tall and branched. Its stems are usually finely-downy, its leaves hairless, linear-lanceolate, 1½ to 3 inches long, and with toothless, rolled-under margins. From their axils sprout short leafy, flowerless branches. The about 1-inch-long, bluish-purple or rarely white or pink flowers are in cylindrical spikes up to 3 inches long. Perhaps not specifically distinct from the last, somewhat bigger ***D. argunense,*** of northeast Asia, has blue-purple or white flowers. Native from southern Europe to the Caucasus, ***D. austriacum*** has erect, branched stems 1 foot to 2 feet tall and linear leaves 1 inch long or a little longer, pinnately-cleft into three to seven lobes with rolled-under margins. Their upper surfaces are hairless, their undersides hairy. The 1¼- to 2-inch-long, blue-violet flowers are in dense spikes.

Annual ***D. moldavica,*** native from Europe to Siberia and naturalized in North America and 1 foot to 2 feet tall, has branched stems and linear to lanceolate leaves ¾ inch to 1¾ inches long that are coarsely-lobed or toothed, the teeth sometimes bristle-tipped. The ¾- to 1½-inch-long blue-purple to purple flowers are in slender, erect, interrupted, terminal spikes 6 to 8 inches in length. Variety *D. m. album* has white flowers.

Dracocephalum moldavica album

Chinese ***D. rupestre*** is often grown under the name *D. grandiflorum.* From that species, which is probably not in cultivation, it is distinguishable by its bristly flower clusters, the calyx teeth of which are spiny. This forms dense clumps of sparingly-hairy foliage and has flowering stalks up to about 1 foot tall that tend to flop with age. The long-stalked leaves have broad-heart-shaped, round-toothed

Dracocephalum rupestre

blades 1½ to 3 inches long. The dark bluish-violet flowers are in short, terminal spikes and occasionally also in whorls from the upper leaf axils. They are 1½ to 2 inches long. Closely related ***D. purdomii,*** of China, has narrower heart-shaped basal leaves 1 inch to 1½ inches long and flowers only 1 inch long. Leaves cleft in finger-fashion almost to their bases into usually seven lobes are characteristic of 1- to 1½-foot-tall ***D. isabellae,*** of China. They are clothed with white hairs. This early-bloomer has spikes 3 to 4 inches long of rich violet flowers 1½ inches long and in whorls of four to six.

Other Asian species include these: ***D. calophyllum,*** of China and Tibet, is from 6 inches to 1½ feet tall or sometimes taller. It has stalkless or short-stalked deeply-pinnately-cleft, linear leaves up to 1¾ inches long. Short, flowerless branches sprout from the leaf axils. The purplish-blue flowers are ¾ to 1 inch long. ***D. c. smithianum,*** in gardens often misidentified as *D. forrestii,* a species probably not in cultivation, differs from *D. calophyllum* in its flower spikes being woolly-hairy and its leaves having mostly seven to eleven instead of five to seven lobes. ***D. nutans,*** of eastern Russia to Siberia, is a graceful perennial or biennial 9 inches to 2 feet tall. It has stalked, coarsely-round-toothed, ovate-oblong leaves and bright blue, ½- to ¾-inch-long flowers with purple calyxes in slender, loose spikes up to 8 inches in length. ***D. peregrinum,*** native of central Asia, is 1½ to 2½ feet tall and freely-branched. Its lower leaves are elliptic, toothed, up to 1½ inches long. Those above are considerably smaller and elliptic to linear-lanceolate. The flowers, on side branches as well as the main stems, 1 inch to 1½ inches long, have purple to blue, rose-red, or white corollas and colorful red-purple calyxes.

Garden and Landscape Uses and Cultivation. Dracocephalums are suitable for flower beds and borders and for planting in naturalistic fashion in semiwild areas. Most grow without difficulty in sunny locations in well-drained ordinary garden soil and are easily increased by seeds, division, and cuttings of young shoots.

DRACONTIUM (Dracón-tium). The thirteen species of *Dracontium* of the arum family ARACEAE occur as wildlings from Mexico to tropical South America. The name, from the Greek *drakon,* alludes to the color of the inflorescences.

Dracontiums are similar to *Amorphophallus* with a usually big tuber from which develops a solitary, deciduous, large, long-stalked leaf with a blade divided into three primary divisions that are again divided or lobed. The flower-like organ (actually an inflorescence), as is common in the arum family, consists of a spikelike spadix with from its base a bract called a spathe. The tiny true flowers, in *Dracontium* bisexual, are tightly crowded along the spadix, which is much shorter than the spathe that envelops it.

A native of Brazil, ***D. asperum*** has a leaf with a mottled green and brownish stalk about 3 feet long or sometimes considerably longer and a blade up to 3 feet in diameter with twice-divided middle lobes. The ultimate segments are oblong to oblong-lanceolate. Appearing when the plant is leafless, the short-stalked to nearly stalkless inflorescence has a somewhat hooded, dark purple-brown spathe, open on one side, and 3 to 6 inches long. The stout, erect spadix is similarly colored.

Garden and Landscape Uses. These are as for *Amorphophallus.*

Dracontium asperum

DRACOPHILUS (Dracó-philus). Consisting of four species of succulent plants of the *Mesembryanthemum* section of the carpetweed family AIZOACEAE, the genus *Dracophilus* is confined in the wild to desert regions in South Africa. Near kin of *Juttadinteria,* it has a name, alluding to the Drakensberg mountains, derived from the Greek *draco,* a dragon, and *philios,* loving.

Dracophiluses form low cushions of tightly clustered, short, leafy shoots. Opposite, with alternate pairs at right angles to each other, and without stem showing between them, the three-angled, very fleshy, more or less oblique leaves are bluish-green. The terminal, white or pink flowers, like others of the *Mesembryanthemum* relationship daisy-like in aspect, but since they are single blooms and not heads of numerous florets, really very different from daisies, have stalks mostly concealed by a basal pair of joined bracts. The fruits are capsules. From *Juttadinteria* this group differs in the calyxes of its flowers being five- instead of four-lobed, and in technical details of its fruits.

Species of *Dracophilus* are neatly divided into kinds with toothed and kinds with toothless leaves. Of the former, ***D. delaetianus*** is one of the best and most satisfactory to grow. It has leaves ¾ inch to 1¼ inches long and from rather less to somewhat more than ½ inch broad and thick. Four or six on each shoot, they have rounded surfaces and keels with chinlike ends, that, like the margins, are broadly notched. The rosy-violet blooms, ¾ inch or slightly more across, have stalks about ¾ inch long. From the last ***D. montis-draconis*** differs in its fleshy leaves being 1¼ to 2 inches long, two of its sepals being fleshy, and its flower stalks only ⅓ inch long. The blunt leaves are slightly upcurved and have one or two flattish protrusions at their tips. White or pale pink, the blooms are 1 inch or slightly more across.

With toothless leaves, commonly four on each shoot, 1 inch to nearly 2 inches long, twice to thrice as long as broad, and thickest near their ends. ***D. dealbatus*** (syn. *D. rheolens*) has pink or sometimes white blooms 1 inch or slightly more across. They usually have ten stigmas. Toothless-leaved ***D. proximus*** is distinguished by its very smooth, glaucous leaves being thickest near their middles and about four times as long as they are wide. Also, its 1-inch-wide, pink flowers have mostly eleven or twelve stigmas.

Garden Uses and Cultivation. To collectors of dwarf, nonhardy succulents dracophiluses have considerable appeal. Not particularly difficult to cultivate, they generally respond more satisfactorily to gardeners' efforts than do juttadinterias. Conditions and care appropriate for *Lithops* and other dwarf, fat plants of the *Mesembryanthemum* relationship are satisfactory. The resting season is winter, during which the soil must be kept dry. Propagation is by seeds and by cuttings. For additional information see Succulents.

DRACOPIS (Dracòp-is)—Coneflower. This genus of one species of the daisy family COMPOSITAE was previously included in *Rudbeckia.* Modern botanists segregate it from that and closely related *Echinacea* and *Ratibida* on technical characteristics of the flower heads. Its colloquial name, coneflower, is shared with *Rudbeckia, Echinacea,* and *Ratibida,* all of which like *Dracopis,* have high, conelike or cylindrical centers to their flower heads. Its botanical name, alluding to the style appendages, comes from the Greek *drakon,* a dragon.

An attractive, hairless, glaucous annual, ***D. amplexicaulis*** (syn. *Rudbeckia amplexicaulis*) attains a height of 1 foot to 2½ feet and is branched above. Its leaves are up to 4 inches long, oblong to ovate, and sometimes slightly-toothed. The solitary flower heads, about 2 inches across, have broad, short, yellow ray florets often with brown-purple or orange bases, and brownish, cylindrical centers that grow 1 inch high or a little higher. The fruits are seedlike achenes. This kind is indigenous from Georgia to Missouri, Texas, and Kansas.

Garden Uses and Cultivation. These are as for annual rudbeckias.

Dracopis amplexicaulis

DRACUNCULUS (Dracún-culus)—Dragon-Arum. The arum family ARACEAE includes, as well as a great array of inoffensive ornamentals such as anthuriums, calla-lilies, philodendrons, and caladiums, a not inconsiderable population of kinds with foul-smelling blooms. The two species of *Dracunculus* belong here. They rate with such notable stinkers as voodoo-lily (*Sauromatum*), twist-arum (*Helicodiceros*), devil's tongue (*Amorphophallus rivieri*), and champion of all, *Amorphophallus titanum.* The name *Dracunculus* is one used by Pliny for some plant with curved rhizomes.

From its near relative the twist-arum, *Dracunculus* differs technically in having male and female flowers not separated by a portion of spadix devoid of fully developed flowers, but it is also markedly different in appearance, most obviously in its inflorescence (commonly called the flower) not being sharply bent, but facing upward, and in its stench being somewhat less intense. The Mediterranean region and the Canary Islands are the homes of this species.

Dragon-arums are deciduous herbaceous perennials. They have rounded, subterranean rhizomes, and clusters of erect leaves, with their stalks enveloping each other at their bases. The leaf blades, unlike those of related *Arum,* which are always arrow-shaped, are divided finger-fashion into spreading lobes that fancifully suggest the claws of a dragon, which is the reason for the vernacular name. The inflorescence or "flower" is typically calla-lily-like. It is erect and has a central, upright spike (spadix) closely furnished with small, unisexual flowers, females toward the base, males above. These blooms are inconspicuous and of no decorative importance. The spadix is enveloped by a large trumpet-shaped bract called the spathe. Comparable parts in the familiar calla-lily are the central, pencil-like, yellow column, the spadix, the white petal-like trumpet, the spathe. The fruits of *Dracunculus* are berries. The plant previously named *D. crinitus* is *Helicodiceros muscivorus.*

Most common in cultivation, ***D. vulgaris*** inhabits the Mediterranean region and is up to 3 feet tall. It has large leaves, streaked with white, with blades with bow-shaped bases and outer portions deeply cleft into nine to fifteen lanceolate-oblong lobes. The wide-mouthed, trumpet-shaped spathes at their tips spread nearly horizontally. Dark, velvety, and reddish-brown, except on the inside at their bases, they may exceed 1½ feet in length and may be about 8 inches in width. The nearly black, tail-like spadix is up to 1 foot or more in length and ½ to ¾ inch in diameter. Less well known and less hardy than the Mediterranean species is ***D. canariensis,*** native of the Canary Islands. It grows about 1½ feet tall and has a tubular, narrow, greenish-yellow spathe about 1 foot long and a yellow spadix that may be twice as long.

Garden and Landscape Uses and Cultivation. As curiosities and conversation pieces dragon-arums are sometimes grown in greenhouses, and they are grown outdoors in warm, essentially frost-free climates. It is well not to locate them where the stench of their blooms will be strongly offensive. They need partial shade and fertile soil with a fairly good humus content, one that during spring and the part of the summer when foliage is present is moist but not saturated. Planting, or for greenhouse culture potting, is done in the fall. Indoors, temperatures at night of about 50°F, and 55 to 60°F by day are suitable. Water should be given sparingly at first, generously after leaf growth is well advanced. When the roots have permeated the soil applications of dilute liquid fertilizer are helpful. After blooming, but not until the foliage begins to die naturally, watering is gradually reduced and finally the tubers are stored quite dry in the soil in which they grew until repotting time in the fall. Propagation is by division of the tubers, and seeds.

DRAGON or DRAGON'S. These words appear as part of the common names of a number of plants including these: dragon-arum (*Dracunculus*), dragon root or green dragon (*Arisaema dracontium*), dragon tree (*Dracaena draco*), dragon's blood (*Daemonorops draco*), dragon's eye pine (*Pinus densiflora oculus-draconis*), and dragon's mouth (*Arethusa*).

DRAGONHEAD. See Dracocephalum. False-dragonhead is *Physostegia.*

DRAINAGE and DRAINING. The majority of plants, excepting those that inhabit bogs, swamps, and other wet places, will not live in poorly drained soil. It is essential for their well-being that their roots be adequately supplied with air as well as moisture. For them, the ideal condition is to have each soil particle or cluster of soil particles surrounded by a film of moisture, with air spaces between the particles or clusters. To achieve this, water must drain by gravity to levels lower than those the roots occupy and the soil must be sufficiently porous to permit such draining of surplus water at a fairly rapid rate. The roots of some plants, cinerarias for example, are killed if submerged for as short a period as twenty-four hours, or even less, with the result that the plants promptly collapse. Many kinds are more tolerant and do not evidence harm as dramatically. Nevertheless they are injured, and

Draining a pot: (a) Place a large crock hollow side down over the drainage hole in the bottom of the pot

(b) Add a few smaller pieces of crock

(c) Cover with dead leaves or similar material

most importantly, it is the fine feeding roots that are killed first as a result of the soil being waterlogged.

Pots, tubs, and other containers for plants should be adequately furnished with drainage holes. To prevent clogging, cover these at potting or planting time with a generous layer of crocks (shards) or other loose drainage material that will last rather than rot, and before filling in soil, surface this with a layer of dead leaves, moss, grass turf turned upside down, straw, hay, or fine plastic mesh as further insurance that fine soil will not wash down to impede surplus water drainage.

Not all poorly drained land is wet enough to qualify as swamp or bog. Only extremes where the level of the water (the water table) is at or near the surface belong in these categories. But if the water table stays for long periods as close as 1½ feet to the surface, it can produce uncomfortably moist conditions for the roots of most plants. Much better that it be 3 to 4 or more feet down in fine-textured soils, possibly a little nearer the surface in coarse sandy or gravelly ones. In many places the depth of the water table varies from time to time, being closer to the surface in spring and after long rainy periods, and lower in summer, especially after long dry spells.

Signs indicative of poor subsurface drainage are patches of ground, usually low lying, that remain conspicuously moist when other parts of the garden are much drier following a dry period. The dampness is caused by moisture rising by capillarity from below. To check further, make test holes in suspected spots. This can be done by excavating with a spade to a depth of 2 feet, but much less effort is needed to bore smaller holes with a soil auger 3 inches or more in diameter. Ring each hole with a dyke of soil and cover it with a watertight cover to prevent surface water entering. Observe the water level in the holes from time to time. If it remains nearer the surface than 2 feet for weeks

Water lying on the ground for several hours or days following heavy rain indicates the need for drainage

together and you expect to cultivate successfully the general run of garden plants, the desirability of artificial drainage is usually indicated. But before proceeding consider the situation further.

Poorly drained or waterlogged ground needs careful study before drainage operations are begun. The first concern is where the water is coming from. Is the wetness chiefly the result of excessive surface flow from higher ground? This is rarely true, but if it is possible relief can be had by constructing a diversion ditch to lead the surplus to an acceptable outlet or some less critical place. Water trapped underground by an impervious layer of clay, hardpan, or rock and fed by rain and perhaps subsurface seepage and flow is a more usual cause.

Waterlogged low land bordering rivers and lakes is generally impossible to improve by installing drain pipes, but if its surface is 1 foot or more above the water table improvement in its ability to support plants may be increased by running parallel ditches through it at intervals of 20 to 30 feet. These may not carry away any appreciable amount of water, but they will assist in aerating the soil above the water table and will make for more uniform and agreeable subsurface conditions. Another sometimes effective method of dealing with situations of this kind is to construct raised beds and grow the plants on these.

Having determined the reason for unsatisfactory drainage ponder the practicability and desirability of taking measures to relieve it. This may be easy or difficult to achieve or impossible. Not infrequently it involves considerable expense and work. Under some circumstances it may be desirable, advisable, or even imperative that you accept the existing situation. You may even be able to take favorable advantage of it by converting the wet area to a bog garden or pond as an alternative to attempting to make it fit for the cultivation of dry land plants.

If you decide to drain, the next step is to find an outlet, where the water is to go. Obviously this must be lower than the land to be drained, so as to allow water from 1½ feet or more beneath the surface to flow away. If the reason for the dampness of the soil is a thin layer of impervious material, such as clay or hardpan, breaking this by deep spading, using a subsoiler plow or, in extreme cases and done only by an expect after permission of local authorities has been obtained, by exploding small charges of dynamite underground. All these methods are designed to fragment the impervious layer so that water can seep downward.

Sometimes instead of draining entire areas gardeners attempt to improve conditions by digging extra deep holes for individual trees and shrubs and partially filling them with stones or other drainage material. Only if the excavations are sufficiently deep to break through the water-holding layer into a porous stratum below does this technique work. Unless they do this all that is accomplished is the creation of water-holding basins partly filled with stones.

Where conditions are propitious the most satisfactory way of draining land is by the installation of a system of agricultural drains. These may be of clay tiles or of composition or plastic drain pipes perfo-

Agricultural (unglazed) tile land drains: (a) Cylindrical type, with, in front, two sections of glazed tile for a main drain

(b) Hexagonal-type agricultural tile land drain

(c) Perforated plastic land drains

Tools for draining. Left to right, narrow-bladed spade for digging ditches, spade for scooping out bottom of ditch, hook for lowering drain tiles into ditches, and ordinary spade

Ditches dug in readiness for drains

rated to permit water to enter. The outlet for the system may be a natural one, such as a stream or pond, or a ditch leading to such a feature. In suburban areas it may be the curb of a street or, if permission can be obtained, a town storm sewer. In any event there must be an outlet and it must be at the lowest point in the system.

Keep the system simple. Drains 15 to 25 feet apart, the closer spacing for very clayey soils, the greater for more porous ones, suffice. For areas big enough to require more than one drain, it is usual to have a main drain with laterals angling into it in herringbone fashion, but land contours may dictate some other plan. It is important that the drains be as straight as possible and that laterals angle into the main drain, usually at about forty-five degrees, in the direction of its flow. Alternate the laterals so that no two come into the main drain at the same point.

The best depth to lay the drains is normally 2 feet. If conditions necessitate, they may be set at 1½ feet, but not shallower. Depths greater than 2½ feet are usually disadvantageous. For the main drain or if there is only one use 6-inch pipes, for laterals use pipes 4 inches in diameter. A minimum pitch of 3 inches in each 100 running feet of drain is needed for the water to flow without clogging the drains with silt. If the ground allows, the pitch may be steeper.

Start excavating the ditch or ditches at the outlet. Keep topsoil and subsoil separate. Do not dig deeper than necessary, otherwise the bottom is loosened and the drains may not slope evenly after they are covered with earth. Agricultural drains are without sleeves or collars so they are not nested as are most drains. When laying them simply butt their ends together. As the work proceeds check at frequent intervals with a spirit level and straightedge or by other means to make sure that an even pitch is maintained. A usual way of doing

Laying drains: (a) In wide ditch

(b) In a narrow ditch, using the hook

(c) Drain covered with gravel or crushed stone before back-filling with soil

(d) Junction of feeder drain into main drain, with one tile lifted out to show water flow

this is to stretch along the center of the ditch and about 2 feet above it, between suitably spaced supports, a taut line sloped as the drains are to slope. The drains are then laid at a uniform measured distance below the line. After the drains are positioned and you are satisfied that they form a true, evenly sloped conduit, cover them with 3 to 6 inches of gravel, crushed stone, or coarse cinders. Then replace the soil, subsoil below, topsoil uppermost.

Installing a drainage system calls for considerable ingenuity and effort, but if adequately done results in a tremendous permanent improvement of the soil that under particular circumstances can be had in no other way. Half-way measures such as digging ditches and partly filling them with stones are not recommended as acceptable substitutes for well-laid agricultural drains.

DREGEA (Drég-ea). This genus, native from South Africa to China, of the milkweed family, ASCLEPIADACEAE consists of a dozen species. The name of the genus *Dregea* commemorates the German botanical collector Johann Franz Drege, who died in 1831.

Dregeas are twining woody vines with opposite, heart-shaped to pointed-ovate leaves and small white, greenish, or yellow flowers in stalked umbels from the leaf axils. The blooms have calyxes of five sepals alternating with small glands, corollas with five overlapping, rather fleshy, spreading lobes (petals) and a star-shaped central crown or corona, and five stamens that press against the tops of the two styles. The fruits are podlike follicles.

Occasionally cultivated, Chinese ***D. corrugata*** (syn. *Wattakaka sinensis*), up to 10 feet tall, is vigorous. It has slender stems, when young, downy, and slender-stalked, pointed-heart-shaped leaves with blades 1½ to 4 inches long by one-half or more as wide, and gray-velvety-downy on their undersides. The downy-stalked clusters, 1½ to 2½ inches wide, of up to twenty-five deliciously fragrant flowers much suggest those of hoyas. Individual flowers are star-shaped, a little over ½ inch across, and white or creamy-white dotted with red around their centers.

Dregea corrugata

Garden and Landscape Uses and Cultivation. These are as for *Hoya*, but the species described is hardier than members of that genus. It survives outdoors in the south of England, in some parts of the South, and elsewhere where winters are mild.

DRIED BLOOD. Less commonly available than formerly, dried blood, a product of slaughterhouses, is an excellent, quick-acting organic fertilizer. Acid in its reaction, it contains 12 to 14 percent nitrogen and much smaller amounts of other nutrients. It is used in spring and summer at the rate of about 2 ounces to 10 square feet. It is also called blood meal.

DRIED FLOWERS, FOLIAGE, AND FRUITS. Flowers and other parts of many plants can be preserved for use in artistic arrangements by drying. Several procedures are available. If two-dimensional effects are satisfactory, as they are for making framed "pictures" and greeting cards, laying the materials to be dried between sheets of newspaper interspersed with blotters and putting them in a plant press such as botanists use for making herbarium specimens, produces satisfactory results. This is the preferred method for drying fern fronds. The press consists of a pair of wooden lattice frames of any convenient size that, after the stack of papers and blotters containing the plants is put between them, can be pulled tightly together with straps, thin rope, or clamps so that sufficient pressure is exerted to press and flatten the plants. Quick drying is necessary to assure the best result and good color preservation. Keep the press in the sun or in a warm dry place and change the blotters frequently, every day or two at first, less often as the specimens become drier.

Arrangement of dried grasses and fruits

Three-dimensional drying of flowers, foliage, and fruits to be displayed in vases may be done in various ways. Light, airy, thin-textured, nonfleshy sorts such as everlasting flowers, grasses, goldenrods, knotweeds (*Polygonum*), and some others respond to air drying. To do this cut the stems or branches before the flowers are fully open, tie them together in small, fairly loose bundles, and hang them upside down in a dry attic, cellar, garage, or similar place to dry slowly. Alternatively, they may be stood in containers without water in similar places. With the latter treatment those with not too rigid stems may arch gracefully and retain the curves when dry. Branches of many berried trees and shrubs can be preserved for considerable periods by treating them similarly after first dipping them in a mixture of equal parts wood alcohol and clear shellac.

Flowers for dried arrangements: (a) Baby's breath (*Gypsophila paniculata*)

(b) Bells of Ireland (*Moluccella*)

(c) Cat-tails (*Typha*)

(d) *Celosia*

(e) Everlasting flower (*Helichrysum*)

(f) Goldenrod (*Solidago*)

(g) Heath (*Erica*)

(h) Knotweed (*Polygonum*)

(i) Lavender (*Lavandula*)

(j) Pampas grass (*Cortaderia*)

Foliage of beeches, oaks, evergreen magnolias, euonymuses, and hollies keep in good condition for long periods if conditioned by standing them with the basal 2 or 3 inches of their stems in a solution of two parts glycerin to one part water for two or three weeks. To encourage absorption of the liquid split and crush the branch ends that will be in the solution, which, at the time of immersion, should be at a temperature of 80 to 85°F.

Drying in fine sand, a mixture of one part sand and two to three parts laundry borax, borax and white cornmeal in proportions of from equal parts of each to three of the former to nine of the latter, or in silica gel (obtainable from florists) are the best ways of achieving three-dimensional results with flowers of soft, moist, or fleshy textures. It is very easily done.

Fruits for dried arrangements: (a) Bittersweet (*Celastrus*)

(b) Chinese lantern (*Physalis*)

(c) Honesty (*Lunaria*)

(d) Maple (*Acer*)

(e) Pine (*Pinus*)

(f) Privet (*Ligustrum*)

Take a box, flower pot, or other container of sufficient size to accommodate the flowers to be dried. Place in it a 2-inch or deeper layer of the drying medium. Next lay the flowers on this, spiky types such as snapdragons horizontal, saucer-shaped and flat blooms such as zinnias face downward with each bloom set on a small, individual mound and their stems sticking up into the air, looser kinds facing upward. Gently sift more of the drying agent over and among the flowers and flower parts so that they are held in position without distortion of their natural forms, making sure that the blooms are covered to a depth of at least a couple of inches. The time it takes to dry the blooms depends upon the character of their tissues. From two to nine days is about normal. As soon as they are dry, and before the flowers become excessively brittle, carefully remove the drying material from around them, blow off or with a soft brush dust off any particles that adhere, and then store the blooms until needed on layers of cotton or shredded wax paper in fairly airtight boxes.

Air drying is successful with ornamental grasses, everlasting flowers of all kinds, and such flowers as amaranths (*Amaranthus*), artemisias, beauty-berries (*Callicarpus*), bells of Ireland (*Moluccella*), black-alder (*Ilex verticillata*), cat-tails (*Typha*), celosias, cupid's dart (*Catananche*), docks (*Rumex*), eryngiums, flowering onions (*Allium*), gilias, globe-amaranths (*Gomphrena*), globe-thistles (*Echinops*), goldenrods (*Solidago*), hawthorns (*Crataegus*), heather (*Calluna*), heaths (*Erica*), horned-poppies (*Glaucium*), hydrangeas, knotweeds (*Polygonum*), lavender (*Lavandula*), perennial baby's breath (*Gypsophila*), pussy willows (*Salix*), and sea-lavender (*Limonium*).

Fruits (including seed pods) that air-dry well are those of bayberry (*Myrica*), bittersweets (*Celastrus*), Chinese lanterns (*Physalis*), gladwin (*Iris foetidissima*), honesty (*Lunaria*), maples (*Acer*), nandinas, oaks (*Quercus*), pines, privets (*Ligustrum*), sumacs (*Rhus*), sweet gum (*Liquidambar*), teasels (*Dipsacus*), and yuccas.

Foliage that responds well to the glycerin and water method includes that of barberries (*Berberis*), beeches (*Fagus*), cannas, English ivy (*Hedera*), eucalyptuses, evergreen euonymuses, evergreen hollies (*Ilex*), evergreen magnolias, leucothoes, mahonias, oaks (*Quercus*), and rhododendrons.

Flowers suitable for drying in sand, borax, silica gel, and various combinations of these include nearly all kinds. Among the most successful are acacias, black-eyed Susans (*Rudbeckia*), blue lace flower (*Trachymene*), butterfly bush (*Buddleia*), calendulas, camellias, carnations, Christmas-roses (*Helleborus*), clematises, cornflowers (*Centaurea*), dahlias, flowering dogwood (*Cornus*), forget-me-nots (*Myosotis*), hollyhocks (*Alcea*), Japanese anemones, larkspurs (*Consolida*), narcissuses, pansies, peonies, Queen Anne's lace (*Daucus*), salvias, and zinnias.

DRILL. Used as a noun a drill is a shallow trench in which seeds are sown, as a verb it alludes to making drills and sowing seeds in them. Drills are commonly made, or to use the gardener's term, drawn, by stretching a garden line tautly across the ground surface and then drawing the corner of the blade of an ordinary draw hoe or the point of the triangular blade of a Warren hoe along the line. The action consists of a series of short pulling strokes. An alternative method is to take a wooden stick about 1½ inches square, cut a deep V-shaped notch in its end, saddle this over the line and then with the stick held at an angle of about forty-five degrees push and pull it in a series of short strokes to make a drill in the loose earth. Yet a third way, useful when considerable seed sowing is to be done, is to use a special attachment to a wheel hoe. It is better not to make

Making drills: (a) With a hoe

(b) With a stick saddled over a tautly stretched line

(c) End of stick showing V-shaped notch

(d) Sowing seeds in drill

drills until one is ready to sow. If made much previously, they are likely to dry out, become sun-baked, or perhaps be washed out by heavy rains.

DRIMIA (Drìm-ia). Little known in American gardens, there are forty-five species of *Drimia,* a genus confined in the wild to Africa. It belongs in the lily family LILIACEAE and is named, because of its irritating, acrid juice, from the Greek *drimys,* acrid. From *Scilla,* to which it is related, *Drimia* differs in having the perianth segments of its flowers united below, and its seeds angular or compressed.

Drimias are bulb plants with lanceolate to narrowly-lanceolate leaves, and bell-shaped, whitish to reddish, often green-tinged flowers in naked-stalked racemes. The blooms have six recurved tepals (commonly called petals), six slender-stalked stamens, and one style ending in a head-like stigma. The fruits are capsules.

Scarcely beautiful, but interesting because of its peculiar bulb, which is composed of the scaly remains of old leaves, and has somewhat the aspect of a lily bulb, ***D. haworthioides*** is occasionally cultivated. Native of South Africa, it usually has up to four lanceolate, conspicuously-veined leaves 2 or 3 inches in length, with hair-fringed margins. Up to about 1 foot long, the flower stalk has its upper 3 or 4 inches loosely furnished with little, greenish-white, short-tubed blooms with the free parts of the petals about ⅓ inch long. The flowers appear before the foliage.

Garden and Landscape Uses and Cultivation. Not hardy in the north, drimias are occasionally grown by bulb fanciers, outdoors in nearly frostless climates, and in greenhouses. They need the conditions and care suitable for ixias, freesias, and most other South African deciduous bulb plants.

DRIMIOPSIS (Drimi-ópsis). The twenty-two species of *Drimiopsis* of the lily family LILIACEAE are endemic to Africa. The name derives from that of closely related *Drimia* and the Greek *opsis,* resembling.

The kinds of *Drimiopsis* are nearly related to *Ornithogalum* and *Scilla.* They have small bulbs, each with two to four oblong to broad-elliptic or ovate-heart-shaped, sometimes keeled, often spotted leaves and spikes of up to thirty small, somewhat bell-shaped, greenish-white to white flowers, each with six separate petals (more properly, tepals), the three inner hooded at their tips, six stamens, and one style tipped with a knoblike stigma. The fruits are capsules.

South African ***D. maculata*** has pointed, ovate-heart-shaped, somewhat fleshy, bright green leaves blotched with darker green and with strongly-channeled stalks. The flowering stalks, 6 inches to 1 foot long, are topped with a close spike of twenty or more little blooms, at first white, but later greenish. Very like the last, and differing only in technical details, is East African ***D. botryoides.***

Drimiopsis maculata

Garden Uses and Cultivation. These interesting, but not showy-flowered plants are hardy only in warm, frostless climates and may be accommodated in rock gardens. In greenhouses they adapt well to pot cultivation, the South African species needing the same care as ornithogalums of the same geographical origin, those from more tropical parts needing warmer environments. Propagation is easy by seeds, offsets, and bulb cuttings.

DRIMYS (Drì-mys)—Winter's Bark. The aromatic evergreen trees and shrubs that compose *Drimys* of the magnolia family MAGNOLIACEAE, or according to those who segregate the families, the drimys family WINTERACEAE, number seventy species. The genus occurs natively in Central America and South America and from the Malay Archipelago to New Caledonia and Australia. Its name, which has also been spelled *Drymis,* is the Greek word for acrid. It alludes to the taste of the bark. The bark of *D. winteri,* called winter's bark, is used medicinally.

This genus has alternate, evergreen leaves with translucent dots. Its chiefly bisexual, rarely unisexual or a mixture of bisexual and unisexual flowers have deciduous calyxes that in the bud stage enclose the petals and then split irregularly as the flowers open. There are four to nine overlapping petals, several thick-stalked stamens, and stalkless stigmas. The berry-like fruits do not cohere. For *D. axillaris* see *Pseudowintera axillaris.*

Winter's bark (***D. winteri***), a native of Chile and Argentina, is a choice and lovely species, in cultivation up to about 25 feet tall, but in the wild sometimes twice that height. It has mahogany-red branches and shoots and bright green, broad-elliptic, leathery leaves 5 to 10 inches long. The jasmine-scented, ivory-white flowers, 1 inch to 1¼ inches wide, are in clusters of four to nine. Untrained, this kind tends to develop several trunks, but can be restrained to one and made more treelike by appropriate pruning. Variety *D. w. andina* grows more slowly and is dwarfer. Even in the wild it does not exceed 15 feet in height. Australian and Tasmanian ***D. lanceolata*** (syn. *D. aromatica*) is a medium- to large-sized shrub. Its lanceolate leaves, 1½ to 3 inches long and one-half as wide, have stalks that, like the young shoots, are rich red. Produced in clusters from the leaf axils and ends of the branchlets, the many ½-inch-wide, narrow-petaled, unisexual flowers are white, the sexes on separate plants. Male blooms have five to eight petals, females usually four.

Drimys winteri in Scotland

Drimys winteri (flowers)

Garden and Landscape Uses and Cultivation. Not hardy in the north, in mild climates such as characterize California and the Pacific Northwest winter's bark is a choice landscape tree or shrub. It prefers lightly shaded locations where the soil is lime-free and well-drained, yet not dry. To maintain a pleasing form, judicious corrective pruning may be needed from time to time. This is best done as soon as flowering is through. The Australian species, considerably more tender to cold than the winter's bark, in nearly frostless climates does well under similar conditions. Propagation is by seeds, by cuttings about 3 inches long made in summer from firm, but not hard shoots and set in a greenhouse propagating bench or under mist, and by layering.

DRIP. Great harm can result in greenhouses and frames when water from leaks or collected condensation drips onto plants. It is important to maintain such structures in leakproof condition. Sash bars that support the glass should have along both sides of their undersides a drip groove running their entire lengths to conduct condensed moisture to the sidewalls of the greenhouse or frame. It is important when painting not to clog these grooves.

The argument is sometimes presented by botanists and others more familiar with plants in the wild or on herbarium sheets than in cultivation that drip may be disregarded because plants in natural environments are subjected to rains by night as well as day, and apparently without harm. Although undoubtedly true, this does not invalidate experienced greenhouse growers convictions that drip in greenhouses and frames causes great harm by packing soil, keeping it too wet, in cold weather subjecting roots to too low temperatures, and encouraging mildews, rots, and other diseases.

DROPWORT is *Filipendula vulgaris.*

DROSANTHEMUM (Dros-ánthemum). Very free-flowering and amenable to cultivation, *Drosanthemum,* of the *Mesembryanthemum* complex of the carpetweed family AIZOACEAE, comprises seventy species. Its members, succulent subshrubs, are often woody toward their bases. They are natives of South Africa. The name, from the Greek *drosos,* dew, and *anthemon,* a flower, acknowledges the glistening papillae (small projections) that roughen the leaf surfaces.

Drosanthemums differ from *Lampranthus* in their blooms having long, slender stigmas, and in the circle formed by the nectary glands not being closed, but separated into five parts. They have stems that, when young, like the foliage, have glistening papillae, but those on the stems later dry and produce a rough surface. The fleshy leaves are opposite, cylindrical or three-angled, or almost globular. The red, purple, pink, yellow, or white blooms, produced freely, have much the appearance of daisies, but are very different structurally, each being a single bloom instead of a head of many florets. The fruits are capsules.

One of the most beautiful species, ***D. hispidum*** is up to 2 feet tall and twice as broad, its lower branches sometimes rooting. Somewhat distantly spaced, and spreading or slightly incurved, the ½- to 1-inch-long, cylindrical leaves, which in the sun tend to become reddish, glisten with large papillae. The gleaming, rich purple blooms, a little over 1 inch wide, are solitary or in twos or threes on stems up to 2 inches long. As free flowering and as beautiful as the last, which it resembles, ***D. floribundum*** forms mounds of slender, prostrate or creeping, freely-branched stems and cylindrical leaves that thicken slightly toward their apexes. The leaves are from a little less to somewhat more than ½ inch long. The flowers are light pink and about ¾ inch wide. Another of this group, ***D. candens*** has bluntish, semicylindrical leaves, thickened toward their apexes, and under ½ inch long. Its solitary, white flowers are nearly ½ inch wide.

Very showy ***D. speciosum*** has greenish-centered, bright orange-red blooms about 2 inches across. They are solitary. This species, up to 2 feet tall and with somewhat spreading branches, has distantly spaced, upcurved, blunt, semicylindrical, fleshy leaves approximately ½ inch long, set with glistening papillae.

Others cultivated include *D. bicolor* and *D. micans,* both up to about a foot high. With up-pointing green or yellowish-green, cylindrical leaves ⅛ inch in diameter, up to ¾ inch long, and recurved at their tips, ***D. bicolor*** has solitary flowers, 1¼ inches or slightly more across, with purple petals with yellow bases. Freely-branched ***D. micans*** has nearly cylindrical leaves ⅛ inch wide, which are somewhat flattened on their upper sides and up to 1 inch long. The flowers, ½ inch wide or somewhat wider, have purple to yellowish petals.

Garden and Landscape Uses and Cultivation. In warm, dry, essentially frost-free climates drosanthemums are excellent for outdoors, for the fronts of borders, covering banks, as edgings, and in rock gardens. They are also satisfactory in pots or pans (shallow pots) and in ground beds in greenhouses with other succulents. Their chief needs are sharp soil drainage and full sun. Given these, they prosper in a variety of soils. They are easily increased by cuttings, by removing rooted portions and starting them as new plants, and from seeds. In greenhouses a winter night temperature of 45 to 50°F with a few degrees increase by day is satisfactory. Little water is needed in winter, more from spring to fall, but the soil should always be permitted to become nearly dry between waterings.

DROSERA (Drós-era)—Sundew. This interesting group of insectivorous plants inhabits many parts of the world including North America. There are more than 100 species, mostly inhabitants of constantly wet soils and bogs, but some Australian kinds have tubers and can survive considerable periods of drought. The group includes a few annuals and biennials, but consists chiefly of deciduous and evergreen herbaceous perennials, mostly of low growth. It belongs to the sundew family DROSERACEAE. Both the scientific and common names allude to the dewlike appearance of the excretions of the gland-tipped hairs on the foliage, the botanical name stemming from the Greek *droseros,* dewy.

Sundews are mostly short-stemmed and have usually basal rosettes of foliage, but some kinds develop erect stems several inches long, with alternate leaves. The leaves vary in shape according to species from narrowly-linear to nearly circular, and are clothed with glandular, generally reddish, usually sensitive hairs that exude a clear, sticky fluid that entraps unwary insects and other tiny creatures that come into contact with it. When this happens the hairs bend inward slowly and the prey is gradually digested and its body nutrients absorbed by the plant. Unlike those of the Venus' fly trap (*Dionaea muscipula*), the leaves of *Drosera* are not triggered to move rapidly when touched. Sundews have slender flower stalks terminating in curving, one-sided racemes, or clusters of flowers, more rarely solitary blooms. The flowers are often of magenta tones, but vary from white to red. They have usually five, but occasionally four or up to eight each sepals, petals, and stamens, and two to five, often divided, styles. The fruits are capsules containing minute seeds.

Northern Hemisphere species that occur as natives in Europe and Asia as well as the United States and Canada are *D. rotundifolia, D. anglica,* and *D. intermedia.* To the two last species the name *D. longifolia* has in the past been applied. White- to pink-flowered ***D. rotundifolia*** forms flat rosettes of leaves with long, flat stalks and nearly round blades under ½ inch long and broader than long. The flowers, up to fifteen on apparently terminal, hairless stalks 3 inches to 1 foot tall, are less than ⅓ inch in diameter. Also in basal rosettes, the leaves of ***D. anglica*** are more erect, long-stalked, and have spoon-shape blades ½ inch to 1½ inches in length. The apparently terminal flower stalks are hairless, 2½ to 10 inches long, and carry up to nine white blooms ¼ to ⅓ inch wide or slightly wider. Similar, but usually with proportionately shorter leaf blades and lateral flower stalks scarcely longer than the leaves, ***D. intermedia*** differs in the technical detail that the stipules (appendages at the bases of the leafstalks) are separate rather than joined. A sterile, intermediate hybrid between *D. anglica* and *D. rotundifolia* is *D. obovata.*

Native of the northeastern United States, ***D. filiformis*** has erect, slender, almost threadlike, all-basal leaves without evident stalks and up to 10 inches long. The flower stalks, up to 9 inches tall, bear up to sixteen purple-pink blooms a little under ½ inch in diameter. A larger, more vigorous variety, *D. f. tracyi* (syn. *D. tracyi*) occurs in the southeastern United States.

South African sundews occasionally cultivated include *D. capensis* and beautiful *D. cistiflora.* The leaves of ***D. capensis*** are all basal, or nearly so, and ascending. They

Drosera filiformis

Drosera cistiflora in its native South Africa

Drosera capensis

Drosera binata

have oblong blades 2 to 4 inches long and under ¼ inch wide and narrowly-winged stalks about as long as the blades. The leafless flower stalks, up to 1 foot long and sometmes branched above, bear up to thirty purplish-red blooms approximately 1 inch wide. Very different is ***D. cistiflora.*** This kind has erect, branchless, leafy stems 4 to 8 inches tall. Its stalkless leaves are ¾ inch to 1¼ inches long by about ⅕ inch wide. Solitary or several together, the white to deep rose-pink flowers, up to 1½ inches in diameter, have orange anthers. This handsome species, rare in cultivation, inhabits dryish soils.

Large-flowered, perennial ***D. binata*** (syn. *D. dichotoma*), of Australia and New Zealand, has erect, all-basal leaves one or two times forked into slender, linear segments. The flowers, in branching clusters, are white and about 1 inch across. Another native of Australia and New Zealand, ***D. spathulata*** is a perennial with short-stalked, all-basal, spoon-shaped leaves and slender, one-sided racemes of purple-pink flowers on slender stems 3 to 8 inches tall.

Garden and Landscape Uses. Sundews can be cultivated in bog gardens and in similar places in rock gardens, but are easier to manage and are more often grown as curiosities and as teaching aids in greenhouses and terrariums. They are extremely interesting. Species confined in the wild to the Southern Hemisphere or to warm regions north of the Equator are not hardy in the north. Others are.

Cultivation. Suitable soil is a sandy peaty mixture topped with live sphagnum moss. It should be sufficiently porous and well drained to permit free passage of water without danger of stagnation. Pans (shallow pots) are suitable receptacles. These are kept standing in saucers of water (preferably rain water) so that moisture seeps up from below. Planting or potting is done just as new growth begins in spring. Kinds that form tubers are kept nearly dry during their dormant season, at other times constantly moist. The soil in which species that do not form tubers are grown is always kept moist. Good light, with only light shade to break the full intensity of strong summer sun, is requisite, and a constantly humid, but not oppressively dank and hot atmosphere. Indoors the winter night temperature should be 50°F, with a five to ten degree increase by day. At other seasons warmer conditions, with fairly free ventilation, is in order. Propagation is by seeds, division of tubers, and in the case of *D. binata* and perhaps some others, by root cuttings planted horizon-

Drosera spathulata

tally and covered to a depth of about ½ inch in a mixture of sand and peat, in a temperature of 65 to 70°F. Feeding sundews with bits of meat is harmful. They may entrap occasional insects, but such sources of nutrients are not essential to their well-being.

DROSERACEAE—Sundew Family. This remarkably interesting family of carnivorous plants contains slightly more than 100 species representing four genera. It has a very wide natural distribution, the genus *Drosera* being nearly cosmopolitan, the others, of which Venus' fly trap (*Dionaea muscipula*) is a well-known example, much more local.

Characteristically, members of this dicotyledonous group are low herbaceous plants inhabiting wet and swampy soils. Mostly their leaves are in basal rosettes. They are conspicuously glandular-hairy, the hairs tipped with drops of a sticky exudate that entrap insects and other tiny creatures. The halves of the leaf blades of Venus' fly trap snap together when trigger hairs are touched. The hairs of droseras are more slowly motile.

The flowers, solitary or in spikes or racemes, have usually five persistent sepals more or less united at their bases, generally five separate petals, four to twenty stamens, and three to five styles. The fruits are capsules. Genera are *Aldrovanda, Dionaea, Drosera,* and *Drosophyllum.*

DROSOPHYLLUM (Drosophýl-lum)—Portuguese-Sundew. Native to Portugal, southern Spain, and Morocco, *Drosophyllum* is a member of the sundew family DROSERACEAE. Its name is from the Greek *drosos,* dew, and *phyllon,* a leaf. It refers to the dewlike droplets that tip the numerous leaf glands. There is only one species.

A nonhardy, subshrubby, carnivorous plant up to 1 foot in height, ***D. lusitanicum*** has a short woody stem and long, slender leaves beset along their lengths with stalked, purple, hairlike glands each tipped with a droplet of clear, viscous fluid that attracts small insects and entraps them as effectively as flypaper. In addition to the stalked glands there are stalkless ones. These, brought into contact with nitrogenous matter, as when an insect is entrapped, secrete a digestive ferment that breaks down the body of the insect into fluids that can be absorbed by the plant. In this way the Portuguese-sundew satisfies some of its nitrogen requirements. The young leaves are coiled at their ends like the tops of a developing fern frond, a most unusual characteristic among flowering plants. The flowers are borne, in summer, on slender stalks. They are 1½ inches in diameter, have five bright sulfur-yellow petals, ten to twenty stamens, and five styles. The fruits are narrow capsules.

Garden Uses and Cultivation. This is a collectors' item of interest for growing in cool greenhouses. It succeeds in sunny, humid conditions in a minimum winter night temperature of 45°F and winter day temperatures up to ten degrees higher in acid, sandy-peaty soil or in milled sphagnum moss kept a little on the dryish side rather than constantly wet. Seeds sown in sphagnum moss or in a sandy-peaty soil in late winter or spring germinate without trouble. Insecticides or other sprays must not be used on this plant. The glands of *Drosophyllum* have no power of movement.

Drosophyllum lusitanicum

Drosophyllum lusitanicum (flowers)

DRUPE. This is a botanical term for a fleshy or pulpy fruit, such as a plum, that does not split to release the seed and has its seed or seeds enclosed separately in hard, bony coverings. Such seeds with their coverings are called stones. Hence, stone fruits is a common name for orchard fruits that are drupes. Many other plants bear drupes very much smaller than those of these familiar kinds.

DRYANDRA (Dry-ándra). The Swedish botanist Jonas Dryander, who died in 1810, is honored by the name of this genus of the protea family PROTEACEAE. It comprises fifty species of beautiful evergreen trees and shrubs endemic to Australia and differing from nearly related *Banksia* in their flower heads having involucres. Few kinds are cultivated. Many are worthy of trial.

Dryandras have alternate, usually pinnately-lobed or prickly-toothed, but sometimes pinnate, leathery leaves. Their yellow to orange flowers are in crowded heads, somewhat suggestive of those of thistles, backed by an involucre (collar) of overlapping, scalelike bracts. They are terminal on the stems or at the ends of short side shoots. The individual blooms that compose the heads are tubular toward their bases. Above they split into four slender segments (petals), each with a stalkless or nearly stalkless anther attached on its upper side close to its tip. There is a slender style without a thickening, enlargement, or brush of hairs at its apex. The fruits are compressed capsules.

Holly-leaf dryandra (***Dryandra floribunda*** syn. *D. sessilis*), 6 to 8 feet tall, has silky hairy shoots and stalkless, obovate, holly-like, prickle-toothed leaves 1 inch to 3 inches long and up to 1 inch wide. Its heads of yellow flowers, as long as the leaves beneath them, are about 1½ inches across. The bracts of their involucres are about one-half as long as the heads.

From 8 to 15 feet tall, and with downy shoots, ***D. formosa*** has leaves, 4 to 8 inches in length and from less than to slightly more than ½ inch in width. They are downy on their undersides and are divided into many little triangular leaflets so that in outline they suggest the sword of a swordfish. The heads of fragrant orange-yellow flowers, with densely-hairy petals, are 2½ inches wide and about 1½ inches long.

Dryandra formosa

Garden and Landscape Uses and Cultivation. Dryandras are for outdoor cultivation in sunny locations and well-drained soils in warm Mediterranean-type climates such as are typical of parts of California. They can also be grown in cool greenhouses. They require the same conditions as banksias, but there is need for trial and experiment to determine their requirements more precisely. Propagation is by seeds and by cuttings.

DRYAS (Drỳ-as)—Mountain-Avens. The number of species that constitute this charming genus of low shrublets of the rose family ROSACEAE is variously interpreted as being two to four, with American botanists tending toward recognition of the larger number. Gardeners find all manifestations of the genus delightful and reasonably easy to manage. The name is from the Latin, *dryas,* a wood nymph. The genus is widely distributed through arctic and alpine regions of the Northern Hemisphere.

Prostrate, wide-spreading evergreens, out of bloom 2 to 3 inches tall or slightly taller, members of the genus *Dryas* have woody stems clothed with the persistent remains of old leaves. Their broad-elliptic, oblong, or ovate leaves have eight or fewer lobes or large teeth along each side, in which case they look like tiny oak leaves, or are smooth-edged. Their upper surfaces are dark green, their undersides heavily clothed with white down. The flowers, on short stalks, are solitary. They have seven to ten sepals, and seven to ten, but usually eight, elliptic to obovate petals. There are many slender-stalked stamens, and numerous pistils with styles that elongate and become plumose in fruit and are then very decorative. The seed heads much resemble those of *Clematis* and certain anemones.

North American species, occurring at high latitudes and altitudes across the continent, are ***D. integrifolia*** and *D. drummondii.* The former is a dense plant with crowded, often toothless leaves up to ½ inch long and one-half as wide and generally with heart-shaped bases. Its upturned white flowers, about ¾ inch across, have lanceolate to linear-oblong sepals, and spreading, white petals. The feathery seed heads are 1 inch to 2 inches wide. From it ***D. drummondii*** differs in being of looser growth and having conspicuously

Dryas drummondii in seed

toothed leaves with wedge-shaped bases. Its nodding yellowish flowers, on stalks 2 to 4½ inches long, have ovate sepals and petals up to ½ inch long that do not spread widely. This species has larger seed heads than the last. American, European, and Asian, ***D. octopetala*** has white, upturned flowers, and usually conspicuously toothed leaves, heart-shaped at their bases, and up to 1¼ inches long by ½ inch wide, but often considerably smaller. The flowers are on stalks 1 inch to 2 inches long. They have linear sepals. This is a highly variable species, of which botanists distinguish several races.

Dryas octopetala

A pretty hybrid between *D. octopetala* and *D. drummondii,* intermediate between its parents, is ***D. suendermannii.*** Yellowish in bud, and when they first open nodding, the flowers of this kind become white when fully expanded.

Garden Uses and Cultivation. Among the easiest to grow of choice alpines, dryases are suited for rock gardens. They succeed suprisingly well even in lowland regions, such as the vicinity of New York City, but are undoubtedly happier where summers are cooler. They need well-drained soil; some limestone content is advantageous. Locations in full sun are to their liking. Each spring a top dressing of sandy soil with some crushed limestone and peat moss mixed in may, with advantage, be worked around the plants. Propagation is easy by seeds, summer cuttings, and layering.

DRYING OFF. Gardeners, especially growers of potted plants, commonly simulate conditions natural in regions where moist and dry seasons alternate by keeping the soil of deciduous plants dry during their periods of dormancy. The drying is not initiated by a sudden cessation of watering nor by giving less water at each application. It is done gradually by allowing progressively longer intervals between soakings until finally, after the foliage has completely died, water is withheld entirely. This procedure, called drying off, is done with deciduous bulb plants such as

Drying off gloriosas laid with their pots on their sides to facilitate drying

certain crinums and hymenocallises, babianas, hippeastrums, and tritonias. It is also practiced with many succulents including stone or pebble plants of South Africa, and with some shrubby and subshrubby plants, Lady Washington geraniums, for example.

DRYMIS. See Drimys.

DRYMOGLOSSUM (Drymo-glóssum). The half dozen secies of *Drymoglossum,* of the polypody family POLYPODIACEAE, are natives of Malagasy (Madagascar), tropical Asia, and Malaysia. Alluding to their natural habitats and the shape of the leaves, the name derives from the Greek *drymos,* a woodland, and *glossa,* a tongue.

These ferns are epiphytes that in their native habitats cling chiefly to the barks of trees. They have slender rhizomes and spaced along them two forms of small undivided leaves (fronds), the sterile ones ovate to broad-elliptic or nearly circular, the fertile ones linear-elliptic. The spore capsules are in a continuous line.

Native to tropical Asia, ***D. piloselloides*** (syn. *D. heterophyllum, Pteropsis piloselloides*) has almost threadlike rhizomes and broad-elliptic or roundish sterile fronds ½ inch to 2 inches long and up to ¾ inch wide. The fertile fronds are 2 to 4 inches long by up to ¼ inch wide. The lines of spore clusters are close to and parallel the frond margins. The plant sometimes named *D. microphyllum* is *Lemmaphyllum microphyllum.*

Garden Uses and Cultivation. These are attractive ferns for cultivating outdoors in the humid tropics and in tropical greenhouses and terrariums. Outdoors, they may be grown on the trunks of trees where they are shaded and receive adequate moisture. Indoors, they succeed attached to slabs of cork bark or tree fern trunk or in pans (shallow pots) in a mixture of peat moss, leaf mold, and coarse sand or grit. Propagation, easy by division, may also be effected by sowing spores. For further information see Ferns.

DRYMONIA (Dry-mònia). The tropical American and West Indian genus *Drymonia* belongs in the gesneria family GESNERIACEAE. It comprises thirty-five species of prostrate, climbing, and more or less scrambling shrubs. They grow both in the ground and perched on trees as epiphytes. The name, alluding to the plants in their homelands growing on trees, comes from the Greek *drymos,* an oak woodland.

Drymonias have evergreen, opposite, undivided, thickish, stalked, toothed leaves. Often solitary, sometimes clustered, their usually whitish or yellowish blooms have a big, five-lobed calyx and a somewhat drooping, swollen, tubular, two-lipped corolla with a spur at its base and with five spreading lobes (petals). There are four fertile stamens and one style capped with a mouth-shaped stigma. The fruits are berries containing many small seeds.

A shrub up to 6 feet tall, ***D. macrophylla*** (syn. *Alloplectus macrophyllus*), of Costa Rica and Guatemala, has purplish stems and grayish, scalloped-edged, elliptic, minutely-hairy leaves that may be 1 foot long or longer and nearly 5 inches wide. Waxy-white or creamy-white on their outsides, within beige-yellow with reddish-purple markings, the flowers, 1 inch to 1¼ inches long, are in clusters from the leaf axils or from the nodes of leafless parts of the stems. Costa Rican ***D. stenophylla*** has more or less vining stems and asymmetrical, lanceolate leaves. Its 1-inch-long blooms are trumpet-shaped, their corolla tubes cream, their lobes (petals) fringed and pale yellow.

Drymonia stenophylla

Garden Uses and Cultivation. The chief appeal of drymonias is to keen collectors of gesneriads (plants of the family to which drymonias belong). They are easy to grow in fairly humid tropical and subtropical environments. They succeed in loose, porous soil containing abundant coarse organic matter and need some shade from strong sun. The soil should be kept moderately moist, not wet. Propagation may be achieved by seeds and by cuttings, and quite probably by leaf cuttings.

DRYMOPHILA (Drymó-phila)—Turquoise Berry. Two species of eastern Australia and Tasmania are the only representatives of *Drymophila,* of the lily family LILIACEAE. The name, from the Greek *drymos,* a forest, and *philios,* loving, alludes to the habitat of the genus.

Drymophilas are nonhardy, herbaceous perennials with bulbous rootstocks and branchless or sparingly-branched, arching stems, leafy only in their upper parts. The thin-textured leaves, in two ranks, are stalkless or nearly so. The flowers, solitary or less often in pairs from the leaf axils, have a perianth of six lanceolate segments (commonly called petals, but more properly tepals) that spread widely or are reflexed, six stamens not longer and often shorter than the perianth, and three slightly flattened, recurved, slender styles The fruits are spherical or egg-shaped, beadlike berries.

Turquoise berry (***D. cyanocarpa***) grows in humid forests. Up to 1 foot tall or somewhat taller, it has arching stems, and leaves 1 inch to 3 inches long by approximately one-third as wide. The starry, white flowers, approximately ½ inch wide, have recurved stalks ¼ to ½ inch long. Bright blue, or rarely white, the globular, fleshy berries are about ⅓ inch or a little more in diameter and contain eight to twenty seeds. From the turquoise berry, ***D. morrei*** differs in having orange-colored berries containing not more than four seeds. It is perhaps no more than a variety of *D. cyanocarpa.*

Garden and Landscape Uses and Cultivation. Drymophilas are adapted for outdoor cultivation in mild, essentially frostless climates. In North America they may be expected to flourish best on the Pacific Coast. Woodland plants, they need cool, shaded locations and well-drained, moist, but not wet soil that contains an abundance of organic matter. Propagation is by division and by seed.

DRYNARIA (Dryn-ària). The ferns that constitute *Drynaria* have in the past been included in *Polypodium.* The group belong in the polypody family POLYPODIAEAE. It numbers twenty species. They are natives of the Old World and the Australian tropics and subtropics. The name derives from that of the dryads (wood nymphs).

From closely related *Aglaomorpha* this genus differs in having two types of fronds (leaves), spore-bearing, fertile ones of normal leafy appearance and shorter, papery, shallowly-lobed, barren ones that trap and retain fallen leaves and other forest debris, which as it decays provides nourishment for the fern.

Drynarias are mostly large and evergreen and in the wild grow as epiphytes, perched on trees, but taking no nourishment from them. They have scaly, fleshy rhizomes, and rigid, pinnately-lobed or pinnate normal fronds. The spore cases are in round clusters.

Native of Queensland, ***D. rigidula*** (syn. *Polypodium diversifolium*) has thick, rabbit's-foot-like rhizomes covered with long, brown, hairlike scales. Its stalkless barren fronds, up to 9 inches long by 4 inches wide, are cleft about halfway to their centers into blunt lobes. The long-stalked fertile fronds, erect at first, but later drooping, have blades 2 to 4 feet long by almost or quite one-half as wide, divided into numerous pointed, short-stalked, round-toothed leaflets up to ⅜ inch wide. The spore-case clusters form a single row each side of the mid-vein of the leaflet and parallel to it. Variety *D. r. whitei* is especially beautiful. It has long, drooping, dark green fronds with deeply-lobed and frilled leaflets that rattle "like tinsel paper when stroked."

Coarser-foliaged ***D. quercifolia*** (syn. *Polypodium quercifolium*), sometimes called the oak leaf fern, is found in the wild from India to Australia perching on trees or sometimes on rocks. Its thick, woody rhizomes are clothed with bright brown scales. The barren fronds are stalkless, brown, and up to 1 foot long by one-half or more as wide. They are lobed, frequently to their middles, with the lobes alternating on either side of the midrib. The fertile fronds have long stalks and are 2 to 3 feet long by 1 foot wide or wider. They are cleft very nearly to their midribs into many leaflet-like lobes ¾ inch to 1½ inches wide, with spore-case clusters in lines paralleling veinlets that spread from the center veins of the lobes. From *D. quercifolia,* closely allied ***D. sparsisora*** differs in its clusters of spore capsules being scattered irregularly instead of forming double rows between the veinlets.

Garden and Landscape Uses and Cultivation. These imposing and beautiful ferns are handsome ornamentals for outdoors and growing in lath-houses in the tropics and warm subtropics and for ornamenting greenhouses. They are not difficult to manage. Where they grow outdoors they can be established in crotches or other suitable places that trees afford, or on stumps or rocks, in a rooting mixture consisting largely of organic debris. Indoors, they may be accommodated in large, very well-drained pots or pans (shallow pots), in hanging baskets, or on wooden rafts. Let the rooting medium be coarse, porous, and mainly organic. One consisting of a mixture of chips of bark of the kind used for potting orchids, peat moss, half-decayed leaves, and sand or perlite, with some dried cow manure, broken pieces of charcoal, and a little turfy loam added suits admirably, but almost any type of material in which tree-perching orchids, bromeliads, and plants of similar habits grow well will give good results. In greenhouses the night temperature in winter should be about 60°F and that by day five to fifteen degrees higher. At other seasons warmer conditions are in order. Maintain at all times a decidedly humid atmosphere and provide sufficient shade to prevent yellowing or scorching of the foliage, but not more. Keep the soil always moderately moist, and at about two-week intervals from spring to fall apply dilute liquid fertilizer to well-rooted specimens. When the rooting medium begins to compact as a result of excessive decay, or when the plants outgrow their containers, transplant or repot them, taking care to keep the rhizomes on the surface of the new rooting compost. This is best done in early spring. Then, too, is the best time to propagate drynarias by division. An alternative means of increase is by spores. For more information see Ferns.

DRYNARIOPSIS (Drynari-ópsis). Formerly included in *Aglaomorpha,* the only species of this genus belongs in the polypody family POLYPODIACEAE. An enormous fern, it is an epiphyte (a plant that grows on trees without taking nourishment from them). Its name is compounded from that of another genus of ferns, *Drynaria,* and the Greek *opsis,* similar to. From *Aglaomorpha* it differs in having minute clusters of spore capsules.

Majestic ***Drynariopsis heraclea*** (syn. *Aglaomorpha heraclea*), native from the Philippine Islands to Java and the Solomon Islands, has rhizomes clothed with rusty scales and leaves up to 8 feet long and 2 feet wide or wider, with broadly-winged stalks. The puckered, thin-leathery blades are deeply divided into segments, with broadly triangular bases, that narrow rapidly to slender points. The leaf bases serve to collect and hold leaves and other organic debris that decays to form humus that nourishes the plants.

Garden and Landscape Uses and Cultivation. These are as for *Aglaomorpha.* For additional information see Ferns.

DRYOPTERIS (Dryóp-teris)—Shield Fern, Male Fern, Wood Fern. A goodly number of ferns formerly included in *Dryopteris* have been transferred to other genera, notably *Ctenitis* and *Thelypteris.* Others once named *Aspidium* and *Nephrodium* are included in the about 150 species now accepted as constituting *Dryopteris.* Of nearly cosmopolitan distribution in temperate, subtropical, and tropical regions, this genus of the aspidium family ASPIDIACEAE has a name derived from the Greek *drys,* an oak, and *pteris,* a fern. It alludes to some species inhabiting oak woodlands.

Mostly medium-sized to large, ferns of the genus *Dryopteris* have thick, erect or short-creeping rhizomes. Their usually evergreen, semileathery fronds (leaves) are generally clustered to form definite crowns rather than being, as are those of many species of *Thelypteris,* scattered along the rhizomes. Their stalks contain seven conductive bundles. The blades, pinnately-divided and pinnately-lobed or twice- or thrice-pinnate, are generally broadest at their bases. Although they may have scales, unlike those of *Thelypteris,* they are without needle-shaped hairs. The usually branched veins, generally enlarged at their tips, stop short of the edges of the frond segments. The round clusters of spore capsules, set in from the margins of the leaf segments, have a round-kidney-shaped cover (indusium). In the wild hybridization between species of *Dryopteris* is fairly common. The resulting progeny generally exhibit characteristics intermediate between those of the parents and fail to produce viable spores. To this group belong some well-known ferns. Since the time of Theophrastus and probably before, the virtues of the male fern as an expellant of tapeworms has been known. It is still employed as a vermifuge.

Crested shield fern (***D. cristata***) inhabits wet woodlands and marshes from Newfoundland to Saskatchewan, Montana, Tennessee, and Arkansas, and also in Europe. It has short-creeping rhizomes and two kinds of fronds. The sterile ones are evergreen and spreading, the somewhat longer spore-bearing ones, deciduous and erect, are up to 2½ feet long. The narrowly-lanceolate blades of the fronds are 3 to 6 inches wide and pinnate, with their rather widely-spaced primary divisions triangular-lanceolate and pinnately-lobed. Each ultimate segment is toothed and has five or six pairs of mostly twice-branched veins. The spore-case clusters, located midway between the mid-veins and the margins of the segments, have persistent, glandless covers.

Fragrant shield fern (***D. fragrans***) is evergreen. Native from New York to Ontario and Minnesota, and in Europe and northern Asia, this often grows on limestone formations, favoring shaded cliffs and hillsides. It has short, thick, erect rhizomes and short-stalked, linear-oblanceolate fronds with blades up to a little over 1 foot in length and about 2 inches in width. They narrow gradually and markedly to their bases and are pinnate with the primary divisions pinnately-lobed. They are glandular on midribs, veins, and margins. Their ultimate segments are oblong, blunt, and round-toothed. Their veins, two to three pairs to each segment, are obscure and without branches. Situated well in from the margins, the clusters of spore cases have large, circular covers with glands along their margins.

Marginal shield fern (***D. marginalis***) is an evergreen native of woodlands and slopes from Nova Scotia to Ontario, Wisconsin, Georgia, Arkansas, and Oklahoma. It has erect, thick rhizomes and numerous leathery, glandless, lanceolate, pinnately-divided fronds up to 3 feet long or longer by 9 inches wide, their primary divisions pinnately-lobed to pinnately-divided, the lowermost only slightly smaller than those above. There are six to nine pairs of branched veins to each ultimate round-toothed or toothless segment. The clusters of spore cases with small shield- to kidney-shaped covers, are borne near the margins of the segments.

Dryopteris marginalis

Spinulose shield fern (***D. austriaca*** syn. *D. dilatata*) is a variable species with several intergrading varieties some deciduous, some evergreen. Indigenous through much of North America as far south as Georgia, Kentucky, and Iowa, as well as in northern Europe and northern Asia, it occurs chiefly in moist woodlands and on moist banks. It has thick rhizomes and clustered ovate-lanceolate fronds with stalks shorter than their blades, up to 3 feet long and 1 foot wide. They are twice- or thrice-pinnately-divided, with the lower primary divisions scarcely smaller than those above. The ultimate segments have three to five branched or branchless veins. Their margins are furnished with tiny, spiny teeth. The persistent spore case coverings are sometimes glandular. Typical *D. austriaca* has deciduous fronds. Those of *D. a. spinulosa* (syn. *D. spinulosa*) are evergreen as are those of *D. a. intermedia* (syn. *D. intermedia*). The spinulose shield fern and its varieties are harvested to supply cut foliage to florists. They are sometimes called florists' fern and fancy fern.

Clinton's fern (***D. clintoniana***) is a hybrid between Goldie's fern and the crested shield fern that evidences intermediate characteristics. Evergreen, it has short-creeping rhizomes and pinnate fronds up to about 4 feet long with lanceolate blades up to about 8 inches wide. They are pinnately divided into pinnately-lobed to nearly pinnate, lanceolate primary divisions. The ultimate segments have five to seven pairs of mostly twice-branched veins. The clusters of spore cases are midway between the margins and mid-veins of the segments. Their covers are persistent and without glands. This fern occurs as a native in swamps and wet woodlands from Quebec to Wisconsin, North Carolina, and Tennessee.

Goldie's fern (***D. goldiana***), sometimes called giant wood fern, is of impressive

Dryopteris austriaca spinulosa

Dryopteris austriaca intermedia

size. It has short, creeping, thick rhizomes and many firm-textured, deciduous or evergreen fronds up to 4 feet long. Their ovate blades, narrowed only slightly toward their bases, are pinnate with the primary divisions deeply-pinnately-lobed and widest some little distance above their bases. The seven to ten pairs of veins of each ultimate segment are mostly forked. The clusters of spore capsules, closer to the mid-veins than to the margins of the segments, have persistent, glandless covers. Goldie's fern is a native of moist woodlands from New Brunswick to Minnesota, Tennessee, and Iowa.

Male fern (***D. filix-mas***) inhabits much of the Northern Hemisphere. In North America it favors woody, rocky hillsides often of limestone formation from Newfoundland to Vermont, Michigan, British Columbia, South Dakota, Texas, and Cal-

Dryopteris filix-mas

ifornia. This has thick, erect rhizomes and deciduous, lanceolate to narrow-elliptic fronds up to 4 feet tall or taller and 1 foot wide. They are pinnate with the primary divisions pinnately-lobed or are nearly twice-pinnate. The primary divisions of the leaf decrease in size downward. There are four to six pairs of branched veins to each utimate segment. The spore-cluster coverings are large, persistent, and without glands. In *D. f. cristata*, a horticultural variety of English origin, the ends of the primary divisions of the fronds are attractively crested.

California or coastal wood fern (***D. arguta***) is native from California to Washington. Up to, but mostly under 3 feet tall, its evergreen leaves form clusters. They are twice- or thrice-pinnate, about one-half as wide as long, and of firm, leathery texture. This withstands dryish conditions, but has not proved easy to establish. Native from California to Mexico and in South America and Hawaii, ***D. parallelogramma*** forms crowns of stiffish leaves up to a little more than 3 feet in height, twice-pinnate and with toothed margins. The dark and conspicuous veins of the segments are once or more times branched. Another native of

Doronicum cordatum

Dracaena sanderana

Drosera intermedia (leaf)

Dudleya brittonii

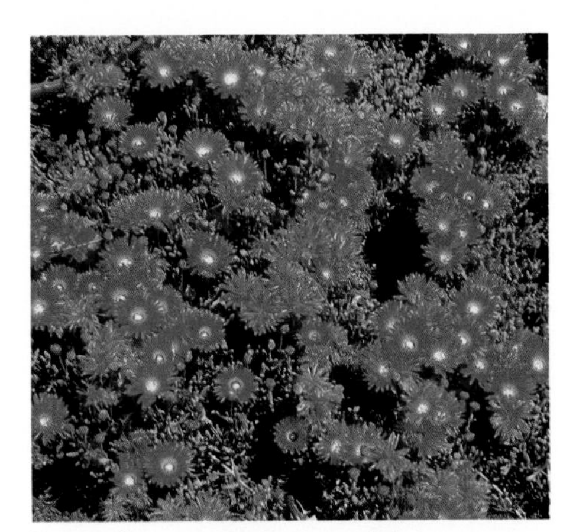
Drosanthemum speciosum variety

Echeveria secunda glauca

Dryas octopetala

Echinocactus grusonii, with flowers

Echinacea purpurea

Echinocereus sciurus

Echium fastuosum

Dryopteris parallelogramma

Hawaii and of many other parts of the tropics and subtropics as well as of New Zealand, ***D. dentata*** bears once-pinnate fronds up to 3 feet long, widest at their middles and there about one-third as wide as long.

Exotic species, kinds not native to North America, include some hardy and some nonhardy species. The more tender, and their degree of hardiness can be inferred from their geographic origins, are suitable for outdoor cultivation in mild climates only. Elsewhere they are strictly greenhouse and conservatory plants. Among exotic kinds likely to be cultivated are these: ***D. atrata,*** of China, has crowns or clusters of lanceolate, once-pinnate fronds up to 2½ feet long, with coarsely-toothed margins. ***D. erythrosora,*** a variable native of Japan, Taiwan, the Philippine Islands,

Dryopteris erythrosora

and China, has erect rhizomes and crowns of 4-feet-tall, broadly-triangular to broadly-ovate, twice-pinnate, coarsely-toothed, evergreen fronds. In their young stages the covers of the clusters of spore cases are red. ***D. sieboldii,*** of Japan and Taiwan, has thick, short-creeping rhizomes and triangular-ovate fronds usually about 1 foot long but sometimes considerably over, each with two to five pairs of undivided, lobeless, toothed primary divisions. The fertile fronds, usually narrower than the sterile ones, have spore clusters mostly toward the margins. ***D. pseudo-mas*** (syn. *D. borreri*) of Europe and southwestern Asia is very variable. It forms sturdy clumps of fronds up to 3 feet long, those of some of its variants attractively crested. From *D. filix-mas,* which it much resembles, this species is most readily distinguished by the shaggy, yellow to dark brown, hairlike scales that clothe its leafstalks and midribs.

Dryopteris pseudo-mas

Garden and Landscape Uses and Cultivation. Dryopterises are excellent furnishings for fern gardens, woodland paths, native plant gardens, naturalistic areas, and other places where suitable conditions of soil, shade, and moisture exist or can be devised. They are splendid underplantings in open woodlands and can be used effectively as backgrounds and complements to flowering plants. Single specimens or small groups of those kinds, such as the marginal shield fern and the spinulose shield fern, which have clumplike crowns instead of creeping rootstocks, can be displayed to good advantage near large boulders and close to the bases of tree stumps.

Hardy dryopterises prosper best in cool locations in shade or filtered sunlight. The most agreeable earth is a fairly loose, slightly acid one containing an abundance of organic matter such as leaf mold, rich compost, or peat moss and not subject to drought. Stony ground is favored by shield ferns and some others. For preference plant in early fall or early spring. Allow 2 to 3 feet between individuals, the greater distance for such massive kinds as Goldie's fern, the lesser for shield ferns. Let the crowns be slightly above the soil surface, but make sure the roots are all covered.

Tropical and subtropical species may in suitable climates be used in the manner of hardy ones in cooler regions. In greenhouses and conservatories they succeed in ground beds, in large pans (shallow pots), and in pots. See also Ferns.

DRYPETES (Dry-pètes)—Guiana-Plum. This horticulturally little known genus of the spurge family EUPHORBIACEAE includes kinds previously named *Hemicyclia.* It comprises about 200 species of the tropics and subtropics of the Old and the New World, extending into southern Florida. The name, from the Greek *drypetes,* ripened on the tree, mature, alludes to the fruits.

The kinds of *Drypetes* are trees and shrubs with alternate, undivided leaves, and small, unisexual, solitary or clustered, petal-less flowers, segregated as to sex on separate plants. The fleshy fruits, which do not split when ripe, have a single stone. Those of some species are edible.

The yellow tulip wood of Australia (***D. australasica*** syn. *Hemicyclia australasica*) is a broad-headed tree 40 or 50 feet tall. It has lustrous, leathery, blunt-ovate, stalked leaves 1½ to 3 inches in length, and flowers, solitary or in clusters or short racemes of few in the leaf axils, with four or five sepals and, the males, four to ten stamens. The red, succulent fruits are about ½ inch long, smooth, and egg-shaped. A rare native of Hawaii, ***D. phyllanthoides*** (syn. *Hemicyclia phyllanthoides*) is about 30 feet tall. It has blunt, ovate, light green leaves, whitish on their undersides and 3 inches long or longer by about one-half as wide. The flowers are in clusters along the shoots. The fruits are red and ¼ inch in diameter.

Guiana-plum (***D. lateriflora***) is a brown-barked shrub or tree up to about 30 feet tall, a native of Florida and the West Indies. It has stalked, elliptic to oblong, toothless leaves 2 to 5 inches in length. The flowers, in dense clusters from the leaf axils, have four sepals and the same number of stamens. The red to dark brown, subspherical fruits are ¼ inch or a little more in diameter. Differing from the last in its flowers having eight each sepals and stamens, and its ovoid fruits being ¾ to 1 inch long, whitewood (***D. diversifolia***) is a native of southern Florida and the West Indies. A white-trunked tree up to 40 feet tall, it has ovate leaves 3 to 4 inches long, of two kinds, the lower sharp-pointed, toothed, and long-stalked, the upper with longer stalks, blunter, and toothless.

Garden and Landscape Uses and Cultivation. The species described here are suitable for the tropics and warm subtropics. They adapt to ordinary conditions and are increased by seed. The Florida natives grow satisfactorily in coastal areas.

DUABANGA (Dua-bánga). One of two genera of the sonneratia family SONNERATIACEAE is named *Duabanga.* Comprising three species of trees, it is wild from the Himalayas to New Guinea. The name is derived from an Asian vernacular one.

One species, ***D. grandiflora*** (syn. *D. sonneratioides*) is rare in cultivation in Florida. A rapid-growing, graceful tree, it has a massive trunk with flaking bark and drooping branches. The leathery leaves are opposite, oblong, and 6 inches to 1 foot long. They have a strong mid-vein and parallel side veins angling from it. Somewhat like those of crape-myrtle, the white flowers are in large clusters at the branch ends. They face downward, emit a foul odor, have a persistent six-lobed calyx, six spreading petals, numerous stamens, and a long-protruding style. They are about 3½ inches in diameter. The fruits are large, pointed-ovate capsules.

Garden and Landscape Uses and Cultivation. Limited experience in southern Florida indicates that the stately tree described here is suitable for wider cultivation in the warmer parts of the southern United States and elsewhere where frost is practically unknown. In Florida a specimen attained a height of 45 feet in a dozen years. It was resistant to pests, diseases, and hurricanes, and is recommended for wider trial. Propagation is by seeds and, probably, by cuttings.

DUCHESNEA (Duchés-nea) — Indian-Strawberry or Mock-Strawberry. Of the possibly six species of this Asian genus only one, the Indian-strawberry or mock-strawberry appears to be known to American gardeners. By some botanists *Duchesnea* is united with *Potentilla,* to which it is obviously closely related. Named after the French botanist Antoine Nicolas Duchesne, who died in 1827, *Duchesnea* is a member of the rose family ROSACEAE.

Duchesneas are low herbaceous perennials with short rhizomes and long-stalked leaves each of three leaflets. They develop long, slender runners after the fashion of those of strawberries. The solitary flowers have five sepals alternating with large leafy bracts, five petals, twenty to twenty-five slender stamens, and numerous pistils each with a slender style. At maturity the receptacle, in that stage commonly, but not with strict botanical correctness called the fruit, becomes swollen and red, but not juicy and edible. As with strawberries, its surface is studded with seed-like achenes that are the true fruits.

The Indian-strawberry (*D. indica*) is a neat trailer, naturalized in moist soils from New York to Missouri and southward, and on the Pacific Coast. It has leafy runners and leaves with coarsely-toothed leaflets sparsely-hairy on their undersides. The flower stalks, as long as or longer than the leaves, arise from the runners rather than from the center of the plant. Each bears a bright yellow flower with a corolla about ½ inch wide and a large, leafy calyx forming a collar beneath and extending beyond the petals. The deep red fruits, resembling strawberries and usually less than ½ inch in diameter, are dry and insipid, reason enough for the colloquial name mock-strawberry.

Duchesnea indica

Duchesnea indica (leaves and fruits)

Garden Uses and Cultivation. The Indian-strawberry has limited uses as a groundcover and makes an attractive specimen for hanging baskets, vases, and strawberry jars. Interestingly, although perfectly hardy quite far north, it succeeds admirably when grown in greenhouses, even those affording tropical environments. Its requirements are simple. It thrives in any ordinary well-drained soil kept fairly moist, where it receives full sun. It is very easily propagated by removing and planting rooted portions of its runners and by seeds.

DUCK'S EYE is *Ardisia humilis.*

DUCKWEED. See Lemna.

DUDLEYA (Dúd-leya). Closely resembling *Echeveria,* this genus of the orpine family CRASSULACEAE is distinguished from it by its broad, stalkless, not readily detached leaves having more or less stem-clasping bases and by its flowers having mostly equal sepals and petals, the latter thinnish, scarcely keeled on their backs, and not deeply hollowed on their faces. As natives, the sixty species of *Dudleya* inhabit California, Baja California, Arizona, and some adjacent territories. The name commemorates William Russel Dudley, first professor of botany at Stanford University, California, who died in 1911.

Dudleyas are hairless, stemmed or stemless, succulent herbaceous perennials with leaves in rosettes. The stems are branched or not. The flower stalks are lateral, coming from the axils of leaves well below the centers of the rosettes. They bear fleshy, stalkless, more or less bract-like leaves that diminish in size upward. Above, the stalks divide into branched or branchless, one-sided racemes of usually pale or dull blooms, less often bright red or yellow ones. The flowers have five sepals, five erect petals more or less joined toward their bases and with sometimes spreading tips. The stamens are in two circles of five. The female parts are five separate or nearly separate, slender, straight carpels, each with one style and with a small, thinnish nectar gland at the base. The fruits are of five little podlike follicles.

Dudleya, unidentified species in Baja California

Kinds with slender leaves, more or less pencil-shaped or sausage-shaped, include ***D. attenuata*** (syns. *Echeveria attenuata, Stylophyllum attenuatum*), native to Baja California. This clump-forming species has branched stems up to 1 foot long. Its rosettes are of glaucous, pointed, linear to linear-oblanceolate leaves 1 inch to 4 inches long by up to ⅕ inch wide. Up to 10 inches tall, the flower stems have slender panicles of upfacing, red-lined, yellow blooms about ⅓ inch long. The petals spread from near their middles. Variety *D. a. orcuttii* (syn. *Echeveria orcuttii*), of southern California and Baja California, has densely-glaucous leaves and flowers with red-lined, white petals, often flushed with rose-pink. Its branched stems up to about 8 inches long, ***D. cultrata,*** of Baja California, has rosettes of green, pointed-oblong leaves 2 to 4½ inches long by from somewhat less to somewhat more than ½ inch wide. The flower stems, up to 8 inches tall, have usually three erect branches. The flowers have pale yellow petals from a little less to a little more than ½ inch long. Endemic to southern California, ***D. densiflora*** (syns. *Echeveria nudicaulis, Stylophyllum densiflorum*) has branched stems up to 4 inches long. Its rosettes are of glistening, white-mealy, linear leaves, round in section and widened at their bases. They are up to 6 inches long by less than ½ inch wide. Up to 1 foot tall, the flower stems have branches again branched. The blooms have white or pinkish petals approximately ⅓ inch in length. Native of southern California and Baja California, ***D. edulis*** (syns. *Echeveria edulis, Stylophyllum edule*) forms clusters of branched stems up to 6 inches tall. Its rosettes are of light green, sharp-pointed leaves 3 to 8 inches long, under ½ inch wide, but broader at their bases. Up to 1½ feet long, the flower stalks have branched branches. The blooms have yellowish to white petals under ½ inch long.

Several broader-foliaged dudleyas with thinner leaves than those discussed above, leaves not nearly round in section, are cultivated. Indigenous to Nevada, Arizona, New Mexico, and California, ***D. arizonica*** has solitary, stalkless rosettes up to 8 inches across of ten to thirty-five pointed-oblong leaves coated with waxy meal. The flowers, on stalks 6 inches to 2 feet tall and from a little less to a little more than ½ inch long, have erect, yellow to red petals. Native of Baja California, ***D. brittonii*** has usually branchless stems up to 4 inches long. Its rosettes are of many densely-white-mealy, pointed-oblong leaves 3½ to 10 inches long by 1½ to 3½ inches wide.

Dudleya brittonii

The flower stalks, up to 4 inches long, with mostly close-set branches, are of blooms with pale yellow petals ½ inch or a little more in length. Variable ***D. caespitosa*** (syns. *Cotyledon caespitosa, Echeveria cotyledon*) of California has usually branched stems and rosettes of fifteen to thirty green to glaucous, oblong leaves 2 to 8 inches long by ½ to 2 inches wide. From 6 inches

Dudleya caespitosa

Dudleya caespitosa (flowers)

to 1½ feet long, the flower stalks carry white to yellow or red flowers up to ½ inch long and with erect petals. Beautiful ***D. candida*** (syn. *Echeveria candida*) is endemic to Baja California and Coronado Islands.

Dudleya candida

lands. Its short, thick stem bears clustered rosettes of usually heavily white-mealy, triangular-ovate to oblong-oblanceolate leaves 2 to 4½ inches long by ½ inch to 2½ inches wide, broader at their bases. From 8 inches to 1½ feet long, the flower stems have three to five closely-set, sometimes forked branches. The blooms have pale greenish-yellow petals ⅓ to ½ inch long. Attractive ***D. farinosa*** inhabits the coasts of central and northern California. Its usually much-branched stems are low. The rosettes are of green to thickly white-mealy, pointed-ovate-oblong leaves 1 inch to 2½ inches long by from nearly ½ inch

Dudleya farinosa

to 1 inch wide. From 4 inches to about 1 foot long, the flowering stalks have three to five primary branches, sometimes branched again. Pale yellow, the petals of the starry blooms are about ½ inch long. Its branchless stem up to 1 foot long, and its foliage completely covered with chalky-white, waxy meal, ***D. pulverulenta,*** of southern California and Baja California, has rosettes of pointed-oblong leaves 3½ to 10 inches long by 1½ to 4 inches wide. The branched flowering stalks, 1 foot to 3 feet long, bear more or less nodding, red blooms ½ to ¾ inch long. Forming large clumps of branched stems, ***D. virens*** (syns. *Echeveria insularis, Stylophyllum virens*) has pointed-strap-shaped, green to glaucous leave, 3½ to 8 inches long by ¾ inch to 2½ inches wide that are often reddish to-

Dudleya virens

ward their apexes. The flowering stalks have branches that again branch. They are up to 2 feet long. The blooms have white to reddish-yellow petals approximately ⅓ inch long.

Garden and Landscape Uses and Cultivation. These are as for *Echeveria,* except that dudleyas when grown indoors seem to prefer slightly higher temperatures than the lowest temperature appropriate for echeverias. For more information see Echeveria, and Succulents.

DUGGENA (Dug-gèna). The twenty-two species of West Indian and South American species that compose *Duggena* belong in the madder family RUBIACEAE. The author of the name, an eighteenth-century Norwegian botanist, says only that it commemorates "Duggen."

Duggenas are small trees and shrubs with opposite leaves, and terminal spikes, racemes, or panicles of small, tubular, salver-shaped flowers with usually four- or sometimes five-lobed calyxes and corollas. There are four or rarely five stamens and a slender style. The fruits are berries.

From 3 to 12 feet tall, ***D. spicata*** (syns. *D. hirsuta, Gonzalagunia spicata*), of the West Indies and South America, has somewhat the aspect of *Clethra.* It has slender branches and short-stalked, broadly-ovate to lanceolate leaves 2½ to 7 inches long and more or less hairy especially on the veins beneath. The ½-inch-long white flowers are in one-sided spikes 3½ inches to 1¼ feet long. The small, fleshy fruits are white or blue.

Garden and Landscape Uses and Cultivation. In warm, frostless climates the species described may be planted as a general purpose shrub. It succeeds under ordinary conditions without special care. Propagation is by seeds and by cuttings.

DUMB CANE. See Dieffenbachia.

DURANTA (Dur-ánta)—Pigeon Berry or Sky Flower or Golden Dewdrop. Thirty-six species of shrubs and trees belonging to the vervain family VERBENACEAE belong in *Duranta.* Their name commemorates Castor Durantes, an Italian botanist and physician, who died about 1500. They are natives from Mexico to South America and the West Indies; one is indigenous in Key West, Florida.

Durantas have opposite or whorled (in circles of three or more), toothed or smooth-edged leaves, and terminal or rarely axillary racemes of small flowers with a five-toothed calyx, five spreading corolla lobes, and a usually curved tube. There are four stamens and an unequally four-lobed stigma. The berry-like, juicy fruits, each containing eight seeds, are enclosed by persistent, enlarged calyxes.

The pigeon berry, sky flower, or golden dewdrop (***D. repens*** syn. *D. plumieri*) has

Duranta repens

Duranta repens (fruits)

a natural range extending from Key West, Florida, to Brazil and the West Indies. It is a shrub or tree up to 20 feet tall with branches often drooping or trailing. Normally in small specimens they are unarmed, but those of large shurb or tree size are mostly spiny. The leaves are ovate-elliptic, obovate or ovate, ½ inch to 2 inches long, and short-stalked. Coarsely-toothed above their middles, they are hairless or finely pubescent. The flowers, under ½ inch across and lilac-blue, are in branched racemes up to 6 inches long. Bright yellow and spherical, the fruits are ¼ to nearly ½ inch wide. The calyx that encloses them has a curved beak. Varieties are *D. r. alba,* with white flowers; *D. r. canescens,* with densely white-hairy calyxes and purple flowers with white centers; *D. r. grandiflora,* with flowers ¾ inch in diameter; and *D. r. variegata,* with variegated foliage. From *D. repens* the Brazilian ***D. stenostachya*** differs in having oblong-lanceolate, toothed or toothless leaves 3 to 8 inches long and pubescent on the veins beneath. The lilac-blue flowers, in slender racemes and under ½ inch across, are succeeded by yellow fruits ⅓ inch in diameter. This species attains a height of about 15 feet.

Garden and Landscape Uses. These are popular in the tropics and warm subtropics, attractive in bloom and even more so in fruit. The fruits are long lasting. Durantas are good as solitary specimens, for grouping, and for hedges and backgrounds. They stand drought well and prosper even in poor soils, although better results are had in fertile ones that do not dry excessively. Full sun is desirable. Durantas are occasionally grown in conservatories, but there they rarely bloom or fruit as freely as outdoors in warm climates.

Cultivation. Little trouble attends the cultivation of these plants. They transplant with ease provided they are pruned hard at the time of moving and are readily increased by seed, layering, and air layering. They may also be propagated by cuttings, but these are sometimes rather difficult to root. An application of fertilizer at the beginning of the growing season promotes good growth, and at that time or a little earlier, pruning to remove dry, dead, crowded, and weak stems and to contain the bush to the size desired should be done. When grown as formal hedges more frequent pruning (shearing) is necessary. Care under greenhouse cultivation is about the same as for outdoor specimens. In the greenhouse they need full sun and a winter night temperature of about 50°F with a rise of five degrees or so by day.

DURIAN is *Durio zibethinus.*

DURIO (Dùr-io)—Durian. The only well-known species of the more than twenty that comprise this genus of Malayan trees is the durian. This is rarely planted in the Western Hemisphere. Although its edible fruits are highly prized in the humid Asian tropics, they scarcely appeal to Western tastes. Visitors to torrid Asian lands, however, are likely to hear of the durian. The genus *Durio* belongs in the bombax family BOMBACACEAE. Its name is a modification of the Malayan one for the fruits, *duryon.*

The durian (***Durio zibethinus***) attains a height of 80 to 100 feet or sometimes more and has short-stalked, obovate-oblong, pointed leaves, clothed on their under sur-

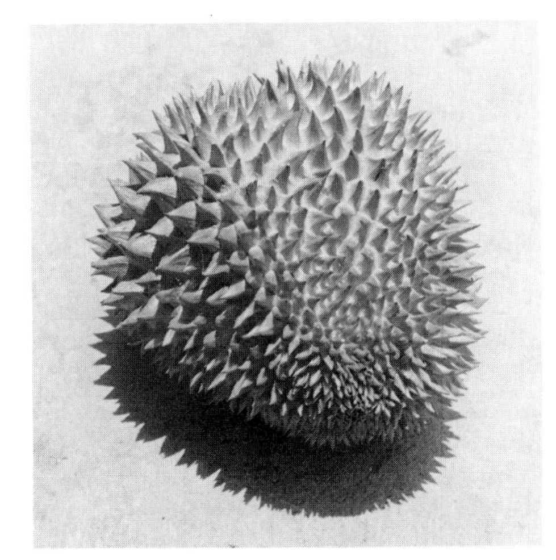
Durio zibethinus: (a) Fruit

(b) Fruit cut open

faces with coppery or silvery scales. They are up to 7 inches long. The flowers, white, cream, yellow, or red and about 2 inches long, are in clusters. They have five petals and numerous stamens united at their bases into a column.

The fruits are remarkable. Spherical to egg-shaped, they are normally 8 to 10 inches long. Exceptionally, they may weigh 100 lbs, usually less. Still, they are weighty enough to seriously injure or even kill a person on whom they fall. They are green or yellowish and studded with hard, sharp spines. Their insides are filled with a few large edible seeds and much creamy, custard-like pulp reputed to be aphrodisiacal. The odor of the pulp may be most generously likened to that of Limburger cheese. It has been described as resembling the smell of rotten onions, and one European who sampled the pulp likened it to "French custard passed through a sewer pipe." Yet not all Westerners are repulsed by it (after all many enjoy Limburger), to Alfred Russell Wallace "the sensation of eating durians is worth a voyage to the East." Like so many of the tastes of man, this seems to be an acquired one. It is unlikely that in the foreseeable future many Americans will develop it.

Garden and Landscape Uses and Cultivation. The durian will not grow outdoors in the continental United States. For its success it requires steamy tropical conditions throughout the year and deep, fertile, moist soil. It is propagated by seed and by budding.

DUSTS and DUSTING. The application of dry dusts instead of wet sprays is an accepted method of combating a wide variety of pests and diseases of plants. Except in rather rare instances, however, experienced gardeners usually prefer spraying. This because properly applied sprays give more even coverage and usually leave less unsightly deposits. The chief advantages of dusting are that if ready-prepared dusts are employed, and they are the only sensible ones for amateur gardeners to use, there is no messy mixing involved and the apparatus need not be cleaned after every use. The material can be left in the duster immediately ready for the next need.

Dusting in a vegetable garden

If you elect to dust, be sure to buy an effective duster. The salt-shaker type cartons and applicators that operate in trombone fashion in which dusts are sometimes sold are generally unsatisfactory. For effective results it is imperative that the material leaves the duster forcibly in an even smokelike cloud of extremely fine particles that will settle as a delicate film on both the upper and undersides of the foliage, without unsightly blobs. Power dusters are made, but a hand one is all the amateur is likely to need. One of the best for small operations is a midget rotary type made of aluminum and fitted with a hopper and a crank.

Dusters that work on the bellows principle, others that are operated with plunger devices, and knapsack types with an arm or lever to be pumped are also available. All kinds should be fitted with an extension making it easy to blow the dust in an upward direction, if needed, from ground level.

Suggestions made in the Encyclopedia entry Sprays and Spraying about the importance of timeliness of application, choice of calm days, and the importance of adequate coverage, apply equally to dusting. Dusting is often most effective if done when the foliage is slightly damp.

DUSTY MILLER. This is a common name applied to the following plants generally notable for their white-hairy or light-gray-hairy stems and foliage: *Artemisia stelleriana, Centaurea gymnocarpa, C. ragusina, C. rutifolia, Lychnis coronaria,* and *Senecio cineraria* and *S. vira-vira.*

DUTCH BULBS. See Bulbs or Bulb Plants.

DUTCH HOE. See Hoes and Hoeing.

DUTCHMAN'S BREECHES is *Dicentra cucullaria.*

DUTCHMAN'S PIPE. See Aristolochia.

DUVALIA (Duvàl-ia). The small leafless succulents that comprise this genus number fifteen or more species, chiefly natives of South Africa, but a few indigenous to other parts of Africa and one to Arabia. They belong in the milkweed family ASCLEPIADACEAE and are related to *Stapelia.* Their name honors the French botanist Henri Auguste Duval, a student of succulent plants, who died in 1814.

Duvalias have brief fleshy stems, prostrate with upturned ends or erect. In their native habitats they are often partly subterranean. The stems have four to six angles, coarsely-toothed and with each tooth terminating in a tiny spinelike tip that is actually a rudimentary leaf. The flowers, much resembling small stapelias, are in small clusters or solitary from the bases or middles of young stems and are somehow reminiscent of starfish. They are distinct in details from those of stapelias and closely related genera. They have two coronas (central crownlike portions). The outer one is disk-shaped and is surrounded by a raised circle formed by a thickening of the corolla tissue. There are five spreading corolla lobes (petals). For usually at least one-half and often the whole of their lengths from their tips downward, those of most kinds are severely folded backward along their centers. The fruits are erect, smooth, podlike, follicles.

Two South African species, *D. elegans* and *D. pubescens,* have stems with four or five angles and the entire inner surfaces of their corollas hairy. In the first the petals are folded backward only at their apexes, in the former along their entire lengths. Their stems are up to 2 inches long and ⅓ to ⅔ inch thick. In ***D. elegans*** the flattish flowers are two to three together from the bases of the stems, ½ to ¾ inch across, and shining dark purple-brown with a pale brown inner corona and a nearly circular outer one. The hairs that cover the petals are purple. The blooms of ***D. pubescens,*** ⅔ to nearly 1 inch in diameter, arise from the middles of the stems and are two to four together. The hairs that cover the corollas are chocolate-brown. The inner corona is light brownish-yellow, the outer one five-sided and deep red-brown.

South African ***D. corderoyi*** is the largest-flowered species. It has six-angled stems up to 1¼ inches long. Its blooms, 1¼ to 2 inches wide and two to four together from the middles or lower parts of the stems, have petals folded backward only at their tips or for up to two-thirds of their lengths, and hairy at their bases only. They are dull olive-green darkening toward their tips. The ring surrounding the central corona is also hairy. The inner corona is buff, the outer dull brick-red.

Kinds without hairs on the faces of their corollas, or at most with some minute ones on the rim of the ring surrounding the central corona, are *D. polita, D. modesta, D. compacta,* and *D. reclinata.* All are indigenous to South Africa. In ***D. polita*** the broad petals do not fold backward, and the stems, about ½ inch wide, are up to 2½ inches long. They are six-angled. Three or four together, the flowers, rich purple-chocolate on their insides and glossy toward their centers, have dull green reverses and are 1 inch to 1¼ inches in diameter. The inner corona is dull orange-red, the outer chocolate-red and five-sided. The petals of ***D. modesta*** are bent backward for one-half their lengths and have their edges fringed with purple hairs. The flowers, two to three together from the middles of the stems, are ½ to ¾ inch across and dark chocolate-brown. The coronas are reddish-brown, the outer one obscurely five-sided. The stems of this species are usually not more than 1 inch in length.

Folded sharply backward for their entire length are the petals of *D. compacta* and *D. reclinata.* The flowers of the former are under, those of the other over, ¾ inch in diameter. The stems of ***D. compacta*** do not exceed and are usually considerably less than 2 inches long, those of ***D. reclinata*** are ¾ inch to 3 inches or, according to some reports, considerably longer. In both they are four- or five-angled. The blooms of *D. compacta,* developed from the centers of the stems, are solitary or up to five together and dark chocolate-brown. Only with the aid of a lens can the few minute hairs at the bases of the petal margins be seen. The inner corona is dull orange-red, the outer brownish-red and somewhat five-sided. The blooms of *D. reclinata* originate from the bases or middles of the stems. They are up to 1¼ inches in diameter and are dark brown with the rim of the ring surrounding the central portion of the flower glossy and greenish. The inner corona is brownish-orange to dull orange, the outer one similar and obscurely five-sided.

Duvalia compacta

A dwarf variety of *Picea abies*

Garden Uses and Cultivation. These easy-to-grow plants chiefly interest collectors of succulents. They are grown indoors and, in climates such as that of southern California, in the open. Their uses and needs are the same as for *Stapelia.*

DWARF and DWARFED. In practice these terms as adjectives are not always applied as precisely as they should be. The best usage limits the first to plants that in comparison to others of the same species or genus are unusually low and generally more compact as is the mugo pine (*Pinus mugo mugo*) in relation to the Swiss mountain pine (*P. mugo*).

Dwarfness is an inherent characteristic of dwarf plants, not the result of cultural practices of manipulation. It may or may not be transmitted to their offspring raised from seed. It normally will be continued in offspring that result from any method of vegetative propagation. Not all dwarfs are completely stable. Some (this often occurs with dwarf varieties of *Picea abies*) after remaining dwarf for many years or even decades develop strong leader shoots of the type normal to the species, and if these are not consistently pruned out will gradually revert to the typical species.

Dwarfed plants are those in which dwarfness has been induced by environmental conditions or cultural manipulations. Naturally dwarfed specimens of many species occur at high elevations in mountain regions, in desert and semidesert areas, and in other inhospitable habitats. If transferred to more favorable environments, such specimens can resume the kind of growth normal to their species.

Bonsai are plants artificially dwarfed by methods, developed by the Chinese and Japanese, that involve controlling the environment by strictly limiting the volume of soil available and its quality and by restricting growth by pruning the tops and roots, by removing leaves, and to some extent by constricting the stems with wires.

A dwarfed (bonsai) *Chamaecyparis obtusa*

Dwarfed fruit trees, the most common of which are apples and pears, are produced by grafting varieties of normal growth onto understocks that have the characteristic of materially limiting the amount of growth made by varieties grafted upon them. If scions from such dwarfed trees are grafted onto nondwarfing understocks, they resume the habit of growth normal for the variety. Dwarfed fruit trees are often called, somewhat incorrectly, dwarf fruit trees.

DWARF-DANDELION. See Krigia.

DWARF-MORNING-GLORY is *Convolvulus tricolor.*

DWARFING STOCKS. The understocks upon which plants are grafted or budded in some instances have marked effects on the growth of scions implanted on them. When this takes the form of restricting the growth of the graft to the extent that the resulting tree is dwarfed as compared with the same kind growing on its own roots or on those of other understocks, the understock responsible is classed as a dwarfing one and is commonly referred to as a dwarfing stock.

Dwarfing stocks are much employed for some fruit trees, notably apples, to a lesser extent for pears, cherries, and certain others. A great deal of investigation and experimental work has been done in isolating and developing dwarfing understocks, abroad and in the United States and Canada. The first in-depth studies were made on apple understocks at the East Malling Agricultural Research Station in England. As a result of this and breeding work carried out there, a series of potentially useful understocks including some dwarfing ones were segregated and named. These understocks are known as East Malling or EM stocks. The most dwarfing is 'East Malling IX'. Other apple understocks developed by the East Malling Station in cooperation with the John Innes Horticultural Institution formerly located at Merton in England are identified as MM stocks. Yet others have been developed in the United States and Canada.

DYCKIA (Dýck-ia). Unlike most genera of the pineapple family BROMELIACEAE, the about eighty species of *Dyckia* grow in the ground instead of perched on trees. They are natives of South America. Their name commemorates the German botanist Prince Salm-Dyck, who died in 1861. Dyckias have close affinities with *Puya,* but differ in their petals being united partway from their bases and in technical details of their fruits.

These are stemless, evergreen perennials, with usually clustered rosettes of many rigid, fibrous, semisucculent, sword-shaped or narrowly-lanceolate, usually spiny-edged leaves ending in sharp points and more or less mealy on their undersides. In usually erect, long-stalked spikes or racemes that develop from the sides of the plants rather than their apexes, the rather small, waxy, yellow, orange, or sometimes reddish flowers are borne, usually in spring. They have three commonly completely separate sepals much shorter than the three petals. The latter are joined in their lower parts into a tube. There are six stamens with their stalks joined at their bases to the petals and a short, three-branched, twisted style. The fruits are three-angled capsules containing winged seeds.

Few kinds are at all commonly cultivated, among them ***D. brevifolia*** (syn. *D. sulphurea*). This native of Brazil has leaves 8 inches to 1 foot long with white scales on their undersides and small spines along their margins. There are about thirty leaves to each rosette. Its branchless flower stalks, up to 1 foot tall or sometimes taller, bear spine-tipped, recurved bracts and several to many tubular, yellow blooms, greenish toward their bases and about ¾ inch long. Another Brazilian, ***D. rariflora*** has spiny-margined leaves up to 6 inches long or somewhat longer. The erect flower stalks, up to 2 feet tall, have short, broad, tapering bracts. The orange flowers, about a dozen to each stalk, are under ½ inch long. Somewhat similar to the last, ***D. remotiflora*** has leaves almost twice as long and lighter-colored blooms more distantly spaced on longer, branchless flower stalks, which are downy toward their apexes. The flowers are orange-yellow. This native of Uruguay is in gardens often misnamed *D. rariflora.*

Dyckia remotiflora

A popular hybrid between *D. brevifolia* and *D. leptostachya* named *D.* 'Lad Cutak' is vigorous and free-flowering. Intermediate between its parents, it has narrow leaves, bronzy-green above and green on their undersides, 6 to 10 inches long. The slender, branchless, erect flowering stalks are 2 to 3 feet tall. They have up-pointing, orange blooms. A very ornamental Brazilian, remarkable for its thick leaves, densely clothed with gray scales, and with closely-set, curved spines along their edges, ***D. marnier-lapostollei*** has stemless rosettes of about ten spreading, triangular leaves up to about 4½ inches long by 1½ inches wide. The flower stalk, from a leaf axil below the center of the rosette, and about 1½ feet tall and branchless, has a few, not showy, small blooms.

Other kinds include these: ***D. altissima*** has rosettes up to 3 feet in diameter of 1-inch-wide, brown-spined leaves. The bright yellow blooms are in ample, branched panicles 3 to 5 feet tall. ***D. coccinea*** forms compact clumps of rosettes of scaly, olive-green, erect leaves 4 to 7 inches long. The spikes of reddish-orange flowers are 1½ feet tall. ***D. fosterana*** is a silvery-gray gem with rosettes 3 to 4 inches wide of sharply

Dyckia fosterana

spined leaves. Its flowers are rich golden-yellow. ***D. frigida*** has recurved, linear-lanceolate, spine-toothed leaves up to 3 feet long, and downy, branched spikes of numerous orange-yellow flowers. ***D. leptostachya,*** with green to plum-colored, channeled leaves up to 1½ feet in length,

Dyckia leptostachya: (a) Flowering spike

(b) Flowers

has rich orange flowers in 1½- to 2-foot-long spikes. ***D. microcalyx*** is medium in size and blooms profusely. Its yellow flowers are on tall, branching stalks. ***D. minarum*** forms fairly compact masses of green rosettes 2 to 3 inches in diameter that produce spikes 6 to 8 inches long of orange flowers. ***D. ursina*** is medium in size and has flower spikes 3 to 4 feet tall that, like the blooms themselves, are thickly covered with brown wool. It is a vigorous grower with 1-foot-tall rosettes of bronzy-green leaves, scaly on their undersides. The orange flowers are on erect stalks.

Garden and Landscape Uses. Dyckias are suitable for growing with agaves, aloes, and other desert succulents. They stand considerable cold, but not much frost. They succeed best in warm, sunny, arid climates. In such places, as in the Southwest, they are splendid for rock gardens and similar informal areas. They are also often included in greenhouse collections of succulents, and the smaller kinds can be grown in windows.

Cultivation. Stony, gravelly, or sandy soil, well-drained, porous, and fairly nourishing is to the liking of dyckias. When well established outdoors their roots strike deeply and watering is usually not necessary. Plants in containers and in greenhouses should be watered moderately freely from spring to fall, much more sparsely in winter. Very well-rooted specimens are benefited by occasional applications of dilute liquid fertilizer. Repotting is needed only at intervals of several years. Propagation is by seed and by division.

DYER'S. This word is incorporated as part of the common names of several plants including dyer's greenweed (*Genista tinctoria*), dyer's-rocket (*Reseda luteola*), dyer's woad (*Isatis tinctoria*), and dyer's woodruff (*Asperula tinctoria*).

DYSCHORISTE (Dyschorís-te). Horticulturally little known, *Dyschoriste*, of the acanthus family ACANTHACEAE, consists of 100 species of annuals, herbaceous perennials, and shrubs. It is related to *Ruellia*. The name, in allusion to the stigma being scarcely lobed, comes from the Greek *dys*, poorly, and *choriste*, divided. This genus is widely distributed in warm regions.

Dyschoristes have opposite, undivided, toothless or almost toothless leaves. Their flowers are clustered in the leaf axils. They have persistent, nearly always five-parted calyxes, and two-lipped, five-lobed, tubular corollas. There are two pairs of stamens and a slender style. The fruits are four-seeded capsules.

Native from Texas to Arizona and Mexico, ***D. linearis*** is an erect herbaceous perennial up to 1½ feet tall. Each of its several to many erect, hairy stems have three or four pairs of linear to spatula-shaped leaves up to 2½ inches long. Somewhat over 1 inch in length, the up-facing blooms,

Dyschoriste thunbergiflora

clustered in the leaf axils, are lavender to violet spotted with purple.

A much more showy species from East Africa may be cultivated. This is ***D. thunbergiflora***, a subshrub or shrub 4 to 10 feet tall that has pink to violet, slender-tubed, bell-shaped blooms 1½ inches to 2 inches long and 1 inch wide or wider. Its elliptic to obovate leaves are ½ to 1 inch long.

Garden and Landscape Uses and Cultivation. In regions where it is native, *D. linearis* is sometimes planted at the fronts of flower beds and in rock gardens and other locations. It responds well to reasonably fertile soil and to sun. Just how far north it is hardy is not determined, but it would be worth trying in climates somewhat colder than those of its natural range. The African species described above succeeds in intermediate-temperature greenhouses under conditions that suit begonias, and in warm, frostless climates outdoors. Its care is that of nonhardy ruellias. Propagation is by seeds and by cuttings, and in the case of *D. linearis*, by division.

DYSOXYLUM (Dy-sóxylum)—Kohekohe. Of the 150 to 200 species of the mahogany family MELIACEAE that compose this genus probably only one is cultivated in North America. It is the kohekohe, of New Zealand, the only member of the group indigenous to that land. The genus, native of tropical Asia, Indonesia, Australia, and New Zealand, takes its name from the Greek *dysodes*, malodorous, and *xylon*, wood, referring to the garlic-like odor of some kinds. It comprises trees with alternate, pinnate leaves, usually with a terminal leaflet, and with flowers in panicles, those of some kinds being very fragrant. Each bloom has four or five each sepals and petals, twice as many stamens joined to form a tube, and one style. The often large fruits split to release the seeds, which are more or less covered with a usually highly colored layer called an aril. The seeds of some kinds dangle from the open fruits on white strands.

The kohekohe (***Dysoxylum spectabile***), one of the loveliest trees of its native land, attains a height of about 50 feet and has stout branches and handsome leaves with an odd number of leaflets. In addition to the terminal one there are usually four pairs of opposite or nearly opposite leaflets, ovate to ovate-oblong, leathery, and mostly 3 to 6 inches long. In pendulous panicles up to 1¼ feet long or slightly longer, from the trunk and branches, the waxy-white, linear-petaled flowers are borne. Each is approximately 1¼ inches wide and short-stalked. The calyx is divided into separate sepals. The petals are spreading. The style has silky hairs and is capped with a disk-shaped stigma. The fruits are egg-shaped to roundish capsules, about 1 inch long, containing seeds with brilliant scarlet coats. The wood of this species is extremely long lasting in contact with the ground and is useful for fence posts.

Garden and Landscape Uses and Cultivation. The kohekohe is suitable for warm, frost-free or nearly frost-free regions and is adaptable to ordinary reasonably fertile soils. It is propagated by seed.

DYSSODIA (Dys-sòdia)—Dahlberg-Daisy or Golden Fleece. Represented in the wild from the southern United States to Central America, this genus of strongly-scented annuals, herbaceous perennials, and subshrubs belongs in the daisy family COMPOSITAE. The name *Dyssodia* is from the Greek *dysodes*, stinking, and alludes to the strong odor of the bruised foliage of some kinds.

These plants have alternate or opposite, pinnately-divided leaves and yellow, daisy-like flower heads with a central eye of disk florets encircled by petal-like ray florets. The seedlike fruits are achenes. There are somewhat more than thirty species.

Dyssodia tenuiloba in a narrow border

Dyssodia tenuiloba (flowers)

The Dahlberg-daisy or golden fleece (***D. tenuiloba*** syn. *Thymophylla tenuiloba*) is most commonly grown. Resembling a small-flowered marigold, it has pleasantly-scented foliage and a multitude of bright golden-yellow, upfacing flower heads ½ inch in diameter. An annual, or possibly sometimes a short-lived perennial, it is 6 inches to 1 foot tall and up to 1 foot or more across. Its ¾-inch-long leaves, the lower ones opposite, those above alternate, are finely dissected into linear segments.

Garden Uses and Cultivation. This gay little, long-blooming plant is delightful for the fronts of flower borders and for edgings and lends itself well to display in rock gardens. A sunny location, and well-drained, moderately fertile, but not excessively rich, soil satisfy its needs. Unlike many annuals, it revels in hot weather and blooms continuously for many weeks. Good results are had by sowing seeds in early spring outdoors where the plants are to remain and thinning out the young plants sufficiently to prevent them from becoming too crowded. If the soil is not disturbed, volunteer seedlings are likely to appear where plants were growing the previous year. Earlier blooming can be had by starting the seeds indoors some eight weeks before the weather is warm enough to set the young plants in the garden, and growing the seedlings in flats or small pots in a sunny, cool greenhouse or under approximately equivalent conditions.

E

EARLY BLOOMER is *Echinomastus intertextus.*

EARPOD is *Enterlobium cyclocarpum.*

EARTH STAR. See Cryptanthus.

EARTHING UP. Chiefly a British term, this is the same as hilling up.

EARTHWORMS. The vast good done by earthworms far outweighs the occasional trouble they cause. They are definitely friends of gardeners, but even friends can sometimes be a bit of a nuisance. This is as true of *Lumbricus terrestris* and most of its kin as of *Homo sapiens.* Surely too familiar to need description, earthworms are soft-bodied, slender, legless, mostly pinkish creatures formed of a number of cylindrical segments or rings. As every boy who has sought them to impale on a bent pin or fishhook knows, they are most plentiful where there are abundant supplies of organic matter and moisture. They thrive and multiply astonishingly well in rotting leaf mold, compost, and well-decayed manure.

Earthworms burrow in soil to great depths. Their tunnels, which may extend to as much as 8 feet below the surface, greatly improve aeration and drainage. Into them the worms drag pieces of dead leaves and other vegetation thus increasing the humus content and making the earth agreeable to plant roots way below regions reached by spade, rototiller, or plow.

Earthworms are bisexual. Each is both male and female. Yet they do not fertilize their own ovules. They mate and as a result produce capsules of many eggs that in due time hatch into young that except in size resemble the parents. The food of these busy creatures is organic matter. This passes through their bodies and is excreted as a very rich and fertile additive to soils. The deposits are called casts. Remaining below ground during the day, earthworms surface at night and scavenge for food. Just how, from the depths of their burrows, they know night from day does not seem to be known. They also come up if their tunnels are flooded.

Valuable manipulators of the soil though earthworms be, little is gained by adding them to ground where there are naturally few. If the earth is favorable to earthworms they will be there aplenty. If not, the introduced worms will neither prosper nor multiply. The best way of promoting the earthworm population is by increasing the organic content of the soil and by seeing that it is not excessively dry for long periods.

The chief objections to earthworms outdoors is that on lawns, golf greens and similar areas their casts are unsightly. Also moles, which more seriously disfigure lawns, seek them as food and are attracted to places where they are plentiful. Sweeping lawns each morning disperses the casts and in doing so fertilizes the grass. It does not of course discourage or eliminate the worms. That objective can be

(b) Disturb roots by tunneling through the soil

Earthworms: (a) Disturb the surface soil of potted plants

achieved by treating the soil with lead arsenate (a poison) at the rate of five to ten pounds to a thousand square feet or by spreading mow-rah meal about.

Despite claims to the contrary by earthworm enthusiasts these creatures harm potted plants by disturbing the roots and clogging drainage. Discourage them. If found, turn the plants out of their containers, pick out the worms and return them to the ground outdoors. Or persuade them to surface by watering the earth with lime water (made by steeping newly slaked lime, at the rate of a pint to two gallons, in water) and dispose of them in the same manner. Lime water can be obtained at drug stores. When plunging pot plants (burying them to the rims of their containers) in soil outdoors always place a piece of slate, tile or flat stone beneath the drainage hole to deny access to worms.

EARWIGS. Rear appendages that resemble a pair of forceps identify the nocturnal beetle-like insects called earwigs, creatures with biting mouth parts that feed on decayed and live vegetation and on insects and other small game. The species that is most pestiferous in gardens and that sometimes invades houses is the European earwig. Reddish-brown, comparatively slender, and up to ⅘ inch in length, this lively pest has forceps about one-quarter as long as its body. In babyhood and youth it feeds on tender shoots and foliage of a wide variety of plants, later it is likely to invade flowers including chrysanthemums, dahlias, sunflowers, and many others to dine on petals, stamens, and other succulent parts. It is also partial to ripening fruits, particularly peaches and nectarines.

Adult earwigs overwinter in debris, under stones and planks, and in crevices in bark and other hiding places. Eggs are laid in spring, and often a second batch is laid later in the season in the soil or beneath stones. The female, a good mother for she sits on her eggs like a broody hen and watches over her young for some time after they hatch, differs from the male in her forceps having nearly straight rather than strongly curved jaws. Young earwigs are without wings. Later they develop them and can fly although in the main they prefer to run. Earwig populations can be reduced by sanitary measures, such as the prompt clearing away of all dead tops of plants and all refuse that can provide hiding places. Earwigs are fairly easily trapped by stuffing flower pots with hay or straw and inverting them over the tops of stakes set among infested plants. The creatures crawl into the pots to hide through the day. Then lift the pots off the stakes and shake any earwigs they contain into a container of kerosene. Bran baits containing sodium fluosilicate are effective, but are poisonous and unless carefully used can injure pets and other creatures. An old-fashioned belief reflected in their name to the contrary, earwigs do not crawl into the ears of sleeping people and bore into their brains.

EAST MALLING STOCKS. These are particular understocks used for grafting apples. See Dwarfing Stocks.

EASTER. The word Easter is a part of the common names of some plants including these: Easter bells (*Stellaria holostea*), Easter cactus (*Rhipsalidopsis gaertneri*), Easter lily (*Lilium longiflorum*), Easter-lily (much less commonly for *Erythronium*), and Easter orchid (*Cattleya mossiae*).

EASTERN-BORAGE is *Trachystemon orientalis.*

EBENACEAE—Ebony Family. Mostly tropical and subtropical, this dicotyledonous family of three genera and 500 species consists prevailingly of very hard-wooded trees and shrubs, several of the former yielding valuable lumber. Members of this group have opposite or whorled (in circles of three or more), mostly leathery undivided leaves. The symmetrical, bisexual or unisexual flowers, solitary or in small groups from the leaf axils, have usually three- to seven-cleft calyxes and corollas, the former persistent, and two to three times as many stamens as petals. There are two to eight styles. The fruits are technically berries. Persimmons are the most familiar sorts. The best-known genera are *Diospyros, Maba,* and *Royena.*

EBENOPSIS. See Pithecellobium.

EBENUS (Éb-enus). Eighteen species are recognized in this genus of the pea family LEGUMINOSAE. Indigenous to the Mediterranean region, Asia Minor, and Baluchistan, these subshrubs and herbaceous plants bear a name applied by Hippocrates to a plant of the same family. They have alternate, pinnate leaves with three or other odd number of leaflets, and pea-like flowers with conspicuously exposed awl-shaped calyx lobes, in axillary, densely-clustered, rounded, clover-like heads or spikes. The fruits, one- or two-seeded pods, are contained in the calyxes.

Inhabiting rocks and cliffs in Crete, ***Ebenus cretica*** is an evergreen subshrub 1½ to 2 feet tall. Its younger parts are densely clothed with silky hairs. The leaves are of three or five leaflets that although technically pinnately arranged appear to spread palmately (like the fingers of an open hand). The leaflets are elliptic-oblong, ½ to slightly over 1 inch long, and have basal appendages (stipules) cleft to form two lobes at their apexes. The stalks of the dense, 2- to 3-inch-long, cylindrical racemes of bright pink or purplish blooms are scarcely longer than the leaves. The flowers are ⅓ to a little over ½ inch long and have side petals and keels of about equal length. A herbaceous perennial, somewhat woody at its base, with leaves of five or seven elliptic-oblong to obovate leaflets ¼ to ¾ inch long, and with the apexes of the stipules three- or four-cleft, ***E. sibthorpii*** inhabits cliffs and rocks in Greece. Its reddish-purple flowers, as large as those of the last species, are in rounded heads 1 inch to 1½ inches in diameter, on stalks at least as long as the leaves. The keel of the bloom is distinctly shorter than the wing petals.

Garden and Landscape Uses. In California and other places with warm, dryish climates these plants may be grown at the fronts of flower beds and in rock gardens and similar places. They thrive in well-drained soil in full sun and are propagated by seed.

EBERLANZIA (Eber-lánzia). About ten species closely related to *Ruschia* compose this genus. All are South African. They belong in the *Mesembryanthemum* complex of the carpetweed family AIZOACEAE and are nonhardy, succulent subshrubs. The name honors F. Eberlanz, of South Africa. From *Ruschia* eberlanzias are distinguished by technical details of their fruits and by the persistent flower clusters that ordinarily include a number of long vicious thorns, which are modified, sterile flower stalks. In their desert homes these serve to protect the plants from browsing animals.

Generally rounded and compact in the wild, eberlanzias in cultivation not uncommonly tend to become straggly. They have woody roots and erect stems. Their opposite, three-angled to nearly cylindrical, gray- or blue-green leaves, joined at their bases, are besprinkled with tiny dark dots. The blooms, in branching clusters, are small to medium in size. Pink, red, or purplish, they are daisy-like in appearance, but unlike daisies, each is one bloom, not a composite head of many florets.

Most likely to be cultivated, ***Eberlanzia spinosa*** (syn. *Mesembryanthemum spinosum*) has forking stems up to 2 feet tall. Its fierce thorns are branched, the branches spreading at right angles. The up-curving or erect, slender leaves are ½ to 1 inch long. The flower clusters are two- or three-times branched, with the central axis represented by a spine ½ to ¾ inch long. The flowers, in pairs, are pink, and ½ to ¾ inch wide.

Garden Uses and Cultivation. Except to collectors of succulents this genus is not likely to have any strong appeal. In warm, desert regions it may be grown outdoors and is suitable for greenhouses devoted to succulent plants. Well-drained earth and full sun are requisites. In greenhouses a winter night temperature of 40 to 50°F, and slightly more by day, is agreeable.

The atmosphere should at all times be dryish, and on all favorable occasions the greenhouse should be ventilated freely. In summer these plants benefit from being placed outdoors. Watering is done sparingly to moderately from spring to fall; very little water is given in winter. Propagation is by seeds and by cuttings. For more information see Succulents.

EBONY. True ebony is *Diospyros ebenum.* Mountain-ebony is *Bauhinia variegata,* Texas-ebony *Pithecellobium flexicaule,* West-Indian-ebony *Brya ebenus.* Ebony spleenwort is *Asplenium platyneuron.*

EBRACTEOLA (Ebract-èola). There are three species of this *Mesembryanthemum* relative, all natives of South Africa. They belong in the carpetweed family AIZOACEAE. Their name derives from the Latin *e,* without, and *bracteola,* a bract. They are dwarf perennial succulents.

Ebracteolas are small, thick-rooted, and have opposite, very fleshy, narrowish, three-angled to slender, sausage-shaped leaves, crowded without spaces between successive pairs along very short shoots. The blooms, which look like the flower heads of daisies, are very different. Each is a single flower, whereas daisy flower heads are conglomerations of numerous little flowers called florets. In *Ebracteola* the blooms are short-stalked, solitary, and at the shoot ends. Red, violet-pink, or white, they come in summer.

Native to southwestern Africa, ***E. montis-molkei*** is clump-forming. It has shoots under 1 inch long with four to eight densely-semitransparent-dotted, gray-green leaves ¾ to a little over 1 inch long, ¼ inch thick, and joined at their bases for about one-third of their lengths. They are more or less boat-shaped, strongly three-angled, and have lanceolate upper sides. The nearly stalkless, bright purple-pink blooms are over ½ inch across.

Garden Uses and Cultivation. Appropriate for collectors of choice succulents and adaptable for outdoor cultivation only in mild desert climates, ebracteolas thrive under conditions that suit most dwarf, nonhardy succulents. Indoors, winter night temperatures of 45 to 50°F are appropriate, with a few degrees increase by day. In summer normal outdoor temperatures suit. At all seasons dry atmospheric conditions are needed, and free ventilation, on all favorable occasions. The soil must be very sharply drained. From spring to fall watering is sparingly moderate and even less generous in winter. Propagation is easy by seed and by division. For further information see Succulents.

ECBALLIUM (Ecbàl-lium)—Squirting-Cucumber. A rampant perennial vine, of the gourd family CUCURBITACEAE, is the sole representative of *Ecballium.* Its name comes from the Greek *ekballein,* to cast out, and alludes to the interesting fruits that, if touched lightly or otherwise disturbed when ripe and turgid, eject their seeds explosively in a stream that carries 6 to 12 feet or more. The squirting-cucumber, a native of the Mediterranean region, is the source of the drug elaterium, a powerful purgative prepared from the fruits.

Echballium elaterium

The squirting-cucumber (***E. elaterium***) is a weak-stemmed, trailing or clambering vine without tendrils or other means of attaching itself to supports. Its grayish leaves, like its stems, are rough-hairy. They are heart-shaped and somewhat three-lobed. The flowers are yellow and unisexual. The females are solitary and mostly arise in the same leaf axils as the males, which are in racemes. The blooms have a five-lobed calyx, a deeply-five-lobed, wheel- or shallowly-bell-shaped corolla, three separate stamens (two represent united pairs), and a short style with three stigmas. The prickly fruits, about 2 inches long, are borne at right angles to their stems. They contain many small, black seeds.

Garden and Landscape Uses and Cultivation The squirting-cucumber thrives in exposed locations and near the sea. It can be used effectively as a summer groundcover and for temporary screening. Its cultivation is similar to that of melons and cucumbers and the same as recommended for *Lagenaria.* Because it is without means of attaching itself to supports it must be tied to them when it is grown other than as a groundcover.

Ecballium elaterium (fruit)

ECCREMOCACTUS (Eccremo-cáctus). Three species by some authorities included in closely related *Epiphyllum* constitute *Eccremocactus,* of the cactus family CACTACEAE. Tree-perchers (epiphytes) rather than ground plants, they have flowers that open at night and close early the next day. The name comes from the Greek *ekkremes,* hanging from, and cactus.

Eccremocactuses have flat, leaflike, jointed stems with more or less spiny areoles along their margins. The flowers are solitary from areoles without spines located toward the ends of the stems. They have short, scaly, but spineless tubes. The withered blooms do not drop, but remain attached to the fruits.

Costa Rican ***E. bradei*** has pendent to upright stems with joints 6 inches to 1 foot long, 2 to 4 inches wide. The center rib is raised. The spines, sometimes absent from cultivated specimens, are solitary or in twos or threes. The short-tubed, funnel-shaped flowers are up to 3 inches long. Their outer petals are pinkish, the inner ones white. Also Costa Rican, ***E. imitans*** has stems up to 3 feet long with branches 3 to 6 inches long by ¾ inch to 1¼ inches wide and usually spineless. When spines are present they are solitary or in twos or threes from each areole; reddish-brown, they are under ⅜ inch long. The flowers are cream and up to 2¾ inches long. Native to Ecuador, ***E. rosei*** has stems up to 3 feet long by up to 3¼ inches wide, and branches with short, semicircular lobes. Mostly in threes, the white to dark brown spines are usually under ⅛ inch long. The greenish-cream flowers are about 2¾ inches long.

Garden Uses and Cultivation. These appeal to cactus fanciers and require the same treatment as epiphyllums. For more information see Epiphyllum, and Cactuses.

ECCREMOCARPUS (Eccremo-cárpus) — Chilean Glory Flower. The five species of *Eccremocarpus,* of the bignonia family BIGNONIACEAE, are natives of Chile and Peru. They are evergreen, more or less woody vines. The name, alluding to the fruits, is from the Greek *ekkremes,* pendent, and *karpos,* a fruit.

The opposite, twice-pinnately-lobed or pinnate leaves of eccremocarpuses terminate in branched tendrils. Their asymmetrical blooms, mostly in racemes, are yellow, orange, or red. They have a five-parted, bell-shaped calyx, a sometimes two-lipped, tubular corolla, four stamens in pairs of two sizes, which do not protrude, and a slender style ending in a two-lobed stigma. The fruits are ovoid capsules.

Best known is ***E. scaber,*** a fast grower that attains heights of 9 to 15 feet and makes pretty displays of foliage and flowers. Its stems are more herbaceous than woody, except that where the plant is perennial they become woody toward the base. The leaves have asymmetrically ovate, usually irregularly lobed or toothed leaflets up to 1¼ inches long. In racemes 4 to 6 inches long, of up to a dozen, the nodding flowers, with 1-inch-long corollas bellied on one side, nearly closed at their throats, and with small, rounded lobes (petals), are typically bright orange-red, but vary in such varieties as *E. s. aureus,* which has golden-yellow blooms, *E. s. carmineus,* in which the flowers are carmine-red, and *E. s. coccineus,* with scarlet blooms. The bladdery seed pods are 1½ inches long by about one-third as wide.

Eccremocarpus scaber

Garden and Landscape Uses. Hardy in mild, nearly frostless regions only, Chilean glory flower can easily be treated as an annual and bloom the first year. It is charming for furnishing supports that afford holds for its tendrils, such as trellises, arbors, wires or strings stretched along the faces of walls, and teepees of branches of brushwood with their sharpened ends thrust into the ground.

Cultivation. Sunny locations and well-drained, rather sandy, fertile soil are most to the liking of this vine. Where hardy, it needs no special attention other than thinning out crowded growths and shortening others. This may be done in fall or early spring. In regions where the tops are killed, but the roots are likely to survive the winter, they may be protected by hilling soil over them in late fall and removing it in spring. To grow this vine as an annual in colder regions sow seeds indoors in a temperature of about 60°F in February or March. Transplant the seedlings to small pots and keep them growing in a greenhouse or under similar conditions until it is safe to set out such frost-tender plants as tomatoes, dahlias, and geraniums. Then plant the eccremocarpuses where they are to bloom.

ECHEVERIA (Echev-èria). The genus *Echeveria* is composed of about 100 species of usually evergreen, succulent, herbaceous perennials and subshrubs of the orpine family CRASSULACEAE. It inhabits warm-temperate and subtropical parts of the Americas from Texas to Argentina, with Mexico the main center of distribution. None is native to California, Baja California, or Arizona. In addition to the species there are numerous hybrids and horticultural varieties. The name commemorates a Mexican botanical artist, Athanasio Echeverriay Godoy, who died in 1797. Plants previously named *Courantia, Oliveranthus,* and *Urbinia* are combined in *Echeveria.* Some species previously included in *Echeveria* belong in *Cremnophylla, Dudleya,* and *Graptopetalum.* The common name hen-and-chickens is applied to some sorts of *Echeveria* and also to kinds of *Sempervivum* and *Bellis.*

Echeverias have thick, branched or branchless stems or are essentially stemless. Their generally fleshy, usually broad, often attractively colored leaves are commonly in compact rosettes, but sometimes are more loosely arranged in spirals around the stems. According to kind, they are from about ¾ inch to 1½ feet long, and from ¹⁄₁₀ inch to 1 foot wide. Their apexes are usually tipped with a short, projecting point. The manner in which they are joined to the stems is distinctive and helpful in identifying the genus. Their broad, thick bases, slightly or more obviously contracted into stalks, fit tightly against the stem, but only a small core of tissue forms the actual attachment. Often red, the leaf margins rarely are wavy or toothed. The branched or branchless flower stalks are lateral, sprouting from the axils of leaves below the centers of the rosettes. They carry fleshy bracts, which are generally similar in shape to the rosette leaves, but much smaller. The flowers are usually in one-sided, branched or branchless racemes, panicles, or spikes; more rarely they are in pairs or solitary. They are urn-shaped to cylindrical, most often reddish to orange, less frequently paler yellow to pinkish. They have a calyx of five often unequal lobes or separate sepals, a strongly-five-angled, fleshy corolla of five erect, strongly-keeled petals deeply hollowed on their faces and joined at their bases to form a short tube, spreading only at their tips. Appendages to the petals, characteristic of related *Pachyphytum,* are rarely present in *Eicheveria.* There are ten separate, nonprotruding stamens, in two circles of five. The female organs are five separate or nearly separate, bottle-shaped carpels, each with a single style, and each with at its base a nectar gland commonly called a scale. The carpels spread as they ripen into follicles, which are the fruits. The preferred natural habitats of echeverias are often cliffs, rocks, lava flows, and steep slopes, but some few are epiphytes (tree-perchers), and some are alpines, growing at high altitudes up to the snowline.

Stemless or very short stemmed species more or less commonly cultivated are now presented in alphabetical sequence. Unless otherwise mentioned all are natives of Mexico. ***E. affinis*** is very short, rarely branched. It has many oblanceolate, short-pointed, brownish-olive-green leaves with green bases, 2 inches long, upturned above their middles, their undersides convex. The flowers are red, in three- to five-branched, flat clusters topping stems 4 to 5 inches tall. ***E. agavoides*** is stemless, rarely clump-forming. Its rigid, apple-green leaves with brown tips and sometimes red margins, are pointed-ovate, 1½ to 3½ inches long and up to 1½ inches wide. They are in compact rosettes of up to twenty-five. The flowers are reddish with yellow apexes, on usually two-branched,

Echeveria agavoides

pink stalks up to 1½ feet tall, each branch with five to eight blooms. ***E. carnicolor,*** of Mexico, is short-stemmed and hairless. In rosettes of twenty to thirty, its green to metallic-bluish, oblanceolate-spoon-shaped leaves are 1 inch to 2 inches long and up to ¾ inch wide. The flowers, in racemes

Echeveria carnicolor

of six to twenty that with their stalks are from 6 to 10 inches long, are flesh-colored and salmon-pink and about ½ inch long. ***E. craigii*** is stemless or short-stemmed. It has many semicylindrical, upcurving, green and brownish leaves up to 4½ inches long by ¾ inch wide, their upper surfaces flat, undersides rounded. The flowers, their outsides pink, their insides red, are in panicles with many short branches that with their stems are about 1½ feet tall. ***E. cuspidata*** is stemless. Its very numerous leaves are thinnish, pointed-triangular-ovate, bronzy-glaucous, up to 3½ inches long by 2 inches wide. The flowers are in paired racemes, with their stalks 9 inches to 1½ feet tall. They are coral-red with orange-buff interiors. ***E. derenbergii*** is short-stemmed. It has tallish rosettes in clusters. The leaves are crowded, thick, obovate-wedge-shaped, green with red edges, glaucous, and up to 2½ inches long by 1 inch wide. The usually branchless flower stems are up to 4 inches tall. The red-tipped flowers are yellow. ***E. elegans*** is stemless. Its globose rosettes of thick, alabaster-white to pale glaucous-green leaves make offsets freely. The leaves, flat to slightly hollowed above and convex beneath, are up to about 2 inches long by ¾ inch wide. The flowers, rose-pink with yellow-orange tips and yellow inside, are in branchless pink-stalked racemes. ***E. halbingeri,*** stemless and eventually with offsets, has thick, obovate, somewhat glaucous, green leaves up to 1 inch long by one-half as wide. The branchless racemes,

(b) Glaucous-green-leaved form

Echeveria elegans: (a) White-leaved form

up to 6 inches long, are of six to nine red-orange flowers with the tips of the petals strongly recurved. ***E. heterosepala*** (syn. *Pachyphytum heterosepalum*) has a very short stem. Its slightly glaucous, pea-green leaves, tinged with purplish-brown, are rhomboid-lanceolate and 1½ to 3 inches long by up to ¾ inch wide. The branchless racemes are of flowers, often with appendages on the petals similar to those of *Pachyphytum,* that are green in bud, later red. ***E. obtusifolia*** (syn. *E. scopulorum*), stemless or short-stemmed, has rather open rosettes of thinnish, round-ended, obovate leaves 3 to 4 inches long. Prevailingly light green, they become reddish as

Echeveria obtusifolia

they age. From 8 inches to over 1 foot tall, the flowering stalks carry three racemes of red blooms ⅓ to ½ inch long. ***E. peacockii,*** stemless or short-stemmed with rosettes of numerous obovate, pale-glaucous, blue-white leaves, often with red edges and apexes, is up to 2½ inches long by 1¼ inches wide and slightly hollowed above, slightly keeled beneath. Up to 1 foot tall, the many-flowered racemes of pendulous, strongly-five-angled, pink blooms, their tips in the bud stage blue, have powder-blue stalks. ***E. pilosa,*** its stem short, but evident, is densely clothed with spreading, mostly white hairs. In loose rosettes, the oblanceolate, thick leaves are up to about 3 inches long by 1¼ inches wide. About 1 foot tall and with white-hairy stalks, the panicles are of orange-red to salmon-orange blooms. ***E. purpusorum,*** very distinct, makes no or few offsets. Stemless or nearly so, this kind has thick, rigid, sharp-pointed, ovate leaves about 1½ inches long by one-half as wide, each with a strongly defined mid-line, and green, with closely mottled brown spots. The reddish-brown flower stalks have rose-pink to red blooms; their apexes and interiors are yellow. ***E. runyonii,*** stemless or short-stemmed, has rosettes of upcurving, spatula- to wedge-shaped, very blue-glaucous leaves, with usually blunt, often notched apexes. They are sometimes over 3 inches

Echeveria peacockii

Echeveria runyonii

long by 1½ inches wide. The two-branched racemes, 6 to 8 inches tall, are of very short-stalked, pink to scarlet blooms. ***E. secunda*** has short, prostrate stems and rosettes of up to sixty 1- to 3-inch-long, more or less spatula-shaped, blunt leaves, about one-half as wide, and ending in a short point. Glaucous, sometimes with a purplish or reddish edge, they have convex upper surfaces. Pink stalks 5 inches to 1 foot tall carry up to fifteen yellow-tipped, old-rose-red to red flowers up to ½ inch long, with spreading, unequal-sized sepals. This species makes offsets very freely. *E. s. byrnesii* (syn. *E. byrnesii*) has smaller rosettes with not glaucous, green rosettes sometimes suffused with red. *E. s. glauca* (syn. *E. glauca*) has very glaucous leaves, broader and perhaps thinner than those of

Echeveria secunda byrnesii

Echeveria secunda glauca

the typical species. *E. s. pumila* (syn. *E. pumila*) has narrower, flatter leaves than the typical species, and the sepals of its flowers are of nearly equal size. ***E. setosa*** is stemless and has numerous offsets. Its foliage and flower stalks are densely clothed with bristly, white hairs. The many leaves, convex on both surfaces and oblanceolate-spatula-shaped, are up to 2 inches long by ¾ inch wide. The flower stalks, branchless or two-branched, may exceed 1 foot in height. The blooms are yellow, their outsides suffused with red. ***E. shaviana*** is stemless. Its numerous, thinnish, sometimes toothed, white-mealy, ovate-spatula-shaped leaves are up to about 2 inches long by one-half as wide. The racemes, up to 1 foot high are not branched. The flowers are rose-pink. ***E. turgida*** is stemless and offset-forming. Its crowded, upcurved, heavily mealy leaves, about 2 inches long by one-half as wide, are oblong-wedge-shaped. Blunt-ended except for a short point, they are flattish with thick edges. The racemes, up to 8 inches tall, are without branches. The blooms are pink outside, deep chrome-yellow within. ***E. whitei,*** native to Bolivia, has a short stem. Its brownish-green, thick leaves, 1½ to 2 inches long or longer and ¼ to ¾ inch wide, are flat above, rounded beneath, and pointed-obovate-oblong. In loose racemes with orange-brown stalks, the flowers are peach-red to bright coral-red.

Echeveria secunda pumila

Echeveria setosa

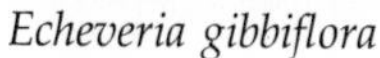

Echeveria gibbiflora

Echeveria gibbiflora carunculata

Shrubby kinds with decidedly long stems include the following, all unless otherwise stated natives of Mexico. ***E. coccinea,*** 1 foot to 2½ feet tall, branched above, and densely-hairy, has pointed, oblanceolate leaves, rounded beneath and concave on their upper sides, and up to 3½ inches long by ¾ inch wide. The older leaves are reddish. Up to a dozen are loosely arranged at the end of each branch. The scarlet, sharply-five-angled blooms are many in slender spikes 6 inches to 1 foot long or longer. ***E. gibbiflora*** has a thick branchless or nearly branchless stem up to about 1 foot tall, topped by a loose rosette of broadly-obovate, purplish-tinged, yellowish-olive-green leaves up to nearly 1 foot long by 6 inches wide, their margins often wavy. Sometimes 3 feet tall, the panicles are of light red flowers. *E. g. carunculata* has conspicuous warty protuberances toward the bottoms of the leaf blades. *E. g. decora* has leaves variegated with white, rose-pink, and green. *E. g. metallica* has purplish-lilac young leaves. Its older leaves are nearly circular, and up to 7½ inches long by 5 inches wide. *E. g.* 'Wavy Leaf' has finely-waved leaf margins. ***E. harmsii*** (syn. *Oliveranthus elegans*), one of the most decorative echeverias, is minutely-softly-hairy on all its parts. It has freely-branched stems with leaves loosely clustered at the branch ends. The leaves are broad-lanceolate to spatula-shaped, thick, flat or slightly hollowed above, keeled beneath, and ¾ inch to nearly 2 inches long. The flowering stalks, forked or not, and 4 to 8 inches long carry one to three scarlet flowers with yellow interiors. Narrow-urn-shaped and five-angled, they are up to 1½ inches long. ***E. leucotricha,*** a subshrub up to 6 inches tall or taller, is much like *E. pulvinata*, but differs in its longer, narrower leaves being clothed with whiter hairs and in its panicles of blooms being more spikelike. The thick leaves, in loose rosettes, are up to 4 inches long by 1 inch wide, but often smaller. The cinnabar-red blooms, ¾ inch long, are orange-yellow inside. ***E. macdougallii*** (syn. *E. sedoides*), up to 1 foot tall, as its synonymous name suggests, has much the appearance of a *Sedum*. Its branches spread or droop. Its club-shaped, slightly-three-angled, green leaves, tipped with maroon-red, are widely spaced rather than in rosettes. They are under ½ inch long by about two-thirds as wide. The flowers have strongly-five-angled, scarlet corollas edged with yellow, on their insides apricot-yellow. They are ½ inch long or longer. ***E. pulvinata*** has its parts covered with fine, soft hairs. Somewhat branched and 4 inches tall or taller, its stems are clothed with rusty hairs. The leaves, loosely arranged near the ends of the branches and ¾ inch to 2¼ inches long, are thick and obovate. Their upper sides are flat or slightly hollowed, their undersides roundish. The flower stems bear ¾-inch-long, red or yellowish-red blooms, which are yellow in their interiors. From rather similar *E. leucotricha* this differs in its hairs becoming reddish in the sun rather than remaining white, and in its more oblong, broader leaves. Variety *E. p.* 'Ruby' differs from *E. pulvinata* in the hairs of its leaves being more definitely rusty-red.

Echeveria gibbiflora (flowers)

Echeveria gibbiflora carunculata variety

Echeveria pulvinata

Echeveria 'Set-Oliver'

Many named and unnamed hybrid echeverias are cultivated. Not infrequently they are misidentified and grown under the names of species. Commonly hybrids exhibit characteristics intermediate between those of their parents. One of the most popular, ***E. imbricata,*** its parents *E. secunda glauca* and *E. metallica,* is often misnamed *E. secunda glauca.* Other favorites are *E.* 'Pulv-Oliver', its parents *E. pulvinata* and *E. harmsii; E.* 'Set-Oliver', its parents *E. secunda* and *E. harmsii;* and ***E. derosa,*** its parents *E. derenbergii* and *E. setosa.* Additional hybrids include *S. graessneri,* its parents *E. derenbergii* and *E. pulvinata;* ***E. gilva,*** its parents *E. agavoides* and *E. elegans,* which resembles its first-named parent, but has less pointed, narrower leaves and forms offsets more freely; and ***E. haageana*** its parents *E. agavoides* and *E. derenbergii.* Beautiful ***E. hoveyi*** has rosettes of 2-inch-long, spatula-shaped, gray-green leaves longitudinally striped with irregular pink markings. This is apparently a variant of ***E. zahnii,*** which has scattered pink spots on its leaves and may be of hybrid origin. There are hybrids between *Echeveria* and allied genera. Here belongs *Graptoveria,* the parents of which are *Graptopetalum* and *Echeveria.* Hybrids between *Cremnophila* and *Echeveria* are named *Cremneria,* those between *Pachyphytum* and *Echeveria* constitute *Sedeveria.*

Echeveria 'Set-Oliver' (flowers)

Echeveria 'Pulv-Oliver'

Garden and Landscape Uses. In climates where they are hardy, that is in frostless or essentially frostless, dry or dryish ones, echeverias are splendid for rock gardens, banks, slopes, dry walls, and other places where soil drainage is rapid and succulents can be displayed advantageously. They are also suitable for window boxes, pots, and other containers. Where winters are too severe for them to live outdoors they can be lifted before the first fall frost and stored in a greenhouse, light cellar, window, or similar place, to be planted outdoors the next spring. Strictly formal flower beds, including the intricate-patterned ones done in the once-fashionable style of carpet bedding are less in favor than formerly. For such beds selected echeverias were commonly employed.

Cultivation. Among the easiest of plants to propagate, echeverias are readily multiplied by offsets, cuttings consisting of

Echeveria gilva

single rosettes, leaf cuttings (these may even be bracts of the flower stalks), grafting (rarely done, but quite possible, using *Echeveria* or *Sedum praealtum* as understocks), and seeds. The latter, collected from hybrids or from plants that have been pollinated by other species or hybrids, will not give plants true to the seed parent. Porous, thoroughly drained soil is a must. It should be reasonably fertile rather than nearly devoid of nutrients, and moist, but surely not wet, from spring through fall. In winter drier conditions are in order. In their native habitats many kinds receive no moisture through the winter, but in cultivation the finest specimens are had where complete dryness is not maintained. This is especially true of pot specimens. Judicious applications of dilute liquid fertilizer to pot plants, especially large-leaved kinds, in summer do much to promote the production of impressive foliage. Echeverias are sun-lovers, but in a bright climate, such as that of California, they tolerate part-day or light shade. For more information see Succulents.

ECHIDNIUM (Echíd-nium). This genus of the arum family ARACEAE, a close relative of *Dracontium*, most likely consists of only one, but perhaps of two closely similar species. It is native to British Guiana. Its name, derived from the Greek *echis*, a viper, undoubtedly alludes to the appearance of the inflorescence.

Echidiums have tubers and erect leaves with three-parted blades and speckled or variegated stalks 1½ to 3 feet long. The scarcely-stalked flowering structures (inflorescences) appear before the leaves, are calla-lily-like in form, and have a central, erect, dull green spikelike spadix on which the insignificant brown flowers are borne. Cupped around the spadix in the manner of the white spathe of a calla-lily, is a pointed, dull greenish-brown spathe 2 to 4 inches long, strongly-ribbed on its almost granular outside. The flowers are not ill-scented. The fruits are berries. Named as species are ***E. dubium*** and ***E. regelianum.*** If these are considered one species, the first name takes precedence.

Echidnium dubium

Garden Uses and Cultivation. A collectors' curiosity, this responds to treatment appropriate for *Amorphophallus rivieri.*

ECHIDNOPSIS (Echid-nópsis). This genus of eight or more species of leafless, succulent-stemmed plants of the *Stapelia* relationship of the milkweed family ASCLEPIADACEAE inhabits Africa, Arabia, and the island of Socotra. It must not be confused with very different and commonly cultivated *Echinopsis* of the cactus family CACTACEAE. The name *Echidnopsis* comes from the Greek *echidna*, a viper, and *opsis*, resembling, and alludes to the angled-cylindrical, often sinuous, snakelike stems that branch freely from their bases and are covered with longitudinal rows of low tubercles (bumps) that give somewhat the appearance of the surface of an ear of corn.

The flowers of *Echidnopsis*, very like those of *Caralluma*, are solitary or clustered in grooves between the tubercles near the ends of the stems. They have a five-parted calyx and a starfish-like or wheel- to bell-shaped, fleshy, five-lobed corolla. In the center of the bloom is an inner corona or crown topping the very short column of stamens. There may or may not be a cup-like or five-lobed outer corona. The fruits are paired, slender, podlike follicles.

Popular among fanciers of succulents, variable ***E. cereiformis*** inhabits both sides of the Red Sea, often favoring limestone soils. Its bluntly-eight-angled stems are ½ inch thick or a little thicker. The bright yellow, very short-stalked, bell-shaped blooms are clustered, ⅓ inch across and without outer coronas. The inner coronas have five fleshy lobes and are yellow. The blooms of *E. c. obscura* are yellowish-brown. Native to Ethiopia, ***E. dammanniana*** differs from *E. cereiformis* in being dark purple with five-lobed outer coronas.

Garden Uses and Cultivation. Succulent fanciers find these interesting asclepiads easy to satisfy. In a warm, dry climate, such as that of southern California, they can be accommodated outdoors in rock gardens. They are also appropriate in greenhouses, and even in window gardens. Sharp drainage and well-aerated, coarse soil is to their liking. The addition of limestone chips is helpful. Indoors, slight shade from the most brilliant summer sun may be advantageous. Water with restraint. Night temperatures in winter of 45 to 50°F, with a daytime increase of five to ten degrees, are adequate. At other seasons higher temperatures are in order, but on all favorable occasions free ventilation must be given. Propagation is by cuttings and by seeds. For additional information see Succulents.

ECHINACEA (Echinàc-ea)—Coneflower. The common name coneflower is applied to this and three other genera, *Dracopis, Ratibida,* and *Rudbeckia,* all once included in *Rudbeckia* and all with flower heads with tall, conelike or cylindrical centers. From these other genera *Echinacea* is distinguished by technical differences in its flower heads and by their ray florets being purple or rose-purple to white. The genus belongs in the daisy family COMPOSITAE. Its name is derived from the Greek *echinos*, a hedgehog, in allusion to the sharp-pointed bracts that protrude from the centers of the flower heads.

There are three species of *Echinacea*, all North American, rather coarse, hardy herbaceous perennials with heavy black roots and sturdy, erect stems. Their leaves are alternate and lobeless, the lower ones usually long-stalked. The flower heads are large and solitary. They terminate long, few-branched stalks. The sterile ray florets dry on the heads without dropping. The fruits are seedlike achenes.

Most commonly seen in gardens, the purple coneflower (***E. purpurea*** syns. *Rudbeckia purpurea, Brauneria purpurea*), is old in cultivation. Known to European gardeners by the end of the seventeenth century, it has been popular ever since. This is a native of open woodlands and prairies from Georgia to Louisiana, Iowa, and Ohio.

Echinacea purpurea

Stout, with hairy stems and foliage, it attains a height 2 to 5 feet. Its stems tend to branch above. The lower leaves are 3 to 8 inches long, long-stalked, broadly-lanceolate to ovate, and coarsely-toothed. Higher, the leaves gradually become narrower, shorter-stalked, and less obviously toothed. The flower heads are up to 6 inches in diameter. In the wild form the rays are typically reddish-purple, varying occasionally to almost white. The centers of the flower heads are dark reddish-brown with conspicuous, protruding, often bright orange-red, pointed bracts. There are improved horticultural varieties that have been selected for their attractive flower colors as well as for the large size and forms of their flower heads. Notable among these are 'The King', with handsome blooms with bright crimson rays, and 'White Lustre', with white ray florets and green centers. The varieties 'White Prince' and 'White King' are closely similar to, if not identical with, 'White Lustre'. Other named horticultural varieties are described in the catalogs of European and American nurseries.

Other kinds occasionally cultivated are ***E. angustifolia*** (syn. *Brauneria angustifolia*), a plains species indigenous from Minnesota to Texas, the Dakotas, Saskatchewan, and in Tennessee, and ***E. pallida,*** a native of prairies from Louisiana, Alabama, and Georgia through the Midwest. By some botanists *E. angustifolia* is considered to be merely a variety of *E. pallida*. The most obvious difference is that the ray florets of the former are spreading, whereas those of the latter are reflexed or drooping. Also, the ray florets of *E. angustifolia* are narrower and shorter than those of *E. pallida,* from 1 inch to 1½ inches long, whereas those of the latter are 1½ to 3½ inches in length. A more robust plant, *E. pallida* sometimes exceeds 3 feet in height, whereas *E. angustifolia* ordinarily is not more than 2 feet tall. Both have toothless, lanceolate to linear leaves, 3 to 8 inches long, narrowing to their bases and the lowermost long-stalked. The ray florets of both range from rose-purple to whitish. Their disk florets are brownish-purple.

Garden Uses. Because they are easy to grow, stand hot weather and humidity well, and provide a good display of blooms useful for garden ornamentation and for cutting, echinaceas merit a place in American gardens. Certainly the improved horticultural varieties of the purple coneflower will find wider acceptance among gardeners than other kinds, but none is to be despised. The garden varieties are admirable for perennial and mixed beds and borders and for inclusion in cut-flower gardens and, because they do not look too "tamed," they as well as the wild species can with propriety be used to decorate informal areas. The natural wild species are excellent for native plant gardens and maintain themselves with minimal attention. Primarily echinaceas are sun-lovers, but they will stand a little part-day shade. They prosper in any ordinary garden soil.

Cultivation. Seeds afford a ready means of increasing the wild species. These may be sown in May in an outoor seedbed or cold frame and the seedlings transplanted as soon as they are big enough to handle comfortably, to nursery beds in rows about 1 foot apart and with 6 inches between plants in the rows. By fall or the following spring the plants will be ready for transferring to their permanent locations. Another means of propagation, and the one practicable for perpetuating selected varieties, is by careful division in early spring or early fall. Ordinarily this should not be done oftener than every three or four years.

In the garden echinaceas may be spaced 1½ to 2 feet apart. Their routine culture is simple. An application of a complete fertilizer each spring is helpful. Weeds must be kept down by shallow surface cultivation or mulching. Faded flower heads should be removed promptly and spent flower stalks cut off. Little or no staking is required and, since the plants stand dry conditions remarkably well, watering is needed only during extended spells of dry weather. In very cold climates a light winter covering of salt hay or branches of evergreens affords protection against root damage caused by alternate freezing and thawing and is advantageous, but not essential. So long as the plants are prospering they may be left undisturbed, but usually it is good practice to dig them up, divide them, and replant them in deeply-spaded and newly-fertilized soil every four or five years.

ECHINOCACTUS (Echino-cáctus)—Barrel Cactus, Golden Barrel. Many plants by some included in *Echinocactus* of the cactus family CACTACEAE are accommodated in *Ancistrocactus, Coryphantha, Echinomastus, Ferocactus, Gymnocalycium, Hamatocactus, Homalocephala, Notocactus, Parodia, Thelocactus,* and other genera. Ten or fewer species, natives of the southwestern United States and Mexico, are retained. Some botanists take a broader view and incorporate additional species. The name, alluding to their extremely spiny character, derives from the Greek *echinos,* a hedgehog, and cactus.

Echinocactuses are impressive and beautiful, fiercely-spiny, spherical to barrel-shaped or massive-columnar plants, mostly of considerable size. They have solitary or clustered stems with few to many often somewhat knobby (tuberculate) ribs. The flowers come from woolly areoles near the centers of the tops of the stems. Short-funnel- to somewhat bell-shaped, they have more or less spreading petals. The oblong to top-shaped, dry fruits have densely-woolly apexes to which the withered blooms remain attached.

The golden barrel (***E. grusonii***) is probably the most familiar species. Few cactus collections of importance are without this very lovely Mexican kind. Well-grown, mature specimens are nearly globular, about 3 feet in diameter, their tops nearly flat. They have twenty to thirty or more sharp ribs, and areoles with white or yellowish wool. In young plants the ribs are deeply notched into separate tubercles, in older ones they are continuous. The awl-shaped spines are in clusters ¾ inch apart or closer, each cluster consisting of eight to ten radials up to 1¼ inches long and three to five centrals 1¼ to 2 inches long. When young the spines are golden-yellow. Later

Echinocactus grusonii, a young specimen

Echinocactus grusonii with flower buds and blooms

An old, cluster-stemmed specimen of *Echinocactus grusonii*

Echinocactus grusonii and columnar cactuses in the Huntington Botanical Gardens, San Marino, California

they become whitish or grayish. From 1½ to a little over 2 inches long, the flowers are yellow with brownish outer petals. Variety *E. g. cristatus* is beautifully crested.

Massive ***E. visnaga*** of central Mexico has a cylindrical to ovoid plant body up to 10 feet tall and 3 feet wide or more with fifteen to thirty pronounced ribs and closely-set clusters of four stout, brown, awl-shaped spines, the upper one up to 2 inches long, the others shorter. The yellow flowers, about 3 inches in diameter, are succeeded by fruits 3 to 4 inches long. Another Mexican, ***E. palmeri,*** is 3 to 6 feet tall and up to 1½ feet or a little more in diameter. Its stems have twelve to twenty-six broad, blunt, slightly-notched ribs with clusters of four to eight radial and four stouter central spines, the longest up to about 3 inches in length. All are yellow except for the brown bases of the centrals. The flowers have fringed yellow petals about ¾ inch in length. The fruits are 1¼ inches long. Mexican ***E. ingens*** has a spherical or short-cylindrical stem up to 4½ feet tall by 4 feet in diameter, very woolly at its apex. Young specimens have five to eight broad, rounded, more or less notched ribs, the old ones more. One inch or a little more apart, the spine clusters have eight radials and one central. The straight, rigid spines are ¾ inch to 1¼ inches long. The flowers, canary-yellow with reddish-yellow outer petals, are 1¼ inches long by up to 2 inches wide. The fruits are woolly, ovoid, and about 1¼ inches long. Nearly spherical with little or no tendency to elongate, a handsome Mexican sort, ***E. platyacanthus*** is up to 2 feet in diameter, almost as tall, and has twenty-one to thirty sharp ribs. Its spines, brownish at first, are later grayish. Each cluster consists of four radials up to ½ inch or slightly more in length and three or four centrals approximately twice as long. The flowers are yellow and about 1¼ inches long. Smaller ***E. horizonthalonius,*** of Texas, Arizona, New Mexico, and Mexico, is globular to short-cylindrical. Rarely exceeding 8 inches in height and diameter, it generally has eight wide, rounded ribs, occasionally fewer or more, which are vertical or spiraled around the stem. Old plants are conspicuously woolly at their tops. The spine clusters, up to ¾ inch apart, are of five to eight radials with, in mature plants, one central. The spines, which overlap and interlock with those of neighbor clusters, are ¾ inch to 1¼ inches long and brownish to reddish fading to gray or nearly black. Brilliant rose-red, the flowers are 2 to 3 inches long and wide. The oblong fruits are up to 1¼ inches in length by approximately ½ inch in diameter. This species, sometimes called eagle claws, seldom makes offsets.

Echinocactus visnaga in the botanical garden, Mexico City

Echinocactus ingens, a young specimen

Garden and Landscape Uses and Cultivation. Echinocactuses are splendid plants for cactus and succulent collections outdoors in suitable warm dry climates and in greenhouses elsewhere. They succeed under conditions and care appropriate for the majority of desert catuses. The golden barrel appreciates a limestone soil and is not averse to a little shade from the fiercest sun. For more information see Cactuses.

ECHINOCEREUS (Echino-cèreus). This genus of the cactus family CACTACEAE is endemic to the southwestern United States and Mexico. The name *Echinocereus,* which alludes to the spiny fruits, comes from the Greek *echinos,* a hedgehog, and the name of the related genus *Cereus.* As is common with genera of the cactus family, botanists differ widely in interpreting it and consequently in their conclusions as to the number of species. There are possibly seventy-five. They include many attractive, easy-to-grow sorts.

Echinocereuses are erect or prostrate low plants with solitary or clustered spherical to cylindrical stems with longitudinal lumpy or smooth ribs. It is rare for the stems to exceed 1¼ feet in length or 3½ inches in diameter. Mostly the flowers are large and showy, generally pink, red, or purple, seldom yellow. The lobes of their stigmas are green. The ovaries and fruits are invariably clothed with spines easily detachable from mature specimens of the usually colored, often edible, thin-skinned fruits. The seeds are black. As a matter of convenience and without representing botanical relationship the kinds described here are grouped according to their general habits as was done by the great cactophile, Professor J. Borg.

Kinds with low, globular to egg-shaped stems include these: ***E. gentryi,*** of Mexico, 2 to 6 inches high, has clusters of 1-inch-wide, five-ribbed stems. The very short spines are in clusters of eight to twelve radials and one central. From 2½ to 3 inches long or a little longer and 2 inches wide, the flowers are bright rose-pink. ***E. knippelianus*** is Mexican. Usually 4 to 8 inches tall, it has solitary, 2-inch-thick, dark green, five- to seven-ribbed stems, which eventually develop weak, yellow spines up to ½ inch long or slightly longer, solitary or in twos or threes. The funnel-shaped blooms, 1 inch to 1½ inches long, have pale magenta-pink to carmine-violet, pointed petals, brown on their outsides. Their stigmas have six or seven lobes. ***E. palmeri*** is a Mexican 2 to 3½ inches tall by ¾ inch to 1¼ inches thick. The spine clusters are of twelve to fifteen spreading, brown-tipped, slender radials and one brown to blackish central ½ to ¾ inch long. ***E. pulchellus,*** of Mexico, has stems at first spherical, but later ovoid or cylindrical. From 1½ to 2½ inches in diameter, branching from their bases, and bluish- or grayish-green, they have eleven to thirteen ribs closely set with clusters of three to five very short, yellow to gray spines. About 1½ inches in length, the blooms, whitish to deep rose-pink and green on their outsides, have toothed inner petals. The stigma has eight or nine lobes. Variety *E. p. amoenus* (syn. *E. amoenus*) has its very short spines in clusters of six to eight. Its flowers are bright magenta-pink. ***E. subinermis,*** also Mexican, has one to few bluish-green to green stems up to about 1 foot long and 4 inches thick with five to nine ribs. The yellowish spines, mostly over $\frac{1}{10}$ inch long, are in clusters of three to eight radials and one central. The blooms, funnel-shaped and up to 3 inches long, are yellow with brown outsides. The stigma has ten lobes. ***E. triglochidiatus melanacanthus*** (syn. *E. coccineus*), of western North America, forms dense, hemispherical clumps up to 6 feet in diameter of eight- to eleven-ribbed stems 2 to 6 inches tall by about 2½ inches thick. They have clusters of five to twelve radial and one to three central, round spines, the former yellowish and ¼ inch to 1¼ inches long, the latter yellow with reddish-brown tips, up to 3 inches long. The orange-scarlet to crimson, funnel-shaped blooms are yellowish at the bases of their petals. They are 1½ to 2 inches long and have stigmas with six to twelve lobes.

Echinocereus gentryi

Sorts with erect, rigid, cylindrical to cylindrical-ellipsoid stems with clusters of small spines that spread in starlike or comblike fashion include these: ***E. adustus*** (syn. *E. caespitosus adustus*), of Mexico, differs from nearly related *E. reichenbachii* in having fewer-ribbed, more slender stems. They are short-cylindrical, not over 4 inches tall and 2 inches thick. The spine clusters are of sixteen to twenty needle-like radials arranged in comblike fashion and none or only one about 1-inch-long central. The purplish flowers are up to 2 inches long. ***E. chloranthus*** of New Mexico, Texas, and Mexico, is quite beautiful. Variable, it has cylindrical, seldom-branched stems up to 10 inches high by 2½ inches in diameter. They have twelve to eighteen low, lumpy ribs. The closely-set spine clusters are of twelve to twenty radials ¼ to ½ inch long and of various colors and up to six centrals of different lengths, the biggest 1¼ inches long. The 1-inch-wide, funnel-shaped flowers, which do not open widely, are yellowish-green with suffusions of brown. ***E. enneacanthus,*** the strawberry cactus of the southwestern United States and Mexico, has clustered flabby stems up to 1 foot tall and 4 inches wide. They have seven to thirteen ribs and rather distantly spaced clusters of usually eight, but from seven to fifteen translucent-white radial spines up to 1 inch long and one to four longer centrals, at first brown, aging to gray. From 2 to 3½ inches wide, the flowers are purplish-red. Variety *E. e. stramineus* forms mounds of very many 2- to 3-inch-wide stems up to 9 inches long and with mostly ten to twelve ribs and closely spaced clusters of straw-colored spines, the radials about ten and up to 1¼ inches long, the two to four centrals up to 3½ inches long. The rather sparingly produced, purple flowers are 4 to 5 inches wide. The stigmas have ten to thirteen lobes. This sort prefers limy soil. ***E. pectinatus,*** a very variable native of the southwestern United States and Mexico, includes several varieties by some authorities recognized as separate species. Typically it has stems up to 1 foot tall and 2½ inches wide, solitary or branched from their bases. The spines, in comblike clusters of about twelve to sixteen radials and usually three to five centrals, are pinkish when young, later white. The 3½-inch-wide blooms are magenta-

Echinocereus enneacanthus

pink with white-hairy, spiny outsides. *E. p. neomexicanus* (syn. *E. dasyacanthus*) of New Mexico, Texas, and Mexico shares with *E. p. rigidissimus* the common name rainbow cactus. It is a robust variety with

Echinocereus pectinatus rigidissimus

cylindrical stems up to 1 foot tall or a little taller and 4 inches thick. They have twelve to twenty-one ribs, seldom branch, and typically become narrower toward their apexes. The comparatively thick spines completely cover the gray-green stems. Each cluster is of fifteen to twenty-five radials, the longest up to slightly over ½ inch in length, and two to nine somewhat shorter centrals. A rather shy bloomer, this has canary-yellow to pink, magenta, or violet flowers up to 5½ inches across. Popular *E. p. rigidissimus* (syn. *E. rigidissimus*), like the last called rainbow cactus is one of the most handsome of the group. It has stems up to nearly 4 inches in diameter and 1 foot tall, densely furnished with bands of different colored spines, white to pink to brown. There are sixteen to twenty spines in each cluster. The pink blooms have white hairs and spines on their outsides. ***E. reichenbachii*** (syn. *E. caespitosus*), of Texas and Mexico, is called the lace cactus and strawberry cactus. The first name alludes to the appearance of the spines, the other to its edible fruits. This kind often has solitary stems spherical when young, but later ovoid and eventually cylindrical. Rarely as much as 1 foot tall by 3½ inches thick, they have ten to nineteen lumpy ribs. The spine clusters consist of twelve to thirty-six various-sized radials arranged in comblike fashion and white to yellowish with darker tips. Usually there are no central spines, but there

Echinocereus reichenbachii

may be up to seven. The flowers are pink to magenta-pink. ***E. sciurus*** is a clump-forming endemic of Baja California. It has slender, densely-spiny stems up to about 8 inches long and with twelve to seventeen warty ribs. The spine clusters are of fifteen to eighteen brown-tipped, slender radials up to ½ inch in length and usually several centrals shorter than the radials. Almost 3 inches long, the flowers are bright magenta-red. ***E. viridiflorus,*** its specific epithet meaning green-flowered, is a well-named native of western North America. It has spherical to more elongated stems 1 inch to 6 inches high, about 2 inches in diameter, with twelve to fourteen ribs. The white to reddish or brownish spines, in clusters of twelve to eighteen radials and occasionally one or two centrals, are mostly in horizontal zones, those of adjacent ones of different hues. The longest are ½ to ¾ inch long. The greenish flowers are about 1 inch long. ***E. weinbergii*** is a Mexican with grayish-green stems at first spherical, but with age becoming longer than their width of about 4½ inches. They branch somewhat from their bases and have fifteen pointed ribs. The reddish-based, white spines are in clusters of about ten short, stout radials all under ½ inch in length. There are no centrals. Coming from near the tops of the stems, the 2-inch-long, bright pink blooms have slender, pointed petals.

Erect to sprawling, more or less rigid, cylindrical stems with usually white to yellowish spines are possessed by these: ***E. berlandieri*** (syn. *E. leonensis*), of Mexico, has light green, upright stems up to 10 inches long and 2 inches thick, with six or seven sharp ribs. They branch freely from their bases. The spines are white and in clusters of about eight awl-shaped radials up to a little over ½ inch long and a solitary straight central about twice as long. The purplish-red blooms have a darker stripe down the center of each blunt-toothed, pointed petal. ***E. brandegeei,*** of Baja California, has clustered, dull green stems 1 foot to 3 feet long or longer and up to 2 inches wide. In place of clearly defined ribs there are numerous tubercles. The spine clusters are of about twelve needle-like radials, ½ to ¾ inch long, and four or fewer centrals, the lowest flattened and up to 3½ inches long. The flowers are purplish-red. ***E. cinerascens,*** of Mexico, forms large patches of erect to sprawling stems 1½ to 2 inches thick, up to 1 foot in length. They have six blunt ribs. The spines are yellowish-white with reddish bases, slender and straight; both radials and centrals are up to ¾ inch long. There are eight to ten of the former, one to four of the latter. Almost 3 inches long, the blooms are purplish-red with brownish-violet outsides. ***E. delaetii,*** of Mexico, is distinct because of the long, white to grayish hairs that clothe its stems. This special attraction gives to this species something of the aspect of the old man cactus (*Cephalocereus senilis*). Usually erect and up to nearly 1 foot in height by 2 to 2½ inches in diameter, the stems which branch freely from their bases, have twenty to twenty-four ribs. Hidden beneath the covering of coarse hairs, which are 2½ to 4 inches long, are clusters of spines. Each is of eighteen to thirty-six straight, yellowish-white radials under ½ inch long and about five bristle- or needle-like centrals ¾ inch to 1¼ inches long. The bright pink to light carmine blooms, almost 3 inches in length, are not plentifully produced. ***E. ehrenbergii,*** of New Mexico, Texas, and Mexico, has long, pure white spines. From erect to nearly prostrate, its green, grayish, or yellowish-white stems are about 6 inches in length and 1 inch in diameter. They have half a dozen ribs. The spines are in clusters of eight to ten radials somewhat over ½ inch long and one central up to 1 inch long. The purple-red to violet-red flowers nearly 3 inches long have toothed petals. ***E. engelmannii*** chiefly inhabits deserts in Arizona and California, but occurs elsewhere including Mexico. It forms mounds of stems 1 foot to 2 feet tall and wider than tall. The stems are green, 6 inches to 2 feet long by 2 to 3 inches wide. They have usually ten to thirteen low ribs. The pale yellow to brownish spines are in closely set clusters of six to twelve awl-shaped radials the largest ½ to 1 inch long, and four to six swollen-based centrals zoned with brown and yellow, often twisted or down-pointing and 1 inch to 2½ inches long. From 2 to over 3 inches long, the toothed-petaled flowers are purple to magenta or lavender. ***E. reichenbachii albispinus*** (syn. *E. baileyi*), of Oklahoma, has stems branched from their bases, 4 to 8 inches tall by up to 2 inches wide, with approxi-

Echinocereus engelmannii

Echinocereus pentalophus with erect-stemmed *Euphorbia* behind

mately fifteen ribs. The spine clusters are without centrals. Each is of sixteen spreading, needle-like white or yellowish radials. A little over 2 inches long, the light pink blooms have toothed petals. ***E. triglochidiatus pacificus*** (syn. *E. pacificus*) inhabits Baja California. A clump-forming kind, this has numerous stems 6 to 10 inches long and 2 inches wide or slightly wider. They have ten to twelve blunt ribs and reddish-tinged, gray spines. The latter are in clusters of ten to twelve radials under ½ inch long and four or five unequal centrals up to 1 inch long. The dark red blooms are only a little over 1 inch in length.

Short ovoid-cylindrical, mostly erect, softish stems are characteristic of these: ***E. maritimus,*** endemic in Baja California, forms dense clumps. It has more or less spherical to cylindrical stems up to about 6 inches long by up to 1 inch thick, branching freely from their bases and green to grayish. The spine clusters are of nine or ten straight, yellowish-white to grayish radials up to 1 inch long and four or fewer stouter, similar centrals up to 1½ inches long. The pale yellow blooms, about 1½ inches in length, come from the tops of the stems. ***E. triglochidiatus mojavensis*** (syn. *E. mojavensis*), ranging from Utah to California and Arizona, forms great clumps of stems 2 to 8 inches long by about 2 inches thick, each with eight to twelve conspicuously warted ribs. Each spine cluster has about ten up to 2-inch-long straight, spreading radials, which at first are reddish, later white, and one central spine not exceeding the centrals in length. The carmine-red blooms have blunt, toothed petals. They are nearly 3 inches long.

Soft, mostly long, sprawling stems are typical of these kinds: ***E. berlandieri angusticeps*** (syn. *E. papillosus*) is a Texan with sprawling, branching stems up to about 1 foot long and 1 to 1½ inches wide. They have six to ten prominently lumpy ribs. The spines are in clusters of about seven radials, under ½ inch long, and a solitary, brown-based, bright yellow central, ¾ inch long. The blooms, about 4 inches wide, have yellow petals with reddish bases and stigmas with nine lobes. ***E. pentalophus*** is a Mexican with mostly prostrate, light green stems up to about 5 inches long by ¾ inch thick, and with five lumpy ribs. The closely set spine clusters are each of three to five short, whitish to yellowish or grayish radials. There are no centrals. The lilac to reddish-violet blooms, 3 to 5 inches long, have wide-spreading, narrow petals, the inner ones toothed. ***E. triglochidiatus neomexicanus*** (syn. *E. polyacanthus*) has dark green, mostly erect, about ten-ribbed stems. They are 4 to 10 inches long and bear spine clusters of eight to twelve spreading radials approximately ¾ inch long, white with dark tips or when young reddish. The three or four centrals in each cluster are bulbous and 1¼ to 2 inches long. Blood-red to deep pink, the funnel-shaped blooms are a little over 2 inches in length. This sort is indigenous to Texas, New Mexico, Arizona, and Mexico. ***E. viridiflorus davisii*** (syn. *E. davisii*) is a native of Texas. It has a spherical to longish-ovoid stem up to 2 inches in length, which rarely exceeds ½ inch in thickness, and with six or seven ribs. The spine clusters are of nine to twelve brown-tipped, white radials. There are no centrals. About 1 inch in length, the flowers are greenish.

Garden and Landscape Uses. These are among the most satisfactory cactuses for planting outdoors in warm, dry climates and often as greenhouse and window plants. Not all are entirely satisfactory for this latter purpose, but many are, and their possibilities are worth exploring by those who enjoy variety among houseplants. In the main, echinocereuses are easy to grow and most bloom freely in cultivation. Their flowers are usually quite large and last longer than those of many cactuses, especially if the plants are in an airy, well-ventilated place. Not only the blooms are attractive, the forms of the plants and their usually dense coverings of sometimes brightly-colored spines have great charm.

Cultivation. This is not basically different from that of most desert cactuses. Echinocereuses need well-drained soil, preferably of fairly high fertility. Water moderately from spring to fall, not at all or with much caution in winter. Cool conditions are needed in winter. Night temperatures of 45 to 50°F are ample, with increases of five to fifteen degrees by day. Propagation is by seeds, offsets, cuttings, and grafting. The last is often the best method with *E. pectinatus* and with allied kinds whose spines form comblike arrangements. For more information see Cactuses.

ECHINOCYSTIS (Echino-cýstis)—Mock-Cucumber or Wild-Cucumber. The approximately twenty species of this genus of the gourd family CUCURBITACEAE are all American. The name comes from the Greek *echinos*, a hedgehog, and *kystis*, a bladder, and alludes to the fruits.

These are annual and perennial vining plants with alternate, undivided, lobed or angular leaves and unisexual flowers; both sexes are on the same plant. The blooms have a small six-lobed calyx, and a flattish, wheel-shaped, deeply-six-lobed corolla. The stamens are joined into a column. The inflated, bladder-like fruits are dry; they open to release the seeds.

Mock-cucumber or wild-cucumber (***Echinocystis lobata***) is a rapid-growing annual vine that, under favorable circumstances, attains a height of 20 feet. Its three- to five-triangular-lobed, or angled leaves are about as broad as long. They are long-stalked and have blades 2 to 5 inches across. The female flowers are solitary or few together and come from the same leaf axils as the more numerous males, which are in long, erect racemes. The blooms are greenish-white to whitish and about 1/3 inch across. The long-stalked spherical to ellipsoid fruits, when ripe, are bladder-like and papery. Furnished with soft spines, they are egg-shaped and 1½ to 2 inches long. This species inhabits thickets and moist soils throughout much of northeastern North America. Native to Texas, ***E. wrightii*** is much like *E. lobata* but its spiny fruits are cylindrical. Other plants previously included in *Echinocystis* now belong in *Marah*.

Echinocystis lobata

Garden and Landscape Uses and Cultivation. One of the fastest growing and most vigorous of annual vines, the mock-cucumber is useful for covering fences, walls, stumps, and other features where a temporary, quickly established cover serves. Unfortunately, it is often affected with a virus disease, which reduces its usefulness. Propagation is easy by seeds sown in early spring where the plants are to remain. These remarks also apply to *E. wrightii*.

Echinocystis wrightii

ECHINODORUS (Echino-dòrus). This genus of the water-plantain family ALISMATACEAE is very like *Sagittaria*. The significant differences are in the fruits, those of *Echinodorus* being merely ribbed or ridged, but not clearly winged. There are about thirty species of *Echinodorus*, natives of temperate, subtropical, and tropical regions of the Americas, including the United States, and of Africa. The name, from the Greek *echinos*, a hedgehog, and *doros*, a leather bottle, presumably alludes to the fruits.

These aquatic and marsh plants have mostly basal, long-stalked leaves, and flower stalks terminating in panicles with whorls (tiers) of three to six branches. The blooms have three persistent sepals and three spreading, white, pink, or rarely yellow petals. There are six to twenty stamens. The seedlike fruits or achenes are in burlike heads.

Native from Ohio, Illinois, and Missouri to Kansas, Texas, California, Mexico, and the West Indies, ***E. berteroi*** (syn. *E. rostratus*) has erect, typically broad-ovate to broad-elliptic leaves with heart-shaped bases, but on dwarfed plants the leaves may be lanceolate. Their blades are up to 8 inches long by 4½ inches wide. The branched or branchless flower stalks are erect and up to 1½ feet tall. The flowers, with broad, white petals, are ¾ inch wide or a little wider. This species has been misidentified as *E. cordifolius*, which name rightly belongs to a kind with prostrate flower stems.

Native of Central and South America and Cuba, ***E. grandiflorus*** has long-stalked, erect leaves with blades up to almost 2 feet in length, and nearly as broad as long. They have shallowly-heart-shaped bases, which, like the tops of their stalks, have stellate (star-shaped) hairs with swollen bases. Branched or branchless, the flower stems carry white or rarely pink blooms. Variety *E. g. aureus*, a native of Cuba, has yellow flowers. White-bloomed *E. g. ovatus* is distinguished by its hairless leaves. A species, ***E. palaefolius***, believed to be native from Mexico southward, has broad, heart-shaped leaves with angled stalks. Its white flowers, in panicles lifted well above the foliage, are large and decorative.

Garden and Landscape Uses and Cultivation. These are the same as for *Sagittaria*. Of the species described here only *E. berteroi* is hardy in the north.

Echinodorus palaefolius

ECHINOFOSSULOCACTUS (Echino-fos-sulocáctus). Endemic to Mexico, this formidably named genus of the cactus family CACTACEAE consists of thirty-two frequently variable species that because of their shapes and spines have often much of the aspect of sea urchins. It was formerly named *Stenocactus.* The name of this group, alluding to the spiny character of the plants and the furrows between the ribs is from the Latin *echinus,* a hedgehog, *fossula,* a little ditch, and the word cactus. From *Echinocactus* this genus differs in not having woolly hairs in the axils of the scales of the ovaries and in the fruits being without hairs or wool. From *Ferocactus* it differs in its prevailingly many more and much thinner ribs of its stems.

These attractive small cactuses, often showy in bloom, frequently have deep taproots. Their stems, usually solitary, are depressed-globular or shortly-cylindrical. They have few to many thin, wavy ribs. The few to many spines are generally flattened, horny, or papery. They may be recurved, but are not hooked. The small blooms, pink to greenish-pink or greenish-yellow, the petals often with a purplish-pink center stripe, are short funnel- to bell-shaped. There are numerous stamens. The spherical to short-cylindrical fruits contain black seeds.

Kinds most likely to be met with in cultivation include those now to be described, others may be grown by specialists. ***E. albatus*** has flattened-spherical to cylindrical, bluish-green usually solitary stems, rarely with offsets. Up to 5 inches in diameter, they have numerous slender ribs and are white-wooly on top. The clusters of whitish spines, up to ¾ inch apart, consist of four central spines up to 1¾ inches long and about ten bristle-like radials ½ inch long. The pure white flowers are about ¾ inch long. ***E. anfractuosus*** has dull green, somewhat cylindrical stems up to 5 inches long by about one-half as thick, with about thirty ribs, each with only a few clusters of somewhat curved, brown-tipped, straw-colored to nearly white spines. Each cluster has about seven stout radials, the three upper much bigger than the others and 1¼ inches long. The radials are slender. There is one brownish, 1-inch-long central spine. The flowers have white-margined, purple petals. ***E. coptonogonus*** is the only species with broad ribs. Spherical with a flattish top and glaucous-olive-green, it is up to 4 inches in diameter. It has up to fourteen ribs with deep notches ¾ inch to 1¼ inches apart in which sit the areoles bearing the clusters of three to five long upcurving spines, the largest about 1¼ inches long. When young the spines are red, later they are grayish. The many flowers, over 1 inch long and nearly 1½ inches wide, have whitish petals each streaked along the center with pinkish-purple. The reddish stamens are tipped with violet anthers. ***E. heteracanthus*** is spherical to short-cylindrical and thickly clothed with clusters of eleven to eighteen spines, the radials white, the four central ones brownish to flesh-colored, slender but flattened. The body of the plant, the stem, is light green. There are thirty to forty, much-compressed, somewhat wavy ribs. The flowers are greenish-yellow. ***E. lloydii*** has a spherical stem 4 or 5 inches in diameter, not flattened on top, and with numerous thin ribs. It is densely covered with spines. Each cluster is of ten to fifteen white, needle-like radials up to ¾ inch long and three thicker centrals 1½ to 3½ inches long and somewhat incurved, their tips touching or overlapping, the upper spine papery. The pale pink flowers are about ½ inch wide. ***E. pentacanthus*** has a flattened-globular to short-cylindrical stem about 3 inches in diameter with twenty-five to forty thin, wavy ribs. The grayish-red, flattened, upcurved spines are in clusters of five. The three upper are longer and wider than the others. They may be 2 inches in length. The flowers have white- or yellowish-edged, violet petals. ***E. violaciflorus*** has a spherical or cylindrical, dull bluish-green stem 3½ to 4 inches wide with about thirty-five wavy ribs with scalloped margins. The spines, all pressed against the plant in clusters of about seven, practically hide the plant body. The three upper, in length from 1¼ inches to twice that and flattened, spread so their tips touch or overlap those of neighbor clusters. The lower, much shorter, thicker spines are spreading. About 1 inch long, the flowers have white petals streaked down their centers with violet, and violet stamens and style. ***E. zacatecasensis,*** pale green, spherical, and very spiny, is up to 4 inches in diameter. It has about fifty-five very thin, wavy ribs. The spine clusters consist of about ten to twelve short, white, spreading, needle-like radials and three brownish centrals, the flattened middle one the longest, erect and curving, and up to almost 1½ inches long. The flowers, up to nearly 1¼ inches across, have petals tipped with lavender-pink.

Garden Uses and Cultivation. Excellent additions to collections of succulents, cactuses of this genus are suitable for rock gardens in dry, warm, desert regions and for greenhouses and other suitable places indoors. Some are a little tricky to satisfy, others flourish with little difficulty. They prefer fertile, slightly acid soil and the general treatment suitable for *Echinocactus* and other groups of more or less globular cactuses. For more information see Cactuses.

ECHINOMASTUS (Echino-mástus). Early Bloomer or White Visnagita. Some authorities include *Echinomastus,* of the cactus family CACTACEAE, in *Echinocactus,* others include it in *Thelocactus.* When kept separate, ten species of northern Mexico and adjacent parts of the United States are recognized. The name comes from the Greek *echinos,* a hedgehog, and *mastos,* a breast. It alludes to the spiny tubercles.

These attractive cactuses have small, usually solitary, spherical to short-cylindrical stems with decidedly lumpy (tubercled), more or less spiraled ribs. The needle-like spines are in clusters of several radials with or without stouter centrals. Blooms are usually pink or reddish. The small fruits become dry at maturity. In aspect these cactuses much resemble *Coryphantha.*

Early bloomer and white visnagita are colloquial names of ***E. intertextus*** (syns. *Neolloydia intertexta, Echinocactus intertextus*), native to Texas, Arizona, New Mexico, and Mexico. The earliest cactus in its region to bloom, this kind has a solitary hemispherical to spherical or short-columnar stem up to 4 inches in diameter, with thirteen broad, low ribs composed of separate tubercles each with a sharply-peaked, chinlike ridge. The spines are in clusters of sixteen to twenty-seven purplish-red radials of different lengths, the longest up to ¾ inch long, and in mature, but not juvenile specimens, three to five somewhat stouter, slightly darker centrals. The spines spread and those of neighbor clusters interlock to such an extent that the blooms often cannot expand fully. Salmon-pink to white, the flowers are 1 inch long and wide. The dry, brown, rotund to oblong fruits are up to ½ inch long. Variety *E. i. dasyacanthus* (syn. *E. dasyacanthus*) is somewhat taller than the species, its stem elongating to 6 inches. Also, the ribs are narrower and the spines are up to nearly 1 inch in length or longer.

Rarely bigger than a golf ball, ***E. mariposensis,*** first discovered in 1945, is endemic to a very limited region in Texas. Its spherical to egg-shaped stem has thirteen to twenty-one ribs, the larger numbers occurring in old specimens. The spine clusters are of twenty-five to thirty-six stiff, white to gray radials, up to ⅜ inch long and sometimes tipped with brown, and four to seven dark-tipped, whitish, pale yellow, or gray centrals, ½ to ¾ inch long or the bottom one shorter. The flowers, ¾ inch to 1¼ inches long and wide, have green-centered, white petals or brown-centered, pinkish petals.

Other kinds are these: ***E. durangensis*** has an ovoid stem approximately 3½ inches long by nearly as thick. Its spine clusters are of fifteen to thirty black-tipped, white, more or less incurved radials, a little over ½ inch long, and three or four slightly longer, needle-like centrals. ***E. erectocentrus*** (syns. *Neolloydia erectocentra, Echinocactus erectocentrus*) has a broad-egg-shaped to short-cylindrical stem up to 6 inches tall and 4 inches in diameter. It has twenty-one low ribs and starry spine clusters of fourteen radials and one or two erect centrals. From 1¼ to 2 inches long, the flowers are pinkish. ***E. macdowellii*** has solitary stems up to about 4 inches tall by a little over 2 inches in diameter. They have twenty to twenty-five ribs. The spine clusters are of fifteen to twenty-five white radials up to ¾ inch long and three or four straw-yellow centrals approximately twice as long. The flowers are rose-pink, about 2 inches long. ***E. unguispinus,*** its stem spherical to short-cylindrical, is up to 4½ inches tall. Native of Mexico, it has clusters of spines consisting of as many as twenty-five radials, usually white except for their darker tips, the biggest up to ¾ inch long, and four to eight stouter centrals, some markedly curved, when young reddish to black, fading to gray later. The reddish flowers are 1 inch long.

Garden and Landscape Uses and Cultivation. In warm desert and semidesert regions these small cactuses are suitable for outdoor cultivation in rock gardens devoted to succulents. They are also treasured items in greenhouse collections of such plants. Because of their sensitiveness to excessive wetness, great care must be taken not to overwater them. They respond to the general conditions and care appropriate for coryphanthas, mammillarias, and other small desert cactuses. For more information see Cactuses.

Echinops exaltatus

ECHINOPANAX. Some species previously included in this genus have been transferred to *Oplopanax* and are discussed under that entry in this Encyclopedia.

ECHINOPS (Echìn-ops)—Globe-Thistle. Coarse thistle-like annuals, biennials, and perennials native from eastern Europe to central Asia and Africa constitute *Echinops,* of the daisy family COMPOSITAE. There are approximately 100 species. The name comes from the Greek *echinos,* a hedgehog, and *ops,* appearance, in allusion to the flower heads.

Globe-thistles have alternate leaves usually two or three times pinnately-cleft or toothed, the lobes or teeth prickly. In cultivated kinds at least their undersides are white-woolly. The tiny florets are technically solitary and each has an involucre of little bracts, but they are crowded into dense spherical heads, usually called flowers, with at the base of each a common involucre (collar) of reflexed bracts. The fruits are seedlike achenes.

Most commonly cultivated, ***E. exaltatus*** is often misidentified as *E. ritro* and *E. sphaerocephalus.* The species to which these last two names rightly belong may not be in cultivation. From the first, *E. exaltatus* can be distinguished by the upper surfaces of its leaves not being rough with short, bristly hairs but practically hairless, and from the second in not being glandular or having the inner bracts of the involucres pubescent on their backs. Native of Siberia, *E. exaltatus* has branchless or branched, white-hairy stems 3 to 10 feet tall or sometimes taller. Its leaves, the lower ones considerably larger than those higher on the stems, are pinnately-cleft into pointed, spiny-toothed, lanceolate lobes. The bluish flower heads are 1½ to 2 inches across. The outer bracts of the involucres are about one-half as long as the inner ones. Differing from the last in its leaves not having cobwebby-hair on their upper surfaces and the plants rarely exceeding 4 feet in height, ***E. humilis*** also hails from Siberia. Its lower leaves are almost spineless. In outline they are fiddle-shaped with sinuate rather than distinctly lobed margins. Those higher on the stems are oblong, pinnately-lobed, and toothed. The outer bracts of the blue flower heads are bristle-like and are one-half as long as the inner ones. A native of eastern Europe and western Asia, ***E. ritro*** is up to 2 feet tall. It has white-woolly, nonglandular stems, and ovate-oblong, deeply pinnately-lobed or twice-pinnately-lobed, spiny-toothed leaves up to 1¼ feet long. They have glossy green upper surfaces and undersides felted with white hairs. The heads of bright blue flowers are up to 1½ inches across.

Echinops exaltatus (head of flowers)

Echinops ritro

Sparingly naturalized in North America, ***E. sphaerocephalus,*** of Europe and western Asia and up to 7 feet tall, has gray-woolly stems glandular-hairy in their upper parts. Also glandular, the spiny-toothed pinnately-lobed, ovate-oblong leaves have green upper surfaces and white-hairy undersides. The flower heads are pale blue to whitish, 2 inches wide or a little wider.

Echinops ritro (heads of flowers)

Echinops sphaerocephalus

Garden and Landscape Uses and Cultivation. The most convenient locations for globe-thistles are toward the backs of perennial borders and in semiwild, naturalistic plantings. They are among the least demanding of plants, hardy and satisfied with any reasonable soil. They do best in full sun, but stand a little part-day shade. Propagation is by division in early fall or early spring and by seed. The flower heads of globe-thistles can be cut and used in dried arrangements.

ECHINOPSIS (Echin-ópsis)—Sea Urchin Cactuses. About forty species of the cactus family CACTACEAE constitute *Echinopsis*. They are natives of southern Brazil, Bolivia, Paraguay, Uruguay, and Argentina. The name comes from the Green *echinos*, a hedgehog, and *opsis*, resembling. As treated here the genus includes kinds by some authors segregated as *Pseudolobivia* and *Pseudoechinopsis*.

Mostly low, these cactuses have solitary or clustered, spherical to short-columnar, strongly-ribbed, spiny stems, the spines sometimes very short. The flowers originate from areoles at the side or near the tops of the stems. Usually white, they open at night. Large, they are funnel-shaped or have a narrow perianth tube and wide-flaring petals. The outside of the tube bears scales with much or little hair in their axils. There are numerous stamens. The fruits, narrowly-ovoid to oblong, contain dullish, black seeds. Echinopsises are not self-fertile. Ordinarily to secure seeds it is necessary to pollinate the flowers with pollen from another seedling (not with that from a plant that has been propagated vegetatively from the same stock).

Species of *Echinopsis* found in contemporary collections include these: ***E. ancistrophora*** (syn. *Pseudolobivia ancistrophora*), of Argentina, has flattened-spherical stems with up to eighteen notched ribs. The somewhat depressed areoles have seven short, spreading and reflexed radial spines and one central, approximately ¾ inch long and curved and reflexed. The narrow-tubed, scentless flowers are up to 6 inches long. ***E. aurea*** (syn. *Pseudolobivia aurea*) is a variable native of Argentina. It stems, spherical or somewhat taller than wide, are approximately 3½ inches in diameter and have fourteen to seventeen sharp, straight ribs. From the woolly areoles clusters of six to ten ¼- to ½-inch-long radial spines sprout and one to four centrals, which are up to 1½ inches in length. Grayish, they are darker at their bases and brown-tipped. Produced from the sides of the stems and curving upward, the flowers, 3 to 4 inches long and approximately 3 inches wide, are yellow with darker yellow centers, and greenish on the outside of the tube. ***E. eyriesii*** seldom appears true to type in collections. Plants so named are usually hybrids. The true species has globular to somewhat elongated stems 4 to 6 inches in diameter, with eleven to eighteen straight, sharp ribs. The spines, ⅛ inch long, are at first reddish-brown, and later darker. They sprout in clusters of about ten radials and four to eight centrals from woolly areoles. Up to 10 inches in length, the blooms are large and white. This is a native of Brazil, Uruguay, and Argentina. ***E. ferox*** (syn. *Pseudolobivia ferox*), of Bolivia, has spherical stems approximately 8 inches in diameter with about thirty ribs. The spine clusters are of ten to twelve brown radials about 1¼ inches long and three or four centrals up to 6 inches long. The white blooms are about 6 inches long. ***E. haematantha*** (syn. *Pseudolobivia haematantha*) usually has a solitary stem up to 6 inches in diameter with about twenty-seven sharp ribs. It is a native of Argentina. The bent, gray spines are in clusters of eight to fifteen, the centrals curved, but not hooked. The white flowers are about 8 inches long. ***E. huottii*** is a native of Bolivia. Globular to ovoid, and 2½ to 4 inches in diameter, its

Echinopsis ancistrophora

stems have about twelve slightly notched ribs. The dark-tipped, light brown spines are in clusters of nine to eleven nearly 1-inch-long radials and about four somewhat longer centrals. The blooms come from near the tops of the stems. From 7 to 8 inches long by 4½ inches wide, they have white inner petals, the outer ones light pink with a greenish midrib. ***E. kermesina*** (syn. *Pseudolobivia kermesina*), of Argentina, except in flower color is much like *E. multiplex.* It has somewhat flattened-spherical, glossy stems 4 to 5 inches in diameter. They have fifteen to thirty-five ribs. The wool of the areoles at first is orange-yellow, later gray. The clusters of needle-like spines are of ten to fifteen brown-tipped radials and four to six centrals up to 1 inch long. Arising from the apex of the stem, the scentless, rich deep-carmine-pink to red blooms are about 7 inches long by one-half as wide. ***E. leucantha*** is a variable native of Argentina. Its spherical to ellipsoid stems, up to 1 foot high, have fourteen to sixteen straight ribs and clusters of spines consisting of nine or ten bent, awl-shaped, yellowish-brown radials and one upcurved, 2-inch-long, brown central. The flowers, fragrant of violets and of large size, are pure white with brownish outsides. Coming from near the tops of the stems, they are 6 or 7 inches long. ***E. multiplex,*** of southern Brazil, is a cluster-forming kind resembling *E. eyriesii.* Plants grown under its name are not infrequently hybrids between these two. True *E. multiplex* is distinguishable from *E. eyriesii* and from the hybrids by its longer, stouter spines the clusters of which are of about ten dark-tipped, yellow-brown, ¾-inch-long radials and about four 1¼-inch-long centrals. The sweet-scented, pale pink blooms are 7 to 8 inches in length. A much-crested, variant is *E. m. cristata.* ***E. pelecyrhachis*** (syn. *Pseudolobivia pelecyrhachis*) is an Argentinian with flattened-spherical stems with about forty ribs. The spine clusters are of about nine radials and one central. The white blooms are about 4 inches long. ***E. rhodotricha*** (syn. *E. forbesii*) is a columnar Argentinian about 2½ feet tall with stems with eight to thirteen notched ribs. The spine clusters are of four to seven radials, about ¾ inch long, and one 1-inch-long, dark-tipped central. The slender-tubed, scentless flowers come from the sides of the stems. Dull white and pinkish on their outsides, they are up to 6 inches long by one-half as wide. ***E. torrecillasensis*** is a Bolivian of curious form. It has, for its size, an enormous turnip-like root some 3 inches long topped by a short, broad, convex stem up to 1 inch tall by up to 2 inches wide. This has about sixteen notched ribs. The spines, curved backward against the stem, are gray. Arising from the circumference of the stem, the narrowly-funnel-shaped, pink to salmon-pink flowers have outer petals greenish on their undersides. The fruits are dark purple. ***E. tubiflora,*** of Brazil and Argentina, is commonly represented in collections by hybrids rather than by the pure species. These usually have yellow spines instead of black ones typical of the true kind. The latter, globular and approximately 4½ inches in diameter, with age becomes cylindrical and up to 2 feet high. It has about twelve slightly wavy ribs with woolly areoles and spines in clusters of fifteen to twenty white-tipped, dark, ½-inch-long radials and three or four stouter, black centrals about ¾ inch long. The white flowers, approximately 8 inches long, have green outer petals tipped with brown. ***E. turbinata,*** of southern Brazil and Argentina, has stems at first globular, later cylindrical to club-shaped, with thirteen to fourteen notched ribs. The spine clusters are of ten to fourteen brown radials under ¼ inch long and about six considerably shorter centrals. Arising from the tops of the stems, the strongly jasmine-scented flowers have white petals with a central greenish stripe along their underside.

Garden and Landscape Uses and Cultivation. Among the most satisfactory cactuses for beginners as well as more experienced growers, echinopsises and their

Echinopsis multiplex

Echinopsis rhodotricha

Echinopsis tubiflora hybrid

hybrids respond well to cultivation and generally bloom with great regularity and freedom even when grown as window plants. They are extremely tolerant of environments that are unacceptable to many of their sister genera and generally multiply aplenty by offsets. Their flowers are among the most beautiful of their family. These plants are much used in dish gardens. In frost-free or nearly frost-free desert climates they are admirable for rock gardens. As young specimens most prefer light shade. Later full sun is satisfactory. For further information see Cactuses.

ECHINOTHAMNUS. See Adenia.

ECHINUS. See Braunsia.

ECHIOIDES (Echiòid-es)—Prophet Flower. Long familiar to rock gardeners as *Arnebia echioides,* the prophet flower has been segregated from *Arnebia* and renamed *Echioides.* From *Arnebia* it is distinguished by its flowers having appendages or glands in their throats and by their stamens being attached to the inside of the corolla tube at various heights from its base. It is the only member of its genus and belongs in the borage family BORAGINACEAE. Its name, derived from that of the genus *Echium* and the Greek *oides,* resembling, alludes to the appearance of the plant.

Echioides longiflorum

The prophet flower (***E. longiflorum***) is a hardy herbaceous perennial, a native of Armenia and the Caucasus. It has a basal rosette of stalkless, narrow, rough-hairy, lobeless, ovate-oblong leaves and smaller similar leaves along its stems, which rise to heights of 9 inches to 1 foot. As is usual in the family, the blooms are in coiled clusters (cymes). They are funnel-shaped and have a five-cleft calyx and a corolla with five spreading, clear lemon-yellow lobes. When the flower opens each lobe has a conspicuous purple-black spot at its base, but as the blooms age these fade and finally disappear. There are five stamens and a divided style. The fruits consist of four small nutlets. The Arabian-primrose (*Arnebia decumbens*), a related plant, has similar blooms, but is an annual.

Garden Uses and Cultivation. This is an interesting species for rock gardens and planted dry walls. For its best success it needs fairly cool summers, and winters not excessively cold, conditions difficult or impossible to provide in many parts of North America. It is perhaps best adapted to the Pacific Northwest. Its environmental requirements are part-shade and a dryish, gritty soil through which water drains readily. Propagation is by seeds, division in early spring or early fall, and root cuttings taken in August or September and planted in a cold frame.

ECHITES (E-chì-tes)—Devil's-Potato or Rubber Vine. A member of the dogbane family APOCYNACEAE, and native from Florida to Mexico, the West Indies, and northern South America, *Echites* consists of six species. The name, the "ch" of which is pronounced as "k," comes from the Greek *echis,* a viper. It refers to the twining habit, or perhaps to poisonous properties of the sap.

Members of this genus are perennial, woody, twining, milky-juiced vines with opposite, undivided, toothless leaves and clustered flowers. The latter have five-lobed calyxes, long, slender, tubular corollas with five lobes (petals) that spread widely in a flat plane, five stamens with anthers that converge around and are partly adherent to the stigma, and one style. The fruits are paired, podlike follicles.

Devil's-potato or rubber vine (***E. umbellata***) inhabits Florida. It has a large, tuberlike root and forms a tangle of intertwined stems. Its short-stalked, ovate to broad-elliptic, hairless leaves are 1½ to 4½ inches long. The flowers, in lateral or subterminal umbels of two to seven that are shorter than the leaves, have white, greenish-white, or yellowish-white corollas with tubes, broader at their middles than at their ends, about 2 inches long, and petals up to 1 inch long. The hairless seed pods are 6 to 10 inches long. This species shares the common name rubber vine with related *Rhabdadenia* and *Cryptostegia.* Occasionally cultivated in greenhouses as ***E. picturata,*** a name without botanical validity, an attractive species of unknown provenance that perhaps is a variety of *Prestonia coalita* has lanceolate, leathery dark green leaves 2½ to 4 inches long decorated with gray veins and with a narrow red line along the mid-vein. For plants previously named *Echites* see Prestonia.

Echites picturata, of gardens

Garden and Landscape Uses and Cultivation. Devil's-potato is suitable for planting in warm, frostless climate in locations where twining veins can be used to advantage. It succeeds in ordinary soils in light or part-day shade or in sun. Increase is by seeds, cuttings, and layering. The other sort discussed needs a humid, tropical environment with some shade from strong sun. It is readily propagated by the same procedures.

ECHIUM (Éch-ium)—Viper's Bugloss, Tower-of-Jewels or Pride of Tenerife. Pronounced as though its "ch" were "k" the name *Echium* is derived from *echion,* the ancient Greek designation for certain of these plants. There are forty species, chiefly natives of the Mediterranean region and the Canary Islands. They belong to the borage family BORAGINACEAE. In their native habitats these plants are a favorite food of goats.

Echiums include annuals, biennials, herbaceous perennials, and subshrubs, some of impressive size and aspect. One, the blue weed or blue devil (*E. vulgare*) is naturalized in North America as a weed of fields, roadsides, and similar places. The stems and foliage of echiums are usually hairy. Their leaves are alternate, the lower ones often in large rosettes. The flowers are in curving or coiled, branchless or branched spikes, sometimes assembled in towering spires. They have calyxes with five slender lobes, tubular, trumpet-shaped, wide-throated, unequally five-lobed corollas, five protruding or included unequal stamens, and a style with a two-lobed or headlike stigma. The fruits are of four seedlike nutlets.

An annual or a biennial 1 foot to 2 feet tall, ***E. lycopsis*** (syn. *E. plantagineum*) is native from the Canary Islands to England, the Mediterranean region, and the Caucasus. Its stems and leaves are furnished with stiff, white hairs. The latter, prominently veined on their undersides, are blunt, broad-elliptic to ovate, narrowed to their stalks, and up to 6 inches long. Those above are smaller and oblong-lanceolate. Approximately 1 inch long, the blooms change from blue through purple to pink. They are in panicles of one-sided, elongating racemes. Their calyxes are one-quarter to one-half as long as their unequally-lobed corollas, which are about equaled in length by the stamens. A showy, bristly-hairy, variable annual from southern Europe, ***E. creticum,*** 1½ to 2 feet tall, differs in its basal leaves not having prominent side veins and in the corollas of its flowers being nearly uniformly hairy all over instead of only on the veins and mar-

Echium creticum

Echiums, unidentified as to species, at Rancho Santa Ana Botanic Garden, California

Echium wildpretii

gins. This has oblongish to lanceolate leaves and open panicles of racemes of brick-red blooms.

Canary Island echiums number possibly twenty-five species, several of truly impressive size. Unfortunately they are not well understood botanically. Sometimes called tower-of-jewels and pride-of-Tenerife, ***E. pininana,*** which attains heights of 6 to 15 feet, blooms only once, at the end of its second year, or later, and then dies. The lower part of its towering, branchless stem is woody. The bristly-hairy leaves, the basal ones in rosettes, are pointed-oblanceolate. The largest are up to nearly 3 feet long by 1 foot wide. In bloom the flowering upper portion of the stem occupies one-half or more of its length. The flowers have pleasing purple-blue, funnel-shaped corollas ½ inch long and approximately ½ inch wide. A hybrid between *E. pininana* and *E. fastuosum* that originated in the San Francisco region is also called tower-of-jewels. Identified botanically as ***E. hybridum,*** this is a perennial with characteristics intermediate between those of its parents. It is more ornamental than either.

Other biennial or short-lived perennials include *E. giganteum* and *E. wildpretii,* of the Canary Islands, and *E. candicans,* also of the Canary Islands and of Madeira. Bushy, branched, and in bloom 6 to 10 feet tall, ***E. giganteum*** has grayish, nonspiny, soft-hairy, stalked leaves, those at the base of the plant up to 1½ feet long. The panicles are of stalked, white flowers on one-sided, short, forked branchlets. The stamens are much protruded. Not branched, 2 to 3 feet tall, and softly-hairy ***E. wildpretii*** develops a rosette of numerous narrow-linear-lanceolate leaves. Its deep rose-red flowers, with long-protruding stamens, are in long terminal spires. Its stalkless or nearly stalkless flowers, white or blue lined with white, bushy, branched ***E. candicans*** is 4 to 6 feet tall. It has pointed-lanceolate leaves and tall, dense to rather loose spires of blooms with corollas scarcely twice as long as the calyxes.

Perennial ***E. fastuosum,*** a native of the Canary Islands, shrubby and branching, is 3 to 6 feet tall. Its stems and evergreen, pointed-lanceolate to oblong-lanceolate leaves are clothed with soft, grayish-green or whitish hairs. The numerous deep purple-blue, bell-shaped blooms are crowded in large ovoid panicles the short branches of which are not branched.

Echium fastuosum

Garden and Landscape Uses. Biennial and perennial echiums are not hardy in the north. They thrive in California and other mild climate areas, especially near the coast. At San Francisco, in severe winters, they are occasionally damaged by frost. Echiums are good flower garden items for use in beds by themselves or in conjunction with other plants, or more informally under seminaturalized conditions. Needing full sun and good soil drainage, they are at their best in poorish, dryish earth. Too much fertility results in rank growth and fewer, poorer-colored blooms than when the plants have to struggle a little for a living. The annuals and more rarely the biennials are sometimes grown in greenhouses.

Cultivation. Echiums come readily from seeds, and the perennials from cuttings. Self-sown seedlings often appear as volunteers in the vicinities of established plants. Seeds of the biennials may be sown in late spring or early summer, the seedlings transplanted 6 to 9 inches apart in a bed or nursery, and in fall or the following early spring transferred to their flowering quarters, 1½ to 2½ feet apart. The annuals are very easily had from seeds sown outdoors in spring, the seedlings thinned to 6 inches to 1 foot apart. Alternatively, young plants, raised indoors from seeds sown eight to nine weeks earlier, and grown in flats or small pots, can be set out after all danger of frost has passed. Temperatures of 60 to 65°F for seed germination, and 50°F at night with a rise of five to fifteen degrees by days during the indoor growing period are appropriate. Harden the plants by standing them in a cold frame or sunny place outdoors for a week or ten days before setting them in the garden. For greenhouse use the annuals may be sown in September and the plants potted successively from small pots to those 5- or 6-inches in diameter for blooming in spring. They respond to porous, nourishing soil and temperatures of 50°F by night and from five to ten degrees more by day. Full sun is needed and an airy rather than highly humid atmosphere, which means that whenever weather permits the greenhouse must be ventilated freely. Biennials for greenhouse use are raised from seeds sown in May or June. The resulting plants are potted into successively bigger pots, giving the last transfer of the first season in August. Throughout the winter they are kept under conditions recommended above for annuals, and in early spring are transferred to their final receptacles, which may be large pots or tubs. Watering is fairly liberal except through the winter months during which the soil is kept dryish, but not dry. After the final pots are filled with roots regular applications of dilute liquid fertilizer stimulate the development of fine spires of bloom.

ECOLOGY AND GARDENS. No living thing exists in isolation. All depend in part upon other plants and animals with which, together with the physical factors of the environment, they form interrelated complexes called ecosystems. The science of ecology is concerned with the study of such systems, and plant ecology is the branch of that science that places particular emphasis on understanding vegetation.

Undisturbed by man, plant communities are likely to remain much the same for untold ages. Only natural catastrophes, such as volcanic activity or lightning-set fires and generally extremely slow, long-term climatic changes, bring variation. But since man became an agriculturist (and incidentally concerned himself with gardens), and since he domesticated animals and adopted the role of herdsman, he has effected vast modifications in the natural ecosystems of the world, many unfortunately, of a destructive nature. Increasing awareness of harmful exploitation of the land and its recourses brought in the twentieth century gradually rising concern among scientists, and broke explosively upon public awareness in the 1960s.

All ecosystems exist in a state of fine balance. Their plants live in harmony, but they are in active competition with other members of their own and other kinds for light, water, and nutrients. If anything disturbs the relationship or alters the physical factors of the environment, species favored by the change prosper and increase in number, those unfavorably affected are gradually dispossessed.

Gardens are very special ecosystems created and maintained artificially by man. For best results it is essential to preserve or improve the natural physical factors of the site and environment and to restrict or eliminate elements likely to unfavorably affect the plants to be cultivated. This is basic to good gardening.

Efforts toward these ends are familiar ones related to soil improvement including the addition of organic matter, fertilizing and liming, and such common garden practices in mulching, watering, shading, and controlling pests and diseases. The single most important factor, however, in maintaining gardens as viable ecosystems is the reduction or elimination of competition between plants. This is achieved by spacing individuals of desired kinds appropriately, which often involves transplanting during early stages of growth, by preventing vigorous kinds from appropriating space allotted to weaker growers, as may happen for example in rock gardens and perennial borders, and by rigorously controlling weeds.

Reduction or elimination of competition alone frequently makes it possible to satisfy plants that otherwise could not survive. For instance, in eastern North America, where without man's interference, forest would prevail, it is possible to grow field crops, lawns, and many sun-loving garden flowers and vegetables by eliminating and preventing the reestablishment of competing trees. Abandoned fields and gardens soon revert to woodland and finally forest. By contrast, on the naturally treeless plains of the West, grasses and other herbaceous plants compete so strongly and successfully that unless they are kept weeded from around trees and unless perhaps the environment is made more agreeable by irrigation, the trees succumb and the grasses take over.

By affording favored plants the opportunity to develop without exposure to the perils of the life-or-death struggle with other plants that practically all species in the wild face, gardeners take a giant step toward successfully cultivating particular kinds in environments often manifestly different from their natural ones.

But suppression of competing vegetation is not always enough. By hand weeding and other means, crab grass, sorrel, and suchlike undesirable elements can be removed from lawns, but by so doing a stand of desirable grasses is by no means assured. The very abundance of the weeds is often a clear indication that soil conditions are more favorable to them than to blue grass and other wanted kinds, and so to achieve a good stand of permanent lawn grasses is is often necessary to improve the soil as well as to destroy the weeds.

The rather common assumptions of rock gardeners, wild gardeners, and others concerned with growing natives and natural species from other parts of the world, as well as of many botanists and others not well acquainted with horticultural methods and achievements, that success in gardening depends almost entirely upon duplicating as nearly as possible the environments under which plants grow in the wild is false. Of course a knowledge of the natural conditions under which particular plants grow is helpful in providing guide lines, and in rare cases close duplication is necessary, but more often attempts at slavish imitation can be self-defeating or at best limiting when applied to cultivated specimens. Numerous examples could be given. Louisiana irises, swamp plants in their natural habitats, survive only under much drier conditions in the north. Many sorts of plants tolerate or even prosper better under cultivation in soils varying considerably in texture, composition, and acidity from those to which they are restricted in the wild. Frequently plants, such as flowering dogwoods and *Paxistima canbyi,* which as natives usually grow in fairly heavy shade, make finer horticultural specimens in sunnier places. It is misleading to assume that the habitats in which plants occur in the wild are necessarily those best suited to them. Some kinds of plants do occupy sites ideal for them. Others grow where they do because they are unable to compete for choicer locations successfully.

Gardeners, then, should direct their efforts toward creating and maintaining environments that they know, from empirical knowledge as well as knowledge from scientific investigation, will most surely promote the satisfactory development of the kinds of plants that to them are the most important elements in the special ecosystems we call gardens.

EDDO is *Colocasia esculenta.*

EDELWEISS. See Leontopodium. For New-Zealand-edelweiss see Leucogenes.

EDEMA. The dropsy-like condition of plants known as edema (sometimes spelled oedema) results from abnormal water relations within tissues manifested by the appearance of small watery pustules or excrecencies, or rust-colored, more or less corky ones easily mistaken for symptoms of a parasitic disease. Such developments are fairly frequent on begonias, geraniums, peperomias, tomatoes, and some other plants grown in greenhouses and elsewhere indoors, are rare on outdoor plants. Excessive atmospheric humidity and a too wet condition of the soil are the causes.

EDGEWORTHIA (Edge-wórthia)—Paper Bush. This genus, which belongs in the mezereum family THYMELAEACEAE and so is related to *Daphne,* comprises three species, and is native from Japan to the Himalayas. Its name commemorates an English botanist, M. P. Edgeworth, and his sister Maria.

Deciduous and evergreen shrubs with few stout branches, crowded toward their ends with alternate, undivided, short-stalked, toothless leaves, edgeworthias are not hardy in the north. Their petal-less flowers have densely-hairy, cylindrical calyx tubes that terminate in four spreading, petal-like lobes. There are eight stamens and a long style tipped with a cylindrical stigma. The fruits are dry drupes.

The paper bush (***Edgeworthia papyrifera*** syn. *E. chrysantha*) is 3 to 6 feet tall and deciduous. It has extraordinarily supple, tough stems, the young parts of which are silky-hairy. They are so flexuous that without breaking they easily can be tied into knots. High quality paper, used for printing currency, is made from their inner bark; this accounts for the vernacular name. This has short-stalked, narrowly-elliptic to narrowly-ovate, dull, dark green leaves 3 to 5½ inches long, with paler undersides, on young leaves are covered with silky-white hairs. Slightly fragrant, the deep yellow flowers, produced in late winter or early spring, are forty to fifty together in rounded short-stalked heads 1 inch to 2 inches in diameter. The individual blooms are ½ to ¾ inch long. This species, native of China, is much cultivated in Japan.

Garden and Landscape Uses and Cultivation. In few parts of North America, notably in the Pacific Northwest, is this daphne relative likely to thrive outdoors. It is very intolerant of hot, dry summers and is best accommodated in well-drained peaty soil in part-shade. Propagation is by seeds, and by summer cuttings under mist or in a greenhouse propagating bench.

EDGING BOX is *Buxus sempervirens suffruticosa.*

EDGING PLANTS. It is often convenient or desirable to emphasize the margins of paths and flower beds with a line or narrow strip of low, compact plants. Those used in this fashion and called edging plants, belong to no particular botanical group. All they have in common are their habits of growth and suitability for the purposes for which they are employed.

Edging plants include such evergreen and deciduous shrubs as edging boxwood, dwarf Japanese barberry, and dwarf lavenders and such shrublets as pachysandra, santolinas, and teucriums. Many low herbaceous perennials are good edging plants, among them ajugas, armerias, aubretias, campanulas, perennial candytuft, dianthuses, echeverias, epimediums, gazanias, heucheras, dwarf irises, lamb's ears (*Stachys byzantina*), liriopes, ophiopogons, dwarf phloxes, low potentillas, sedums, sempervivums, strawberries, thymes, and violas.

Low annuals suitable for this use are available in great variety. They include calendulas, annual carnations and pinks (*Dianthus*), California-poppies (*Eschscholzia*), annual candytuft (*Iberis*), annual lobelias, dwarf marigolds (*Tagetes*), nemophilas, dwarf petunias, annual phlox, sweet alyssum (*Lobularia*), and Torenia.

An informal edging of sweet alyssum (*Lobularia*)

Those listed above are but a few of the numerous plants suitable for edgings. Any low and compact ones or ones that can be kept so by appropriate shearing, that grow with facility and show good foliage and in some cases attractive bloom over long seasons, are worth considering.

EDGINGS, NONLIVING. As a measure of neatness, and to reduce maintenance, it is often desirable to edge flower beds and paths with permanent materials other than living plants. Carefully chosen and well installed, these can enhance the landscape by producing a crisp, well-finished appearance. Near buildings it is often advantageous to employ materials that duplicate or complement those of the structure.

Bricks, especially in association with brick-paved terraces or paths, are highly satisfactory. Set them in cement and leave an expansion joint ½ inch or so wide about every 10 feet. The bricks may be slanted diagonally to give a saw-tooth top to the edging, but a better effect usually results from laying them flat. Other kinds of masonry edgings include paving stones set on edge with only their top 2 to 3 inches above the surface, and tiles. Generally less attractive, but usable under some circumstances are edges made by pouring concrete between wooden or metal forms which are removed after the concrete has dried. The concrete may be tinted gray, brown, or some other, preferably earthy color. Cement coloring materials are obtainable from dealers in building supplies.

Wooden edgings formed of boards set on edge fastened to stakes driven below frost level can be attractive, but have the great disadvantage of not being permanent. If you elect to use wood be sure to choose cypress, redwood, or some other kind noted for its resistance to decay and treat it with a wood preservative other than creosote, which is toxic to plants, before installation.

Metal strips about 6 inches wide fixed to metal stakes driven into the ground are among the most satisfactory edgings. Unfortunately good ones made of heavy-gage rolled steel or other suitable metal are rather costly, and those of flimsy manufacture often bought by amateur gardeners rarely do an adequate job or look as well as sturdier types.

For some uses effective, long-lasting edgings can be made by sawing corrugated sheets of fiberglass or other weather-resistant compositions such as are used for roofing into strips 6 inches wide. Saw across the corrugations so that the strips are wavy and fix them to wooden or metal stakes.

EDITHCOLEA (Edith-còlea). Native to East Africa and the island of Socotra, *Edithcolea* belongs in the milkweed family ASCLEPIADACEAE, or according to those who make the distinction, to the stapelia family STAPELIACEAE. It is nearly related to *Caralluma*, from which it differs in the appearance of its blooms and technical details relating to their inner coronas. The name commemorates Miss Edith Cole, who discovered *E. grandis* about 1885. There are two species.

Edithcoleas are more or less branched, leafless herbaceous perennials. They have fleshy stems with spiny teeth along their angles and flowers from near their apexes. The blooms have five-lobed calyxes and large, wheel- or saucer-shaped corollas cleft halfway to their centers into five lobes. The outer corona or crown is of five short, erect or pouchlike lobes, the inner of the same number of erect, fleshy lobes expanded above, bent nearly horizontally over the anthers, and with small spines or tubercles on its upper side. The styles are not longer than the anthers. The fruits are podlike follicles.

The most attractive of the two species, ***E. grandis*** (syn. *E. sordida*), of East Africa, has five-angled stems up to 1 foot tall and 1 inch in diameter. The briefly-stalked blooms are 4 to 5 inches in diameter and whitish or yellowish freely spotted with purple-brown. Five rows of hairs spread from the center to the bases of the clefts between the corolla lobes and extend as arcs across the ends of the middles of the lobes.

Garden Uses and Cultivation. Little is reported about the cultural needs of this choice, rare plant. Presumably they are similar to those of *Caralluma* and *Stapelia*. For further information see Succulents.

EDRAIANTHUS (Edrai-ánthus). This genus, the name of which has sometimes been spelled *Hedraeanthus*, belongs to the bellflower family CAMPANULACEAE. Most of its kinds are natives of the Balkans, but it occurs wild from Italy to the Caucasus. It is closely allied to *Wahlenbergia*, a group that has a much wider natural distribution and is most strongly represented in the Southern Hemisphere. From it, *Edraianthus* differs in the way its seed capsules open. Also, its flowers are most often in compact heads with an involucre (collar of bracts) below them, or if solitary have bracts on the stalk immediately beneath the bloom. The flowers of *Wahlenbergia* are nearly always solitary or if more than one are quite separate, not clustered, and are without bracts just below them on the stalk. There are ten species of *Edraianthus*. The name comes from the Greek *hedraios*, sitting, and *anthos*, a flower. It alludes to the blooms of some kinds being stalkless in crowded clusters.

Edraianthuses are low, compact, more or less tufted, hardy herbaceous perennials with alternate, undivided, narrow leaves. They have bell- or funnel-shaped blooms, clustered or solitary, usually in shades of blue-purple, rarely white. The calyx is five-lobed, its tube joined to the ovary. The corolla has five spreading lobes (petals). There are five stamens and one style. The fruits are capsules that open irregularly instead of by valves at their apexes to disperse their seeds.

Cluster-headed species are typified by variable ***E. graminifolius*** (syns. *E. caricinus*, *E. kitaibelii*). Tufted, compact, and 2 to 5 inches tall, this has hairy stems and leaves, the latter linear, usually toothed and ½ inch to 1½ inches long. The upper leaves are bristle-tipped. The flower heads, of up to six somewhat hairy, rather narrowly-bell-shaped, purple blooms, are on stems up to 3 inches in length. Much like the last, but less compact and not as attractive, ***E. tenuifolius*** has hairy stems up to 6 inches long, leaves usually toothed and bristly along their edges, and white-based, violet-blue flowers in clusters of six to ten. Erect, and up to 8 inches tall, ***E. serbicus,*** with much the aspect of a bigger, essentially hairless, *E. graminifolius*, is perhaps only a variety of that variable species. Its flowers are purple. Leaves 1 inch to 2 inches long, without teeth, the lower ones forming rosettes, the upper linear-lanceolate, are typical of ***E. dalmaticus*** (syn. *E. caudatus*). This has 3-inch-long, nearly hairless, semiprostrate stems, gray-green, broadly-linear leaves without teeth, and funnel-shaped, violet-blue flowers in clusters of six to ten. The outer bracts are longer than the blooms.

Kinds with flowers distinctly separate and often solitary on the stems instead of being clustered in heads, are now to be described. In gardens they are often misidentified as wahlenbergias, from which they can be distinguished by the presence of bracts on the stems just below the blooms and by the manner in which the seed capsules open. Dwarf and tufted ***E. dinaricus*** (syn. *Wahlenbergia dinarica*) has very slender, grassy, silvery-gray-green leaves up to 1½ inches long, with hairy upper surfaces. The up-to-3-inch-long leafy stems end with a single funnel-shaped, violet bloom about ¾ inch long. The sepals, twice as long as the hairless ovary, are hairy on their outsides. The bracts narrow to slender points. From the last, ***E. wettsteinii*** differs in its floral bracts not narrowing to long points and in the sepals not being noticeably longer than the hairy ovaries. The stalked, violet-blue, funnel-shaped flowers are solitary or in heads of few. The extremely slender, grassy leaves, rarely exceeding ½ inch in length and having hairy upper surfaces, of ***E. pumilio*** (syn. *Wahlenbergia pumilio*) form low, dense cushions upon which sit the stalkless blooms. The latter, slightly over ½ inch long, are funnel-shaped and violet. Their sepals are hairy on both surfaces. Mat-forming ***E. serpyllifolius*** (syn. *Wahlenbergia serpyllifolia*) has oblanceolate leaves, up to about ¾ inch long by ⅒ inch wide, with hair-fringed edges. The upturned, bell-shaped, purple-violet blooms terminate semiprostrate, leafy stems 1 inch to 6 inches long. In the bud stage they are red. Variety *E. s. major* has bigger flowers.

Edraianthus tenuifolius

Garden Uses and Cultivation. These little *Campanula* relatives are choice for rock gardens and alpine frames and greenhouses. They need porous, gritty, not-too-rich soil and full sun, or better in hot climates, a little shade in summer during the middle of the day. In general they do best in limestone soils. Adequate moisture, without stagnation, is needed from spring to fall. Excessively wet soil in winter is disastrous. Propagation is easy by division in spring and by seed.

EEL-GRASS. See Vallisneria.

EGG FRUIT is *Pouteria campechiana*.

EGGPLANT. Popular as a home garden crop and grown to a considerable extent commercially, eggplants are botanical allies of tomatoes inasmuch as both belong to the nightshade family SOLANACEAE. They are even more closely related to potatoes, with which they share membership in the genus *Solanum*. Botanically *S. melongena esculentum*, eggplants are varieties of a species native to tropical Asia. They have been cultivated for several centuries at least. In Europe they are called aubergines.

Eggplants are a warm-weather crop, more tender to cold even than tomatoes and needing a longer growing season to mature their edible fruits, which are prepared for eating by cooking. Not particular as to soil so long as it is well drained and fertile, they succeed best perhaps in deep, sandy loams that have been prepared by deep spading, rototilling, or plowing in the process of which fairly liberal amounts of compost, well-rotted manure, or other organic material has been turned under. Fork in a dressing of a complete garden fertilizer before planting.

As with tomatoes, start seeds of eggplants early indoors to give sturdy young plants to set in the garden after the weather is quite warm and settled, in the vicinity of New York City toward the end of May. To have plants of a size suitable for planting, sow in a temperature of 65 to 70°F eight or nine weeks before the plants are to be set in the garden. Transplant the seedlings to individual small pots or, spaced 2 inches apart, in flats. Later move them to 4-inch pots. Keep them growing steadily in a sunny place where the night temperature is 60 to 65°F. Do not let the plants develop hard woody stems as they will if they become too crowded or if allowed to

Edraianthus tenuifolius (flowers)

Eggplants raised in pots in a greenhouse for planting outdoors when weather permits

Eggplant (flower)

Harvesting eggplants

suffer from lack of water or are subjected to low temperatures. Gradually accustom them to outdoor conditions before setting them in the garden by standing them in a cold frame or sheltered place outside for a week or so.

In the garden space the plants, according to vigor of variety, 2 to 2½ feet apart in rows 2½ to 3 feet apart. Subsequent care is mainly concerned with keeping down weeds. After the ground has warmed and hot weather arrives this is best done by mulching. Until then, and later if mulch is not used, repeated shallow surface cultivations will do the job. Make sure they are shallow. Eggplants are vigorous surface rooters and deep disturbance of the soil does serious harm. In dry weather periodic deep soakings at intervals to keep the foliage from wilting are needed. Harvest the fruits anytime after they have attained one-third their mature size until maturity. The largest fruits are obtained if the number each plant is permitted to carry is restricted. To do this pinch out flowers that come in the latter part of the summer after as many fruits as it is judged the plants can develop to satisfactory size have set.

Varieties, their fruits varying in shape, size, and color, are fairly numerous. Favorites are those with deep purple-black fruits. These include 'Black Beauty', 'Black Magic', 'Early Long Purple', and 'New York Spineless'. White-fruited 'White Beauty' is grown as something of a novelty.

Pests and Diseases. Eggplants are very susceptible to a leaf and fruit blight and to verticillium wilt disease, which causes the foliage to dry and the fruits to be abnormal or rot. Early evidence of infection is the foliage wilting by day and recovering at night. To avoid these diseases use treated seeds and do not plant on ground where eggplants, potatoes, tomatoes or other plants subject to this disease have been previously. Insect pests of eggplants are aphids, caterpillars, Colorado potato beetles, and flea beetles.

EGLANTINE is *Rosa eglanteria.*

EGYPTIAN LOTUS. See Nymphaea.

EHRETIA (Eh-rètia)—Philippine-Tea, Heliotrope-Tree, Sugarberry or Knockaway or Anacua. By some botanists segregated, along with *Cordia* and a few other genera, as the ehretia family EHRETIACEAE, this group of about fifty species of mostly tropical and subtropical, chiefly Old World, evergreen and deciduous trees and shrubs is more traditionally included in the borage family BORAGINACEAE. The name *Ehretia* honors the German botanical artist George Dionysius Ehret, who died in 1770.

Ehretias have alternate, toothed or toothless leaves, often furnished with short, rough hairs, and flowers in terminal clusters or occasionally solitary. The blooms are small, white or whitish, and have five-parted or five-cleft calyxes, short funnel- or wheel-shaped, five-lobed corollas, five usually protruding stamens, and a two-lobed or two-branched stigma. The fruits are small, dry, berry-like drupes (fruits structured like plums).

The hardiest kind, the most cold-resistant woody representative of the borage family, is the heliotrope-tree (***E. acuminata*** syns. *E. thyrsiflora, Cordia thyrsiflora*), which survives outdoors at least as far north as Boston, Massachusetts. Native of China, Korea, Japan, and Taiwan, it is a deciduous tree 45 feet or so tall, with upright, rather slender branches, and oblong-obovate to obovate, stalked, toothed leaves, without hairs except for tufts on the undersides in the axils of the veins. The pointed leaves are 2 to 7 inches long. In summer, in many-flowered panicles 4 to 10 inches long, the nearly stalkless, wheel-shaped blooms, each ¼ inch in diameter, are borne. Turning as they ripen from orange to brownish-black, the spherical fruits are about ¼ inch across. Another deciduous species, probably hardy as far north as Virginia, ***E. dicksonii,*** of China, Taiwan, and the Ryukyu Islands, is about 30 feet tall. It has broad-elliptic, toothed leaves up to 8 inches long, with short, bristly hairs on their upper sides, and hairy beneath. Its fragrant flowers, over ⅓ inch across, are shortly-funnel-shaped and have spreading lobes. In flattish panicles 2 to 4 inches long and almost as wide, they come in late spring or early summer. The sub-

Ehretia acuminata

spherical fruits, yellowish when mature, are from a little under to a little over ½ inch in diameter. This species in cultivation has been misidentified as ***E. macrophylla,*** a much less hardy tree native of India that has rhomboidal, hairy leaves, toothed and up to 8 inches in length. Its small blooms, in panicles, are succeeded by fruits ½ inch or a little more in diameter.

The sugarberry, knockaway, or anacua (***E. anacua*** syn. *E. elliptica*), native to Texas and Mexico, is evergreen or toward its northern limits partially deciduous. It is round-topped, 15 to 50 feet tall, and has broad-elliptic or oblong, dull leaves about 2 inches long, but sometimes longer, that have wavy or more rarely toothed margins, and the tips often spiny. They are roughly hairy or hairless. The ¼-inch-long, narrower-than-long, fragrant flowers, in panicles or clusters, are succeeded by pea-sized ovoid to ellipsoid, orange-yellow fruits with sweet edible flesh. This species is planted for ornament as far north as Dallas, but in severe winters there is often killed back. It is long-lived and grows well in alkaline as well as other type soils.

Native of Mexico and the West Indies, ***E. tinifolia*** is a shrub or tree up to 80 feet in height. Hairless or nearly so, it has broad, stalked, oblong to ovate, lustrous leaves 2 to 4½ inches long and at their apexes rounded to somewhat pointed. The crowded or loose panicles of numerous tiny white blooms are longer than the leaves. The edible fruits, under ¼ inch in diameter, are red. In the tropics and subtropics this kind is planted as a shade tree.

Widely distributed in South Africa, where it is one of the earliest and showiest spring-blooming shrubs, ***E. hottentotica*** has numerous wide-spreading branches, and on short branchlets obovate leaves up to 1 inch long by almost one-half as wide. Its pretty, starry, lavender-blue flowers are in clusters a little over 1 inch long. They are about ⅓ inch across, and have calyxes with five or occasionally six triangular lobes and five, or less often six, spreading petals. There are five or six stamens and a two-branched style. The spherical fruits are ⅕ inch in diameter. This kind succeeds only in frostless or nearly frostless climates.

Ehretia microphylla

Philippine-tea (***E. microphylla***) is native from India and Malaya to the Philippine Islands. A shrub up to 12 feet tall, it has clustered, toothed, obovate leaves up to 2½ inches long and flowers in groups of four or fewer, or solitary. The fruits are ½ inch or a little more in diameter.

Garden and Landscape Uses and Cultivation. Hardy ehretias are rarely planted except in botanical collections. They are of interest because they can be grown where no other trees or shrubs of the borage family survive. Although not showy in flower, *E. acuminata* commands some attention because of its late season of bloom. It and *E. dicksonii* do especially well in limestone soils. In the Deep South and California some of the more tender kinds are satisfactory for shade and ornament. Especially suitable for dry soils and exposed locations, *E. tinifolia* and *E. anacua* are recommended. Ehretias are quite adaptable and need no special care. No pruning other than any needed to shape the trees is required. Propagation is by seeds and cuttings of firm shoots taken in summer.

EICHHORNIA (Eich-hórnia)—Water-Hyacinth, Water-Orchid. In some warm regions, including Egypt, Australia, and the Deep South of the United States, water-hyacinths are among the most pestiferous weeds. Literally millions of dollars have been spent in Florida, and elsewhere, in attempting to control them, mostly with indifferent success. They proliferate so rapidly that they block navigable rivers, streams, bayous, and other waterways. Those concerned with maintaining such facilities see no virtue in water-hyacinths. Yet they are among the most beautiful of aquatics. So lovely are they that northerners, who do not have to cope with their population explosion, sometimes, understandably, call them water-orchids. Six species are recognized. The genus, native to the American tropics and subtropics, commemorates by its name a German clergyman, J. A. F. Eichhorn, who died in 1856.

Belonging in the pickerel weed family PONTEDERIACEAE, the genus *Eichhornia* consists of frost-tender, generally free-floating plants that in shallow water may root into mud bottoms. In addition to being floating plants, they differ from *Pontederia* in having larger blooms. The flowers are somewhat two-lipped, with tubular bases and six wide-spreading perianth lobes (petals). There are six stamens, the three lower ones protruding, and the upper three included in the corolla. The fruits are many-seeded capsules.

The water-hyacinth (***E. crassipes***) forms broad, rosette-like clusters of leaves with much-swollen leafstalks and great masses of long, dangling, freely-branched roots.

Eichhornia crassipes (floating)

When it roots in mud its leafstalks are less markedly inflated. The smooth, glossy leaf blades, round to kidney-shaped, are 2 to 4½ inches across. The spikelike flower panicles are erect and arise from the centers of the plants. Two to 6 inches long, they have several blooms 2 to 3 inches in

Eichhornia crassipes (rooted in mud)

Eichhornia azurea

Eichhornia azurea (flowers)

diameter that are lilac-blue with, on the upper petal, a violet blotch with a yellow center. This tropical American species is abundantly naturalized in parts of the southeastern United States and west to Texas, and in Hawaii. In many places it seriously obstructs waterways. Rosy-lilac flowers distinguish *E. c. major*.

Native of Brazil, ***E. azurea*** differs in not having markedly inflated leafstalks. Its hairless leaves, variable in shape and size, are mostly broadly-ovate to round and up to 4½ inches in diameter. Those under water are quite different; their blades are pinnately-divided into many slender lobes, which have a ferny look. The flower spikes develop from hooded spathes (leafy bracts) and have bright lavender-blue blooms with deep purple centers and a distinct yellow spot.

Brazilian ***E. paniculata,*** about 1 foot high, has short, fleshy rhizomes. It has basal leaves with long-pointed, heart-shaped ovate blades up to 4 inches long and smaller stem leaves. In panicles of five

Eichhornia paniculata

to fifteen, up to 4 inches long, the 1-inch-wide flowers are rosy-lavender to blue with a yellow blotch on one petal.

Garden Uses and Cultivation. Provided the water contains nourishment in the form of minerals that leach from a soil bottom, from tubs of soil containing other plants, or other sources, and is warm, no problems attend the cultivation of water-hyacinths. They are simply floated on water where they will receive full sun. Comparatively quiet conditions are needed, water constantly agitated by splashing fountains or other disturbances is not to their liking. In general these plants are at their best in water not over 1 foot deep, and hence where the soil that supplies needed nutrients is not too far away. They increase rapidly by natural offsets. In cold climates they must be wintered indoors.

ELAEAGNACEAE—Oleaster Family. This dicotyledonous family consists of fifty species of Northern Hemisphere small trees and much-branched shrubs contained in three genera. Many of its members inhabit wind-swept sea coasts, steppes, or plains. They have scaly shoots and alternate or opposite, undivided, lobeless leaves clothed especially on their undersides with silvery or yellow scales. The small or unisexual flowers are in racemes or panicles or are otherwise clustered. Each has a tubular calyx with two to four lobes, no petals, as many or twice as many stamens as calyx lobes, and one style. The scaly, berry-like fruits are technically achenes or small nuts surrounded by a fleshy receptacle. The genera are *Elaeagnus, Hippophae,* and *Shepherdia.*

ELAEAGNUS (Elae-ágnus) — Oleaster or Russian-Olive or Trebizonde-Date, Gumi, Silverberry, Lingaro. Evergreen and deciduous, hardy and nonhardy, sometimes spiny trees and shrubs, including a number of variations in addition to forty-five species, belong here. Horticulturally, they constitute the most important genus of the oleaster family ELAEAGNACEAE. The name comes from the Greek *elaia,* the olive, and *agnos,* the chaste tree (Vitex agnus-cactus).

Elaeagnuses, wild in North America, southern Europe, and Asia, have alternate, short-stalked, undivided leaves, which, like their young shoots, are more or less thickly clothed with silvery or brownish scales. Solitary or in clusters, the inconspicuous, often fragrant little flowers come from the leaf axils. They are tubular or bell-shaped, without petals, but with calyxes with four petal-like, deciduous lobes, white or yellow on their upper surfaces. There are four short stamens, and a much longer style. The berrylike or more correctly drupelike fruits consist of a nutlet surrounded by the persistent fleshy tube of the perianth. Those of some kinds are edible.

Oleaster, Russian-olive, or Trebizonde-date (***Elaeagnus angustifolia***) is a beautiful, extremely hardy, deciduous shrub or tree up to 20 feet tall. Usually crooked-trunked, and sometimes spiny, it has very silvery shoots and willow-like, gray-green, silvery-under-surfaced, lanceolate to oblong-lanceolate leaves 1½ to 3½ inches long. The outsides of its flowers and of the ½-inch-long yellow fruits are clothed with silvery scales. The blooms, in clusters of

Elaeagnus angustifolia

two or three or sometimes solitary, are yellowish and fragrant. Variety *E. a. spinosa* is more spiny than the species. The leaves of *E. a. orientalis* are broadly-elliptic to oblong. Its fruits are 1 inch in length.

From the Russian-olive, the shoots and leaves of which have silver scales only, the deciduous kinds now considered differ in those parts being clothed with mixtures of silver and brown scales. Up to about 10 feet tall, the gumi (***E. multiflora*** syn. *E. longipes*), of China and Japan, has mostly reddish-brown scales on its branchlets. Ovate, ovate-oblong, or elliptic, and 1 inch to 2½ inches long, the leaves are silvery

Elaeagnus multiflora

Elaeagnus multiflora (flowers)

Elaeagnus umbellata (flowers)

Elaeagnus umbellata (fruits)

beneath with a sprinkling of brown scales. Their upper surfaces at first have stellate (star-shaped) hairs, but these soon fall. Solitary or in pairs, the fragrant flowers, with silver and brown scales, and with perianths with tubes as long as the lobes, are succeeded by pendulous, long-stalked, scarlet, acid fruits about ¾ inch in length. Flowers with perianth tubes much longer than the perianth lobes, and short-stalked fruits distinguish *E. umbellata* and *E. commutata* from *E. multiflora*. Native of Japan, Korea, and China, ***E. umbellata*** is a broad, often spiny shrub up to 12 feet tall or sometimes taller, with yellowish-brown to silvery shoots. Elliptic to ovate-oblong and 1 inch to 3 inches long, its leaves, often crinkled at their edges, usually have on their undersides brown scales mixed with much more numerous silver ones. The upper surfaces of young leaves are silver-scaled. Yellowish-white and fragrant, the flowers are in twos or threes or are solitary. The short-stalked, scarlet fruits, when young clothed with silver and brown scales, and ¼ to ½ inch long, are carried erectly. The silverberry (***E. commutata***), native from eastern Canada to Minnesota and Utah, is a suckering, spineless shrub up to 12 feet tall. Its has brownish-scaled shoots and ovate to oblong leaves 1 inch to 4 inches long, with both surfaces covered with silvery scales, and the lower sides sometimes with a few brown ones. Fragrant, yellow, and silvery on their outsides, the flowers are in groups of two or three or are solitary. The very short-stalked, roundish, ½-inch-long fruits are clothed with silvery scales. From the buffalo berry (*Shepherdia argentea*), with which it is sometimes confused, *E. commutata* is easily separable because it has alternate instead of opposite leaves.

Evergreen kinds are much less hardy than the deciduous ones discussed above. They are not generally reliable much north of Washington, D.C.; nevertheless, in well-sheltered locations near New York City *E. pungens*, or perhaps *E. p. simonii*, which appears to be slightly hardier, has prospered outdoors for more than two decades. This species, native of China and Japan, and its varieties are favorites in southern gardens and in mild-climate regions elsewhere. Broad, dense, usually spiny, and up to 15 feet in height, ***E. pungens*** has blunt to somewhat pointed, leathery, elliptic to oblong leaves 1½ to 4 inches long and generally crisped along their margins. Beneath, they are covered

Elaeagnus pungens at The New York Botanical Garden

Elaeagnus pungens (leaves and flowers)

with dull whitish scales dotted with brown ones. Their upper sides are scaly at first, but soon become smooth and lustrous. Solitary or in twos or threes, the pendulous, silvery-white flowers, under ½ inch in length, are gardenia-scented. Brown at first, the ellipsoid fruits ripen red. They are ½ inch long. Varieties are *E. p. aurea*, the leaves of which are edged with yellow, *E. p. fredericii*, with smallish, narrow leaves with pale yellow centers and narrow, dark green margins, *E. p. maculata*, with yellow centers to its leaves, *E. p. simonii*, which has large leaves with few or no brown

Elaeagnus pungens maculata

scales on their undersides, *E. p. tricolor*, its leaves variegated with yellow and pinkish-white, and *E. p. variegata*, which has leaves with cream margins. Claimed to be a hybrid, with one parent *E. macrophylla*, but perhaps only a variety of *E. pungens*, the kind named *E. p. fruitlandii* has wavy-edged, rounded leaves. Reported to be of the same parentage, spineless ***E. ebbingei*** has leaves silvery beneath and 2 to 4 inches long. Yet another possible hybrid, this between *E. pungens* and *E. glabra*, is *E. p. reflexa*. It has pointed ovate-lanceolate leaves, lustrous on their upper sides, rusty-brown beneath.

Other evergreens sometimes cultivated, both natives of Japan, are *E. macrophylla* and *E. glabra*, neither hardy in the north. The largest-leaved species ***E. macrophylla***

Elaeagnus ebbingei (foliage and fruits)

is a generally spiny, broad shrub up to 12 feet tall. Its shoots and the undersides of its comparatively longish-stalked, broad-ovate to broad-elliptic leaves, 2 to 4½ inches long, are silvery-white. The upper surfaces of the leaves, scaly at first, are eventually smooth. The flowers, usually in clusters of four to six, have bell-shaped calyxes with tubes as long as the lobes. A more or less climbing, spineless shrub, when supported, up to 20 or even 40 feet tall, ***E. glabra*** has brown branches and leaves, lustrous on their upper sides, covered beneath with somewhat glossy, yellowish-white and brown scales. The calyxes of its flowers have tubes twice as long as their lobes.

The lingaro (***E. philippensis*** syn. *E. latifolia*), native of the Philippine Islands, is a good ornamental for warm regions. It bears edible fruits of considerable merit and adapts well to a variety of soils, including limestone ones. This kind may be grown as a vine on a trellis, pergola, or wall or as a

Elaeagnus philippensis in southern Florida

wide-spreading, low bush. Its evergreen leaves are elliptic and covered with silvery scales on their undersides. The flowers are small and agreeably scented. The subacid fruits, ¾ to 1 inch long, are dull pink with silvery spots. To have the best fruits it is necessary to thin out the shoots from the center of the bush and so prevent crowding.

Garden and Landscape Uses. Elaeagnuses thrive in a wide variety of soils including those of limestone origin, and in sunny places or those with part-day shade. Deciduous kinds are admired for their forms and attractive foliage, and a few, notably *E. umbellata* and *E. multiflora*, for their decorative fruits. The Russian-olive, hardy far north, withstands exposure to wind and sun and is one of the best shrubs for seaside planting and, in New York City, for roof gardens. Not less hardy, *E. commutata* also withstands exposure, and like the Russian-olive, sheared or unsheared, forms an effective windbreak. Scarcely less hardy, enduring in northern New England and southern Canada, *E. multiflora* and *E. umbellata* are likewise highly tolerant of difficult, exposed sites.

The evergreen kinds, particularly *E. pungens* and its varieties, are highly satisfactory garden furnishings. Admired for their often variegated, always beautiful foliage, they serve splendidly as lawn specimens and in shrub plantings and can be used with good effect as informal hedges and more strictly sheared formal ones. Other uses they serve admirably are as ornamentals for planters, tubs, and other large containers and, in smaller sizes, as window plants in cool rooms and greenhouses.

Cultivation. Elaeagnuses are easily raised from leafy summer cuttings, hardwood cuttings, and root cuttings, and *E. commutata* by means of rooted suckers dug from about the bases of established specimens. Seeds germinate satisfactorily if removed from the surrounding pulp and sown at once in a cold frame or under similar conditions, or they may be stored for up to a year in a cool place in airtight containers and then, before sowing, be mixed with slightly damp peat moss or vermiculite, placed in a polyethylene plastic bag, and stored for three months more at 40°F. The seeds of *E. multiflora* give the best results if, after mixing with the peat moss in the bags, they are kept for five months at normal room temperature before being chilled.

These shrubs and trees require no special care. Unless they are being grown as hedges or in other formal fashion, no regular pruning is necessary, yet there need be no hesitation about cutting back if this is desirable, for elaeagnuses recover quickly from the effects of the saw and the pruning shear, and sometimes it is wise to cut back old specimens, which have become straggly, with enthusiasm and severity. Late winter is the preferred time to prune.

ELAEIS (Elaè-is)—Oil Palm. The genus *Elaeis*, of the palm family PALMAE, consists of two feather-leaved species, occurs natively in tropical Africa and tropical America. The name, derived from the Greek *elaia*, the olive, alludes to the oily character of the fruits.

One of the most important commercial plants of the world, the African oil palm is the source of vegetable oil of prime importance for making margarine, vegetable shortening, soap, and lubricants, and is used for treating metals in the tin plate and iron fabricating industries. Following the British ban on slave trading in 1807, exploitation of this oil replaced traffic in human beings as the most important activity of Europeans in West Africa, but the operation did not reach its greatest development there until after World War I. In the meantime the palm had been introduced to other tropical regions, notably Malaya, Sumatra, and South America, and is cultivated in those lands for its oil.

The African oil palm (***E. guineensis***) reaches an eventual height of 60 feet or more, but grows slowly and takes many years before the maximum is attained. An attractive ornamental, in its early stages it looks much like a date palm (*Phoenix dactylifera*). Its trunk is then covered with the remains of old leafstalks, but with age these fall from the lower part of the trunk to reveal a dark gray surface, smooth except for the rings or scars that mark the spots from which leaves have dropped. The widely-arching leaves, 10 to 15 feet long and with spiny stalks, are numerous and form a dense head. The flowers are in tight masses grouped in the center of the crown among the leaf bases. They first appear on trees five or six years old and annually thereafter. Each cluster or spike contains flowers of one sex only, with both sexes on the same tree. The ovoid or globose clusters of female flowers are about 1 foot long, the cylindrical clusters of male flowers shorter. The fruits, the size of plums, are dry and golden to reddish-brown. Each contains one to three seeds. A yellow-brown oil is extracted from the fibrous pulp surrounding the seeds, and a white oil called palm kernel oil from the seeds themselves. It is this latter that is used in margarine. The residue, palm kernel cake, is nourishing cattle feed.

Elaeis guineensis, Jamaica, West Indies

The American oil palm (***C. oleifera*** syn. *Corozo oleifera*) is similar in appearance to the African species but is much smaller and differs in botanical details. As a producer of commercial oil it is of little importance. Even in the American tropics the palm planted for oil is the African kind. Nevertheless, the red fruits of the American species, about 1 inch long, contain an oil very similar to that of the African oil palm. In early colonial days it was used by Europeans to make candles. Indian women in some parts of its native range use it to dress their hair. This species has a trunk with its ascending part up to 6 feet tall, and usually its lower end reclines like a log on the ground, emitting roots and often curved. Its pinnate leaves, up to 24 feet long, consist of numerous slender, drooping leaflets. Their stalks are spiny and spread at all angles like great plumes to form a fountain-like crown. The flower clusters, dense and crowded, are borne close to the crown of the tree.

Garden and Landscape Uses. The African oil palm has much decorative merit and is well suited for ornamental plantings in the humid tropics. In the continental United States it can be expected to survive outdoors only in southern Florida. It is graceful as a solitary specimen, or grouped. It is also appropriate for growing in conservatories, especially those devoted to plants useful to man. The American species is also attractive, but is little known in cultivation. It has been grown in southern Florida.

Cultivation. These palms thrive in any fairly good soil, reasonably well drained and not excessively dry, in full sun or light shade. In greenhouses they need a minimum winter night temperature of 60 to 70°F and a day temperature five or ten degrees higher. From spring through fall both day and night temperatures can with advantage be higher than winter minimums. At all times a humid atmosphere is necessary and shade from strong sun must be provided. Watering should be done freely from spring through fall, somewhat less generously in winter. Well-rooted specimens benefit from biweekly applications of dilute liquid fertilizer from spring through fall. For additional information see Palms.

ELAEOCARPACEAE—Elaeocarpus Family. Comprising a dozen genera and possibly 350 species of tropical and subtropical dicotyledons, this family consists of trees and shrubs with alternate or opposite, undivided leaves. In racemes or panicles, its flowers have calyxes of four or five lobes or petals, corollas of four or five rarely united, often much apically-divided petals, or corollas may be lacking. There are many stamens and one style. The fruits are capsules or drupes. Genera cultivated are *Aristotelia, Crinodendron, Elaeocarpus,* and *Muntingia*.

ELAEOCARPUS (Elaeo-cárpus). Approximately 100 species of this, the type genus of the elaeocarpus family ELAEOCARPACEAE, inhabit eastern Asia, Indomalaysia, Australia, New Zealand, and islands of the Pacific. They are trees and shrubs, mostly tropical. None is hardy in the north. The name, in reference to the fruits, stems from the Greek *elaia*, an olive, and *karpos*, a fruit.

Elaeocarpuses have usually alternate, more rarely opposite, undivided, toothed or toothless leaves, and axillary racemes of bisexual or sometimes mixed bisexual and unisexual flowers, with four or five sepals and the same number of most commonly fringed petals. There are generally numerous, more rarely from eight to twelve stamens, and one style. The fruits are drupes (fruits of plumlike construction). Buttons and beads are made from the seeds of *E. sphaericus*.

Nearly hairless ***Elaeocarpus reticulatus*** (syn. *E. cyaneus*), native only in Australia, is a shrub, small tree, or rarely is up to 60 feet tall. It has pointed, elliptic-oblong to lanceolate-oblong, strongly-veined, shallowly-toothed leaves 2 to 5 inches long. Its creamy-white, fringed flowers have numerous stamens and are in loose racemes shorter than the leaves. The fruits are blue and spherical.

From the last, ***E. dentatus*** an endemic of New Zealand, differs in its shoots being silky-hairy, and its leaves being linear-lanceolate and having sinuously-toothed, recurved margins. This is a tree 45 feet tall or taller. Its leaves have stalks 1 inch long, and blades that may vary on the same plant from narrow- to ovate-oblong, blunt or pointed. Their undersides are more or less white-hairy. From eight to twelve together, the flowers are in racemes 4 to 6½ inches long. They have white petals usually between ⅓ and ½ inch long. The purplish, egg-shaped fruits are about ½ inch long. Variety *E. d. obovatus* has broad-obovate leaves narrowing to long stalks.

Native of India, ***E. sphaericus*** (syn. *E. ganitrus*) is a tree with elliptic, toothed leaves 4 to 6 inches long by up to 2 inches wide. Its white, about ½-inch-wide flowers are in drooping clusters from old wood. The purple fruits are spherical, and up to 1 inch in diameter.

Garden and Landscape Uses and Cultivation. In mild, nearly frostless climates, the quite handsome species described above are planted as ornamentals. They succeed in ordinary soils and situations and may be used effectively as lawn specimens and in mixed plantings. They are propagated by cuttings and by seeds.

ELAEODENDRON. See Cassine.

ELAEOPHORBIA (Elaeo-phòrbia). From *Euphorbia* to which it is very closely related, *Elaeophorbia* differs only in its thick, fleshy fruits, which unlike the usually nonfleshy ones of *Euphorbia,* do not separate into lobes. Five species are accepted in *Elaeophorbia,* of the spurge family EUPHORBIACEAE, natives of tropical Africa and South Africa. The name, from the Greek *elaia,* the olive, and a part of the name *Euphorbia,* presumably alludes to the appearance of the fruits.

Elaeophorbia drupifera

Elaeophorbias are succulent, milky-juiced trees with angled branches and alternate, fleshy, undivided leaves, each with a pair of spines at its base. The floral structure is identical with the rather complicated one of *Euphorbia.*

Native of tropical West Africa, ***E. drupifera*** (syn. *Euphorbia drupifera*) has a slightly five-angled, nearly cylindrical trunk gradually narrowing toward its top and with erect branches. The conical spines are not over ⅕ inch long. The long-persistent leaves are long-obovate, from 6 inches to 1 foot or more in length and narrowed gradually to their stalks. They are dark green and thick. Neither flowers nor fruits are of any decorative importance.

Garden and Landscape Uses and Cultivation. The species described is a handsome ornamental for outdoor planting in warm, frost-free climates and in large greenhouses. Its requirements are those of large, succulent euphorbias except that more moisture in the soil and in the atmosphere are to the liking of elaeophorbias. Some shade from strongest sun is advantageous in preserving the rich green of the foliage. Increase is easily had from cuttings and seeds.

ELAPHOGLOSSUM (Elapho-glóssum)—Elephant-Ear Fern. This botanically imperfectly understood genus, believed to consist of about 400 species, chiefly inhabits warm parts of South America, but occurs elsewhere in the tropics and subtropics. Especially prolific of species in the Andes, it consists of terrestrial and epiphytic (perching on trees without abstracting nourishment from them) ferns of the aspidium family ASPIDIACEAE. The name comes from the Greek *elaphos,* a stag, and *glossa,* a tongue, and refers to the shape of the leaves.

Considered in geological time very old, and with only two, not cultivated, near, living relatives, the sorts of *Elaphoglossum* are mostly moderate-sized, occasionally small or large plants with creeping rhizomes, usually short, but in some species long and slender, and sometimes climbing. The leathery fronds (leaves) are stalked, and have, except in one species, undivided blades sometimes with thickened, horny margins. Commonly, the fertile fronds are longer-stalked, smaller and narrower than barren ones; their clusters of spore capsules are distributed over their entire surfaces.

Most frequent in cultivation is the elephant-ear fern (***E. crinitum*** syn. *Acrostichum crinitum*). This native of the West Indies, Mexico, and Central America, has sterile leaves up to about 2 feet tall, including their shaggy-hairy stalks that may account for one-third of their height. The blunt blades, broadly-oblong and up to 9 inches wide, have a prominent midrib, margins fringed with whisker-like hairs, and both surfaces with long scattered ones. The spore-bearing fronds are shaped similarly, but are smaller and have longer stalks. Long trailing rhizomes are a feature of ***E. lingua,*** of the West Indies. Its ovate, leathery, sterile leaves taper at their bases. They are 1 foot to 1½ feet long and without the prominent hairs so characteristic of the previously discussed kind. As its name signifies, ***E. longifolium,*** of tropical America and the West Indies, has proportionately longer leaves than the common elephant-ear fern. In clusters from stout creeping rhizomes, they are linear-lanceolate and 1 foot to 1½ feet in length. Thick of texture, they are slightly wavy, and have prominent midribs. They are hairless.

Much smaller and quite distinct from those discussed above, ***E. pilosum*** produces in crowded tufts or clusters, pointed-linear, erect and spreading, flexuous leaves 6 to 8 inches long and about ¾ inch wide. Both sides of the leaves have bright brown scales. This species is indigenous from Mexico to Colombia.

Garden and Landscape Uses and Cultivation. In the humid tropics and warm subtropics these ferns are attractive for planting outdoors in shaded places where the ground is well drained, but not dry. They appreciate rooting mediums that contain a goodly proportion of organic matter. Elaphoglossums are handsome greenhouse plants. They show to better advantage in pans (shallow pots) than in deeper containers. Coarse soil that remains porous and drains rapidly is needed. It may consist largely of leaf mold or peat moss, with a little turfy loam, a generous dash of large-grained sand or perlite, and possibly some fir bark, such as is used for orchids, and crushed charcoal. It is a mistake to overpot; these ferns grow best when their roots are slightly restricted. They succeed in greenhouses shaded from strong sun and moderately humid, where the winter night temperature is about 60°F and day temperatures at that season are from five to fifteen degrees higher. At other seasons higher temperatures are appreciated. Watering must always be done with care. Constantly wet soil is detrimental. It should be allowed to become dryish, but not to the extent that the leaves wilt, between waterings. An occasional application of dilute liquid fertilizer is helpful with well-rooted specimens. Propagation is by careful division and by spores. For further information see Ferns.

ELATINACEAE—Elatine Family. Consisting of forty species in two genera, the ELATINACEAE is a family of dicotyledons the members of which are undershrubs and herbaceous plants including some aquatics. They have undivided, toothed or toothless leaves, opposite or in circles of more than two. The solitary or grouped flowers have two to six each sepals and petals, as many or twice as many each stamens and stigmas. The fruits are capsules. Only *Elatine* is cultivated. The family is practically cosmopolitan in its natural distribution.

ELATINE (E-lát-in-e)—Waterwort. The name of the genus *Elatine* is based on an ancient Greek one for some unknown plant. The genus belongs in the waterwort family ELATINACEAE and enjoys a wide natural distribution, chiefly in the Northern Hemisphere. Opinions vary regarding the number of species, but twenty is probably a fair estimate. Precise identification is often difficult.

Elatines are small, wet-soil or aquatic, annual and perennial creeping herbaceous plants with rooting stems. They have leaves opposite or in whorls (circles of more than two) and minute flowers with two to four sepals, the same number of petals, and as many or twice as many stamens. There are two to four styles or stigmas. The flowers are in the leaf axils or are terminal. The fruits are capsules.

Common throughout much of North America, Europe, and parts of Asia is variable ***E. triandra.*** A branching annual, it grows in mats on wet soils or under shallow water. Its stems, up to about 2 inches long, have linear to narrowly-obovate leaves about 1/6 inch long. The parts of the flowers are usually in threes. Confined as a wildling to the northeastern United States and adjacent Canada, ***E. minima*** is similar, but its flower parts are usually in twos. Mediterranean region ***E. macropoda*** has short-stalked, spatula-shaped, opposite leaves, up to 3/4 inch long, and flowers with four sepals, four petals, and eight stamens. It is perennial.

Garden Uses and Cultivation. Horticulturally waterworts are of minor importance. They are cultivated to some extent, especially *E. macropoda,* to carpet the bottoms of aquariums and shallow pools and sometimes in bog gardens. They thrive best in soft water and on sandy or clayey bottoms. For *E. macropoda* a minimum water temperature of 65°F is needed to prevent the plant hibernating.

ELATOSTEMA (Elato-stéma). A genus of 200 species, *Elatostema* of the nettle family URTICACEAE is endemic to the Old World. Its name is probably derived from the Greek *elatos,* driving (or elastic). It alludes to a springing action of the stamens.

Elatostemas are annual or perennial herbaceous plants related to *Pilea.* They have alternate, very asymmetrical, coarsely-toothed leaves and minute, unisexual

Elatostema sessile

Elatostema sessile (foliage)

flowers aggregated in crowded cushions in the leaf axils.

The only sort known to be cultivated, ***E. sessile*** is a variable, widespread native of warm parts of Asia, Africa, and islands of the Pacific. It has stems up to 1 foot long or longer and rich green, pointed-ovate leaves up to 5 inches long by 2½ inches wide, with glossy, wrinkled surfaces and well-marked, depressed veins.

Garden Uses and Cultivation. An attractive foliage plant for warm, humid, shady environments, this is a good groundcover for gardens in the tropics and an ornamental in tropical greenhouses. It is very easily increased by division and cuttings.

ELDER. See Sambucus. Box-elder is *Acer negundo.* Yellow-elder is *Tecoma stans.*

ELDERBERRY. See Sambucus.

ELECAMPANE is *Inula helenium.*

ELEOCHARIS (Eleóch-aris)—Chinese Water-Chestnut. Most species of this genus of the sedge family CYPERACEAE are rushlike plants with distinctly angled stems. Eleocharises characteristically grow in marshy and boggy soils. Few are cultivated. One, the Chinese water-chestnut, is cultivated in China and Japan for its edible tubers. Another plant, *Trapa natans,* is also called water-chestnut. There are 150 species of *Eleocharis* distributed throughout the world; most are perennials, a few annuals. The name, sometimes spelled *Heleocharis,* is derived from the Greek *helos,* a marsh, and *charis,* grace, and refers to the habitats and appearances of the plants. They are tufted or solitary and often have long rhizomes. Their spreading or erect, branchless stems bear leaves usually reduced to scales (the green stems perform the work of photosynthesis or food building usually associated with leaves) and terminate in a solitary spikelet of inconspicuous flowers.

Cultivated in aquariums, pools, and bog gardens, ***E. acicularis,*** sometimes called hair-grass, in the wild inhabits muddy watersides from Greenland to Alaska, Florida, and Mexico, and in Europe and Asia. It forms dense clumps or mats of hairlike stems rarely exceeding 6 inches in length, each terminating in a tiny, reddish-brown or purplish, ovate or lanceolate spikelet. A form called *E. a. submersa* grows entirely under water. Another, *E. a. fluitans,* has its roots under water and the upper parts of its stems floating on the surface. Probably these forms are not stable, but vary in response to environment. A rather similar species, ***E. parvula,*** inhab-

Eleocharis dulcis

its wet, saline soils along the coasts of North America, Cuba, Europe, Asia, and Africa. From *E. acicularis* it is distinguished by its subterranean rhizomes or stolons ending in small tubers. Distinct in habit is ***E. vivipara,*** which occurs along peaty watersides from Virginia to Florida. From rhizomes it produces tufts of slender, erect or reclining stems 4 inches to 1 foot in length, and most usually branching freely at their tops, in much the manner as those of papyrus (*Cyperus papyrus*), but considerably smaller. When the stems bend and touch the water or wet soil, they develop new plants at their tips. Flowers and seeds are rarely produced by this species.

The Chinese water-chestnut (***E. dulcis***) attains a height of 3 to 4 feet and forms clumps of longitudinally-grooved, green stems that have interior horizontal partitions; most or all of the stems terminate in a cylindrical flower spikelet 1 inch to 2 inches long. Its tubers are up to 2 inches in length by about 1 inch in diameter and are white with a thin brown or blackish skin. They have a sweet, nutty flavor and are used raw and cooked in Chinese foods. For this purpose they are grown in Hawaii and other warm climates. In some places the reedlike stems are made into baskets, mats, and hats.

Garden Uses. Except for the Chinese water-chestnut, which is raised for food, the cultivated sorts of *Eleocharis,* all of which are perennials, are chiefly grown in aquariums, usually submersed. They are also appropriate for bog gardens.

Cultivation. These plants grow without difficulty in mud or in unwashed river sand mixed with some clay and organic debris. Slightly acid water suits them best, that of a brackish character is appropriate for *E. parvula,* although it succeeds perfectly well in water that is not salty. The Chinese water-chestnut is cultivated under conditions that suit rice. Propagation is by division and by seeds and, for *E. vivipara,* by separating the upper branched portions of the stem, which root freely. The Chinese water-chestnut is grown from tubers.

ELEPHANT or ELEPHANT'S. These words are incorporated in various common names of plants including elephant-apple (*Dillenia indica*), elephant ear fern (*Elaphoglossum crinitum* and *Platycerium angolense*), elephant tree (*Bursera microphylla* and *Pachycormus*), elephant's ear (*Colocasia antiquorum* and *Enterolobium cyclocarpum*), elephant's foot (*Dioscorea elephantipes*), and elephant's head or little red elephants (*Pedicularis groenlandica*).

ELETTARIA (Elet-tària)—Cardamon. Tropical Asia is home to the seven species of this genus of the ginger family ZINGIBERACEAE. Its name is an East Indian one. The sorts are related to, and in cultivation are confused with, *Amomum.* They differ in the technicality that the connectives between the anthers are without appendages. Elettarias are ginger-like, perennial herbaceous plants with rhizomes from which rise erect leafy stems and leafless flower stalks. The species cultivated is the cardamon (*Elettaria cardamomum*). This is the chief source of the spice cardamom, which is used in curry powder, as a flavoring, and for chewing. An oil expressed from it is employed in perfumery. Cardamom of inferior quality is obtained from related genera including *Amomum.*

The cardamon (*E. cardamomum*) has fleshy rhizomes and jointed, somewhat canelike stems 6 to 12 feet tall, with alternate, near-ly stalkless, lanceolate leaves 1 foot to 2½ feet long. Above, they are dark green. Their under surfaces are paler and covered with fine silky hairs. The small white flowers, with blue and yellow lips, are in long, loose, racemes. The fruits, triangular capsules, contain the small light colored seeds (cardamom), which keep their flavor better if they are retained in the fruits until use.

Garden Uses and Cultivation. In addition to its use as a spice, the cardamon is an attractive foliage plant for partially shaded locations in tropical gardens and greenhouses. It has the same cultural needs and is propagated in the same way as *Alpinia.*

ELEUSINE (Eleusìn-e)—African-Millet, Wire Grass or Yard Grass. Rarely some of the about eight species of these warm region grasses, chiefly of the Old World, of the grass family GRAMINEAE, are cultivated for garden ornament and to provide flower spikes for fresh and dried arrangements. Tufted or spreading by stolons, they have mostly flattened stems, and narrow, folded or flat leaf blades. The spikelets of flowers are in crowded, one-sided spikes that are rarely solitary, more commonly in clusters of two to several at the tops of the stems. The spikelets are flattened and are without stalks or awns (bristles). They are in two rows and are much overlapped. The name of the genus *Eleusine* comes from that of the Greek town Eleusis, where the goddess of harvests Demeter was worshipped.

African-millet (***E. coracana***) is an agricultural derivative of some African or Asian species. It is a stout annual 4 to 5 feet in height with leaf blades up to 1 foot long and up to ⅓ inch wide or slightly wider. The finger-like spikes of more or less persistent spikelets may be 1½ inches long by ¼ to ¾ inch wide. The spherical seeds are used as food in Africa and India.

Similar to the last and perhaps the species from which it was derived, wire or yard grass (***E. indica***), of the tropics of the Old World, is naturalized in North America and occurs as a weed in lawns and elsewhere. It is less robust than African-millet. This annual, from under 1 foot to 3 feet tall, has leaf blades up to 1 foot long by ⅓ inch wide. The flower spikes, 1 inch to 7 inches in length by under ¼ inch wide, and spreading, are usually in groups of two to five at the tops of the stem, with often one or two spikes lower on the stem. This species has much the appearance of, and is often confused with, crab grass (*Digitaria sanguinalis*), but whereas the spikelets of crab grass are one-flowered, those of *Eleusine* have several.

A perennial, ***E. tristachya*** (syn. *E. oligostachya*), native of tropical and subtropical America and naturalized in parts of North America, is tufted and up to 1 foot tall. It has leaf blades up to 8 inches long by up to ¼ inch wide. The one to four flower spikes are 1 inch to 4 inches long and under ½ inch broad.

Garden Uses and Cultivation. As ornamentals and for cut flowers eleusines may be grown in ordinary garden soil in full sun. They are raised from seeds, the annuals by sowing them in spring where they are to remain and thinning out the seedlings so that they are not unduly crowded.

ELEUTHERINE (Eleùtherin-e). This genus of the iris family IRIDACEAE has a rather unusual natural distribution. Two species occur in tropical Central America and South America and in the West Indies, and two in Indochina. Also remarkable is that, atypical for a family in which the number of stamens is almost universally three, its flowers sometimes have six stamens. The name comes from the Greek *eleutheros,* free, in allusion to the stamens not being united.

Eleutherines are bulb plants related to *Trimezia* and *Cipura.* Their parallel-veined leaves are all basal. The naked flower stalks carry at their tops several white, tubeless flowers with six spreading perianth segments (petals) of equal size. The three or sometimes six stamens are quite separate. There is a short, three-branched style. The fruits are capsules.

The kind usually cultivated, ***Eleutherine bulbosa*** (syns. *E. palmifolia, E. plicata*), is indigenus to tropical Central and South America and the West Indies. It has a few longitudinally pleated leaves up to 1½ feet long and about 1 inch wide. The leafless flower stalks, 6 inches to 1 foot long, produce from their tops several short-lived blooms that open in succession, and are 1 inch across. In the wild they have three stamens, but a cultivated form has six.

Garden Uses and Cultivation. In the tropics *Eleutherine* is adapted for flower gardens and rock gardens. It can also be grown in greenhouses and succeeds where the winter night temperature is about 50°F and by day a few degrees higher. At other seasons higher temperatures are appreciated. It succeeds in well-drained, fertile, sandy peaty soil in sun. When the plants are in active growth the soil is kept fairly moist, at other times drier. Applications of dilute liquid fertilizer benefit well-rooted specimens in full growth. Repotting is done at the beginning of the growing season. Increase is by seeds and by offsets.

ELISENA. See Hymenocallis.

ELLANTHERA. This is the name of bigeneric orchid hybrids the parents of which are *Renanthera* and *Renantherella.*

ELLEANTHUS (El-leánthus). This genus of the orchid family ORCHIDACEAE comprises about fifty species, natives of Central America and South America and the West Indies. Its members closely resemble in growth their near relative *Sobralia,* but are very different in bloom. The name comes from the Greek *elio,* I shut in, and *anthos,* a flower. It alludes to the blooms being enclosed by the bracts.

Elleanthuses, mostly tree-perchers (epiphytes), more rarely grow on rocks or in the ground. They have clusters of reedlike stems with few to many ribbed leaves. Some kinds are dwarf and grasslike, others several feet tall. Unlike those of *Sobralia,* the flowers of *Elleanthus* are tiny and are many together in spikes or heads furnished with bracts usually bigger than the blooms. Also, the flowers have a pouch at the base of the lip containing a pair of protuberances. This is not true of *Sobralia.*

Said to attain heights of up to 8 feet in the wild, ***E. capitatus*** in cultivation is usually 1½ to 2 feet tall. It occurs from Mexico to South America and the West Indies. The leaves have elliptic-lanceolate, spreading blades up to 8 inches long by approximately one-half as wide and with tubular basal parts that sheathe and almost hide the stems. The tubular, rose-purple flowers, up to ½ inch long, are in dense, spherical heads nested among usually purple-edged, overlapping, green bracts. From the last, ***E. trilobatus,*** of Costa Rica and Panama, differs in having leaves not over 1 inch wide and up to 6 inches long, and flower spikes up to 3½ inches long that by their weight cause the stems to arch or droop. The blooms, like the floral bracts, are golden-yellow. Also from Costa Rica and Panama, ***E. hymenophorus*** has stems 9 inches to 2 feet tall with pointed-elliptic leaves 4 to 6 inches long by one-third to nearly one-half as wide. They terminate in spikes 2 to 4 inches long of brilliant orange to yellow flowers with the lip sometimes edged with red.

Garden Uses and Cultivation. These are easy-to-grow orchids. They respond to conditions and care that suit sobralias. For further information see Orchids.

Elleanthus hymenophorus

ELLIOTTIA (Elliótt-ia). There is only one species of *Elliottia,* a little-known deciduous shrub or rarely small tree of the heath family ERICACEAE. It occurs as a wildling at a few stations in South Carolina and Georgia only and unfortunately is rare in cultivation. Choice and beautiful, it is said to have been cultivated as early as 1813 and was described by Stephen Elliott, after whom it was named, in his "A Sketch of the Botany of South Carolina and Georgia," which appeared in 1821.

Only in very favorable locations and moist soils does ***E. racemosa*** attain its maximum height of 30 feet, elsewhere it is shrubby and usually 4 to 12 feet tall. It has slender, upright branches and alternate, pointed, elliptic to elliptic-oblong, toothless leaves 3 to 4 inches long and with hairy stalks, that remain late in the fall and turn rich crimson before they are shed. From most members of the heath family, *Elliottia* differs in having flowers with separate and distinct petals and from kinds that do have flowers of this type (*Ledum, Leiophyllum,* and *Cladothamnus*), by its blooms being in racemes. It is closely related to Asian *Tripetaleia,* but from that is easily distinguished by its flowers having four rather than three petals. The racemes, carried well above the foliage, are 6 to 10 inches long, often branched below, and with slender-stalked, slightly fragrant flowers loosely arranged. Each flower is 1 inch or slightly more in diameter; the petals are pure white and narrowly-oblong. They have eight short-filamented stamens with yellow anthers and a much longer, slender style. The flowers remain attractive for a long time. The fruits are capsules. This shrub is hardy as far north as Philadelphia.

Garden Uses and Cultivation. This is such a rare species that it can be considered at present only as a choice plant for collectors, to be located wherever it would seem most likely to succeed. It has been thought a difficult plant to grow, but this does not seem to be the case if it is accommodated in a peaty, sandy, acid soil always fairly moist, but not waterlogged. It should be transplanted with care to preserve as many as possible of its fine hairlike roots, and it is advisable to keep the ground surface around it mulched with peat moss or other acid organic material. It grows in sun or part-shade. Because seed is rarely available propagation must usually be by vegetative means and this has in the past proved neither rapid nor easy. The plants occasionally produce suckers, which can be carefully removed and used to establish new individuals. The production of suckers from roots left in the ground is stimulated if, after a specimen is dug for transplanting, the crater is left unfilled. From this it would seem probable that the same results would follow judicious root pruning accomplished by thrusting a spade into the soil not too distant from the trunk. Another method that has been successfully employed is that of planting pieces of root, 4 or 5 inches long and about ⅜ inch wide or wider, horizontally in flats (shallow boxes) of sandy soil, placing these in a greenhouse in late winter and using the shoots that arise from them as cuttings, which after treating with a root-inducing preparation, are rooted in a greenhouse propagating bench. This method of propagation was developed at the Arnold Arboretum and reported in 1969.

ELM. See Ulmus. The water-elm is *Planera aquatica*.

ELMERA (El-mèra). The only species of this genus of the saxifrage family SAXIFRAGACEAE is named in honor of its discoverer, the American botanist Adolph Daniel Edward Elmer, who died in 1942. A close relative of *Heuchera* and *Tellima*, it is endemic to subalpine and alpine altitudes in Washington. The most obvious differences between *Elmera* and *Heuchera* are that the stems of the former are leafy, the flowers in branchless racemes, and the petals cleft. Flowers with five instead of ten stamens distinguish *Elmera* from *Tellima*.

A more or less hairy and glandular, hardy herbaceous perennial with slender, horizontal rootstocks and flowering stems 4 to 10 inches tall, ***E. racemosa*** (syn. *Heuchera racemosa*) has long-stalked, kidney-shaped, double-toothed basal leaves up to 2 inches wide and considerably shorter than their lengths, and usually two or three stalked similar, but smaller stem leaves. The flowers have greenish-yellow, cup-shaped calyxes about ⅓ inch long, with triangular, usually erect sepals. The linear to spatula-shaped petals, up to one half as long as the sepals, may be partly or completely lacking. There are five small, white, three- to seven-cleft petals, five stamens, and two thick, persistent styles. The fruits are many-seeded capsules.

Garden and Landscape Uses and Cultivation. These are as for *Heuchera*.

ELODEA (Elo-dèa)—Water Weed. Sometimes known by its alternate name *Anacharis*, the genus *Elodea* of the frog's bit family HYDROCHARITACEAE consists of about a dozen species of underwater aquatic perennials inhabiting temperate and tropical parts of the Americas. The name, of obvious application, derives from the Greek *helodes*, bred in marshes.

The botany of elodeas is complicated and identification to species difficult. Some kinds are unisexual, others have unisexual and bisexual flowers on the same plant. The blooms originate from two-lipped spathes (leafy bracts) in the leaf axils. The solitary females have slender stalklike perianth tubes up to 1 foot in length, which raise the perianth lobes (petals) and stigmas to the water surface. There are three two-lobed stigmas. The males reach the surface either at the ends of filament-like stalks or, when their pollen is ready to be transferred to the females, they separate from the parent, rise to the surface, and float free. The fruits ripen under water.

A common North American, ***E. canadensis*** (syn. *Anacharis canadensis*) is native from Quebec to Saskatchewan, New York, and Kentucky. It inhabits still waters, often forming large colonies. Its branching stems are thickly furnished with blunt leaves 1/16 to ⅙ inch wide and in whorls (circles) of three. Female plants commonly have their leaves more crowded toward the ends of the branches than males. The blooms are about ⅓ inch in diameter. The males do not separate from the parent plant.

In about 1835 *E. canadensis* was introduced to Ireland. It spread from there with remarkable swiftness to other parts of the British Isles and continental Europe, and became a serious pest that clogged rivers and other waters. In the British Isles and other parts of Europe only male plants have been found, which indicates that the rapid propagation of the plant was entirely by vegetative means, largely, undoubtedly, by pieces carried by water birds from place to place, as well as by currents.

Another North American of widespread natural distribution and naturalized in Europe, ***E. nuttallii*** (syns. *E. occidentalis*, *Anacharis nuttallii*) differs from *E. canadensis* in having narrower leaves and male flowers that separate from the parent plant and float free.

South American ***E. callitrichoides*** (syn. *Anacharis callitrichoides*) has much-branched stems up to several feet in length, and opposite or in circles of three, finely-toothed, linear to ovate leaves up to ¾ inch long. The flowers are solitary and unisexual, the females with corolla tubes up to 9 inches long. A South American naturalized in North America, ***E. densa*** (syn. *Anacharis densa*) is much stouter than *E. canadensis* and has its circles of four to six leaves more distantly spaced along the stems, which may be many feet long. The leaves are linear, up to 1½ inches long and under ¼ inch wide. The white flowers are about ¾ inch across. Also South American, ***E. naias*** (syn. *Egeria naias*) has slender stems and extremely narrow linear leaves up to 1 inch long in circles of four to eight.

Garden Uses and Cultivation. These submersed aquatics are grown in aquariums and ponds. Under favorable conditions they increase rapidly and in mud-bottomed water may be troublesome because of their excessive growth. They are quite satisfactory in tubs or other containers in concrete and other solid-bottomed pools. They provide hiding attractive to young fish. They should be planted in mud or washed river sand that contains some mud content and are increased with great ease by means of pieces broken off and pushed into mud or sand. They need good light.

ELSHOLTZIA (Elshólt-zia)—Mint Bush. This genus must not be confused with the California-poppy (*Eschscholzia*), a member of the poppy family PAPAVERACEAE. Thirty-five Asian, European, and Abyssinian annuals and subshrubs are included in *Elsholtzia*, which belongs in the mint family LABIATAE, and is named after Johann Sigismund Elsholtz, a Prussian physician and botanist, who died in 1688.

Elsholtzias have short-stalked, opposite, toothed leaves, often glandular-puncate and aromatic, and numerous white, blue, lilac, or purplish-pink, small flowers in usually one-sided terminal spikes. The blooms are slightly two-lipped with the upper lip concave and notched and the lower one three-lobed. The four usually protruding stamens are in two pairs, one of which is longer than the other. The slender style has two stigmas. The fruits consist of four small nutlets.

Two quite different kinds are cultivated to a limited extent, one a shrub commonly called mint-bush (*E. stauntonii*), the other an annual, *E. ciliata* (syn. *E. cristata*). The mint-bush is by no means showy. Its chief merits are its late season of flowering, its long season of bloom, and its reliability. From 3 to 5 feet tall, ***E. stauntonii*** has round stems (unusual in the mint family). Its light green leaves, 2½ to 4 inches long, ovate to oblong-lanceolate and toothed, when crushed or bruised have a pleasant minty fragrance. The flowers, muted bluish-pink or pinkish-lavender, are in spikes 4 to 8 inches long mostly displayed in panicles at the termination of the branches. The stamens and style protrude. This modest little shrub succeeds in any ordinary soil not excessively dry, and in full sun. It is a

Elsholtzia stauntonii

native of northern China and is hardy throughout most of New England, but in the north its stems die down in winter and are replaced by new ones in spring. Another species, ***E. fruticosa*** of temperate Asia, is described as being a subshrub 6 feet in height with white flowers. Annual ***E. ciliata,*** of Europe and temperate Asia, has upright stems 1½ to 2½ feet in height, pale green, oval, toothed leaves about 2 inches long, which smell strongly of mint when bruised, and spikes of pleasantly fragrant, light lavender-blue flowers in 3-inch-long spikes. The display rarely lasts more than two weeks, but a succession of blooming plants may be had by making more than one sowing at intervals of about three weeks. It is reported that the Ainu, of northern Japan, use this as a palliative for the aftereffects of intoxication.

Garden Uses. Both cultivated species can be appropriately included in herb gardens. The annual is also worth growing to a limited extent for interest at the fronts of flower borders, and the perennial in shrub borders and mixed flower beds.

Cultivation. The annual kind is grown from seeds sown outdoors where the plants are to remain. In spring the seeds are raked into the surface of the ground, and the young plants are thinned to stand 6 inches apart. No special care is required. Weeds must be kept down, and in very dry weather watering is needed.

The mint-bush (*E. stauntonii*) thrives in any ordinary garden soil in full sun, but grows best in one that is deep and fertile. It is very easily raised from seeds, which it produces freely, sown in spring either indoors or outdoors, and from leafy cuttings of youngish shoots in summer. Pruning consists of cutting the stems back to ground level every late winter or early spring.

ELYMUS (Ély-mus). Mostly hardy, annual and perennial grasses indigenous to temperate and cold parts of North America, Europe, Asia, and South America constitute this genus of some fifty species. It belongs to the grass family GRAMINEAE and has a name based on the ancient Greek *elymos,* a kind of grain.

Tufted or with rhizomes, elymuses have usually upright, slender to stout stems, and leaves with flat or rolled linear blades. The slender to stout flower spikes are of stalkless spikelets, each two- to seven-flowered, and generally paired, more rarely solitary or in threes, fours, or sixes from each joint, and arranged alternately on opposite sides of the axis that carries them. In some kinds, long, projecting bristle-like awns give a whiskery appearance to the spikes.

Several species have decorative merit, despite which they are not widely cultivated. One of the most attractive, ***Elymus glaucus,*** is native from Ontario to Michigan and Arkansas, and from Alaska to California. From 2 to 4 feet tall, this kind has leaves up to 1 foot long, with soft blades, up to ½ inch wide. The slender, whiskered flower spikes, straight or sometimes nodding, are often purplish. Another glaucous-leaved kind, ***E. canadensis glaucifolius*** differs from the last in the bristle-like awns of its flower heads being much curved and spreading instead of erect and nearly straight. From 3 to 5 feet tall, ***E. canadensis*** is wild from Quebec to Alaska, North Carolina, Texas, and California. Its leaves, 1 foot long or longer, have flat blades ½ to ¾ inch wide.

Elymus glaucus

The largest North American species, ***E. condensatus*** in the wild occurs from Minnesota to the Pacific Coast. Densely tufted, and 3 to 9 feet in height, it has stout stems and very long leaf blades ¼ to ¾ inch in width. The dense, sometimes interrupted flower spikes, 4 inches to 1 foot in length, are without bristles.

Sea lyme grass or dune grass (***E. arenarius mollis*** syns. *E. mollis, E. arenarius villosus*) is planted to stabilize shifting sands. Seeds of this American variety of a species native in Europe and Asia were used as food by some tribes of Indians. Native on beaches and in sandy soils from Greenland and arctic North America to Massachusetts, the Great Lakes, and California, the variety differs from the typical species in parts of its flower spikes being hairy. Sea lyme grass has extensively creeping, cordlike rootstocks and stout stems up to 5 feet tall. The leaves are long, glaucous, and up to ¾ inch wide. The stiff, dense flower spikes are 4 inches to 1 foot long and without projecting bristles.

Elymus racemosus

Attractive ***E. racemosus*** (syn. *E. giganteus*), of eastern Europe and northern Asia, 2 to 4 feet tall, has long, creeping rhizomes. Its mostly basal, glaucous-blue-gray leaves have pointed blades up to 2½ feet long by ¼ to ⅔ inch wide. The crowded, narrow flower spikes are 6 to 10 inches long. This prospers near the sea.

Garden and Landscape Uses and Cultivation. All the grasses discussed above are hardy perennials. The chief use of sea lyme grass is as a sand binder. Other kinds, especially those with glaucous foliage, are planted for ornament in beds, as informal groups, and on banks. They grow without difficulty in a variety of soils, moist or dryish, and in full sun, which is generally most to their liking, or part-day shade. Little routine care other than trimming back dead tops in fall or spring is needed. A yearly spring application of a complete fertilizer stimulates vigorous growth. Propagation is easy by division in early fall or spring. Increase can also be had from seed.

ELYTRARIA (Ely-trària). Seven species of warm-temperate and subtropical regions constitute *Elytraria,* of the acanthus family ACANTHACEAE. The name, alluding to the bracts, comes from the Greek *elytron,* a sheath.

These are nonhardy herbaceous perennials with leaves chiefly in basal rosettes. The flowering stalks, clothed with rigid, sheathing, scalelike bracts, terminate in spikes of small, slightly two-lipped, white or blue flowers that have a five-lobed calyx, a tubular corolla with five spreading

Elytraria carolinensis

lobes (petals), two stamens, and one style. The fruits are capsules.

Native from South Carolina to Florida, ***E. carolinensis*** (syn. *Tubiflora carolinensis*) has a basal rosette of broad-elliptic to spatula-shaped leaves 2 to 8 inches long and slender, erect flowering stalks 6 inches to 1½ feet tall. The white or blue flowers are about ⅓ inch long.

Garden Uses and Cultivation. The species described is occasionally grown in botanical collections and is an interesting window garden plant. It succeeds in ordinary soil in sun or light shade, and is easily raised from seed.

EMBLIC. See *Phyllanthus emblica.*

EMBOTHRIUM (Embòth-rium). Eight species of evergreen trees or shrubs of Andean South America constitute *Embothrium.* Belonging in the protea family PROTEACEAE, this genus takes its name from the Greek *en,* in, and *bothrion,* a small pit, in allusion to the anthers. One species is numbered among the most beautiful and decorative of flowering trees. Unfortunately neither it, *E. coccineum,* nor its somewhat more cold-resistant variety *E. c. longifolium* are hardy except in very mild climates. They are grown successfully in Pacific Coast gardens and in the warmest parts of Great Britain and Ireland. The plant sometimes named *E. wickhamii* and *E. pinnata* is *Oreocallis pinnata.*

Embothrium coccineum (flowers)

In bloom ***E. coccineum,*** native of Chile, and *E. c. longifolium,* indigenous to Tierra del Fuego, are superb. The latter has longer and narrower leaves than the former. A suckering tree up to 40 feet tall or sometimes taller, *E. coccineum* has alternate, deep green, undivided, leathery leaves 2½ to 4½ inches long and ¾ inch to 1½ inches wide, and ovate-lanceolate to narrowly-ovate. In spring, in terminal and axillary racemes 3 to 4 inches long and as wide, the crimson-scarlet flowers contrast brilliantly with the foliage. The blooms have slender stalks ½ to ¾ inch long. Their corolla tubes, 1 inch to 1½ inches long, are divided deeply into four recurved and twisted, strap-shaped lobes that enclose the anthers. The long style, tipped with a yellow stigma, protrudes. The fruits are many-seeded follicles.

Garden and Landscape Uses and Cultivation. These trees are spectacular as specimens and conversation pieces and show to especially fine advantage against a background of tall evergreens. For their successful cultivation embothriums require full sun and reasonably fertile, not alkaline soil. One of a peaty nature is to their liking. They are benefited by a mulch of compost, peat moss, or similar material. What little pruning is necessary is only enough to keep them shapely. It should be done as soon as flowering is through. Propagation is by seeds, cuttings, and suckers.

EMILIA (Emíl-ia)—Tassel Flower, Flora's Paint Brush. To the daisy family COMPOSITAE belong the thirty species of annuals and herbaceous perennials, chiefly natives of the Old World tropics, that constitute the genus *Emilia.* The derivation of the name of the genus is obscure. It is, perhaps, a commemorative one.

The solitary or clustered flower heads of emilias are without ray florets, the disk ones are all fertile. The fruits are seedlike achenes. The leaves are in basal clusters and are alternate on the stems. Two species, both grown under the names of tassel flower and sometimes Flora's paint brush, and which are commonly confused as to identification, are cultivated.

Best known, ***E. javanica*** (syn. *E. sagittata*) is an annual about 2 feet tall with erect stems and crowded lower leaves that are ovate and toothed and taper gradually to narrowly-winged stalks. The upper stem leaves are smaller and more widely spaced. The scarlet or orange-scarlet flower heads, about ½ inch in diameter, are borne in loose clusters throughout the summer. Variety *E. j. lutea* has golden-yellow flower heads. Another tropical annual, 2-foot-tall ***E. sonchifolia*** has toothed or lobed, obovate leaves, and rose-pink to purple or more rarely white flowers in heads about ½ inch across.

Garden Uses and Cultivation. Although not abundantly showy, the emilias described here are sufficiently attractive to be given places in flower beds, borders, and rock gardens. They are of some use as cut flowers, lasting quite well in water. Propagation is easy from seeds sown outdoors in ordinary garden soil in full sun in early spring. All that is necessary is to rake the seeds into the surface of the ground and thin the seedlings so that they are 5 to 6 inches apart. Weeds, of course, must be kept down. Staking is not needed. If the soil near colonies of these plants is undisturbed, it is not unusual for self-sown seedlings to appear in abundance the following year. An alternative method is to sow seeds indoors about eight weeks before it is safe to set the young plants in the garden. A temperature of 60 to 65°F is appropriate for germination. Transplant the seedlings about 2 inches apart in fertile soil in well-drained flats and grow them in full sun where the temperature is about 55°F at night and by day is five to ten degrees higher. Water moderately. When there is no longer danger of frost and it is safe to set out tomatoes, plant the emilias where they are to bloom, spacing them about 6 inches apart.

EMMENANTHE (Emmen-ánthe) — Whispering Bells or Golden Bells or Yellow Bells. This agreeably fragrant western North American annual is related to and closely resembles *Phacelia.* It belongs in the water leaf family HYDROPHYLLACEAE and consists of one species, called whispering bells, golden bells, or yellow bells. From *Phacelia* it differs in having yellow or cream, pendulous flowers with persistent corollas. The name is from the Greek *emmeno,* to abide, and *anthos,* a flower, and alludes to the corollas remaining attached after they wither.

Whispering bells (***E. penduliflora***) is a native of dry and desert places from California to Utah, Arizona, and Baja California. Erect, freely-branched, and ½ to 1½ feet tall, it is sticky-pubescent and has linear-oblong, pinnately-divided leaves 1½ to 4 inches long, with toothed or toothless lobes. The tubular-bell-shaped flowers are in panicle-like clusters and are up to ½ inch long; their stalks are recurved and slender. They have a five-lobed calyx, five rounded corolla lobes, and five stamens, which do not protrude from the corolla. The short style is slightly two-cleft. The fruits are capsules. In variety *E. p. rosea* the flowers are pink.

Garden and Landscape Uses. In the West this is a pleasing plant for inclusion in native plant gardens and, in suitable places, for naturalizing. It is also attractive for flower garden decoration, being useful for such purposes as edging paths and planting in drifts at the fronts of flower borders. If cut as soon as the blooms are fully open, tied in bundles, and hung upside down in a dry, airy place out of direct sun, the flowers may be used as everlastings in dried flower arrangements.

Cultivation. Full sun and porous, well-drained soil are best to the liking of whispering bells. Seeds are sown in early spring where the plants are to remain and the young plants are thinned to stand about 6 inches apart. Alternatively, and this is the best procedure in regions of hot, humid summers, seeds may be sown in a greenhouse about eight weeks earlier than it is safe to transplant the seedlings to the garden, which may be done as soon as all danger of frost is passed. Later care is minimal and routine; it consists chiefly of keeping down weeds.

EMMENOPTERYS (Emmen-ópterys). A member of the madder family RUBIACEAE and related to the American genus *Pinckneya,* the only species of *Emmenopterys* is a deciduous tree of China, which is not hardy in the north. Its name, from the Greek *emmeno,* persisting, and *pteryx,* a wing, alludes to the long-lasting, winglike calyx lobes.

Scarce in cultivation, ***E. henryi,*** 30 to 80 feet tall, is one of the most beautiful forest trees of its native land. It has opposite, pointed-ovate to broad-elliptic, rather leathery, toothless leaves, the largest up to 9 inches long by nearly one-half as wide, but commonly about 6 inches long and 3 inches wide. They are dullish green, without hairs on their upper sides, paler and hairy along the veins beneath. The reddish leafstalks are ½ inch to 2 inches long. Borne in many-flowered, terminal panicles 4 to 8 inches long, the white or perhaps sometimes yellow flowers make handsome displays in summer. They are about 1 inch long and wide and have bell-shaped corollas, downy inside and out, and expanding above into five spreading, rounded lobes (petals). Neither the five stamens nor the style are longer than the corolla. The calyx is five-lobed. As in more familiar, related *Mussaenda* and unrelated *Schizophragma,* one calyx lobe of some of the flowers is petal-like and greatly enlarged. These big lobes are ovate to oblong, their blades up to 2½ inches long by 1½ inches wide. The fruits, woody, spindle-shaped, ribbed capsules up to 2 inches long, contain winged seeds.

Garden and Landscape Uses and Cultivation. Here is a tree that should receive more attention from gardeners in regions of mild winters. It is hardy in southern England, but it has rarely if ever flowered there. In eastern North America it may survive perhaps about as far north as Washington, D.C. It succeeds in ordinary, fertile, well-drained soil. Propagation is easy by summer cuttings 2 to 4 inches in length made of firm, but not hard shoots, planted under mist or in a greenhouse propagating bench, preferably with slight bottom heat. No pruning other than occasional cutting that may be desirable to ensure shapeliness is needed. This is done in late winter.

EMPETRACEAE—Crowberry Family. Low, heathlike, evergreen, dicotyledonous shrubs numbering ten species or more and representing three genera constitute this family, which occurs natively in north temperate regions, the Andes, the Falkland Islands, and Tristan da Cunha.

Family characteristics are crowded, alternate, linear leaves with turned-under margins that leave a deep lengthwise groove along the underside of the leaf and small, symmetrical, usually unisexual, petal-less flowers in terminal heads or from the leaf axils. They have two to six sometimes petal-like sepals, or are occasionally without sepals. There are two or four stamens and a very short, variously-branched style, the branches often toothed or fringed. The fruits are fleshy or dry, berry-like drupes. Genera cultivated are *Corema* and *Empetrum.*

EMPETRUM (Ém-petrum)—Crowberry. Three or four species of heathlike shrubs or shrublets of the crowberry family EMPETRACEAE constitute *Empetrum,* which inhabits arctic and subarctic regions around the world and in North America extends southward to California, northern Michigan, and New York. The name, from the Greek *en,* and *petros,* rock, alludes to a common habitat of these plants.

Crowberries are low or procumbent evergreens with undivided narrow leaves in whorls (circles of more than two). The small, unisexual or bisexual, greenish to purplish, petal-less flowers, which arise in the leaf axils, have three each sepals and stamens, and one style. The fruits are berry-like drupes.

The black crowberry (***E. nigrum***), indigenous throughout the natural range of the genus, is spreading and procumbent. Up to 10 inches high, its branches are crowded with linear-oblong, dark green leaves ¼ inch long or slightly longer, those on young shoots glandular. The flowers are unisexual, the sexes on the same or different plants. The fruits are black or rarely dark purple, edible berries under ¼ inch in diameter.

The rockberry (***E. eamesii*** syn. *E. rubrum eamesii*), of Newfoundland and the nearby mainland, has white-hairy, but not glandular branchlets and very glossy elliptic-oblong to oblong-linear leaves ⅜ inch long. The flowers are bisexual or unisexual with the sexes on the same plant. The fruits are pink or red, about 3/16 inch in diameter, and edible. Variety *E. e. hermaphrodidum,* of northern North America, Europe, and Asia, has branches that are hairless or have a fuzz of brown hairs, and black fruits. The plant sometimes named *E. nigrum purpureum* probably belongs here.

Garden Uses and Cultivation. These are attractive ground covers for rock gardens and native plant gardens and for associating with heathers, heaths, and other ericaceous plants. They form dense mats of good-looking foliage and are ornamental in fruit. They flourish in fairly moist, sandy peaty acid soil in full sun. Their berries are edible. Empetrums are easily increased by division, layering, summer cuttings, and by seeds freed from the pulp of the berries and sown in sandy peaty soil or milled sphagnum moss kept evenly moist. Established plants need little or no care.

EMPRESS TREE is *Paulownia tomentosa.*

EMU BUSH is *Eremophila maculata.*

ENCELIA (En-cèlia)—Brittle Bush or Incienso. Shrubs, subshrubs, and herbaceous perennials of the daisy family COMPOSITAE compose *Encelia,* a genus of about fifteen species that occurs in the wild from the western United States to Chile. Its name commemorates Christopher Encel, a botanist, who died in 1577.

Encelias are often strongly scented and have alternate, generally pubescent, nearly or quite toothless leaves, and small to medium-sized flower heads, usually of disk and ray florets, but sometimes of only the former. The flower heads are solitary or in panicles. The ray florets (those equivalent to the petal-like ones of daisies) are yellow, those of the disk (comparable to those composing a daisy's eye) are yellow, purplish, or blackish, and fertile. The seedlike fruits are flat achenes.

Brittle bush or incienso (***E. farinosa***) in bloom is one of the showiest inhabitants of deserts and semideserts of the western United States and adjacent Mexico. It favors gravely and rocky slopes, and in bloom is a dense, slightly woody bush 1½ to 2½ feet tall. Its gray-green leaves, ¾ inch to 2 inches long by up to 1 inch wide are thickly clothed with crooked, whitish, short hairs. Lifted well above the foliage on long, leafless stalks, the loose clusters of flower heads are displayed in profusion. About ¾ inch in diameter, they have showy, yellow rays and yellow or purple centers. The name incienso alludes to the use made in parts of Mexico of a clear resin obtained from the wood that is burned as incense. The Indians used this resin as glue.

Inhabiting coastal regions in California and Baja California, ***E. californica*** differs from the last in its leaves not being white-hairy, and its flower heads being solitary. From 2 to 5 feet in height, it forms rounded clumps. Its pointed, lanceolate to ovate leaves are pubescent and up to 1¼ inches long. The showy, dark-centered yellow flower heads are about 1 inch in diameter.

Garden and Landscape Uses and Cultivation. In regions where they are native and in similar desert and semidesert places encelias are occasionally planted and are effective in informal landscapes. They need full sun and sharp soil drainage. Propagation is by seeds, and young plants should be grown in cans or other containers until they are big enough to set out in their permanent locations.

ENCEPHALARTOS (Encepha-lártos). Endemic to Africa, except for the solitary species of *Stangeria,* this is the only genus of the cycad family CYCADACEAE native there. It comprises thirty species. The name, which alludes to the practice of African aboriginals of preparing an edible meal from the starchy seeds, is from the Greek *en,* within, *kephale,* a head, and *artos,* bread.

Usually without branches, members of this genus have trunks from entirely subterranean up to heights of 30 feet, according to species. They are rough with scales or scars that mark the breaking-off points of fallen leaves. Clustered in palmlike fashion as a crown at the apex of the trunk are the leathery, stiff, pinnate leaves. They have linear to lanceolate, mostly more or less spiny-toothed leaflets. Male and female cones are on different plants. From *Cycas,* which *Encephalartos* generally resembles, the latter is readily distinguished by its leaflets being without distinct midribs.

Encephalartos horridus, South Africa

Encephalartos villosus with cone, Pretoria, South Africa

Two kinds with subterranean or at most short-above-ground trunks are *E. horridus* and *E. villosus,* both natives of South Africa. Very distinct ***E. horridus*** has silvery-blue-glaucous leaves 2 to 4 feet in length, with recurved tips, and four-angled stalks. Their forked and twisted, opposite or alternate leaflets are asymmetrically ovate-lanceolate, very spiny-margined, and up to 4 inches long by one-half as wide. The cylindrical cones of male plants are about 1 foot long by 2½ inches wide, those of females oblong-ovoid, shorter, and squatter. By contrast the erect to spreading, slightly arching leaves of woolly-trunked ***E. villosus*** are 5 to 9 feet long and the broadest up to 1½ feet wide. Less rigid than those of many cycads, and densely-woolly-hairy when young, they have on each side of the midrib sixty to ninety alternate or opposite leaflets, those at the base mere spines, those above linear-lanceolate, up to 8 inches long by ¾ to 1 inch wide, and with two spines near their apexes. From 1 foot to 2 feet in length and cylindrical, the male cones are pale yellow. Those of female plants are greenish-yellow to apricot-yellow and about 1½ feet long by a little over one-third as wide. Another trunkless or practically trunkless species, rare *E. kosiensis* of South Africa has green leaves about 3 feet long. Its cones are orange.

Very glaucous foliage, the leaves about 3 feet long, topping stems up to 9 feet high is typical of South American ***E. lehmannii.*** The spine-pointed leaflets, which taper to both ends and are up to 8 inches long by ¾ inch at their widest, have one or two spiny lobes along each side, or are sometimes without these. Male cones of this kind are cylindrical, 9 inches long by 2 inches wide, and yellow. Cones of female plants are reddish-brown and up to 1½ feet long by 1 foot wide. Of unrecorded nativity, ***E. latifrons*** is up to 8 feet tall. Its 1 to 3 foot-long leaves have about thirty pairs of ovate to ovate-lanceolate, spreading, mostly overlapping, spine-tipped leaflets each with three or four coarse lobes or teeth and the biggest about 5 inches long by 2 inches wide. The brownish-yellow male cones are about 2 feet long by 6 inches in diameter. Female cones, which weigh up to fifty pounds, may be nearly 2 feet long by 8 inches wide.

Encephalartos lehmannii, South Africa

Encephalartos latifrons at the Royal Botanic Gardens, Kew, England

Encephalartos latifrons (cone)

From 15 to 20 feet tall, South African ***E. altensteinii*** has a trunk up to 2½ feet thick crowned with circlets of leaves 5 to 6 feet long; these have numerous leaflets about 6 inches long by 1 inch wide, with three to five spiny teeth on each margin. The leaflets stand up from the midrib at quite a sharp angle. The cones of male plants are 1 foot to 1¼ feet long by 4 inches wide,

Encephalartos woodii at the botanic garden, Durban, South Africa

Encephalartos woodii with cones, Longwood Gardens, Kennett Square, Pennsylvania

those of female plants are ovoid and up to 1½ feet long by 10 inches wide.

Extremely rare (male specimens have been found only three times in the wild, females never) is ***E. woodii,*** of South Africa. Up to 18 feet tall, this remarkable and handsome species has numerous arching

Encephalartos transvenosus at the botanic garden, Durban, South Africa

leaves 5 to 6 feet long, with leaflets at and above their centers up to 8 inches long and 1½ inches wide, with a few spiny teeth along their edges, and lower leaflets, except the very lowest which are reduced essentially to spines, ovate, with spine-tipped lobes. Male cones, up to 4 feet long, are slender and cylindrical.

One of the tallest species is ***E. transvenosus,*** of South Africa. Up to 30 feet in height, this has a splendid head of handsome foliage. Its trunk, up to almost 3 feet in diameter, sometimes branches. Somewhat similar to *E. altensteinii,* it differs from that kind in its leaflets curving back slightly from the midrib of the leaf instead of being raised to form a stiff v pattern. When young its leaves are hairy. Other species that much resemble *E. altensteinii* are *E. natalensis,* and *E. longifolius.*

A giant of the genus, ***E. laurentianus*** comes from tropical Africa. In the wild it is up to 30 feet or more in height and has a trunk 2 to 2½ feet or more in diameter. Sometimes exceeding 20 feet in length, the leaves have the leaflets toward their middles 1¼ feet long and 2 inches wide. The leaflets are lanceolate and spiny-toothed. Both male and female cones are about 10 inches long. The latter are red. Another sometimes large species from tropical Africa is ***E. hildebrandtii.*** This, up to 20 feet tall, but often lower, has a trunk about 1 foot in diameter. Its leaves, up to 9 feet long, have fifty to seventy pairs of linear-lanceolate leaflets 6 to 9 inches long by ¾ to 1 inch wide, with two or three spines along each margin. From 8 inches to 1½ feet long, the male cones are 2½ to 4 inches thick. The cylindrical female cones are up to 2 feet long by 7 inches in diameter.

Largely through the efforts of the Fairchild Tropical Garden, Miami, Florida, a number of other *Encephalartos* species have in recent years been introduced to American gardens. These include ***E. bubalinus,*** which has a thick trunk 2 to 3 feet tall and grayish leaves with thorny leaflets. Its greenish-yellow fruiting cones may be 1 foot long. ***E. caffer,*** nearly trunkless, has

Encephalartos natalensis, Durban, South Africa

Encephalartos natalensis with female cone

Encephalartos natalensis with male cones, the Royal Botanic Garden, Edinburgh, Scotland

leaves about 2 feet long. ***E. cycadifolius,*** as its name implies, has much the aspect of a *Cycas.* Low and handsome, it has leaves with narrow leaflets. ***E. ferox*** is short-trunked. Its dark green leaves have sharply-spiny, holly-like leaflets. The fruiting cones are 1 foot to 1½ feet long and bright orange. ***E. gratus*** has trunks sometimes over 4 feet wide and leaves 12 to 18 feet long by up to 3 feet wide. ***E. laevifolius*** has leaves with slender leaflets much resembling those of cycases. ***E. lebomboensis*** is up to 6 feet tall, and has 4-foot-long leaves with spiny leaflets. ***E. manikensis*** is a handsome, thick-trunked kind with ample crowns of 4-foot-long leaves the leaflets of which have few marginal spines, but are spine-tipped.

Garden and Landscape Uses and Cultivation. Encephalartoses have landscape values similar to other cycads. They are choice and handsome show pieces for plant connoisseurs and, especially in large sizes, cannot fail to attract favorable attention from even the less horticulturally knowledgeable. As specimens standing alone or if near other plants at least not crowded, they are superb. They associate fittingly with architectural features. As a group encephalartoses are lovers of summer heat. Their degree of winter cold resistance varies somewhat according to kind, those native to the cooler parts of South Africa being hardier than species that in the wild inhabit warmer areas. None withstands much frost. These are plants for outdoor cultivation in the tropics and warm subtropics only. They prefer fertile, loamy soils that are satisfactorily drained and sunny locations, although they will stand part-day shade. Propagation, by seeds and offsets, is slow. For more information see Cycads.

ENCEPHALOCARPUS (Encephalo-càrpus). A genus of one species, *Encephalocarpus*, of the cactus family CACTACEAE, is a native of Mexico. Closely related to *Pelecyphora* and by some authorities included there, it has a generic name derived from the Greek *en*, in, *kephale*, a head, and *karpos*, a fruit, alludes to the manner in which fruits are borne. It is known colloquially as pine cone cactus.

A thick taproot and nearly spherical, greenish-gray, flattish-topped plant bodies about 2 inches in diameter are characteristic of ***E. strobiliformis.*** They have flat-topped, triangular tubercles regularly arranged in spirals in much the manner of the scales of some pine cones (the *strobiliformis* part of the name means conelike). From the top of each sprout a few soft, white spines, but these are soon shed. The 1-inch-wide, purple flowers come from the woolly axils of the uppermost tubercles. Usually they open about noon and last only a few hours.

Garden Uses and Cultivation. Not considered easy to grow, this choice and charming species is best treated as advised for *Ariocarpus*. For additional information see Cactuses.

ENCHANTER'S-NIGHTSHADE. See Circaea.

ENCHYLAENA (Enchy-laèna). Half a dozen species constitute this Australian genus of the goosefoot family CHENOPODIACEAE. The name, given in allusion to the soft, berry-like fruiting parts that enclose the seed, is from the Greek *en*, in, and *chlaina*, a garment.

A dense, spreading or procumbent, much-branched shrub up to 3 feet tall or sometimes taller, ***Enchylaena tomentosa*** has slender, silvery or gray, finely-hairy stems, and alternate, linear or cylindrical leaves generally under ½ inch long. Solitary in the leaf axils, the tiny flowers have urn-shaped to spherical, five-toothed persistent perianths, and five stamens. The fruits are enclosed by the perianths, which become fleshy and red or yellow when ripe. In its native land this species is to some extent grazed by sheep. Its fruits are eaten by Australian aborigines.

Garden and Landscape Uses and Cultivation. Not hardy in the north, *E. tomentosa* is suitable for mild, dry climates. It succeeds in ordinary soil in sunny locations. Propagation is usually by seed.

ENCYCLIA. See Epidendrum.

ENCYCLIPEDIUM. This is the name of orchid hybrids the parents of which are *Cypripedium* and *Encyclia.*

ENDEMIC. An endemic plant is a kind native only to a specified particular region. Flowering dogwood (*Cornus florida*) is endemic to eastern North America, redwood (*Sequoia sempervirens*) to western North America. Both are endemic to North America. Endemic plants of one region are often cultivated elsewhere. Compare Indigenous and Exotic.

ENDIVE OR ESCAROLE. This must not be confused with Belgian or French endive, which is more correcty whitloof chicory and is dealt with under Chicory. True endive is lettuce-like *Cichorium endivia*, of the daisy family COMPOSITAE. Probably native to India, it is cultivated as a salad and for cooking as greens. It withstands heat better than lettuce, but yet may bolt prematurely to flower and seed in very hot weather.

Soil for endive should be fertile and fairly moist. Medium-heavy loams are better than sandier earths. Prepare the ground by turning under with a spade, spading fork, rototiller, or plow manure, compost, or other suitable organic matter and mix in the surface few inches a dressing of a complete garden fertilizer.

The first sowing may be made as early as the land is in fit condition in spring, succeeding ones at intervals. The best results are often had from plants that mature in fall from summer seedings. Sow thinly or drop three or four seeds every foot along rows 1½ feet apart. Cover the seeds to a depth of ½ to ¾ inch. When the plants are 1½ to 2 inches tall, thin them to 1 foot apart. If carefully removed, the thinnings can be transplanted.

Routine care consists of restricting weed growth by cultivating shallowly at frequent intervals or by mulching, and by a certain amount of hand pulling along the rows. Watering is necessary in dry weather. Endive must never be allowed to suffer for lack of moisture. Keep the ground damp. Like all leaf vegetables this requires adequate supplies of nitrogen. When the plants are half-grown apply a light dressing of a fertilizer, such as nitrate of soda or urea, rich in a quickly available form of this element. If the plants are to be used as salads blanching is needed to ensure the leaves being tender, crisp, and not unpleasantly bitter. This will turn the centers of the plants creamy-white. To accomplish this, about three weeks before harvesting, on a dry day when no moisture is on the plants, lift the outer leaves upward and over the centers of the plants until they touch, wrap a piece of waterproof paper around to hold the leaves in place, and secure it with an elastic band or a string tie. An alternative method is to position a 6-inch-wide board over the plants at such a height that it nearly touches the foliage. After rain or heavy fog lift the board and if the plants are wet leave it off until they have dried. If moisture remains for long periods it rots the hearts of the plants.

Harvest as the plants mature by cutting them as you would lettuce just below the lower leaves. Endive withstands a touch of frost, but not hard freezing. Its season can be prolonged by digging the plants in fall with ample roots and planting them in a cold frame or brightly lighted cool cellar or garage.

ENDYMION (Endỳm-ion)—Spanish Bluebell, English Bluebell. At one time included in the related genus *Scilla*, the about ten species of *Endymion* differ from the group as now conceived by well-marked characteristics. Among the most definite are two of which many successful cultivators of Spanish and English bluebells are not aware, that their bulbs are completely renewed each year, and their roots are of only annual duration. The genus *Endymion* inhabits western Europe and northwest Africa, and belongs in the lily family LILIACEAE. Its name is that of a mythological character.

Endymions are deciduous bulb plants with narrow basal leaves and erect, branchless, leafless, succulent stalks bearing racemes of six-parted flowers. Each bloom has at the base of its individual stalk a bract and a bracteole (tiny bract). There are normally six perianth lobes (commonly called petals, but more correctly tepals), six stamens, and a single pistil and a knob-like stigma. The fruits are angled capsules.

The Spanish bluebell (***E. hispanicus*** syns. *Scilla hispanica, S. campanulata*) is most familiar to American gardeners. It is grateful

Endymion hispanicus

in cultivation and is offered in a number of flower color forms by dealers in hardy bulbs. The Spanish bluebell has egg-shaped, whitish bulbs ¾ to 1½ inches long. Its smooth, glossy, somewhat fleshy foliage appears with the flowers in spring, at about the time late tulips are in bloom, and dies in early summer. Strong bulbs produce six to nine narrowly-strap-shaped leaves, tapered at their ends, 1 foot long or longer, and generally ⅓ inch to 1¼ inches wide. One erect, not arching flower stalk comes from each bulb and is 1 foot to 2 feet tall. Along its upper part are displayed ten to thirty nodding, bell-shaped flowers slightly over ½ inch long and as wide. Their stamens do not protrude and their petals are not strongly recurved. In the typical wild type the blooms are blue-lavender with an indefinite central line of blue-lilac down each petal. Unlike those of the English bluebell, the flowers of the Spanish bluebell have no strong, pleasing fragrance.

The English bluebell (***E. nonscriptus*** syns. *Scilla nonscripta, S. nutans*) brings nostalgic memories to those familiar with the woodlands of England and other parts of northwestern Europe. There, in spring, it carpets the ground with seas of dainty, delightfully scented, slender "wild hyacinths" that children pick in vast quantities to decorate homes and schoolrooms. Because little or no foliage is taken, such annual harvesting does not limit future blooming. The English bluebell differs from the Spanish in being of frailer appearance, in the tops of its flowering stalks being usually arched, in the stalks mostly carrying no more than a dozen blooms, in the stalks of the lower flowers not exceeding ½ inch in length, in its blooms being slightly smaller, with the ends of the petals strongly recurved, and, most pleasing of all, in having a wonderful fragrance.

Endymion nonscriptus

Charming, rather rare ***E. italicus*** (syn. *Scillia italica*), native to southern France and Italy, has ovoid bulbs under 1 inch in diameter. There are four to six spreading, strongly-keeled, fleshy leaves 4 to 8 inches long and ¼ to ½ inch wide. The slender flower stalks, 6 inches to almost 1 foot long, end in conical racemes up to 2 inches long by one-half as wide of blue, not nodding blooms nearly ½ inch wide.

Endymion italicus

Hybrids between Spanish and English bluebells are common and likely to occur in gardens, as they do in the wild, wherever the two species grow near each other. Such hybrids are intermediate between the parents and at least some of the named garden varieties of *Endymion* are of this origin. Named garden varieties or hybrids of *E. hispanicus* are 'Blue Queen', light blue, 'Excelsior', deep blue, 'Sky Blue', dark blue, 'La Grandesse', excellent white, and 'Rose Queen' and 'Queen of the Pinks', pink.

Garden and Landscape Uses. Spanish and English bluebells are among the loveliest and most delightful of hardy spring bulbs. With little attention they remain indefinitely, gradually increasing in number and enlivening the garden each May with a wealth of pretty spikes remindful of those of slender, untamed hyacinths. The blue and white ones are most admired. The pinks are less commendable, invariably they have that suggestion of bluishness that renders certain pinks unlovely. It is better that they be planted more sparingly and with more careful thought to their placement than those of other colors.

These plants will grow in full sun, but are grateful for a certain amount of shade, and show to better advantage where they receive it. They are woodlanders adapted for planting beneath trees, at the fringes of shrub borders, along shaded paths, and in similar locations. The English bluebell stands denser shade than the other. This liking for a tempering of the sun's rays does not mean, of course, that they are happy planted where they must compete with the vigorous, soil-depleting roots of some old maple or other surface-rooting species. Deep, fertile soil abundant in humus and, although never unduly dry, certainly not waterlogged, is best to the liking of these bluebells. Yet even under pine trees they flourish, and their foliage and flowers show to excellent effect against a carpet of fallen brown needles. In addition to their usefulness as garden decorations, the flowers are delightful for cutting. These are also attractive plants for forcing into early bloom in greenhouses. The third species discussed above, *E. italicus,* does well in sun or light shade and is suitable for rock gardens and similar places.

Cultivation. Under the environmental conditions detailed above endymions prosper with little attention. They are propagated by natural multiplication of the bulbs and by seeds. A mulch of compost, leaf mold, or other suitable organic material spread each fall, and a spring application of a complete fertilizer, do much to keep them vigorous. By natural increase of the bulbs the plants form clumps that gradually increase in size and tend to become crowded. When this occurs to the extent that the amount of bloom produced is impaired, which is at long intervals only, the remedy is to dig them up as soon as the foliage has browned, sort the bulbs to size, spade over the ground and improve it by mixing in decayed organic material, and replant immediately.

Normal planting time for purchased bulbs is as early in fall as they can be obtained. They are set 4 to 5 inches deep and 3 to 5 inches apart, not exactly spaced as hyacinths or tulips so often are, but more irregularly and casually to produce naturalistic effects.

For early bloom the bulbs are potted in early fall in pots or pans (shallow pots) 5 to 7 inches in diameter. They are set in porous, fertile earth with their tops just covered, and spaced so that there is about as much distance between them as one-half the diameter of the bulb. After planting, they are buried outdoors or in a cold frame under 6 to 8 inches of sand or peat moss or are stood in a cellar or other place where the temperature is about 40°F. Early in the new year they are brought indoors and forced in the same manner as tulips.

ENGELMANNIA (Engel-mánnia). There is only one species of this genus. It belongs to the daisy family COMPOSITAE and is a spring- to summer-bloomer that inhabits dry, often limestone soils from Kansas to New Mexico and Texas. The name honors Dr. Georg Engelmann, American physician and botanist, who died in 1884.

Rough-hairy and 9 inches to 2 feet tall or sometimes taller, ***Engelmannia pinnatifida*** is a taprooted, herbaceous perennial. Its stems branch above and it has deeply-pinnately-lobed leaves, with the lobes usually toothed. The basal leaves may be 1 foot long, but are usually shorter, those of the stems are rather smaller. Each of the branches terminates in a daisy-type flower head 1 inch to 2 inches wide, with a central eye of yellow, five-toothed disk florets surrounded by about eight, spreading, petal-like, yellow ray florets that are elliptic, and minutely-three-toothed at their apexes. Only the ray florets produce fertile, seedlike fruits (achenes).

Garden and Landscape Uses and Cultivation. Not widely grown, but planted to some extent where it is native and in similar climatic regions, *Engelmannia* is hardy and suitable for flower borders and informal landscapes. It succeeds in ordinary soils, including those of limestone derivation, and withstands dryish conditions. Full sun is needed. Propagation is easy by seed and by division.

ENGLISH. The word English as part of their common names is definitive of these plants: English bluebell (*Endymion non-scriptus*), English boxwood (*Buxus sempervirens*), English hawthorn (*Crataegus laevigata* and *C. monogyna*), English holly (*Ilex aquifolium* and *I. altaclarensis*), English ivy (*Hedera helix*), English primrose (*Primula vulgaris*), English wallflower (*Cheiranthus cheiri*), and English yew (*Taxus baccata*).

ENGLISHMAN'S FOOT is *Plantago major*.

ENKIANTHUS (Enk-iánthus). This group deserves more attention from American gardeners. It consists of spring-blooming, deciduous, or rarely evergreen shrubs or small trees of great charm and quality. Those discussed below are deciduous. Of neat appearance, they have small, stalked leaves and nodding blooms somewhat like those of blueberries (*Vaccinium*), that make pleasing, but not blatant displays. In fall their foliage is handsomely colored. There are ten species or more, members of the heath family ERICACEAE. They are natives from Japan to the Himalayas. Of Greek derivation, the name comes from *enkyos*, pregnant, and *anthos*, a flower, in allusion, it is supposed, to the pouched corollas of some kinds.

Enkianthuses have branches in whorls (circles of more than two). Their leaves, undivided and usually toothed, are mostly clustered near the ends of the twigs. The bell- or urn-shaped flowers are in terminal racemes or umbels. They have five persistent small calyx lobes (sepals), a usually five-toothed corolla, ten stamens, and a slender style longer than the stamens. The fruits are small capsules containing seeds with wings or angles, a seed characteristic that distinguishes *Enkianthus* from nearly related *Pieris*.

The kind most frequently cultivated ***E. campanulatus***, native to mountains in Japan, and hardy throughout most of New England, may attain a height of 12 to 15 feet, but is sometimes rather lower in cultivation. Of fairly erect habit, it has dullish green, pointed, obovate-elliptic, toothed leaves 1½ to 3 inches long, usually sparsely-hairy above, and on their undersides brown-hairy in the axils of the veins. The bell-shaped flowers are not pouched at the bases of the corollas. They are in short racemes of five to fifteen that appear after the foliage. They have pubescent stalks ½ to ¾ inch in length. Yellowish to orange-yellow and streaked with red, the blooms are ⅓ to ½ inch long. They have broad-lanceolate sepals. The stalks of the stamens are hairy, but the ovary and style are without hairs. In fall the foliage turns brilliant red. In *E. c. albiflorus* the flowers are white or nearly so. The deep toothing of the corolla, to one-third to one-half its length, and the elliptic leaves distinguish *E. c. longilobus*. The blooms of *E. c. palibinii* are red with the corollas toothed to one-third their length. Its leaves are obovate to oblong. Obovate leaves, narrow calyx lobes, and rose-pink flowers with slightly protruding styles are characteristic of *E. c. sikokianus*.

Generally similar to *E. campanulatus*, but with larger blooms, which may measure ¾ inch across the mouths of their broadly-bell-shaped corollas, ***E. deflexus*** is less hardy. It is native to western China and the Himalayas. The Chinese phase survives in southern New England, but the Himalayan one is not hardy in the north. A desirable species, *E. deflexus* attains a

Enkianthus campanulatus

Enkianthus perulatus

maximum height of about 20 feet. It has elliptic, obovate, or oblong-lanceolate leaves 1 inch to 3 inches long, hairy on the midribs beneath, and usually with scattered hairs on both surfaces. The flowers, many together in umbels, are yellowish-red with dark veinings. The stalks of the stamens and the anthers are hairy, as are the lower parts of the style and the ovary.

One of the most attractive enkianthuses is the Japanese ***E. perulatus*** (syn. *E. japonicus*). A much-branched shrub hardy in southern New England, and up to about 6 feet in height, it has umbels of three to ten green-tinted white flowers, displayed to great advantage because they appear before or together with the leaves. Unlike the blooms of *E. campanulatus*, those of this species are urn-shaped and distinctly pouched at their bases. They are ⅓ inch long and have the stalks of their stamens hairy below their middles. The pointed-elliptic-ovate to obovate, toothed leaves, which in fall turn red, are bright green, ¾ inch to 2 inches long, hairless on their upper surfaces except on the midribs, and white-hairy on the veins beneath.

From the last, upright-growing, much-branched ***E. subsessilis***, 3 to 10 feet tall and a native of Japanese mountains, differs in having flowers that come after the foliage and are in drooping, five- to ten-flowered, pubescent-stalked racemes. They are almost white, roundish urn-shaped, and under ¼ inch long. The bases of the corollas are pouched, and the lower parts of the stalks of the stamens pubescent. From ¾ inch to 1½ inches long, the elliptic to obovate, toothed, pointed leaves are brownish-hairy beneath, and white-hairy on the midribs above. Differing from *E. subsessilis* chiefly in that the branches of its racemes are hairless, ***E. nudipes*** is also of the mountains of Japan.

The irregularly fringed teeth of the corolla lobes readily distinguish the blooms of the much-branched ***E. cernuus*** from those of other species discussed. This 15-foot-tall, native of Japan has pointed- or bluntish-toothed, obovate, lively green leaves ¾ inch to 2 inches long, with hairs on their upper sides, and brown-hairy on the veins beneath, especially toward their bases. The slender-stalked, urn-shaped, white flowers, ¼ to ⅓ inch long, are five to twelve together in pubescent-stalked racemes. Their styles sometimes protrude slightly. Variety *E. c. rubens* (syn. *E. c. matsudai*) has rich red blooms.

Garden and Landscape Uses. Enkianthuses are gracious shrubs that fit pleasingly into landscapes without any suspicion of loudness or of being insistent. They compose wonderfully well with azaleas, pierises, vacciniums, and others of the health family and with nearly all acid-soil plants. They are happiest in full sun, but tolerate part-day or light shade. Their soil should be agreeably moist, but not wet, and contain appreciable amounts of such organic matter as leaf mold or peat moss. Many are the places where enkianthuses can be used advantageously—in shrub borders, foundation plantings, rock gardens, heather gardens, and near watersides. They are cherished for their neat appearance, good foliage, pleasing blooms, and beautiful fall coloring.

Cultivation. These shrubs give little trouble. They need no pruning or other special care, but are grateful if the earth around them is kept mulched with such organic material as peat moss, compost, or pine needles. In dry weather soakings at intervals of ten days or so are of great benefit. Planting may be done in early fall or spring. Enkianthuses transplant easily provided a good ball of soil is kept intact about their roots. Seeds afford a good way of securing increase. They should be sown in spring in containers of sandy peaty soil in a cool greenhouse or cold frame and be barely covered with soil. The soil in the seed pots must never be permitted to dry. Alternative methods of propagation are by summer cuttings under mist or in a greenhouse propagating bench, and by layering.

ENSETE (En-sét-e)—Abyssinian Banana. The seven species of *Ensete*, the name of which is pronounced with three syllables and is derived from the Amharic name, *anset*, belong to the banana family MUSACEAE. Once included in the genus *Musa*, they are now segregated in *Ensete* chiefly because their seeds are much larger than those of *Musa*, and because of technical differences in the pollen. Indigenous to Africa, Madagascar, southern China, and other parts of southeast Asia and Malaysia, ensetes are imposing large-leaved plants that look like bananas (*Musa*).

The Abyssinian banana (***E. ventricosum*** syn. *Musa ensete*), the largest of the banana family, is native to Ethiopia. Under favorable conditions it may be 30 to 40 feet in height, but often is not over one-half as tall. It has great paddle-shaped leaves up to 20 feet long by 3 feet wide, with red mid-veins. As with *Musa*, its apparent trunk, which is about one-half as tall as the plant, is composed of sheathing leafstalks and therefore the entire aboveground structure is that of a gigantic herbaceous plant. It is not a tree. Each plant, when it has gained sufficient strength, develops a flower stem that pushes upward inside the false trunk and emerges from the top to present an erect, globose flower cluster with claret-brown bracts, up to 1 foot long, and with up to twenty whitish flowers 1½ to 2 inches long. Each bloom has a three-lobed calyx, and one short three-lobed petal. The banana-like fruits of this species are 2 to 3 inches long, dry, and inedible. The shining black seeds are almost 1 inch in diameter.

Other species sometimes cultivated are *E. gilletii* (syn. *Musa religiosa*), of Africa, and Indian *E. superbum* (syn. *Musa superba*). With much the aspect of a canna, ***E. gilletii*** is smaller than the Abyssinian banana. The seeds are up to slightly more than ½ inch in diameter. Up to 12 feet high and at its base 2½ feet in diameter, ***E. superbum*** has leaves that may be 5 feet in length by 1½ feet wide. The about 3-inch-long fruits contain ½-inch-wide seeds.

Garden and Landscape Uses. The Abyssinian banana can be strikingly effective in tropical, subtropical, and warm-temperate regions. It provides a distinctly exotic atmosphere and can be used to good effect near buildings and other architectural features. It is satisfactory in southern Florida and California. It also makes a handsome tub plant for big containers for ornamenting patios, terraces, and similar places. In large conservatories it can be accommodated in ground beds or tubs. The other sorts have similar uses.

For summer effects in temperate climates, the Abyssinian banana enjoyed considerable popularity when subtropical bedding (planting ornamental beds for temporary outdoor display with plants that suggest the tropics) was popular. This is little practiced now and the cultivation of the Abyssinian banana is mostly limited to tropical, subtropical, and warm-temperate regions. The other sorts described can be used effectively in similar ways.

Cultivation. These plants produce no natural offsets, and propagation is most usually by seed. In recent years methods of inducing shoots to develop from stumps of old cut-back specimens have been devised, but they are not likely to prove of great practical importance. Seeds sown in sandy peaty soil in a temperature of 70 to 80°F germinate well, and the seedlings are potted and grown under humid atmospheric conditions in temperatures of 60°F and higher. To give of their best ensetes need deep, fertile, satisfactorily drained soil, and generous amounts of water. Well-established specimens benefit from liberal fertilization.

When needed for temporary summer bedding, plants raised from seeds sown in January or February in a tropical greenhouse, and grown indoors under similar conditions until the weather is thoroughly settled and warm, are satisfactory and are likely to attain heights of 6 to 8 feet or more in one season.

ENTELEA (Entelè-a)—Corkwood or Whau. This endemic genus of New Zealand contains only one species. A member of the linden family TILIACEAE, its name derives from the Greek *enteles*, perfect, and alludes to all its stamens being functional. It is closely related to *Sparmannia* and *Grewia*.

In its native land the corkwood or whau (***Entelea arborescens***) is a large shrub or tree up to 20 feet in height that bears attractive flowers and foliage. Its softly-hairy leaves, heart-shaped, often slightly three-lobed, and with toothed margins, are among the largest of any borne by native trees of New Zealand. They are 9 to 10 inches long. The flowers, in pendulous clusters, are 1 inch in diameter, pure white, and have four or five crinkled, pointed petals, a central cluster of yellow-tipped stamens, and one style. The fruits are roundish capsules, about 1 inch in diameter, and covered with long, weak bristles, at first pale green, but dark brown at maturity. The wood of the corkwood weighs about one-half as much as cork. It was used by the Maoris as floats for fishing nets and for rafts. Its foliage is browsed by cattle and horses. It is unfortunate that in its natural state this handsome plant is becoming scarce. The plant sometimes called *E. palmata* is *Sparmannia palmata*.

Entelea arborescens

Garden and Landscape Uses and Cultivation. Well adapted for outdoor cultivation as an ornamental in mild climates, such as that of southern California, the corkwood is also attractive for cultivating in large pots or tubs in cool greenhouses. It blooms in spring. Any ordinary, well-drained garden soil suits this species, which thrives in sunny locations. Needed pruning should be done after the flowers fade. Container-grown specimens are repotted or top-dressed in late winter or early spring. Those that have ample healthy roots benefit from regular applications of liquid fertilizer from spring through fall. At no time should the soil be excessively dry, but in winter less frequent watering is needed than in summer. During the warmer months greenhouse-grown specimens benefit from being placed outdoors, but they must be brought inside before the advent of fall frosts. A winter minimum night temperature of 50 to 55°F is adequate. This may be raised about five degrees in late winter or early spring. Day temperatures should be a few degrees higher. Whenever outside conditions admit the greenhouse should be ventilated moderately. Propagation is by seeds and by cuttings.

ENTEROLOBIUM (Enterolòbium)—Elephant's Ear or Earpod. This is a tropical American and West Indian genus of ten species of trees of the pea family LEGUMINOSAE. Its name alluding to the fruits, is derived from the Greek *enteron*, intestine, and *lobos*, a lobe.

Enterolobiums have fernlike, twice-pinnate leaves with numerous paired leaflets. The small, greenish, mostly bisexual flowers are in solitary or clustered globular heads. They have a five-toothed calyx, five petals joined to their middles, many stamens joined at their bases into a tube, and a slender style. The fruits are broad, flat, leathery pods that do not split to discharge the seeds.

Elephant's ear or earpod (***Enterolobium cyclocarpum***) is a quick-growing, wide-spreading deciduous tree up to 125 feet in

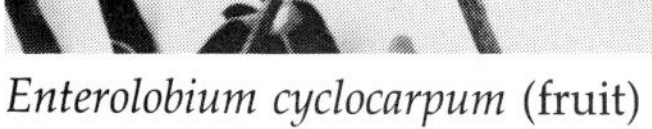

Enterolobium cyclocarpum (fruit)

Enterolobium contortisiliquum

height. Native of Central America, tropical South America, and the West Indies, it is planted in Hawaii and other parts of the tropics. Its leaves have four to nine pairs of primary divisions, each with twenty to thirty pairs of asymmetrical, narrow leaflets about ½ inch long, with glaucous undersides. The small heads of whitish flowers are in loose panicles. The very distinctive seed pods, some of which remain on the tree after the foliage falls, are polished-brown. They are broad, flat, and bent in one plane, or coiled until their blunt ends almost meet to form a circle 3 to 4 inches in diameter. This species is a source of good lumber. In the tropics its pods are used for stock feed, and in a young state as human food. They and the bark are used for tanning. Lopped branches are employed as green manure. In Hawaii the seeds are used in leis. Brazilian ***E. contortisiliquum*** (syn. *E. timbouva*) has leaves with primary divisions of ten to fifteen pairs of leaflets up to about 1 inch long, and seed pods bent to form not more than two-thirds of a circle.

Garden and Landscape Uses and Cultivation. As a lofty shade tree the elephant's ear has much to recommend it. It should not, however, be planted close to dwellings or other buildings because its very large surface roots are likely to do damage, sometimes to the extent of lifting or tilting the structures. These trees succeed in ordinary soils in open locations and are propagated by seed.

ENVIRONMENT. Every plant is the product of two sets of conditions, its heredity or genetic capabilities and its environment. No amount of excellence in one can compensate for shortcoming in the other. Wise gardeners, by selecting the best possible varieties and strains of seeds and plants, do their best to assure heredity factors favorable to their objectives. From then on, gardening consists of providing and maintaining by selection and modification environments most favorable to the sorts of plants being grown.

Environmental factors that affect plants are physical and biological. The physical are soil, water, air, light, and temperature, the biological ones other plants and animals. Let us consider each in turn.

The soil admits of much manipulation. Its fertile upper layer, the topsoil, can be deepened and otherwise improved by spading, rototilling, or plowing into it humus-forming materials. Its porosity can be increased by similar means, by the use of additives such as sand or grit, and by draining. Its ability to supply plant nutrients can be boosted by the use of manures and fertilizers and its acidity decreased or increased by liming and by mixing in sulfur or other acidifiers. For container-grown plants and sometimes for plants in outdoor beds soil substitutes, usually based on peat moss and a nonorganic light-weight material such as perlite, are frequently used. In the practice of nutriculture or hydroponics plants are grown in media containing no organic material.

That water be available in adequate supply whenever plants need it and in many cases be withheld when they do not are matters of prime concern. Outdoors natural rainfall is frequently supplemented by irrigation or by watering with hoses or cans. Indoors the amounts of water plants are given and when is entirely at the discretion of the gardener. But the volume and frequency of application are not the only concerns. The quality of the water may greatly influence plant growth. Notable examples are when it contains lime or other substances that cause it to be alkaline and unsuitable for such acid-soil plants such as rhododendrons, including azaleas, and most other members of the heath family ERICACEAE. Saline water such as sometimes pervades the ground in coastal areas and in some desert regions, and in tidal reaches affects river water, is detrimental to all but a few species especially adapted to such environments.

Air quality greatly affects the growth of plants as gardeners and others who have attempted or must engage in their cultivation in numerous urban regions know to their dismay. Usually not a great deal can be done to improve the quality of the air in individual gardens, but fortunately public concern, greatly increased from the middle of the twentieth century on, has resulted in less pollution of the air in many areas. Gardeners exercise certain controls over the air. They employ windbreaks to direct or slow its movement and in greenhouses fans to ensure its circulation. In greenhouses too, and sometimes outdoors, they increase its relative humidity by spraying or misting water on the ground, floors, and other surfaces and often on the plants too. Unfortunately, there are no practical means available for lowering relative humidity on occasions when that is desirable.

For light, necessary for all chlorophyll-containing plants, which means practically all cultivated sorts except mushrooms, chief dependence is upon that received directly from the sun. In greenhouses and elsewhere indoors this may be supplemented by electric light, and hobbyists and experimental scientists sometimes grow plants with artificial light only.

Except for experimental purposes the quality of light is not substantially changed from that of natural sunlight. The chief modifications employed relate to its intensity and duration. The first is often diminished with advantage to many kinds of plants both outdoors and in greenhouses by shading. Indoors, increasing light intensity by using electric lights as supplementary sources is a common practice.

The duration of light allowed during each twenty-four hour period is frequently controlled for plants such as chrysanthemums and poinsettias the growth and flowering habits of which are affected by day length (or more accurately by the length of the periods of darkness). This manipulation consists of lengthening the nights by shading, or shortening them by the use of artificial light.

Temperature sensitivity is much more acute in plants than animals, at least warm-blooded ones. The latter have the ability to maintain nearly constant body levels in nearly all environments, but plants for all practical considerations assume the temperatures of the air and soil around them. Minimums, optimums, and maximums vary widely for different kinds and frequently for the same sort at different seasons and different stages of growth. Night temperatures are commonly more critical and for the plants' well-being are within a narrower range than those acceptable during the day. Tropical plants commonly killed by freezing are likely to be harmed at temperatures considerably above that level. Many alpine and other cool-climate plants do not prosper if temperatures much above 60°F are experienced over extended periods.

Outdoor temperatures are not much subject to manipulation. Chief reliance is had upon selecting locations most favorable to the plants to be grown. Choices are influenced by such considerations as avoidance of low-lying frost pockets where cold air collects and cannot drain away, the selection of south-facing, north-facing or other-oriented slopes according to the need for warm or cool growing conditions, and the taking advantage of sheltered sites and sometimes of the temperature-modifying effects of the proximity of fairly large bodies of water or walls or buildings. More positive, but practicable only under special circumstances, is the employment of smudge pots and other devices used to protect citrus orchards.

Some protection from damage by low temperatures may be afforded near-hardy plants by covering them in winter or by mulching the soil around them to prevent too-deep frost penetration. Sheets of black plastic mulch are sometimes spread between rows of the early sowings of vegetables as an aid in warming the soil by absorbed sun heat. Mulching is also effective in keeping the ground relatively cool in summer.

Indoors, temperatures are commonly raised, in greenhouses and hotbeds to meet the needs of plants, in dwellings, offices, stores, and some other places where plants are grown, to assure the comfort of humans. In these latter places cooling in hot weather is frequently practiced, in greenhouses it is less common, but not unusual. Because air-conditioning equipment for dwellings and like places dehumidifies as well as cools, its use is not conducive to the best growth of many plants and it is impractical for greenhouses. Instead, evaporative cooling systems are generally employed. These suck air into the structure at one end, pass it through a thick web of excelsior or similar material kept moist by dripping cold water, and after it has picked up surplus heat discharge it at the other end of the greenhouse.

Biological environmental factors are susceptible to various manipulations and many garden operations consist of these. Chiefly such activities are directed toward the suppression or elimination of unwanted and harmful plant life as exemplified by weeds and by fungi and bacteria that cause diseases, and to limiting by various pest control measures animal life deemed deleterious. Although perhaps not living organisms, viruses behave as though they were, and gardeners must do everything within their power to keep those that cause plant diseases under control.

Undoubtedly humans are often the most influential animal element in the biological environments of plants. It is they in the persons of gardeners who decide where cultivated plants are grown and the methods of husbandry employed, they who determine what, how, and when gardening operations are performed. Even under adverse circumstances enthusiastic, perceptive, knowledgable gardeners, alert to the possibilities of manipulating environmental factors in favor of plants, are much more likely to enjoy success than poorer gardeners operating where conditions are basically more favorable.

EOMECON (Eo-mècon)—Snow Poppy. One species constitutes this genus of the poppy family PAPAVERACEAE. It is the snow-poppy of China, a late spring- and summer-blooming deciduous herbaceous perennial that has lived outdoors at The New York Botanical Garden for several decades. The generic name derives from the Greek *heos*, eastern, and *mekon*, a poppy, and refers to its habitat and appearance.

The snow-poppy (*Eomecon chionantha*) has wide-spreading underground rhizomes from which are produced many broadly heart- to kidney-shaped leaves about 4 inches in diameter and with undulated margins. They are distinctly fleshy, palmately-veined, bluish-green, and have dull undersurfaces. Rising to a height of 8 inches to 1 foot or more, the flower stalks bear several short-lived blooms in loose panicles. The flowers are about 1½ inches across and face upward. They have a pair of joined sepals that soon fall, four pure white roundish elliptic petals, and numerous yellow stamens. The style is two-lobed. All parts of this plant contain an abundant red-orange juice. The fruits are capsules.

Eomecon chionantha

Garden Uses and Cultivation. For covering slopes and other places where the slightly invasive habit of its rhizomes is not objectionable, the snow-poppy has merit and is interesting because of its comparative rarity. It grows in any ordinary well-drained and reasonably fertile soil, but seems to favor those that are fairly deep and contain abundant organic matter. It thrives in full sun. In late summer and fall the foliage discolors and becomes a little shabby. No special care is needed to grow the snow-poppy. Where it prospers it may remain undisturbed for many years. A spring application of a complete garden fertilizer is helpful as is regular watering in dry weather. Propagation is easy by division in early spring. New plants can also be raised from seed.

EPACRIDACEAE — Epacris Family. This family of dicotyledons, most plentiful of species in the native floras of Australia and Tasmania, occurs also in New Zealand, South America, Hawaii, and Indochina. It is closely related to the heath family ERICACEAE and contains some 400 species alloted among thirty genera. Most are decidedly *Erica*-like in aspect and are mostly shrubs and small trees of wet ground. Their alternate small, slender, rigid leaves are often crowded. The often fragrant, usually bisexual flowers have generally four- or five-lobed, persistant calyxes, a five-lobed corolla, five or less commonly four stamens, and one style tipped with a headlike stigma. The fruits are capsules or drupes. Genera in cultivation are *Cyathodes, Epacris, Leucopogon,* and *Richea.*

EPACRIS (Épac-ris). The epacris family EPACRIDACEAE, to which this genus belongs, is closely related to the heath family ERICACEAE. Epacrises have much the aspect of heaths (*Erica*), and in their native lands are often known by the same common name. They total nearly forty species, mostly natives of Australia, but with two species in New Zealand and one in New Caledonia. The name, from the Greek *epi,* on, and *akris,* a hilltop, refers to the natural habitats of some kinds.

Epacrises are alternate-leaved, evergreen shrubs with solitary, axillary flowers often so numerous and crowded that they form showy, columnar spikes. The leaves are small, narrow, sharp-pointed, and stalked or stalkless. The flowers, cylindrical to bell-shaped, have five short corolla lobes (petals), five stamens, and a single style tipped with a pinhead-like stigma. The fruits are little capsules.

Epacris, unidentified species

Epacris longiflora

The floral emblem of the State of Victoria, Australia, ***Epacris impressa*** is there called heath. One of the commonest and most admired wild flowers of the southeastern part of the Australian continent and Tasmania, it is 1½ to 3 feet tall, and variable. Its slender, erect branches are minutely-hairy, its stalkless, linear-lanceolate, sharp-pointed leaves ⅓ inch long or slightly longer. The flowers, spreading horizontally or nodding, and crowded in long spikes, vary from pure white through pink to deep red. They are ⅓ to ¾ inch in length, showy, and long lasting. From all other species of epacrises this differs in having five small pits at the bottom of the corolla tube. Variety *E. i. grandiflora* is distinguished by its coarser and broader, gray-downy leaves, and wider, rich rosy-crimson blooms. Double-flowered forms of this species are known. From 3 to 5 feet tall, ***E. longiflora*** has almost stalkless, sharp-pointed, ovate to lanceolate leaves ¼ to ⅔ inch long. Strung along the usually downy shoots, the hanging, tubular flowers, except for the upper end and spreading lobes of the corolla, which are white, are rosy-crimson. They are ¾ inch to 1¼ inches long, ¼ to ⅓ inch wide across the mouth.

Garden and Landscape Uses. Epacrises are splendid for mass planting outdoors in parts of California and elsewhere where the climate is to their liking. In such regions they are also excellent for rock gardens. In addition, they are beautiful plants for cool greenhouses, although their successful cultivation in pots, like that of ericas which they resemble, is a little challenging. They prefer well-drained, sandy-peaty soil of reasonable fertility, never so dry that they suffer from lack of moisture, but without suspicion of ever being overwet and stagnant. Full sun and free air circulation are needed.

Cultivation. This is essentially the same as for nonhardy heaths (*Erica*). Propagation is by cuttings and by seeds. Lax, pendent kinds need little pruning, but those of erect growth are cut back fairly severely as soon as blooming is through, and pot-grown specimens are repotted shortly afterwards. When potting, it is important not to disturb the roots more than absolutely necessary and, following the operation, the plants should be kept a little warmer than usual, and in a more humid atmosphere, until new growth is well started. Great care must be taken at this time not to water excessively. During summer epacrises in pots benefit from being put outdoors in a sunny place with their containers buried to their rims in sand, ashes, peat moss, or other material to keep them evenly cool and moist. Before frosts, they are removed to a greenhouse and wintered in a night temperature of 45 to 50°F with a daytime increase of five to ten degrees permitted. On all favorable occasions the greenhouse is ventilated freely. Watering at all times is done to keep the soil moderately moist. Discreet applications of dilute liquid fertilizer benefit pot-bound specimens.

EPAULETTE TREE is *Pterostyrax hispidus.*

EPHEDRA (Éph-edra)—Mexican Tea, Mormon-Tea-Bush, Joint-Fir. These curious plants are of greater interest and significance to botanists than to gardeners, although some are sometimes cultivated. They belong to a small order of plants called the *Gnetales,* which, apparently, are maverick relatives of the conifers. By some authorities the order has been thought to represent a connecting link between conifers and other gymnosperms and the more typical flowering plants called angiosperms, but this is generally doubted. In any case, ephedras are very distinct from even their nearest relatives. They alone constitute the ephedra family EPHEDRACEAE or, by some botanists, are included in the gnetum family GNETACEAE. There are approximately forty species, chiefly inhabitants of warm, dry regions in many parts of the world, including North America. The American species, of which there are about fifteen, are most abundant in the southwestern United States and Mexico. The name is one used by Pliny for horsetails (*Equisetum*).

Ephedras are freely-branching shrubs of distinctive appearance, erect, climbing, or trailing, and ordinarily not more than 6 or 7 feet tall; often they are lower. They have much the aspect of brooms (*Cytisus*) or horsetails (*Equisetum*). Because they are essentially leafless (the leaves are ordinarily reduced to thin sheaths, which grow in whorls or circles on the stems), the jointed, green stems function in food building (photosynthesis) that in most plants is done by leaves. The primitive flowers are unisexual and minute. The males on tiny stalks form what technically are compound cones, the females, usually paired, are on short shoots. After pollination, in most species the females develop brightly colored, berry-like, fleshy fruits, but in some desert kinds the fruits are dry.

The alkaloid ephedrine, used to treat colds, hay fever, asthma, and similar ills, and the virtues of which have been known to the Chinese since prehistoric times, is contained in all species of *Ephedra.* Until fairly recently the drug was obtained from kinds native to Asia, but it is now made synthetically. It is a common constituent of nosedrops. Medicinal teas are prepared from native kinds by Mexicans and by North American Indians.

American species, to which the names Mexican-tea and Morman-tea-bush are applied, that may be cultivated include *E. viridis* and *E. trifurca,* both natives of dry regions in the western United States and Mexico. From 1½ to 3 feet tall, ***E. viridis*** has many bright or pale green, broomlike stems with scalelike leaves in pairs. Its fruits are usually two, but sometimes one to each catkin. Attaining heights of 1½ to 5 feet tall and with pale or yellowish-green branches, ***E. trifurca*** has its scale leaves in threes and its fruits one to each catkin. This species is used by Mexicans and Indians to make a tea.

The European ***E. distachya,*** 1 foot to 2 feet tall, has underground rhizomes and male cones of four to eight pairs of flowers. The female flowers are solitary or in pairs. The fruits are red and ¼ inch in diameter. This is a native of the Mediterranean region. Its subspecies *E. d. helvetica* has erect, dark green twigs. The tall-growing ***E. fragilis*** (syn. *E. altissima*), of southern Europe and North Africa, scrambles to heights of 15 feet or more or is procumbent. Its fruits, ⅓ inch in diameter, are red.

Garden and Landscape Uses and Cultivation. These shrubs are occasionally planted as sand binders and are grown in botanical collections, outdoors in warm, dry climates, and in greenhouses elsewhere, for their interest and for study. They are content in porous, well-drained soil kept dryish, in full sun. In greenhouses winter minimum temperatures of 40 to 50°F are adequate. They are propagated by division, suckers, layers, and seeds.

EPHEDRACEAE — Ephedra Family. The characteristics of this family of gymnosperms are that of its only genus, *Ephedra.*

EPICATTLEYA. This is the name of bigeneric orchid hybrids the parents of which are *Epidendrum* and *Cattleya.*

Epicattleya orpetii

EPIDENDRUM (Epi-déndrum). Estimates of the number of species of *Epidendrum* of the orchid family ORCHIDACEAE range from 400 to 800 or more. Except for *E. nocturnum,* which is indigenous in West Africa as well as the New World, all are restricted in the wild to warm parts of the Americas, where the genus occurs from North Carolina to Mexico, the West Indies, Central America, the Galapagos Islands, and Argentina.

As their name suggests (it comes from the Greek *epi,* upon, and *dendron,* a tree), epidendrums perch on trees. They are epiphytes, which means they take no nourishment from their hosts. They sometimes have pseudobulbs, but often only reedlike branching stems. Their leaves, mostly short and thick, are less commonly grasslike. The flowering stalks, terminal at the ends of the stems, bear comparatively small blooms in spikes, racemes, or sometimes panicles. The blooms, individually short-stalked, have three sepals, the dorsal one ordinarily bigger than the lateral ones, two petals similar to the sepals, but sometimes narrower, and a usually deeply-three-lobed, comparatively broad lip with a basal shaft or claw more or less united with the column. There are typically four pollen masses (pollinia), more in three-anthered varieties of *E. cochleatum* and *E. boothianum.*

Eleven species inhabit Florida, nearly all restricted in that state to the southerly part and the Keys, but also found in the West Indies and the great majority also in Central and South America. One occurs as far north as the Carolinas. The majority of the Floridians are not sufficiently attractive to appeal to other than avid orchid fanciers and those interested in native plants.

Most alluring and among the commonest and most beautiful tree-perching or-

Epicattleya 'Winter Sunshine'

chids indigenous to the United States is ***E. tampense*** (syn. *Encyclia tampensis*). Endemic to the Florida Keys and the Bahamas, this is plentiful in sun and shade, in humid and dry habitats. It forms large mats of clustered, somewhat compressed, ovoid to nearly spherical pseudobulbs 1 inch to 2¾ inches long. Its linear to linear-lanceolate, rigid leaves may be 1 foot long or longer by ⅓ to ¾ inch wide. The racemes or panicles, slender-stalked and up to 2½ feet long, are of few to many fragrant blooms with yellowish-brown to yellowish-green or green sepals and petals, sometimes stained with purple. The sepals are oblong-spatula-shaped, up to nearly 1 inch long. The petals are slightly shorter and oblanceolate. The deeply-three-lobed lip, ½ to ¾ inch long, is white with a blotch or streaks of rosy-purple on the large, nearly circular middle lobe and has side lobes veined with the same color.

The clam shell orchid (***E. cochleatum*** syn. *Encyclia cochleata*) is very different from the last. In Florida it is represented by variety *E. c. triandrum*, which differs from the typical species only in its blooms

Epidendrum cochleatum

having three instead of two anthers. Indigenous from Mexico to Brazil and the West Indies, *E. cochleatum* has somewhat compressed, ovoid to spindle-shaped pseudobulbs 2 to 8 inches long or longer by about 1½ inches wide, each with usually two lustrous, oblong-lanceolate to linear, rigid leaves up to 1 foot long or longer and about ¾ inch wide. The seldom-branched flowering stalks, up to about 1¼ feet long, terminate in loose racemes of up to ten blooms with slender sepals and petals 1 inch to 1½ inches long, incurved, long-pointed, twisted, and greenish-yellow with a few purple spots near their bases. The heart-shaped, concave lip, ½ to ¾ inch long and as wide or wider than long, is dark purple to black-purple with a white center with radiating purple lines. Surprisingly, the Florida population of ***E. boothianum*** (syn. *Encyclia boothiana*), identified as *E. b. erythronioides*, like that state's population of *E. cochleatum*, has flowers that differ from those of the typical species in having three instead of two anthers. Typical *E. boothianum* is native from Mexico to northern Central America and the West Indies. It has crowded, compressed, roundish pseudobulbs up to 1¼ inches long, and rigid, often twisted leaves 3 to 6 inches or a little more in length. The loosely few-flowered racemes have slender stalks and are up to about 1 foot long. About ¾ inch across, the blooms have reddish-brown- or magenta-blotched sepals and petals. The lip is small, greenish-yellow and white and sometimes marked with magenta. In Florida found only in the southermost part, ***E. nocturnum*** is widely distributed in the West Indies, from Mexico to Peru, Ecuador, and Amazonian Brazil, and in West Africa, where it was probably introduced in the distant past from the Americas. This has erect, reedy stems up to 3 feet tall, compressed in their upper parts. The two-ranked, lustrous leaves, mostly toward the apexes of the stems, are elliptic to lanceolate and up to 7 inches long by 1 inch wide. The spidery blooms are in racemes, often branched, of few to several, of which only one is usually expanded at any one time. Pungently fragrant, especially at night, they have very slender sepals and petals, the former yellowish, 2 inches long or somewhat longer, the petals whitish and nearly as long. The center lobe of the somewhat shorter, deeply-three-lobed lip is very slender and almost threadlike, the side lobes are much broader. The lip is white with a small yellow center. Another Floridian native only of the southern tip, ***E. difforme*** (syn. *E. latilabre*) is a variable species found also from southern Mexico to South America. It has cylindrical, usually zigzag, leafy, cylindrical stems 6 inches to 1 foot long, at first erect, later drooping. The usually glossy leaves are elliptic to oblong-lanceolate, up to more than 4 inches long and 1¼ inches wide. From 1 inch to 1¼ inches across, its fragrant, green to yellowish blooms are in nearly stalkless, crowded racemes of few to many. They have narrow sepals and petals and an indistinctly three-lobed, more or less kidney-shaped lip. Native from southern Florida to Bolivia, Brazil, and Peru, ***E. pygmaeum*** forms mats of branched rhizomes and, somewhat distantly spaced along them, erect, ellipsoid, spindle-shaped or cylindrical pseudobulbs 1 to 4 inches long. Each pseudobulb has usually two elliptic to linear leaves up to 5 inches long, often smaller. The flowers, in groups of few to several or solitary, are up to ½ inch long. The sepals and petals are greenish to brownish-green, sometimes tinted with lavender. The lip is white with a central purple blotch.

Epidendrum cochleatum (flowers)

Flowers with lobeless lips illuminated with radiating dark lines are characteristic of these exotic epidendrums: ***E. fragrans*** is a variable native from Mexico to Peru, Brazil, and the West Indies. This sort has generally spindle-shaped pseudobulbs up to 3 inches long or longer, each with one strap-shaped to broad-elliptic leaf up to about 6 inches long. The short-stalked clusters of up to eight or fewer fragrant, waxy blooms, about 1½ inches across,

Epidendrum nocturnum

Epidendrum fragrans

Epidendrum lancifolium

Epidendrum ciliare

have twisted, narrowly-linear, creamy-white sepals and petals and a pointed-heart-shaped, white lip lined with red or purple. ***E. chacaoense*** (syn. *E. ionophlebium*), much like the last, differs chiefly in each pseudobulb having two leaves up to 1 foot long or longer and the blooms having at the base of the lip a fuzzy, nearly square callus. The blooms are inverted, up to 2 inches wide, and fragrant. Their sepals and petals are cream- to yellow, suffused with green. They have concave, shell-like lips with radiating lines of dull purple. This is native from Mexico to Panama. ***E. lancifolium*** is an attractive, fragrant-flowered Mexican with two-leaved, ovoid pseudobulbs. The leaves are about 9 inches long by 1½ inches wide. In erect, rather loosely-flowered racemes up to about 9 inches tall, the flowers have creamy-white sepals and petals and a lip of similar color tinged with brown and lined with violet. ***E. pentotis,*** indigenous from Mexico to Brazil, has slender-cylindrical to rather spindle-shaped pseudobulbs 4 inches to 1 foot in length with mostly two linear-elliptic to lanceolate, leathery leaves up to 1 foot in length. The very fragrant, upside-down flowers are usually two to three together in short racemes. About 2 to 3 inches across, they have light greenish-yellow to white sepals and petals and a usually white lip with purplish radiating lines in its lower part. ***E. radiatum*** is sometimes confused with *E. fragrans,* but it lacks the long point to the lip that is characteristic of that species. Native from Mexico to northern South America, *E. radiatum* has more or less egg-shaped, stalked, clustered pseudobulbs up to 5 inches tall, each with two or three rigid, narrowly-strap-shaped leaves up to 1 foot long. The fragrant, fleshy, upside-down blooms are close together in few- to many-flowered racemes with erect or arching stalks. About 1 inch wide, they have creamy-white to yellowish-green sepals and petals and a white lip with purple radiating lines.

Epidendrum radiatum

Conspicuously fringed, three-lobed lips are a characteristic of the flowers of these exotics: ***E. ciliare*** hails from Mexico to Brazil and the West Indies. This variable, vigorous kind has generally oblong, flattened, usually stalked, spindle-shaped to cylindrical pseudobulbs 4 to 5½ inches tall. Each carries one or two blunt, elliptic-oblong, rigid, lustrous leaves up to 1 foot long by 3½ inches wide. Produced before the new pseudobulbs are mature, the loose racemes, up to 1 foot in length, are of a few rather spidery blooms 3 to 7 inches wide, with a heliotrope-like fragrance. They have narrow, pale green, yellowish-green, or rarely yellow sepals and petals and a three-lobed, lacy-fringed, white lip often with a blotch of bright yellow. The center lobe is very slender. ***E. medusae,*** of Ecuador, is without pseudobulbs. It has crowded, branched, drooping stems up to 1 foot long furnished along their lengths with glaucous-green, two-ranked leaves. From the axils of the uppermost leaves emerge the fleshy blooms. They have narrow, brown-tinged, yellowish-green sepals and petals, a nearly circular, prominently fringed lip some 2 inches in

Epidendrum medusae

Epidendrum o'brienianum

diameter, deep maroon with a green center. ***E. o'brienianum*** is a hybrid with slender, reedlike, leafy stems in place of pseudobulbs. Its parents are *E. evectum,* of Colombia, and *E. ibaguense.* The hybrid, a spring and summer bloomer and one of the easiest to grow of orchids, most resembles its last-named parent. Its long, erect, slender flowering stalks terminate in rounded clusters of small, bright carmine blooms with some yellow on their fringed lips. Native from Mexico to South America, variable ***E. ibaguense*** (syn. *E. radicans*) is common throughout its range. It has exceptionally long, slender, branching stems, reportedly sometimes 30 feet in length, that root from the nodes. This species is practically everblooming.

Epidendrum alatum

Flowers with lobed, but not fringed lips are borne by these exotic species: ***E. alatum*** (syn. *Encyclia alata*) of Mexico and Central America has pear-shaped pseudobulbs 3 to 5 inches long, each with two or three leathery leaves up to 1¼ feet long. The fragrant flowers, 1½ to 2 inches across, are in loose racemes on erect stalks about 1½ feet long that sprout from the tops of the pseudobulbs. Their linear-spatula-shaped sepals and petals are greenish at their bases, and above are brown suffused with purple. Shorter than the sepals and petals, the three-lobed lip is white to pale yellow with crimson streaks and spots. ***E. atropurpureum*** is an attractive native from Mexico to Brazil, Peru, and the West Indies. It has pear-shaped to cylindrical pseudobulbs up to about 5 inches long by one-half as wide, each with two or three stiff, linear-lanceolate leaves that may attain 1½ feet in length by 2 inches in width. In erect racemes up to 2 feet long, of two to ten, the 2- to 3-inch-wide, fragrant blooms may vary considerably in color patterns. Typically their sepals and petals, incurved at their purplish-tinged, green apexes, are chocolate-brown, and the large, deeply-three-lobed lip is white with a magenta or magenta-crimson center. A number of variants have been given identifying

Epidendrum atropurpureum

Epidendrum atropurpureum roseum

Epidendrum cnemidophorum

Epidendrum cnemidophorum (flowers)

names. Noteworthy is *E. a. roseum,* the flowers of which have pink lips. ***E. cnemidophorum*** (syn. *E. pfavii*), native to Costa Rica, has canelike, leafy stems up to 3 feet long or longer furnished with pointed-elliptic leaves. The long-stalked, arching racemes have many rather loosely-arranged, flowers with slender tubes about 1½ inches long and strongly-reflexed sepals and petals. They are deep old-rose with some white on the strongly three-lobed lip. ***E. dichromum*** is a Brazilian of much merit. It has ovoid pseudobulbs 2 to 2½ inches tall each with two or three leaves. The erect flowering stalks, 1½ to 3 feet tall, bear blooms about 2 inches across with broad-spatula-shaped sepals and petals of good texture, pure white to pale lavender with a few purple markings, and a white-edged, rosy-violet, three-lobed lip. ***E. endresii*** is a dainty, low native of Costa Rica. Its tightly-clustered, slender stems under 1 foot long have two-ranked, rigid, 1-inch-long or a little longer, blunt-elliptic leaves. The flowers are in erect, short-stalked, termi-

Eggplant

Elmera racemosa

Elsholtzia stauntonii

Emilia javanica

Encephalocarpus strobiliformis

Endymion nonscriptus

Epidendrum ibaguense

Epidendrum ibaguense variety

Epilobium latifolium

Epimedium youngianum niveum

Epidendrum dichromum

Epidendrum endresii

Epidendrum parkinsonianum

nal racemes of up to ten. About 1 inch across and fragrant, they are pure white tinged with pink on their outsides and with the lip marked with lavender and orange. ***E. parkinsonianum*** (syn. *E. falcatum*) ranges in the wild from Mexico to Panama. This is a pendulous, clump-forming species with clustered, cylindrical pseudobulbs up to 4 inches in length, each with one thick, flaccid, linear, often purple-tinged leaf 1 foot to 1½ feet long. The very fragrant flowers are borne in threes or fewer from the ends of the new growths. Reportedly sometimes up to 6 inches across, but often considerably smaller, they have greenish-yellow to white sepals and petals tinged pinkish on their outsides and a big, three-lobed, white lip with a yellow or orange blotch and a pair of flaplike keels in front of the column. ***E. prismatocarpum,*** of Costa Rica and Panama, has 2-inch-wide, fragrant blooms with pointed, broad-linear sepals and petals, usually sulfur-yellow, and abundantly spotted, especially the petals, with dark brown or dull magenta. The side lobes of the fleshy, creamy-yellow to rosy-red lips are small and earlike. The middle lobe is sharply-pointed. This pleasing kind has its blooms, loosely arranged or somewhat crowded, in erect racemes of few to many. It has more or less clustered, narrowly-ovoid pseudobulbs up to about 6 inches long by one-third as broad. Each has two or three, blunt, strap-shaped to narrowly-ovate leaves up to 6 inches long by up to 2¼ inches wide. ***E. rhynchorphorum,*** of Mexico, has short, spindle-shaped pseudobulbs each with two broad-elliptic leaves. The spidery flowers, in short clusters, have pale greenish sepals and petals and a strongly three-lobed, yellow-centered, white lip, the center lobe slender, the side lobes wing-shaped. ***E. schumannianum*** has fragrant blooms heavily spotted with chocolate-brown to brownish-red. Native of Costa Rica and Panama, this has slender to fairly thick, clustered, reedy stems 6 or 7 inches long, leafy in their upper parts. The leaves are lanceolate to elliptic-oblong, up to 4 inches long and over 1 inch wide. Erect, branched or branchless, the 1- to 2-foot-tall flowering stalks carry usually many up to 1-inch-wide blooms. The inside base color of the obovate to oblanceolate sepals and petals is sulfur-yellow to tan. Their backs are amethyst-purple. The quite imposing, deeply-three-lobed, lavender to blue-lavender or rose-pink lip has a notched or cleft center lobe. ***E. stamfordianum,*** of Central and northern South America, is one of the handsomest species. It has spindle- to club-shaped pseudobulbs up to 1 foot long and with three or four leaves that may be nearly as long. The many-flowered, arching to pendulous racemes, up to 2 feet long, come from the bases of the pseudobulbs. About 1 inch wide and fragrant, the blooms have red-spotted, yellow, or greenish-yellow sepals and petals and a yellow, deeply-three-lobed lip,

Epidendrum prismatocarpum

Epidendrum rhynchorphorum

Epidendrum stamfordianum

Epidendrum stamfordianum (flowers)

the center lobe with a serrated margin. ***E. odoratissimum*** (syn. *Encyclia odoratissima*), native to Guyana and Brazil, has slender to ovoid pseudobulbs each with two or three strap-shaped leaves up to 1¼ feet long. The approximately 1-inch-wide, very fragrant flowers, in loose panicles up to about 1 foot long, have greenish sepals and petals and a three-lobed, yellowish lip veined with purple.

Lobeless, fringeless lips not decorated with radiating dark lines are characteristic of these exotic kinds: ***E. brassavolae*** is a Central American with ovoid to more or less pear-shaped, clustered pseudobulbs up to 7 inches tall each with a pair of leaves up to 9 inches long by 1¾ inches broad. The spidery flowers, in racemes of six to nine, up to 1½ feet long, are 3 to 4 inches across. The narrow sepals and petals, their upper parts turned backward, are yellowish-brown to greenish. The lip, with a long basal shaft or claw, has a white, pointed-heart-shaped blade stained, as though dipped in grape juice, with purple-red or magenta at its apex. ***E. coriifolium*** of Central America and northwestern South America forms dense clumps of erect stems 6 inches to 1 foot tall and clasped at the bases by erect, oblong to oblong-elliptic, stiff leathery leaves 6 inches to 1 foot long. The sometimes purple-tinged, green to yellowish, fleshy flowers, ¾ to 1 inch long or a little longer, are solitary in the axils of strictly two-ranked, large, triangular, bright green bracts that partly conceal them. ***E. mariae,*** of Mexico, one of the most beautiful of the genus, has pear-shaped pseudobulbs 1½ to 2 inches long, commonly with one, or more often two, leaves up to about 7 inches long. The arching racemes are of up to five blooms 2 to 3 inches across. The sepals and petals are green. The large frilled, slightly-notched, but not fringed lip is pure white with a spot of green or yellow at the base. ***E. polybulbon*** (syn. *Dinema polybulbon*) is a diminutive native of Mexico, Guatemala, Honduras, and the West Indies. It has long, creeping rhizomes and rather distantly-spaced pseudobulbs 1 inch to 1¼ inches long. One, or more commonly two, lustrous, ovate-elliptic to elliptic-oblong leaves 1 inch to 3 inches long sprout from each pseudobulb. The flowers, solitary or rarely two together, with stalks shorter than the leaves, have narrow, greenish-yellow or brownish, similar sepals and petals and a broadly-spade-shaped, white or creamy-white lip touched with yellow at its base. There are two small lateral horns on the column. The blooms are about 1¼ inches across. ***E. porpax*** is a creeping, mat-forming native from Mexico to Venezuela. It has no pseudobulbs. Its 1-inch-long or shorter leaves are fleshy with

Epidendrum mariae

Epidendrum porpax

Epidendrum coriifolium

Epidendrum polybulbon

Epidendrum pseudepidendrum

Epidendrum skinneri

a conspicuous mid-vein. Approximately ¾ inch across, the solitary flowers are green with a large, notched, but scarcely lobed, convex, glossy, brown to maroon lip. ***E. pseudepidendrum*** of Central America has slender, erect stems 1½ to 3 foot tall and two-ranked, narrow-elliptic leaves. The flowers, few together in terminal racemes, are about 2½ inches across. They have green sepals and petals and a bright-orange-red, obscurely-lobed, roundish lip with a recurved apex. The upper part of the column is red. ***E. skinneri*** (syn. *Barkeria skinneri*) is a native of Guatemala. It has club-shaped-cylindrical, leafy stems 6 to 8 inches tall. The leaves are about 3½ inches long by ¾ inch wide. The blooms 1 inch to 1¾ inches across are dark rose-pink to purple with a white blotch and yellow ridges on the lip. They are in few- to many-flowered, erect racemes. ***E. campylostalix*** has clusters of flattened pseudobulbs each with a single, glaucous, oblong to elliptic leaf up to 1 foot long. The slightly drooping flowers, in erect racemes or panicles up to 1 foot long sprouting from the apexes of the pseudobulbs, have narrow, dull purplish-red sepals and petals lined with yellow and about ¾ inch long. The lip is ivory-white, and the column is red. Its trailing stems hidden by two rows of overlapping, stem-clasping, fleshy, pointed-elliptic leaves up to about

Epidendrum campylostalix

Epidendrum schlechteranum

1 inch long, ***E. schlechteranum*** of Central and South America has stalkless flowers, solitary or sometimes in twos at the ends of the stems. From ¾ inch to 1½ inches across, the flowers have linear-lanceolate to strap-shaped, greenish to dull brown sepals and petals and a bluish ovate-heart-shaped lip with the column for most of its length united to it.

Garden Uses and Cultivation. Among epidendrums are many kinds well suited for the beginner collector with a small greenhouse or somewhat similar environment at command. Certain of the smaller species can be bloomed in terrariums and under artificial light setups. In the warm, humid tropics and subtropics, these orchids are satisfactory in lath houses, outdoors perched on the trunks of palms and other trees, and the taller, reed-stemmed ones in raised ground beds. Mainly they are of interest to collectors, but some of the taller kinds are good garden ornamentals and supply useful cut flowers.

In a genus as vast as *Epidendrum* it is to be expected that differences will be reflected in their needs, and to some extent this is true. Yet on the whole most kinds respond remarkably well to similar environments, at least the commonly cultivated kinds. Conditions and care that suit cattleyas are appropriate for epidendrums. Indoors most are satisfied with winter night temperatures of 55 to 60°F, rising five to ten or in sunny weather as much as fifteen degrees by day. In favorable weather ventilate the greenhouse fairly generously. Keep kinds with fat pseudobulbs drier in winter than at other times and even those with reedy stems a little drier, but do not overdo this. Do not fertilize in winter.

When new growth is developing water generously. Gradually decrease the amounts given, or rather lengthen the periods between applications, when the blooming begins. As a group, these orchids need fairly high light intensity, just short of that which would scorch the foliage. A few kinds need quite bright light and fail to bloom without it. A few others, *E. medusae,* for example, prefer comparatively dim light. A little trial and error will enable you to determine the most suitable intensities for species with which you are not entirely familiar.

Potting mixes for epidendrums may be any of those agreeable to the run of epiphytic orchids, tree fern fiber, osmunda fiber, or bark chips. Some few kinds are suitable for attaching to slabs of cork bark or tree fern bark. For further information see Orchids.

EPIDIACRIUM. This is the name of bigeneric orchids the parents of which are *Diacrium* and *Epidendrum.*

EPIGAEA (Epi-gaèa)—Trailing-Arbutus or Mayflower. These charming low plants of the heath family Ericaceae are notoriously balky to grow. They are, as British rock gardeners say, miffy, difficult to establish, often sulky in residence, and given to dying unaccountably despite the best efforts of experienced gardeners. Yet they are by no means impossible. In some gardens they flourish, but more of that later.

Choice and horticulturally challenging, *Epigaea* comprises two species, one native of eastern North American, the other of Japan. Thus they are part of that extensive flora that points so convincingly to the one-time existence of an ancient land connection between North America and the Orient. Without it how can we account for the many plant genera that have species native in the eastern United States and Canada and closely related ones in Japan and China, and nowhere else? The name *Epigaea* derives from the Greek *epi*, upon, and *gaea*, the earth, and alludes to the low, trailing habit.

Epigaeas have prostrate stems and alternate, stalked, lobeless leaves. They are normally evergreen, but under some conditions may lose their foliage in winter and renew it in spring. The stalkless clusters of very fragrant small flowers nestle in the axils of leafy bracts toward the ends of the stems and branches. They have five green sepals, a tubular corolla with five spreading lobes (petals), ten stamens, and a hairy ovary. The fleshy fruits open to reveal numerous tiny seeds. They are favorite foods of birds, ants, and snails, and can be eaten by humans.

Admired by all, and especially beloved by New Englanders, the trailing-arbutus or mayflower (***E. repens***) is a fairly common inhabitant of poor, acid soils from Massachusetts to Ohio, Georgia, and Tennessee. It has wiry, hairy stems and slightly hairy, leathery leaves, ovate to nearly round, up to 3 inches in length, and fringed with eyelashes of bristly hairs. Ranging in color from nearly white to quite deep pink, its flowers, about ⅓ inch across and up to twice as long, are a special delight of spring. Variety *E. r. plena* has double flowers. According to some, the pilgrims named *E. repens* mayflower as a tribute to the ship that brought them safely to American shores, but it seems more probable that the name refers to the time the plant blooms in Massachusetts.

Epigaea repens

The Japanese species ***E. asiatica*** is said to be easier to grow than the American, but this is by no means certain; it is little known in American gardens. It is a prostrate shrub with ascending branches similar to its American relative, but with calyx lobes one-half instead of one-third as long as the corolla tubes.

Garden Uses. Rock gardens, open woodlands, and gardens in which native plants are exclusively or predominantly grown are obvious choices of places to try the American trailing-arbutus. In gardens limited to American natives only, the Japanese species is not, of course, appropriate, otherwise it can be tried under conditions considered likely to suit its American relative. Poor, gravely or sandy, acid soil and partial, but not heavy shade seems best to suit these plants in cultivation. In the wild they not infrequently grow in full sun.

Cultivation. One way of *not* succeeding with trailing-arbutus is to grub up mats from the wild with little or no soil attached, and transplant them to gardens. Such transplants invariably perish, sometimes slowly, more often quickly. Even carefully dug sods with goodly amounts of soil attached to their roots rarely move well, although if small plants are found and taken in this way the chances of success are somewhat increased. By far the best plan is to set out pot-grown plants. And, strangely enough, these are not too difficult to come by. Trailing-arbutus is more easily raised from seeds, or even from cuttings, than most gardeners imagine, and in a cool greenhouse or cold frame they are easier to grow than they usually are outdoors. True, this takes a little time, normally about three years from seed, until specimens of suitable size for transplanting to gardens are achieved, but it is surer than moving plants from the wild and certainly more in accordance with the principles of conservation. Only from areas where wild plants face certain extermination as a result of road-making or building operations or some similar disaster, is it suggested that attempts be made to transplant them. The first necessity to success is suitable soil and location. Their needs in these respects are indicated above under garden uses. Planting is best done in early spring or early fall and should be followed by mulching with partly decayed pine needles, rotten wood, or acid leaf mold or peat moss. It is advantageous to keep for a while a light covering of branches of pines or other evergreens over the newly moved epigaeas; this provides a little shade and aids in checking too rapid water loss from the foliage. The soil must not be permitted to dry out, and care must be taken that the plants are not disturbed by mice or other rodents. Yearly renewal of the mulch should be routine.

In pots or pans (shallow pots) in cold frames and cool greenhouses, trailing-arbutus is easier to grow than it often is outdoors. At The New York Botanical Garden it succeeded for years in pans in cold frames without any great fuss or attention. The containers were sunk to their rims in a bed of coal ashes. Watering was with New York City water. In summer the frames were shaded and through the coldest part of the winter, too, the plants were afforded light shade as protection against the dehydrating effect of strong winter sun.

Epigaea repens in a pan

Propagation is by seeds sown as soon after they are ripe as possible on the surface of acid sandy-peaty soil. The seeds are not covered with soil, but a sheet of shaded glass is kept over the pot in which they are sown to ensure uniform humidity. Water by immersing the containers part way into water so that moisture reaches the surface soil from below. A shaded cool greenhouse or cold frame affords a suitable germinating place for the seeds and for growing on the young plants. The water used should not be alkaline. Cuttings of firm young shoots inserted in a mixture of sand or perlite and acid peat moss in August and kept, preferably under mist, in a frame or cool greenhouse root without too much difficulty. Layering in June affords a comparatively sure means of increase and is available to those without greenhouse or cold frame propagating facilities. The young layers will be ready for taking from the parent plants the following spring. Then they should be potted individually and kept in a cold frame for at least a year. The rare, double-flowered trailing-arbutus can be increased only by cuttings and layering.

EPIGENIUM (Epi-gènium). The about thirty-five species of *Epigenium* of the orchid family ORCHIDACEAE are natives of tropical Asia, Indonesia, and the Philippine Islands. The name, from the Greek *epi*, upon, and *genion*, a chin, alludes to the position of the petals and sepals on the column foot.

Epigeniums are tree-perchers (epiphytes) with one or two-leaved pseudobulbs, often distantly spaced along the creeping rhizomes. The flowers are in terminal or apparently terminal racemes from the tops of the pseudobulbs.

Sorts cultivated include these: ***E. amplum*** (syns. *Dendrobium amplum, Sarcopodium amplum*), of tropical parts of the Himalayan region, has long rhizomes, pseudobulbs 1½ to 2 inches long, leaves up to 6 inches long by about one-third as wide, and 2½-inch-wide flowers that are white to greenish-brown with darker brown spotting and with a lip having a dark purple middle lobe. ***E. coelogyne*** (syns. *Dendrobium coelogyne, Sarcopodium coelogyne*), of Burma, has distantly spaced along long rhizomes 2-inch-tall pseudobulbs each with a pair of glossy, oblongish leaves 4 to 6 inches long, mostly 2 inches or more in diameter. The flowers, except for their lips, which are dark purple, are yellowish-green heavily spotted with purple. ***E. cymbidioides*** (syns. *Dendrobium cymbidioides, Sarcopodium cymbidioides*), of the Philippine Islands, Java, and Sumatra, has short rhizomes on which are clustered 1½- to 2-inch-long, four- or five-angled pseudobulbs, each with two elliptic-oblong leaves 4 to 6 inches long. In racemes up to 10 inches long, five to ten or sometimes more 1¼- to 1½-inch-wide flowers are displayed. They have light yellow to creamy-yellow sepals and petals and a short, purple-lined, white lip. ***E. lyonii*** (syns. *Dendrobium lyonii, Sarcopodium lyonii*), of the Philippine Islands, has spaced along

Epigenium lyonii

its rhizomes ovoid pseudobulbs 2 to 3 inches long, each with a pair of oblong-elliptic leaves. The flowers, ten to fifteen on an arching stalk, are up to 5 inches across. They are rose-carmine with a wine-red lip.

Garden Uses and Cultivation. These orchids are best suited for growing in pans (shallow pots) or hanging baskets that assure adequate room for their rhizomes to spread. They need a rooting medium that drains freely so there is no danger of stagnation. Bark chips and shredded tree fern fiber are satisfactory, and the plants do well when attached to slabs of tree fern trunk. These orchids have no season of complete dormancy so water must be supplied throughout the year. For the sorts described here winter night temperatures of 60 to 65°F are satisfactory with a daytime rise of five to ten degrees permitted. Good light, but not so strong that the foliage is scorched, is essential for good results. For more information see Orchids.

EPIGOA. This is the name of bigeneric orchid hybrids the parents of which are *Domingoa* and *Epidendrum*.

EPILAELIA. This is the name of bigeneric orchid hybrids the parents of which are *Epidendrum* and *Laelia*.

EPILAELIOCATTLEYA. This is the name of trigeneric orchid hybrids the parents of which include *Cattleya, Epidendrum,* and *Laelia*.

EPILAELIOPSIS. This is the name of bigeneric orchid hybrids the parents of which include *Epidendrum* and *Laeliopsis*.

EPILOBIUM (Epi-lòbium) — Fireweed, Willow Herb. This genus belongs in the evening-primrose family ONAGRACEAE. It comprises a few more than 200 species, nearly one-quarter natives to New Zealand, the others scattered throughout temperate regions of the Northern and the Southern Hemispheres, with a few intrusions into warmer parts. The name, alluding to the placement of the flowers, comes from the Greek *epi*, upon, and *lobos*, a pod.

Probably the best known *Epilobium* in North America is fireweed. The colloquial name of this testifies to the frequency and enthusiasm with which it invades and colonizes forest areas destroyed by fire. It behaved similarly in parts of London and other European cities devastated by bombing in World War II. In fall it paints vast areas of the Northern Hemisphere rose-purple with its attractive spires of blooms. It is quite lovely, but many consider it too dedicated to perpetuating itself by self-sown seeds, or express their horticultural snobbery by tagging it as too common, to be admitted to gardens. Yet there are places where it can be used effectively. Less lovely plants by far are treasured by gardeners. In addition to the fireweed and other tall species there are several very dwarf epilobiums worth considering.

There are annual, herbaceous perennial, and shrub epilobiums, ranging from low creepers to kinds erect and tall. They have alternate or opposite, undivided, toothed or toothless leaves, frequently narrow and willow-like, from whence derives the common name willow herb. The flowers, in terminal spikes or racemes, or solitary from the leaf axils, are prevailingly purple-pink, magenta-pink, pink, or more rarely white or yellow. They have four sepals, four broad, spreading or erect petals, usually notched at their apexes, eight stamens of varied lengths, and one style capped with a four-lobed stigma. The fruits are slender capsules containing many seeds each with a tuft of long silky hairs that acts like a parachute, enabling the seeds to be air-borne for long distances and incidentally to colonize vacant territories. The kinds discussed here are all hardy herbaceous perennials.

Fireweed (***E. angustifolium***) is ordinarily 3 to 6 feet tall. It has upright, usually branched stems amply supplied with short-stalked, spirally-arranged leaves up to 6 inches long. From ¾ inch to 1¼ inches wide, the rosy-purple blooms are in long, terminal, spikelike, not conspicuously leafy

Epilobium angustifolium in seed

racemes. Variety *E. a. album* has white flowers. Similarly widely deployed in the wild through temperate parts of the Northern Hemisphere, ***E. latifolium***, 6 inches to 2 feet tall, has ovate to lanceolate-ovate leaves up to about 2 inches long and rather short, terminal, leafy spikes of beautiful rosy-purple blooms about 2 inches across.

A lovely western North American mountain species about 3 inches tall, ***E. obcordatum*** has underground or prostrate, spreading stems up to 6 inches long or longer, and opposite or densely-crowded, nearly stalkless, ovate to obovate, obscurely-toothed leaves approxi-

Epilobium latifolium

mately ½ to 1 inch long. Upfacing, and solitary in the axils of the upper leaves, the glowing magenta-pink blooms, 1 inch to 2 inches wide, have petals cleft so deeply that at a glance the blooms appear to be eight-petaled. This is probably the finest of dwarf willow herbs.

New Zealand natives include several low kinds, among them ***E. hectori,*** a mountain species with stems 2 to 6 inches long or sometimes longer, postrate below, then more or less erect. Its oblong or linear leaves, from ½ to 1 inch long, are opposite except beneath the flowering parts, where they are alternate. The flowers are white and about ¼ inch wide. From the last, ***E. alsinoides*** differs in having thicker stems up to 1 foot long and broad-ovate to nearly circular leaves. This is wild at lower altitudes than *E. hectori.* Delightfully dwarf ***E. nummularifolium*** is a real treasure. It forms ground-hiding mats 4 to 6 inches in diameter, of slender, creeping, rooting stems, and russet-brown to bronzy foliage, in sun about ½ inch high, in partial shade up to twice that. The leaves are opposite, very briefly stalked, round-bladed, and ½ to under ⅓ inch wide. White or delicate pink, the cross-shaped flowers come from the axils of the middle leaves. Similar ***E. nerterioides*** has smoother leaves and less hairy stems and seed capsules. Variable ***E. chloraefolium,*** usually much branched from the base, is 6 inches to 1½ feet tall and nearly hairless. It has opposite, rather distantly-spaced, ovate to oblong, short-stalked, toothed leaves ⅓ to ¾ inch long. About ⅓ inch in diameter, the white or pink, up-facing flowers sprout from the axils of the upper leaves. Variety *E. c. kaikourense,* a superior garden plant, has flowers twice as large as those of the species. The most handsome New Zealand epilobium, it breeds true from seed.

Epilobium chloraefolium kaikourense

Garden and Landscape Uses and Cultivation. Epilobiums are moist-soil plants. Located where self-sown seedlings will not cause undue trouble, the tall ones supply abundant and welcome herbage and fall bloom at watersides and in other damp places. These need sunshine. They can be encouraged to be bushier and lower than if left to their own devices by cutting them about halfway to the ground in June. This results in more, but smaller flower spikes. The low epilobiums discussed above are fit furnishings for rock gardens and other intimate plantings. They grow well in moraines and moist screes, needing little in the way of nourishment and prospering where their roots and underground stems can ramble among small stones or grit. In hot climates these benefit from a little shade from the fiercest sun of summer. Epilobiums are easily raised from seed, and the low ones by division.

EPIMEDIUM (Epi-mèdium) — Barrenwort. The unattractive common name barrenwort, which fortunately is less often used than the botanical *Epimedium,* does ill justice to the charming plants considered here. Belonging to the barberry family BERBERIDACEAE, although only the informed botanist would be likely at first acquaintance to guess that, *Epimedium* consists of twenty-one species. It occurs wild in southern Europe, North Africa, and temperate Asia, and is kin to *Vancouveria* of western North America. The name is derived from the Greek *epimedion* which was used by Dioscorides for a quite different plant.

Epimediums are hardy deciduous and evergreen herbaceous perennials with short, more or less woody, creeping rhizomes. The foliage of some kinds is all basal, that of others chiefly basal, with, in addition, a leaf on each flowering stem. The leaves are once- to three-times-pinnately-divided and have a rather coarse-fernlike aspect. Their leaflets are smooth-edged or spine-margined and when mature of firm tex-

Epimedium grandiflorum

ture. In slender-stalked, branchless or sparsely-branched terminal racemes, the graceful flowers are displayed concurrently with the growth of the new foliage. They are white, yellow, pink, red, or purplish, and of a curious, attractive form. They have eight sepals in two circles of four, the outer ones small, the inner petal-like, four petals that function as nectaries, which are hooded or often have prominent spurs, four stamens, and one style. The fruits are slender, many-seeded capsules. One species, *E. sagittatum*, has long been cultivated in China for medicinal use.

Japanese ***E. grandiflorum*** (syn. *E. macranthum*) is deciduous, and with both basal and stem foliage. It attains heights of 8 inches to somewhat over 1 foot, and has leaves three or four times thrice-divided into ovate to triangular-ovate, spine-edged leaflets 1¼ to 2¼ inches long by two-thirds-as wide. They have heart-shaped bases. Their upper surfaces are hairless, their undersides nearly so and glaucous. The white to light purple flowers, in sprays decidedly longer than the stem leaves, have long-spurred petals, and are 1 inch to 2 inches wide. A white-flowered variety, *E. g. album*, and one named 'Rose Queen,' with large red flowers, are cultivated.

Evergreen ***E. diphyllum*** (syn. *Aceranthus diphyllus*) is a Japanese endemic 8 inches to 1 foot tall. In addition to basal foliage, there is one leaf on each stem. The leaves are forked once or twice into two divisions, or sometimes the final time into three. The spineless or sparsely-spine-margined leaflets are ovate to broad-ovate, and 1 inch to 2 inches long by two-thirds as wide. Their undersides have a few short hairs. White, and nearly spurless, the flowers, under ½ inch wide, are in racemes of four to fifteen.

Believed to be a hybrid between the two above-mentioned species, ***E. youngianum***, about 1 foot tall, has basal and stem leaves, usually three times thrice-divided into pointed-ovate leaflets up to 3 inches in length. The white flowers are in racemes, longer than the stem leaves, of up to eight. Especially beautiful is *E. y. niveum* (syn. *E. musschianum*), which is a lower plant with larger, white blooms. Another

Epimedium youngianum niveum

Epimedium rubrum

variant of this hybrid, *E. y. roseum* (syn. *E. y. lilacinum*) is about 8 inches tall and has deep violet-rose blooms.

Another hybrid of *E. grandiflorum*, this with *E. alpinum* as the other parent, is ***E. rubrum.*** One of the showiest epimediums, it is about 1¼ feet tall. Its leaves have red-edged, sometimes red-mottled leaflets. The flowers, up to 1 inch across, in racemes longer than the stem leaves, in total effect are red, although the petals, except for red margins around their open ends, are white or yellowish. Their spurs are as long as the inner sepals and produce the visual effect of a white cross on a background of red. The flowers of ***E. alpinum*** are dull red, except for the yellow petals. This kind is similar to the last, but less robust. It commonly is under 1 foot tall and differs also in the stem leaf being usually longer than the branched sprays of flowers. A native of central and southern Europe, its leaves are usually twice- or thrice-divided into pointed-ovate, spine-edged leaflets. Quite often *E.rubrum* is cultivated as *E. alpinum.*

Yellow-flowered ***E. pinnatum,*** native to the Caucasus and Iran, is about 1 foot tall and without stem leaves. Its basal leaves are once- or twice-divided into six or nine spine-margined leaflets that have heart-shaped bases and are up to 3 inches long. The yellow flowers, nearly ¾ inch across, have red-tinged petals and protruding stamens. More commonly cultivated, *E. p. colchicum* of Transcaucasia has leaves with usually five or fewer leaflets.

The name ***E. versicolor*** belongs to a hybrid between *E. grandiflorum* and *E. pinnatum* that has both basal and stem leaves mostly twice divided into threes. The spine-edged leaflets are ovate and when young are clearly mottled with red. The racemes of blooms equal or exceed the stem leaves in length. The flowers, about ¾ inch wide, have old-rose-red inner sepals and yellow petals with red-tinged spurs. In the excellent variety *E. v. sulphureum* (syns. *E. luteum, E. ochroleucum, E. sulphureum*) the sepals are pale yellow, and the petals bright yellow. The young leaves of this are not mottled with red. From the last, *E. v. neosulphureum* differs in its stem leaves usually having only three leaflets and the spurs of its flowers being shorter.

Very distinct, white-flowered ***E. pubigerum*** has pale pink inner segments to its blooms. This native from the Balkan Peninsula to Asia Minor in flower possesses an airy grace resulting from its small, loosely arranged blooms being carried much higher above the foliage than those of any other kind here described. The branched flower stems are about 1¼ feet tall, the foliage about one-half as tall, and usually white-pubescent on its undersides.

Garden and Landscape Uses. Epimediums rank especially high among easy-to-grow, choice garden plants and surely deserve more attention from American gardeners. They have durable, beautiful foliage, that of some kinds pleasingly bronzed or variegated with red when young, and make charming spring displays of bloom. They need partial shade, and reasonably moist soil that contains a liberal proportion of decayed organic matter, such as compost, leaf mold, or peat moss. Planted in broad reaches they are

Epimediums as a groundcover in a woodland garden

among the choicest of low groundcovers, and in narrower bands form excellent edgings to lightly shaded paths. They are splendid in smaller groups and as solitary specimens in rock gardens, woodland gardens, and tucked into suitable places at the fronts of foundation and other shrub plantings.

Cultivation. Planting distances of 9 inches to 1 foot are about right for these plants. Once established they need hardly any care, beyond the cutting away of dead foliage. This in harsh climates is best delayed until late winter instead of done at the time of the fall cleanup of the garden. A mulch of leaf mold, compost, or peat moss spread around the plants in spring and left to decay is beneficial, as is watering during long periods of dry weather. Increase can be had by division in spring, but because the plants do not spread by runners in the manner of pachysandra, nor increase in size rapidly, this is rather a slow process. A faster means of propagation is by cuttings taken in July and planted in a soil, sand, and peat moss mixture in a shaded cold frame that is kept covered with glass sash or polyethylene plastic film, except for sufficient ventilation to prevent the temperature inside from rising to harmful levels.

EPIPACTIS (Epi-páctis)—Helleborine. This genus of twenty-four species occurs in north-temperate regions and southward to Mexico, Thailand, and South Africa. It belongs in the orchid family ORCHIDACEAE and has a name derived from the Greek *epipegnuo,* meaning coagulating milk.

Epipactises are ground orchids of minor horticultural importance. They have creeping rootstocks and branchless, leafy stems. Their longitudinally corrugated leaves are ovate or lanceolate. Purplish-brown, reddish, or nearly white, the blooms have three spreading sepals, two slightly larger, but otherwise similar petals, and a pouched lip that broadens at its apex into a three-lobed blade. For other species previously included in *Epipactus* see Goodyera.

The only European orchid naturalized in North America the helleborine (***Epipactis helleborine***) occurs from Quebec and Ontario to New Jersey, Washington, D.C., and Missouri. From 1 foot to 4 feet tall, it has stem-clasping or short-stalked, pointed, ovate to lanceolate leaves up to 7 inches long. Its madder-purple- or pink-tinged green flowers, solitary in the axils of long, narrow bracts, have sepals and petals about ½ inch long. The lip has a strongly recurved triangular blade.

Giant helleborine (***E. gigantea***) sometimes attains a height of 5 feet. Native in moist soils and swamps from South Dakota to British Columbia, Texas, California, and Mexico, it has variously shaped leaves up to about 8 inches long and flowers approximately 1 inch across. They come from the axils of bracts up to 6 inches long.

Epipactis gigantea

Epipactis gigantea (flowers)

Garden Uses and Cultivation. The species described here may be grown in native plant gardens and like places where environments similar to those in which the plants grow in the wild can be closely duplicated. They may be increased by carefully dividing them. For further information see Orchids.

EPIPHRONITELLA. This is the name of bigeneric orchid hybrids the parents of which are *Epidendrum* and *Sophronitella.*

EPIPHRONITIS (Epi-phronìtis). This is the name of orchid hybrids the parents of which are *Epidendrum* and *Sophronitis.* Similar to *Epidendrum radicans* in habit of growth and foliage and responding to the same cultural conditions, ***Epiphronitis veitchii*** blooms freely. In flower it is 1¼ to 1½ feet tall. Its erect, slender-stalked racemes have up to about ten loosely arranged blooms about 1¾ inches across, and deep cinnabar-red with a red-spotted, yellow lip.

Epiphronitis veitchii

Epiphronitis veitchii (flowers)

Garden Uses and Cultivation. These are as for *Epidendrum.*

EPIPHYLLANTHUS (Epi-phyllànthus). The genus *Epiphyllanthus* of the cactus family CACTACEAE consists of three Brazilian species closely related to and by conservative botanists included in *Schlumbergera.* The name, alluding to the similarity in the appearance of the blooms to those of *Epiphyllum,* is derived from the name of the last and the Greek *anthos,* a flower.

These are tree-perching (epiphytic) cactuses with stems of broad-oval and flat, more elongated and flat, or slender-cylindrical joints. Their areoles, sprouting woolly felt and a few short bristles, are in spirals. From 1 inch to 2 inches long, the tubular, decidedly asymmetrical flowers open by day. The fruits are small.

Uncommon in cultivation, ***E. obtusangulus*** (syn. *E. microsphaericus*) has stems at first upright, later more or less prostrate, of slender, cylindrical or obscurely angled, somewhat spiny to spineless joints. The flowers, terminal on the stems and branches, are rose-pink to purple.

Garden Uses and Cultivation. These plants are suitable for cultivation in fairly humid, lightly shaded greenhouses and outdoors in the tropics. Of slow, somewhat uncertain growth, they are usually grafted onto such related cactuses as *Schlumbergera* (syn. *Zygocactus*). They respond to environments and care suitable for those genera, *Rhipsalis,* and other epiphytic cactuses. For more information see Cactuses.

EPIPHYLLUM (Epi-phýllum)—Orchid Cactus. This genus of the cactus family CACTACEAE and hybrids between it and related genera are fascinating plants of easy culture, adaptable to the needs of window gardeners, greenhouse gardeners, and in mild climates outdoor gardeners. They include kinds that bloom freely and bear stunning flowers, those of the hybrids in many sizes and beautiful colors on plants of varying degrees of compactness and vigor. From among their vast number selections can be had to satisfy all tastes. In addition to these bigeneric hybrids many natural species are well worth cultivating and some are popular. Their flowers are predominantly white to cream-colored.

There are sixteen to twenty species of *Epiphyllum,* natives of Mexico, Central and South America, and the West Indies. The name, from the Greek *epi,* upon, and *phyllon,* a leaf, alludes to the flowers developing from what appear to be leaves. Closely related *Disocactus, Eccremocactus,* and *Nopalxochia* are by some authorities included in *Epiphyllum* but that is not done in this Encyclopedia. Occasionally epiphyllums are cultivated under the now discarded name of *Phyllocactus.*

Epiphyllums are sometimes known as leaf cactuses. This is rather misleading because the parts that look like and serve the functions of leaves are really broad, flattened, and for cactuses thin stems. Other genera containing species with similar stems and also called leaf cactuses are *Rhipsalis* and *Schlumbergera.* It should be noted that there is a small group of cactuses, *Pereskia, Pereskiopsis,* and *Quiabentia,* that produce normal leaves similar to those of most plants, but *Epiphyllum* does not belong with those.

In the wild epiphyllums are mostly epiphytes. They grow on trees after the fashion of many orchids and bromeliads and like those do not take nourishment from their hosts; they are not parasites. They have flat, leaflike, two-, or sometimes three-angled, rounded-toothed stems. The areoles (specialized areas from which spines, branches, and flowers arise) are small and are restricted to the stem margins. On adult plants they are generally spineless, but are conspicuously spiny on young seedlings. The flowers of epiphyllums are commonly, but not always large and showy. Those of most kinds open by day, some only at night. They have long perianth tubes (in this they differ from nearly related *Nopalxochia*) and many spreading, narrow perianth lobes (petals). The pulpy fruits, spherical to egg-shaped and containing many seeds, are red or purple.

The species most commonly grown include *E. anguliger* and *E. crenatum,* both day-bloomers. Mexican ***E. anguliger,*** its lower stems rounded, has flat branches with very deep round lobes that do not point forward. The flowers, white with brownish outer petals and a white style, and 4 to 5 inches in diameter, have perianth tubes 3 inches long. The plant grown as *E. darrahii* probably belongs here. Na-

Epiphyllum darrahii of gardens

Epiphyllum crenatum

tive of Honduras and Guatemala, day-blooming ***E. crenatum*** has stems 2 to 3 feet long that not infrequently root from their tips. The older parts are rounded, the branches flat, glaucous, about 2 inches wide, and coarsely-round-toothed. About 4 inches or somewhat more in diameter and with a perianth tube approximately as long, the scented flowers, white to cream with yellow stamens, remain expanded for a few days. They are approximately 8 inches long by 4 to 6 inches wide. The style is white.

Night-blooming ***E. oxypetalum*** is native from Mexico to Brazil. It has stems up to 20 feet long or longer, with thin, flat, more or less lanceolate, deeply-round-toothed branches up to 3 feet long and 4 to 5 inches wide. The richly fragrant, white flowers are about 10 inches long by one-half as wide. The outsides of the perianth tube is reddish. The style is white or red. Another night-bloomer, ***E. phyllanthus,*** native from Panama to Brazil and Peru, has stems up to about 10 feet long, with blunt, oblong branches up to 2 feet long by 1½ to 3 inches wide, often with purplish edges. The fragrant flowers are white, 8 inches to 1 foot long, and about 2 inches wide. They have a white or pinkish style.

Other species cultivated include these: ***E. caudatum*** is Mexican. It has slender, cylindrical stems with round-toothed, lanceolate branches up to 8 inches long by 1½ inches wide and white flowers up to 8 inches long. ***E. chrysocardium,*** of Mexico, has flattened branches up to 6 inches long by 1½ inches wide pinnately-cleft to their midribs. Its 8-inch-wide, 1-foot-long flowers are white. ***E. hookeri,*** of northern South America and Trinidad, climbs to heights up to 30 feet and has blunt-toothed branches 1½ to 3 inches wide and 8-inch-long white flowers that open at night. ***E. lepidocarpum,*** a native of Costa Rica, has flat or three-angled branches approximately 9 inches long and night-opening, white flowers about 8 inches long. ***E. macropterum,*** also Costa Rican, and up to 6 feet tall, has shallowly-round-toothed branches up to 1½ feet long by 2 to 5 inches wide. The white flowers are up to 1 foot long by 8 inches across. ***E. pittieri,*** of Costa Rica, up to 10 feet tall, has blunt-toothed branches 1 inch to 2 inches wide and fragrant white flowers about 3½ inches wide by up to 5 inches long that open at night and remain open through the next day. ***E. pumilum,*** of Guatemala, is 18 feet or so tall, has pointed, shallowly-round-toothed branches up to 2 feet long by 1 inch to 3½ inches wide and 2-inch-wide, 4- to 5-inch-long white flowers that open at night. ***E. stenopetalum*** is Mexican. About 10 feet tall, it has linear to lanceolate round-toothed branches up to 3 feet long and 5 inches wide and night-opening, fragrant, white flowers 8 to 10 inches long. ***E. strictum,*** wild from Mexico to Panama, and about 10 feet tall, has thick, rigid, linear, round-toothed branches 2 to 3 inches wide. Opening at night, its white flowers are about 9 inches long by 4 to 7 inches wide.

Epiphyllum hybrid 'Imperator'

Orchid cactus is the common name of a group of popular, often astonishingly beautiful hybrids between *Epihyllum* and related genera including *Heliocereus, Nopalxochia,* and *Selenicereus.* For this hybrid swarm, the name *Phyllocereus* has been proposed. It seems best, however, to admit them here under *Epiphyllum* for that is the name under which they are commonly grown and under which they will undoubtedly continue to be known. It is next to impossible to sort out the parentage of most with any degree of certainty, and those of multi-generic ancestry merge almost imperceptibly into those of less complicated parentage. These hybrids of *Epiphyllum* covered by the name *Phyllocereus* chiefly owe their origin to the work of German breeders from the latter part of the nineteenth century on. The later introductions are comparatively compact plants that produce large flowers freely and in a considerable range of colors. They are listed and described under their horticultural names and are often well illustrated in catalogs of specialist nurseries. An early hybrid between *E. crenatum* and *Selenicereus grandiflorus,* now correctly named *Seleniphyllum cooperi,* exists in more than one variety.

For a long time one of the most puzzling plants to which the name *Epiphyllum* was applied was a favorite and easily grown sort popularly, but wrongly called *E. ackermannii.* True *E. ackermannii,* which now is correctly named *Nopalxochia ackermannii,* is quite different. The misidentified plant is now known to be of hybrid origin, to be one of a group of offspring raised early in the nineteenth century by crossing *Heliocereus speciosus* and *Nopalxochia phyllanthoides.* The correct group name for plants of this parentage is *Heliochia* and the sort usually offered in the American trade as *Epiphyllum* hybrid *ackermannii* is *H. violacea.*

Garden Uses and Cultivation. Orchid cactuses are wonderfully showy plants for growing in pots or hanging baskets in greenhouse, and as houseplants, and outdoors attached to the trunks of palms or other trees or in lath houses in warm climates. Unlike the majority of cactuses they are not terrestrial desert plants geared to withstand baking sun, very low humidity, and soils deficient in organic matter and moisture. On the contrary, they grow in shaded places where the available rooting material is likely to be entirely organic

Epiphyllum hybrid 'Memories'

and adequately moist for most of the year. Likewise, humidity for long periods is high rather than very low. They live under conditions that suit many orchids and bromelaids. Most perch on trees or rocks in environments generally congenial to plant growth. An understanding of this is helpful in caring for them in cultivation.

The first point to remember is that their roots need a loose, coarse, organic soil through which they can ramify, one that will not become compact and remain saturated for long periods after watering. Of just what the soil consists is of less importance than its texture and permeability. A good mixture might be one part each of coarse peat moss, crocks (broken flower pot) or broken soft brick, coarse sand or perlite, and cow or sheep manure, two parts coarse leaf mold, and one-half part chopped charcoal. The addition of a pint of bonemeal to each bushel of the mixture is helpful.

Epiphyllums need good light, but not direct strong sun. From early fall until spring full exposure is advantageous. Too much shade inhibits blooming and results in weak growth. In summer they may be accommodated under a tree or an arbor outdoors or in some similar place in part-shade, where they are not subjected to strong wind. They must, of course, be brought indoors before frost. Moderate temperatures in winter are best. At night 50 to 55°F is adequate, by day this may be increased five to ten degrees. At that season too, the soil should be kept fairly dry. When new growth begins the frequency of watering is gradually increased and soon higher outdoor temperatures make it possible to ventilate the greenhouse more freely. Frequent light spraying with water promotes good growth. Flower buds set in winter and develop obviously with the coming of spring.

Epiphyllums can be accommodated in pots, tubs, hanging baskets, and in warm, dry climates in cans with holes in their sides. Potting and repotting (applying this term no matter what type of container is used), is done in summer, immediately following blooming. For young plants, until they occupy containers 8 to 10 inches in diameter, annual repotting is in order. From then on repotting every second or even third year is satisfactory. It is a mistake to put them in containers much too large for their roots. They seem to enjoy being slightly crowded. Plants that have filled their containers with roots benefit greatly from applications of dilute liquid fertilizer regularly throughout their growing season, but not in winter.

Neat staking and tying are needed for plants not in hanging baskets. Care must be taken to use a soft tying material in such a way that the stems are not injured. Individual branches bloom for many successive years and little pruning is needed, but branches that are really old and that obviously will not bloom again may be cut out low down to make way for younger shoots. Any long, thin, weak, rounded stems that may develop during the winter season of partial rest may also be removed with impunity. It is sometimes advisable to cut off the tips of growing branches that are becoming too long and so concentrate their energy on forming flower buds below.

As houseplants epiphyllums can be very satisfactory, but unless a cool, light room can be provided for their winter semihibernation difficulty arises in giving them the rest they need at that time. If they must be wintered in a warm room it becomes a matter of having them survive with minimum damage. This necessitates watering oftener than under more ideal conditions. If temperatures are high and water is not given, the stems shrivel.

Propagation is chiefly by cuttings, which root with great ease. The best are made from mature branches that have bloomed for two or three seasons rather than from younger shoots. Late spring is a good time to make cuttings. Plants from them are well rooted before winter. After the cuttings are made allow the cut surfaces to dry and heal in a shady cool place for two to three weeks before inserting them in the propagating medium, which may be a mixture of sand and peat moss, perlite and peat moss, or plain sand or perlite. The cuttings are set with their bases 1½ inches beneath the surface. In order to achieve this without toppling it may be advisable to tie each cutting to a stake, part of which extends below the bottom of the cutting.

Even easier than rooting cuttings is layering. This is done by bending a stem over, pegging its tip upon the soil or other rooting medium, and severing it after roots and new shoots develop from the point of contact.

Seeds of epiphyllums germinate most readily when they are six to nine months or more old. They are sown on the surface of coarse, sandy-peaty soil and covered with a ¼-inch layer of grit-sized charcoal. They are kept shaded and moist in a temperature of 70 to 90°F. Germination takes place in about two weeks. The sturdy little plants that result are transplanted individually to small pots. Seeds of hybrid plants usually give a highly variable progeny. Some growers graft the seedlings, as soon as they are large enough to conveniently handle onto understocks of *Selenicereus macdonaldiae* or other species of that genus and later regraft onto *Opuntia*. This speeds the growth of the epiphyllums, which eventually makes stems strong enough to be rooted as cuttings. For further information see Cactuses.

EPIPHYTE or AIR PLANT. Derived from the Greek *epi*, upon, and *phyton*, a plant, this term encompasses all plants that perch on other living plants, most commonly trees or shrubs, without extracting nourishment from their hosts. Compare with parasite, the name for organisms that live on or in other living organisms, depending upon them for sustenance, and saprophyte, the term for plants that live on dead plants, deriving nourishment from decaying tissues. The term lithophyte is sometimes used for rock-perching plants, but these are generally individuals of epiphytic species. It is not unusual to find some epiphytes, notably certain bromeliads, flourishing on such unlikely perches as telegraph wires and roofs.

Epiphytes, very common in the humid tropics and subtropics, are rare or absent in temperate regions. They predominate in the orchid and bromeliad families. In

Epiphytes, bromeliads, and rhipsalises on a tree, Castleton Gardens, Jamaica, West Indies

Epiphytic *Tillandsia recurvata* growing on telegraph wires, Tegucigalpa, Honduras

addition, many ferns, aroids, nepenthes, and other warm-climate plants are epiphytes. Most commonly epiphytes root into small accumulations of moss or debris that collect in crotches and other holds or into rough bark. Very rarely do they attach themselves to green leaves or stems. Some plants, notably strangler figs (*Ficus*), begin life as epiphytes, but later send down roots that anchor in the ground. Eventually these types are likely to kill their hosts by constricting their trunks and cutting off supplies of water and nutrients, and by denying them light. Certain species that usually are epiphytes, may occasionally be ground-inhabiting; the reverse sometimes occurs.

In cultivating epiphytes the conditions they favor in the wild are duplicated to the extent of seeing that they have extremely porous, organic, rooting mediums and for most kinds a humid atmosphere. Often they are grown in hanging baskets or other suspended containers. Sometimes in greenhouses and lath houses, frequently in the humid tropics and subtropics, they are attached to slabs of tree fern trunk, bark, or to the trunks and branches of living trees.

EPIPREMNOPSIS (Epipremn-ópsis). The only species of *Epipremnopsis* of the arum family ARACEAE is a native of the Malay Peninsula and from the Philippine Islands to the Molucca Islands. The name is derived from that of the closely related genus *Epipremnum,* and the Greek *opsis,* resembling.

A wiry-stemmed vine up to 40 feet tall or taller, ***E. media*** (syn. *Epipremnum medium*) has rich green, parchment-textured ovate leaves up to nearly 1 foot long, with puckered surfaces obscurely mottled with yellow and a single row of elliptic holes flanking each side of the midrib.

Garden and Landscape Uses and Cultivation. These are as for *Monstera* and *Philodendron.*

EPIPREMNUM (Epi-prémnum) — Pothos-vine or Ivy-arum or Devil's Ivy, Taro Vine. The tropical Asian genus *Epipremnum* consists of about ten species. Belonging to the arum family, ARACEAE, it is so closely allied to Asian *Raphidophora* and American *Monstera* that only after microscopic examination of the ovaries of their tiny blooms can species with certainty be assigned to one or other of these genera. The name, from the Greek *epi,* upon, and *premnon,* a trunk, alludes to the mode of growth. The common name taro vine is used also for *Raphidophora.*

Epipremnums are high-climbing vines that cling to trees and other supports by aerial roots. They have alternate leaves, those of a juvenile character commonly different from and decidedly smaller than adult-type foliage. The calla-lily-like floral structures, commonly thought of as flowers, are properly inflorescences, which resemble except in recondite details those of *Monstera* and *Raphidophora.* Their tiny true flowers are bisexual. The fruits consist of berries crowded along the spadix, and angled as a result of compression by neighbor berries.

Commonly cultivated in southern Florida, Hawaii, and other humid tropical and subtropical parts, ***E. pinnatum*** is a vigorous climber with stems up to 1½ inches in diameter. It has glossy, leathery, pendulous leaves with stalks up to 1½ feet in length. Their oblong blades, up to 2 feet in length, are mostly deeply-cleft into narrow, closely-spaced, essentially opposite lobes and may be perforated with tiny holes that appear as silvery dots along the midribs. Smaller, juvenile-type, lobeless, ovate leaves are often borne by the same plants as the larger, divided ones.

Ivy-arum, pothos-vine, or devil's ivy, (***E. aureum*** syns. *Pothos aureus, Scindapsus aureus, Raphidophora aurea*), a native of the Solomon Islands, is a popular tall, yellow-

Epipremnum aureum

variegated, heavily foliaged vine long cultivated under the names *Pothos aureus* and *Scindapsus aureus.* It very seldom blooms, and not until botanists had the opportunity to observe its flowers was it possible to classify it. Following an examination of specimens that bloomed in Puerto Rico in 1956 and in Florida in 1962 it became evident that the plant was neither *Pothos* nor *Scindapsus.* It was then named *Raphidophora aurea,* but later investigation determined that it belongs in the genus *Epipremnum.* As a young specimen this kind has broad-ovate leaves very much smaller than the similar-shaped adult ones, which have blades up to 2½ feet long by almost one-half as wide, often lobed or cleft. Both juvenile and adult foliage is smooth and dark green freely variegated with light yellow. Variety *E. a.* 'Marble Queen', a less robust grower, has leaves with almost pure white streaking and marbling. In *E. a. tricolor* the leaves are marbled and spotted with light green, yellow, and creamy-white against a background of dark green. Those of *E. a. wilcoxii* have more clearly defined variegation, and the stalks often white.

Cultivated to a lesser extent are *E. elegans,* of New Guinea, and *E. falcifolium,* of Java. With thick, nearly woody stems, ***E. elegans*** has glossy, pinnately-lobed leaves

Epipremnum aureum 'Marble Queen'

Epipremnum aureum 'Marble Queen' (foliage)

up to 4 feet in length, with long, rigid, angled stalks. The leaves of ***E. falcifolium*** are glossy, narrowly-lanceolate, and have winged stalks.

Garden and Landscape Uses and Cultivation. These bold-foliaged vines are striking ornamentals for clothing tree trunks and other supports in the humid tropics and for displaying in large tropical greenhouses. They respond to the same environments and care as *Monstera* and *Philodendron* and are propagated in the same ways.

EPISCIA (Epís-cia)—Flame-Violet. As subjects of interest to fanciers of gesneriads (plants of the gesneria family GESNERIACEAE), flame-violets or episcias are beginning to rival African-violets (*Saintpaulia*) in popularity. Beginning is the key word here, it is highly improbable that they will ever seriously challenge the supremacy in popular appeal of America's favorite gesneriad. Yet more and more amateur gardeners are surrendering to their lure each year, and the number of varieties constantly increases.

The genus *Episcia* is restricted in the wild to Central and South America and the West Indies. It consists of forty species of chiefly forest-floor plants. The most familiar kinds have stolons or strawberry-like runners that bear new plantlets. Often their foliage is highly colored. A less well known group of species are upright, rather than creeping, and are without runners. Because of the wide disparity of the plants presently included in *Episcia* it seems not improbable that sooner or later the genus will be split into two entities or more, or that some of its elements will be transferred to other genera. The name, from the Greek *episkios*, shaded, alludes to the preferred habitats of the group.

Episcias have opposite, stalked leaves with ovate to broad-elliptic or lanceolate, toothed blades, often rough-surfaced or quilted with tiny bumps. Despite the name flame-violet, their blooms very much less resemble violets, even discounting color, than do those of African-violets. Solitary or in groups of up to four, they have a calyx of five separate or briefly-united, hairy sepals often unequal in size. The corolla is long-tubular with a small, but distinct pouch or spur on its upper side at the base. The tube is cylindrical or funnel-shaped. It terminates in five spreading lobes (petals) that form an asymmetrical face to the flower, and which have minutely-toothed or fringed margins. The blooms range from bright red to yellow, lavender, yellowish-white spotted with purple, or white. They have four or very occasionally five nonprotruding stamens united at their bases, the pairs joined by their stalks. The style ends in a two-lobed or mouth-shaped stigma. A distinctive feature, shared with some other American genera of gesneriads, is the presence at one side and at the base of the ovary of a large, conspicuous gland. The fruits are fleshy capsules.

Red-flowered, trailing, runner-producing species in cultivation are *E. cupreata* and *E. reptans* (syns. *E. coccinea, E. fulgida*). These are the sources of numerous lovely varieties and hybrids, mostly with red flowers and frequently with beautifully variegated foliage. Their leaves are broad-elliptic to ovate, 2 to 5 inches long by approximately one-half as wide. The runners, naked of leaves, bear new plantlets at their ends. Native of Colombia and Venezuela, ***E. cupreata*** in its most typical form has coppery or deep bronzy-green foliage, but there are numerous variants many with leaves handsomely variegated with green, silver, white, or pink. The bright red flowers, about 1 inch long, have inconspicuously toothed or toothless sepals. Inside, the corollas are yellow spotted with red. Their faces are set at an oblique angle to the corolla tube, the upper petals slanting backward, the lower forward. The leaves of *E. c. viridifolia* are smoother than those of the typical species and are bright green or green variegated

Episcia cupreata

Episcia cupreata varieties (a) 'Acajou'

(b) 'Emerald Queen'

(c) 'Frosty'

with paler green and silver. From *E. cupreata* and its numerous varieties ***E. reptans*** differs in the calyx lobes of its 1½-inch-long, bright red blooms usually being conspicuously toothed at their apexes, in the interiors of the corollas being pink, lined, but unspotted, and in the faces of its blooms being at about right angles with the corolla tube. Also, the corolla tubes are essentially cylindrical, whereas those of *E. cupreata* are enlarged above so that they are narrowly-bell-shaped. The foliage of *E. reptans* is deep bronzy-green with pale green or silvery veins. In variety 'Lady Lou', they are irregularly blotched with pink. Hybrids between *E. cupreata* and *E. reptans* are given the group name ***E. variabilis.*** The first recorded were produced at Cornell University in 1955.

Episcia reptans

Bright yellow blooms distinguish *E.* 'Tropical Topaz'. In other respects this notable kind closely resembles *E. cupreata viridifolia,* of which it is believed to be a variant. It occurs wild in Panama and does not produce seeds. It was introduced to cultivation in North America in 1956. Flowers cream at first that become yellowish later are borne by 'Sun Gold'. This dark-bronze-green-foliaged variety originated as a sport of a pink-flowered *E. wilsonii* hybrid.

Lavender-colored blooms of delicate hue are typical, as known in cultivation of ***E. lilacina*** (syns. *E. chontalensis, E. fendleriana*), but there is reason to believe that other flower colors are to be found in the wild. A hairy species, this has leafless runners with plantlets at their ends and leaves of sizes and shapes comparable to the red-flowered kinds discussed above. They are deep bronze-green to reddish-green or plain green, with or without a herringbone pattern of emerald-green above and purple-pink beneath. Solitary or in groups of up to four, the slender-tubed flowers, 1 inch to nearly 1½ inches long, have slightly toothed or toothless calyx lobes and a white-hairy corolla tube, prominently spurred at its base, gradually widening and compressed at the sides above. There is a pale yellow patch in the throat of the bloom.

Episcia lilacina

Hybrids between *E. lilacina* and *E. cupreata* are grouped as ***E. wilsonii***. The first were raised at Cornell University and at the Fairchild Tropical Garden in Florida. Among the best known of these hybrids and their variants are 'Pinkiscia', with rich coppery-brown leaves and pink blooms, and its sport 'Ember Lace', which has foliage variegated with white and pink.

White-flowered ***E. dianthiflora*** is distinctive and well named. Its botanical specific, meaning with flowers like a dianthus, is very descriptive. This compact kind has slender, hairy runners, furnished along their lengths with pairs of leaves and with plantlets at their ends. Its leaves, thickish, dull olive-green- and green- or red-veined on their paler undersides, have 1-inch-long stalks and ovate to elliptic blades 1 inch to 1½ inches long. The short-stalked, solitary flowers are 1¼ inches long and approximately 1 inch wide. The petals have deeply-fringed margins. Another with fringed petals, ***E. punctata,*** a native of Mexico and Guatemala, is much coarser and more upright than the last. It develops many runners with pairs of leaves spaced along them and plantlets at their ends. Its coarsely-toothed, ovate to ovate-elliptic, white-hairy leaves, sometimes suffused with red at their edges, are ¾ inch to 3

Episcia dianthiflora

Episcia punctata

inches long. The very short-stalked, solitary, creamy-white flowers have purple-spotted, hairy corolla tubes 1 inch to 1¼ inches long, that gradually widen upward from their bases. The petals, up to ⅜ inch long, are thickly spotted with purple or violet. Intermediate between its parents, *E. dianthiflora* and *E. punctata*, is the hybrid 'Cygnet'. This is more graceful than *E. punctata* and blooms more freely than *E. dianthiflora*. It has large, purple-spotted, white flowers. Other procumbent episcias with runners that are in cultivation are *E. fimbriata* and *E. hirsuta*. Native of South America, ***E. fimbriata*** has pointed-elliptic, round-toothed hairy leaves with blades 2 to 4 inches long and one-third to one-half as wide. They are green sometimes relieved by silver variegation. The fringed, white or white and violet flowers are about 1¼ inches long and approximately 1 inch across. Native to Venezuela, ***E. hirsuta*** has elliptic to obovate or copper leaves, sometimes silvery down their middles, with blades 2 to 5 inches long. The white to pale purple flowers have slender corolla tubes, striped with deep purple in their throats.

Kinds without runners, quite different from those dealt with previously, form a group of more upright species, many with a tendency to develop tubers, and none with variegated or highly colored foliage. For convenience these are reviewed here in alphabetical sequence. ***E. adenosiphon,*** of Venezuela, has rather weak, semierect or more or less sprawling stems, and small, softly-hairy, ovate leaves. The small flowers are white tinged with pink. ***E. lurida*** is taller and more graceful than the last and upright. Its arching leaves, up to more than 1 foot long by 1½ inches wide, have somewhat recurved margins, and brown stalks and midribs. The flowers are white, with maroon spots on the petals and yellow throats. ***E. panamensis,*** wild from Mexico to Panama, is tuber-bearing. It has white-hairy leaves some 7 inches long by 4 inches wide, and 1¾-inch-long, 1½-inch-wide, white blooms, thinly lined with red in their throats, with a hooked spur at the bottom of the corolla tube. They are in clusters of three. Native to Panama and Costa Rica, ***E. lineata*** is an epiphyte (a plant that perches on trees but does not derive nourishment from the host). Short-stemmed, its erect to spreading leaves have oblanceolate blades 6 to 9 inches long with a prominent mid-vein from which parallel side veins angle out-

Episcia hirsuta

Episcia lineata

ward and upward. Their undersides are grayish with reddish veins. Small and short-stalked, the flowers are white to cream with the petals spotted or lined with red.

Garden and Landscape Uses. In the tropics episcias can be used in rock gardens and as groundcovers in shaded places, but most Americans know them as delightful and accommodating greenhouse plants and houseplants. There are collectors who avidly assemble and grow as many kinds as possible, there are amateurs and professionals who hybridize to generate new varieties, there are a few enthusiasts who seek them in their native haunts, and there are the many who pick up flame-violets from a florist, garden center, or even dime store counter, or who get a start from a neighbor or friend. However obtained, and wherever grown, these lovely tropicals give pleasure and supply beauty. Best of all, they are easy to grow and to multiply. Indoors, they do especially well in artificial light gardens, but can also be grown with natural light only.

Cultivation. To succeed with episcias certain conditions must be met. You need a warm place for them, where the winter night temperature never falls below 65°F and daytime temperatures are five to fifteen degrees higher. At other seasons higher temperatures are in order. These plants simply revel in summer heat. Just a few kinds tolerate temperatures five or possibly even ten degrees lower than those just suggested. These include *E. dianthiflora* and *E. punctata*. A fairly moist atmosphere is a must. The relative humidity should be at least 50%. Higher is better. Give episcias good light without exposing them to more than weak direct sun. Under a fluorescent light station them near the center of the tubes and as close as 6 inches beneath those of 40-watt output. With artificial light only, a day length of about sixteen hours is advisable. In greenhouses supply shade from spring to fall. The general idea is to give the plants as much illumination, natural or artificial, as they will take without the foliage being harmed. This encourages best blooming. If your objective is foliage rather than flowers your episcias will do nicely with less light.

Soil for episcias must be fairly loose, porous, and contain a moderate proportion of organic matter, or the plants can be grown in soilless mixes. A suitable soil consists of about equal parts good top soil, leaf mold or peat moss, and coarse sand or perlite, to which has been added some crushed charcoal, a generous sprinkling of dried cow or sheep manure, and bonemeal at the rate of a teaspoonful to a pint of the mixture. Make sure that the containers are well drained, and water to keep the roots always moderately moist, but not in a soggy, waterlogged condition. Do not use water much below room temperature. It does no harm, indeed it is beneficial, to allow the earth to become a little dry between soakings, but not to the extent that the foliage even begins to wilt. Episcias are heavy feeders. Give plants that have filled their containers with roots, and that are otherwise in good health, weekly or biweekly applications of dilute liquid fertilizer from spring through fall, more widely spaced ones in winter. Plants in soilless mixes need more frequent fertilizing. You will probably grow most of your plants in pots or pans (shallow pots), but if you have a place for them do not forget to try a few in hanging baskets. Repotting is needed from time to time. Late winter or spring is a good time to attend to this. Not a great many plants are as easily propagated as episcias. With those that make runners, the plantlets they produce afford obvious means of increase. Cuttings and leaf cuttings root with ease, and seeds sprout readily. Except in the case of species, seeds do not give plants identical with the seed parent, nevertheless the results are interesting. The seeds are fine and should be sown and cared for in the same way as those of gloxinias, African-violets, and other gesneriads.

Pests and Diseases. These are practically the same as for *Saintpaulia* (African-violet).

EPITHELANTHA (Epithel-ántha)—Button Cactus. There are three species of *Epithelantha,* a genus of the cactus family CACTACEAE, endemic to the southwestern United States and Mexico. The name is from the Greek *epi,* upon, and *thele,* a nipple. It refers to the location of the spine clusters. From *Mammillaria,* which button cactuses superficially resemble, *Epithelantha* differs in its flowers arising from the ends of the tubercles instead of from depressions between them.

Epithelanthas have branched or branchless, spherical or nearly spherical plant bodies or stems that form clumps and are 1 inch to 1½ inches in diameter. They are covered with spirals of bumplike tubercles with at the top of each a cluster of twenty to one-hundred needle-like spines. The flowers, from the tops of the stems, come from areoles (spine-producing areas) that develop wool as well as spines. The blooms are ⅛ to ½ inch in diameter and funnel-shaped. The small fruits are usually red and fleshy and edible.

Most commonly cultivated, ***E. micromeris,*** of Arizona, New Mexico, Texas, and Mexico, has usually solitary, occasionally paired or tripled, typically flattened-spherical plant bodies ½ to 1 inch wide, less often up to 1¾ inches across, that in aspect resemble rounded buttons or marbles. They have twenty to thirty-five rows of tubercles and are practically hidden beneath clusters of twenty to forty or sometimes more slender, rigid, white spines. Usually hidden among the spines, and the wool that accompanies them, the flowers, whitish to pale pink, are about ¼ inch in diameter. The red fruits, club-shaped and from under ½ to ¾ inch long, are very decorative. Mexican *E. m. greggii* differs in being larger, its plant bodies sometimes attaining 3 inches in diameter, and in producing offsets freely from the bases of its plant bodies. *E. m. rufispina* has all its spines tipped with reddish-brown. Others cultivated include white-flowered ***E. pachyrhiza,*** of Mexico, which has cylindrical plant bodies with a strong constriction between their bases and their tuberous roots, and ***E. polycephala,*** of Mexico, which has very small spherical plant bodies that form loose cushions and have coarse, brown-tipped, gray spines.

Epithelantha micromeris

Garden Uses and Cultivation. Button cactuses are easy to grow under conditions that suit most desert members of their family, outdoors in warm dry climates, and in greenhouses. It is important to pay special attention to soil drainage and watering so that there is no danger of harm resulting from excess moisture to which they are very sensitive. For further information see Cactuses.

EPITHET. The name of every species consists of two words, the first of which is that of the genus, and thus the generic name. Contrary to somewhat common loose usage, the second word is not the specific name, but the specific epithet. The specific name or name of the species consists of the generic name followed by the specific epithet.

EPITONIA. This is the name of bigeneric orchid hybrids the parents of which are *Epidendrum* and *Broughtonia.*

EQUISETACEAE — Horsetail Family. The characteristics of this family are that of its only genus, *Equisetum.*

EQUISETUM (Equi-sètum) — Horsetail, Scouring-Rush. The about twenty species of horsetails extant are the remnants of a much larger number of the same and allied genera that inhabited the earth eons ago. Not only are present horsetails fewer in number, they are dwarfs compared with the giant treelike kinds abundant in remote geological periods. Today *Equisetum* is the only genus in the horsetail family EQUISETACEAE. It is geographically widespread, but does not occur in Australia or New Zealand. The horsetails probably best known to gardeners are the noxious weedy species, such as the common deciduous *E. arvense,* that invades meadows, woods, roadsides, and cultivated and waste places throughout North America and elsewhere. The name comes from the Latin *equus,* a horse, and *seta,* a bristle, and alludes to the appearance of these fern allies.

Horsetails include evergreen and deciduous kinds. They do not have flowers, but like ferns produce spores. In horsetails these are in cones on separate shoots without chlorophyll or on the regular green ones. They have deep underground, jointed rhizomes that bear roots, and branched or branchless, generally hollow, corrugated stems with minute, bractlike, point-tipped leaves joined several to many at each node into an encircling sheath. Because they contain considerable silica equisetums can be used as abrasives and once were important polishing and scouring agents. This is reason for the name scouring-rush for some kinds.

Giant scouring-rush (***E. hyemale***) is a widely creeping, evergreen species that produces one kind of shoot only, and those hollow, usually branchless, and with fourteen to forty longitudinal ridges. Up to 5 feet in height, this kind is indigenous in North America, Europe, and Asia. Eastern North American *E. h. affine* (syn. *E. prealtum*) is 9 inches to 3 feet tall and drops the tips of its tiny leaves early; *E. h. robustum* (syn. *E. robustum*) of the same provenance drops the tips of its leaves tardily.

Dwarf scouring-rush (***E. scirpoides***), an evergreen sort of North America, Europe, and northern Asia, has wide-spreading rhizomes and prostrate or ascending, branchless or few-branched shoots, all similar. From 3 to 6 inches long, they are not hollow. Another wide-spreading evergreen with only one type of shoot, ***E. variegatum,*** up to 1 foot tall, has hollow stems branched near their bases, but without whorls (circles) of branches along their lengths. The tips of the leaves are persistent. This is native to wet soils and sandy shores through much of North America, Europe, and Asia.

Giant horsetail (***E. telmateia***), a deciduous native of Europe and North Africa, is represented in Pacific North America and Michigan by its scarcely distinguishable variety *E. t. braunii.* This has two distinct types of stems, the sterile 2 to 8 feet tall, with dense whorls (circles) of slender, spreading branches. The white to brownish, short-lived, cone-tipped fertile stems are 9 inches to 2 feet high. Occurring as a native around the world in north-temperate regions, deciduous ***E. fluviatile*** (syn. *E. limosum*) has all nearly similar stems 2 to 3 feet tall or sometimes taller, and branchless or with whorls of few to many four- to six-angled branches.

Equisetum fluviatile

Garden Uses and Cultivation. Only rarely are horsetails cultivated, for instance when needed for botanical instruction, to add variety to collections of native plants, and sometimes in water gardens and pools. When selecting locations for them keep in mind their tendency to spread vigorously and sometimes troublesomely. In water gardens they may be accommodated in containers with the soil surface at or slightly below the surface of the water. Moist or wet soil and full exposure to sun is to the liking of these plants. Propagation by division is very easily accomplished.

ERAGROSTIS. (Era-gróstis)—Love Grass. Teff. Grasses, mostly warm-temperate, subtropical, and tropical in origin, but some from cooler regions, and many native to North America, compose *Eragrostis* of the grass family, GRAMINEAE. A few of the possibly more than 300 species are of some horicultural interest. The name, derived from the Greek *eros,* love, and *agrostis,* grass, is without particular application.

Love grasses are annuals and perennials. They have flat or rolled leaf blades, and few- to many-branched, very dense to very loose flower panicles with strongly compressed spikelets. The panicles are of very diverse forms, sizes, and aspects in different species.

Annual love grasses sometimes grown for ornament include teff (***E. tef*** syn. *E. abyssinica*), which is an important food grain in Ethiopia and is grown there and elsewhere for fodder. Native of northern Africa and Arabia, teff has tufted or solitary stems 1 foot to 5 feet long. Its pointed, flat leaf blades are 6 inches to 1½ feet long by under ½ inch wide. The slender-branched, nodding flower panicles, up to 2 feet long, but often smaller, are of green, grayish-green, whitish, pink, red, or purple spikelets, loosely or densely aggregated. From 1 foot to 2 feet tall, ***E. japonica,*** of Japan, Korea, and Taiwan, forms tufts of slender stems. Its leaf blades are hairless, flat, pointed, and up to 8 inches long by under ⅛ inch wide. The stiffish, open panicles of pale or purplish spikelets have short, spreading branches and are up to 10 inches long. From the last, with which it has been confused, ***E. amabilis*** (syn. *E. tenella plumosa*), of warm parts of the Old World, differs in being scarcely as tall and in having leaf blades up to 4 inches long and flower panicles not over 5 inches long. Its spikelets are purplish or green.

Mexican love grass (***E. mexicana***) ranges in the wild from the southwestern United States to Mexico. It is 1 foot to 3 feet tall and loosely tufted. Rough on their upper surfaces, the pointed, flat leaf blades are up to 9 inches long by ¼ inch wide. The very open, decorative flower panicles vary much in size, the largest being over 1 foot long by one-half as broad. Ovate to cylindrical, they are composed of grayish or purplish spikelets.

Garden and Landscape Uses and Cultivation. Like other annual grasses used as decoratives, the species described above are pleasing in patches in sunny flower beds. They are also grown for cutting for use in fresh and dried flower arrangements. Growing readily in ordinary well-drained soils in sunny locations, they are raised from seed sown in spring where the plants are to remain. The seedlings are thinned just enough to prevent overcrowding.

ERANTHEMUM (Erán-themum). Some plants that previously belonged in this genus, and are often still grown under this name, properly belong in the closely related genus *Pseuderanthemum.* From pseuderanthemums true eranthemums differ most obviously in having large, conspicuous floral bracts. They belong in the acanthus family ACANTHACEAE. There are thirty species, natives of tropical Asia. Their name is from the Greek *erranos,* loving, and *anthemon,* a flower; it alludes to the beauty of the blooms.

Eranthemums are evergreen shrubs and herbaceous plants with opposite, smooth-edged or toothed, undivided leaves. Blue, purple, or pink, the flowers are in branched or branchless spikes with showy bracts that hide the five-sepaled or five-lobed calyxes. The corollas are slender-tubular and have five spreading lobes (petals) of equal or about equal size, which are symmetrically disposed. They are not two-lipped. There are two fertile stamens included in the corolla tube. The fruits are capsules with up to four seeds.

A popular kind is ***Eranthemum pulchellum*** (syns. *E. nervosum, Daedalacanthus nervosus*). Native of India, this is remarkable because of its beautiful blue flowers, profusely borne. A hairless or nearly hairless shrub 2 to 4 feet or more in height, it has long-stalked, ovate to elliptic leaves 4 to 8 inches long, markedly pinnately-veined and sometimes with rounded teeth. The overlapping bracts of the axillary and terminal spikes of bloom are white conspicuously feathered with green. The flower spikes are up to 3 inches long. The rich blue blooms are about ¾ inch in diameter and ¾ to 1 inch in length.

Eranthemum pulchellum

A smaller plant, ***E. wattii,*** of India, up to 1½ feet tall, has rich deep purple, phlox-like blooms 1 inch to 1½ inches wide with corolla tubes 1 inch to 1½ inches long, and paler on the undersides of the petals. Its broad-ovate, stalked leaves are 3 to 6 inches long and have hairs on their veins beneath. The plant known as *E. tricolor* is *Pseuderanthemum atrosanguineum tricolor.*

Garden and Landscape Uses. For southern Florida, Hawaii, and other places with warm, moist climates these are lovely plants for flower beds, the fronts of flower borders, and similar locations. They are attractive pot plants for winter bloom in tropical greenhouses.

Eranthemum wattii

Cultivation. Eranthemums prefer fertile, well-drained soil that never is saturated for long, but that does not want for moisture. They grow in sun or part-day shade and need little attention other than the removal of spent flower spikes, pruning to shape at the beginning of each growing season, and if the ground is poorish, an application of a slow-acting complete fertilizer at the same time.

In greenhouses they are usually best raised each year from cuttings taken from late winter to early summer. These root readily in a greenhouse propagating bench. Depending upon the time when they are started, plants of *E. pulchellum* will finally occupy 5-inch, 6-inch, or larger pots; those of *E. wattii* will not need containers more than 5 inches in diameter. The plants are ready for transferring to their final pots in August or early September. The soil should be coarse, fertile, and porous. A minimum winter night temperature of 60°F, or while the plants are in bloom, 55°F is satisfactory. Day temperatures may rise five to ten or fifteen degrees above those maintained at night. At other seasons both night and day temperatures may be higher. Good light with a little shade from strong summer sun is needed. The atmosphere should be moderately humid. An excessively moist atmosphere, like a constantly too wet soil, causes edema, the development on the leaves of whitish, sugary-looking particles, that result from liquid being excreted through the leaf pores. To encourage branching and to promote shapeliness, the shoot tips are pinched out two or three times before mid-July, the first time when the young plants have developed their third or fourth pair of leaves. Repotting is done as growth demands and when the final containers are filled with roots regular applications of dilute liquid fertilizer are desirable. After flowering, the plants are rested briefly by keeping them somewhat drier, then they are cut back, top dressed with rich soil, placed in a temperature of 70°F, and misted frequently with water to encourage the production of new shoots suitable for use as cuttings.

ERANTHIS (Erán-this) — Winter-Aconite. About six species of low, hardy, herbaceous perennial, early spring-flowering plants of the buttercup family RANUNCULACEAE belong here. They are natives of Europe and temperate Asia. The name is from the Greek *er,* spring, and *anthos,* a flower, and alludes to their season of bloom.

Winter-aconites have small tuberous rootstocks and stalked basal leaves, deeply-divided in the manner of a hand with spreading fingers. The flower stems are naked of foliage except for one dissected leaf just beneath the flower. This forms a collar of greenery that cushions a solitary golden or white bloom. Its showy parts, which look like petals, are actually sepals. The true petals are two-lipped, tubular, nectar-producing, shorter organs, sometimes called nectaries. There are many stamens and few to many pistils. The fruits are many-seeded follicles.

The common winter-aconite (***Eranthis hyemalis***), of western Europe, is 2 to 6 inches tall, the height depending largely upon the fertility and moistness of the soil. Its long-stalked, nearly round basal leaves are cleft almost to their centers into seven segments, which are again divided. The stem leaf is unequally five- to eight-cleft, but not to its base. The five to eight bright yellow sepals, ¼ inch wide or very slightly wider, spread to form a bloom 1 inch to 1½ inches in diameter. Similar, but stouter, ***E. cilicica,*** native from Greece to Asia Minor and Syria, has stem leaves of equal-sized, usually more numerous segments than those of *E. hyemalis,* with the segments cut almost to the base of the leaf. Its sepals are almost ½ inch wide.

Eranthis hyemalis among English ivy

The finest garden winter-aconite, ***E. tubergenii,*** a hybrid between the species discussed above, was raised in Holland and named in 1924. This beautiful plant has shining, golden-yellow blooms, 2 to 3 inches in diameter, and stem leaves intermediate between those of its parents. Being sterile, it produces no seeds, and because of this, the flowers remain in good condition very much longer than those of the fertile species. A selection of this hybrid with bronze-tinged, deep golden-yellow flowers and bronzy foliage is named *E. t.* 'Guinea Gold'.

The white-flowered ***E. pinnatifida*** is a rather rare woodland native of Japan. It has roundish, kidney-shaped, somewhat

Eranthis hyemalis (flowers)

five-angled, three-cleft leaves with the divisions deeply-pinnately-lobed. The blooms are about ¾ inch wide. This species is 2 to 6 inches tall.

Garden and Landscape Uses. These very hardy spring-bloomers are admirable for planting in groups at the fronts of perennial borders and in clumps or drifts in rock gardens, and to carpet and naturalize beneath deep-rooting deciduous trees and shrubs. They appreciate light shade and can be used for very much the same purposes as crocuses. They begin blooming before the last snowdrops fade and often while snow still lingers. The flowering of early scillas and *Muscari azureum* overlaps the season of winter-aconites and so, if they are suitably placed one may enjoy the white of snowdrops, the blues of scillas and *Muscari azureum,* and the gold of winter-aconites at the one time. A very pleasing effect results from planting *E. tubergenii* close by *Puschkinia scilloides;* the pale grayish-blue of the puschkinia compliments beautifully the yellow of the eranthis. For their best satisfaction they need a not excessively compact soil that contains a generous measure of organic matter; it should be moist, at least through spring, but not saturated. Later, the moisture content is of small importance, since the growing season of winter-aconites is short and their foliage dies very early. After that, dryness is not harmful. The early disappearance of all above-ground parts means, of course, that these are not long-season groundcovers. Because of this, it is often desirable to oversow the spaces they occupy with such quick-developing annuals as sweet alyssum, Dahlberg-daisy, and *Sanvitalia.*

Cultivation. The tubers are best planted in early fall, setting them 2 to 3 inches deep and 3 to 4 inches apart. In naturalistic plantings the spacing should be irregular to give the casual effect of self-sown seedlings. Established plants benefit from light mulching with well-rotted compost or some similar decayed organic material in fall and, if the ground is at all dry, from generous watering during the period that their foliage is in evidence. Kinds that produce seeds can be increased readily by these if sown, as soon as they are ripe, in woodsy soil in a shaded cold frame or well-protected spot outdoors. Under favorable circumstances self-sown seedlings spring up near old plants. The hybrid *E. tubergenii* can only be multiplied vegetatively. This is done, and the method is equally as applicable to other kinds, by dividing the tubers. The best time to do this is immediately after the foliage has died, but before it has disappeared. Then, the bulbs can be located easily. It is much better to replant immediately than to keep the tubers out of the ground until fall, although if it is essential to do that, it is possible to store them in slightly damp peat moss or sand in a cool, shady place. So long as winter-aconites are doing well they should not be disturbed except to propagate them.

ERCILLA (Ercíl-la). Two representatives of the pokeweed family PHYTOLACCACEAE constitute *Ercilla.* Climbing evergreen shrubs, they are natives of South America. The name commemorates Don Alonso de Ercille, of Madrid, who died in 1595.

Chilean and Peruvian ***E. volubilis*** is occasionally cultivated and is useful for clothing walls and other supports to which its aerial rootlets can cling. It attains a height of 15 to 20 feet and has slender stems closely clothed with alternate, stalked, ovate-heart-shaped to oblong, fleshy leaves 1 inch to 2 inches long and dark shining green. The inconspicuous flowers, in evidence in spring, are bisexual. In spikes 1 inch to 1¾ inches long, they are without petals, but have five purplish or greenish sepals and eight to ten white stamens. The blooms are succeeded by dark purple berries. This plant, hardy only in mild climates where little or no frost occurs, succeeds outdoors in sheltered locations in the south of England and in the Pacific Northwest.

Garden and Landscape Uses and Cultivation. This rare vine is likely to appeal to lovers of the unusual. Any ordinary well-drained garden soil suits it. Pruning consists of the removal of some of the older stems when the plant becomes crowded and may be done in early spring or immediately following flowering. Although a self-clinger, its aerial rootlets are often not sufficient to keep it securely attached to a vertical surface, and it is desirable to tie the main branches to nails or other devices driven into the masonry or other surface to be covered by the vine. Propagation is easy by cuttings of young shoots taken in late summer and planted in a propagating bed in a greenhouse or cold frame.

Ercilla volubilis

Ercilla volubilis (flowers and foliage)

ERDISIA (Erdís-ia). The genus *Erdisia* of the cactus family CACTACEAE is by some authorities included in *Corryocactus.* It comprises ten species in the wild confined to Peru and Chile. Few are cultivated. The name honors Ellwood C. Erdis, a member of a Yale University expedition to Peru in 1914.

Erdisias are much-branched, spiny plants with erect or prostrate, slender branches arising from partly or wholly underground stems. The areoles (areas on the stems from which spines arise) are depressed. Broadly-bell-shaped, the blooms are small and have stamens much shorter than the perianth segments (petals). Their warty ovaries bear scales and spines. Very tiny seeds are contained in small, spherical to egg-shaped fruits.

Most likely to be met with in cultivation are *E. squarrosa, E. spiniflora,* and *E. meyenii.* Native of Peru, ***E. squarrosa*** has cylindrical stems 3 to 6 feet long, up to 1 inch wide or a little wider, and with seven or eight ribs. The spines from each areole number about fifteen and are unequal in length with the longest about 1¾ inches. About 1¾ inches long and as wide, the bright pink to red blooms are borne near the branch ends. The juicy fruits have whitish wool, small scales, and short spines. Quite different is Chilean ***E. spiniflora.***

Its narrowly-club-shaped, above-ground branches arise from subterranean stems and are up to 8 inches long. They are about eight-ribbed and have slender spines. The 2-inch-long flowers are purple, the yellowish fruits up to 2 inches long. With underground stems and above-ground narrowly-club-shaped branches with slender spines is ***E. meyenii,*** a native of Peru. It differs from the last in having yellow flowers. Its stems have five to eight ribs. The red fruits are ¾ inch in diameter.

Garden and Landscape Uses and Cultivation. These are as for most small terrestrial cactuses. See Cactuses.

EREMOCHLOA (Eremó-chloa)—Centipede Grass. Except for centipede grass, which in warm regions is used for lawns, *Eremochloa,* of the grass family GRAMINEAE is without importance to gardeners. It comprises eight Asian and Australian species. The name comes from the Greek *eremos,* solitary, and *chloa,* grass.

Centipede grass (***E. ophiuroides***) is low, has thick, rooting stolons, and forms a close, coarse turf. Its bluish-green leaves are about 3 inches long by ¼ inch wide. The arching, narrow, one-sided flower spikes are of many small spikelets. For more information see Lawns, Their Making and Renovation.

EREMOCITRUS (Eremo-cítrus) — Australian Desert-Lime or Australian Desert-Kumquat. The only species of this Australian relative of the orange and other citrus fruits belongs in the rue family RUTACEAE. Its name, derived from the Greek *eremia,* a desert, and *kitron,* the citron, alludes to the plants' relationship and natural habitats.

The Australian desert-lime or Australian desert-kumquat (***Eremocitrus glauca***) is a shrub or bushy tree mostly 8 to 12 feet tall, with sharp, stout spines ½ inch to 2 inches long from the leaf axils of the lower branches, and narrowly-linear or wedge-shaped, stalkless or nearly stalkless, evergreen leaves up to 1½ inches long by under ½ inch wide, and gray-scurfy and minutely-hairy. The fragrant flowers appear in twos or threes in the leaf axils. They are about ½ inch in diameter and have three to five, usually four each sepals and petals, nine to twelve stamens with yellow anthers, and a thick style. The globular to egg-shaped fruits, technically berries, with usually four compartments, are slightly over ½ inch across and have pale yellow rinds sprinkled with tiny oil glands. They are mildly acid, edible, and in Australia are made into preserves.

Garden and Landscape Uses and Cultivation. Not hardy in the north, this species is well suited for dryish, mild climates and withstands more cold than the orange. It does well under semidesert conditions in full sun and is satisfied with well-drained soil of ordinary fertility. It may be planted singly, in groups, or as informal barrier hedges. It is propagated by seeds and by grafting onto other citrus fruit trees. It may also be used as an understock to graft citruses upon.

EREMOPHILA (Eremó-phila)—Emu Bush. The Australian genus *Eremophila* belongs to the myoporum family MYOPORACEAE. Its name, unclear as to significance, is derived from the Greek *eremos,* solitary, and *phileo,* to love. It has forty-five species.

Eremophilas are small trees and shrubs with alternate, undivided, often shiny and sticky leaves. The white, pink, red, orange, or purple flowers, from the leaf axils, have persistent calyxes, five-lobed or of five sepals, and asymmetrical, two-lipped, five-lobed corollas with tubular bases. There are four fertile stamens in pairs, a staminode (nonfunctional stamen), and one style. The dry or fleshy fruits are drupes (fruits of the structure of plums).

An erect, evergreen shrub 4 to 8 feet tall, the emu bush ***E. maculata*** has stiff, downy shoots, and nearly stalkless, pointed-lanceolate to wedge-shaped, erect leaves ½ inch to 2 inches long, and usually with toothless margins. The solitary flowers, 1 inch to 1¼ inches long, have a five-lobed calyx and a red-spotted-with-yellow corolla with four short, erect lobes (petals) and one much larger, down-pointed one. The fruits are drupes.

Garden and Landscape Uses and Cultivation. In frostless or nearly frostless, Mediterranean-type climates, such as that of California, the species described can be grown outdoors as an interesting and rather rare item and may be expected to thrive in ordinary soils and locations. Propagation is usually by cuttings. Seeds are often difficult to germinate.

EREMOSTACHYS (Eremó-stachys). This group of possibly sixty species of temperate western and central Asian herbaceous perennials, of the mint family LABIATAE, is closely related to *Phlomis.* The name, derived from the Greek *eremos,* solitary, and *stachys,* a spike, alludes to the way of blooming.

These plants have pinnate or pinnately-lobed leaves. Their foliage is chiefly basal, the stem leaves becoming progressively smaller above. The flower spikes are composed of whorls (circles) of many two-lipped blooms, with five-toothed calyxes that surround the corolla tubes, and long, somewhat compressed, hooded upper lips. Their lower lips are three-lobed with the middle lobe the largest. There are four stamens. The fruits consist of four seedlike nutlets.

Best known is ***Eremostachys laciniata,*** native to Asia Minor. This hairy plant has branchless or scarcely branched stems 1 foot to 2 feet tall, and deeply-twice-lobed leaves up to about 6 inches long. Its yel-

Eremostachys laciniata nuda

Eremostachys labiosa

low or creamy flowers are in massive spikes of whorls of ten to twenty blooms. The upper whorls are close together, the lower ones more widely spaced. Variety *E. l. nuda* is not hairy. A more recent introduction into cultivation is Siberian ***E. labiosa,*** an acanthus-like species with deep green, pinnately-cut leaves, and dense spikes, about 1½ feet in height, of large, golden-yellow flowers.

Garden Uses and Cultivation. These are hardy summer-blooming perennials adapted for flower beds. They grow well in ordinary, fertile garden soil and are propagated by seed and by division. They need full sun.

EREMURUS (Er-emùrus) — Foxtail-Lily, Desert Candle. For some unexplainable reason these stately herbaceous perennials are less common in American gardens than their merits deserve. Yet given an understanding of their needs, they are not difficult to grow. There are some fifty species, natives of steppes and mountains in temperate Asia, and belonging to the lily family LILIACEAE. Their name, given in allusion to the appearance of the flower spikes, is from the Greek *eremos,* solitary, and *oura,* a tail.

The rootstocks of eremuruses consist of a central crown from which spread horizontally in starfish-like manner a number of thonglike, fleshy roots. The foliage is all basal; the parallel-veined leaves are strap-shaped. Towering high above the foliage, the strictly erect, cylindrical or tapering flower spikes from a little distance can indeed give the illusion of being giant candles. They consist of usually hundreds of small, white, pink, yellow, or orange-yellow blooms. Bell- to saucer-shaped, these have six perianth segments (properly tepals, but commonly called petals) slightly joined at their bases, six stamens often longer than the petals, and a slender style. The fruits are capsules.

Eremurus robustus with irises

A vigorous kind with bluish-green, hairless leaves about 2 feet long by 1 inch to 2 inches wide, ***Eremurus robustus,*** of central Asia, has impressive, rigid, cylindrical spikes 4 to 9 feet tall of pink flowers with brown keels to their petals. The blooms are 1 inch to 1½ inches wide. Closely related ***E. elwesii,*** of unknown nativity, possibly of hybrid origin, has longer leaves, which are hairy at their apexes. There are both pink- and white-flowered forms of this.

The white flowers, about 1 inch in diameter and with brown keels to the petals, of ***E. himalaicus*** are in spikes 2 to 3 feet tall. This native of the Himalayan region has bright green, broadly-strap-shaped leaves, up to 2 feet long or longer, with their margins fringed with tiny hairs. Clear yellow flowers in crowded spikes up to 4 feet tall are displayed by ***E. stenophyllus*** (syn. *E. bungei*), of Iran and Russia. The leaves of this are finely-hairy or hairless and under ½ inch wide. Similar, but with pink or white blooms, is ***E. olgae,*** of central Asia.

Hybrid eremuruses are handsome. One betweeen *E. himalaicus* and *E. robustus* is named *E.* 'Himrob'. Intermediate between its parents, this has blush-pink blooms.

Eremurus himalaicus

Eremurus isabellinus

Spikes of lovely pale yellow flowers are displayed by ***E. tubergenii,*** the parents of which are *E. stenophyllus* and *E. himalaicus.* A series of lovely hybrids between *E. olgae* and *E. stenopetalus* ranging in flower color from white to light yellow, coppery-yellow, orange, and pink are grouped as ***E. isabellinus*** (syns. *E. shelfordii, E. warei*). The first of these, the result of a cross made in 1902, was named *E.* 'Shelford'. Among the best of those extant are *E. i.* 'White Beauty', with white flowers, *E. i.* 'Isobel' and *E. i.* 'Rosalind', with pink flowers, and *E. i.* 'Moonlight', with pale yellow flowers.

Garden and Landscape Uses. Because of the strong vertical lines of their impressive flower spikes, eremuruses in bloom are telling elements in the landscape. They need locations sheltered from wind, and to be seen to best advantage, a dark background. This last can fittingly be supplied by evergreens at a sufficient distance that their roots do not interfere with the eremuruses. These are good plants for the rears of mixed flower beds, and for use alone in beds. Cut, the flower spikes last well in water and are useful in large arrangements.

Eremurus isabellinus (flowers)

Cultivation. Eremuruses are generally root-hardy in southern New England, but because their shoot and foliage growth begins so early in spring special measures should be taken to protect this from late freezes. In fall, after the ground is frozen to a depth of an inch or two, mound over each plant several inches of sand, peat moss, or sawdust. When, in spring, the shoots begin to grow out of the mounds, if danger of frost still exists give additional protection by laying branches of evergreens over the mounds or by covering them each night with large boxes. Once threat of frost is over remove the mounds.

Eremuruses are long-term perennials. Once established do not transplant them if they are doing well unless you absolutely must. In preparation for planting make ready a bed of perfectly drained, nutritious, porous, loamy soil 1½ to 2 feet deep. Set the roots, their spokes spreading horizontally, on an inch or two of sand or grit at a depth of 4 to 6 inches. Handle the roots with care for they are very brittle and easily broken. Early fall is the best time to plant and transplant. Propagation is by very careful division of the root clumps in early fall, and by seeds, which are slow to germinate. The seeds are sown in fall in pots of sandy peaty soil. After sowing, the pots are buried nearly to their rims in a bed of sand or peat moss in a cold frame where they remain from fall until late winter. Then they are transferred to a greenhouse to germinate. An alternative method is to mix the seeds with slightly damp peat moss or sand, put them in a polyethylene plastic bag, and store them in a temperature of 35 to 40°F for several weeks before sowing. Established specimens benefit from a spring application of a complete garden fertilizer. Take care when working around eremuruses not to spade or fork near them. If you do, you are likely to injure their roots.

EREPSIA (Erép-sia). This genus of the *Mesembryanthemum* complex of the carpetweed family AIZOACEAE comprises forty-five species of erect, succulent, nonhardy small shrubs and subshrubs, in the wild restricted to South Africa. The name alludes to the covering or roofing over of the short stamens by the staminodes. It is from the Greek *erepsis,* a roof. The plant sometimes named *Erepsia haworthii* is *Lampranthus haworthii.*

Erepsias have erect, flattened, two-angled, branched stems and opposite fleshy leaves, triangular in section, the bases of each pair slightly united, and with the keel horny and sometimes toothed. The flowers, commonly rose-purple, but sometimes yellow or whitish, unlike those of many of the *Mesembryanthemum* group, do not close when light is poor or at night. Daisy-like in aspect, they differ markedly from daisies in structure, being single blooms rather than heads of many florets. They have nonfunctional stamens (staminodes) as well as fertile ones. The fruits are capsules.

Closely related ***E. aspera*** (syn. *Mesembryanthemum asperum*) and ***E. compressa*** (syn. *Mesembryanthemum compressum*) are differentiated by the leaves of the first being longer and rougher than those of the other. These rigidly branched plants attain heights of about 2 feet. From 1¼ to 1¾ inches long, the rough-dotted leaves of *E. aspera* are incurved. They have recurved, hooklike apexes, and rough keels. Those of *E. compressa* are ¾ inch to 1¼ inches in length and are less coarsely roughened with smaller dots. Solitary or paired, the whitish-centered, rose-purple to red flowers of both kinds are about 2 inches wide. The spreading, ¾- to 1-inch-long, somewhat sickle-shaped leaves of ***E. heteropetala*** (syn. *Mesembryanthemum heteropetalum*) are laterally compressed and triangular in section. They are bluish-gray, with tiny translucent dots, and have slightly furrowed, flat upper surfaces, and horny, toothed angles. A little over ½ inch in diameter, the blooms are reddish to whitish.

Upright or prostrate, the branches of ***E. inclaudens*** (syn. *Mesembryanthemum inclaudens*) have rather distant pairs of smooth, glossy, somewhat reddish, dark green leaves with conspicuous translucent dots. Broadly sickle-shaped and ½ to 1 inch long, they are slightly toothed only along the keels near their pointed apexes. On 2-inch-long stalks, the beautiful, broad-petaled blooms are reddish-purple.

Solitary pink flowers with light yellow centers are characteristic of ***E. mutabilis*** (syn. *Mesembryanthemum mutabile*). This kind has upright stems and branches, and laterally flattened, pointed, incurved, gray-green leaves ½ to ¾ inch long, with three narrow, horny, sometimes wavy angles, and a short, bristle-like apex. Usually there are short shoots in the leaf axils. Also commonly developing shoots in its leaf axils, slender-stemmed ***E. gracilis*** (syn. *Mesembryanthemum gracile*) is up to 2 feet tall. It has erect, tapering, somewhat incurved leaves, green with tiny translucent dots, and about ¾ inch long. They are triangular in section and are tipped with a little recurved hook. In twos or threes, the 1¼-inch-wide blooms are rose-purple.

Garden and Landscape Uses and Cultivation. In arid climates erepsias are appropriate for rock gardens and other places outdoors. They are also suitable for greenhouses and sunny windows. They make no special demands, accommodating to any reasonably good soil so long as it is thoroughly porous and well drained and is not kept too wet. Full exposure to sun is needed, and indoors, night temperatures in winter of 40 to 50°F with an increase of up to about fifteen degrees by day. Pot specimens can with advantage be summered outdoors or in cold frames. Propagation is extremely easy by cuttings and by seeds.

ERIA (Èr-ia). This large and diverse genus of the orchid family ORCHIDACEAE is represented in contemporary American collections by a few species only. Its natural geographical range embraces warm parts of Asia, Polynesia, and Australia. There are about 375 species. The name, from the Greek *erion,* wool, alludes to the flower stalks and flowers often being woolly. From *Dendrobium,* to which it is most closely allied, *Eria* differs in its flowers have eight instead of four pollen masses (pollinia).

Erias include deciduous and evergreen species of very diverse growth habits. Some have long, cylindrical, leafy stems, others more or less creeping rhizomes and roundish pseudobulbs. Most are epiphytes (tree-perchers that take no nourishment from their hosts), more rarely they live on cliffs and occasionally in the ground. The flowers are solitary or more often in rather loose to dense, erect or drooping, lateral or apparently terminal racemes, sometimes with the added attraction of brightly colored bracts. Of small to medium size, the flowers have three sepals, two generally slightly smaller petals, and a three-lobed or lobeless lip usually with longitudinal ridges or thickenings.

Among erias occasionally cultivated by orchid fanciers is ***E. cymbiformis,*** of Borneo. This, about 2 feet tall, has erect, slender, stemlike pseudobulbs topped with recurved, pointed-strap-shaped leaves about 1 foot long by 1½ to 2 inches wide. The numerous small pale yellow flowers are in dense, pendulous, cylindrical spikes some 8 inches long. Himalayan ***E. ferruginea*** has creeping rhizomes with spaced along them slender, cylindrical pseudobulbs up to 8 inches long. At the apexes of the pseudobulbs are up to four leathery, elliptic-oblong, spreading or drooping leaves up to about 6 inches long. Arising from beneath the foliage, the erect, bracted flower spikes have up to fifteen blooms ¾ inch to 1½ inches wide. Their sepals and petals are olive-brown, or the petals may be white blushed with rose-red. The lip and column are white with pink to purple-red markings. The flowering stalks and backs of the flowers are clothed with short, soft, brown hairs. Slightly compressed conical to nearly cylindrical pseudobulbs up to 4 inches tall are characteristic of ***E. hyacinthoides,*** of Malaya and Indonesia. Each has two or sometimes three more or less erect, stalked, lanceolate leaves with blades 8 inches to over 1 foot long, woolly-hairy or hairless on their undersides. The racemes are, as the specific epithet suggests, hyacinth-like. Two or sometimes more are produced from the upper part of the pseudobulb. They are erect and about 6 inches tall. The individual blooms, ½ inch wide and fragrant, are white with the side lobes of the lip violet-purple. Native from southern China to Indochina, Malaya, Indonesia, and New Guinea, ***E. javanica*** has 2- to 3½-inch-long, egg-shaped pseudobulbs spaced 1 inch to 3 inches apart along creeping rhizomes. From near the top of each come a pair of fleshy, lanceolate-strap-shaped leaves up to 1½ feet long and 4 inches wide. The fragrant blooms, 1¼ to 1¾ inches across, are rather loosely

Eria javanica

arranged in erect racemes up to about 2 feet long that come from near the tops of the pseudobulbs. They are white to creamy-yellow, usually with purple veins or stripes, and are hairy on the backs of their sepals and on the ovary. The center of the three lobes of the ridged lip is narrowly edged with violet. Somewhat resembling the last, ***E. merrillii,*** of the Philippine Islands, has conspicuously-angled pseudobulbs, up to

Eria cymbiformis

Eria ferruginea

4 inches long, with two oblong-lanceolate leaves up to 1 foot long. Its short-lived, faintly fragrant, 1-inch-wide flowers are in racemes up to 1¼ feet long. They are creamy-white, sometimes lined or suffused with purple and with the lip marked with yellow on its disk. Native from the Himalayas to southeast Asia, ***E. spicata*** (syn. *E. convallarioides*) has rather slender pseudobulbs 6 to 9 inches long with three or four pointed-lanceolate-elliptic leaves up to 7 inches long. In crowded, usually drooping racemes 4 to 6 inches long, the very fragrant flowers, which do not open fully, are approximately ½ inch in diameter. Their broad-ovate sepals and much smaller obovate petals are white to yellowish. The lip is spoon-shaped and at its front golden-yellow.

Garden Uses and Cultivation. Erias are appropriate for inclusion in collections of orchids. Their temperature needs vary somewhat, depending upon the native habitats of the various kinds, those from higher elevations in the tropics naturally preferring somewhat cooler conditions than those from lower ones. In general they respond to environments that suit dendrobiums. For more information see Orchids.

ERIANTHUS (Eriánth-us)—Plume Grass, Ravenna Grass. By some botanists this genus of the grass family GRAMINEAE is united with that of the sugar cane *Saccharum.* Usually it is considered separately on the basis that its spikelets have awns (terminal bristles). It consists of up to twenty-five species of the tropics, subtropics, and temperate regions of the Old and the New Worlds. Several are natives of North America. The name is from the Greek *erion,* wool, and *anthos,* a flower. It alludes to the conspicuous hairiness of the flower panicles of most kinds.

These are tall, reedlike perennials. They form bold clumps of slender to stout stems with very narrow to linear, flat or rolled leaf blades. The mostly dense, long and slender to broad-cylindrical, fragile flower panicles are composed of spikelike racemes of paired, similar spikelets, one of each pair stalked, the other not.

Ravenna grass (***E. ravennae***), native from the Mediterranean region to India, attains a height of 4 to 12 feet and has leaves up to 3 feet long and ¼ to ½ inch wide. They are rough on both surfaces and have stout, pale mid-veins. The grayish or purplish, loose or compact, flower plumes are silky-hairy and 1 foot to 3 feet long.

Two American species worth considering are *E. alopecuroides* and *E. giganteus* (syn. *E. saccharoides*). Native from New Jersey to Indiana, Missouri, Florida, and Texas, ***E. alopecuroides*** is a moist soil plant 3 to 9 feet tall, with leaf blades ½ to 1 inch wide and densely-hairy toward their bases. The silvery to tawny flower heads, 6 to 9 inches long by about one-third as wide, have many long hairs that almost hide the spikelets. The latter have twisted or spiraled bristles (awns). About the same height, ***E. giganteus*** differs in the bristles of its spikelets not being twisted or spiraled. It is indigenous from New Jersey to Kentucky, Arkansas, Florida, and Texas, favoring moist soils. Tawny or purplish, its panicles of blooms are 6 inches to 2 feet in length and fairly narrow. The hairs are as long or often much longer than the spikelets.

Garden and Landscape Uses. The most decorative species is the ravenna grass, but the others discussed are worth growing and are particularly appropriate for native plant gardens and naturalizing. Although less magnificent than the pampas grass (*Cortaderia selloana*), the ravenna grass serves very much the same purpose in the landscape and is much hardier. It survives at least as far north as New York City and favors well-drained, somewhat sandy soils. It makes great purplish clumps that when well established produce numerous feathery flower heads. This grass may be accommodated in lawn beds, borders, or at the fronts of tree and shrub plantings.

Cultivation. Once established these grasses need no particular care. If they show signs of losing vigor a spring application of a complete garden fertilizer may act as a restorative. They are easily increased by division in early spring or early fall and by seed.

ERIASTRUM (Eri-ástrum). Differing from closely related *Gilia* in the lobes of the calyxes of its flowers being unequal, *Eriastrum,* of the phlox family POLEMONIACEAE, comprises fourteen species, natives of western North America. Its name, in allusion to the woolly-hairiness of the plants is derived from the Greek *erios,* wool.

Mostly annuals, less often herbaceous perennials or shrubs, eriastrums are clothed with cobwebby hairs. They have alternate, smooth-edged to toothed or pinnately-lobed leaves. Usually in heads furnished with woolly hairs, but sometimes solitary, the blue, pink, yellow, or white flowers have a five-parted calyx, a more or less funnel-shaped, five-lobed corolla, five stamens, one style, and a three-lobed stigma. The fruits are capsules.

Native to California and Baja California, ***E. densifolium*** (syn. *Gilia densifolia*) is herbaceous with a woody base or is a subshrub or shrub up to 2 feet tall or taller. Its leaves are linear, lobed or lobeless, and often stiff. The flowers have corollas with white or yellow tubes and yellow lobes (petals). Variety *E. d. austromontanum* is smaller and less hairy. Variety *E. d. elongatum* has less woody stems and more rigid leaves and is usually white-woolly.

Garden Uses and Cultivation. These are as for *Gilia.*

ERICA (Erì-ca)—Heath, Bell-Heather. Ericas or heaths are by amateurs often confused with callunas or heathers. Although both belong to the heath family ERICACEAE, they are quite different. In *Erica* the tubular, very briefly-lobed corolla forms the showy part of the flower and is much bigger than the calyx. In *Calluna* the showy part of the flower is the calyx. This is deeply-cleft into petal-like sepals and is much bigger than the corolla. Also, heathers have the leaves of their nonflowering shoots opposite and closely overlapping, whereas those of *Erica* are most commonly in whorls (circles) of three to six. The genus is estimated to contain 500 species, natives of Europe, North Africa, islands of the Atlantic, and South Africa. They are evergreen shrubs, subshrubs, or rarely small trees.

Along with heathers, which in many places grow with them, heaths have for centuries, often out of want of more suitable materials, been used in northern Europe for building, thatching, and making brooms and as bedding for animals and man. The most important modern use is that of the roots of *E. arborea* to make brier pipes. The word brier in this context is a corruption of the French *bruyere,* heath.

Heaths have needle-like leaves and usually nodding, small- to medium-sized flowers, solitary or few to many in axillary or terminal spikes or racemes. They have brief, four-lobed calyxes, cylindrical-tubular to urn-shaped corollas with four small lobes (petals), usually eight stamens, and one each style and stigma. The fruit capsules contain numerous very small seeds. The name *Erica* derives from the Latin *erice* and the Greek *ereike,* names for these plants. For convenience cultivated heaths may be considered as two groups, those of South African origin, and the rest. None of the first group is hardy. Many, but not all of the others are.

Hardy heaths, those native of Europe that survive outdoors north of the Mason-Dixon line, are mainly excellent low plants deserving of more attention from gardeners than they receive. Prominent here is the spring heath or winter heath (***E. carnea*** syn. *E. herbacea*). Not exceeding 1 foot in height, this very hardy, friendly species, native chiefly in mountain woods and stony slopes through central and southern Europe, has given rise to numerous beautiful garden varieties. It is a hummocky plant with procumbent, more or less hairless stems and erect branches that in its typical form bear ¼-inch-long, rosy-red, fragrant, ovoid to urn-shaped flowers in leafy, terminal, one-sided racemes 1 inch to 3 inches long. They are displayed in winter or early spring, or sometimes in fall depending upon climate and weather. The stigma and usually the anthers protrude beyond the corolla. The stems between the closely-set leaves have conspicuous ridges

Erica carnea

Erica darleyensis

that, unlike those of *E. mediterranea,* do not diminish in height or width downward. The leaves, mostly in whorls of four and dark, glossy green above and more or less spreading, are linear and ¼ to ⅓ inch long and have rolled-under margins that hide the underside of the leaf. Horticultural varieties with red, pink, and white flowers are offered by specialist nurserymen.

Except for its larger size much resembling the spring heath, and like it spring-blooming, ***E. mediterranea*** (syn. *E. erigena*) is native of Spain, Portugal, southwestern France, and western Ireland. Erect and from 2 to 10 feet tall, this has distinctly short-hairy young shoots that become hairless later. A distinction between it and *E. carnea* is that the ridges that extend from the leaf bases down the stems of *E. mediterranea* rapidly narrow and become less conspicuous downward. The more or less bluntish, about ¼-inch-long, linear leaves, dark green above and in whorls of four or sometimes five, have rolled-under margins that do not meet, but leave exposed a narrow central strip of white along the underside of the leaf. The honey-scented, rosy-red flowers are in leafy racemes often assembled as panicles at the ends of shoots of the previous year's growth. Their stigmas and sometimes anthers protrude.

A hybrid swarm intermediate between its parents *E. carnea* and *E. mediterranea* is identified as ***E. darleyensis.*** It includes a number of varieties that have been given horticultural names. They approximate 1½ feet in height and have blooms ranging in color from magenta-red through pink to white. They bloom in late fall, winter, and spring.

Among European summer-flowering heaths hardy in the north, the bell-heather or twisted heath (***E. cinerea***) ranks highly. Of rather loose habit, this native of western and northern Europe forms mats of wiry stems and ascending branches, their young shoots pubescent. From 6 inches to 2 feet tall, this has hairless, somewhat spreading to recurved, linear leaves approximately ¼ inch in length in whorls of three. Their strongly rolled-under margins essentially conceal the lower sides of the leaves. The stems that bear the 1- to 3-inch-long terminal umbels or short racemes of typically reddish-purple flowers have below the blooming portion many short, leafy side shoots. The ovoid to urn-shaped flowers, from a little under to a little over ¼ inch long, have hairless, semitransparent sepals and four-toothed corollas. The anthers do not protrude. Many horticultural varieties, varying in height, habit of growth, color of flowers and foliage, and other details, are grown by specialists. Flower colors include purple, lilac-pink, pink, and white.

Cornish heath (***E. vagans***), a native of western Europe, attains heights up to 2 or exceptionally 3 feet. Blooming in summer and fall, this sort has flexuous, decumbent to erect stems, the young twigs hairless or finely-hairy. In whorls of four or five, the hairless or slightly hairy, spreading or deflexed leaves are linear to oblong. Their rolled-under margins cover the entire undersides of the leaves. In small axillary clusters grouped as dense, raceme-like panicles up to about 4 inches long, the broadly-bell-shaped, purplish-pink to lilac-pink blooms are ¼ to nearly ½ inch long. Their anthers and stigmas protrude. Variety *E. v. alba* has white blooms. Garden varieties in a number of improved flower colors are numerous.

Dorset heath or fringed heath (***E. ciliaris***), of western Europe, comparatively large-flowered and 6 inches to approximately 1 foot tall, has semiprocumbent stems, spreading branches, and glandular-downy young shoots. Not over ¼ inch long, and often shorter, its spreading, ovate-lanceolate to oblanceolate leaves are in whorls of three or, more rarely, four. At least when young they are hairy on their upper surfaces and are fringed with gland-tipped hairs. Their rolled-under margins hide only about one-third of the whitish undersides of the leaves. The anthers of the urn- to pitcher-shaped, rosy-pink flowers, the latter ⅜ inch long, in short, terminal racemes 1 inch to 5 inches long, do not protrude. This blooms in summer and fall. Horticultural varieties with white and with red flowers are cultivated. From the cross-leaved heath the Dorset heath differs in its leaves mostly being in threes and its flowers in racemes.

The cross-leaved heath (***E. tetralix***), of western and northern Europe, is a straggling shrub 9 inches to 2 feet tall with few-branched, spreading to prostrate older stems and ascending younger ones. The young shoots are glandular-hairy. In whorls of four spread to form a cross, the linear-

Erica tetralix

oblong to lanceolate, often hair-fringed leaves, are up to ⅛ inch long. Their margins may be more or less rolled-under, but not so that the under leaf surface is completely hidden. The light pink, urn-shaped flowers are displayed in summer and fall. From about ¼ to over ⅓ inch long, they are in umbels of from four to twelve at the branch ends. They have lobes (petals) that spread or roll backward. The anthers do not protrude. There are numerous horticultural varieties, some with white flowers.

Hybrids of the cross-leaved heath include ***E. praegeri,*** the other parent of which is *E. tetralix.* It does not produce fertile seeds. Another sterile hybrid, this between the cross-leaved heath and *E. ciliaris,* named ***E. watsonii,*** is intermediate between its parents. Yet a third *E. tetralix* hybrid, an intermediate between that species and *E. vagans,* is named ***E. williamsii.***

A rare native of western Ireland and native also in Spain, ***E. mackaiana*** in many details resembles the cross-leaved heath. From 6 inches to 2 feet tall and compact, it has spreading to erect stems, suberect branches, and glandular-pubescent young shoots. In whorls of four the obong-lanceolate leaves, including those below the flowers, are spreading and the spaces between them on the stems are all approximately the same length. Their margins are less rolled backward than those of *E. tetralix.* The bright purplish-pink, urn-shaped flowers are in terminal umbels. Their corolla lobes (petals) spread or recurve. Their anthers do not protrude. Variety *E. m. plena* has double flowers.

European and Mediterranean-region heaths, generally less hardy than those discussed above, include a number, the taller ones sometimes called tree heaths, of higher shrubs or rarely small trees. The tallest, ***E. arborea,*** exceptionally attains a height of 20 feet, but more commonly not more than 12 feet. Native of the Mediterranean region, this spring-bloomer has young shoots densely clothed with branched hairs. Its linear, erect to spreading, hairless leaves, up to ¼ inch long, are arranged closely, usually in whorls of four, sometimes of three. Their margins are rolled under to the extent that they completely hide the under leaf surfaces. Produced in spring, the broadly-bell-shaped to nearly globular, fragrant, grayish-white to pinkish flowers, in racemes toward the ends of lateral twigs usually assembled in terminal or subterminal panicles, are up to ⅙ inch long. The anthers do not protrude. The stigmas, broadly-head-shaped and white, are exerted. Variety *E. a. alpina,* decidedly hardier than the species, is 6 to 8 feet tall. Resembling *E. arborea,* but not exceeding about 12 feet in height and often lower and blooming earlier, the Portugal heath (***E. lusitanica***) has slightly bigger, usually pink-tinged flowers with nonprotruding stamens and exerted or included, obconical, red stigmas. One of the showiest species, the Spanish heath (***E. australis***), of Spain and Portugal, is a slender, elegant shrub that attains a maximum height of 8 feet, but does not usually exceed 3 to 4 feet. From *E. arborea* it differs in its paler foliage and by the hairs on the young shoots not being branched. It has suberect branches. When young its shoots, like its young leaves, are downy. At maturity the latter are without hairs. Erect or spreading, linear, and ¼ inch long, they are in whorls of four. Their margins are rolled under to hide completely the undersides of the leaves. In clusters of four to eight in spring at the ends of short branchlets from growths of the previous year, the fragrant, rose-purple, cylindrical to narrowly-bell-shaped flowers with recurved lobes (petals) and a small red stigma are approximately ⅓ inch long. Their anthers may or may not partly protrude. A beautiful variety with white flowers is *E. a.* 'Mr. Robert'. Other horticultural varieties, with blooms of varying shades of pink, are grown.

Other heaths of the Mediterranean region include the besom heath (***E. scoparia***), which ranges from Spain to Italy and central France. Slender and up to 10 feet tall or sometimes taller, this irregularly-shaped shrub has hairless or very slightly hairy young shoots and erect to spreading, pointed-linear, glossy dark green leaves up to or scarcely over ¼ inch long in whorls of three or four. Except for a center line about one-third of their widths, the undersides of the leaves are hidden by the rolled-under leaf margins. Of minor decorative importance, the broadly-bell-shaped, red-tinged, greenish flowers in terminal, usually interrupted, slender racemes, which sometimes form loose panicles, are displayed in late spring. They have nonprotruding stamens and a reddish-purple, head-shaped stigma that protrudes scarcely or not at all. Variety *E. s. azorica,* up to 18 feet tall and endemic to the Azores, has smaller blooms with conspicuously projecting stigmas. The horticultural variety *E. s. pumila* does not exceed 1½ feet in height. The Corsican heath (***E. terminalis*** syn. *E. corsica*) that flowers in late summer is native from Spain to Italy and Corsica. Erect and bushy, 3 to 8 feet tall, this has suberect branches, and slightly hairy young shoots. In whorls of four, rarely five or six, the spreading, glossy dark green, finely-hair-fringed, linear leaves are up to ⅜ inch long. Their margins are rolled under without hiding all the lower surfaces of the leaves. The fragrant, bright pink blooms, approximately ¼ inch long and urn-shaped with recurved lobes (petals), are in umbels of up to eight at the branch ends. Their anthers do not protrude. Attractive ***E. umbellata,*** of Portugal, Spain, and North Africa, erect or semiprostrate and 9 inches to 3 feet tall, has downy young shoots and blunt, linear leaves ⅙ inch long in whorls of three. Their rolled-under margins completely hide the leaf undersides. The

Erica canaliculata

broadly-bell-shaped to subspherical flowers, produced in late spring, are in terminal umbels of up to eight. They are up to ¼ inch long and cerise-pink. Their anthers do not protrude.

South African heaths are very numerous. None is hardy in the north. A few are cultivated outdoors in California and other places with mild, dryish climates, and in greenhouses. Many others would be worth introducing. Some have very much larger blooms than the European and Mediterranean-region kinds, and they come in a wider range of colors. The most widely grown ***E. canaliculata,*** often misidentified as *E. melanthera,* is very popular in the florists' trade as a winter-flowering pot plant, which is frequently sold as Christmas-heather. An erect, bushy shrub 2 feet or so tall, this has densely-hairy young shoots and recurved, linear leaves up to ¼ inch long and usually in whorls of three. They are in great numbers in clusters at the ends of short lateral shoots that form large panicle-like arrangements. From white to pinkish and with protruding conspicuous black anthers, the broadly-bell-shaped blooms are about ⅛ inch long. Varieties most commonly grown are *E. c. boscaweniana,* with white blooms, and *E. c. rosea,* with pink ones. Attractive ***E. peziza,*** 1 foot to 1½ feet tall, has downy shoots and, in whorls of three, lustrous, linear leaves up to ⅙ inch long. Its abundant, cup-shaped to obconic, densely-downy, white flowers, mostly in groups of three and not constricted at their throats, are about ⅛ inch long. About as tall as the last, *E. quadrangularis* has minutely-downy shoots and linear leaves up to ⅛ inch long in whorls of four. The numerous white to rosy-pink flowers, about ⅛ inch long and in groups of four, are cup-shaped to reverse-conical with the lobes (petals) commonly erect rather than spreading. A pleasing plant cultivated as *E. tricolor* is probably misidentified. The species to which that name properly belongs has distinctly hair-fringed leaves, but the leaves of the plant grown in gardens as *E. tricolor* are hairless. Moreover, its flowers are in elongated spikes of many whereas that of true *E. tricolor* are few together at the shoot ends. The cultivated plant attains a height of about 2 feet. It has narrow-linear leaves that spread at right angles from the shoots. The cylindrical to narrow-conical flowers, crowded in terminal spikes, are white at their bases, deep pink at their apexes, and paler pink between. The styles protrude slightly.

Other South African species and hybrids of South African species of special merit include the following. ***E. abietina*** (syn. *E. patersonii*), 2 feet or more tall, has linear leaves up to a little over ½ inch long in fours, and solitary, rich yellow, ¾- to 1-inch-long flowers in leafy racemes. ***E. baccans,*** hairless, 3 to 5 feet tall, with linear leaves up to ⅓ inch long, has numerous spherical, ¼-inch-wide, reddish-purple flowers mostly in terminal clusters of four. ***E. cavendishiana,*** its parents *E. abietina* and *E. depressa,* 1½ to 2 feet tall, has awl-shaped leaves ⅜ inch long and beautiful rich-yellow, cylindrical blooms nearly 1 inch long. ***E. chamissonis,*** 1 foot to 3 feet

Erica abietina

Erica baccans

tall, has downy shoots densely crowded with leaves ¼ to ½ inch long in whorls of three. The broadly-bell-shaped, rose-pink flowers are in clusters of up to four at the ends of short lateral branches that form cylindrical spires about 6 inches in length. ***E. cruenta,*** 2 to 3 feet tall, has linear leaves

Erica peziza

Erica quadrangularis

Erica tricolor

Erica cruenta

Erica curviflora

up to ⅓ inch long and 1-inch-long red, pink, or white flowers widening from their bases to their mouths and in racemes of twos or threes near the shoot ends. ***E. curviflora*** is 2 to 5 feet tall. Its leaves, linear and up to ⅓ inch long, are in fours. The

Erica doliiformis

Erica foliacea

solitary flowers, terminal on short shoots, are in long racemes. They have curved, downy, red to orange corollas 1 inch to 1¼ inches long. Variety *E. c. sulphurea* has yellow blooms. ***E. doliiformis*** (syn. *E. blanda*) is a low shrub with tiny leaves and umbels of flask-shaped, purplish-red flowers. ***E. foliacea,*** up to about 3 feet tall, erect and practically hairless, has branches crowded with more or less erect leaves ½ to ⅓ inch long and overlapping. Its tubular-inflated green to yellow-green flowers, a little more than ½ inch long, are in small close clusters at the ends of short branches. ***E. glauca*** is a hairless shrub up to about 6 feet tall with glaucous, mostly incurved leaves up to ½ inch long. The pretty pink flowers in nodding clusters of usually eight together with pink bracts are ⅓ to ½ inch long. Variety *E. g. elegans* has somewhat larger blooms. ***E. hirtiflora,*** 2 to 4 feet tall, has hairy young shoots and linear, hairy, blunt leaves up to ¼ inch long in whorls

Erica glauca elegans

of four. Its nodding clusters of light purple, broadly-ovoid, purse-mouthed flowers, up to ⅙ inch long, are in clusters of seven or fewer at the ends of short, leafy twigs. ***E. hyemalis,*** possibly of hybrid origin, is an erect shrub up to 2 feet tall or taller with linear leaves in whorls of three or four. The cylindrical, ¾-inch-long, pink-tinted, white flowers are in long racemes. *E. h.* 'Professor Diels' has blue-tinged, scarlet blooms. ***E. mammosa,*** up to 5 feet tall, has linear leaves ¼ to ¾ inch long in whorls of four or scattered singly along the twigs. In crowded clusters 2 to 2½ inches long, the ¾-inch-long, cylindrical flowers, contracted at their mouths, are red, reddish-purple, or white. A selected variety is named *E. m.* 'John McLaren'. ***E. persoluta*** is 1 foot to 3 feet tall. This has downy young shoots and leaves up to ¼ inch long, slightly hairy or hairless. In clusters of four at the ends of short branchlets arranged to form leafy inflorescences 4 to 7 inches long, the cupped,

Erica mammosa

Erica perspicua

Erica verticillata

⅙-inch-wide flowers are rosy-red. Variety *E. p. alba* has white blooms. ***E. perspicua*** is 2 feet tall or taller. It has linear leaves up to ¼ inch long in threes or fours. In spire-like panicles, its flowers, deep pink or purplish toward their bases, and paler above, are single or in twos or threes at the ends of short twigs. They have dark anthers and are ¾ to 1 inch long. ***E. ventricosa,*** 2 to 6 feet tall, has linear-awl-shaped leaves ½ inch long or a little longer in crowded whorls of four. In terminal umbels, the very glossy, white to rosy-pink, ovoid flowers, constricted at their mouths, are ½ to ¾ inch long, with small, reflexed petals. Many varieties of this fine heath have been raised, some with flowers larger than the maximum given above and in a range of pink and red and combinations of these colors and white. ***E. verticillata*** (syn. *E. concinna*), up to 5 feet tall, has few erect branches bearing whorls of shorter branches. The spreading leaves up to ¼ inch long are mostly in fours to sixes. The nearly cylindrical, rosy-pink flowers about ¾ inch long are clustered slightly below the apexes of the branches.

Garden and Landscape Uses. Heaths have many garden and landscape uses. Hardy kinds are useful as groundcovers, in rock gardens, at the fringes of shrub beds and fronts of flower beds, and in beds by themselves. A border of heaths planted with something of the casualness of pattern suggested as appropriate for a heath garden (see Heath or Heather Garden), located in front of a bed of shrubs can be highy effective. In Europe heath or heather gardens, often of considerable extent, and planted almost entirely with species and varieties of *Erica, Calluna,* and near relatives, are popular. Such features, although practicable where these plants grow well, are less common in North America. Nonhardy heaths, many considerably taller than hardy kinds, are splendid for mild climates. They prosper in the same types of locations and soils as other kinds and serve well as general purpose, fine-textured, flowering shrubs. The lower kinds, such as *E. canaliculata,* are excellent greenhouse pot plants.

Heaths are basically sun-lovers that succeed best in open, breezy places where air circulation is good, but the atmosphere is not arid. Some endure a little part-day shade. They adapt well to coastal areas. All heaths grow satisfactorily in acid soil, many do as well in neutral ones, a very few, including *E. cinerea, E. erigena, E. herbacea, E. darleyensis,* and *E. tetralix,* survive in limestone earths, especially where a layer of chiefly organic debris overlies a limestone substratum. The chief requirements are that subsurface drainage be good and that the texture be such as to encourage the fine roots of these plants. An excellent earth for heaths is a sandy, gritty, gravely, or even stony loam with which has been mixed generous amounts of peat moss, leaf mold, or compost. Cold, wet, clayey soils are unsuitable. It is inadvisable to fertilize. Heaths do better in a lean soil loose enough for their roots to ramify through in search of nutrients than on a fatter diet that even if the plants live causes them to become gross and atypical.

Cultivation. Caring for hardy heaths, and in climates adapted to their cultivation nonhardy kinds, is not onerous. The time to plant is early spring or early fall, in cold climates the first. Make sure the roots do not dry while the plants are out of the ground. As soon as they are planted water thoroughly. Then mulch with peat moss or other organic material. One of the most helpful practices is to maintain around them at all times afterwards such a mulch. It conserves soil moisture and aids in keeping the roots cool in summer and in tempering the severity of winter freezing.

Pruning should not involve severe cutting-back. Shearing to remove faded blooms is desirable. Do this to winter- and spring-blooming varieties as soon as they are through blooming. In cold climates defer pruning kinds that bloom in summer until the following late winter. Left on the plants the tops afford some winter protection. Additional protection against severe cold and drying winter winds may be given by a loose cover of branches of pine or other evergreen or by salt hay or similar loose material that gives some shade while permitting free air circulation. This winter cover is removed early in spring.

New plants of heaths are raised from cuttings, layers, and seeds. The latter are not reliable for the multiplication of horticultural varieties, but are satisfactory for species. Sow in pots or flats of sandy peaty soil, the top layer finely sifted, in a cold frame or cool greenhouse. Make sure the soil is kept evenly moist. Cuttings root readily. Make them from terminal shoots or branches at any time in summer after the new growth has become firm or in fall before hard freezing. Even shoots that have flowered are satisfactory as propagating material. Plant the cuttings in a propagating bed in a cold frame or cool greenhouse in a mixture of peat moss and sand or of peat moss and perlite. Layering may be done in spring or summer.

As greenhouse plants, South African species need cool, airy conditions. A winter night temperature of 40 to 50°F is adequate. In relation to the brightness of the weather, day temperatures may be from five to fifteen degrees above the night ones. Full sun is needed. The plants benefit from being outside during the summer with their pots buried to the rims in a bed of sand, peat moss, or similar material. A rather fine, sandy peaty soil, kept always evenly moist, is suitable. Shear the plants to shape after flowering. Repotting, when this attention is necessary, is best done in late winter or in September. When attending to this take great care not to disturb the roots more than absolutely necessary. Drain the pots adequately. After potting, keep the atmosphere somewhat more humid than usual for three or four weeks and on bright days spray the foliage with a fine mist of water. Pack the soil rather firmly. Well-rooted specimens benefit from regular applications of dilute liquid fertilizer from spring to fall. In their early stages heaths to be grown as pot plants need some training. This is done by occasional pinching out the tips of the shoots to induce branching. With those kinds that make strong, erect branches it is advisable, while they are yet young, to tie some down into a horizontal position to form a low framework from which erect shoots can develop. If this is not done straggly specimens bare of foliage below are likely to result.

ERICACEAE—Heath Family. As a source of ornamentals the heath family is extremely important. A few of its members produce edible fruits, among the best known, blueberries, cranberries, huckleberries, and lingon berries. The roots of *Erica arborea* supply wood used for brier pipes. From *Gaultheria* oil of wintergreen is obtained. Some members of the *Rhododendron* relationship are poisonous to livestock and produce poisonous pollen. By some authorities *Vaccinium* and allied genera are segregated as the *Vacciniaceae,* but that is not done in this Encyclopedia.

In the wild widely distributed throughout most parts of the world, the heath family consists of fifty genera totaling 1350 species of subshrubs, shrubs, and small trees with evergreen or deciduous, undivided leaves that are alternate, opposite, or in whorls (circles of more than two). The flowers, symmetrical or slightly asymmetrical, are solitary or in racemes or panicles from the leaf axils or shoot ends. They have usually persistent calyxes and corollas with four or five lobes or separate sepals or petals, as many or twice as many stamens as corolla lobes or petals, and one style and stigma. The fruits are capsules, berries, or drupes. The vast majority of the sorts of this family of dicotyledons are acid-soil plants that will not thrive in limestone or other alkaline soils.

Among the best known genera of the heath family are *Agapetes, Andromeda, Arbutus, Arctostaphylos, Bejaria, Bruckenthalia, Bryanthus, Calluna, Cassiope, Chamaedaphne, Chimaphila, Cladothamnus, Daboecia, Elliottia, Enkianthus, Epigaea, Erica, Gaultheria, Gaulthettya, Gaylussacia, Kalmia, Kalmiopsis, Ledum, Leiophyllum, Leucothoe, Loiseleuria, Lyonia, Macleania, Menziesia, Oxydendrum, Pentapera, Pernettya, Phyllodoce, Pieris, Pterospora, Pyrola, Rhododendron, Rhodathamnus, Tripetaleia, Tsusiophyllum, Vaccinium, Xylococcus,* and *Zenobia.*

ERICAMERIA. See Haplopappus.

ERIGENIA (Eri-genìa)—Harbinger-of-Spring. The common name of the only species of this genus of the carrot family UMBELLIFERAE, indicates its early-flowering propensities. Its botanical identification does also, being derived from the Greek *erigeneia,* early born.

Native from New York to Michigan, Minnesota, Alabama, and Arkansas, harbinger-of-spring (***Erigenia bulbosa***) is a tuberous, herbaceous, hairless perennial of delicate aspect, 2 to 6 inches tall, an inhabitant of rich, deciduous woodlands. In outline its leaves are broadly-ovate. At maturity they are 4 to 8 inches long, shorter at flowering time. The leaves are once-, twice-, or thrice-divided into linear or spatula-shaped leaflets. The flowers are in solitary, terminal, stalkless umbels, each associated with a single leafy bract and composed of two to four smaller, stalked umbels. Individual blooms are about ⅓ inch in diameter and have spreading white petals. The ellipsoid, seedlike fruits are small and flattened.

Garden Uses and Cultivation. This modest spring flower is suitable for wild gardens and shaded spots in rock gardens. Its needs are few, a woodland soil, broken shade, and care to prevent incursions of more vigorous neighbors. It is propagated by seed.

ERIGERON (Eríg-eron) — Fleabane, Orange-Daisy. Erigerons or fleabanes are much like asters, not the annuals called China-asters, which belong in the genus *Callistephus,* but the plants of the genus *Aster,* of which the kinds known as Michaelmas-daisies are the most familiar. As natives fleabanes are widely dispersed throughout much of the world being especially plentiful in temperate and mountain regions. There are about 200 species, belonging in the daisy family COMPOSITAE. The name, from the Greek *eri,* early, and *geron,* old man, perhaps alludes to the early development of the seed heads of some kinds of *Erigeron.*

Fleabanes include annuals, biennials, and herbaceous perennials, the cultivated ones hardy. Their foliage may be all basal or along the stems and branches. The leaves, rarely more or less finely-cut, are more usually at most toothed. Those of the stems are alternate. The solitary or clustered flower heads of cultivated kinds are daisy-like, having two or more circles of ray (petal-like) florets and a center disk floret. The rays are white, pink, lavender, purple, blue, or more rarely yellow or orange. Hemispherical or bell-shaped, the involucre (collar at the back of the flower head) is of one or two rows of nearly equal, narrow bracts. The fruits are seedlike achenes. The plant sometimes called *E. dubius* is *Hysterionica montevidensis.*

Native to western North America, ***E. speciosus*** is a handsome perennial that, except for the margins of its leaves is hairless or nearly so. From 1½ to 2½ feet tall,

Erigeron speciosus

it has leafy stems branched above and toothless leaves 3 to 6 inches long, the basal ones spatula-shaped, those above lanceolate and slightly stem-clasping. The flower heads, 1 inch to 1½ inches wide and at the ends of branches of the flower clusters, have very many narrow, violet to purple ray florets. Variety *E. s. macranthus* (syns. *E. macranthus, E. grandiflorus*) is less hairy and has broader leaves, those of the stem ovate.

A hybrid group, the parents of which are *E. s. macranthus* and some unspecified North American species, are in gardens commonly referred to as *E. hybridus,* although that name correctly belongs to a

Erigeron hybridus 'Sincerity'

Erigeron karvinskianus

South American species not in cultivation. Notable among these perennials are such splendid varieties as 'Charity', soft pink, 'Dignity', violet-mauve, 'Felicity', clear pink, 'Festivity', lilac-mauve, and 'Sincerity', mauve-blue.

Himalayan fleabane (***E. multiradiatus***) is a handsome, hairy perennial 1 foot to 2 feet tall. It has long-stalked, often few-toothed, oblanceolate to lanceolate leaves 1½ to 3 inches long. Solitary or few together on long stalks, the flower heads are 2 to 3 inches in diameter. They have thickly-hairy involucres and very many purple ray florets. As its name implies, this is native to the Himalayas.

Endemic from Maine to Ontario, Minnesota, Georgia, Mississippi, and Texas, ***E. pulchellus*** is a biennial or short-lived, variable perennial 1 foot to 2 feet tall. In gardens forms of apparently *E. speciosus* travel under its name. True *E. pulchellus* is distinct in having stolons and offsets. It has short-stalked, toothed, spatula-shaped basal leaves and stem ones stalkless and usually without teeth. The flower heads, 1 inch to 1½ inches across and one to six on slender stalks, have bluish-purple, pink, or rarely white ray florets about ⅓ inch long.

Dainty, trailing ***E. karvinskianus*** (syn. *E. mucronatus*), in gardens is not infrequently misidentified as *Vittadinia triloba*. Native from Mexico to Panama, this fleabane is a perennial that blooms the first year from seed. It has much-branched, slender stems 1 foot long or longer, and 1-inch-long, or smaller, wedge-shaped-ovate to linear leaves, the lower often few-toothed at their apexes, and hairless except along the margins at their bases. About ¾ inch wide, the slender-stalked, daisy-like flower heads are solitary. They have white ray florets purplish on their undersides.

The orange-daisy (***E. aurantiacus***), native to Turkestan, a velvety, often not long-lived perennial 6 inches to 1 foot tall, has toothless, spatula-shaped basal leaves and narrower, oblong, toothless, stalkless ones on the stems. The solitary, 1- to 2-inch-wide, semidouble flower heads terminate erect stalks. They have five or six rows of bright orange ray florets, with between them and the disk florets some rows of tubular female florets. In *E. a. sulphureus* the rays are paler. Another in this color range, its ray florets clear yellow, is perennial ***E. aureus,*** native at high altitudes in the northwestern United States and adjacent Canada. This has a woody, branched rootstock and is 2 to 6 inches tall and finely-hairy. It has stalked, spatula-shaped to elliptic to almost round basal leaves with blades about 1 inch long. The few stem leaves are smaller. Solitary, the flower heads are ¾ to 1 inch wide or a little wider.

Erigeron aureus

Erigeron compositus

Erigeron simplex

Alpine ***E. compositus*** ranges as a wildling from the Rocky Mountains to California and Alaska. A perennial 2 to 8 inches tall, it has a short-branched, woody rootstock and mostly basal foliage. The hairy, long-stalked leaves 3 inches or usually much less in length are twice- or thrice-cleft into linear or spatula-shaped lobes. Solitary on erect stalks and ¾ inch to 1¼ inches in diameter, the flower heads have white or purplish ray florets. Variety *E. c. discoideus* (syn. *E. trifidus*) has leaves usually only once-cleft into three lobes and flower heads ½ inch wide. Also native of the Rocky Mountains, ***E. leiomerus*** is a perennial up to 6 inches in height. It has a branched, woody rootstock. Its essentially hairless, toothless leaves are spatula-shaped to oblanceolate or obovate, up to 3½ inches long by almost ¾ inch wide. The stem leaves are smaller. With deep blue to nearly white ray florets, the flower heads are ¾ to 1 inch in diameter. A taller alpine species, ***E. formosissimus*** is a native of the Rocky Mountains and Black Hills. A perennial with slender, erect flowering stems 1 foot to 1½ feet tall, with chiefly basal foliage, this has oblanceolate to spatula-shaped or elliptic leaves up to 6 inches long, gradually diminishing in size upward. The blue-, pink-, or rarely white-rayed flower heads are 1 inch to 1½ inches wide. A delightful western North American high-altitude alpine, perennial ***E. simplex*** forms tufts of hairless to somewhat hairy, oblanceolate to spatula-shaped basal leaves 2 to 3 inches long and with hair-fringed edges. There are a few smaller stem leaves. The hairy, more or less sticky stems, from 3 to 6 inches tall, each terminate in a solitary flower head about ½ inch wide. Their ray florets are blue or pink or more rarely white.

Garden Uses. The taller fleabanes serve well at the fronts of perennial borders and have flowers useful for cutting. Except for

Erigeron leiomerus

Erigeron formosissimus

Epiphyllum darrahii of gardens

Episcia cupreata variety

Erysimum allionii of gardens

Eryngium, garden variety

Eriophorum species

Erythrina princeps

Erythrina species

Erythronium americanum

Erysimum capitatum

Erigeron karvinskianus in a dry wall

Erinacea pungens

the orange-daisy, which prefers full sun, fleabanes do best where they receive just a little shade from the strongest sun and succeed in ordinary garden earth of reasonable fertility. They are impatient of poorly drained locations. Trailing *E. karvinskianus* is delightful tucked into crevices in masonry steps and for poorish soils in sunny rock gardens. It blooms through most of the summer and fall. Of tidier growth, *E. compositus* and the other mountain species discussed above are charming in gritty screes or moraines in rock gardens and for growing in pots or pans (shallow pots) in alpine greenhouses. There are other species native in North America worth seeking out and trying.

Cultivation. With the possible exception of the orange-daisy, which is inconsiderate enough to expire on occasion without apparent cause, fleabanes other than high alpine ones, which can be tricky to grow in regions of hot summers, give no trouble. Spring is the preferred season to plant. Summer care is that needed by the run of easy-to-grow perennials. It involves weeding, watering, and staking of taller kinds. Propagation of horticultural varieties is easy by division, of species, by that method and by seed.

ERINACEA (Erin-àcea)—Hedgehog-Broom. One not very hardy low deciduous shrub or shrublet is the only species. It belongs in the pea family LEGUMINOSAE and is native of the western Mediterranean region of Europe and Africa. Its name, from the Latin, *erinaceus*, a hedgehog, was applied in recognition of its compact and spiny character.

The hedgehog-broom (***Erinacea pungens***) attains a height of 1 foot or a little more and forms a dense, rounded tussock, often wider than high, of rigid, much-branched, green or grayish stems each ending in a sharp spine. The young shoots are clothed with silky hairs. The leaves soon drop. They are silky-hairy, linear or narrowly-spoon-shaped, and ¼ to ½ inch long. The lower ones are opposite, the upper often alternate. Carried two to four together on short stalks from the uppermost leaf axils, the pea-shaped blooms are about 1 inch long and are lilac- to violet-blue. They have a persistent, swollen, silky-hairy, tubular calyx with five small teeth, and a corolla with a notched, ovate standard petal larger than the wing petals or the keel. The pointed seed pods are silky-hairy and ¾ inch long.

Garden Uses and Cultivation. Primarily a rock garden plant, this species is not reliably hardy in the north, but may be grown in parts of the south and on the West Coast. It needs perfect soil drainage and a warm, sunny location. Because it resents transplanting young specimens should be planted from pots to minimize root disturbance. They are often slow at becoming established and coming into bloom. New plants are obtained from seeds and by cuttings.

ERINUS (Erì-nus). One alpine species of the mountains of Europe is the only member of *Erinus* of the figwort family SCROPHULARIACEAE. It is a pretty, little, hardy plant of the easiest culture. The name is an adaption of the Greek *erinos,* used by Dioscorides for basil.

A tufted plant 2 to 4 inches tall, ***E. alpinus*** forms crowded basal clusters of toothed, spatula-shaped leaves up to ½ inch long and in spring has racemes of magenta-purple flowers up to 2½ inches long, erect and with stems furnished with scattered, alternate, toothed leaves. Each flower has a five-lobed corolla about ½ inch across and four stamens included in the throat of the bloom. The fruits are capsules containing numerous small seeds. In addition to the typical kind, varieties based on flower color have been named, thus we have *E. a. albus,* with white flowers, *E. a. roseus,* with pink blooms, and *E. a. carmineus,* with flowers of deeper pink. For the most part these color variations come true from seeds. A distinct variety, *E. a. hirsutus* is a common native of southern Europe. It is a hairier and softer plant than the typical species and has paler pink flowers lined with crimson in their throats.

Garden Uses and Cultivation. Seeds, which germinate readily, provide a satisfactory means of propagation. Plants can also be increased by division in spring or early fall. This species prospers in light, sandy soil, moderately moist, but not wet, in locations where it receives a little shade from the most intense sun. It is especially well suited for planting in deep vertical or near-vertical crevices between rocks where its roots can ramble into cool moist earth

Erinus alpinus (flowers)

Eriobotrya japonica

and its flower racemes display themselves to fullest advantage held vertically against the cliff face. In like fashion they can be used effectively in planted dry walls. Steep slopes in rock gardens not subject to erosion also are appropriate planting places. Because erinuses do not tolerate wet soil and because the flowers are not seen to full advantage under such circumstances, it is better not to plant them on level ground. When happily located *Erinus* often spreads by self-sown seedlings.

ERIOBOTRYA (Erio-bótrya)—Loquat. Belonging in the rose family ROSACEAE, there are about ten species in this genus of eastern Asian evergreen shrubs and small trees. None is hardy in the north. The name *Eriobotrya* is apt. Derived from the Greek *erion,* wool, and *botrys,* a cluster, it alludes to the pubescent flower clusters.

Eriobotryas have large, distinctly pinnately-veined, toothed, lobeless, short-stalked or almost stalkless leaves. Their white or creamy-white, five-sepaled, five-petaled flowers, in terminal panicles, have broad, clawed (narrowed to their bases) petals, about twenty stamens, and one style. They are succeeded by firm-fleshed fruits that contain one to few seeds and are tipped with the withered remains of the calyx.

The loquat (*E. japonica*) is cultivated in frost-free and nearly frost-free climates for ornament and for its edible fruits. The latter are pear-shaped to globose, about 1½ inches long, and pleasantly acid and have a few large seeds. They are in woolly-stalked clusters. The tree, symmetrical in outline, attains a height of 20 to 25 feet and has shoots clothed with rusty-brown hairs. Its firm, obovate to elliptic, pointed leaves, 8 inches to 1 foot in length, have conspicuous parallel veins running from midrib to margin, each ending in a pointed tooth. Their upper sides are rich glossy green, beneath they are whitish- or rusty-hairy. The fragrant, dingy-white flowers ⅜ to ¾

Eriobotrya japonica (flowers)

inch across develop in fall or early winter and are succeeded by pear-shaped to roundish fruits as large as small plums that ripen in winter or spring. They are pale yellow to orange-yellow and have white or yellowish flesh. Their outsides are covered with whitish down. The loquat is a native of China. There are superior horticultural varieties and only these should be planted for fruit production. Seedlings are likely to have inferior fruits or not to crop regularly and abundantly. Among recommended improved varieties are 'Advance', 'Champagne', 'Early Red', 'Oliver', 'Pineapple', 'Premier', 'Tanaka', and 'Thales'.

Native of Taiwan, ***E. deflexa*** (syn. *Photinia deflexa*) has coarsely-toothed, oblong-obovate to elliptic leaves 5 to 10 inches long and 2 inches wide, with parallel veins angling from midrib to margin. When young rusty-hairy, later they become more or less hairless. The flowers are white, a little over ½ inch wide, the subspherical fruits are about ¾ inch in diameter. Variety *E. d. buisanensis* (syn. *E. buisanensis*) has 4- to 6-inch-long, ½- to 1-inch-wide, oblong to lanceolate leaves with a pointed apex and tapered base. Those of *E. d. grandiflora* (syn. *E. grandiflora*), 4 to 6 inches long by 1½ to 2 inches wide, have round-toothed margins, rounded apexes, and wedge-shaped bases. From 3 to 6 inches long, the leaves of *E. d. koshunensis* (syn. *E. koshunensis*) are obovate-oblong, 3 to 6 inches long by 1 inch to 2½ inches wide, blunt at the apexes, and tapered at the base.

Garden and Landscape Uses. The loquat is a very handsome ornamental evergreen with bold distinctive foliage that lends character to landscape plantings and contrasts pleasingly with finer, more delicate-leaved items. It associates well with architectural features and can be placed attractively near houses and other buildings. It is a good courtyard and patio plant that accommodates well to cultivation in large tubs and other containers. As an ornamental it is worth growing in large greenhouses and conservatories. The fruits are eaten fresh and in preserves and jellies. For fruit production loquats can be grown only where winters are extremely mild; the fruits are damaged by slight freezing.

Cultivation. Provided drainage is good, eriobotryas thrive in a wide variety of

Eriobotrya japonica (fruits)

soils, in sun or part-day shade, and are quite tolerant of dryish conditions. They are best propagated by grafting, but budding and air-layering may be practiced. Where specimens are required for ornament only, seeds afford a satisfactory and ready means of increase. No systematic pruning is required, but if it is desirable to limit the plants to size, needed cutting to accomplish this may be done just before new growth begins. Ordinarily the trees are symmetrical and shapely. For the best fruit production annual fertilization, and irrigation during dry periods, is helpful.

When grown in tubs or other containers good drainage is essential, and coarse, fertile soil is suitable. Repotting or retubbing is needed at intervals of several years only, but at the beginning of each growing season some surface soil may be forked away and replaced with a rich top dressing. Watering, sufficient to keep the soil evenly moist, but not constantly saturated, requires attention throughout the year. Specimens that have filled their containers with roots should be given regular applications of dilute liquid fertilizer. In greenhouses a winter night temperature of 40 to 50°F is satisfactory and at that season the day temperatures should be only a few degrees higher. At all seasons airy conditions should be maintained.

Pests and Diseases. Mealybugs are partial to these plants and, in greenhouses, are especially given to invading the flower clusters. Fire blight disease sometimes kills branches and flowers of outdoor trees. The best control is had by carefully pruning away and burning affected parts.

ERIOCACTUS (Erio-cáctus). Closely related to *Notocactus* and by conservative botanists included there, two species those preferring finer generic splitting segregate as *Eriocactus* are natives of South America. They belong in the cactus family CACTACEAE. The name, derived from the Greek *erion,* wool, and cactus, alludes to the woolly character of the tops of the stems.

Eriocactuses have long columnar stems that eventually become prostrate, that have more than thirty ribs, and that are furnished with yellow spines and at their tops a felt of white hairs. The short, funnel-shaped, yellow blooms from near the ends of the stems open by day. They have spreading petals. The fruits are spherical berries containing reddish-brown seeds.

Native to southern Brazil, ***E. leninghausii*** (syn. *Notocactus leninghausii*) has stems cylindrical from their earliest stages, and up to 3 feet tall by about 4 inches wide. The spine clusters are of up to fifteen bristle-like radials under ¼ inch long and three or four centrals about 1½ inches long. The flowers are 2 inches wide. From the last, ***E. schumannianus*** (syns. *Notocactus schumannianus, N. grossei*) differs in its stems at first being spherical and later cylindrical and up to 3 feet long, and its approximately 1-inch-long spines being needle-like and in clusters of mostly four to seven, but up to ten. The flowers are 1¼ inches in diameter.

Eriocactus leninghausii

Garden and Landscape Uses and Cultivation. These are attractive, fast-growing, free-blooming plants well suited for outdoors in warm, semidesert regions such as much of the Southwest and for including in indoor cactus collections. They do well in full sun and, so long as the soil drains rapidly, appreciate more frequent watering than many cactuses. For more information see Cactuses.

ERIOCAULACEAE—Pipewort Family. Of minor horticultural importance, this family of monocotyledons includes thirteen genera totaling 1150 species, mostly of the tropics and subtropics, and especially South America. Plants of wet soils, they usually are more or less grasslike, stemless or nearly stemless herbaceous perennials with small to minute, nearly always unisexual flowers in heads with involucres (collars) of bracts, usually terminating erect, slender stalks. The perianths have usually two rows of two or three segments. The fruits are capsules. Only *Eriocaulon* seems to be occasionally cultivated.

ERIOCAULON (Erio-caùlon)—Pipewort. Of scant horticultural interest, except perhaps to fanciers of native plants and bog garden enthusiasts, the pipeworts are estimated as comprising from 200 to 400 species of perennial herbaceous plants. The genus is widely distributed in tropical, subtropical, and temperate regions in many parts of the world. It belongs in the pipewort family ERIOCAULACEAE and bears a name derived from the Greek *erion,* wool, and *kaulos,* a stem, that alludes to the woolly flower stalks of some species.

These plants form basal tufts of grasslike foliage. They have erect, leafless flower stalks terminating in solitary, more or less spherical, woolly heads of tiny, inconspicuous, unisexual flowers that have two or three sepals and petals, and twice as many stamens. The petals of the female flowers are separate to their bases, those of the males are joined below into a tube. The fruits are tiny capsules. Any of several North American species may occasionally be cultivated. They are described in floras of the regions in which they occur. All grow in wet soils or shallow water and may be propagated by seed and by division.

ERIOCEPHALUS (Erio-céphalus). This South African genus of thirty species of evergreen, aromatic shrubs belongs in the daisy family COMPOSITAE. Its name, alluding to the flower heads becoming woolly as they pass into seed, is from the Greek *erion,* wool, and *kephale,* a head.

Usually linear and more or less silvery, the leaves of *Eriocephalus* are small, alternate or opposite, and often clustered. Sometimes they are three-toothed at their apexes. The subspherical flower heads are solitary, or in umbels or racemes. They have disk florets that, like those of daisies, form a central eye, in *Eriocephalus* yellow or purple, and encircled by a single row of whitish, petal-like ray florets. The collar (involucre) at the back of the flower head is of two rows of bracts, the inner ones densely-woolly. The fruits are achenes.

The species most likely to be cultivated ***E. africanus*** has opposite or clustered, linear or three-lobed, thickish, silky-hairy leaves ½ to 1 inch long. Its white flower heads are in umbels at the branch ends.

Eriocephalus africanus at home in South Africa

Garden and Landscape Uses and Cultivation. Only in frost-free or essentially frost-free Mediterranean-type climates, such as that of California, will these shrubs thrive outdoors. They need sunny locations and well-drained soil and are raised from cuttings and seeds.

ERIOCEREUS (Erio-cèreus). By conservative botanists this genus of the cactus family CACTACEAE is included in *Harrisia*. Those who favor segregation do so on the basis that the bright red or carmine fruits of *Eriocereus*, of South America, burst open when ripe, whereas the yellow to orange or dull red ones, of West Indian and North American *Harrisia*, do not. Accepted in this way, *Eriocereus* comprises seven species. Its name, from the Greek *erion*, wool, and *Cereus*, the name of an allied genus, alludes to woolly hairs of the areoles (specialized parts of the stems of cactuses from which spines come).

These are vigorous, slender-stemmed, erect, clambering, or prostrate cactuses, usually conspicuously spiny and with ridged stems. The long-tubed, funnel-shaped, predominantly white, large flowers, borne near the tops of the stems, open in the afternoon or evening or at night. The fruits, often warty, usually with deciduous scales, and rarely spiny, are fleshy and more or less spherical.

Species cultivated include these: ***E. bonplandii*** (syn. *Harrisia bonplandii*), of Argentina, Uruguay, and Brazil, has scrambling or trailing, bluish-green to grayish stems up to about 10 feet long and 1¼ to 2¼ inches in diameter, with four to six flattish ribs that tend to be obscured as the stems age. The clusters of usually eight spines are ¾ to 1 inch apart. The biggest are about ¾ inch long. Pure white with greenish-brown outside petals, the blooms, about 10 inches long, are conspicuously woolly in the axils of the scales of their perianth tubes. They open in the afternoon and close at nightfall. The fruits are spherical and warty. ***E. jusbertii*** (syns. *Harrisia jusbertii, H. bonplandii brevispina*), believed by some to be wrongly identified as an *Eriocereus* and instead to represent a hybrid between an *Echinopsis* and a *Cereus*, is not known as a wild plant. Reports that it breeds true from seeds seem to belie the possibility of its hybrid origin. A free-bloomer, this has stems ¾ to 1 inch thick with four to six low ribs. Except those of young seedlings the stems are erect. Clusters of dark spines are spaced about ¾ inch apart. Greenish toward their outsides, the otherwise pure white blooms are about 7 inches long. ***E. martinii*** (syn. *Harrisia martinii*), very vigorous, has much-branched, cylindrical, scrambling, green to grayish stems up to 6 feet long by ¾ to 1 inch thick, with four or five lumpy ribs. Each spine cluster has one long central spine and a few shorter radials. About 8 inches in length, the white flowers have reddish-tipped, pale green outer petals. They are produced freely. The spherical, red fruits are warty and spiny. This is a native of Argentina. ***E. pomanensis*** (syn. *Harrisia pomanensis*), of Argentina, has arching or prostrate, bluish-green, glaucous stems with four to six blunt, rounded ribs. The spine clusters have each one central up to ¾ inch long and five to seven shorter radials. Up to 6 inches in length, the flowers are white. ***E. regelii*** (syn. *E. martinii regelii*), probably of Argentina, is much like *E. martinii*, but has stouter stems with longer spines and smaller bright red fruits. The blooms are white with light pink outer petals. ***E. tortuosus*** (syn. *Harrisia tortuosa*) is native to Argentina. It has slender, arching, bright green stems ¾ inch to 1½

Eriocereus tortuosus

inches thick, with usually seven rounded, sometimes lumpy ribs. The clusters of six to ten spines are of one central and shorter radials. The 5- to 6-inch-long, freely-produced blooms are white to pink with dull greenish-brown outer petals. The red fruits are globular, warty, and spiny and up to 1½ inches in diameter.

Garden and Landscape Uses and Cultivation. Where reasonable space is available these cactuses are well suited for outdoor collections in warm climates and for greenhouses. In greenhouses they lend themselves for training up pillars or to wires strung beneath the roof glass. Often they bloom while fairly small. They are very suitable for use as understocks upon which to graft weaker-growing cactuses. Eriocereuses are of very easy cultivation. They thrive with minimum attention under conditions that suit the majority of columnar *Cereus*-type cactuses. For more information see Cactuses.

ERIODENDRON ANFRACTUOSUM. See Ceiba.

ERIODICTYON (Erio-díctyon) — Yerba Santa. Referring to the undersides of the leaves, the name of this group of the water leaf family HYDROPHYLLACEAE comes from the Greek *erion*, wool, and *diktuon*, a net. There are eight species, natives of the southwestern United States and Mexico. Sometimes they are planted in their home regions. The plant previously known as *Eriodictyon parryi* is *Turricula parryi*.

Eriodictyons are loose, aromatic, evergreen shrubs with underground, spreading rootstocks and foliage mostly at the ends of the branches. The leaves are alternate, undivided, leathery, and toothed or not. White to purple, the small flowers are many together in branching panicled clusters. They have deeply-lobed calyxes and funnel- to bell-shaped corollas. The stamens do not protrude. The style is divided to its base. The fruits are capsules.

Yerba santa (***E. californicum***) was used medicinally by the Indians and early settlers in the treatments of colds, influenza, and asthma. Native to California, it is from 2 to 7 feet tall and has sticky, hairless or sparsely-hairy stems, and sharp-stalked, lanceolate to oblong, usually toothed or more rarely smooth-edged leaves 2 to 6 inches in length, sticky and hairless on their upper sides and with almost a varnished appearance, and shortly-hairy underneath. The lavender to white, narrowly-funnel-shaped blooms, from under to a little over ½ inch long, are in clusters 2 to 8 inches in length, with lanceolate-linear, sparsely-hairy calyx lobes and corollas sparsely-hairy above on their outsides. Very similar, ***E. trichocalyx*** differs chiefly in its calyx lobes and corollas being smaller and densely-hairy. This kind is also Californian. From the above kinds ***E. crassifolium*** differs in its leaves not being sticky on their upper sides. In the typical species they are hoary, sometimes white-hairy above as well as on their undersides, but hair is absent from the upper surfaces of the leaves of *E. c. denudatum*. A shrub 3 to 10 feet in height, *E. crassifolium* has hairy shoots and short-stalked, lanceolate to elliptic or ovate, toothed leaves 2 to 6 inches long. Its broadly-funnel-shaped blooms, in clusters 2 to 3 inches across, are from a little under to a little over ½ inch long, and not glandular, but have hoary-hairy, lanceolate-linear calyx lobes and densely-pubescent corollas. By some botanists treated as a variety of the last, ***E. tomentosum*** has smaller flowers constricted at their throats and with glandular calyxes and corollas.

Garden and Landscape Uses and Cultivation. These shrubs have limited landscape uses in dry, warm climates where they may serve as general purpose shrubbery in sunny, well-drained locations. They are propagated by seed.

ERIOGONUM (Erióg-onum)—St.-Catherine's-Lace, Wild-Buckwheat or California-Buckwheat. The genus *Eriogonum* of the buckwheat family POLYGONACEAE is endemic to North America, the tremendously greater number of its approximately 200 species to the western part of the continent. It includes annuals, biennials, herbaceous perennials, subshrubs, and shrubs, some of considerable beauty. Few are cultivated. The name is from the Greek *erion*, wool, and *gony*, a joint or knee. It alludes to the jointed stems and to the heavy pubescence of some kinds.

Eriogonums frequently have deep, woody taproots and chiefly basal foliage. Their leaves vary much in shape and size according to species. They are alternate, opposite, or in whorls (circles of more than two). Often they are more or less hairy, sometimes glandular, sometimes hairless. The small flowers are individually inconspicuous, but are assembled in many-flowered clusters with at the base of each a cuplike involucre (collar) of bracts and frequently slender bractlets. The clusters form often showy umbels or heads or are sometimes solitary along the branches. Each flower has six petal-like sepals, no petals, nine stamens, and three styles. The fruits are three-angled achenes.

The only eastern North American species ***E. allenii*** is endemic to shale barrens of Virginia and West Virginia. An excellent hardy, herbaceous perennial, it is well worth cultivating in gardens and thrives without special care in the vicinity of New York City and undoubtedly further north. It blooms in late summer. With a strong taproot and heavy stems and foliage, this is from 1 foot to 2 feet tall. It has long-stalked, ovate to oblong basal leaves and smaller ones in whorls of three to five on the stems. These become progressively smaller above. The flattish-topped, broad, branched clusters of clear yellow flowers are displayed over a long period in late summer and fall.

Eriogonum allenii

St.-Catherine's-lace (***E. giganteum***) is an attractive, nonhardy, rounded, coarse shrub 3 to 8 feet tall or occasionally taller, native to islands off the coast of California. It has hairy young shoots and, mostly toward the ends of the branches, leathery, oblong to oblong-ovate leaves 1 inch to 2½ inches or rarely 4 inches long, grayish-white on their upper surfaces, white-hairy beneath. The loosely branched clusters of blooms, often 1 foot or more across, have twice- or thrice-forked stalks. Their sepals are white. From the last, ***E. arborescens,*** also an inhabitant of islands off the Californian coast, differs in branching less freely and in having linear to narrowly-oblong leaves not over 1½ inches in length, with rolled-under edges. They are crowded in terminal tufts, are densely-white-hairy beneath, and nearly hairless on their upper surfaces. The flowers, in crowded, leafy clusters 2 to 6 inches across, have whitish to pale pink or rose-pink sepals. This nonhardy shrub is mostly 3 to 4, but sometimes up to 8 feet tall, and broader than high.

Wild- or California-buckwheat (***E. fasciculatum***) is a nonhardy shrub about 3 feet tall. It has leafy, branched stems 2 to 4 feet long and numerous leaves in tufts or clusters. They are ¼ to a little over ½ inch long, and oblong-linear to linear-oblanceolate. Their upper surfaces are green, their underneaths white-woolly. Their edges are strongly rolled under. The flower clusters, at the branch tips, are of many loosely arranged, rounded heads. The sepals are white or pinkish. This sort inhabits deserts from Utah and Nevada to California.

A variable species of which botanists recognize several varieties, ***E. latifolium*** is native from Colorado to Oregon and California. A low shrub or shrublet, often much branched and densely leafy, this kind has ovate to oblong or nearly round leaves, often wavy-edged and densely-white-hairy beneath. They have blades 1 inch to 2 inches long and woolly stalks broadened at their bases. From 9 inches to 2 feet tall, the leafless, more or less hairy flowering stalks, branchless or once or

Eriogonum latifolium

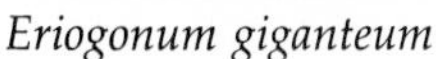

Eriogonum giganteum

Eriogonum giganteum (flowers)

twice forked, have blooms with white to rose-pink sepals. Variety *E. l. rubescens* (syn. *E. rubescens*) has branches that tend to sprawl, but have erect tips about 1 foot high. The flowers are rosy-red. Variety *E. l. sulphureum* (syn. *E. sulphureum*) has more numerous, smaller flower heads of blooms with sometimes yellow sepals.

Common in southwest Canada and the western United States, ***E. umbellatum*** is a variable herbaceous perennial or subshrub often low and spreading, rarely erect, and

Eriogonum umbellatum

exceeding 3 feet in height and as much in width. Its leaves, in tufts or rosettes, have obovate to broad-elliptic blades ½ to 1 inch long and stalks approximately as long. At first hairy on both surfaces, the green upper sides soon become hairless or nearly so, the undersides remain white-hairy. The yellow to cream flowers are in usually five- to ten-rayed umbels, the rays up to 2 inches long. Felty-white-foliaged ***E. crocatum,*** of California, has much the aspect of *E. umbellatum,* but its flowers have much narrower sepals. Attaining a height of about 8 inches, it has stems that sprawl somewhat from their bases. The leaves are many, bluntish-broad-elliptic to nearly round and 1 inch to 1½ inches in length. Like the stems they are densely felted with

Eriogonum crocatum

Eriogonum ovalifolium

white hairs. The flowers are crowded in clusters 1 inch to 1½ inches across on once- or twice-forked stalks. They have sulfur-yellow sepals.

Low, compact eriogonums suitable for trial in rock gardens include several mountain species. At The New York Botanical Garden beautiful ***E. ovalifolium*** has prospered for several years. This silvery-white-hairy gem forms dense mats or cushions about 3 inches tall of broad-oblanceolate to nearly round leaves with blades up to ¼ inch long, above which rise on slender stalks 6 to 10 inches tall crowded spherical heads ¾ inch in diameter of cream blooms that as they age become purplish. This species is a native at high altitudes of western North America.

Garden and Landscape Uses. With the exception of *E. allenii,* most eriogonums do not take kindly to the generally humid conditions and wet winters of eastern and central North America. The vast majority are adapted only to the drier climates to which they are accustomed in their natural environments. They need full sun and for preference well-drained, loose, gravelly soil. They are useful for planting on slopes and banks, and the lower ones in rock gardens. They generally do well in places exposed to wind. The flowers of the taller kinds are useful for cutting. They may be used fresh or dried.

Cultivation. Because of their deep taproots most eriogonums resent transplanting. Once established, it is better to leave sizable specimens undisturbed. Even in dry regions little or no watering is needed, at most an occasional deep soaking during summer. The shrub kinds may be kept shapely for a long time by a little periodical pruning, beginning when the plants are small, but even so with age they may tend to become lanky and ungainly. This occurs sooner if early pruning is omitted. Once specimens have become leggy, restoration by hard pruning is not practicable. It is better then to replace with young plants. Eriogonums are easily raised from seed. In gardens in climates congenial to their cultivation if the old flower clusters are not cut off volunteer seedlings are likely to appear near established plants.

ERIOPHORUM (Eri-óphorum) — Cotton-Grass. Natives of bogs, swamps, and wet soils, cotton-grasses (*Eriophorum*) are suitable for bog gardens, watersides, and similar locations. There are about twenty species, all natives of the Northern Hemisphere, except one that occurs in South Africa. They belong in the sedge family CYPERACEAE. The name derives from the Greek *erion,* wool, and *phoros,* bearing, and alludes to the conspicuous tufts of persistent hairs or bristles that accompany the flowers and fruits.

These sedgelike or rushlike herbaceous perennials have triangular stems, slender, flat leaves, and bisexual flowers that, except for the long cotton-like hairs or bristles that are the perianth parts, are minute. There are one to three stamens and a three-cleft style. The flowers are many together in spikelets that may be single and form solitary tufts of "cotton" or that may occur in umbel-like clusters or heads of several cottony tufts. The seedlike fruits or achenes have bristles attached.

The kinds with more than one tuft of cotton on each stem are the more ornamental. They include the closely similar ***E. viridi-carinatum,*** of North America, ***E. angustifolium,*** of North America, Europe, and northern Asia, and ***E. latifolium,*** of Europe. With stems in tufts or solitary, these are 1 foot to 2 feet tall and carry near the tops of their stems few to several soft,

Eriophorum angustifolium

Eriophyllum lanatum

Eriophorum latifolium

brushlike, downy, white flower heads that, when the seeds mature, are 1½ to 2½ inches wide.

Garden Uses and Cultivation. Possible garden uses for cotton-grasses are given at the beginning of this entry. Provided the soil is wet or constantly moist and contains abundant organic matter they grow with little care, and are hardy. They prefer somewhat acid soils, and need full sun. Propagation is easy by division, and by seed sown in soil kept wet.

ERIOPHYLLUM (Erio-phýllum). Endemic to western North America, *Eriophyllum,* of the daisy family COMPOSITAE, consists of eleven species of annuals, biennials, herbaceous perennials, and subshrubs. Its name, alluding to the woolly-hairiness of the foliage, comes from the Greek *erion,* wool, and *phyllon,* a leaf.

Eriophyllums have mostly alternate, usually divided, lobed, or toothed leaves. Their flower heads, solitary or clustered at the branch ends, are prevailingly daisy-like in structure. They have a central eye of disk florets usually encircled by petal-like ray florets. Usually both disks and rays are yellow. Two kinds have white rays. The fruits are angled, seedlike achenes.

Most commonly cultivated, ***E. lanatum*** (syn. *E. caespitosum*) is a herbaceous species of which botanists recognize several varieties, including some previously accepted as distinct species. Woody-based and with few to many sprawling to erect stems, it is a few inches to 2 feet tall and woolly-hairy. Usually pinnately-divided or toothed, its leaves are variable. They are from ½ inch to 3½ inches long. The naked-stalked, showy solitary or very loosely clustered flower heads are 1 inch to 1¼ inches wide and bright yellow. Variety *E. l. integrifolium* (syn. *E. leucophyllum*) has leaves usually dissected into narrow, pointed lobes and smaller flower heads than the typical species. From *E. lanatum,* highly variable 1- to 2-foot-tall ***E. confertiflorum*** differs in having crowded clusters of ¼-inch-wide, short-stalked or stalkless flower heads. Densely-white-hairy, this has stems woody at their bases, and wedge-shaped to obovate, deeply-three- to -five-lobed leaves, sometimes with their lobes cleft into linear segments.

A much-branched, shrubby herbaceous seaside perennial, ***E. staechadifolium*** is 1 foot to 5 feet tall, with sprawling to erect stems. Its lobeless or few-lobed, linear to linear-oblanceolate, somewhat leathery leaves, hairless on their upper surfaces and hairy beneath, have rolled-under margins. The many short-stalked flower heads, in rather dense clusters, are yellow and under ½ inch wide.

Annual ***E. wallacei,*** a beautiful little desert species, is indigenous from Utah to Arizona, California, and Baja California. Well-branched, densely-white-felted, 3 to 6 inches tall, and often of tufted appearance, it has spatula-shaped to obovate leaves, sometimes three-lobed at their apexes and up to ½ inch long. The short-stalked, yellow flower heads, about ⅓ inch wide, are scattered rather than clustered. Variety *E. w. rubellum* has pinkish, purplish, or rarely whitish ray florets.

Garden and Landscape Uses and Cultivation. Although eriophyllums are less satisfactory in eastern North American gardens and elsewhere where climates are humid than in more arid regions, it is possible to grow them there and have surprisingly good displays of bloom. To do this, it is important to give them a sun-baked location where the soil is sandy, low in nutrients, and very well drained. In dry climates they succeed with little trouble and produce quite glorious displays of bloom. They are suitable for sunny rock gardens, wild gardens, and similar informal plantings and, the taller ones, for the fronts of borders. Well-drained, porous soils are to their liking. Established specimens do not transplant readily. Propagation is easy by seed sown in sandy soil.

ERIOPSIS (Eri-ópsis). Tropical Central and South American orchids numbering half a dozen species make up *Eriopsis.* They belong to the orchid family ORCHIDACEAE. The name, from that of another genus of orchids, *Eria,* and the Greek *opsis,* resembling, is of obvious application.

Here belong epiphytic species (they perch on trees without taking nourishment from their hosts) and terrestrial species (they grow in the ground). They have pseudobulbs with two to four lanceolate, ribbed leaves, and in long, graceful racemes from the bases of the pseudobulbs, yellowish to brownish flowers with similar spreading sepals and petals, and with a three-lobed lip that does not closely surround the quite prominent column.

Almost black, finely wrinkled pseudobulbs 2 to 6 inches long or longer are typical of ***E. rutidobulbon.*** They are topped by two or three prominently veined, broadly-pointed-lanceolate leaves. Arching or erect, and up to 1½ feet long, the stalked racemes are of many fairly closely arranged, fragrant blooms about 1½ inches wide. Their sepals and petals are tawny-orange-yellow to bronzy-yellow margined with reddish-purple to chocolate-brown. The lip is dull orange-red with a dark-purple spotted, white terminal lobe. The erect column is greenish. Summer to autumn is the blooming season. This appears to be an epiphytic species.

The long-egg-shaped pseudobulbs of terrestrial or sometimes tree-perching ***E. biloba*** are much wrinkled at maturity. They are 3 to 8 inches long and surmounted by a pair of elliptic leaves up to 1½ feet long by 4 inches wide. The crowded, almost cylindrical racemes of fragrant blooms are erect and are carried to heights of up to 2½ feet. About 1 inch in diameter, the flowers have golden-yellow sepals and petals with brownish-red edges, and a lip of two kidney-shaped portions, the lower much the larger and yellowish with brownish suffusions, the other yellow spotted with red. This blooms in fall.

Garden Uses and Cultivation. Cool- or intermediate-temperature greenhouses with shade from all strong sun are suitable for these orchids. They are best grown in well-drained pots, and prosper in any of several rooting mediums, such as osmunda fiber, tree fern fiber, and fir bark, so long as water drains through rapidly. Water must be supplied liberally through most of the year, but is withheld completely for about a month after the new pseudobulbs have reached maturity. For more information see Orchids.

ERIOSTEMON (Erios-tèmon)—Wax Flower. Common throughout Australia, and one kind native of New Caledonia, the genus *Eriostemon* consists of over thirty species. It belongs in the rue family RUTACEAE and is named from the Greek *erion,* wool, and *stemon,* a stamen, in allusion to its stamens being hairy.

Eriostemons are small to fairly big evergreen shrubs with alternate, undivided, toothless leaves bespeckled with tiny translucent dots. Axillary or terminal, the flowers are white, pink, or bluish. They have usually persistent, bell-shaped, five-lobed calyxes and five waxy petals arranged in starlike fashion or more rarely forming a cupped bloom. There are ten flat, hairy-edged stamens that form a sort of cage around the pistil. The fruits are of five divisions or carpels, with a solitary, glossy, black seed in some or all of them.

Most likely to be cultivated is ***E. myoporoides,*** which has dark green, lanceolate, hairless leaves and white flowers. In Australia called pink wax flower, although its blooms are sometimes white, ***E. lanceolatus*** is an upright species with comparatively large, mostly pale pink flowers and narrow-lanceolate, gray-green leaves. A spreading shrub, ***E. spicatus*** has narrow-linear leaves and crowded terminal spikes of small pinkish-mauve blooms.

Garden and Landscape Uses and Cultivation. For outdoor cultivation these shrubs are adapted only to mild climates, such as that of California. In greenhouses they may be grown under conditions that suit tender heaths (*Erica*). They need well-drained, sandy peaty soil that does not dry excessively and a sunny location. Any needed pruning is done in spring. The seeds of some kinds, including those of *E. myoporoides* and *E. spicatus,* germinate readily, but those of many are notoriously difficult to start. They should be sown in sandy peaty soil in a temperature of about 60°F. Cuttings, taken in spring, may be used to secure increase. Another method once favored in Great Britain was to graft eriostemons onto young plants of *Crowea alba.* This induced more rapid growth of the eriostemons.

ERIOSYCE (Erio-sŷce). One somewhat variable species is the only kind of this genus. It belongs in the cactus family CACTACEAE and grows at high altitudes in Chile. The name, alluding to the fruits, is from the Greek *erios,* wool, and *syke,* the fig.

The bluish plant body of this fiercely-spiny cactus (***Eriosyce ceratistes*** syn. *E. sandillon*) is solitary, spherical to shortly-columnar, and has many (up to thirty-five or sometimes more) ribs. It may be 3 feet in height and 1 foot or more in diameter. The areoles (cushions from which spines sprout) are approximately 1 inch apart. They have eleven to twenty nearly equal, awl-shaped spines 1 inch to 1½ inches long. Yellowish-red to red, the bell-shaped flowers arise from the top of the plant. From 1¼ to 1½ inches long, they have narrow petals slightly over ½ inch in length, and ovaries, which later become the fruits, thickly covered with matted wool. Bristles as well as wool are usually borne in the axils of the scales of the perianth tubes. The blooms begin to fade after being open for only a few hours. The fruits are 1½ inches long and spherical to ellipsoid. Their tops are spiny.

Garden and Landscape Uses and Cultivation. This quite handsome species has much the aspect, uses, and needs of the larger echinocactuses. It provides a distinctive element in succulent collections and can be displayed to good advantage among rocks or in other surroundings outdoors in warm, essentially frost-free climates and in greenhouses. It needs very well-drained, dryish soil and full sun and is propagated by seed. For more information see Cactuses.

ERITRICHIUM (Eritrích-ium). This enticing genus of the borage family BORAGINACEAE, the name of which is derived from the Greek *erion,* wool, and *trichos,* a hair, and refers to the hairiness of the plant, consists of four species. Its interpretation, however, is difficult and some botanists regard these species and their variants as being greater in number. Plants now belonging in other genera, including *Amblynotus, Anchusa, Cryptantha, Cynoglossum, Hackelia, Krynitzkia, Myosotis, Omphalodes,* and *Plagiobothyrus,* once were named *Eritrichium* and may appear in the literature as such. Occasionally seeds of some of the less exciting of these are invested with the magic name *Eritrichium* and thus sold. As now delimited *Eritrichium* is a strictly Northern Hemisphere alpine and arctic genus. It consists of perennials, in effect, tiny forget-me-nots, bearing flowers of the purest cerulean blue or more rarely white. They are among the choicest and most brilliant denizens of the high mountains and northlands.

Eritrichiums are low and sometimes woody at their bases. They have alternate, undivided leaves and flowers in terminal cymes (spraylike clusters). The blooms have a five-lobed calyx, a short-tubed corolla with five spreading lobes (petals), five nonprotruding stamens, and a nonprotruding style ending in a knoblike stigma. The fruits are tiny nutlets.

Best known, at least by reputation, is ***E. nanum,*** a variable kind of the three northern continents. In North America it ranges from Alaska to the Yukon and down the spine of the Rockies to Colorado. With silvery-fuzzy foliage, often it is not over an inch or two in diameter, with the flowers resting on the 1-inch-high foliage; sometimes it grows taller and looser. Its blue or rarely white flowers are ¼ inch across. The strictly American ***E. howardii*** mostly inhabits moraines in the Rockies and other mountains in the West. A low, compact species sometimes 1 foot wide, it frequently forms vast turfs of hundreds of thousands of plants. Its leaves are spoon-shaped and silvery-hairy, its flowers brilliant blue and somewhat larger than those of *E. nanum.*

Garden Uses and Cultivation. Only the most intrepid gardeners, or the most abysmally ignorant, attempt *Eritrichium.* Few alpine plants are as difficult and perhaps none more so than these. Yet success, even if only for a season or two, gives the rock gardener a well-merited sense of accomplishment, and thrill. Taming *Eritrichium* is not easy. In their homelands these plants live under deep snow for one-half the year, where at all times soil and air are cool. Ample water is available during their brief season of growth. Their roothold consists of mineral detritus often almost devoid of organic matter and usually, but by no means always, of limestone derivation. In view of the almost complete lack of success that has attended efforts to establish these plants in gardens, it is futile to offer recommendations for their cultivation. Every gardener who attempts this must pioneer. Guidance must come from what is known of their native habitats. It is well, too, for such adventurers to consult the writings of those who have tried, and almost always failed; these are to be found in specialist books on rock gardening. Some day this Mount Everest of alpine plant cultivation may be conquered; perhaps the answer will come when we refrigerate beds of soil in which such plants are attempted. In any case, good luck to those who try.

ERLANGEA (Erlán-gea). One or two of this group of sixty species of the daisy family COMPOSITAE are cultivated. The genus consists of African shrubs and herbaceous plants that differ from *Vernonia* in technical details of the achenes (fruits). The name *Erlangea* honors the University of Erlangen, Bavaria.

A nonhardy shrub up to 5 feet tall, ***E. tomentosa*** as ordinarily grown in greenhouses does not develop its shrubby characteristics and appears to be no more woody than a chrysanthemum. Its fragrant leaves, up to 5 inches in length, are ovate-lanceolate to oblong, toothed, and on their undersides white-hairy. The flower heads, each about ⅔ inch in diameter, are in flat-topped clusters. They are lilac-mauve or powder-blue and without ray florets. A nonhardy herbaceous perennial, ***E. cordifolia*** has ovate, nearly stalkless, heart-shaped, toothed leaves up to 3 inches long. The rayless, purple flower heads are about ⅓ inch across.

Garden Uses and Cultivation. Both kinds mentioned may be and occasionally are grown outdoors for garden embellishment in ordinary well-drained, moderately moist soil in frost-free or nearly frost-free climates. More often, but not commonly *E. tomentosa* is cultivated in greenhouses chiefly for cut flowers. For this use it is especially welcome because its blooms are of a color unusual in midwinter among kinds suitable for cutting and are light and dainty in appearance so that they can be used effectively with more massive ones such as chrysanthemums. The plants begin blooming in late November and continue all winter. The flowers, which last well in water, can be cut with stems up to 1½ feet long.

When grown in greenhouses this species is raised from cuttings each year. Made in spring, these root readily in a propagating bench in a cool greenhouse. The rooted cuttings are potted singly into 2½-inch pots and are pinched when 3 or 4 inches high to encourage branching. They may be planted 1 foot apart in ground beds or benches in moderately fertile, porous soil in July or August or be transferred to successively larger pots until they occupy containers 7 or 8 inches in diameter. They need support, which may be supplied by staking or by erecting over the bench or bed a grid of wires and strings, as is done for carnations. This is an easy crop to grow in a sunny, airy greenhouse with the night temperature in winter held at 50°F, and the day temperature allowed to rise only five to ten degrees above that before the greenhouse ventilators are opened. Erlangeas must not be allowed to dry out, but the soil should not be constantly wet. Moderate fertilizing is appropriate after the available soil is well filled with roots. Whitefly and red spider mites are the most likely pests.

ERNESTARA. This is the name of hybrid orchids the parents of which include *Phalaenopsis, Renanthera,* and *Vandopsis.*

ERODIUM (Erò-dium)—Heronsbill. Widely distributed in temperate and tropical regions, the ninety species of *Erodium* belong to the geranium family GERANIACEAE. They are predominantly low annuals and herbaceous perennials, the cultivated kinds hardy or nearly hardy. Their name, in allusion to the beaked fruits, is from the Greek *erodios,* a heron. From closely related *Geranium* this genus differs in having only five anther-bearing stamens. Like *Geranium* it differs from *Pelargonium* in its flowers being without spurs and in having glands alternating with the petals.

Erodiums have mostly opposite leaves with blades usually longer than wide, and nearly always pinnately-divided or lobed, sometimes with comparatively large leaflets or lobes alternating with very much smaller ones. Rarely solitary, generally in umbels of more than one, with at the bases of the umbels a pair of usually papery bracts, the flowers are symmetrical or slightly asymmetrical. They have five each sepals and petals, five fertile stamens, and five staminodes (nonfunctional stamens). The five styles, united at flowering time, sometimes separate later. The fruits are composed of five carpels (capsule-like parts), each containing one seed.

A convenient way of classifying cultivated heronsbills is as those in which the leaves are only lobed, with perhaps one or two pairs of separate leaflets at their base, and those whose leaves for the greater parts of their lengths are separated into distinct leaflets, with perhaps a small section at the apex lobed. All kinds described here are perennials.

Noteworthy among the group with undivided leaves, ***E. chamaedryoides*** (syn. *E. reichardii*) is a delightful native of damp rocky places in the Balearic Islands. Stemless, tufted, and 1 inch to 2 inches high, this has blunt-heart-shaped, round-toothed, sparsely-hairy, green leaves about ½ inch long, and short-stalked, solitary blooms,

Erodium chamaedryoides

Erodium chamaedryoides flore-plenum

their petals white, veined with pink or purplish-pink. Especially lovely *E. c. roseum* has pink flowers. Double pink blooms are borne by *E. c. flore-plenum.* Native on rocks near the sea in Sardinia and Corsica, ***E. corsicum*** is much like *E. chamaedryoides,* from which it differs in having distinct stems and in its leaves being gray-hairy. Its flowers are pink with darker veinings. Those of *E. c. album* are white.

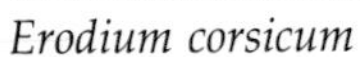

Erodium corsicum

Erodium pelargoniflorum

Two others with undivided leaves are *E. guttatum* and *E. pelargoniflorum.* Inhabiting sandy and rocky places in southern Spain and North Africa, ***E. guttatum*** is from 3 inches to a foot tall. It has triangular to ovate leaves, heart-shaped at their bases and ½ to 1 inch long, the upper ones three-lobed, the others lobeless. In umbels of two or three, the rosy-violet flowers, ½ to ¾ inch wide, have a nearly black spot at the base of each petal. Quite lovely ***E. pelargoniflorum,*** of Asia Minor, up to 1 foot tall, has ovate-heart-shaped, hairy leaves with shallowly-scalloped and toothed margins. Its flowers, in longish-stalked clusters of up to ten, are white or delicate pink with purple veining and the two upper petals blotched toward their bases with brownish-purple.

Heronsbills with ferny foliage divided into separate leaflets, the alternate leaflets of markedly different sizes, include ***E. petraeum,*** a mostly mountain species of the Pyrenees. This has a woody rhizome and is stemless and generally hairy. Its ill-scented, ovate-oblong leaves ½ inch to 2½ inches long have lobed leaflets. The flowers, ¾ to 1 inch wide and in umbels of up to five or solitary, are bright pink without dark blotches at the bases of the two upper petals.

So closely akin to *E. petraeum* that some authorities regard them as subspecies rather than distinct species, *E. cheilanthifolium* (syn. *E. petraeum crispum*) and *E. macradenum* are natives of Spain. Perennial ***E. cheilanthifolium*** is 3 to 4 inches tall with ill-scented, all-basal foliage. Its twice-pinnate leaves are densely-white-hairy, but usually not glandular. Three or four in a cluster, the ¾-inch-wide flowers are white or delicate pink veined with rose-pink. Larger than the others, the two upper petals have a basal red blotch. From the last, ***E. macradenum*** (syn. *E. petraeum glandulosum*) differs in having usually rather densely-glandular-hairy, ill-scented leaves and violet to purple flowers with a dark patch at the bottom of each of the two upper petals.

An inhabitant of dry, limestone cliffs and rocks in Spain, ***E. supracanum*** (syn. *E. rupestre*) is stemless. Silvery-white-hairy on their upper sides and green and much less hairy beneath, its leaves, ½ inch to 1¼ inches long, have lobed leaflets. Soli-

Erodium petraeum

Erodium supracanum

Erodium absinthoides

tary or in umbels of two or three, the darker-veined, pale pink blooms are approximately ½ inch wide.

Unisexual flowers with the sexes on different plants are characteristic of ***E. absinthoides,*** a native of rocky and stony places in Macedonia. Up to 8 inches tall, this has sparsely-glandular-hairy, green leaves up to 2 inches long that have lobed leaflets or are twice-pinnate with the ultimate segments linear or linear-lanceolate. In umbels of two to eight, the violet flowers are about ¾ inch across. From the last, ***E. guicciardii,*** an inhabitant of high mountains in Greece and the Balkan Peninsula, differs in its foliage being silvery-hairy on both surfaces. The leaves are up to 4 inches long. The pink to violet-purple flowers ¾ inch to 1½ inches across are in umbels of two to six. Males and females are on separate plants. Yellow or creamy-yellow blooms ½ to ¾ inch wide in few-flowered umbels distinguish ***E. chrysanthum*** from others of this group. Native of mountains in Greece, this has thick, branching, woody stems up to 6 inches tall, and silvery- or gray-hairy leaves, rarely exceeding 1¼ inches in length, that have lobed leaflets or are twice-pinnate. The blooms are unisexual with the sexes on different plants.

Erodium manescavii

Ferny-leaved kinds with the alternate leaflets of the leaves not conspicuously smaller than the others include several cultivated heronsbills. Here belongs ***E. manescavii,*** a stemless species of meadows and pastures in the Pyrenees. Its lanceolate leaves, up to 1 foot long, have deeply-lobed, ovate leaflets. The umbels of five to twenty rosy-purple flowers, each bloom 1 inch to 1½ inches in diameter top stalks up to 1½ feet long. Also in this group is ***E. daucoides,*** a stemless kind with hairy leaves up to 3½ inches long, the lanceolate to linear-lanceolate leaflets of which are lobed or toothed. This has flowers, in umbels of up to seven or solitary, ½ to ¾ inch wide, and pale lilac to purplish with a dark purple blotch at the bases of the two upper petals. Also without small and large leaflets alternating, the short-stalked leaves of ***E. romanum*** (syn. *E. acaule*) are all basal, and sparsely-hairy. The ½-inch-wide or somewhat broader flowers, without darker patches at the bottoms of the upper petals, are lilac to purplish. This is a native of dry soils in the Mediterranean region.

A few hybrid heronsbills are cultivated. The offspring of *E. manescavii* and *E. daucoides,* named ***E. hybridum,*** except for more finely-divided foliage, and smaller, paler blooms, resembles its first-named parent. Pale pink flowers with darker veins are borne by ***E. kolbianum,*** an intermediate hybrid of *E. supracanum* and *E. macradenum.* The intermediate hybrid between *E. cheilanthifolium* and *E. macradenum* is ***E. wilkommianum.***

Erodium kolbianum

Garden and Landscape Uses. The taller, more vigorous heronsbills, such as *E. manescavii* and *E. hybridum,* are useful for the fronts of flower beds and less formal locations. They can also be effective in rock gardens. Dwarfer kinds are suited only for rock gardens and other intimate environments, and some, especially high alpine kinds, are best accommodated in pots or pans in an alpine greenhouse or a cold frame.

Cultivation. As a group heronsbills need sunny locations, and dryish, very well-drained, nonacid soils. Most appreciate limestone. In cold climates it is advisable to protect the plants over winter with a covering of branches of evergreens or other material that permits a free circulation of air and to carry a small stock of young plants in cold frames to replace any lost in winter. Propagation is generally easily achieved by division, in early spring, and by seed.

ERPETION. See *Viola hederacea.*

ERUCA (Er-ùca)—Rocket Salad, Roquette. This little known, mustard-like annual is occasionally grown for its leaves for salads. Unless they have developed quickly and are comparatively young, their flavor is too strong for most American palates. Rapid, unchecked growth is needed to produce acceptable salad makings. Rocket salad is the only sort cultivated. Like the others, it is a native of Europe and western Asia. It was well known to Dioscorides who, with typical directness, pronounced it as being "good for the belly."

A member of the mustard family CRUCIFERAE, this genus bears the ancient name of some unknown plant of the same family. It consists of about half a dozen annuals, biennials, and perennials with markedly-pinnately-lobed or toothed leaves and racemes of comparatively large, whitish, yellow, or purplish, four-petaled flowers, each with four long and two shorter stamens and a prominent style. The fruits are prominently beaked pods containing two rows of wingless seeds.

Rocket salad (***Eruca vesicaria sativa***) is an annual, 1 foot to 2½ feet tall, with scattered hairs and irregularly-pinnately-lobed leaves. The flowers, up to 1 inch long, are

Eruca vesicaria sativa

whitish or cream, with purplish veins. The seed pods are erect, pressed against the stems, and about 1 inch in length.

Cultivation. For best success rocket salad must be grown in the cooler seasons of spring and fall rather than in high summer. Good results can be had by starting the plants early indoors and transplanting them to the garden after danger from frost is passed, as is done to secure early lettuce. Alternatively, the seeds may be sown as early in spring as possible where the plants are to remain and the seedlings thinned to about 1 foot apart. The rows of plants may be about 1½ feet apart. A later sowing may be made as soon as the most intense heat of summer is passed. The first picking of leaves may be done in six to eight weeks from the time of sowing and continue as long as tender young leaves are produced. To further the development of quickly grown foliage judicious applications of rapidly available nitrogenous fertilizer are helpful and irrigation should be used to prevent the soil from ever becoming excessively dry.

ERVATAMIA. See Tabernaemontana.

ERYNGIUM (Erýn-gium) — Sea-Holly, Eryngo, Button Snakeroot or Rattlesnake Master. Eryngiums are unlikely looking members of the carrot family UMBELLIFERAE. Annuals, biennials, and herbaceous or rarely subshrubby perennials, some hardy, some not, they number approximately 230 species. Widely dispersed as natives of temperate and warm areas, they are most abundant as to species in the Mediterranean region. The name *Eryngium* is derived from *eryggion,* the ancient Greek one of one species. For many centuries the roots of certain species were employed medicinally, those of the sea-holly being especially esteemed because of their supposed virtue as an aphrodisiac.

Eryngiums are admired for their showy, decorative upper leaves, bracts, and flower heads, in cultivated kinds prevailingly steely-blue to metallic-purplish. They are stemmed or stemless and have alternate, usually rigid, leathery leaves, commonly lobed or variously cleft, less often without indentations. Nearly always their margins are spiny. Small, short-stalked or stalkless, the little white, blue, or greenish flowers are crowded in hemispherical or spherical heads of cylindrical spikes, with at their bases an involucre (collar) of three or more spiny bracts. The individual flowers have conspicuous, commonly spine-tipped, usually stiff sepals, five erect petals, generally shorter than the petals, and two styles. The usually scaly fruits are cylindrical, subspherical, or ovoid seedlike carpels. All kinds discussed here, except *E. leavenworthii,* are herbaceous perennials.

Sea-holly (***E. maritimum***), a native of sandy coasts of Europe, has stems, branched above, and 1 foot to 2 feet tall. Its roundish, long-stalked basal and lower leaves are thick and leathery, the basal ones roundish, 4 to 6 inches long, three- to five-lobed, and with coarse, spreading, spiny teeth. The upper leaves are similar, but stalkless. The many flower heads, ¾ inch to 1¼ inches in diameter and bluish, have four to seven ovate to lanceolate bracts 1 inch to 1½ inches long, with one to three pairs of spiny teeth. The fruits are densely-scaly. From 1½ to 3 feet in height, herbaceous perennial ***E. planum,*** a native of dry places in central and southeastern Europe, has persistent, slightly leathery, toothed basal leaves. They are ovate to ovate-oblong, heart-shaped at their bases and have wingless stalks approximately as long as the blades. The upper parts of the stems are branched and like the uppermost leaves and flower heads are usually bluish. The flower heads are broad-ovoid and rarely over ½ inch long. They have six to eight linear-lanceolate bracts, ½ to 1 inch long, with one to four pairs of spiny teeth. The fruits are densely-scaly. In gardens *E. planum* not infrequently masquerades as *E. amethystinum.* From it true ***E. amethystinum,*** a herbaceous perennial not over 1½ feet tall, native to dry places in Italy, Sicily, and the Balkans, differs in its persistent, leathery basal leaves being deeply-pinnately-lobed or divided instead of only toothed, and in its fruits being sparsely-scaly. The basal leaves have obovate blades 4 to 6 inches in length, their segments spiny-toothed. The upper parts of the stems, their leaves, and many flower heads are usually bluish to amethyst, the heads up to ¾ inch long. There are five to nine bracts, each with one to four pairs of spines, ¾ inch to 2 inches long.

Easily distinguished from other kinds here considered by its flower heads having at least twenty-five bracts, many pinnately-dissected in comblike fashion into long, soft spines, ***E. alpinum*** inhabits mountains in Europe, favoring meadows

Eryngium planum

Eryngium amethystinum

Eryngium alpinum

and other grassy places. A herbaceous perennial up to 2½ feet tall, this has soft, persistent, basal foliage, the long-stalked leaves having irregularly-toothed, heart-shaped to triangular-ovate blades up to 6 inches long and wingless stalks. The uppermost leaves and stems are bluish or whitish. The cylindrical-ovoid flower heads are ¾ inch to 1½ inches long. Another low kind, ***E. bourgatii*** inhabits dry, stony places in Spain and the Pyrenees. Up to 1½ feet tall, this has persistent, slightly leathery, long-stalked basal leaves with roundish blades deeply-cleft into three or five once- or twice-pinnately-lobed, spine-toothed segments. The upper parts of the plants, including the rather few flower heads, are usually bluish. The spherical-ovoid heads, up to 1 inch long, have ten to fifteen linear-lanceolate bracts ¾ inch to 2 inches long, with one to rarely three pairs of spiny teeth. The fruits have few scales.

Eryngium bourgatii

Robust and up to 6 feet tall, ***E. giganteum*** comes from Asia Minor and the Caucasus. Its stout stems branch above. The leathery, spine-toothed, stalked leaves have blades up to 6 inches long by two-thirds as wide. The basal ones are heart- to triangular-heart-shaped, the stem leaves more or less three-lobed. The many cylindrical to ovoid-cylindrical flower heads, 3 to 4 inches long, have very stiff, much dissected, lanceolate to obovate bracts 3 to 4 inches long. The upper parts of this bold species are bluish to sea-green.

Eryngium giganteum

One of the finest of the genus, ***E. protaefolium*** when out of bloom has much the aspect of a bromeliad. Native at high altitudes in Mexico, this remarkable species has rosettes of arching, silvery-green, pointed-linear leaves 1½ to 2 feet long and furnished along both margins with ¾-inch-long white spines. Approximately 3 feet tall, the leafy flowering stalks are much branched. Each branch ends in one or two heads, 2 to 3 inches long and about 1½ inches wide, of light blue flowers backed by a collar 6 to 8 inches across of silvery-white bracts.

Eryngium protaefolium

Button snakeroot or rattlesnake master (***E. yuccifolium***) differs much in appearance from other kinds discussed here. Native from Virginia to Minnesota and southward, this inhabits damp and dry soils in open woodlands and prairies. Except in its flowering portion usually without branches, and 3 feet tall or taller, rattlesnake master has strictly longitudinally-parallel-veined leaves, the lower ones up to 2½ feet long and ¾ inch wide, the others gradually reduced in size upward. The flower heads are up to ¾ inch long and spherical-ovoid. The ovate-lanceolate bracts rarely protrude from beneath the head. There has been considerable confusion between this species and ***E. aquaticum,*** which does not seem to be cultivated. This last has leaves, with a single mid-vein, and side veins angling from it in pinnate-fashion.

Eryngium yuccifolium

Hybrid ***E. oliveranum,*** a strong grower of uncertain parentage, is about 3 feet tall. Its long-stalked, spiny-toothed basal leaves heart-shaped-ovate, are obscurely three-lobed at their tips. The leaves of the stem, also spiny-toothed, are cleft into three lobes, which are again lobed or angled.

Cylindrical-ovoid to ovoid, the blue flower heads have ten to fifteen, linear to awl-shaped, somewhat spiny bracts.

An annual or a biennial, hairless, rather slender, 1 foot to 3 feet tall, and its upper parts purple, ***E. leavenworthii*** inhabits dry plains and prairies from Kansas to Texas. Very spiny, its leaves clothe the stems throughout. The lower stem ones up to 2½ inches long, are broadly-oblanceolate to nearly round, and deeply-palmately (in hand-fashion) -cleft into pinnately-cleft lobes. The many violet, ovoid to ovoid-cylindrical flower heads, up to 2 inches long, have about eight spiny bracts up to 1½ inches long, some or all of the same hue as the heads.

Garden and Landscape Uses. Eryngiums are highly decorative and of considerable worth for perennial beds and more informal plantings. Once established they need little care, but they resent root disturbance and so should not be transplanted more often than is quite necessary. They are sun-lovers, and with few exceptions, need well-drained soil. In addition to providing long-lasting displays, outdoors eryngiums are useful as sources of cut blooms. If taken before the flower heads are fully expanded, and dried by tying them in bundles and hanging them upside down in a shaded, airy place, they make effective winter decorations.

Cultivation. Treatment that suffices for the general run of popular hardy herbaceous perennials suits perennial eryngiums. In cold climates winter protection of a covering of branches of evergreens, salt hay, or other loose material is advantageous. Fertilizing in spring stimulates strong growth. Propagation is by seeds, root cuttings, and careful division in early fall or early spring. The last method is rather tricky, and not always successful. At best it is a slow way of increasing stock. Seeds of annual *E. leavenworthii* may be sown in fall or early spring, in well-drained porous soil where the plants are to bloom, and the seedlings thinned just enough to prevent overcrowding.

ERYNGO. See Eryngium.

ERYSIMUM (Erýs-imum) — Blister-Cress, Siberian-Wallflower. This genus is very closely related to that of the wallflowers (*Cheiranthus*), and some of its members have the word wallflower as part of their colloquial names, the coast-wallflower and beach-wallflower, indigenous to the Pacific Coast, for example, and the Siberian-wallflower, which despite its name is also North American. They belong to the mustard family, CRUCIFERAE. The generic name is derived from the Greek *erysimon,* that of a related plant, the hedge-mustard. There are 100 species of *Erysimum* very widely distributed, mostly in temperate regions. From *Cheiranthus* they differ in having flowers with median as well as lateral nectary glands at the bases of the stamens.

Erysimum allionii of gardens

Erysimums are low, leafy-stemmed, usually much-branched annuals, biennials, and evergreen perennials. They have undivided, sometimes scallop-toothed leaves and terminal racemes of yellow, orange, lilac, or bluish, four-petaled flowers, each with six stamens, two of which are shorter than the others, and one style. The racemes lengthen markedly as the lower flowers pass out of bloom and the buds above expand. The seed pods are slender and four-angled.

The Siberian-wallflower, commonly cultivated as *E. allionii* and *Cheiranthus allionii,* in all probability is a horticultural derivative of ***E. hieraciifolium***. A biennial or short-lived perennial, it is bushy, freely-branched, and up to 1 foot or a little more in height. It has narrow-oblanceolate or spatula-shaped, sparingly-hairy leaves up to 4 inches long, and brilliant, almost dazzling, orange or yellow flowers in spring. Other American species that may occasionally be cultivated include the very variable biennial coast-wallflower (***E. capitatum***), a branched or branchless kind about 1½ feet tall with linear, toothed or toothless leaves up to 3 inches long and white or yellowish flowers. It is native from California to British Columbia and Idaho. The plant formerly called *E. elatum* is now included in *E. capitatum.* The beach-wall-

Erysimum allionii of gardens (flowers)

Erysimum kotschyanum

flower (*E. suffrutescens*) is a somewhat woody, subshrubby plant of coastal California. It is rather succulent, freely-branched, and has narrow, linear-lanceolate leaves up to 3 inches long and bright yellow flowers. It is cultivated as a biennial.

Among perennial erysimums not native to North America one of the prettiest is the alpine-wallflower (***E. linifolium,*** syn. *Cheiranthus linifolius*), indigenous to Spain, of branching, somewhat decumbent habit, and with very narrow-linear, sometimes toothed, grayish leaves and racemes of lavender or lilac blooms. This species forms a broadish mound about 1 foot in height. In the high mountains of Asia Minor, ***E. kotschyanum*** is at home. It is a tufted plant not over 5 inches in height with narrow-linear or awl-like, fine-toothed, much crowded leaves and yellow-orange blooms. Another low grower, also from Asia Minor and about 6 inches tall, is ***E. purpureum,*** which has narrow leaves, the lower of which are pinnately-lobed, the upper toothed or not, and purple flowers. A native of Asia Minor and Greece, ***E. pulchellum*** forms much-branched mounds from a few inches up to 2 feet in height. Its oblong to spatula-shaped leaves are deeply toothed on the upper parts of the plant. Its flowers are orange. Native to the European mountains, ***E. pumilum*** has mostly basal leaves, linear-lanceolate and without teeth, and sulfur-yellow, slightly fragrant flowers. It is about 6 inches in height. A native of Norway and Lapland, ***E. alpinum*** (syn. *Cheiranthus alpinus*) is about 1 foot in height and has lanceolate, pubescent leaves and lemon-yellow, sweetly fragrant flowers. A variety, *E. a.* 'Moonlight', is distinguished by its dark flower buds and pale yellow blooms, another, *E. a.* 'Pamela Pershouse', has deep golden-yellow flowers. From Greece comes ***E. senoneri,*** a subshrubby much-branched kind with a more or less woody base, ovate-oblong, sometimes toothed leaves, and fragrant orange flowers. ***Erysimum dubium*** (syn. *E. ochroleucum*), an attractive native of the European Alps, has decumbent, branched stems, oblong-lanceolate toothed leaves, and pale yellow flowers. Its variety *E. d. helveticum* has narrow leaves.

Biennial species of garden merit other than the Siberian-wallflower include ***E. canescens,*** a much-branched native of eastern Europe that has long, narrow, grayish-green, more or less lanceolate leaves and becomes 3 feet tall; ***E. cuspidatum,*** of Europe and southwestern Asia, a somewhat hairy plant 2 feet or sometimes more in height, often with lobed leaves and with bright yellow blooms; robust ***E. hieraciifolium*** of Europe, which is 3 feet or more in height and has oblong-lanceolate leaves and flowers that are usually pale yellow and more than ½ inch across; quite attractive ***E. suffruticosum*** (syn. *E. murale*), of uncertain provenance but probably of Mediterranean origin, leafy, compact, about 9 to 20 inches in height, and with short, narrow leaves and golden-yellow blooms; and western American ***E. torulosum,*** a quite attractive plant, 1½ feet tall, mostly without branches, with spatula-shaped, often conspicuously toothed leaves and yellow flowers exceeding ½ inch in diameter.

Annual erysimums include ***E. perofskianum,*** a pretty grayish pubescent plant from the Himalayas. This sort, little-branched and up to 2 feet tall, has toothed or plain-edged, narrowly-lanceolate or linear leaves and showy, yellow or orange flowers. Less showy ***E. repandum,*** which ranges as a native from southeast Europe to India, is 1 foot or sometimes more in height and has lanceolate, sometimes distantly-toothed leaves and small yellow flowers. It should be noted that the plant frequently cultivated as *E. perofskianum* is not that species, but the Siberian-wallflower.

Garden Uses. Erysimums are easy to grow. By making use of their simple charms, the gardener can with little trouble add bright color to his or her domain in spring and summer. The annual kinds are suitable for use in flower borders and semiformal areas, and *E. repandum* can, with good grace, be admitted to rock gardens. The biennials include some of the more attractive kinds, none showier or more useful as a garden plant than the Siberian-wallflower. This is one of the best plants for spring bedding, ranking for that purpose with the more popular pansies, forget-me-nots, English daisies, and polyanthus primroses and introducing colors that are approached only in the primroses and even then not attained in the dazzling purity of the Siberian-wallflower. With much the appearance of the English wallflower, but not exhibiting the wide color range of that species, the Siberian-wallflower has the distinct advantage of being much hardier to cold. It winters outdoors at New York City with little or no protection. Used in conjunction with deep maroon or black-purple tulips the typical orange-flowered form of this erysimum is quite stunning, and its paler yellow variety offers wider possibilities for combining with tulips of more subtle hues. The blooms of the Siberian-wallflower last well in water and are attractive as cut flowers. The other biennials are useful for planting in patches in flower beds and borders and can be planted in naturalistic fashion in semi-wild parts of the garden; for the latter purpose the taller growers commend themselves. The biennial *E. torulosum* and the annual *E. repandum* planted in the lean soil of a rock garden are more limited in height than in more nourishing flower garden earth and are entirely appropriate there. The true *E. perofskianum* lends itself well to the decoration of flower beds and borders, and its flowers are useful for cutting.

Cultivation. From seeds erysimums are raised with greatest ease. Those of the annuals are broadcast in early spring where the plants are to bloom and the seedlings are thinned while quite small to appropriate spacing, which is about 9 inches between individuals. Seeds of biennial species are sown in June in cold frames or in outdoor seed beds, and as soon as the seedlings are large enough to handle they are transplanted to outdoor nursery beds or to cold frames, about 6 inches apart. In fall, in mild climates, or in spring, the plants are transplanted to the stations where they are to bloom; final spacing depends upon the species and the actual size of the plants, but 9 inches to 1 foot between individuals is generally satisfactory.

Perennial erysimums come satisfactorily from seeds sown any time in spring or early summer, and the young plants are handled in the same manner as those of the biennials. The perennials can also be increased by cuttings made of nonflowering shoots in late spring or early summer and inserted in a propagating bed in a humid, shaded cold frame or under mist. The rooted cuttings are potted into gritty, porous soil and, when they are established, are planted out in their permanent locations or in nursery beds. Routine care of perennial kinds is not exacting, the removal of spent flowers and a little trimming to shape after they are through blooming and an annual top dressing of fresh soil applied in spring about meet their requirements.

All erysimums are sun-lovers, and all give the best results in porous, well-drained soil that is not excessively rich in nitrogen and that is neutral or nearly so. If the soil is decidedly acid the addition to it of lime is generally beneficial.

Diseases and Pests. Diseases that sometimes infect erysimums are gray mold or botrytis, mildews, white rust, bacterial wilt, and club root. They may be harmed by aphids, caterpillars, flea-beetles, and nematodes.

ERYTHEA. See Brahea.

ERYTHRAEA. See Centaurium.

Erythrina crista-galli

ERYTHRINA (Ery-thrìna) — Coral Tree. Mostly trees and shrubs, but including a few herbaceous perennials, the 100 species of *Erythrina*, of the pea family LEGUMINOSAE, are inhabitants of tropical, subtropical, and occasionally warm-temperate regions of the Old and the New Worlds. The name, alluding to the prevailing color of the blooms, is aptly derived from the Greek *erythros*, red. The vernacular designation coral tree refers to the same feature.

Erythrinas have usually thick, generally spiny stems and mostly coarse, deciduous or partially evergreen, stalked leaves with three broad leaflets. Their quite large, commonly showy pairs or clusters of few blooms of pea-flower-like construction are in terminal or axillary leafy or leafless racemes. They have five-toothed or toothless calyxes, often two-lipped or splitting down one side to give a spathelike effect. The sometimes stalked standard or banner petal is erect or reflexed or folded and enclosing the other floral parts. The wing petals are short or lacking. The petals of the keel, separate or united, are markedly shorter than the standard, and may exceed, equal, or be shorter than the wings. There are ten stamens, of which either nine are joined and one is separate, or all are more or less united. The incurved, hairless style is tipped with a small stigma. The fruits are pods constricted between the often brightly colored seeds. Coral trees are not hardy in the north. In warm countries, in addition to their employment as ornamentals, some kinds are used to shade plantations of chocolate and coffee trees and to serve as supports for vines.

The wiliwili of Hawaii and Tahiti (***E. tahitensis*** syn. *E. monosperma*) inhabits hot, dry sites and has pale red, orange, greenish or white flowers. Its extremely light wood is used for surfboards, floats, and similar purposes.

The cockspur coral tree (***E. crista-galli***) is probably the most commonly cultivated in the United States. Native to South America, it is a bushy, more or less thorny, tree 20 to 30 feet tall or sometimes taller. Its lustrous, dark green, deciduous leaves have broad-elliptic leaflets up to 4 inches long by 3 inches wide. Appearing with the leaves, the brilliant crimson blooms, in terminal racemes leafy in their lower parts and up to 2 feet long, have a toothless or short-toothed calyx, a reflexed standard petal 1½ to 2 inches long by a little over one-half as wide, a narrow keel one-half to two-thirds as long as the standard petal, and wings one-third as long as the keel. About 6 inches in length, the curved pods have slight constrictions between the dark brown to black seeds. Somewhat resembling the cockspur coral tree, ***E. falcata,*** also South American and bearing flowers and foliage together, is a large tree that differs from *E. crista-galli* in having leafless terminal or axillary racemes 6 inches to 1 foot long of smaller flowers. The racemes are often in clusters of two or more from near the branch ends.

A hybrid between the cockspur coral tree and *E. herbacea*, ***E. bidwillii*** is intermediate between its parents. A deciduous shrub with stems that tend to die back after blooming, and blooms that appear with the foliage, this has leaves with ovate leaflets up to 3 inches long. Its flowers have spreading standard petals, broad and up to 2¼ inches long, and joined keel petals and wings about one-half as long as the standard.

Kinds with calyxes toothless or short-toothed, but not markedly two-lipped or split down one side, more or less erect standard petals, keels not over one-half as long as the standard, and wings as long or slightly longer than the keel, include *E. humeana*, *E. arborescens*, *E. acanthocarpa*, and *E. zeyheri*. The first two, except for a variety of *E. humeana*, are trees, the others shrubs. Native of South Africa, ***E. humeana*** is a small tree that has leaves with short-pointed leaflets, red blooms with wings and separated keels about ⅕ inch long, and pods without prickles, containing red seeds. Variety *E. h. raja*, a shrub up to 6 feet tall, has longer-pointed leaflets up to 4½ inches long by 3 inches wide. Flowers with both wings and a united keel over ⅓ inch long, that are succeeded by pods containing black seeds, distinguish Asian ***E. arborescens*** from *E. humeana*. From all others discussed in this paragraph shrubby ***E. acanthocarpa,*** of Africa, is distinguishable by its flowers having the standard petal tipped with green, and the

Erythrina acanthocarpa

seed pods prickly. Its seeds are red. An African herbaceous species not over about 1½ feet in height, ***E. zeyheri*** has underground stems from which spring the leaves and flower spikes. The leaves, with prickly undersides and stalks, have blunt-pointed leaflets 2 to 4½ inches long by about two-thirds as wide. Calyxes split down one side to the base and linear-lobed at their apexes are characteristic of South African *E. latissima* (sometimes misidentified as *E. abyssinica*) and tropical Asian and Polynesian *E. variegata* (syn. *E. indica*). The bark of ***E. latissima*** is thick and corky; its leaves are leathery at maturity and densely-hairy when young. It blooms when in leaf. The bark of ***E. variegata*** is not thick and corky, the leaves do not become leathery, and the flowers come before the leaves. Variety *E. v. alba* has white flowers, *E. v. orientalis* has leaves veined with yellow.

Kinds with the standard petal folded to enclose the other flower parts so that individual blooms are long and slender, without the other petals or stamens displayed, are *E. coralloides* (which in gardens has been misnamed *E. polianthes*), *E. speciosa* (syn. *E. polianthes*), *E. americana* (frequently misidentified as *E. corallodendrum*), *E. macrophylla, E. princeps* (often wrongly identified as *E. caffra*), and *E. herbacea*. In the first two the densely crowded flowers appear when the trees are leafless. Those of ***E. coralloides,*** a shrub or small tree of Mexico, have wing petals and keels of the same length and one-fifth as long as the standard petal. Also, its solid-stalked leaves have leaflets 2 to 4½ inches long and usually not as wide as their lengths. The leaves of ***E. speciosa,*** a small Brazilian tree, have hollow stalks and leaflets usually as wide or wider than long. The flowers have a banner petal two or three times as long as the keel, which is three times as long as the wings.

Flowers that appear with the foliage and are less crowded than those of the two kinds last discussed, are typical of Mexican *E. americana*, Central American *E. macrophylla*, and *E. herbacea*. The first two are small- to medium-sized trees, the last a shrub 2 to 3 feet tall, the stems of which die back each year, and which has leaves with long-pointed leaflets. The leaflets of the other two are blunt or pointed, and ovate. The slender blooms of ***E. herbacea,*** native from North Carolina to Florida and Texas, are brilliant red and 2 inches long. The flower stalks and calyxes of ***E. americana*** are lightly pubescent, those of ***E. macrophylla*** densely so. Their blooms are scarlet to light rose-pink. A large, thorny, nearly evergreen African tree, ***E. princeps*** (syn. *E. lysistemon*) has bright red flowers with two-lipped calyxes without linear lobes, and corollas 1¾ to 2 inches long. They are crowded, drooping, and lie rather close to the stalk, in contrast to those of *E. caffra*, with which the species is commonly confused, which are more spreading.

Erythrina princeps, Pretoria, South Africa

Erythrina princeps (flowers)

Erythrina caffra

True ***E. caffra*** (syn. *E. constantiana*), also African, and a large, almost evergreen, thorny tree, differs from *E. princeps* in its blooms spreading well out from the stalk and in having standard petals not folded, but reflexed so that the other flower parts are clearly visible. In this kind the calyx is two-lipped and not linear-lobed.

Garden and Landscape Uses. In the tropics, subtropics, and warm-temperate areas including Florida and California, coral trees are commonly cultivated for their often astonishingly showy blooms. Frequently somewhat ungainly, they mostly have coarse foliage. Their fruits are without decorative merit. They are effective as individual specimens and often are planted as hedges. Coral trees thrive in well-drained, fertile soil, in full sun, and do well near the sea.

As far north as Virginia, *E. crista-galli,* which, even if killed to the ground in winter, so long as the roots survive will produce new shoots that bloom the next summer, and *E. herbacea* will live permanently outdoors in sheltered places if protected over winter by a heavy mulch. These, and other kinds such as *E. speciosa,* which tend to die back or may be pruned back in fall, have thick roots that in cold climates can be dug up and stored in a cellar or other frost-free place until spring. They can then be planted outdoors to provide summer bloom. Coral trees are also sometimes grown in conservatories and greenhouses in ground beds and large containers.

Cultivation. Where hardy, coral trees are easy to handle. The herbaceous kinds such as *E. herbacea* and the kinds such as *E. crista-galli* and *E. speciosa,* the flowering branches of which die back after the blooms have faded, bear flowers on shoots of the current season's growth. Prune these kinds either by cutting back the shoots that have died or by more severe pruning of all shoots, in fall or early spring. Kinds that bloom on shoots developed the previous year may, if desired, be pruned immediately after flowering finishes.

In greenhouses a night temperature in winter of 50 to 55°F is satisfactory. By day an increase of five to fifteen degrees is permissible. From spring to fall temperatures considerably higher are in order, and then the soil should be kept moist and the atmosphere humid. In winter the soil is kept dry or nearly dry. Generous fertilization in spring and summer is very helpful, and exposure to full sun is necessary. Propagation is by seeds and cuttings, and of herbaceous kinds, by division.

ERYTHRODES (Ery-thròdes). About 100 species of the orchid family ORCHIDACEAE constitute *Erythrodes,* a genus that inhabits the tropics and subtropics around the world. The name is presumably derived from the Greek *erythros,* red. Its application is not clear.

The sorts of *Erythrodes* are ground orchids. They have erect to prostrate stems and ovate to lanceolate, short-stalked leaves, frequently attractively variegated. The small, nearly stalkless flowers are in loose or crowded, spikelike racemes.

Erythrodes nobilis, in bloom

Erythrodes nobilis, foliage

Brazilian ***E. nobilis*** in bloom is about 10 inches tall. Its nearly stalkless, pointed-ovate leaves, the lower ones the largest, are dark green clearly veined with silver. The ¼-inch-long, greenish-white flowers are in crowded spikes. The lip has three lobes, the middle one fringed.

Garden Uses and Cultivation. The sorts of *Erythrodes* appeal to orchid collectors. Their chief attractions are the rarity and the beauty of their foliage. They succeed under conditions and care encouraging to *Haemaria.*

ERYTHRONIUM (Ery-thrònium)—Adder's Tongue, Dog's-Tooth-Violet, Trout-Lily, Avalanche-Lily. The vernacular names listed above are not the only ones given erythroniums. Others, often of local application, are fawn-lily, Easter-lily, mariposa-lily, and star-lily. By whichever name known, these are charming plants well deserving the attentions of gardeners. The greatest number of the twenty-five species of *Erythronium* are natives of western North America, a few occur elsewhere on that continent, and four inhabit Europe and temperate Asia. They belong to the lily family LILIACEAE. The name, alluding to the flowers of the European dog's-tooth-violet, comes from the Greek *erythros,* red.

Erythroniums are deciduous, hardy, herbaceous perennials. They have corms (bulblike organs) deep in the ground and branchless, mostly subterranean stems terminated at or near ground level with two or occasionally three beautifully brown-mottled or plain green leaves of unequal size. The leaves, broad or narrow, and undivided, are without lobes or teeth. The flowers are usually solitary or few to several in leafless, stalked racemes. They are showy, nodding, lily-like, and gracefully disposed. Each has six lanceolate to oblanceolate perianth segments, in some species strongly reflexed (bent backward), commonly called petals, but more correctly designated as tepals, six stamens shorter than the petals, and a style that may be three-cleft, three-lobed, or entire and lobeless. In some species the three inner petals have appendages or ears on either side at their bases. The fruits are somewhat triangular capsules.

Species native in North America west of the Rocky Mountains may be divided into two groups, those with plain green leaves and those with foliage handsomely mottled. To the first category belongs ***E. grandiflorum,*** a somewhat variable mountain species with beautiful golden-yellow flowers, often streaked with green on their outsides, paler toward their centers within, with white, yellow, or red anthers, and a distinctly three-lobed style. The petals are ¾ inch to 1¼ inches in length, the inner ones eared. There are one to several blooms on each 6-inch- to 1-foot-tall stalk. The oblong-elliptic leaves are 4 to 8 inches long. Beautiful ***E. tuolumnense,*** one of the most distinct of erythroniums, is known only from a very limited area in Tuolumne County, California. Its lanceolate to broadly-lanceolate, not mottled leaves are up to 1 foot long. The flower stalks about equal them. The blooms, as large as those of *E. grandiflorum,* have petals with greenish-yellow bases, the inner three with ears. The stigma is not distinctly three-lobed.

White to cream flowers are borne by *E. klamathense* and *E. purpurascens,* both high mountain species with plain green leaves. The lovely avalanche-lily (***E. montanum***),

Erythronium tuolumnense

of Washington and Oregon, also belongs with this group that has foliage that is not mottled and blooms with eared inner petals. Its flowers are pure white with orange bases to the petals. Unfortunately, in contrast to other erythroniums, this beautiful denizen of mountain peaks has proved untamable in gardens. The yellowish-green leaves of ***E. klamathense*** are up to 6 inches long by up to 1 inch wide. The flower stalks, 3 to 8 inches tall, carry one to several blooms with very narrow petals, yellow at their bases, and ¾ to slightly over 1 inch long. The three inner petals are eared. From the last, ***E. purpurascens*** differs in having smaller flowers with slenderer petals, none of which has ears. The petals, not over ½ inch long, are often shorter. As the flowers age, they become tinged with purple. The narrow leaves have wavy margins.

Mottle-leaved western North American species, chiefly natives of lower elevations than the kinds dealt with above, can be subdivided on the basis of other characteristics. Notable is ***E. revolutum*** and its many varieties. These favor wet and moist, fertile soils, and except for flower color are similar to each other. They have blooms with distinctly lobed stigmas, and stamens with stalks dilated at their bases. The leaves of *E. revolutum,* crisped at their margins, lanceolate-ovate, and mostly pointed, are 6 to 8 inches long, and from a little over 1 inch to 2 inches wide. The blooms, one to several on a stalk, have pointed, lanceolate to linear petals 1¼ to 1¾ inches long, often with inrolled margins. Typically they are pink with cross bands of yellow near the base, and deeper rose-pink backs, but there are many variants. One called 'Pink Beauty' has soft lavender-pink blooms, 'Rose Beauty' has deep pink flowers and more darkly mottled foliage than *E. revolutum,* another variant has white blooms tinted with lavender. The flowers of *E. r. johnsonii* are rose-pink with white centers. Less robust and less erect than *E. revolutum,* variable ***E. oregonum*** (syn. *E. revolutum watsonii*) has blooms that range in color from white to rich cream, with the petals banded or not at their bases with maroon. It commonly has large leaves, up to 3 inches wide. Pure white flowers with maroon centers are borne by variety 'Purdy's White'. One named *E. o. praecox* has brown-zoned, rich creamy-white flowers.

A mottle-leaved species that among western kinds has the peculiarity of developing offset bulbs on long, threadlike, subterranean stems, and differs from other erythroniums in having blooms, when more than one, which occasionally happens, in umbels rather than racemes, ***E. multiscapoideum*** (syn. *E. hartwegii*) inhabits hot, dry foothills where there is little shade. Its leaves are oblanceolate and up to 4 inches long by 1 inch wide. The flower stalks are 4 to 8 inches tall. The blooms have white to whitish petals 1 inch to 1½ inches long, greenish at their bases. There are no sizable ears at the bottoms of the inner petals, but sometimes four small glands.

Mottle-leaved species that favor well-drained moist to wet soils in cool woodlands are *E. californicum, E. citrinum, E. howellii,* and *E. hendersonii.* With lanceolate to oblongish leaves 4 to 6 inches long and 1 to 2 inches broad, ***E. californicum*** has one to several flowers, on stalks up to approximately 1 foot tall, with white to cream petals with bases of light greenish-yellow with cross bands of yellow, orange, or

Erythronium californicum

brown. The anthers are white. The stigma has three stubby lobes. A variant with the upper parts of the petals pure white and the lower portions orange-yellow is named *E. c. bicolor.* The lovely variety 'White Beauty' has slightly creamy-white blooms usually with a basal zone of maroon to the petals. The other species considered here differ from *E. californicum* in their flowers having lobeless or essentially lobeless stigmas. Those of ***E. citrinum,*** one to seven on stalks up to 1 foot tall, have the upper halves of their petals white to creamy-white, the lower portions citron-yellow, and their backs sometimes pinkish. The petals, the inner ones eared, are 1 inch to 1½ inches long. The anthers are white. Up to 6 inches long by 2 inches wide, the leaves are wavy-margined. From the last, ***E. howellii*** differs in its inner petals being without earlike appendages. Lavender blooms with petals with rich brown-maroon to dark purple bases margined with a yellowish or whitish zone distinguish ***E. hendersonii.*** Its purplish flower stalks, up to 1 foot tall, carry one to four blooms that have strongly recurved petals 1 inch to 1½ inches in length. The inner three have basal ears. The anthers are white.

Erythronium americanum (flowers)

American species native east of the Rocky Mountains have solitary, nodding flowers. They inhabit moist woodlands or, *E. albidum mesochoreum,* dry woodlands and prairies, and often form extensive colonies of many single-leaved, nonblooming plants with scatterings of two-leaved, flowering individuals intermixed. Their leaves are usually mottled. Ranging from Nova Scotia to Minnesota, Florida, and Alabama, ***E. americanum,*** 4 to 8 inches tall, normally has yellow flowers, often darker on their outsides, with petals 1 inch to 2 inches long. There is a glandular spot near the bottoms of the margins of each of the three inner petals. The stigma is not lobed. Native chiefly from Ontario to Minnesota, Kentucky, Arkansas, and Oklahoma, more rarely to the east, ***E. albidum*** differs from *E. americanum* in having bluish-white or sometimes pale pink blooms with the petals yellow at their bases and often tinged on their outsides with blue or green. Their stigmas are strongly three-lobed. Variety *E. a. mesochoreum* (syn. *E. mesochoreum*) does not develop offshoots as does *E. albidum,* and the leaves of nonflowering specimens come later rather than along with those of blooming plants. Rich woodlands in Minnesota and perhaps Ontario are home to ***E. propullans,*** which has pink blooms with petals up to scarcely over ½ inch long.

Erythronium americanum

Variable ***E. dens-canis,*** of Europe and Asia, has beautifully mottled, pointed, broad-elliptic leaves, and solitary, nodding, rose-pink, rose-purple, lilac, or more rarely flesh-pink or white flowers on stalks up to 6 inches tall. They have long-pointed, strongly reflexed petals. This kind does not increase rapidly by offsets as do eastern North American species. Several named varieties are cultivated including 'Charmer', with light pink flowers, 'Rose Queen', with rose-pink flowers, and 'Snowflake', with pure white blooms. Japanese *E. d. japonicum* (syn. *E. japonicum*), with broad-elliptic to nearly ovate leaves, occurs also in Korea. On stalks 8 inches to 1 foot long, it has solitary rose-purple blooms with

Erythronium albidum

slender petals up to 2 inches or slightly more in length, with a three-lobed, dark purple spot at the base of each.

Garden and Landscape Uses. Among the most charming flowers of spring, these miniature lily relatives, except for the untamable avalanche-lily, are not difficult to grow, although in eastern North American gardens there is a tendency for western American species to gradually deteriorate with passing years. In the main erythroniums accommodate to dappled shade and deep, loose, dampish earth full of well-decayed organic debris, such as leaf mold, compost, or peat moss. Drier soil conditions and more sun are to the liking of *E. multiscapoideum.* Too heavy shade inhibits blooming.

Rock gardens, woodland gardens, and similar informal areas are by far the best locations for erythroniums, but it is not impossible to fit the more vigorous western North American species into the fronts of perennial borders. The Westerners are also delightful when forced gently into bloom in pots in cool greenhouses.

Cultivation. The corms (bulbs), available for planting in early fall, are set vertically with their thin ends upward and covered to a depth of 2½ to 3½ inches. Under no circumstances must they, except those of *E. multiscapoideum,* which are not harmed by such treatment, be allowed to dry while out of the ground. To prevent this, they may be packed in slightly damp peat moss, vermiculite, sand, or other loose medium and kept in a cool, shaded place, but they should not be stored for long; the sooner they are planted the better. A mulch of peat moss, leaves, or other litter helps to maintain equable conditions of moisture and temperature. When the plants become crowded to a degree that flower production diminishes, they should be dug up, separated, and replanted as soon as the foliage has died down. Some gardeners are of the opinion that *E. americanum* and other eastern North American kinds can be stimulated to bloom more freely by fertilizing them with bonemeal and unleached wood ashes, and some contend that it is helpful when planting to place stones beneath the corms to prevent them from making their way to deeper levels. Increase of some erythroniums is by offsets, but except *E. multiscapoideum,* the Westerners do not multiply in that way. Seeds must be relied upon. Under favorable circumstances self-sown seedlings may appear. Seeds sown as soon as they are ripe in a cold frame may germinate the following spring, but more often take an additional year, and a further three or four years is needed for the plants to attain flowering size.

Erythrorhipsalis pilocarpa

ERYTHRORHIPSALIS (Erythro-rhípsalis). The only species of *Erythrorhipsalis* of the cactus family CACTACEAE is one of the minor genera into which *Rhipsalis* conceived broadly has been dismembered. Its distinguishing characteristic is that its flowers have a definite, though extremely short, perianth tube, which is lacking in *Rhipsalis.* Also, the seeds are bigger than those of *Rhipsalis.* The name, alluding to the color of the fruits, comes from the Greek *erythros,* red, and the name of the related genus.

Native of Brazil, ***E. pilocarpa*** (syn. *Rhipsalis pilocarpa*) is an epiphyte (tree-percher) with slender, at first upright, eventually drooping stems up to 1½ feet long. These fork or bear branches in whorls (circles of more than two). They have bristly areoles and from their ends bear small, fragrant, white to pink flowers ½ to ¾ inch wide that remain open for several days. These have many stamens and a style longer than the fifteen to twenty perianth segments (sepals and petals). The wine-red, berry-like fruits are almost ½ inch in diameter.

Garden Uses and Cultivation. These are as for *Rhipsalis.* For additional information see Cactuses.

ERYTHROXYLACEAE—Coca Family. Two tropical genera of dicotyledons totaling approximately 250 species of trees and shrubs compose this family. The most notable member, *Erythroxylum coca* is the source of cocaine. Family characteristics are alternate, undivided, lobeless, toothless leaves, and small flowers in clusters from the leaf axils with persistent calyxes with five lobes or sepals, five petals often with appendages, ten persistent stamens united at their bases, and three styles. The fruits are small drupes. Only *Erythroxylum* is cultivated.

ERYTHROXYLUM (Ery-thróxylum) — Cocaine Plant, Coca. One other genus and this, the name of which is sometimes spelled *Erythroxylon,* constitute the coca family ERYTHROXYLACEAE. Of the 250 species of tropical trees and shrubs that compose *Erythroxylum* only one has any slight claim to horticultural interest, and that is far more renowned for other reasons. It is the cocaine or coca plant, which must not be confused with the cocoa or cacao tree (*Theobroma*).

The name *Erythroxylum* comes from the Greek *erythros,* red, and *xylon,* wood. The wood of some species is red. Erythroxylums are chifly natives of tropical America and Madagascar. They have alternate, undivided, toothless leaves and tiny flowers, mostly in clusters from the leaf axils, each with five or six sepals and petals and twice as many stamens. The berry-like fruits contain a single stone, usually called a seed.

The cocaine or coca plant has served man both as the source of one of the most beneficial of drugs and as a corrupting influence. Some twenty-four years after 1860, when the alkaloid cocaine was first isolated from it, the discovery was made that this could be used to desensitize the eye in preparation for operations. Before long, doctors found that, injected hypodermically, cocaine deadened pain. It was the

Erythroxylum coca

Escallonia, unidentified sort showing flowers

first anesthetic, and the most important one until synthetic substitutes became available.

Centuries before white man learned of the coca plant, South American Indians had discovered certain properties of this shrub, to which the Incas ascribed magic powers and permitted only the ruling class to chew its leaves. After the Spanish conquest, the practice of chewing coca leaves spread. This habit made it possible for its devotees to exercise great feats of strength and endurance and to go without food and sleep for long periods. But its use exacted a toll. The drug is devastatingly habit forming, and when used regularly it weakens mind and body and may result in death. Sniffing cocaine can destroy nasal cartilage. These are the penalties paid by chewers of coca leaves and "sniffers" and other habitual users of cocaine.

The cocaine plant (*E. coca*) is a twiggy shrub 5 or 6 feet tall with slender branches and alternate elliptic to obovate leaves 1 inch to 2½ inches long, clearly defined on their undersides with two veins paralleling the mid-vein. Its yellowish flowers, barely ¼ inch across, in clusters of three to five, have five each sepals and petals, and ten stamens.

Garden Uses and Cultivation. Occasionally cultivated as an ornamental in humid, essentially tropical climates, and in greenhouse collections of plants of importance to man, the cocaine plant is easy to grow. It responds to any ordinary, well-drained soil kept fairly moist and succeeds in sun or part-shade. It may be pruned or sheared to shape. This is best done at the beginning of a new season of growth. It reproduces readily from cuttings and seeds. In greenhouses a minimum winter night temperature of 50 to 60°F is adequate. Day temperatures may be five to ten degrees above those maintained at night.

ESCALLONIA (Escal-lònia). Popular in European and western North American gardens, this genus is composed of sixty species and many hybrids and horticultural varieties. It belongs to the saxifrage family SAXIFRAGACEAE or, if the division of that group is adopted, to the rather poorly defined escallonia family ESCALLONIACEAE. Endemic to South America, *Escallonia* is most numerous as to species in the Andean region. Its name commemorates Señor Escallon, a Spaniard who traveled in South America.

Escallonias are handsome, mostly evergreen shrubs and trees, unfortunately not hardy in the north. They have alternate or clustered, undivided, toothed or toothless leaves, often glandular and with resinous dots. The often fragrant fowers are in terminal racemes or panicles of usually little clusters of white, pink, or red blooms that resemble those of some *Ribes.* Each little bloom has a bell- or top-shaped receptacle, usually thought of as the calyx tube. This encloses the ovary, which is capped by a disk that may be shaped as that term suggests or more or less conical. There are five each sepals, petals, and stamens, and one style ending in a two- to five-lobed stigma. The petals are not joined, but often stand so closely together that their lower parts seem to form a tube. The fruits are many-seeded capsules tipped with the persistent calyx and style.

The only deciduous species cultivated and the most cold-hardy, ***E. virgata*** is a vigorous, erect or spreading shrub 3 to about 8 feet in height. Native of Chile, this has erect or arching, leafy branches and pure white flowers up to ½ inch wide. Its leaves are hairless, obovate, ½ to ¾ inch long by up to ¼ inch broad and toothed. Unlike those of other kinds dealt with here, the flowers, which have much the aspect of those of *Leptospermum,* have petals that spread widely without their lower parts seemingly forming tubes. This escallonia is hardy approximately as far north as Washington, D.C. Hybrids between *E. virgata* and evergreen *E. leucantha* are named ***E. stricta*** (syn. *E. bellidifolia*).

Extremely variable ***E. rubra,*** native of Chile and adjacent Argentina is up to 15 feet tall. It has usually sticky-glandular, hairy shoots, and leaves likely to vary considerably in size and shape even on the same plant. Mostly they are pointed and long-obovate to oblanceolate or ovate, toothed or double-toothed in their upper parts, 1 inch to 2 inches long by ½ to 1 inch broad, with stalked glands toward the base. Their undersides, often thickly bespeckled with resin glands, are prominently veined. The veins on the upper surfaces are depressed. Pink to deep crimson, the flowers are in loose, 1- to 4-inch-long panicles. The lower parts of the petals form an apparent tube up to ½ inch long. Their upper parts spead. The disk is narrow and conical. Variety *E. r. glutinosa* (syn. *E. glutinosa*) is a red-flowered low shrub with leaves furnished thickly on their lower sides with resin glands. Variety *E. r. macrantha* (syn. *E. macrantha*), up to 10 feet tall, has sticky shoots, and is downy and with erect glands. Its broad-elliptic to obovate leaves are 1 inch to 3 inches long by up to 1¾ inches wide. The bright rose-red, downy-stalked flowers, ⅝ inch long and wide, are in racemes or panicles 2 to 4 inches long. Other varieties, or possibly hybrids, of *E. rubra* are *E. r. compacta*, columnar and dense, with winged stems, without resin dots on the undersides of its leaves and with rose-red flowers; *E. r.* 'Glasnevin Hybrid', with dull leaves thickly dotted with resin glands on their undersides and bright rose-red flowers; *E. r.* 'C. F. Ball', a robust kind that produces large, rich red blooms in profusion; *E. r. ingramii*, which resembles *E. r. macrantha* but has smaller leaves and flowers.

A hybrid swarm developed from *E. virgata* and *E. rubra* or *E. r. macrantha* is named ***E. langleyensis.*** The first of these, all of which have large red or pink flowers, raised at the end of the nineteenth century in England, is evergreen, up to 8 feet or more in height, with arching branches and shoots with plentiful stalked glands. From ½ to 1 inch in length and approximately one-half as broad as long, its stalkless, glossy, toothed leaves have tiny resin glands on their undersides. Notable among this group of hybrids are several raised by the Donard Nursery Company of Northern Ireland. The names of many of these contain the word Donard, for example, 'Donard Beauty', 'Donard White', and 'Pride of Donard'. Others of similar aspect and lineage are *E. edinensis*,

Escallonia langleyensis 'Apple Blossom'

'Apple Blossom', 'Peach Blossom', 'William Watson', and 'Gwendolyn Anley', the last much like the original *E. langleyensis*, but deciduous.

Native of Brazil and Uruguay, ***E. bifida*** (syn. *E. montevidensis*) is the most beautiful white-flowered escallonia. A shrub up to 10 feet tall or taller, or sometimes a small tree up to 25 feet high, this kind has hairless, sometimes slightly sticky shoots and elliptic to obovate or spatula-shaped leaves, generally conspicuously toothed near their apexes, 1½ to 3 inches long, and sprinkled on their undersides with small, resinous dots. The rounded, terminal panicles of ½- to ¾-inch-wide, pure white flowers rarely are 9 inches long by more than one-half as wide. Commonly they are considerably smaller. A presumed hybrid between *E. bifida* and the hybrid *E. exoniensis* originated in Scotland as a chance seedling beneath plants of these that grew close together. Named ***E. iveyi,*** this esteemed evergreen has angled young shoots with dark hairs and lustrous, round-apexed, broad-elliptic, finely toothed, hairless leaves 1 inch to 2¾ inches long. The ½-inch-wide white flowers are in terminal panicles up to 6 inches long by 4 inches wide at their bases.

Another Brazilian, ***E. laevis*** (syn. *E. organensis*) hails from the Organ Mountains near Rio de Janeiro. A densely leafy shrub, this is 10 to 15 feet tall. It has slightly glandular-resinous shoots and bronzy-green, narrowly-obovate to elliptic leaves, toothed except near their bases, up to 3 inches long by 1 inch wide. From ⅓ to ½ inches wide and in the bud stage red to pink, the flowers when expanded are pink to white. They are in short, broad, terminal panicles. This species is ill-suited for planting close to the sea. In such places its foliage is likely to be scorched.

Often misidentified in gardens and nurseries as *E. montevidensis,* Chilean ***E. illinita*** is up to about 10 feet tall. A characteristic feature is its strong resinous odor. It has smaller flowers than *E. bifida* (syn. *E. montevidensis*). One of the hardiest sorts, *E. illinita* has glandular-sticky young shoots and clammy or sticky, obovate to broad-elliptic leaves ¾ inch to 2½ inches in length, almost or quite one-half as broad as long. The white flowers, in panicles 3 to 4 inches long, are glandular and sparsely-hairy. They have cylindrical corollas about ½ inch long by ⅜ inch wide across their mouths. Hybrid ***E. franciscana,*** its parents *E. illinita* and *E. rubra macrantha* is an evergreen with the odor of *E. illinita* and, like it, with resin glands. Often misidentified as *E. rosea* this has chocolate-brown bark and dark green leaves ½ to 1 inch long. Of open habit, up to about 10 feet tall and often developing long shoots, it bears rose-pink flowers over a long season.

Chilean ***E. rosea*** is up to 15 feet tall. It has angled, downy shoots and very short-stalked, narrowly-obovate leaves up to 1 inch long by ¼ inch wide, often smaller. They are lustrous above, paler on their undersides, toothed and, except along the midrib above, hairless. The fragrant, cylindrical, short-petaled flowers, white and the lower parts of their ½-inch-long petals forming an apparent tube, have top-shaped receptacles. An evergreen hybrid between *E. rosea* and *E. rubra* or *E. r. macrantha* named ***E. exoniensis*** is a good-looking evergreen shrub or small tree 10 to 20 feet in height. It has slightly-glandular, ribbed, downy shoots and terminal panicles up to 3 inches long of white or pink-tinged blooms about ½ inch long. The lower parts of the petals form a tube. The double-toothed leaves, lustrous above, paler beneath, are ½ inch to 1½ inches long, not more than one-half as wide as long. Except for the midribs on their upper surfaces, they are hairless. Raised in Scotland, *E. e. balfourii* (syn. *E. balfourii*), up to 10 feet tall and of graceful habit, has drooping branchlets and blush-white, cylindrical blooms much resembling those of *E. rosea.*

Another Chilean, this with white flowers, ***E. pulverulenta*** attains heights of 10 to 12 feet. Its shoots and finely toothed leaves are very downy and sticky. The oblongish, round-apexed leaves taper to their bases. They are 2 to 4 inches long by ¾ to 1½ inches broad. From other cultivated escallonias that have white flowers this differs in that they are in slender, cylindrical racemes, branchless or sometimes branched at their bases, and 4 to 9 inches long by up to 1 inch wide. Native of Argentina, ***E. tucumanensis,*** a shrub or tree up to 15 feet tall and first introduced to cultivation in

Escallonia tucumanensis

1961, has finely toothed oblong to elliptic leaves about 2 inches long. Its ½-inch-long and -wide, white flowers are in loose, slender panicles, leafy at their bases.

Garden and Landscape Uses. Escallonias are splendid garden and landscape furnishings in mild climate regions such as California. The hardier kinds succeed as far north as Washington, D.C. All lend themselves for use as general purpose shrubs. They can be espaliered against walls or, with the sacrifice of some bloom, be sheared as hedges. They withstand exposure to wind and most sorts do well in seaside environments. Full sun or a little part-day shade is acceptable. Ordinary

Escallonia, unidentified sort, espaliered against a wall

garden soils, except highly alkaline ones, are satisfactory. Escallonias are reasonably drought-tolerant but do best with fairly generous supplies of moisture.

Cultivation. These beautiful, clean-looking shrubs, most of them evergreens, present no serious problems to gardeners. Pruning to shape the plants, prevent overcrowding, or limit them to size consists of cutting out low down about one-third of the branches of taller kinds each year and of cutting out obviously crowded branches and removing the tips of others of low kinds. This is done as soon as flowering is through or in late winter or spring. Hedges may be sheared whenever necessary, usually after flowering. Propagation is easy by cuttings of firm, but not hard side shoots taken in summer and planted under mist or in a greenhouse or cold frame propagating bed. Species, but not hybrids or varieties, can be raised true to type from seeds.

ESCAROLE. See Endive or Escarole.

ESCHSCHOLZIA (Eschschólz-ia)—California-Poppy. Ten variable species of the poppy family PAPAVERACEAE are included here, all natives of western North America. The name commemorates the naturalist and surgeon Dr. Johann Friedrich Eschscholz, who accompanied Russian expeditions to the American Pacific Coast and also went around the world. He died in 1831. The name has also been spelled *Eschscholtzia.*

The genus *Eschscholzia* consists of annual and perennial herbaceous plants that unlike true poppies (*Papaver*) contain colorless sap. The finely dissected, usually hairless leaves are alternate. The flowers are stalked, poppy-like, and have two united sepals that form a cap, pushed off as the bloom opens. In this they differ from *Hunnemannia,* the sepals of which fall separately. The petals in *Eschscholzia* are usually four, occasionally six or eight. In dull weather they fold longitudinally enclosing some of the sixteen or more stamens. The style is short. The fruits are slender capsules that when ripe rupture explosively. This genus is indigenous from the Columbia River to New Mexico.

Eschscholzia californica

Of chief horticultural importance, ***E. californica*** is a freely-branched annual or short-lived perennial with glaucous, blue-gray stems and foliage. It has produced many horticultural varieties with cream-white, yellow, bronze-yellow, orange, orange-red, rose-pink, and carmine-pink flowers. The blooms, on long slender stalks, are up to 3 inches across. Those of some varieties are double. The plants range in height from 10 inches to 1¾ feet. Another cultivated species, ***E. caespitosa*** (syn. *E. tenuifolia*) is an annual with several stems and usually thrice-ternate, somewhat glaucous leaves with blunt lobes. It seldom exceeds 6 inches in height and has bright yellow flowers about 2 inches across on slender stems well above the foliage.

Garden and Landscape Uses. California-poppies are among the gayest and most brilliant annuals, esteemed for flower beds and borders and window and porch boxes and as carpets for sunny slopes. They succeed especially well near the sea and make gorgeous displays of color over an extended season, continuing even after the first frosts. Although when cut they last for only a few days, the flowers can be used effectively in arrangements located in bright light. For such purpose they should be cut when the buds are well advanced, but before the petals unfurl. Cultivated in pots, California-poppies are attractive for

Eschscholzia californica (flowers)

keeping greenhouses colorful in late winter and spring.

Cultivation. California-poppies, even the more or less perennial *E. californica*, are grown as annuals. They give of their best in rather poor, sandy soil. Too rich a diet encourages coarse, vigorous growth instead of prolific bloom. Exposure to full sun is essential. Like so many of their relatives in the poppy family, they do not transplant well, therefore sow where the plants are to remain and thin out the seedlings to about 6 inches apart. Sow outdoors in spring or where winters are mild in fall. Plants from fall-sown seeds bloom earlier than those from spring seeding. Cover seeds with soil to a depth of about ⅛ inch.

To have plants in bloom in pots in greenhouses in winter and spring, sow in September. Fill 2½- or 3-inch pots with sandy soil and sow about three seeds in each. When the young seedlings are well up, pull out all except the strongest one in each pot. When the small pots are filled with roots repot into 4-inch pots and later if desired into pots or pans 5 inches in diameter. Use a porous, moderately fertile soil. Grow the plants in full sun in a greenhouse where the night temperature is about 50°F and day temperature five to ten degrees higher. Water moderately and after the plants are well rooted in their final containers, give them dilute liquid fertilizer once a week. Prevent seeding and encourage flower production by picking off all faded blooms. This prolongs the flowering season.

ESCOBARIA (Esco-bària). The genus *Escobaria*, by conservative authorities included in *Mammillaria*, belongs in the cactus family CACTACEAE. Its name commemorates the work of two Mexican brothers, Romulo and Numa Escobar. The group comprises twenty species, natives of the southwestern United States and Mexico. From *Coryphantha* this genus differs in the outer segments (petals) of its flowers being usually fringed with hairs and its seeds being minutely pitted.

Escobarias are small plants with globose to cylindrical, clustered or sometimes solitary stems, the older, lower parts of old specimens knobby and naked because the lower tubercles shed their spines and are presented as lumps of brownish, dead-appearing, corky tissue. The tubercles are grooved along their upper sides. The spines are always straight, never hooked. The small, symmetrical, bell- to funnel-shaped flowers originate in the grooves of young tubercles near the tops of the plants. Neither stamens nor style protrude. Except sometimes for one scale, the red, spherical to oblong fruits are naked. They do not split open and are tipped with the withered remains of the perianth. The black or brown seeds have pitted coats. In fruit these plants are decidedly attractive.

Of kinds cultivated, *E. dasyacantha* (syn. *Mammillaria dasyacantha*) and *E. tuberculosa* (syn. *E. strobiliformis, Mammillaria tuberculosa*) much resemble each other and are frequently confused in cultivation. Native of Texas and adjacent Mexico, they are spherical to ovoid when young and become cylindrical as they age. Their stems are covered with lumpy tubercles. Their flowers are white to lavender-pink. The bright, egg-shaped, red fruits are about ¾ inch long. The smaller of these species, ***E. tuberculosa***, 1 inch to 2 inches in diameter, attains heights of 5 to rarely 7 inches or more. The other may exceed this maximum by about 1 inch and has stems up to 2¾ inches wide. A chief difference between these species is that ***E. dasyacantha***, apparently the commonest in cultivation, has spine clusters with twenty or more radials and about nine slender, reddish-brown-tipped centrals that spread in all directions, whereas those of *E. tuberculosa* are of twenty to thirty radials and five to nine purplish-tipped centrals four of which, especially the lowermost, are much stouter than the radials. Also, this last species has bell-shaped blooms ¾ to 1 inch long by ¾ inch to 1¼ inches wide, whereas those of *E. dasyacantha* are 1 inch tall by only ½ to ¾ inch wide. Mexican ***E. lloydii*** much resembles *E. tuberculosa*, differing mainly in its spine clusters having stouter centrals and its flowers being greenish-white. Of the same relationship is yellow-flowered ***E. zilziana.*** Until quite large the species described above have solitary stems, but when old they may produce offsets. Their surfaces are covered with numerous tubercles.

Other species of *Escobaria* cultivated include these: ***E. albicolumnaria*** (syn. *Mammillaria albicolumnaria*), endemic to Texas, and by some authorities included in *E. tuberculosa*, in cultivation is often misnamed *E. dasyacantha*. In appearance it strongly resembles this last species, but has more translucent, more brittle and easily broken, white spines. Also, its seeds are brown, not black. It has usually solitary stems. ***E. chaffeyi***, Mexican, has a short-cylindrical stem up to 4½ inches long by 2¼ inches wide nearly covered with white spines. Each spine cluster is composed of a large number of dark-tipped radials and several shorter centrals. The cream to purplish flowers are approximately ½ inch in length. The crimson fruits are ¾ inch long. ***E. chihuahuensis***, of Mexico, has often solitary, spherical to short-cylindrical stems up to 4½ inches long by 2¼ inches wide practically hidden by the spines that are in clusters of many spreading radials and fewer, longer, usually dark-tipped centrals. The small, purple blooms, mostly under ½ inch long, have broad outer petals. ***E. leei*** (syn. *Mammillaria leei*), one of the rarest native cactuses of the United States, inhabits a few canyons in New Mexico. A curious species, this has a stout taproot and tightly-packed clusters of sometimes several hundred spherical to club-shaped or sometimes cylindrical stems mostly not over 1½ inches long by one-half as thick, sometimes twice as big. The tiny, white radial spines, pressed tightly against the plant body, are from forty to ninety in each cluster along with six or seven stouter centrals. About ¾ inch long by ½ inch wide, the flowers are deep pink suffused with brown. The oblong to club-shaped fruits, green tinged with pink, are about ½ inch long. ***E. runyonii*** (syn. *Mammillaria robertii*), native of the Rio Grande Valley, forms crowded clumps up to 1 foot in diameter, 2 or 3 inches high, of spherical to oblong stems about 1¼ inches thick. They are covered with small tubercles and clusters of twenty to thirty slender, usually brown-tipped, white radial spines and five to ten white-based, dark brown centrals up to a little over ½ inch long. The tan or buff flowers are ¾ inch in diameter. Egg-shaped and scarlet, the fruits are ¼ inch long or somewhat longer. ***E. sneedii*** (syn. *Mammillaria sneedii*), almost eliminated from its native haunts in Texas and New Mexico by ruthless collectors, forms patches of crowded stems 2 inches long by ¾ inch wide covered with tiny tubercles almost hidden beneath clusters of pinkish to white spines consisting of twenty-five to forty-five radials about ⅛ inch long and thirteen to seventeen somewhat bigger, spreading centrals. The flowers are pink, about ½ inch long and wide, with their inner petals fringed in their lower parts. The nearly spherical fruits, ¼ inch in diameter, are deep pink. ***E. varicolor*** (syn. *Mammillaria varicolor*), an endemic of Texas, has solitary stems that branch only if their tops are damaged. Egg-shaped, they are up to 5 inches tall by almost as wide and covered with small tubercles. The clusters of spines are of fifteen to twenty very slender, white radials, bristle-like or thicker, up to ¼ inch long. There are four or more rarely five stouter, semitranslucent centrals about ½ inch in length, brownish with yellowish bases. The white to pink flowers are ¾ inch long by 1¼ inches wide. Their outer petals have fringed margins. The inner petals vary in color from white to light rose-pink. The rose-red fruits are ½ inch or a little more in length, ellipsoid to club-shaped.

Garden and Landscape Uses and Cultivation. Great caution must be exercised not to give these plants too much water. The uses and general cultivation of escobarias are those of mammillarias. For more information see Mammillaria, and Cactuses.

ESCONTRIA (Escón-tria). One to three species are accepted as composing *Escontria* of the cactus family CACTACEAE. A native of Mexico, the genus is separated from nearly related *Cereus* because its ovaries, instead of being practically naked, bear persistent, papery scales. The genus is dedicated to the distinguished Mexican Don Blas Escontria, who died in 1906.

Large and treelike *Escontria* has many erect and ascending, few-ribbed branches, with spines in comblike groups. The small, solitary, somewhat bell-shaped blooms, which open by day, have erect perianth lobes (petals) longer than the stamens and style. The fleshy, globular, purple fruits contain many black seeds.

The first species described by botanists, ***E. chiotilla*** (syn. *Cereus chiotilla*) is 15 to 20 feet or somewhat more tall. It has a very short trunk and numerous, easily broken, bright green branches each with seven or eight longitudinal, sharp ribs. The closely-set, sometimes joined areoles (specialized regions from which spines and blooms arise) have ten to fifteen, short, often reflexed, light-colored spines in addition to several centrals of which one much exceeds the others in length and may be fully 3 inches long. Borne near the branch ends, the yellow flowers are slightly over 2 inches long. The hairless, scaly fruits, about 2 inches in diameter, are edible and are sold in Mexican markets.

Garden and Landscape Uses and Cultivation. In warm desert and semidesert regions *Escontria* is an agreeable addition to outdoor collections of cactuses. It can be grown also in greenhouses wih other succulents. For further information see Cactuses.

A young pear tree espaliered against a wall

Older pear trees espaliered against a wall

Apples espaliered with branches extending horizontally: (a) A fence of young trees

(b) A fence of older trees

ESENBECKIA (Esen-béckia)—Jopoy. Two dozen or more species of New World trees and shrubs constitute *Esenbeckia* of the rue family RUTACEAE. Most are natives of South America. The kind described below was once native of Cameron County, Texas, but perhaps is now extinct in the wild. The name commemorates Christian Gottfried Daniel Nees von Esenbeck, a German professor of botany, who died in 1858.

The jopoy (***E. runyonii*** syn. *E. belandieri*) is 10 to 30 feet tall, whitish-barked, and round-topped. It has alternate, hairless, glossy leaves that remain all winter. They have three elliptic to oblong leaflets 2½ to 6 inches long by ½ inch to 2 inches broad. The white flowers, about ¼ inch wide, are in broad, terminal panicles. They first come in late spring and are succeeded by fruits that mature in August. In September more flowers appear, with a second crop of fruit in December or January. The blooms have four or five sepals and the same number of spreading petals and stamens. The broad fruits are thick-skinned, 1¼ to 2¼ inches long, and deeply four- or five-lobed and ridged. When green they are strongly orange-scented. At maturity the seeds are ejected with considerable force.

Garden and Landscape Uses and Cultivation. Showy in bloom, this rare orange relative is worth growing for floral display as well as for its interest of being, or at least having been, a native of the United States. It can be used in locations appropriate for a small flowering tree, and succeeds in ordinary soils in sun. Experience in Florida suggests that it grows slowly. It is propagated by seeds and cuttings.

ESKIMO-POTATO is *Fritillaria camschatcensis.*

ESPALIER. In its original senses an espalier is a trellis against which a tree or shrub is trained so that its branches lie in one plane or is a tree or shrub grown in this way on a trellis. Commonly the definition is expanded to include all trees and shrubs trained flat (in one plane) whether against trellis, wall, or other support. As a verb, espalier denotes the training of espaliers.

Espaliered trees have long been, and still are, popular in Europe as practical means of raising home and even commercial crops of apples, pears, peaches, figs, and other fruits. Their chief advantages are economy of space (an espalier can be accommodated against the wall of a house or a garage or several can be used to fence a garden) and the protection walls afford, which may reduce the danger of late spring frosts damaging blossoms and may hasten the ripening and improve the quality of the fruits. In North America, except as ornamentals, fruit trees are rarely espaliered. There are three chief reasons for this. Land for fruit growing is more abundant than in many parts of Europe. Espaliering is time-consuming and calls for considerable skill. In hot American summers espaliers against walls are likely to suffer from reflected heat. Remember, too, that because of the need for periodic painting or otherwise treating wooden surface espaliers against them are likely to be troublesome. Espaliering for ornament is not uncommon in North America and orchard fruits as well as other kinds of trees and shrubs, both deciduous and evergreen, are used. When well done the results can be highly decorative. If the specimens are to be against walls remember the danger of reflected heat. South-facing walls are less satisfactory than others. Instead of training the branches flat against the masonry as is commonly done in Europe, establish, spaced about 1 foot apart and at a distance of 4 to 6 inches from its face, a trellis or series of tautly stretched horizontal wires passing through long-stemmed screw eyes fixed into the wall and tie the branches to these.

Rather few kinds of trees and shrubs are commonly espaliered as ornamentals. Among them are the various fruit trees

Apples espaliered in double-U pattern, in bloom

Pear trees, espaliered in grid-iron pattern

Young espaliered peach tree

Fan-trained espaliered peach tree

Pyracantha crenulata as a fan-trained espalier

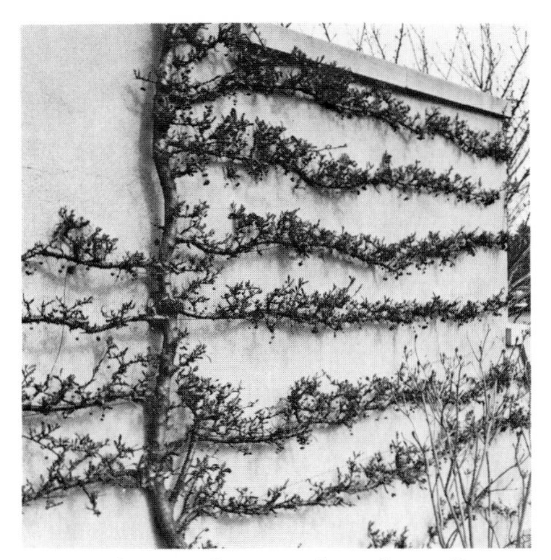

Pyracantha crenulata as a formal espalier

Pyracantha coccinea as an informal espalier

Espaliered English hawthorn (*Crataegus monogyna*)

traditionally grown in this fashion, and pyracanthas, yews, *Cotoneaster horizontalis, Magnolia grandiflora*, forsythias, geraniums, and fuchsias. Others as suitable include various deciduous and evergreen viburnums, flowering quinces, ceanothuses, English hawthorns, *Acanthopanax sieboldianus, Buddleia alternifolia, Caragana arborescens, Ilex crenata, Poncirus trifoliata, Prunus triloba, Streptosolen*, and *Tamarix*.

Patterns in which espaliering can be done are numerous. Choose those appropriate to the space to be filled and the plants to be used. Basically they can be grouped as formal and informal. The former emphasize man's dominance over nature and appeal especially to those whose tidy minds lean to the mechanically and mathematically precise. Informal or free-style espaliers permit a full display of the individual characters of the plants, more individual artistic expression on the part of the gardener.

Espaliered wisteria

Formal espaliering savors of the Old World and its painstaking traditions. Free-style espaliering, less precise, but more modern, suggests the New. But do not be influenced too much by that. There has been a considerable revival in the use of formal espaliers to ornament rather stark wall surfaces of many contemporary style buildings, as witness the splendid use made of espaliered fruit trees in the gardens of the United Nations building in New York City.

The basic designs for formal espaliers, better illustrated in pictures than conveyed in words, are upright, diagonal, and horizontal cordons and a variant called the horizontal-T, U-shaped espaliers and variations of it known as the double-U and triple-U, palmette verniers, Belgian fence, lozenge, and arcure espaliers, and fans. Other patterns can be devised. As their name suggests, informal espaliers conform to no rigid pattern and permit the placement of the stems and branches as natural growth suggests and fancy determines. One great advantage of the informal style is that if something goes wrong during the training period and branches do not develop just where you want them, or if a branch is inadvertently lost later, the pattern is not so obviously marred.

To train plants as espaliers begin with young ones. Attempts to fashion older specimens of pyracanthas, yews, Japanese holly, and some other kinds into presentable espaliers may meet with some success, although rarely as much as when a beginning is made with younger and smaller individuals, but to achieve strictly formal specimens and to espalier such rigid-branched trees as apples and pears an early start must be made. It is best to begin with single-stemmed whips, one-year-old trees that have not yet developed branches. With apples, pears, plums, and cherries they should be grafted on dwarfing understocks as a check to too rampant growth.

After planting, training begins. This consists of pruning and tying branches into place. Pruning usually needs attention in late winter or spring and again in summer. Its purpose is to restrict branch development to the production of limbs necessary to form the permanent design and the removal or drastic shortening of "breast wood," shoots that grow out from the vertical plane of the espalier. The extent and management of these operations depends to a large extent upon the flowering habits of the subject. Where fruits are the objective the best subjects for formal espaliers are kinds that lend themselves to spur pruning, such as apples, pears, vinifera-type grapes, and red currants. Peaches, plums, and cherries that bloom along shoots of the previous year's growth are best suited for training as fans. Ornamentals grown for foliage or flowers only can be effectively trained formally or informally.

To train a formal espalier beginning with a one-year-old whip (a young tree without branches) and supposing you wish it to be a double horizontal cordon, a specimen with a pair of limbs spreading in opposite directions parallel to the ground and about 1½ feet above it, begin by cutting the whip back in late winter to a height of 1½ feet. Permit only two shoots to grow and from the beginning keep these tied to trellis or wires in the positions they are to assume. Rub off as they appear any shoots that grow from the stem below the two branches

An orange tree espaliered informally

Viburnum plicatum espaliered informally

Formal espaliered fruit trees in the gardens of the United Nations, New York City: (a) A fan-trained pear flanked by espaliers with tiers of horizontal branches

(b) Espaliered apple with horizontal branches

(c) Apples espaliered in Belgian fence pattern

and any that come from the branches except those spaced 5 to 6 inches apart that are to be kept to develop into spurs. In late June or early July cut all retained side shoots, but not the terminal or leading shoot, back to within three or four leaves of their bases. If secondary shoots develop as a result of this, pinch them back just above the first leaf. The following late winter prune the leading shoots to a length of 2 feet and all side shoots that were shortened the previous summer to 1 inch. Side shoots so treated will become the spurs from which flowers and fruits as well as leafy shoots grow each year. When new growth begins select one strong shoot from the end of each cut-back leader and treat them as the original leaders were the year before, tying them into position without shortening them and allowing side shoots to develop only at desired intervals. In summer prune side shoots along the whole length of the branches partway back as was done the previous year. Repeat this pattern of pruning and tying-in extensions of the leading shoots annually until the espalier has filled the space you wish, then vary it only by at the late winter pruning cutting back the leading shoots as well as the laterals to within 1 inch of their bases. Manage trees to be trained in more complicated formal patterns than double horizontal cordons similarly, but over a period of two years or more select and tie into place as many more permanent branches as are needed to complete the pattern.

Fan-trained specimens of fruiting trees that bloom all along shoots of the previous year, peaches, for example, call for a somewhat different procedure. The whips are cut back to about 1½ feet. From each is allowed to grow a number of shoots to become permanent branches. These are tied to the support so that they spread fanwise. Some side shoots are allowed to grow and are tied in as permanent limbs to fill the widening spaces between the main branches as they fan outward. Pruning is done to encourage shoots to develop one season that will fruit the next. With this in mind the only summer pruning carried out is any necessary to remove surplus shoots. Those to be retained for fruiting the following season, and these should be selected from as low down on the previous year's shoots as possible, are tied to the support beside and parallel to the previous year's shoots that bear the current season's flowers and fruits. In late winter, shoots older than those of the previous year's development are cut back as far as possible without removing the shoots of the last season.

Trees and shrubs not grown for fruits, but only for foliage and floral effects are espaliered in accordance with the same principles as are applied to those that bear comestible or decorative fruits. Often the permanent framework branches are spaced less precisely with the result that greater informality is achieved. With the same purpose in mind current season's shoots of kinds such as forsythias, tamarixes and *Buddleia alternifolia* may be allowed to develop without tying to the supports. In this way a looser, cascading effect is achieved. Be very careful to delay pruning spring-blooming kinds until flowering is through.

ESPOSTOA (Es-postòa). Of the six species of the Peruvian genus *Espostoa* of the cactus family CACTACEAE, the most commonly cultivated is *E. lanata,* by connoisseurs rated one of the most strikingly beautiful cactuses. The genus is dedicated to a Nicolas E. Esposto, of Lima, Peru, a keen twentieth-century botanist. Some authorities include in this genus *Pseudoespostoa* and *Thrixanthocereus.*

Espostoas are branching cactuses with cylindrical or columnar, many-ribbed stems closely set with areoles (small cushions from which spines come) bearing long hairs and spines. The flowers, usually rather small, open at night. They develop from areoles, different from ordinary spine-producing ones, that form patches called pseudocephaliums of closely set, long bristles and hairs. The pseudocephaliums are located near the tops of the stems. The symmetrical, tubular-bell-shaped blooms have short, scaly calyx tubes, densely-hairy, in the axils of the scales, short, spreading or recurved perianth segments (petals), and many stamens. The spherical to obovoid fruits may have tufts of hairs or be more or less naked.

Handsome and popular ***E. lanata,*** 4 feet tall or taller and branching like a candelabrum, has stems 2 to 5 inches in diameter with twenty to twenty-five ribs. From the areoles sprout long white hairs that wrap themselves around the stems and form a belt through which protrude the yellowish or reddish spines, the central ones of each cluster 1 inch to 3 inches long, the ten or twelve radials shorter. The flowers are white, pinkish, or reddish, 2 to 2½ inches long. They are succeeded by white to pink, edible, juicy fruits containing black seeds. Variety *E. l. sericata* has thicker stems than the typical species and a denser covering of slightly yellowish hairs. Treelike and up to 12 feet tall, ***E. ritteri*** of Peru is broader than it is high. Its stems have eighteen to twenty-two ribs furnished with white areoles, spine clusters of about twenty-five radials and one central, and many long white hairs. The cephalium is yellowish. The flowers are white and up to 1¾ inches long.

Euadenia eminens

Espostoa ritteri, a young grafted specimen

Garden and Landscape Uses and Cultivation. Espostoas are excellent for outdoors in warm, semidesert climates and for including in indoor collections of succulents. They succeed in rather poor soils, so long as they are porous and well drained, and need the same general conditions and care as other columnar cactuses. For more information see Cactuses.

ESTRAGON is *Artemisia dracunculus.* See Tarragon.

ETHROG or ETROG. This is a form of the citron (*Citrus medica*).

EUADENIA (Euadè-nia). The tropical African genus *Euadenia,* of the caper family CAPPARIDACEAE, consists of two species. Its name is derived from the Greek *eu,* well, and *aden,* a gland, and refers to the glands on the flowers.

Euadenias are hairless shrubs with alternate, leathery, evergreen leaves, each of three leaflets. The flowers, in terminal clusters or racemes, have four each sepals and petals, five to seven fertile stamens, two, one, or no staminodes (abortive stamens), and one stigma. The fruits are capsules.

Sometimes cultivated, ***E. eminens*** has long-stalked leaves the central leaflets of which are obovate and the side leaflets unequal-sided. The pale, sulfur-yellow flowers have narrow petals the two upper of which are 3½ to 4 inches long and hooded. The others are much shorter. The five fertile stamens are curved and pale green.

Garden Uses and Cultivation. This is attractive for outdoor planting in the tropics and for growing in tropical greenhouses. Indoors it succeeds in a humid atmosphere and a minimum winter night temperature of 65 to 70°F with an increase of five to ten degrees by day. From spring through fall it needs shade from strong direct sun. Repotting or top dressing should receive attention in late winter just before new growth begins. Usually repotting is needed only every two or three years. In the intervening years top dressing is satisfactory. Specimens that are well rooted benefit from fertilizing regularly from spring through fall. Cuttings taken in spring or late summer root readily in a humid propagating case in a tropical greenhouse. They respond favorably to mild bottom heat.

EUANTHE. See Vanda.

EUBOTRYS. See Leucothoe.

EUCALYPTUS (Eucalýpt-us)—Gum Tree. Eucalypts (a useful group name for members of the genus *Eucalyptus*) are characteristically Australian and Tasmanian. Of the approximately 500 species only two or three occur elsewhere, in Indo-Malaysia. The group belongs in the myrtle family MYRTACEAE. Its name, derived from the Greek *eu,* well, and *kalyptos,* covered, refers to the flower buds being capped with a lidlike structure called an operculum. This is the most distinctive distinguishing feature of the genus.

The great majority of eucalypts are evergreens. They range in size from some of the tallest trees on earth to kinds barely 6 feet high. Their leaves are neither lobed nor toothed, but except in very few cases, at various stages in the development of individual trees they assume markedly different forms. The different types are classed as cotyledons, seedling leaves, juvenile leaves, intermediate leaves, and adult leaves. The last three, especially the juvenile and adult types, are most obvious. Almost always they are markedly dissimilar, with the intermediate type transitional between the other two.

The foliage of all eucalypts contains volatile oil and so does the bark, flowers, fruits, and other parts of some. Usually juvenile leaves are opposite with each successive pair set at right angles to the pairs immediately below and above. Generally without stalks, they spread horizontally with one surface facing upward, the other downward. This is in marked contrast to the adult leaves, which, in most cases, hang with tips pointing earthward and, unlike typical juvenile leaves, show no marked difference in appearance between their two surfaces. Adult leaves are usually alternate and stalked. In the vast majority of eucalypts they are green, whereas the juvenile foliage of most has a bluish, glaucous appearance. Juvenile foliage is produced by some species for several years before the changeover to the development of adult leaves begins. Other kinds make the transition early, sometimes when only a few months old.

The flowers of eucalypts are in umbel-like groups of three or more or rarely are solitary. The groups may be solitary or clustered. Each flower bud is separated into two parts by a distinct horizontal line. The upper part, the operculum, at a certain stage of development falls away to reveal the stamens and style that are attached to a platform called the receptacle that often merges imperceptibly into the flower stalk. There are no sepals or petals (the operculum is believed to represent these). The many stamens are responsible for the showiness of the blooms. They are most commonly white or creamy-white, more rarely yellow, pink, or red. The fruits are capsules, in almost all cases joined to the receptacles and partly buried within them. Each capsule ordinarily matures a few fertile seeds.

Eucalypts constitute about three-fourths of the vegetation of Australia and are one of the most important, if not the most important hardwood timber resource of the world. Among them is the world's tallest nonconiferous tree, the Australian-mountain-ash (*E. regnans*), the largest specimen of which comes within a few feet of equaling in height the tallest redwood of California. Joseph Banks, the British botanist who accompanied Captain Cook on his first voyage to the Pacific, collected the first eucalypt brought to the attention of scientists at Botany Bay in 1770. It was *E. gummifera*, but it was nineteen years later before the genus was described and named by the French botanist L'Heritier de Brutelle.

As ornamentals and shade trees several eucalypts rate highly in California and other parts of the Southwest as well as in Australia and many parts of the world

Eucalyptus, unidentified species in California

where similar climatic conditions prevail, but the importance of the genus as a source of decorative subjects is far overshadowed by their immense value as producers of lumber and other commercial products and by their value for reforestation. The introduction of the blue gum (*E. globulus*) into Ethiopia, beginning in the 1860s, is credited with the survival of the people in that country, with its development, and with the creation of its capital Addis Ababa. At the time, the destruction of the original forest was nearly complete, and the inhabitants were on the verge of being forced through lack of building material and fuel, quite essential to survival in that harsh climate, to push southward and westward. Had they done this, what forests remained would soon have been denuded and inevitably wars of annihilation with other peoples would have followed. The introduction of the blue gum to Ethiopia stands as one of the most successful reforestation projects ever undertaken. Some 15,000,000 small landowners cultivate the land and every village has its own grove or groves of these useful trees. In Argentina, Brazil, Chile, and other parts of South America, as well as in many other parts of the world, Australian eucalypts are planted for their useful products as they are in their native continent. About sixty species are exploited commercially for their lumber. Their woods include some of the heaviest, hardest, and most durable known and are used for a wide variety of purposes. Many are handsomely grained and figured and find favor with furniture manufacturers. Others supply timbers for construction, mining, railroad ties, wharves, paving blocks, plywood, veneers, agricultural implements, musical instruments, sporting goods, paper pulp, fiberboard, and other purposes. The wood of most eucalypts is excellent fuel. From the foliage of about twenty species essential oils are distilled on a commercial scale. These are used in medicines, perfumes, disinfectants, and deodorants. Eucalypts produce large amounts of nectar, which is the base of excellent honey. They are the most important bee pasture of Australia. Another useful product of this amazingly versatile genus is tanning material. The foliage of some kinds is used as fodder and that of some forms the exclusive diet of the Australian koala-bear.

The aborigines and early settlers and travelers in Australia discovered what might be considered unexpected uses for certain eucalypts. They first ground and ate the roots of several of the smaller growing kinds (mallees), and they used for food and for the preparation of beverages a sugary secretion, known as lerp or manna, of scale insects that infest certain eucalypts. But the most critical value of some of the mallees of desert areas was as sources of drinking water. The aborigines were well aware that large amounts of slightly astringent water were stored in the roots and they often were dependent upon it for survival when other sources failed. The water was obtained by digging up long lengths of root, blowing vigorously into one end, and collecting in a container the liquid that bubbled out at the opposite end. It was not unusual for white travelers, unfamiliar with this resource, to die from thirst in regions where mallees abounded and where life-saving water was to be had in quantities sufficient to ensure survival had they been exploited. The Australian song "Waltzing Matilda" made known to many outside Australia the name of the coolabah tree (*E. microtheca*).

The blue gum (***E. globulus***) is probably the most widely cultivated eucalypt. A rapid-growing species, it has been introduced to all regions where eucalypts can be expected to persist outdoors, and is also commonly cultivated as a pot plant and for temporary summer display in the open in many places where winters are too harsh for its survival in the open. Native of Tasmania, where it grows on land well supplied with moisture, at its best the blue gum is 180 feet in height and may

Eucalyptus globulus: (a) In California

(b) In Ireland

(c) In Buenos Aires

have a trunk 7 feet in diameter. It has smooth, blue-gray bark that peels above, but on old specimens is persistent at the base of the trunk. Its adult leaves are stalked, narrowly-sickle-shaped, dark lustrous green, and 4 inches to 1 foot long. Especially beautiful is the juvenile foliage. This consists of broad, opposite, stalkless leaves set along quadrangular stems. They are a lovely glaucous-blue and distinctly waxy to the feel. The flowers of the blue gum are large, cream, and solitary or occasionally in threes in the leaf axils. In southern California this species may reach a height of 45 or 50 feet in five years, and specimens 30 years old and 150 feet in height are not uncommon. The wood of this eucalypt is used commercially, but is not as durable as that of some. Felled trees sucker freely from their stumps and soon renew themselves. The blue gum is certainly one of the most decorative of eucalypts but its shedding bark and leaves are likely to be messy and its brittle branches are subject to storm damage. Variety *E. g. compacta* does not attain a height of more than 70 feet and is of more bushy habit.

The manna gum (***E. viminalis***), of southeastern Australia and Tasmania, is another fast grower. It attains a height of 150 feet, and forms a spreading, open crown with drooping branchlets. The bark on the lower part of its trunk is rough and gray, but elsewhere it peels in long strips to expose white or yellowish underbark. The stemless juvenile leaves are green, narrowly-lanceolate to ovate-lanceolate, and in opposite pairs. The adult ones are stalked, sickle-shaped, pale green, and 4 to 8 inches long. The flowers are white. In outline the manna gum resembles the blue gum, but its juvenile foliage is not bluish. This species is inferior as a timber tree. Its common name derives from considerable quantities of manna that frequently occur on its leaves and buds. The manna gum succeeds at moderate altitudes in the interior valleys of California and withstands more frost than most eucalypts.

The red ironbark (***E. sideroxylon***), of eastern Australia, is a tall, slender, graceful species with black, furrowed, persistent bark and opposite, narrow, slightly glaucous, short-stalked juvenile leaves. The adult leaves, 2½ to 6 inches long, are alternate, green or slightly glaucous, and similar in shape to the juvenile ones, but larger. The flowers, three to seven together in axillary clusters, are white or yellowish or, in variety *E. s. rosea,* pink. This species,

Eucalyptus sideroxylon rosea

Eucalyptus globulus (flowers)

sometimes 100 feet tall and one of the most ornamental eucalypts, is not at all particular as to type of soil and is adaptable to a quite wide range of climatic conditions. It grows rapidly and is a fine shade tree and street tree that endures considerable cold. The red ironbark is a good bee tree and is exploited for lumber and oil.

The sugar gum (*E. cladocalyx*) is another rapid grower. An inhabitant of Victoria and South Australia, it attains a height of 100 feet and forms a picturesque head with the foliage concentrated toward the ends of the branches. This has yellowish-brown bark persistent on the lower part of the trunk, but deciduous and mottled with white and light gray above. Its juvenile leaves are opposite, stalked, elliptic, and light green. The adult ones are alternate, lanceolate to broad-lanceolate, stalked, and 3½ to 6 inches long. The flowers are in axillary or terminal clusters. The sugar gum is less tolerant of cold than some eucalypts. In California it is injured by temperatures of 20 to 25°F. This is a good highway tree and a commercial source of lumber. It provides pasture for bees, and in Australia its leaves are used as fodder.

The desert gum (*E. rudis*), which in the interior valleys of southern California survives desert heat and cold down to at least 17°F, is highly resistant to drought. Native to Western Australia, this has a short trunk and spreading branches. It rarely exceeds a height of 50 feet. The bark of its trunk is rough and persistent, but that of the branches peels. The slightly glaucous, juvenile leaves are short-stalked and ovate or orbicular. The lanceolate, sickle-shaped adult leaves are alternate and bright green. They are 4 to 6 inches long and up to 1½ inches wide. Not especially showy, the flowers are white. The desert gum is a good shade and shelter tree. The chief use of its wood is as fuel.

The red gum (*E. camaldulensis* syn. *E. rostrata*) like the last has remarkable powers of resisting high temperatures and drought. Native to Australia, it is the most widely naturally dispersed of all eucalypts. It attains a height of 150 feet and most frequently inhabits river banks and flat lands subject to periodic flooding. Usually it has a short trunk and spreading crown. The red gum has peeling, gray and buff bark, opposite, broadly-lanceolate, glaucous juvenile leaves, and alternate, lanceolate, green or sometimes reddish adult leaves 2½ inches to 1 foot long. The flowers are in axillary clusters of five to ten. This is more cold tolerant than the blue gum (*E. globulus*) and succeeds even in alkaline soils. It grows well even in the extremely high temperatures of the Imperial Valley of California. The red gum is a good shade tree and windbreak and is a commercial lumber species. Its foliage is used as feed for livestock. A product called kino gum obtained from it is used medicinally. It occurs in several rather distinct forms.

The spotted gum (*E. maculata*), of New South Wales and Queensland, with a tall shaftlike trunk and a shapely crown, may attain a height of 150 feet. Its bark peels, producing a smooth, mottled pinkish and whitish effect. The juvenile leaves are opposite and have short stalks. The adult ones are alternate, lanceolate to broadly-lanceolate, dark green, and 2½ to 4 inches long. The flowers are in axillary clusters. The lemon-scented gum (*E. citriodora*) is closely related to the spotted gum and by some considered to be a variety of it. A native of Queensland, it is distinguished by the mouth-watering lemon odor that its leaves emit when crushed or bruised. Its adult leaves are 2½ to 6 inches long. This sort is the source of commercial lumber and oil.

The white ironbark (*E. leucoxylon*) grows up to 90 feet tall and has on the upper parts of its trunk and its branches smooth, shedding bark mottled white and bluish-gray. The base of the trunk is covered with persistent, rough bark. The opposite, stalkless, juvenile leaves are broadly-lanceolate to nearly round and slightly glaucous. The adult leaves are alternate, narrowly- to broadly-lanceolate, grayish or dull green, and 3 to 6 inches long. The flowers are in axillary, three- to eleven-flowered clusters. This species, which grows well near the sea, is esteemed for its lumber, oil, and honey.

The swamp-mahogany (*E. robusta*), native of Queensland and New South Wales, attains a height of about 90 feet. It favors waterlogged, often saline soils, but grows well under drier conditions, and is rather tender to frost. Its juvenile leaves are opposite, short-stalked, and broadly-lanceolate to ovate. The adult leaves are lanceolate to ovate and 4 to 6½ inches long. The flowers are in six- to ten-flowered clusters. This handsome tree is a source of commercial lumber and is a good bee tree. It is satisfactory in coastal regions.

Eucalyptus maculata (trunk)

Eucalyptus ficifolia (foliage and flower)

Among smaller ornamental eucalypts the scarlet-flowering gum (*E. ficifolia*) is outstanding. It has a low, spreading crown and ordinarily does not exceed 30 feet in height. This is a native of a very limited coastal area in Western Australia. Its bark is persistent, rough, and flaky. The juvenile leaves are ovate to broadly-lanceolate, the adult, alternate, lanceolate to ovate, and 3 to 6 inches long. The flowers of individual trees vary greatly in color from almost white to brilliant scarlet. The latter are the most desirable. In California this kind blooms twice each year, somewhat

Eucalyptus cornuta (flower)

sparsely in spring and very much more generously in late summer and fall. It is rather tender to cold, but stands heat and drought well. The yate tree (*E. cornuta*) reaches a maximum height of about 65 feet, but is usually lower. Its trunk and larger branches are covered with deeply-furrowed, brown to nearly black bark, its upper branches with smooth, white, shedding bark. The juvenile leaves are stalked, nearly round to broadly-lanceolate, dark green, and opposite or alternate. The adult leaves are lanceolate, alternate, stalked, and 3 to 4½ inches long. The flowers, in the leaf axils, are in clusters of from five to fifteen. This kind, a native of southwestern Australia, is tolerant of saline and alkaline soil and is a good shade tree.

One of the most ornamental eucalypts is the very distinct red-box or Australian-beech (***E. polyanthemos***) of Victoria and New South Wales, the cut juvenile foliage of which is much used in indoor decorations. Slower growing than most, this attains an eventual height of 80 feet or more and develops an irregular spreading crown. The juvenile leaves, opposite or alternate, 1½ to 4½ inches long and almost circular, are stalked, and a beautiful silver-gray. The adult leaves, 2 to 6 inches long, are alternate, broadly-lanceolate to ovate, slightly glaucous-grayish-green. The flowers, profusely borne, are in terminal panicles. For a considerable distance from its base the bark of the red-box is rough and persistent, above it is smooth and deciduous. The red-box endures high temperatures and drought well. A small species of much interest as an ornamental and for screening is ***E. erythronema,*** of southwestern Australia. Attaining a height of up to 20 feet, it is a tree or large shrub with rough, reddish bark. Its juvenile leaves are stalked and narrowly-lanceolate, the adult ones are similar in shape, light green, and 1½ to 3½ inches long. The flowers are pinkish or reddish and in clusters.

Other kinds of eucalypts cultivated include these: ***E. amygdalina,*** the peppermint gum, about 50 feet tall, is native of Tasmania. It has gray, persistent bark, and alternate, linear to lanceolate adult leaves 2¼ to 4½ inches long that when bruised smell strongly of peppermint. ***E. astringens,*** the brown mallet, about 80 feet tall, has smooth, brown bark and slightly curved, lanceolate, lustrous, dark green adult leaves 3½ to 4½ inches long. It is a native of southwestern Australia. ***E. botryoides,*** the bangalay or bastard-mahogany of from New South Wales to Victoria often grows in saline soils. Up to 140 feet in height, it has dark brown bark and long-pointed, lanceolate adult leaves 4 to 6 inches long and mostly persistent. ***E. calophylla,*** the marri or red gum of southwestern Australia, has decorative creamy-white or pink flowers in large terminal clusters. A good ornamental and shade tree, it is up to 150 feet in height and has dark green, lanceolate adult leaves 2½ to 6½ inches long. ***E. cinerea,*** the Argyle-apple, is an ornamental species of southeastern Australia that grows on poor soils and is about 50 feet tall. It has silvery-blue, rounded or ovate adult leaves up to 6 inches long, and brown, fibrous bark. ***E. coccifera,*** the Tasmanian snow gum, is a tree up to 70 feet or more tall, but sometimes much lower. Native of Tasmania, and one of the hardiest eucalypts, this has peeling bark, white when first exposed, and alternate, green or glaucous, lanceolate to oblong adult leaves 2 to 2½ inches long and ending in a hooked point. ***E. diversicolor,*** the kauri gum, is the tallest native tree of Western Australia. Up to 250 feet in height, it has shedding, whitish, yellowish and grayish bark. Its adult leaves are lanceolate and 4 to 5½ inches long. ***E. dives,*** the broad-leaved peppermint gum native of Victoria and New South Wales, is a handsome shade tree up to 80 feet in height. The chief commercial source of piperitone used to produce synthetic menthol and thymol, this eucalypt has lanceolate to ovate, dark lustrous green adult leaves 4 to 6 inches long. Its bark is persistent on the trunk and main branches. ***E. gomphocephala,*** the tuart, is native to a very restricted region in Western Australia. Sometimes 130 feet tall, its bark is light gray, deeply fissured, and persistent. Its adult leaves, up to 6½ inches long, are thick, gray-green, narrowly-lanceolate, pointed, and usually curved. ***E. gummifera,*** the red bloodwood, is a native of coastal New South Wales, Queensland, and Victoria that becomes 120 feet tall and has brown, persistent bark. It has a dense, compact head and makes a good shade tree. Its adult leaves are dark, lustrous green above, paler beneath, and 3½ to 6 inches long. ***E. gunnii,*** the cider gum, is one of the hardiest kinds. It succeeds outdoors in parts of Scotland. About 100 feet in height, it has

Eucalyptus coccifera in Scotland

smooth green and white bark and is deciduous. Its adult leaves, lanceolate or ovate, are green or gray-green and 1½ to 3 inches long. It is a good ornamental. A good quality cider-like beverage is made from its sap. ***E. longifolia,*** the woollybutt, a native of coastal Victoria and New South Wales that attains a height of 130 feet, has an irregularly-branched crown and fissured, gray, persistent bark on the trunk and large branches. Above, the bark sheds and exposes smooth pale undersurfaces. The adult leaves are narrow-lanceolate to lanceolate-sickle-shaped, and 6 to 9 inches long. ***E. marginata,*** the jarrah of southwestern Australia, up to 130 feet in height, forms an open crown and has persistent brown bark. Its adult leaves are lanceolate and darker green above than on their lower surfaces. ***E. melliodora,*** the yellow-box or honey-scented gum is native from Victoria to New South Wales. It forms a large, rounded crown and may be 100 feet tall. Its bark, more or less persistent, is yellowish or brown. This is a handsome, quick-growing ornamental with bluish or grayish-green, lanceolate adult leaves 3 to 6½ inches long. It is the best bee tree of all eucalypts. ***E. microtheca,*** the coolabah tree, rarely exceeds 50 feet in height and usually has a crooked trunk and spreading branches. Its bark is gray, more or less persistent, and usually is furrowed at the base of the trunk. Its dull green adult leaves, 2½ to 8 inches long, are lanceolate and often curved. This grows along watercourses from southern Queensland northward. ***E. niphophila,*** the snow gum of New South Wales, is one of the most beautiful and possibly the hardiest eucalypt. A shrub or tree in the wild up to 20 feet tall, it will probably attain greater height in cultivation. Closely related to *E. pauciflora,* this has smooth, white, shedding bark. Its sickle-shaped adult leaves are alternate, glossy, 3 to 5 inches long. ***E. pauciflora,*** the cabbage gum of Australia and Tasmania, is a tree usually under 60 feet high that sometimes attains a height of 100 feet. One of the hardiest sorts, this has peeling, white bark and alternate, stalked, broadly-lanceolate to sickle-shaped, glossy, green adult leaves 2½ to 6 inches long. ***E. pilularis,*** the blackbutt, forms a straight trunk and an open, spreading crown. Its bark is persistent and blackish-gray. Its adult leaves are lanceolate and 4 to 4½ inches long. This species attains a height of 200 feet. It is a native of eastern Australia. ***E. pulverulenta*** rarely exceeds 30 feet in height. Very ornamental, it has greenish-white bark and opposite, stalkless, nearly orbicular, glaucous or mealy-white leaves, the juvenile and adult phases quite similar. It is a native of eastern Australia. ***E. regnans,*** the Australian-mountain-ash, is the tallest nonconiferous tree in the world. It is exceeded in height only by the California redwood (*Sequoia sempervirens*), a conifer. Normally it is from 175 to 250 feet, or more rarely up to 300 feet tall when fully mature, but the largest authentically recorded specimen was one measured in 1918 as having attained a height of 374 feet. At the base the bark of this species is rough and persistent, above it sheds and is smooth and whitish or greenish-gray. The broadly-lanceolate adult leaves are lustrous green and 4 to 6½ inches long. ***E. resinifera,*** the red-mahogany, sometimes 150 feet in height, is a native of coastal parts of eastern Australia. It has reddish, rough, persistent bark and lanceolate adult leaves, paler beneath than on their upper surfaces. ***E. rhodantha,*** a hybrid between *E. macrocarpa* and *E. pyriformis,* is a loose shrub about 8 feet tall with nearly circular to pointed-ovate, silvery leaves and showy, red flowers. ***E. salmonophloia,*** the salmon gum, is a native of southwestern Australia sometimes 100 feet in height. Its bark sheds to show salmon-pink and grayish undersurfaces. Its adult leaves, 2½ to 4 inches long, are yellowish-green and ovate to ovate-lanceolate. This is a fine ornamental for arid areas. ***E. tereticornis*** is a native of eastern

Eucalyptus pauciflora

Eucalyptus pauciflora (trunk)

Eucalyptus niphophila (trunk)

Eucalyptus rhodantha: (a) Emerging flowers (b) Flowers emerged

Eucalyptus bicostata

Eucalyptus falcata

Eucalyptus paniculata

Australia and New Guinea. It grows up to 150 feet in height and has smooth, mottled bark above and persistent rough bark on its lower trunk. Its adult leaves are lanceolate, lustrous green, and 4 to 8 inches long. This is a good shade tree.

Additional to the sorts just dealt with, these are in cultivation in California: ***E. bicostata*** of southern Australia and Tasmania is a close relative of *E. globulus* from which it differs in its buds and fruits having four instead of two ribs. It attains a maximum height of 100 to 140 feet but is often shorter. The lower part of its trunk is clothed with persistent, rough bark; the upper part and the branches are smooth and blue-green to gray-green. The juvenile leaves are opposite, stalkless, glaucous, and up to 5½ inches long by two-thirds as broad. The adult leaves, alternate, stalked, lanceolate, and somewhat sickle-shaped, are glossy dark green on both surfaces and 4 inches to 1 foot long. ***E. falcata*** of Western Australia is either a tree up to 35 feet high or a shrub. It has white to gray or mottled, smooth bark. Its rather conspicuously veined, juvenile leaves, up to 3 inches long by 2 inches wide, are ovate to broadly-ovate or elliptic, and slightly glaucous. The adult leaves are alternate, pointed-linear, and slightly curved. ***E. paniculata*** of New South Wales may attain a height of 80 to 125 feet but often is lower. It has persistent, deeply-furrowed, dark gray bark. Its juvenile leaves are mostly opposite, broad-lanceolate, and stalked. Its adult leaves are alternate, stalkless, narrow- to long-lanceolate, slight curved to somewhat sickle-shaped, and 2½ to 6 inches long. ***E. reduncta*** of Western Australia is up to 6 feet tall or in cultivation may be taller. Its bark is smooth and white. The juvenile leaves are alternate, conspicuously-stalked, heart-shaped, and grayish. The adult leaves, 4 to 6 inches long, lanceolate to almost ovate or nearly linear, are dull green. ***E. trabutii,*** its reputed parents *E. rostrata* and *E. botryoides,* often has fissured bark and is densely-foliaged. Its adult leaves, elliptic to lanceolate, slightly curved and markedly pointed, are 6 to 8 inches long.

Garden and Landscape Uses. Eucalypts are among the most familiar exotics in Cal-

Eucalyptus paniculata (bark)

Eucalyptus reduncta

Eucalyptus traubii

Pot plants: (a) *Eucalyptus globulus*

(b) *Eucalyptus citriodora*

ifornia and there and in other mild, dry climates they afford the landscape planner a wide choice of kinds suitable for many different purposes. They are useful as street and highway trees, windbreaks, seaside planting, for small groups and groves and, the smaller ones particularly, trees for home gardens. The blue gum (*E. globulus*) and the lemon-scented gum (*E. citriodora*) are also useful as pot plants and for temporary summer use in outdoor beds.

Cultivation. Eucalypts are raised from seeds that shed readily if the capsules are collected just before they open naturally and are stored in a dry place exposed to the sun. Spring or early summer are good times to sow. The soil or other medium should be kept uniformly moist, but excessive wetness is to be avoided or damping off of the seedlings may result. For the same reason a good circulation of air should be maintained about the young plants. As soon as the seedlings are large enough to handle with ease they are transplanted about 2 inches apart to other flats where they remain until they are large enough to be transferred to a nursery bed outdoors. These trees transplant fairly easily and are adaptable to a wide variety of soils. They grow most quickly and attain their largest size where they have access to adequate supplies of moisture, but many withstand quite dry conditions.

When cultivated as pot plants and for summer outdoor beds in climates where they will not survive the winter, it is usual to sow seeds in January indoors or to sow the previous September and carry the young plants through the winter in a

Eucharis grandiflora

sunny, cool greenhouse or window. They thrive in any ordinary well-drained soil and are ready for planting in outdoor beds as soon as it is warm enough to set out tomatoes, peppers, geraniums, and other tender plants. At all times they need full sun. In the greenhouse a night temperature of 50°F, with an increase of five to ten degrees in the day, is satisfactory.

Diseases and Pests. Leaf spots and crown gall are the chief diseases, and in greenhouses if they are grown in poorly drained soils and are subjected to excessive watering and a humid atmosphere they are apt to suffer from a physiological edema that causes the leaves to blister. Among pests of these plants are aphids, mealybugs, scales, mites, caterpillars, and borers.

EUCHARIDIUM. See Clarkia.

EUCHARIS (Eùch-aris)—Amazon-Lily. Ten species of the amaryllis family AMARYLLIDACEAE belong in *Eucharis.* The common name of the most familiar, the Amazon-lily, is something of a misnomer, since the genus is confined in the wild to the Colombian Andes. The generic name is from the Greek *eu,* well or good, and *charis,* an attraction. It alludes to the beauty and grace of the flowers.

Eucharises are bulb plants that, unlike most members of the amaryllis family, have broad-bladed, evergreen leaves that narrow to slender leafstalks. The handsome, large, pristine-white, sweet-scented flowers are few together in umbels atop leafless stalks. They have slightly curved, cylindrical perianth tubes, six spreading lobes (petals) almost as broad as long, and six stamens with their filaments (stalks) broadened at their bases to form a central corona or cup to the bloom. The stamens are shorter than the perianth lobes. The stigma is three-lobed. The fruits are large-seeded capsules.

The Amazon-lily (***E. grandiflora*** syn. *E. amazonica*) has tapered bulbs about 2 inches in diameter and two to four ovate, elliptic, or oblong leaves with blades up to 8 inches long and about one-half as wide as long. The leafstalks are as long or longer than the blades.

The nodding, exquisitely fragrant blooms, three to six together on erect stalks 1 foot to 2 feet tall, have prominent cups at their centers and petals 1 inch to 1½ inches broad and only slightly longer. The corolla tube is 2 inches long and slender. Each bloom is carried on a short stalk from the top of the common flower stem. The three-angled fruits usually have many seeds.

Garden and Landscape Uses. These choice and beautiful plants are successful only where high temperatures and high humidity prevail and where they are shaded from strong sun. Out-of-doors these conditions prevail only in the tropics, indoors usually only in greenhouses. They can be grown in ground beds and in large pots or tubs. Unlike many bulbous plants they do not do well in small pots; several bulbs accommodated in containers 8 inches or more in diameter give better results than solitary bulbs in those of lesser size. Container cultivation is advantageous because watering can be better controlled and periods of comparative dry-

Eucharis grandiflora (flowers)

ness can induce them to bloom two or three times a year; if the soil is constantly moist they usually flower once a year. Furthermore, they are well adapted for container cultivation and are handsome both in and out of flower. For good effects several bulbs are planted closely together, whether in containers or ground beds. Amazon-lilies are choice subjects for beds and borders for greenhouse ornament and temporary decorations in homes and buildings. Their flowers are lovely and last well in water when cut.

Cultivation. Requirements for the successful cultivation of these desirable plants are high temperatures, high humidity, coarse, fertile, well-drained soil, and nondisturbance of their roots if they are flourishing. Transplanting or repotting that involves much root disturbance often causes them to omit blooming for a year or more. So partial are Amazon-lilies to warmth and shade that they often respond well when planted in well-drained beds of rich soil beneath greenhouse benches where they get plenty of side light, near hot water pipes. They may also be grown in greenhouse benches. Greenhouses in which they are accommodated should have a minimum night temperature of 65 to 70°F, the temperature may be ten degrees to fifteen degrees higher by day and allowed to run up to 90°F or more in summer. Because Amazon-lilies are evergreens they should at no time be deprived of water to the extent that their leaves wither, but periodically, when all the leaves are full size and no flush of young ones is developing, they should be partially dried and rested by withholding water until the leaves begin to wilt, and then soaked. This procedure is repeated for about six weeks before regular watering is resumed. Then, water is applied copiously and dilute liquid fertilizer is given at weekly intervals to well-rooted specimens. Propagation is most commonly by offsets carefully removed in spring and potted or planted individually. Because the removal of these inevitably disturbs the roots, unless the most rapid possible increase is desired it should be undertaken only at intervals of several years. Seeds sown at 75 to 85°F in sandy peaty soil are also a possible means of increase, but it takes several years for plants so raised to reach blooming size.

EUCNIDE (Eucnì-de). Not as well known in cultivation as it deserves, this genus of the loasa family LOASACEAE is closely related to more familiar *Mentzelia*. It consists of eleven species, natives from the southwestern United States to Guatemala. The name comes from the Greek *eu*, pretty, and *knide*, a nettle, and alludes to the appearance and stinging properties of the foliage. Eucnides are annuals, biennials, or herbaceous perennials, often somewhat woody toward their bases. They have alternate, stalked leaves, broadly-ovate to nearly round or, rarely, obovate. Like the stems the foliage is furnished with barbed, needle-like, stinging hairs. The stalked blooms are in terminal or axillary panicles, or solitary. They have five persistent sepals, white or greenish or reddish-orange, few to many stamens joined at their bases, and a style with five stigmas. The usually nodding, stalked, spherical to obovoid fruits are capsules with numerous tiny seeds.

Native of southwestern Texas and Mexico, ***Eucnide bartonioides*** is a much-branched, biennial or annual rarely over 1 foot tall. Its leaves are ovate, and toothed, lobed, or cleft. Few together at the branch ends, or from the leaf axils, the bright yellow, hypericum-like blooms, with showy brushes of stamens, open only in full sun; then their petals spread widely. They are on stalks about 1½ inches long that lengthen considerably as seeds are formed. In flower size this species is variable, the maximum width of the blooms being about 3½ inches. Closely related ***E. grandiflora,*** of Oaxaca and Guatemala, has flowers up to six inches across.

Garden Uses and Cultivation. These are delightful plants for the fronts of flower borders, rock gardens, and other places where they are assured full sun and thoroughly well-drained, preferably sandy, soil. They are summer bloomers and are cultivated by sowing seeds outdoors, in spring in cold climates, elsewhere in spring or fall, and, except for pulling out any surplus to avoid overcrowding, allowing the plants to remain undisturbed. Alternatively, seeds may be sown indoors in a temperature of 60 to 65°F some eight weeks before it is expected that it will be warm enough to transplant the youngsters to the garden. In the interim the seedlings are transplanted 2 inches apart in flats of porous soil, or individually to 3-inch pots, and are grown in a greenhouse in a night temperature of 50 to 55°F, with a daytime rise of a few degrees permitted until one week or ten days before they are to be set in the garden. Then they are moved to a cold frame or sheltered location outdoors to harden. The plants may be planted 9 inches to 1 foot apart. Usually no staking is required. Avoid excessive watering.

EUCODONOPSIS. See Achimenantha.

EUCOMIS (Eù-comis)—Pineapple Flower. Pineapple flowers are quite extraordinary summer-blooming bulb plants, suitable for outdoors in mild climates, for patio and terrace decoration, and for greenhouses and window gardens. They also supply unusual, long-lasting cut flowers. All except one of the about fifteen species are endemic to South Africa, the exception is native to tropical Africa. The genus belongs to the lily family LILIACEAE. Its name, from the Greek *eukomes*, beautiful headed, alludes to the tufts of bracts that crown the flower spikes.

Eucomises have spreading or vaselike rosettes of fresh green, sometimes wavy-edged, strap-shaped to strap-shaped-lanceolate leaves. From the centers of the rosettes come stout cylindrical or club-shaped flower stalks with thick, cylindrical, spike-like racemes of starry or somewhat cupped, side-facing or down-facing blooms. The spikes are capped, in the manner of crown imperials (*Fritillaria imperialis*), with a tuft of leafy bracts remindful of the topknot of a pineapple fruit, hence the common name pineapple flower applied to *Eucomis comosa* and sometimes other kinds. The bulbs are 3 to 6 inches in diameter. The flowers have six persistent petals (more correctly, tepals) united only briefly at their bases, six stamens, and a style with a rounded or minutely three-lobed stigma. The fruits are capsules with ovoid seeds.

Common pineapple flower (***E. comosa*** syn. *E. punctata*) is the most attractive in bloom. Its flower spikes, raised on cylindrical stalks with purple-spotted bases to

Eucomis comosa

heights of 2 or 3 feet, are more elegant than those of *E. undulata.* The white, pink- or wine-purple-tinged, blooms, on individual stalks 1 inch long, have spreading petals about one-half as long. The maturing seed pods become attractively wine-colored. The leaves of *E. comosa,* up to 2 feet long by 2 to 3 inches broad and not wavy-edged, are more or less spotted on their undersides with purple. Much like *E. comosa,* but with shorter-stalked flowers in more crowded, longer racemes, and flower stalks without spots, ***E. zambesiaca*** is a native of East Africa.

Leaves with markedly wavy margins, ovate-oblong, up to 2 feet long by 3 inches wide, somewhat longitudinally channeled, and not purple-spotted beneath are char-

Eucomis zambesiaca

Eucomis undulata

acteristic of ***E. undulata*** (syn. *E. autumnalis*). Its green or sometimes whitish, individually very short-stalked blooms have petals a little over ½ inch long, and are in dense, cylindrical-stalked spikes lifted to heights of 1 foot to 2 feet or sometimes more. Very similar ***E. bicolor*** is distinguished by its longer-stalked, green flowers having petals slightly over ½ inch long, clearly outlined with purple.

Broadly-strap-shaped leaves, up to 1½ feet in length, and club-shaped flower stalks are typical of ***E. regia.*** Its short-stalked, greenish-white or sometimes pink-tinged, sweetly-scented blooms, those toward the bottoms of the spikes nodding and those above horizontal, have petals about ½ inch long. The spikes attain a height of about 1½ feet. Their topknots are small and of membranous bracts.

Outstanding because of its size, but otherwise much resembling *E. undulata,* is ***E. pole-evansii.*** On stalks that raise them usually about 3 feet but sometimes 5 or 6 feet into the air, the racemes of blooms of this kind are up to 2 feet long. The flowers, larger than those of *E. undulata,* shallowly-bell-shaped, and whitish-yellow, are on individual stalks up to 2 inches long. The deeply channeled leaves, up to 2 feet long by 6 inches wide, have toward their bases deeply crisped margins. The green seed capsules are decorative.

Garden and Landscape Uses and Cultivation. Outdoors eucomises are suitable for flower beds and for colonizing at the fronts of shrub borders and elsewhere. In pots they can be used on porches, terraces, and similar locations, and in greenhouses and windows. Although they can-

Eucomis bicolor

not be regarded as hardy in the north, eucomises were successfully grown outdoors at the base of a southfacing greenhouse wall for many years at the New York Botanical Garden. In winter they were protected with a very heavy mulch of salt hay or dry leaves covered with roofing paper or polyethylene plastic. In mild climates eucomises are accommodating bulbs that succeed with little care in partial shade or sun where the ground is porous, well-drained, and nourishing. The bulbs are planted about 6 inches apart with their apexes 3 to 4 inches beneath the surface. Once established they remain undisturbed for many years. Transplanting can be done as soon as the foliage had died or in very early fall. Then, too, separation of offset bulbs can receive attention. During the period when they are in leaf fairly generous supplies of water are needed, at other times dry or dryish conditions satisfy. A yearly spring application of a complete garden fertilizer stimulates growth.

Pots and other containers for eucomises must be well drained. Repotting, needed by mature specimens at intervals of several years only, is done in late winter or spring just before new growth begins. Rich, porous soil is requisite, and the bulbs are planted, one in a 5-inch pot or three together in larger containers, with their tips slightly beneath or just showing at the surface. During the season of dormancy the earth is kept dry, at other times watering is done freely, and weekly or biweekly applications of dilute liquid fertilizer are supplied. Temperatures of 40 to 50°F are appropriate during the period of winter rest. After growth begins, 50°F by night and a few degrees more by day is adequate. Light shade from the strongest sun only is needed. Airy conditions sustained by ventilating the greenhouse freely on all favorable occasions are requisite. Increase is most commonly by removal of offset bulbs, but can also be accomplished by seeds, by bulb cuttings, and by using the leafy crowns of the flower spikes as cuttings.

EUCOMMIA (Eu-cómmia). A hardy tree that contains good quality rubber in all its parts would seem to be one that should be carefully investigated and perhaps commercially exploited. Investigated thoroughly *Eucommia ulmoides* has been, adapted to commercial rubber production, it has not been. First, because its rubber content is too low and difficult to extract, and second because it would be expecting too much to suppose that a tree growing in regions of short summers could equal the annual production of the great tropical rubber tree *Hevea brasiliensis,* which is active throughout the year. That *Eucommia* contains rubber is easily demonstrated, and this fascinates those shown it for the first time. It may be done by breaking a leaf or a fruit in two and pulling the parts slowly away from each other; fine strands of rubber latex are easily seen stretching from one part to the other. The same result may be had from a piece of inner bark. In its native China the bark of *Eucommia* is used medicinally, and for this the tree is planted; it is not known to exist in the wild. The name is from the Greek *eu,* well, and *kommi,* gum. It alludes to the rubber content of the plant.

The eucommia family EUCOMMIACEAE contains one species, ***Eucommia ulmoides.*** Its botanical relationship is confusing. It is generally considered to be most closely allied to the rose family ROSACEAE and to the witch-hazel family HAMAMELIDACEAE, but some botanists relate it to the elm family ULMACEAE. At one time it was considered to be a botanical ally of the magnolias. In appearance it rather resembles an elm, being a deciduous tree 60 feet or more in height with a branching trunk and a broad head of ascending branches. The pith of its twigs is not continuous, but consists of widely, distinctly spaced thin horizontal plates. Its glossy leaves are alternate, stalked, and sharply-toothed. They are pointed, elliptic to ovate or oblong-ovate and about 3 inches long. Their upper sides are hairless and rather crinkled, their under surfaces pubescent, at least along the veins when young. The flowers make no appreciable display. They are solitary and unisexual and are without sepals or petals. Males and females are on separate trees. The males have four to ten reddish brown stamens, the females consist of a two-lipped ovary with stigmatic surfaces on the inner sides of the lobes. The fruits are winged, one-seeded nutlets, 1 inch to 1¾ inches long and at maturity dark brown.

Garden and Landscape Uses. This vigorous tree, chiefly cultivated as a curiosity in botanical collections and hardy in southern New England, has merit as an ornamental because of its foliage, yet it has no special virtues to recommend it to the planter other than those possessed by many other kinds. It is most effective when standing alone and is suitable for a lawn specimen.

Cultivation. This tree thrives in a variety of soils, but makes its most satisfactory growth in fairly moist loam. It is easily raised from seeds and may be increased by summer cuttings planted in a greenhouse propagating bed or under mist.

Eucommia ulmoides (leaves and young fruits)

EUCOMMIACEAE—Eucommia Family. The only species that belongs here, *Eucommia ulmoides,* is a dicotyledon discussed under Eucommia.

EUCROSIA (Eu-cròsia). Four species of nonhardy deciduous bulb plants of Peru and Ecuador constitute *Eucrosia,* of the amaryllis family AMARYLLIDACEAE. Its name, derived from the Greek *eu,* good, and *krossos,* a fringe, alludes to the blooms.

Eucrosias have more or less paddle-shaped leaves and erect flowering stalks topped by umbels of pendulous, greenish to orange, narrowly-bell-shaped flowers with six petals or more properly tepals, six long-protruding, bent stamens, and a slender style. The fruits are capsules.

Native of Ecuador, ***E. morleyana*** has oblanceolate leaves 1 foot long or longer that develop when the bulbs are without leaves. The umbels of up to about ten blooms are on stalks 2 to 3 feet long.

Garden Uses and Cultivation. The rather peculiar-looking species described is of interest to collectors. Its needs are those of hippeastrums.

EUCRYPHIA (Eucrýph-ia). Americans who visit British gardens in late summer are often greatly and understandably impressed by the beauty of eucryphias. Although not very like and certainly not related to, these handsome trees are likely to conjure nostalgic thoughts of flowering dogwoods, chiefly because of the size and whiteness of their usually four-petaled blooms. Unhappily, eucryphias are little known in North America, over much of the continent because they are not winter-hardy or because summers are too hot for them. Even in the Pacific Northwest, where they flourish, they are rare. In the wild confined to temperate South America, Australia, and Tasmania, *Eucryphia* constitutes a family of its own, the EUCRYPHIACEAE, comprising five species. It is most closely related to the cunonia family CUNONIACEAE and has been included in the saxifrage family SAXIFRAGACEAE and the rose family ROSACEAE. The name comes from the Greek *eu,* well, and *kryphia,* hidden, in allusion to the manner in which in the bud stage the sepals cover the corolla.

Eucryphias are evergreen trees or shrubs, but in climates colder than that of its native habitat *E. glutinosa* loses most or all of its foliage in fall and renews it in spring. The leaves are opposite, leathery, undivided,

Eucryphia glutinosa

or pinnate. In their axils the solitary or sometimes paired blooms, looking like large single roses, stewartias, or white-flowered hypericums, are produced. The fruits are capsules that, when ripe, split into several segments, each with a few winged seeds.

Eucryphias with undivided leaves are *E. cordifolia*, *E. lucida* (syn. *E. billardieri*), and *E. milliganii*, and a hybrid between the last two. The first is distinguished by its leaves being toothed, and pubescent on their undersides. One of the least hardy, ***E. cordifolia,*** of Chile, attains heights of 130 feet, but cultivated specimens are much smaller. Its dullish dark green leaves, grayish-tan on their under surfaces, are ovate-oblong and 1 inch to 1½ inches long. About 2 inches across, the flowers sometimes have five or six, but more usually four petals, and numerous stamens with terra-cotta anthers. Tasmanian ***E. lucida*** is a slender tree sometimes 100 feet in height, but much smaller in cultivation. It has very short-stalked, narrowly-oblong leaves 1½ to 3 inches long by from a little under to a little over ½ inch wide; like the young shoots, they are resinous. Their upper sides are rich glossy green, beneath they are glaucous. Sweetly fragrant, pure white, and 1 inch to 2 inches wide, the nodding, four-petaled flowers are on ½-inch-long stalks. They have many yellow-anthered stamens. So closely related to *E. lucida* that some authorities regard it as merely a variety of it, ***E. milliganii,*** which in the main occurs at higher altitudes in Tasmania, differs chiefly in being smaller in all its parts and in its overall dimensions. Natural hybrids, probably not cultivated, between *E. lucida* and *E. milliganii* are named *E. hybrida*.

Pinnate leaves of three or more leaflets are typical of *E. glutinosa* (syn. *E. pinnatifolia*), *E. moorei*, and the hybrid *E. nymansensis*. The hardiest and largest flowered eucryphia, Chilean ***E. glutinosa,*** never exceeds 35 feet in height. Its leaves, clustered toward the shoot ends, have mostly five, sometimes three, lustrous green, elliptic to ovate, sharply-toothed leaflets 1½ to 2½ inches long and, at least when young, pubescent, especially on their undersides. Singly or in twos from near the ends of the shoots the 2½-inch-wide, four-petaled blooms are borne in considerable profusion. They have yellow anthers. Native of warm, humid regions in New South Wales, ***E. moorei*** attains heights up to 80 feet or more. It has leaves of usually five to eleven, but up to fifteen scarcely-stalked, bristle-tipped, narrowly-oblong, toothless, glossy leaflets ½ inch to 3 inches long and up to ⅝ inch wide. Their margins, and to a greater or lesser extent, their undersides are downy. The solitary four-petaled blooms are about 1 inch in diameter.

Hybrid eucryphias are very beautiful. Those between *E. glutinosa* and *E. cordifolia* are named ***E. nymansensis.*** Especially noteworthy is *E. n.* 'Nymansay', one of the original seedlings raised about 1915 at Nymans gardens in England with *E. glutinosa* as its seed parent, and originally designated Nymans A to distinguish it from other seedlings of the same cross referred to as Nymans B, and so on. Considerable variation, especially in freedom of flowering, is shown by forms of this hybrid. It is vigorous and evergreen, and has undivided leaves and leaves of three leaflets intermixed. Similar, but with mostly undivided leaves, *E. n.* 'Mount Usher' has as its female parent *E. cordifolia*. Other hybrids are *E. intermedia*, between *E. glutinosa* and *E. lucida*, and *E. hillieri* between *E. lucida* and *E. moorei*. Varieties of the last are *E. h.* 'Winton' and *E. h.* 'Penwith'.

Garden and Landscape Uses. Eucryphias are remarkably beautiful trees, or in cultivation tall shrubs, of refined appearance and great ornamental merit. They are suitable for planting as individual specimens, and in groups, and for mixing with other woody plants. They grow well in full sun or part-day shade, where the soil is well-drained, but not dry. In general they need acid soil, but *E. cordifolia* and *E. nymansensis* do quite well in limestone earths. It is desirable to test these plants more widely in parts of the United States where winters are not excessively severe. It seems likely that *E. moorei* hybridized with cooler climate kinds may give progeny adapted for growing in the south.

Cultivation. Little need be said here, because in congenial environments, once established, eucryphias grow with little trouble. Rather special care is needed in transplanting, since they resent undue root disturbance; this seems to be especially true of *E. glutinosa*. A mulch of organic material maintained around them is helpful. Any old, worn-out shoots and very weak ones should be pruned out in late winter or early spring. Propagation is by spring-sown seeds, by summer cuttings under mist or in a greenhouse propagating bench, preferably with slight bottom heat, and by layering.

EUCRYPHIACEAE—Eucryphia Family. The characteristics of this family of dicotyledons are those of its only genus, *Eucryphia*.

EUGENIA (Eugèn-ia)—Cherry-of-the-Rio Grande, Grumichama, Pitomba, Pitanga or Surinam-Cherry. Botanists are not in agreement as to how the vast number of species the more conservative consign to this genus of the myrtle family MYRTACEAE should be classified. Some treat them as belonging to several genera, and as the trend seems to be in that direction it is followed in this Encyclopedia with the result that certain species previously included in *Eugenia* are referred to *Acmena*, *Myrciaria*, and *Syzygium*. Even after this pruning, *Eugenia*, widely distributed in tropical and subtropical regions, consists of possibly 1,000 species. The name commemorates Prince Eugene of Savoy, a patron of gardening and botany, who died in 1736.

Eugenia uniflora (flowers)

Eugenias are nonhardy evergreen trees and shrubs, the vast majority natives of the Americas. They have opposite, short-stalked, undivided, generally glossy, feather-veined, almost always more or less hairy leaves. The solitary racemes of flowers usually come from the leaf axils. The blooms are white, cream, or red. They have turban-shaped calyxes with four lobes (sepals), four separate petals, numerous stamens, and a branchless style. The fruits are berries, with the calyx lobes attached to their apexes, containing smooth seeds.

Cherry-of-the-Rio Grande (***E. aggregata***) was brought to the United States early in the twentieth century mistakenly identified as *Myrciaria edulis*. A shrub or tree up to about 15 feet tall, it has glossy, narrow-elliptic leaves 2½ to 3 inches long, with short, grooved stalks. The white flowers, borne in pairs, have two ½-inch-long, ovate to heart-shaped bracts immediately beneath the blooms which usually fall when the fruits are picked. The thin-skinned fruits, which change from orange-red to purple-red as they ripen, are pleasantly flavored when eaten out of hand.

The grumichama (***E. brasiliensis*** syn. *E. dombeyi*), of Brazil, is a compact tree with glossy, leathery, ovate-oblong to obovate leaves 4 to 5 inches long by one-half as wide. In the wild it attains heights of 50 feet, in cultivation it is generally lower. The flowers are solitary and white. The long-stalked, soft-fleshed, pendent fruits, ½ to 1 inch in diameter, scarlet to purplish-black, are flattened-spherical to spherical, with at their apexes four big, persistent sepals. They contain one or two, sometimes more seeds. They are sweet and pleasantly flavored. This attractive species does not prosper in alkaline soil.

The pitomba (***E. luschnathiana***) is a good-looking Brazilian tree, especially handsome in fruit. At home attaining a height of about 30 feet, but often lower, it is compact in habit. About 3 inches in length, the elliptic-lanceolate, glossy leaves are paler on their under surfaces than above. The thin-skinned, soft, juicy, orange-fleshed, orange fruits are broadly-obovoid. They have four or five persistent sepals at their flattened apexes. They contain one to several seeds, are tartly aromatic, and are esteemed for jellies, jams, and sherbets. In Florida the pitomba has proved somewhat more tolerant of limestone than the grumichama. Like the last, it appreciates earth having a generous organic content.

The pitanga or Surinam-cherry (***E. uniflora***), of Brazil, is a heavily foliaged ornamental shrub or tree up to 25 feet tall that bears heavy crops of long-stalked, thin-skinned, juicy fruits esteemed for eating out of hand, for use in salads, and for jellies and sherbets. The fruits, shaped like tiny Chinese lanterns, are about 1 inch in diameter, light red to nearly black-red, subspherical, and more or less fluted with eight ribs. They are solitary or clustered, in flavor sweet to disagreeably resinous. The somewhat glossy leaves of this species, wine-red when young, are 1 inch to 2 inches long and ovate to ovate-lanceolate. The fragrant, white flowers, ½ inch in diameter, are solitary or in clusters of few.

West Indian and Floridian ***E. foetida*** (syn. *E. buxifolia*) is an erect shrub with closely set elliptic to broad-elliptic, pointed leaves ½ inch to 2 inches long, besprinkled with fine black dots on their undersides. The small white flowers, clustered on the leafless parts of the branches and on short, leafy branchlets, are succeeded by aromatic, black, ellipsoid to nearly spherical fruits. Another Floridian, ***E. confusa,*** the largest tree of the myrtle family

Eugenia uniflora (fruits)

Eugenia rubicunda

native to the United States, is becoming increasingly rare in the wild. At the beginning of the twentieth century specimens 50 to 60 feet tall existed in Florida. Now 30 to 45 feet seems to be about the maximum. This species has branches angled upward. The pointed leaves, considerably larger on young specimens than older ones, are at most about 3½ inches long, ovate to narrowly-elliptic or on young trees nearly round. The young foliage is red. The densely-clustered, minute white blooms are succeeded by spherical scarlet fruits up to ½ inch in diameter.

Native to India, ***E. rubicunda*** is a tall shrub or small tree with four-angled branchlets and nearly stalkless, privet-like, narrow-oblong to elliptic somewhat hairy leaves 1½ to 3 inches long. In clusters at the shoot ends, the flowers, a little over ¼ inch wide, have fleshy petals, convex and silvery-gray above, concave and brown on their undersides. The stamens are red, the style curiously zigzagged. The ovoid fruits are about ⅝ inch long. A medium-sized, round-headed tree native from South Africa to Uganda and Kenya, ***E. cordata*** has leathery, nearly stalkless, broad-elliptic to oblongish, hairless, somewhat glaucous leaves 2 to 4 inches long by about one-half as wide. Its white flowers are in many-flowered terminal panicles. The fruits are usually one-seeded and ½ inch long.

Garden and Landscape Uses and Cultivation. In frostless and essentially frostless climates, such as those of Florida, California, and more tropical places, eugenias are planted for ornament, some kinds for their edible fruits. Some make excellent sheared hedges. They are admired variously, according to species, for their habits, foliage, flowers, and decorative fruits. They are generally easy to grow in ordinary soil in sun or part-day shade. Propagation is by seeds and cuttings.

EULALIA. See Miscanthus.

EULOPHIA (Eu-lòphia). A considerable and variable genus of orchids, *Eulophia* is widespread nearly throughout the tropics and subtropics, the greatest concentration of its more than 200 species being in Africa. At least two there, *E. gigantea* and *E. horsfallii,* have flowering stalks up to 15 feet tall. Most sorts grow in the ground, a few perch on trees or rocks. Belonging to the orchid family ORCHIDACEAE, this genus has a name derived from the Greek *eu,* good, and *lophos,* a crest, in allusion to the crested lip of the flower.

Eulophias have fleshy corms, thickened rhizomes, or more or less prominent pseudobulbs. Their leaves are leathery and pleated. The erect flowering stalks bear few to many often showy blooms that open in succession from below upward. They have three spreading sepals, two similar petals, and a spurred, three-lobed lip nearly always with a prominent crest and with the side lobes encircling the column. Nearly related *Lissochilus* differs in its flowers having petals considerably bigger, broader, and differently colored than the sepals.

Natives of tropical Africa, *E. guineensis* and *E. quartiniana* are much alike in vegetative characteristics, but their flowers differ. They have clustered, ovoid to conical

Eugenia cordata

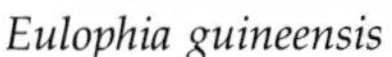

Eulophia guineensis

Eulophidium mackenii

pseudobulbs 1 inch to 1½ or perhaps 2 inches tall each with two to five pointed, narrow-elliptic leaves 1 foot to 1½ feet long and up to 4½ inches wide. The erect flowering stalks are 1 foot to 2 feet tall. Those of ***E. guineensis*** have greenish or brownish, fragrant blooms with three-lobed, white lips with red in the throat, or the middle lobe entirely red, and a prominent spur. The side lobes of the lip are rolled to form an open tube. The flowers of ***E. quartiniana,*** from ten to twenty on each stalk, have greenish-pink sepals and petals and a bright rosy-pink lip. They are 3 to 4 inches across. They have a larger, broader, more flaring lip than those of *E. guineensis.*

Native in damp soils from Florida to the West Indies, Mexico, and South America, ***E. alta*** has a large rutabaga-like corm from the end of which new roots and leaves develop. It has linear-lanceolate, pleated leaves up to 3 feet long by 4 inches wide. Irregularly spaced in quite impressive spikes 1½ to 5 feet tall, the green and maroon-red flowers, each, to use the imaginative imagery of one authority, with "the alert expression of a German shepherd with erect ears and panting tongue." Another Floridian, ***E. ecristata*** occurs also somewhat to the north and west of Florida and in Cuba. It favors drier, sandier soils than *E. alta.* It has a corm up to about 2 inches across and up to four linear-lanceolate, leaves up to 2½ feet long by 1½ inches wide. The flowering stalk, 1½ to 6 feet tall, terminates in a slender raceme of ten to thirty blooms, the yellow sepals and petals of which form a hood over the dark purple, deeply-three-lobed lip.

Eulophia quartiniana

Garden Uses and Cultivation. These are suitable for collectors of tropical orchids. The kinds here described and other terrestrial sorts are most likely to respond to conditions and care that suit *Phaius,* and *Calanthe;* skilled orchid growers, however, have reported difficulty in maintaining eulophias in cultivation. For more information see Orchids.

EULOPHIDIUM (Eulo-phídium). Native of the tropics of the Old and the New Worlds, *Eulophidium,* of the orchid family ORCHIDACEAE, comprises about thirty species. Its name is derived from that of the allied genus *Eulophia* and the Greek diminutive suffix *idium.*

Eulophidiums grow in the ground or rarely perch on trees. They have pseudobulbs from the apexes of which sprout a single, fleshy leaf. The erect stalks bearing few- to many-flowered racemes come from the bases of the pseudobulbs. Small, and with sepals and petals nearly similar, the flowers have a short-spurred, three-lobed lip and a column with an expanded base.

Native of Brazil and Paraguay, ***E. maculatum*** (syns. *E. ledienii, Eulophia maculata*) has ovoid pseudobubs 1 inch to 1½ inches long, and leathery, dark green leaves cross-banded with silvery-white or white

6 inches to 1 foot in length and 2 inches wide or somewhat wider. The flowers, which do not open widely, are about ¾ inch across, have pinkish-brown to wine-red sepals and petals usually prettily marked with white, and a pink to white lip with a short, nearly spherical spur. This species grows in the ground or on trees. South African *E. mackenii* inhabits heavily shaded forest floors. It has procumbent rhizomes with spaced along them nearly spherical to pear-shaped pseudobulbs about ½ inch in diameter. The variegated, nearly stalkless leaves spread horizontally. The green and yellow flowers are marked on the lip with brown lines.

Garden Uses and Cultivation. These are as for *Eulophia.* For further information see Orchids.

EULOPHIELLA (Eulo-phiélla). Endemic to Madagascar, *Eulophiella,* of the orchid family ORCHIDACEAE, consists of four species. Its name is a diminutive of that of another genus of orchids, *Eulophia.*

Eulophiellas are epiphytes, peculiar in that the species usually are highly selective of the trees they perch on. For instance, *E. roempleriana* it seems, occurs naturally only on a species of *Pandanus.* They do not take nourishment from the trees they inhabit.

Eulophiellas have vigorous, creeping rhizomes with rather distantly spaced pseudobulbs and long, narrow, pleated leaves. The sometimes branched, erect to horizontally inclined flowering stalks, from the bottoms of the pseudobulbs, carry many beautifully colored, fleshy blooms with rounded sepals and petals and a three-lobed, spurless lip.

Extraordinarily vigorous ***E. roempleriana*** (syn. *E. peetersiana*) has very thick rhizomes clothed with sheaths that fray into fibers as they age. The cylindrical to ovoid-spindle-shaped pseudobulbs, spaced along the rhizomes, are 3 inches to 1 foot long. They have up to eight leaves, narrowed to both ends, 2 to 4 feet long by up to 4 inches wide. The racemes or panicles together with their stalks are 3 to 5 feet long. Ordinarily they have fifteen to thirty flowers that open in succession. The fragrant, waxy blooms remain in good condition for a long time. From 3 to 4 inches across, they have blunt, broad-elliptic, concave, light pink to purple sepals and petals, the sepals paling toward their bases, and a rose-purple lip with a whitish throat and bright yellow disk with several erect crests.

Garden Uses and Cultivation. Eulophiellas are choice and rare. Considered somewhat difficult to grow, they need tropical, humid environments with good light, but sufficient shade to prevent the foliage from being scorched. They succeed in a mixture of tree fern fiber and sphagnum moss kept moist throughout the year and are best accommodated in large baskets made of wooden slats. Repotting should not be done oftener than necessary. A hybrid between *E. roempleriana* and *E. elisabethae,* named *E. rolfei,* is perhaps more amenable to cultivation than its parents. For more information see Orchids.

EULYCHNIA (Eulých-nia). Chile is homeland to the five species of *Eulychnia,* of the cactus family CACTACEAE. The genus name comes from the Greek *eu,* well or good, and *lychnouchos,* a lampstand (by analogy a candlestick, which the form of the plant suggests).

Eulychnias are shrubby or treelike. They have procumbent or erect, ribbed, spiny stems, with one spine of each cluster very much longer than the others. The broadly-bell-shaped flowers have a short calyx tube with many scales, in the axils of which are bristle-like spines or long hairs, and erect or somewhat spreading short perianth segments. There are many stamens and a stout, short style. The fruits are fleshy and spherical.

Branching freely and 6 to 15 feet tall, ***E. spinibarbis*** has 3-inch-thick stems with twelve or thirteen ribs. The spines are in clusters of about twenty, with the longest one up to 6 inches, the others of varying lengths with most very much shorter than the middle one. The whitish to pinkish flowers are 1¼ to 2 inches long. From the last, ***E. acida*** differs in the brown, woolly hairs in the axils of the bracts of its flowers and in the fruits being very short instead of long. Up to 10 feet tall, ***E. ritteri*** has stems 2 to 3½ inches thick clothed with spines sprouting from tufts of woolly hair in clusters of about twelve brownish radials ⅓ to ⅔ inch long and one to four black centrals 1¼ to 2¾ inches long. The pink flowers are ¾ inch long and nearly as wide.

Eulychnia ritteri

Garden and Landscape Uses and Cultivation. Eulychnias respond to conditions that suit most columnar cactuses. For additional information see Cactuses.

EUODIA. See Evodia.

EUONYMUS (Eu-ónymus)—Strawberry Bush, Wahoo, Spindle Tree. The name of this genus is sometimes spelled *Evonymus.* It designates a nearly cosmopolitan group of about 175 species, the greatest number natives of temperate Asia. It belongs to the staff tree family CELASTRACEAE. The name *Euonymus* is an ancient Greek one.

Euonymuses are deciduous and evergreen small trees, shrubs, or more rarely stem-rooting trailers or climbers. The foliage of some kinds colors brilliantly in fall. Most have four-angled stems, in a few kinds the angles are furnished with conspicuous corky wings. Undivided, toothed or toothless, the leaves are opposite, very rarely whorled, or as in *E. nana,* alternate. The small, inconspicuous flowers are pale green to purplish. They are in stalked clusters of three, seven, or fifteen arising from the leaf axils. Each has four or five sepals, petals, and short-stalked stamens. The three- to five-lobed stigma may top a short style, or the style may be lacking. The more or less pendulous fruits are three- to five-lobed capsules that split longitudinally to reveal the white, red, or black seeds, one to each lobe, each surrounded by a bright orange-red or scarlet covering called an aril.

The strawberry bush, burning bush, or skewerwood (***E. americana***) is a sparsely-branched, deciduous shrub native from New York to Florida and Texas, and hardy in southern New England. Up to about 8 feet tall, it has smooth shoots and pointed, round-toothed, ovate-lanceolate to lanceolate leaves 1 inch to 3½ inches long that turn red in fall, but less intensely so than those of *E. alata.* Solitary, in pairs, or in threes, on slender stalks, the greenish to reddish-green flowers have yellow anthers. The ½-inch-wide, warty, pinkish to scarlet fruits contain brown seeds with scarlet coats. Running strawberry bush (***E. obovata***), native from Ontario to Kentucky, has procumbent or prostrate rooting stems and smooth shoots. About 1 foot tall, it has deciduous, finely toothed, pointed or bluntish, obovate leaves 1½ to 3 inches long. One to three on slender stalks, the flowers are greenish-purple. Their anthers are yellow. The crimson, warty, three-lobed fruits are ½ inch across. They contain seeds enclosed in scarlet arils.

Euonymus americana, in fruit

The wahoo or burning ash (***E. atropurpurea***), another native American, is a not very shapely deciduous shrub or tree up to 25 feet tall. Native from New York to Florida and Texas, and hardy through most of New England, this has pointed, finely toothed, elliptic leaves, 2 to 5 inches long, that become pale yellow in fall and are hairy on their under surfaces. The purple-anthered, purple flowers are seven or more together in slender-stalked clusters. About ½ inch wide, the deeply-four-lobed, pink to scarlet fruits, somewhat over ½ inch across, enclose scarlet-coated, brown seeds.

The European spindle tree (***E. europaea***), native of Europe and western Asia and naturalized to some extent in eastern North America, is hardy as far north as southern Canada. Long a garden favorite, especially in its brighter-colored fruiting varieties, this is an erect, deciduous shrub or bushy-headed small tree up to about 20 feet tall. It has pointed, elliptic-ovate to lanceolate-oblong, round-toothed leaves 1 inch to 3½ inches long that become reddish in fall. Yellowish-green, a little over ⅓ inch across and with yellow anthers, the flowers are in slender-stalked clusters of three to seven. The deeply-four-lobed, bright red to pink fruits, ½ to ¾ inch wide, contain white seeds with orange coats. Varieties *E. e. aldenhamensis* and *E. e. intermedia* have showier, more highly colored fruits than the typical species. Those of the first are brilliant pink, those of the other bright red. The fruits of *E. e. alba* are white. Not over 3 or 4 feet tall, *E. e. pumila* is of compact habit.

Two deciduous endemics of southern Europe and Asia are *E. verrucosa* and *E. latifolia*. The first is as hardy as *E. eruopaea*, the other not reliable in climates harsher than that of southern New England. Erect and up to about 6 feet tall, ***E. verrucosa*** has conspicuously warty shoots. Its short-stalked, ovate to lanceolate-ovate leaves are 1 inch to 2¼ inches long, pointed, with round-toothed margins. The ¼-inch-wide, brownish flowers are usually in threes. Deeply-four-lobed and ¼ inch wide, the fruits are yellowish-red. The black seeds are only partly enclosed by an orange coat. One of the earliest euonymuses to leaf out in spring, ***E. latifolia*** is a shrub or small tree up to 20 feet tall. Its toothed, obovate-oblong to elliptic leaves, 2¼ to 4½ inches in length, turn red in fall. Greenish and a little in excess of ⅓ inch wide, the flowers are in clusters of seven to fifteen. About ¾ inch across, the bright red, pendulous, five- or less often four-winged fruits contain white seeds surrounded by an orange aril.

The cork-winged euonymus (***E. alata***) shares with *E. americana* and *E. atropurpurea* the common name burning bush. Its brilliant crimson and scarlet fall foliage well entitles it to the designation. Native to China and other parts of northeast Asia, this has the welcome distinction of being immune to scale insects. A deciduous, stiff-branched shrub up to 10 feet tall, it has branches with two or four usually broad, corky wings. Its elliptic to obovate, finely-sharp-toothed leaves are 1 inch to 2 inches long. Yellowish and about ¼ inch wide, the flowers are usually in short-stalked clusters of three. They are succeeded by fruits divided almost to their bases into usually four, sometimes fewer, purplish pods up to ⅓ inch long containing brown seeds covered with an orange-red aril. Variety *E. a. aptera* has wingless or slightly winged branches. Intermediates between it and the typical species occur. Most excellent *E. a. compacta* forms a dense, rounded bush, its branches with less conspicuous

Euonymus alata showing corky winged branchlets

Euonymus europaea aldenhamensis

A hedge of *Euonymus alata compacta*

corky wings than those of typical *E. alata.*

A deciduous shrub or tree up to 40 feet tall, ***E. hamiltoniana*** (syn. *E. sieboldiana*) is a native of the Himalayas. It has elliptic, oblong, or ovate-elliptic, finely-toothed leaves 2½ to 4½ inches long. The flowers have four petals and purple anthers, the fruits are yellow. More common in cultivation than the species are some of its varieties: *E. h. lanceifolia* (syn. *E. lanceifolia*), a shrub or tree up to 30 feet tall, hardy in southern New England, has elliptic, elliptic-oblong, or lanceolate-oblong, pointed leaves 3½ to 5 inches long, with round-toothed margins. The flowers, ½ to ¾ inch across, are in clusters of seven to fifteen. Pale pink and four-lobed, the fruits contain crimson seeds with orange coats open at their apexes. *E. h. maackii* (syn. *E. maackii*), of China, Korea, and Manchuria, hardy through most of New England, has elliptic-oblong to lanceolate-oblong leaves 2 to 3½ inches long and yellowish-white blooms, ⅓ inch or a little more in diameter, in six- or seven-flowered clusters. The deeply-four-lobed, ⅓-inch-wide, pink fruits have red seeds with orange coats. *E. h. nikoensis* (syn. *E. nikoensis*), endemic in Japan, is distinguished by its leaves being broader than those of the species and its bright red fruits having orange-coated, green seeds. *E. h. yedoensis* (syn. *E. yedoensis*), of China and Japan, is a flattish-topped shrub up to 12 feet tall with short-pointed, obovate to obovate-oblong or elliptic leaves 2¼ to 4½ inches in length, with finely-round-toothed margins, that in fall turn brilliant red. In rather crowded, many-flowered clusters, the blooms are yellowish to whitish. The pinkish-purple fruits, under ½ inch wide, are deeply-four-lobed. The arils surrounding the seeds are orange.

Two more deciduous east Asian species of merit are *E. bungeana* and *E. sanguinea.* A fast-growing, slender branched shrub of loose habit or tree up to 20 feet tall, ***E. bungeana*** is hardy through most of New England. It has pointed, elliptic-ovate to elliptic-lanceolate, toothed, light green leaves up to 4½ inches long. In clusters of three to seven, the purple-anthered, yellowish flowers are succeeded by long-lasting, deeply-four-lobed, yellowish to pinkish-white fruits with white or pinkish seeds that have orange coverings usually open at their apexes. Variety *E. b. semipersistens,* where climates are not very severe, is semievergreen. Its fruits remain attractive for a long time. Variety *E. b. pendula* has drooping branches. Its leaves appearing as early in spring as those of *e. latifolia,* purplish at first becoming green as they develop, ***E. sanguinea*** is a handsome-foliaged shrub or small tree up to 20 feet high, and hardy in southern New England. It has nearly cylindrical branchlets, broad-elliptic to ovate or oblong-ovate, pointed leaves up to 4½ inches long, often reddish on their undersides and with frilly-toothed margins. The flowers, their parts in fours or sometimes fives, in loose, slender-stalked clusters of three to fifteen, are purplish.The scarcely lobed, winged, 1-inch-wide, red fruits contain black seeds surrounded by orange arils. Unfortunately they are usually not abundant nor do they last long before falling.

Other sorts cultivated include these: ***E. nana*** is a deciduous, low, usually procumbent shrub with branches up to 3 feet tall. It differs from all other cultivated euonymuses in having alternate or whorled (in circles of more than three), very rarely opposite, leaves. Native from the Caucasus to western China, and extremely hardy, this sort has linear to linear-oblong leaves ¾ to 1½ inches long, toothless or remotely-toothed and with rolled-back margins. The brownish-purple, slender-stalked blooms, solitary or in threes, are succeeded by pink, four-lobed fruits containing brown seeds with orange coats. ***E. oxyphylla,*** hardy in southern New England, is a shrub or small tree, native of Japan. Its leaves, up to 3½ inches long, are pointed, rounded to broadly-wedge-shaped, margined with incurved teeth. The brown-tinged, purplish, ⅓-inch-wide flowers are many together in loose clusters. Subspherical and pendulous, the fruits have five or four ribs. Red, and about ⅓ inch across, they contain red-coated seeds. ***E. sachalinensis*** (syn. *E. planipes*), a northeastern Asian shrub up to 12 feet tall, is hardy in southern New England. It has short-pointed, round-toothed, obovate leaves up to 4½ inches long. Its yellowish flowers are many together in slender-stalked, loose clusters. The five-angled fruits, without obvious wings, are showy. About ¾ inch across, they are borne in profusion. ***E. wilsonii*** is a western Chinese climbing shrub up to 18 feet in height, which is not hardy in the north. It has toothed, lanceolate leaves 2½ to 6 inches long. The flowers, yellowish, are in clusters of many. The four-lobed fruits, covered with short spines, are ¾ inch across. Their seeds have yellow coverings.

An evergreen, erect, leafy shrub or small tree up to 15 feet tall or occasionally taller, Japanese ***E. japonica*** is represented in gardens by several excellent varieties. Not hardy in the north, they are popular in the south and are sometimes grown indoors in pots. They have very slightly angled branches and glossy, opposite, broadly-elliptic to obovate or narrowly-obovate, blunt-toothed, blunt-ended to pointed leaves, those of the typical species 1¼ to 2¾ inches long. In short-stalked clusters of five or more, the small, greenish-white blooms come in high summer.

Euonymus sachalinensis

Euonymus japonica albomarginata

Euonymus kiautschovica (leaves and flowers)

The pinkish capsules, spherical and about 1/3 inch wide, contain orange-coated seeds. The following are among the most popular varieties: *E. j. albomarginata* (syn. 'Pearl Edge') has narrowly-white-edged leaves. This may be the plant grown as 'Silver Queen'. *E. j. aureo-marginata* has leaves edged with yellow. This may be the plant grown as 'Giltedge'. *E. j. aureo-variegata* (syn. 'Gold Spot') has leaves blotched with yellow. *E. j. fastigiata*, columnar, has narrow-elliptic leaves. *E. j. macrophylla* has exceptionally large leaves. The sort grown as *E. j. grandifolia* may be this. *E. j. microphylla* has boxwood-like, narrow-oblong to lanceolate leaves from under 1/2 to 1 inch long. *E. j. m. variegata* is like the last, but its leaves are splashed with white. *E. j. pyramidata*, much more compact than *E. j. fastigiata*, is also columnar. *E. j. viridi-variegata* (syn. *E. j.* 'Duc d'Anjou') has large bright green leaves, their centers relieved with green and yellow.

Euonymus kiautschovica with *Taxus cuspidata* and *Vinca minor* in front

Evergreen and hardy about as far north as New York City, although it does not fruit there and its foliage usually suffers some winter damage, ***E. kiautschovica*** (syn. *E. patens*) is popular in milder regions. A shrub up to 10 feet tall, native to China, this has thinner leaves than *E. japonica* and its varieties. They are broad-elliptic to obovate and 2 to 3 inches long and have roundish-toothed margins. In loose clusters, the greenish-white flowers are displayed in late summer or early fall. The spherical fruits, 1/2 inch wide, are pink. They contain pinkish seeds with orange coats. Reported to be hybrids of *E. kiautschovica*, varieties 'Dupont' and 'Manhattan' are probably a little less hardy than *E. fortunei*, but survive in southern New England. They have excellent glossy green foliage. Their leaves are about 2 1/2 inches long.

An evergreen climber or trailer, or in some varieties a bushy shrub, Chinese ***E. fortunei*** has densely-minutely-warted, rooting stems that may climb to a height of 20 feet. Its shoots are nearly round. Pointed and thicker than those of *E. kiautschovica*, its roundish to elliptic leaves are 1 inch to 2 inches long. Their veins are whitish. Greenish and with four each sepals and petals, the flowers, in clusters of five or more, are rarely borne except by bushy forms. The pinkish capsules, about 1/2 inch across, are spherical. The pink seeds have orange coats. The following

Euonymus fortunei

Euonymus fortunei (flowers)

Euonymus fortunei colorata as a groundcover

Euonymus fortunei colorata clothing a bank

Euonymus fortunei gracilis

Euonymus fortunei minima climbing a low wall

Euonymus fortunei kewensis

Euonymus fortunei vegeta

varieties are recognized: *E. f. carrierei* is a handsome, freely-fruiting, broad shrub that, if support is available, tends to climb. It has glossy, somewhat pointed, elliptic-oblong leaves 1½ to 2 inches long. *E. f.* 'Berryhill' is an upright variety with leaves 1½ to 2 inches long. *E. f. colorata,* called purple-leaf wintercreeper, is a vigorous trailer with erect branches up to 1 foot or so tall and lustrous, toothed, pale-veined ovate to elliptic leaves 1½ to 2 inches long that change in fall to shades of purple or pinkish-purple. *E. f. gracilis* is a variable trailing kind with leaves edged and variegated with white or yellow, sometimes with pink at the margins. *E. f. minima* is a creeper or sometimes low climber with leaves only ¼ to slightly over ½ inch wide. A particularly small-leaved expression of this variety known as *E. f. kewensis* is cultivated. *E. f. radicans,* native to Japan and Korea, is a trailer or climber with pointed to bluntish, ovate to broad-elliptic leaves quite evidently toothed, ½ inch to 1¼ inches long. *E. f. reticulata* is a trailer or climber with white-veined foliage. *E. f.* 'Sarcoxie', about 4 feet tall, is an upright shrub with lustrous, 1-inch-long leaves. *E. f. vegeta* is a bushy shrub up to 4½ feet tall that, if supported, climbs. It has dull green, broad-elliptic to nearly round leaves 1 to 1¾ inches long. It fruits freely.

Less hardy than the evergreen sorts just described, ***E. lucida*** of the Himalayan region is a large shrub or tree up to nearly 50 feet tall with opposite, elliptic to lanceolate, finely-toothed, glossy, leathery leaves 2 to 5 inches long by ¾ inch to 1½ inches wide, bright crimson when young, later coppery-salmon, and finally dark green. This sort is hardy in mild climates only.

Garden and Landscape Uses. According to kind, these accommodating, easy-to-grow plants, not choosy about soil or location, serve many useful purposes. This despite the fact that most kinds, the winged-bark euonymus an outstanding exception, are subject to severe, if uncontrolled often devastating infestations of scale insects. Euonymuses are esteemed for their forms, foliage, and fruits. Their flowers make no appreciable show. They grow well near the sea and in city gardens. The deciduous shrubs and small trees, several of which color brilliantly in fall, are general purpose landscape furnishings, well adapted for shrub borders, screens, and similar usages. Low-growing *E. alata compacta* is a splendid hedge plant. It needs shearing only every other year to maintain it as a broad, dense divider or background. In fall its foliage turns fiery red. The upright evergreens are attractive foliage shrubs for beds, borders, foundation plantings, and hedges, as single specimens, and for container plants. Other evergreen kinds lend themselves to espaliering against walls and fences. In selecting euonymuses, consideration must be given not only to their forms, foliage, and similar characteristics, but also to the relative hardiness of the various kinds.

Cultivation. But little care is needed to maintain euonymuses. Chief concern is to keep under control scale insects, which if allowed to flourish unchecked, encrust the branches and twigs as well as infest the foliage. A dormant spray applied before new growth begins in spring is most effective. Neglected specimens, heavily infested, are best treated by cutting out and burning the most infested branches and

Euonymus fortunei vegeta (fruits)

Euonymus lucida

then spraying. In extreme cases destruction of the entire plant may be the only practicable procedure. When grown as hedges, espaliers, or formal specimens in containers, periodic shearing is necessary. Container specimens need well-drained, nourishing soil kept evenly moist, but not saturated. If they are well rooted applications of dilute liquid fertilizer made at two- or three-week intervals from spring to fall are beneficial. In the north container specimens of *E. japonica* and its varieties can be wintered in a cool, light, frostproof cellar or similar place.

EUPATORIUM (Eupat-òrium)—Mist Flower, Boneset, Snakeroot, Joe Pye Weed. This immense genus of well over 1,000 species belongs to the daisy family COMPOSITAE. Most of its members are hardy or nonhardy herbaceous perennials or subshrubs or nonhardy shrubs or small trees. A few are annuals. The chief centers of distribution of *Eupatorium* in the wild are Mexico, the West Indies, and tropical South America, but some kinds are indigenous in the United States and Canada, and a few in Europe, Asia, and North Africa. The name commemorates an ancient King of Pontus, Mithridates Eupator, credited with discovering that one kind of eupatorium was an antidote for poison.

Closely related to *Ageratum,* from which it differs in the pappus (attachments to the ovary and achenes) being of bristles instead of scales or absent, *Eupatorium* most commonly has opposite leaves. Its small flower heads are of all disk-type florets (similar to those that form the central eyes of daisies). They are without petal-like ray florets. The heads are crowded in flat- or round-topped, more or less panicled clusters, often of large size. Most often they are white, but not infrequently they are pink, rosy-purple to nearly red, mauve, or blue. Each flower head generally consists of five or more bisexual florets. The fruits are seedlike achenes.

Mist flower (***E. coelestinum***) is a hardy herbacous perennial with beautiful clear blue to heliotrope or rarely white ageratum-like flowers in large, dense, showy clusters. Wild in moistish soils in woodlands, meadows, fields, and along stream banks from New Jersey to Illinois, Florida, and Texas, this late summer and fall bloomer is 1 foot to 3 feet tall. Somewhat pubescent, it has opposite, stalked, bluntish-triangular-ovate, coarsely-toothed, thinnish leaves, somewhat heart-shaped at their bases, and 1½ to 3 inches long. Nevada to Washington and California is home to ***E. occidentale*** a finely-hairy to hairless species up to 1½ feet tall. This has more or less toothed, ovate leaves up to 2 inches long, with three chief veins. Its clusters of blooms are of pink or sometimes white flower heads.

Three of the best hardy, usually white-flowered herbaceous perennial kinds are common boneset (*E. perfoliatum*), upland boneset (*E. sessilifolium*), and white snake root (*E. rugosum,* syns. *E. urticaefolium, E. ageratoides, E. fraseri*). Distinguished by the bases of its pairs of stalkless leaves being united and encircling the stems, ***E. perfoliatum*** is 2 to 5 feet tall, branched above, with wrinkled, long-pointed, lanceolate, toothed leaves up to 8 inches long. Its grayish-white to sometimes bluish-purple flower heads, of ten to sixteen florets, are in crowded clusters. This is native from New Brunswick to Florida and Texas. Found mostly in fertile, limestone soils, from Vermont to Indiana, Georgia, Alabama, and Missouri, ***E. sessilifolium*** differs from *E. perfoliatum* in its flower heads being of about five florets. From 2 to 6 feet tall, this sort has usually opposite, stalkless or nearly stalkless, thinnish, long-pointed, lanceolate, bright green, toothed leaves 4 to 8 inches long and up to 1½ inches wide. The flowers are in dense clusters. Hairy to hairless and 2 to 4 feet tall, ***E. rugosum*** has branching stems, and pointed-ovate, sharp-toothed, stalked leaves 3 to 6 inches long. Its pure white flower heads of six to eight florets are in ample clusters. This is native from Quebec to Florida and Louisiana.

Other white-flowered hardy herbaceous perennials include ***E. album,*** which prefers dry, open woodlands and occurs wild from southern New York to Florida and Arkansas. From 1½ to 3 feet tall, this kind has nearly stalkless, opposite, elliptic to ellip-

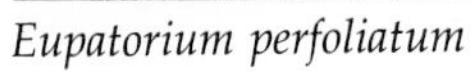
Eupatorium perfoliatum

Eupatorium rugosum (flowers)

tic-lanceolate, toothed, glandular-hairy leaves up to 4½ inches long and flattish clusters of flower heads, each of five florets. It grows in poor, sandy soil. Similarly adapted to sandy soils of low fertility, ***E. aromaticum*** resembles white snakeroot, from which it can be distinguished by its shorter-stalked, less pointed, blunter-toothed leaves. It chiefly inhabits dry woodlands from Massachusetts to Ohio, Florida, and Mississippi. Variety *E. a. melisoides* has smaller, nearly stalkless leaves.

Joe Pye weed is used as a common name for several related eastern North American species, particularly *E. maculatum, E. dubium,* and *E. purpureum.* These are coarse, hardy herbaceous perennials 6 to 10 feet tall. The two first favor swamps and other moist soils, ***E. maculatum*** especially likes those of a limestone character. Drier soils in open woodlands and similar places are the preferred habitats of ***E. purpureum.*** These kinds ordinarily have relatively broad, lanceolate to ovate or elliptic leaves

Eupatorium rugosum

Eupatorium purpureum

Eupatorium ligustrinum

Eupatorium atrorubens

in whorls (circles of more than two), and rose-purple to pale purple or occasionally nearly white flowers in flattish or rounder-topped panicles. Strongly-three-veined leaves up to 6 inches long are characteristic of ***E. dubium.*** The other two have leaves with one prominent mid-vein, those of *E. maculatum* up to 8 inches, those of *E. purpureum* up to 1 foot long. The flower heads of *E. maculatum* are of nine to twenty-two, those of *E. purpureum* of five to seven, florets. The hemp-agrimony (*E. cannabinum*) of Europe is similar to, but less ornamental than American Joe Pye weeds.

Shrubs usually with white flowers and not hardy in the north will now be considered. Native from Mexico to Central America, ***E. araliaefolium,*** which has oblong, leathery, pinnately-veined leaves with stalks up to 2 inches long and blades up to 8 inches long by 3½ inches wide is, except among its panicles of blooms, hairless. The white to pink or purple flower heads are of about twenty-five florets. In the wild this often is epiphytic (perching on trees rather than rooting in the ground, but taking no nourishment from the host). Variable ***E. ligustrinum*** (syns. *E. micranthum, E. weinmannianum*) is handsome and native from Mexico to Costa Rica. From 3½ to 15 feet tall, it has thickish, short-stalked, elliptic-lanceolate, few-toothed, pinnately-veined leaves 1½ to 3½ inches long by ¾ inch to 1½ inches wide, usually with tiny glistening glands on their undersides. Its large, round-topped clusters of white flowers have pink or pink-tinged pappuses (tufts of hairs attached to the ovaries). A much-branched, pure white-flowered shrub 6 to 8 feet tall, ***E. glandulosum,*** of Mexico, has opposite, thin, long-stalked leaves, scarcely-hairy above, and sticky-hairy beneath, especially along the veins. They are coarsely-toothed and triangular with wedge-shaped bases. The flower heads, loosely clustered on glandular-pubescent stalks, are fragrant.

Purple-, violet-, blue-, or pink-flowered shrubs or subshrubs not hardy in the north are *E. sordidum* (syn. *E. ianthinum*), *E. atrorubens* (syn. *E. raffillii*), *E. glabratum,* and *E. macrophyllum.* All are handsome. Its young shoots densely clothed with tawny to reddish, woolly hairs, Mexican ***E. sordidum*** has toothed, ovate to ovate-oblong leaves with five main veins originating well above their bases. They have blades up to 4 inches long, and nearly as broad, and stalks up to 2 inches long. The crowded clusters of fragrant, violet flower heads are up to 3½ inches in diameter. Closely related and one of the most splendid eupatoriums is ***E. atrorubens,*** of Mexico. Stout-stemmed, but not very woody, this kind is 2 to 3 feet tall. Its stems and to a lesser extent its foliage are covered with red or purple, spreading, velvety hairs. Up to 1 foot long and more than two-thirds as wide as long, the stalked, opposite leaves are pointed-ovate, and purple on their undersides. Its numerous reddish-purple to lavender-blue or pinkish-lavender flower heads are in crowded, terminal clusters up to 1 foot across. A nearly hairless shrub with sticky shoots and foliage, ***E. glabratum,*** of Mexico, is 3 to 8 feet tall. It has more or less angled, erect branches, and ovate to oblong-lozenge-shaped, finely-toothed or toothless leaves, wedge-shaped at their bases. They are paler beneath than on their upper surfaces and 1 inch to 2½ inches long by about one-half as wide. The flower heads, of fifteen to eighteen florets, are light pink. Native of the West Indies and South America, ***E. macrophyllum*** is a sub-shrub 3 to 5 feet tall. It has tawny-pubescent, pointed-ovate to broad-heart-shaped leaves with blades

Eupatorium macrophyllum

4 to 9 inches long with round-toothed margins. The purplish, bluish-lilac, whitish, or greenish flowers are in large, showy heads.

White-flowered nonhardy subshrubs, their lower parts more or less woody, their upper parts herbaceous, include the following kinds. Up to 6 feet in height, and wild from Mexico to Guatemala, is ***E. conspicuum*** (syn. *E. grandifolium*). This sort has six-sided branchlets and thin, pointed-triangular-ovate, toothed leaves, with stalks, winged at their summits, up to 4½ inches long and blades 3½ to 5 inches long and wide. The almond-scented flower heads of about forty florets are in loose terminal and subterminal clusters often shorter than the surrounding leaves. From Chile, Ecuador, and other temperate parts of South America comes ***E. glechonophyllum,*** a subshrub or herbaceous perennial with flexuous stems and opposite, slender-stalked, triangular-ovate leaves of varying sizes, mostly ½ to 1 inch long, but sometimes longer. They are slightly pubescent and coarsely-toothed. The loose clusters of bloom are of slightly pink-tinged, white flower heads. Mexican ***E. riparium*** is desirable. It has flexuous, cylindrical, reddish, finely-hairy, scarcely woody stems up to about 2 feet tall and opposite leaves pointed toward both ends, toothed above their middles, and with three main longitudinal veins. They are 2 to 4 inches long by up to 1 inch wide, and slightly hairy along the veins. The numerous small flower heads are in freely produced panicles. Also believed to be Mexican, ***E. vernale*** is a subshrub or perhaps sometimes a shrub. Its stems and foliage are clothed with loose hairs. The leaves are coarsely-toothed, oblong-ovate, and up to 5 inches long. Their under surfaces are grayish-hairy. Its pure white flower heads are in good-sized clusters.

Garden and Landscape Uses. Of the hardy eupatoriums described here the most commonly used in gardens are the mist flower and white snakeroot. These well deserve places in flower beds and borders and are useful as cut blooms. The other hardy kinds for the most part are better suited for naturalistic areas, wild gardens, and native plant gardens. None is difficult to grow. Outdoor cultivation of the nonhardy shrubs and subshrubs is limited to regions where little or no frost occurs. Most prefer dryish to highly humid climates. These shrubs and subshrubs also make very satisfactory plants for growing in pots in greenhouses, where they thrive with minimum care.

Cultivation. Eupatoriums are satisfied with any fairly good soil, damp or dryish, depending upon the needs of different species. The hardy herbaceous perennials are easily propagated by division in spring or early fall, and by seeds. The shrubs and subshrubs may be raised from seeds, but cuttings afford a more usual means of propagation. Outdoors in suitable climates these kinds are useful in flower beds, shrub borders, and the larger ones, as lawn specimens.

To have good specimens in pots in greenhouses, take cuttings in late winter, spring, or summer depending upon the size of plants desired, and root them in a propagating bench in sand, vermiculite, or perlite. Pot the rooted cuttings individually in small pots and successively into bigger ones as root growth makes necessary. As finals they may occupy containers 5 to 8 inches in diameter. During their early stages pinch out two or three times, at intervals, the tips of the shoots of such slender-stemmed kinds as *E. ligustrinum, E. conspicuum, E. riparium,* and *E. vernale,* but not of stout-stalked *E. atrorubens* and *E. sordidum.* It is better to let these grow without pinching, and allow each stem to carry one large cluster of bloom. Use rich, porous soil and water generously. After the pots in which the plants are to be bloomed are filled with roots supply dilute liquid fertilizer at weekly intervals. In summer the plants can be stood in a cold frame or outdoors with their pots buried almost to their rims in a bed of sand or similar material. Bring them indoors before frost. Fall and winter temperatures indoors, should for most kinds be at night 45 to 50°F, but 55°F better suits *E. atrorubens* and *E. sordidum.* Day temperatures five to ten degrees higher than those maintained at night suffice. On all favorable occasions ventilate the greenhouse freely, and keep the air moderately humid. After flowering is through, greenhouse specimens are cut lightly back and rested somewhat by keeping them cooler and rather drier until about a month before cuttings are needed. Then they are pruned further back, fertilized, put into a warmer place, and watered more freely. Plants of *E. atrorubens, E. ligustrinum,* and *E. sordidum* can be grown on a second year to make big specimens. To do this, cut them back fairly severely in late winter, shake some of the old soil from their roots, repot them, and restart the growth cycle as with new plants raised from cuttings. Good standard (tree-shaped) specimens of *E. ligustrinum* can be had by allowing a rooted cutting to grow without pinching until it is 2 to 3 feet high, removing all side shoots as soon as they appear. When the trunk-to-be is of the desired height pinch out its tip and let branches develop from near its top. Pinch the tips out of these and of the secondary branches that develop when they are 5 or 6 inches long, and repeat this until a good head has formed. Be sure to cease pinching long enough before flowering time to allow flowering branches to develop.

EUPHORBIA (Eu-phòrbia)—Poinsettia, Crown-Of-Thorns, Snow-On-The-Mountain, Spurge. The spurge family EUPHORBIACEAE takes its name from this remarkable genus of plants, which belongs to it and accounts for about 2,000 of its 5,000 species. Euphorbias occur as natives throughout most of the world, but are most abundant as to species in the tropics and subtropics. They include some exceedingly familiar garden plants, many lesser known ones, and a vast array of kinds not cultivated, many of which are of no ornamental merit and some of which are more or less weeds, although rarely pestiferous ones. The name is believed to be based on that of Euphorbus, physician to the King of Mauretania.

The genus *Euphorbia* is extremely diverse in the forms its species assume, perhaps more so than any other genus of cultivated plants. To such an extent is this true that it is difficult for anyone not botanically oriented to believe that all can logically be classified in the same genus. That poinsettias and snow-on-the-mountain belong together is acceptable, and the imagination is not too strained to accept crown-of-thorns as an ally of these. But credence boggles somewhat at the conception that the insignificant, prostrate, tiny-leaved, milky-juiced weeds that invade lawns and waste grounds, which botanists tag as *E. preslii* and *E. maculata,* are close relatives of the gorgeous poinsettia. Even less acceptable on the basis of gross appearance is that the great succulent trees and shrubs in African deserts that so closely resemble giant cactuses, and that the uninitiated invariably accept as such, also belong in *Euphorbia.* Yet they do, and so does another species of a shape and size that almost mimics a baseball. That is South African *E. obesa.* How come, one asks, these very different looking plants are all euphorbias? The answer is simple.

Botanical classification is based primarily on floral structure, and the flowers and their arrangement throughout the entire genus *Euphorbia* are so similar that, although attempts have been made, no generally acceptable basis has been found for dividing that vast and admittedly unwieldy group into a greater number of smaller genera. And so, as gardeners, we must content ourselves with grouping the kinds we grow in categories that reflect their general uses by us and their cultural needs. Thus, we can consider annual kinds, leafy herbaceous perennials, the succulent or cactus-like group, and quite different in its needs, the ever-popular poinsettia.

The flowers of *Euphorbia* can, perhaps, be most conveniently understood by a careful examination of a poinsettia in bloom. The brightly colored red, pink, or creamy-white petal-like parts that are its chief attraction are colored leaves (bracts), not parts of the flower proper. For our present purpose they can be regarded as though they were plain green or as though they were missing. If we strip off these showy organs we have left the true

Euphorbia pulcherrima variety

Euphorbia fulgens

blooms. But even then the casual observer can be misled. The yellow-green, curiously-shaped structures less than ½ inch across that might logically be mistaken for individual flowers are really clusters of several. Each is called a cyathium (plural cyathia). The cyathium has an involucre (collar) of five bracts that look like little petals. The center of the simulated flower is occupied by a pistil with a stalked ovary and three styles cleft at their tips. This is the female flower. It is surrounded by a number of stalked stamens, each technically a male flower, in groups of five. In addition, there is a large nectar-producing gland attractive to pollinating insects. So the true flower cluster or cyathium of a poinsettia consists of one female and several male flowers. In poinsettias all the cyathia are bisexual, but in some species of *Euphorbia* some contain only male flowers and others only female, with never more than one of the latter to a cyathium. The number of lobes of the involucre is commonly five, but in some species there are only four. The glands may number five, four, or more rarely, as in poinsettia, fewer. In some species, but not in the poinsettia, the glands are furnished with colored, petal-like attachments. The placement and arrangement of the cyathia vary greatly according to species. Sometimes they are in branched clusters as in poinsettia, sometimes they form racemes, and often they are solitary or in small groups spaced along the branches or near their tops. The fruits of euphorbias are three-lobed capsules each containing three seeds that when ripe are discharged explosively.

The poinsettia (***E. pulcherrima*** syn. *Poinsettia pulcherrima*), highly prized as a winter-flowering pot plant, is grown literally by millions in greenhouses to supply the Christmas trade. There are many horticultural varieties. In California, Florida, and other warm, frost-free or nearly frost-free places poinsettias are planted permanently outdoors. Attaining a height of 3 to 10 feet or more, this stout-stemmed, nonhardy Mexican shrub has ovate-elliptic to lanceolate leaves 3 to 8 inches long, pubescent on their undersides, and often lobed or toothed. The greenish cyathia are in branched, terminal clusters accompanied by large, showy, red, pink, or creamy-white bracts. Double-flowered poinsettias have many more bracts to each cluster of cyathia than single-flowered ones. For the cultivation of poinsettias see Poinsettia.

Another greenhouse favorite for growing for cut flowers and as a decorative pot plant is the scarlet plume (***E. fulgens*** syn. *E. jacquinaeflora*). Native of Mexico, this is a hairless shrub with slender, arching stems 3 to 4 feet tall, which in winter are festooned with small axillary clusters of "flowers" (cyathia), the showy parts of which are small brilliant orange-scarlet petal-like appendages to the nectar glands. The long-stalked, narrowly-lanceolate, dark green leaves, are 2 to 4 inches in length.

Crown-of-thorns, commonly grown in greenhouses, windows, and in the dry and dryish tropics and subtropics outdoors, is a variety of *E. milii*, of Madagascar, distinguished from that species by its more closely-spaced, stouter, broad-based rather than needle-like spines and its larger "flowers." Possibly of hybrid origin, it is correctly identified as ***E. m. splendens*** (syn. *E. splendens*). A thorny shrub 3 to 6 feet tall, this has intricately-branched, sometimes more or less vining, shallowly-furrowed, slightly more than pencil-thick

Euphorbia milii splendens

stems very freely furnished with spines about ¾ inch long. Its elliptic to long-obovate, undivided, toothless, bright green, deciduous leaves up to 2 inches long by ½ to ¾ inch wide are on the upper parts of the stems and branches. The "flowers," a little under ½ inch across and on repeatedly branched, sticky stalks, are in up-facing clusters borne freely and intermittently throughout the year, especially in spring. Their showy parts are a pair of brilliant red, nearly round petal-like bracts. Variants of the crown-of-thorns are grown. They include kinds offered commercially under these names: *E. splendens*

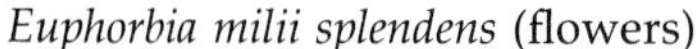

Euphorbia milii splendens (flowers)

Euphorbia keysii

bojeri, upright, freely-branching, with dark green, persistent leaves and red-bracted "flowers"; *E. s. breonii* has 1-inch-thick, spiny stems, bronzy, lanceolate leaves 4 to 6 inches long, and brilliant red "flowers"; *E. s. hislopii* has stout stems with strong, brown to gray spines, long, thin leaves and "flowers" with large salmon-pink bracts with paler bases; *E. s. lutea* has bluish-gray leaves and "flowers" with pink-edged, lemon-yellow bracts. Sometimes called giant crown-of-thorns ***E. keysii*** (syn. *E.* 'Flamingo') is a hybrid between *E. milii splendens* and *E. lophogona.* In habit this resembles crown-of-thorns, but has much stouter branches and ovate leaves up to 6 inches long, which when the plant is grown in full sun become coppery red. The "flowers," somewhat bigger than those of crown-of-thorns, are rosy-carmine to red. Very different, West Indian ***E. punicea*** is a thornless shrub or tree 10 to 30 feet tall with oblanceolate to obovate, stalkless leaves 3 to 5 inches long. The flowers, at the ends of the branches and branchlets, are nested in a collar of three to five brilliant red, spreading, broad-elliptic bracts 1 inch to 1¾ inches long.

Two nonhardy nonsucculent, thin-leaved shrubs or small trees cultivated in southern Florida, Hawaii, and other warm regions are *E. leucocephala* and *E. cotinifolia,* both natives of Mexico and Guatemala. Up to 15 feet tall and showy in bloom, ***E. leucocephala*** has broad-elliptic to lanceolate leaves, opposite or in whorls (circles of three or more). They have blades 1 inch to 3 inches long and are finely-hairy along the veins. The many large, loose, flattish flower clusters are prettied by abundant ½-inch-long, spoon-shaped white bracts. From 10 to 30 feet in height, ***E. cotinifolia,*** has long-stalked, broad-elliptic to nearly round leaves with thin blades up to 4½ inches long by 3½ inches wide. Its abundant loose panicles from the branch ends are of numerous small flowers with many little white or creamy-white bracts. A very lovely

Euphorbia punicea

Euphorbia cotinifolia

Euphorbia marginata

variety with rich copper-purple foliage is cultivated. It would seem appropriate to distinguish this as *E. c. atropurpurea.*

Annual euphorbias and biennials commonly grown as annuals include a few familiar kinds and some less well known. Among the most popular are Mexican fire plant or annual poinsettia, snow-on-the-mountain or ghost weed, and caper spurge, mole plant, or gopher plant.

Mexican fire plant, painted leaf, fire-on-the-mountain, or annual-poinsettia (***E. heterophylla***) is a showy annual 1½ to 3 feet tall. Its red or partially red floral bracts suggest those of the poinsettia, but are smaller and generally much less splendid. Native from Illinois to Florida, Texas, and tropical America, this has slender-stalked, very variably shaped leaves, in outline ovate to linear, smooth-edged, wavy-toothed, or lobed, those near the flowers splashed with crimson, scarlet, white, or yellow. The cyathia, in terminal clusters, have small involucres.

Snow-on-the-mountain or ghost weed (***E. marginata*** syn. *E. variegata*) is a vigorous, branching annual, native of prairies and plains from Minnesota to Colorado and Texas. From 1 foot to 3 feet tall, broad and bushy, this has reddish stems and oblongish leaves 1 inch to 3 inches in length, the lower ones green, those above clearly white-margined. The little "flowers," in umbels, have glands with showy, white, petal-like appendages.

Caper spurge, mole plant, or gopher plant (***E. lathyrus***) is a handsome, hairless, glaucous biennial or sometimes annual native of Europe, naturalized in eastern North

Euphorbia heterophylla

Euphorbia marginata (flowers)

Euphorbia lathyrus

America and California. Its vernacular names allude to its fruits having been used as substitutes for capers and to the aversions moles and gophers are said to have for this species, which is sometimes planted in the hope of driving those pests away. Caper spurge is 2½ to 5 feet tall. It has erect stems with, disposed symmetrically along their lengths, mostly opposite leaves with alternate pairs set at right angles to form four vertical rows. The stem leaves are stalkless, lanceolate, 1½ to 5 inches long, and without teeth. The leaves accompanying the cyathia are ovate to ovate-lanceolate. The cyathia are in umbels with three or four rays. The seed capsules are approximately ½ inch in diameter.

Hardy herbaceous perennial spurges that die to the ground in winter include a few known in gardens. Flowering spurge (***E. corollata***) is one such. It has slender stems up to about 3 feet tall, branched in their upper parts. Its lobeless, toothless leaves, 1 inch to 2 inches long and stalkless or with very short stalks, are oblong to oblong-spatula-shaped. The lower ones are alternate, those above in whorls (circles of more than two). The numerous cyathia are in umbels. Their glands have white appendages with very much the appearance of petals. This is native from Ontario to Florida and Texas. Less well known ***E. griffithii,*** of the Himalayan region, has creeping rhizomes, slender stems 2 feet or so tall, and lanceolate leaves with pink midribs. Its broad, loose, headlike clusters of bloom have yellowish-red to red floral leaves. In variety 'Fire-fly' they are bright orange-scarlet. Cypress spurge (***E. cyparissias***), is a European naturalized in waste places in North America. Spreading rapidly by underground rhizomes, this forms broad patches of upright stems with many erect branches up to about 1 foot high and attractively feathered with slender-linear leaves ½ to 1 inch long. In summer it bears "flowers" in many-rayed umbels. They have yellowish or reddish-purple bracts, but rarely develop seed. Attractive ***E. epithymoides*** (syn. *E. polychroma*) makes a considerably better floral display than the last and does not spread. Native of eastern Europe, this is of tidy habit. It has many erect, leafy stems that form hemispherical 1- to 2-foot-tall mounds somewhat broader than high. Approximately 2 inches long, the sometimes purplish-tinged leaves are thin and oblong. The uppermost, muted yellow when the plants are in bloom, make an at-

Euphorbia griffithii

Euphorbia griffithii 'Fire-fly'

Euphorbia lathyrus (fruits)

Euphorbia epithymoides

Euphorbia epithymoides (flowers)

tractive, by no means gaudy, display. The "flowers" are in five-rayed umbels. Native to Europe and temperate Asia, ***E. palustris*** is a wet-soil plant with erect stems, 3 to 6 feet tall, clothed with stalkless, elliptic leaves 1 to 3 inches long. Its yellow-bracted flowers are in showy, terminal heads 4 to 6 inches in diameter. *E. palustris* is suitable for bog gardens.

Euphorbia palustris

Evergreen herbaceous perennial and subshrubby spurges of garden merit, except for *E. myrsinites* not hardy in the vicinity of New York City, but satisfactorily permanent and easy to satisfy in milder climates, include these few. Native of dry, open sites in the western Mediterranean region, ***E. characias*** is a usually thickly-hairy herbaceous perennial with dull purple

Euphorbia characias

stems up to 2½ feet tall and bluish-green, lobeless, toothless, linear to oblanceolate leaves 1½ to 5 inches long by ¼ to a little over ½ inch wide. The bracts of the flower clusters are united in pairs to form circular, leaflike structures, the uppermost chartreuse and forming cup-shaped collars around the cyathia. The glands of the cyathia are purplish-brown. Closely similar ***E. wulfenii*** (syn. *E. venata*), of the eastern Mediterranean region, apparently hybridizes freely with the last in the wild and in cultivation. Up to 6 feet in height, it has larger leaves and bigger flower heads than *E. characias* and the glands of its cyathia are yellowish-brown. Southern European ***E. myrsinites*** has more the appearance of a warm-country, desert succulent than the hardy perennial or sometimes biennial it

Euphorbia wulfenii

Euphorbia wulfenii (flowers)

Euphorbia myrsinites

Euphorbia myrsinites (flowers)

cylindrical and pencil-thick, often form broomlike clusters at the branch ends. The cyathia are in stalkless clusters at the tops of the branchlets. In Africa and Asia this species is used to poison fish, as an insecticide, and medicinally.

Tree euphorbias of decidedly cactus-like appearance include these: ***E. abyssinica,*** of Ethiopia and Eritrea, is a sizeable tree. Its stems are erect, olive-green marked with lighter and darker green. They have eight almost winglike, slim, wavy ribs with closely-set pairs of spines under ½ inch long. About 2 inches long by one-quarter as wide, the linear-lanceolate leaves arc toward the branch ends. They are deciduous. ***E. angularis,*** of Portuguese East Africa, is poorly defined botanically. As known in cultivation it is a tree or shrub 10 to 15 feet tall with a stout trunk from the upper part of which radiate up-curving branches, secondary branches, and branchlets. The stems are deeply constricted into joints or segments 2 to 4½ inches wide with three or four winglike angles with continuous horny edges. The tiny spines, divergent and in pairs, are gray. The leaves are scalelike and rudimentary. The fruits are deeply-three-lobed. ***E. canariensis,*** native of the Canary Islands, is a large shrub or tree up to 40 feet tall. Its fresh green, flattish-sided

is. This sort has prostrate stems that snake in various directions from a common center and for their lengths of up to about 1 foot are clothed with close spirals of evergreen, broad, strikingly blue-green leaves with hollowed upper surfaces. The cyathia are in terminal umbels of seven to twelve rays. The floral bracts are yellow. Subshrubby ***E. biglandulosa*** (syn. *E. rigida*) is closely related to *E. myrsinites* from which it differs chiefly in its stems being erect or ascending, up to 2 feet high instead of prostrate, and its fleshy gray-green, 1½-inch-long evergreen leaves being lanceolate. The chartreuse-yellow "flowers" are in domed heads. This is a native of Europe and southwest Asia. First described in 1953, ***E. robbiae*** inhabits Asia Minor. A subshrub of good habit and handsome foliage, but without floral merit, it forms broad clumps about 2 feet tall. It has very dark green, leathery, evergreen leaves.

Tree euphorbias of massive size and mostly cactus-like aspect inhabit dry parts of Africa, Asia, and the Canary Islands. In southern Africa there are fourteen species, a few more elsewhere. Often they are difficult for nonbotanists to identify as to species. The possibilities of exploiting some for the rubber they contain has been quite thoroughly explored, but the comparatively poor quality of the product and the high percentage of resin contained in the latex has precluded commercial success. Even so, *E. tetragona* and *E. triangularis* are

Euphorbia robbiae

still sometimes called rubber trees, although in America this designation is more commonly reserved for *Ficus elastica* and for *Hevea brasiliensis,* the last the source of commercial natural rubber.

Milk bush or Indian tree spurge (***E. tirucalli***) is a tree widely distributed as a native of Africa and a naturalized introduction in warm parts of Asia. The least cactus-like of the tree spurges, this has much more slender shoots than others and is without spines. From 15 to 30 feet high, it develops a heavy trunk, thick main limbs, and much slimmer branches. Its dense head is irregularly-rounded. The shoots, alternate, opposite, or clustered,

Euphorbia canariensis

stems usually have five, sometimes four or six, wavy-toothed, sharp angles with pairs of short spines a little over ½ inch apart. The juice of this is reportedly very poisonous. ***E. candelabrum,*** native from the Sudan southward, attains a maximum height of about 30 feet. Its four-ribbed stems and branches are constricted into joints. The ribs, wavy and toothed, have pairs of spines under ¼ inch long spaced about ¾ inch apart. By some *E. ingens* is considered identical with this. ***E. cooperi,*** of South Africa, is somewhat similar to and some-

Euphorbia cooperi

times confused with *E. ingens.* It differs in being lower, 15 to 25 feet tall, and less massive. Also, its lower branches die annually as new ones are made above so that a longer clean trunk is developed, and its branches curve in their lower parts, being normally five- or six-angled and having continuous horny edges. Their segments or joints are distinctly pear-shaped, with the broadest part a little above their bases. The edges of their angles have pairs of black-tipped, gray spines up to ⅓ inch long spaced ½ to ¾ inch apart. The fruits are three-angled. ***E. curvirama*** is a spiny South African usually not over 15 feet tall, but sometimes taller. It has one to three trunklike stems, naked below, each surmounted by a head of markedly upcurved, ascending branches up to 6 feet long. They are 2 to 3 inches thick and usually four-, but sometimes three- or five-angled. The spines, those of each pair widely diverging, are ⅕ inch to slightly over ½ inch long. ***E. excelsa*** is a spiny tree of South Africa, at maturity 20 to 30 feet tall, occasionally more. It has a trunk up to 1 foot in diameter, branchless below, topped with a considerable crown of spreading-erect branches. When young, the stems are bluish-glaucous, slightly six- or seven-angled. The branches, more or less in whorls

Euphorbia curvirama in the Karroo Garden, South Africa

(circles of more than two), slightly segmented or jointed, and most commonly four-angled, are up to approximately 3 feet long, 1 inch or a little more thick. The thinner secondary branches have shorter joints. ***E. grandidens*** is a loose-headed, slender, spiny tree 30 to 50 feet tall with a cylindrical trunk and few to several erect branches each with numerous secondary branches and branchlets clustered toward their ends. The secondary branches are usually three-, occasionally two- or four-angled, ½ to ¾ inch thick. The gray spines, in pairs, are about ¼ inch long or sometimes are wanting. There is often a pair of minute prickles above each pair. Of graceful aspect, this kind is South African. The plant cultivated as *E. alcicornis,* not the spineless, smooth-branched, Madagascan shrub to which that name properly belongs, may be a variety of *E. grandidens.* It differs from the typical species in having more uniformly two-angled branches. ***E. ingens,*** most massive, but not tallest of tree euphorbias, is remarkable for the huge, dense head it develops as a result of its lower branches not dying off annually. Under favorable circumstances, this usually short-trunked tree may be 30 feet high and have a head broader than long composed of hundreds of crowded, erect, four-angled, leafless and usually spineless, thick branches not markedly curved at their bases, less strongly segmented into joints than those of *E. cooperi,* and the joints more oblong than pear-shaped and with wings without continuous horny margins. The fruits are not three-angled. This is native from South Africa to perhaps as far north as Kenya. Variety *E. i. monstrosa,* sometimes called totem pole, has curiously contorted, knobby stems. ***E. neriifolia*** becomes tree-like when old. Native to the East Indies, it has slightly five-ribbed stems and branches, the ribs spiraled and with short, black spines. From 3 to 4 inches long, the obovate, fleshy-leathery, deciduous leaves give reason for the vernacular name of oleander euphorbia sometimes applied to this species. ***E. tetragona,*** up to about 40 feet in height, is a South African with a naked

Euphorbia tetragona in South Africa

trunk up to 6 inches in diameter, sometimes with five or fewer main branches from near the base. These, like the main trunk, terminate in crowns of upright-spreading lesser branches and branchlets. The latter, mostly in whorls (circles of more than two), are four-, five-, or less commonly three-angled and are slightly tubercled. The spines, sometimes absent on old specimens, are widely divergent. Under ½ inch long, they are in pairs. ***E. triangularis,*** a robust tree 30 to 50 feet in height, is South African. It develops a clear trunk surmounted by a rounded crown of branches. The latter, in whorls around the trunk (which with age becomes cylindrical) are usually three-, sometimes two-, four-, or five-ribbed. The winglike ribs have horny, sometimes discontinuous margins with rather closely set pairs of widely-divergent spines ¼ to ½ inch long.

Shrubby, succulent euphorbias mostly of branching, more or less cactus-like aspect and commonly from 2 to 10 feet or sometimes more tall, or exceptionally becoming small trees include these: ***E. angularis*** (syn. *E. lemaireana*), of tropical Africa, up to 15 feet tall, has stems and branches of joints or segments up to 9 inches in length with three broadly-winged angles and with pairs of spines ¼ inch long or somewhat longer. The scalelike leaves are gray. When they fall they leave between the base of the two spines two stiff, persistent, hooklike remains. The cyathea are in branched clusters. ***E. antiquorum,*** of India, is spiny, 10 to 12 feet tall. It has jointed, four-, to five-angled stems up to about 2 inches thick, their branches three-angled. The angles have pairs of about ¼-inch-long spines spaced ¾ inch to 1¼ inches apart. ***E. antisyphilitica,*** the candelilla of Mexico, is 2 to 3 feet high. From its base it has many erect, slender, cylindrical, leafless, spineless, waxy-coated stems. The cyathia have five small white bracts and in appearance suggest little individual flowers. The wax of this species is used for candles and in soaps, polishes, and other products. ***E. aphylla*** is a thicket-forming, low shrub of the Canary Islands. It has slender, jointed, gray-green stems of pencil-thickness spotted with conspicuous, rounded leaf scars. They are forked or have whorls of up-curving branches 2 to 3½ inches long. The few leaves are linear. The cyathia are solitary or in twos or threes at the stem ends. ***E. avasmontana*** is a spiny, stout-stemmed, cactus-like shrub up to 6 feet in height,

Euphorbia avasmontana

much branched from the base but not above. Native of South Africa, it has five- to eight-ribbed, jointed stems, 2 to 3 inches thick, greenish-yellow to bluish. Their ribs have continuous horny edges with closely-spaced pairs of wide-spreading sharp spines ¾ inch or less long. ***E. balsamifera*** is a spineless native coastal shrub of the Canary Islands, much-branched and broader than high and with gray stems with, near their ends, conspicuous rosettes of up to 1-inch-long, linear to narrow-elliptic leaves. The inconspicuous cyathia are solitary. ***E. bubalina*** is pretty well-foliaged, branched or branchless, up to about 4 feet tall. It has tubercled stems about ¾ inch thick and stalkless, oblanceolate to oblong-oblanceolate leaves up to 4 inches long by one-quarter as wide. There are no true spines, but the old flowering stalks are persistent and spinelike. The cyathia are decorative, in long-stalked, umbel-like arrangements. Their bracts are green edged with red or are all red, ⅓ to 1 inch long. A favorite food of goats in its native South Africa, this adapts readily to cultivation and sometimes self-sows. Crested variants are not uncommon. ***E. clava*** is a South African with a spineless, succulent, club-shaped to cylindrical, generally branchless main stem 1 foot to 4 feet in height and with tubercles spiraled around it. Toward its apex it has deciduous, linear leaves. The long, persistent flowering stalks harden and become somewhat spinelike. ***E. coerulescens*** is a stiff-stemmed, spiny, cactus-like shrub of South Africa. It spreads below ground from

Euphorbia coerulescens

the base to form broad bushes up to 4½ feet tall. Often branched above, the roundish-jointed, bluish-gray stems up to 2 inches thick have four to six horny-edged, sinuate ribs with deep furrows between. The spines, rigid and ¼ to ½ inch long, are in pairs. ***E. dregeana*** is a spineless native of South Africa that forms clumps 6 to 9

Euphorbia dregeana

feet tall and frequently broader than wide of erect and sometimes sprawling cylindrical stems. The main stems are 1 inch to 2 inches thick, the branches up to about ½ inch in diameter. The young parts are velvety-hairy. This is the tallest South African species with slender cylindrical stems. Crested variants are not uncommon. ***E. echinus,*** native of Morocco, is 1 foot to 3 feet tall or taller. It forms broad, dense, hemispherical clumps of crowded, erect

Euphorbia echinus

stems and branches 1½ to 2 inches thick, with five to eight or occasionally more, scarcely-toothed, continuously-horny-edged ribs. Up to ¾ inch in length, the whitish to grayish or reddish spines are in closely set pairs. ***E. grandicornis,*** one of the most handsome shrub euphorbias, is remarkable

Euphorbia grandicornis

for its effective armament of pairs of strong spines pointing in all directions. White and up to 3 inches long, they ornament the three wavy, winglike ribs of the irregularly sized joints into which the 1½- to 3½-inch-wide branches are deeply segmented. Each pair of spines may be accompanied by a pair of small prickles. This very distinctive native of southern and East Africa is 3 to 6 feet tall. It has a very short main stem and many erect branches. ***E. handiensis,*** of the Canary Islands, attains heights of 3 feet or more. It has a woody, eight- to twelve-angled main stem. Its leafless branches have pairs of red to white spines up to 1¼ inches long. ***E. heptagona*** is a South African up to 3 feet tall or taller, often lower. Unisexual, from its base it has many erect branches that develop clusters of short secondary branches near their tops. The stems, 1 inch to 1½ inches thick, have from five to ten ribs with solitary spines 1 inch to 1¼ inches

Euphorbia grandicornis

Euphorbia grandicornis (fruit)

long. ***E. horrida*** is a spiny South African. Up to 3 feet tall or taller, it branches from the base to form irregular clumps. Its club-shaped to cylindrical stems, 4 to 6 inches thick, have many deep, wing-like ridges furnished with straight spines, solitary or in groups of five or fewer, the longest ½ inch to 1¼ inches long. ***E. hottentota,*** a broad shrub up to 6 feet high, is South African. It has numerous erect stems that curve outward from their bases and when old sometimes branch above. About 1½ inches thick and not or only slightly constricted into joints, they have five or less

Euphorbia horrida in South Africa

commonly six angles. The short secondary branches are usually four-angled. Not over ⅕ inch long, the spines are in pairs. ***E. intisy*** of Madagascar, is a spineless shrub or sometimes tree up to 20 feet tall the latex of which contains rubber. It has much-branched, slender, cylindrical, forking, gray stems furnished with small, whitish protuberances. ***E. lactea*** is a commonly grown, bushy shrub or small tree with three- or more usually four-ribbed stems 1½ to 2 inches thick. They are dark green

Euphorbia lactea

illuminated with conspicuous milky-white bands and broadly-V-shaped blotches along the channels between the ribs. The latter are wavy and have pairs of short, dark brown spines. This is native to Ceylon or the East Indies. Variety *E. l. cristata* has crested stems. ***E. ledienii,*** sometimes con-

Euphorbia lactea cristata

fused with *E. coerulescens,* which it resembles, is a South African branched from the base and above, and 3 to 6 feet tall. Its four- to seven-angled, green stems, slightly constricted into joints, are up to a little over 2 inches in diameter. The horny, brown margins of the more or less tubercled angles are continuous or interrupted. The spines, occasionally rudimentary or missing, are up to ¼ inch long and in pairs. The yellow cyathia, abundantly produced, are quite decorative. This is generally taller than *E. coerulescens,* its stems less conspicuously jointed. ***E. mauritanica*** is a dense, rounded, spineless shrub of South Africa, mostly 3 to 5 feet tall, often broader than high, and leafless except on the youngest shoots. It has slender stems, their branches alternate,

A crested specimen of *Euphorbia mauritanica*

the ultimate branchlets about ⅕ inch in diameter. The leaves are alternate, stalkless, pointed-lanceolate, up to ½ inch long by one-third as wide. The cyathia are in terminal umbels. Several varieties of this variable species have been recognized. A crested variant is cultivated. ***E. pentagona,*** of South Africa, has erect stems with five, six, occasionally up to eight shallow angles furnished with downward-pointing teeth. The branches are in whorls (circles of more than two). The spines are solitary. ***E. polygona*** is a South African 2 to 5 feet tall that suckers to form clumps of erect stems 3 to

Euphorbia polygona

4 inches thick of unequal lengths which sometimes but not commonly branch again. The young stems have seven, older stems up to twenty, slightly wavy deep angles (ridges). These are furnished with spines less than ½ inch long and solitary or in twos, threes, or rarely fives, or the spines may be nearly lacking. The minute leaves are soon deciduous. Male and female flowers are on separate plants. ***E. pseudocactus*** is spiny, broader than tall, much-branched from the base, and up to 3 feet in height. Its branches, erect to somewhat spreading, are rigid, sometimes three-, more commonly four- or five-angled, the angles

Euphorbia pseudocactus

horny-margined. The stems are constricted into joints or segments 2 to 6 inches long by 1 inch to 2 inches thick. They are generally marked with yellowish-green, V-shaped markings. The spines, up to ½ inch long, are in pairs. This species is native of South Africa. Variety *E. p. lyttoniana* is without spines. ***E. schoenlandii*** is very like *E. fasciculata,* but more robust and spiny. When very old this South African is up to 4 feet tall and 9 inches thick. It has a cylindrical stem, very rarely branched, covered with prominent spirally-arranged tubercles each ending in a slender, teatlike, spiny tip. The withered flower stalks remain for several seasons. ***E. trigona*** (syn. *E. hermentiana*) is indigenous to tropical southwest Africa. A shrub or sometimes a small tree, this has erect branches 1½ to 2 inches thick. With four or five winglike ribs, they are dark green mottled with grayish-white, broad, V-shaped markings. The small, spatula-shaped leaves end in brief points. Although deciduous they usually do not fall quickly. ***E. xylophylloides*** is a sometimes treelike, much-branched, spineless shrub up to 6 feet or more tall, native to Madagascar. It has flat, two-angled, slightly-toothed stems about ½ inch wide. The tiny, roundish leaves are early deciduous.

Usually from 1 foot to 2 feet tall at maturity, some sorts perhaps occasionally slightly higher, are the succulents we shall

Euphorbia trigona

Euphorbia trigona

now describe: ***E. fasciculata*** is not spiny. It develops a partly buried, rarely branched, club-shaped erect stem, on very old specimens up to 2 feet long by about 6 inches thick, surfaced with spirally-arranged tubercles ending in short, conical points. The narrow leaves, up to 1½ inches long, are dark green. The spines are solitary. The withered flower stalks persist for several seasons. Closely related to and sometimes confused with *E. schoenlandii,* this is South African. ***E. griseola*** is a spiny South African shrub with a short main stem and many erect branches that again branch to form dense clumps 1 foot to 2 feet tall or sometimes slightly taller. From four- to six-angled, the branches are up to or ½ inch thick, horny-edged, and tubercled. They have pairs of tiny, slender, gray spines, sometimes accompanied by a pair of prickles. ***E. leuconeura*** bears a general resemblance to *E. lophogona.* A shrub with four-angled stems, the angles fringed in comblike fashion with hairs, this is a native of Madagascar. It attains a height of about 1½ feet and has tapering, obovate-elongate leaves 3½ to 4 inches long by one-third or more as wide, crowded toward the apexes of the stems. ***E, lophogona*** is remarkable for its spines being modified into longitudinal fringes of almost hairlike fineness. These decorate the angles of the slightly club-shaped stems. A native of Madagascar and 1 foot to 2 feet tall, this has white- or pink-bracted cyathia. ***E. polyacantha,*** a cactus-like native of North Africa and Ethiopia, is 1 foot to 2 feet tall. Its thick, gray-green stems are four- or five-angled. The angles are furnished with pairs of long, needle-like spines, those of each pair spreading widely. ***E. pteroneura*** is a Mexican, 1 foot to 2 feet tall. It has pencil-thick or slimmer, branched, five- or six-angled stems and alternate, short-stalked, ovate-lanceolate leaves ¾ inch to 1¼ inches long and approximately one-half as wide. They are deciduous. ***E. resinifera,*** native of Morocco, is a freely-branched, dense shrub 1 foot to 1½ feet in height and forming broad mounds. It has very numerous, erect, gray-green stems with four slightly-toothed angles with closely set pairs of spines up to ¼ inch long. It is the source of the drug euphorbium. ***E. stellaespina*** is akin to *E. pillansii.* A South African, it is remarkable

Euphorbia polyacantha

Euphorbia pteroneura

Euphorbia lophogona

Euphorbia lophogona (flowers)

Euphorbia resinifera

for its strikingly starry, three- to five-branched spines up to nearly ½ inch long that decorate its corncob-like stems. The latter are erect, up to 1½ feet tall by 3 inches in diameter, and branch freely from their bases; rarely they are taller. They have ten to sixteen lumpy ribs and form beautifully-shaped, large mounded clumps. The plants are unisexual. ***E. woodii*** is spineless, low, and succulent. Native to South Africa, it has a thick root and stem, only the upper part of the latter above ground. The exposed portion, 4 to 6 inches in diameter, sprouts in two or more circles around its perimeter up to forty tuberculate, cylindrical branches 4 to 8 inches long and up to ½ inch thick.

Kinds generally under 1 foot tall, many consistently much lower, are now to be considered: ***E. aeruginosa,*** of South Africa, much-branched from the base, is about 6 inches tall. Slightly four- or five-angled, often spirally-twisted, its coppery stems, 2 to 3 inches thick, have ¾-inch-long spines mostly in pairs, often with smaller spines above. ***E. anoplia,*** of South Africa, is spineless, sparingly-branched from the base, and 5 to 7 inches tall. Its gray-green, seven- to nine-angled stems have closely spaced, white tubercles along the angles. The shorter branches have five-angled, less prominent tubercles. ***E. bergeri,*** of South Africa, is without true spines, but retains short, spreading, spinelike remnants of old flowering stalks. It forms conical clumps up to 10 inches wide of branched, slightly lumpy, cylindrical stems up to 6 inches long by about ½ inch thick. The deciduous, folded, linear-lanceolate leaves mostly not over ¼ inch long are clustered at the branch ends. ***E. bupleurifolia*** is a spineless native of South Africa. Its solitary, thick-ovoid, rarely-branched stem, up to 4½ inches tall by 3½ inches in diameter, is covered with a grater-like pattern of spiraled small tubercles. The deciduous, narrow-oblanceolate leaves sprout from near the tops of the stems. ***E. caput-medusae*** is a curious South African with a short, thick, obovoid or club-shaped stem up to 8 inches across, from the top of which sprouts a medusa-head of numerous slender, snaky, gray-green branches up to 2 feet long by ¾ inch to 2 inches in diameter. The leaves are deciduous, linear-lanceolate, ½ to 1 inch long. Their stalks persist and become somewhat spinelike. The numerous greenish-yellow flowers are at the ends of the young branches. ***E. clavarioides*** is a fascinating

Euphorbia bupleurifolia

Euphorbia caput-medusae

Euphorbia woodii

spineless South African that forms compact green to reddish cushions up to 1 foot across of densely-crowded, short, warty branches ¼ to ½ inch thick that sprout from a short, thick stem. At first nearly spherical, the branches elongate to become club-shaped. The tiny, folded leaves are lanceolate. ***E. davyi*** has a more or less globose to egg-shaped main stem crowned with a dense, open-centered cluster of many branches 1½ to 6 inches long by ½ to ¾ inch thick. They are cylindrical and surfaced with closely-set little bumps. Toward their tips are hardened, spinelike remains of old flowering stalks and, in season, pointed, linear to linear-lanceolate, longitudinally folded leaves up to a little over 1 inch long. This is an attractive South African. ***E. ferox,*** about 6 inches tall, branches from below and at ground level to form low mounds up to 2 feet across. Its deeply nine- to twelve-ridged gray, but not glaucous stems, up to about 2 inches thick, have solitary, straight or curved, mostly wide-spreading, brown, later gray, glaucous spines ½ inch to 1¼ inches long. The plants are unisexual. This is a native of South Africa. ***E. fimbriata*** is a close ally of *E. mammillaris* and *E. submammillaris.* Like the former it is called, in allusion to the appearance of its stems, corncob euphorbia. A unisexual South African, *E. fimbriata* has branched or branchless stems usually not over 1 foot tall, in shaded locations sometimes considerably higher. They are erect or more or less procumbent, about 1½ inches thick, with seven to ten lumpy ribs. The solitary spines, sometimes few or missing, usually in encircling bands, are up to ¾ inch long. ***E. flanaganii,*** of South Africa, so closely resembles *E. woodii* that it should probably be considered a variant of that species. *E.*

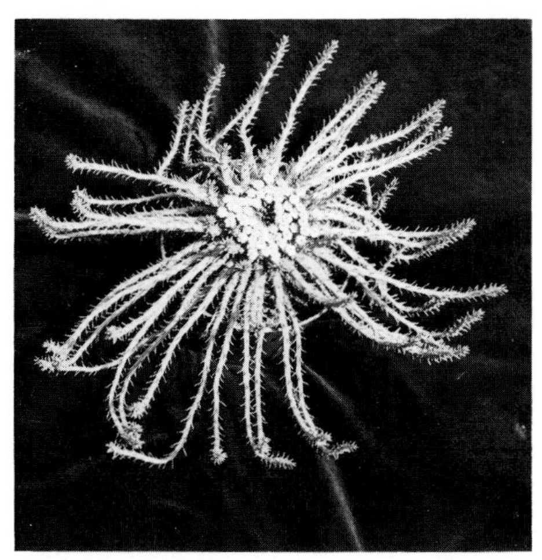

Euphorbia flanaganii

f. cristata is a crested variety. ***E. franksiae,*** of South Africa, a near relative of *E. woodii,* has its chief stem joined to and similar to its carrot-like root. Mostly underground, its apex has a crown of two or three circles of branches up to 2 inches long with a center area free of tubercles. The branches are cylindrical, about ⅛ inch thick. There are no spines. ***E. globosa,*** of South Africa, is spineless and 1 inch to 2 inches tall. Branching from its base, it forms irregular cushions

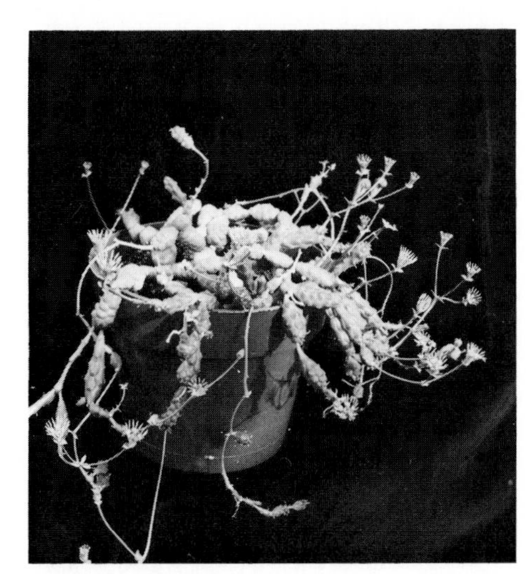

Euphorbia globosa

of branches composed of spherical to ovoid segments 1 inch or so in diameter from which sprout more or less persistent flowering stems up to 2¼ inches long. ***E. gorgonis*** is a spineless, low South African with a main stem, a continuation of its thick taproot, partly below ground, and its exposed, flat top, 2 to 4 inches across, crowned with three to five circles of green or purplish, radiating branches ½ inch to 2 inches long, with a central circular area free of branches. The tuberculate branches are from a little under to a little over ¼ inch thick. ***E. jansenvillensis*** has underground stems from which come clusters of erect, five-angled, glaucous-green, spineless branches mostly not over 6 inches, but sometimes up to 1 foot tall by up to ¾ inch thick. Along their angles they have short, sharp tubercles. From their upper parts they often produce short, egg-shaped branches. The plants are unisexual. This sort is South African. ***E. juglans*** is a South African with a large tuberous root from the apex of which comes a brief main stem and one to five erect, club-shaped, grayish-green or brown-tinted branches up to 2 inches long by 1 inch thick. They have six to nine blunt angles without spines. The plants are unisexual. ***E. knuthii,*** native of East Africa, has tuberous roots, rhizomes, and strongly-spiny above ground stems up to 6 inches long by ¼ to ½ inch thick, narrowed to their bases, and occasionally branched above. They have three or four deep, wavy angles and are light green clearly decorated with greenish-gray. The divergent, needle-like spines up to ⅓ inch long, are in pairs. There may be a pair of small prickles with each pair of spines. ***E. mammillaris,*** because of the appearance of its stems called corncob euphorbia, is a South African about 8 inches tall, sometimes a little taller. It forms clusters of erect

Euphorbia mammillaris

stems that in cultivated plants may branch above. Cylindrical and 1½ to 2¼ inches in diameter, they have seven to seventeen tubercled angles. The solitary spines, ¼ to nearly ½ inch long, are rather distantly scattered or are in encircling bands. The plants are unisexual. ***E. meloformis*** is a variable South African as remarkable as *E. obesa,* from which it is readily distinguishable by the much more widely spaced tubercles along the usually eight, sometimes more, ribs or angles of its spherical to more elongated stems. The ribs are generally more prominent than those of *E. obesa* and the flowering stalks are persistent, becoming spinelike. Commonly, the stems are marked with light green and purplish-brown or darker green transverse bands. They may or may not produce numerous branches or offshoots. ***E. obesa*** as a plant is nearly unbelievable. At sight it suggests a baseball finely chiseled out of smooth, purplish-streaked, light gray stone. Its sol-

Euphorbia obesa

itary stem spherical or when old twice as high as wide is 3½ to 4½ inches in diameter. There are eight extremely shallow, vertical, finely-toothed ribs that, like the bottoms of the slight valleys between them, have the appearance of seams. Individuals are unisexual. This South African can only be increased by seeds. These germinate readily. Hybrids between it and some other species are known. Scarcely different from *E. obesa* except in technical detail, and probably in cultivation sometimes masquerading as it, South African ***E. symmetrica*** has a more strongly developed taproot and a stem or plant body usually broader than tall. ***E. ornithopus*** is a South African 2 to 3 inches in height with a turnip-shaped stem 1 inch to 2 inches wide capping a tuberous root. From this sprout many short-cylindrical to nearly globose, usually more or less pointed stems up to 2¼ inches long by under ½ inch thick. Often they produce jointlike branches, which sometimes tend to flop. They have up to five spiraled rows of conical tubercles. There are no spines. ***E. pillansii,*** not exceeding 1 foot in height, is a somewhat variable, unisexual South African that branches from the base and above. Its stems, up to 2 inches in diameter, have seven to nine, or when young sometimes only five, shallow, blunt, slightly lumpy ribs. They are banded horizontally with light and dark green. The spines are solitary, sometimes branched, and up to about ½ inch long. ***E. pseudoglobosa*** from the top of its subterranean, tuberous stem produces a cluster of spineless branches at first globose, later developing as strings 8 to 10 inches long of two or three nearly spherical to short-cylindrical joints ½ to ¾ inch thick and bluntly five- to seven-ribbed. It is South African. ***E. pugniformis,*** of South Africa, has a short main stem that is a continuation of the stout taproot and is partly subterranean. Its rounded top, 2 to 3½ inches across, supports a crown of two or three circles of radiating branches with a central branchless area. The branches, pencil-thick at the base, taper toward their apexes. They are ½ inch to 1½ inches long and covered with tiny tubercles. This South African differs from nearly allied *E. woodii* in being smaller and having tapered rather than nearly evenly-cylindrical branches. ***E. schinzii*** of South Africa and East Africa, has a stout, underground, slightly-thickened main root and stem from which sprout dark green, spiny, sometimes branched, usually four-angled branches 4 to 6 inches long by ⅓ inch thick or slightly thicker. The angles are markedly tubercled. The nearly ½-inch-long spines are in pairs with often two spiraled prickles associated with each pair. This forms cushion-like clumps. ***E. submammillaris*** is similar to and often confused with *E. mammillaris.* Its geographical origin unknown, it differs from *E. mammillaris* in having thinner stems that branch more freely and that have fewer, more prominent angles, and spines up to ¾ inch long less distinctly in horizontal bands. Unisexual, this forms a dense mound 4 to 8 inches tall. The plant sometimes named ***E. pfersdorfii,*** which has a great number of small, few-angled branchlets, is a variant of this species. ***E. susannae*** is a distinct and very attractive South African that in its natural habitat grows with most of its plant body buried in the soil, but in cultivation should not be so embedded. A unisexual, spineless, taprooted species, this when young has a single bluish-green stem later encircled with many globose to globose-cylindrical branches. The stem is 1 inch to 1½ inches in diameter, twelve- to sixteen-ribbed, their angles with prominent, nipple-like, usually down-pointing tubercles. ***E. tortirama*** is a curious dwarf South African with an enormous carrot-like root and main stem the top of which may be 6 inches in diameter and from which sprout up to fifty branches or more 2½ inches to 1 foot long up to 1¾ inches thick. They have three or rarely two prominently-tubercled, spirally-twisted ribs. The spines, in pairs without attendant prickles, are about ¾ inch long. ***E. valida,*** rarely 1 foot tall and commonly not exceeding 8 inches, is South African. As a young plant nearly spherical, with age it may become about twice as tall as its maximum diameter of generally not over 4½ inches. The stem usually solitary, rarely branched from the base, is eight-angled and dull green sometimes marked with paler bands. The angles, vertical or spiraled, are obscurely round-toothed. The persistent, hardened remains of old flowering stalks are spinelike, but this species, related to *E. meloformis,* is technically spineless. ***E. vandermerwei,*** of South Africa, has a short subterranean stem, which is a continuation of the main root. From its apex sprout few to several branches up to 1 inch thick, occasionally 1 foot long, generally shorter. These branch toward their apexes. The stems and branches have four or five straight or spiraled angles, constricted slightly into jointlike segments and tubercled. The spines, a little under ½ inch long, are in pairs with two small prickles accompanying each pair.

Garden and Landscape Uses. The many and varied garden and landscape uses of euphorbias are closely related to their habits of growth and to the colorfulness or absence of colorfulness of their foliage and floral parts. The annuals cultivated, with the exception of the caper spurge which is appreciated chiefly for its foliage and reputed power to drive away moles and gophers, make quite gay displays and are useful in patches in flower beds, in ribbon borders, and in window and porch boxes. They are also effective as cut flowers. For this last use their lives are prolonged if, immediately after cutting, the ends of the stems are dipped in boiling water for a minute or are seared with a flame to seal them and prevent too great a loss of sap.

Herbaceous perennial spurges and subshrubby, nondesert kinds are suitable for sunny perennial and mixed flower beds where the soil is thoroughly drained and preferably dryish rather than damp; *E. myrsinites* is preferred for rock gardens and similar areas. The flowers of *E. corollata* and *E. griffithii* are useful for cutting for indoor decoration. Their stems should be treated as recommended for annual kinds used as cut flowers. Evergreen *E. robbiae,* serves well as a groundcover among shrubs where shade is not excessive.

Euphorbia susannae

Desert-type euphorbias are admirably adapted for outdoor cultivation in mild, dry regions. In the main, they serve the same purposes and thrive under similar conditions as cactuses. Their range of sizes and forms makes it possible to make selections appropriate for many different locations. These are good subjects for desert rock gardens. Collectors of succulents esteem euphorbias for outdoor plantings where they are hardy and for growing in greenhouses. Some make satisfactory window plants. But it must be remembered that the sap of many kinds is extremely irritating to the skin and can cause serious harm if it enters the eye.

Cultivation. Annual euphorbias are easy to grow. Choose well-drained soil of ordinary quality and sunny locations. Sow the seeds in spring where the plants are to remain and thin the seedlings to appropriate distances, about 1 foot apart for snow-on-the-mountain and Mexican fire plant, 6 to 8 inches for caper spurge. Tall sorts may need sufficient staking to give just a little support, otherwise no special care is needed. If desired, plants can be started early indoors and transplanted to the garden after the weather is warm and settled, but there is little advantage in this.

Herbaceous perennial euphorbias and subshrubby types such as *E. robbiae,* which are not desert plants, prosper with minimum care in ordinary, well-drained garden soil, even in those somewhat infertile. The need for excellent drainage particularly applies to *E. myrsinites.* These spurges are increased by seeds and in the cases of some that lend themselves to the procedure by division.

Succulent desert-type euphorbias require conditions and care much like those satisfactory for cactuses, although most seem to prefer less arid conditions than are acceptable to many cactuses. In the main they are not natives of very dry deserts, but of semidesert regions with somewhat more rainfall. They succeed in full sun, but in greenhouses and indeed outdoors many are more richly colored when afforded a little shade.

Kinds such as crown-of-thorns and *E. tirucalli,* as well as many of more cactus-like appearance, are generally easy to manage although, as to be expected in a group so large and diverse, some are more adaptable than others and respond more posily to either careful attention or neglect.

Soil that drains rapidly and admits air readily, that is nourishing yet without a high organic content, is a first essential. For container specimens let the basis of the soil be a mixture of coarse loam and a fairly liberal amount of peat moss or leaf mold. Add enough coarse sand, perlite, and some finely broken brick, broken oyster or clam shells, or leached coal cinders to keep it open, and a generous dash of bonemeal and perhaps one of dried manure.

Repotting, not needed every year, is best done at the beginning of the growing season. It is a mistake to provide over-big receptacles. Quite large specimens will live successfully for many years in pots or tubs that seem small in comparison to their above-ground parts. Such specimens are greatly aided if from spring to fall they are given monthly applications of dilute liquid fertilizer.

A dry or dryish atmosphere with free air circulation is requisite, so for the best results with most kinds is exposure to full sun. The commonest cause of trouble with succulent euphorbias is excessive watering. Strictures that apply to cactuses are as applicable here. In greenhouses, a winter night temperature of 50°F suits most sorts. A few, especially those from fairly warm regions, are likely to do better with five degrees higher. For all, day temperatures may exceed those maintained at night by five to fifteen degrees.

Propagation of succulent euphorbias is by cuttings, offsets, and seeds and sometimes by grafting. Cuttings of most root readily enough, but those of a few, notably of *E. caput-medusae* and kinds of similar habit that produce slender branches from a stout, thick head, do not develop fully the characteristics of the plants from which they are taken. The cuttings made from branches root and live, but fail to develop a new head. After removal from the parent plant and basal trimming, lay cuttings of succulent euphorbias in a dry, shaded, airy place for a week or to to give the wounds time to dry and partially heal. Then plant them in sand or perlite, kept barely moist, in a light place shaded from strong sun where the temperature ranges from 60 to 70°F. Under these conditions roots soon sprout. When a goodly bunch has developed transplant to well-drained pots of sandy soil. Seeds of these succulents germinate readily under similar environmental conditions. Sow in sandy soil, barely covering the seeds. For more information see Succulents.

EUPHORBIACEAE—Spurge Family. This vast and horticulturally important dicotyledonous family is native in all regions where vegetation occurs except the arctic. Among its most familiar sorts are castor-bean, crown-of-thorns, poinsettias, and many cactus-like desert kinds esteemed by fanciers of succulents. Commercial products of the family include cassava, castor oil, rubber, tapioca, and tung oil. Many kinds are poisonous. The majority of the about 5,000 species, organized in some 300 genera, are shrubs or trees, a few are woody vines, others are herbaceous perennials or annuals. A minority have stinging hairs. A very common feature is the possession of milky sap.

The leaves of euphorbs, as members of this family are called, are generally alternate, more rarely opposite; in some kinds they are alternate on the lower parts of the plant, opposite above. In succulent sorts the leaves are often few, greatly reduced in size, or are vestigial. Such kinds commonly have green, gray-green, or bluish-green, chlorophyll-containing stems that perform the function of photosynthesis.

The flowers are unisexual with both sexes or only one on individual plants. They are variously arranged in often complex inflorescences, those of some sorts with colorful, showy bracts and, as is true of poinsettias, having much the appearance of large and conspicuous single blooms. The true flowers may have usually five-parted calyxes and corollas, or one or both may be lacking. Male blooms have only one stamen or as many as twice as many as the sepals or more.

If more than one, the stamens may be separate, united, or partly joined. Often rudimentary female organs are present in male blooms. Female flowers frequently display staminodes (non-functional stamens). They have three styles, which may be united at their bases and often are two-lobed. There are three or six stigmas. The fruits are usually three-lobed capsules, less often drupelike or berrylike.

Cultivated genera include *Acalypha, Aleurites, Andrachne, Antidesma, Bischofia, Breynia, Codiaeum, Dalechampia, Daphniphyllum, Drypetes, Elaeophorbia, Euphorbia, Glochidion, Hevea, Homalanthus, Hura, Jatropha, Mallotus, Manihot, Mercurialis, Monadenium, Pedilanthus, Phyllanthus, Ricinus, Sapium, Securinega, Stillingia,* and *Synadenium.*

EUPHORIA (Euph-òria)—Longan or Lungan. One species of this genus is cultivated in warm climates for its edible fruits, for shade, and for ornament. It is the longan or lungan. The genus belongs in the sapodilla family SAPOTACEAE. Its name is derived from the Greek *eu,* well, and *phore,* bearing, and alludes to its fruit-producing habits. Euphorias are indigenous in tropical Asia and Indonesia. There are fifteen species. At one time they were included in *Nephelium,* but the trees and shrubs that now constitute *Euphoria* are segregated because they have overlapping calyx lobes and petals. The leaves of euphorias are pinnate, with their leaflets in opposite pairs. The flowers are tiny and either male or bisexual. They have cup-shaped calyxes with four or more segments and, usually, the same number of petals. The fruits do not split open. The seeds are surrounded by a thin layer of pulp encased in a more or less hard skin usually covered with small projections.

The longan (***E. longana*** syn. *Nephelium longana*) has very much the appearance of its close relative the litchi (*Litchi chinensis*), but, unlike it, its flowers have petals. Thirty to 40 feet in height, it has spreading branches and glossy, hairless, evergreen, leathery leaves with two to five nearly opposite pairs of short-stalked, elliptic to lanceolate, narrow leaflets up to 1 foot in length. When young they are reddish. Borne in terminal and axillary panicles, the yellowish-white flowers, male and bisexual ones on the same tree, are about 1⁄10 inch in diameter. They have deeply-five- or sometimes six-lobed, hairy calyxes and the same number of spoon-shaped petals about as long as the sepals. There are six to ten stamens. The fruits, which hang in grapelike clusters, are spherical, 3⁄4 inch in diameter, and are yellowish-red to brown. The pulp surrounding the solitary seeds separates easily from them. It is white, juicy, and slightly sweet. In the Orient the raw, dried, and preserved fruits are esteemed for eating. They are inferior to those of the litchi and in general do not appeal to American taste. They are reputed to possess medicinal qualities. The longan is indigenous to India, Burma, and Ceylon.

Garden and Landscape Uses and Cultivation. In sheltered places in southern Florida and southern California, and in Hawaii and other warm countries, the longan thrives and is cultivated for its fruits and as an ornamental. It stands slight frost, but not heavy freezes. It is hardier than the litchi and less exacting in its requirements. In southern California it appreciates a little shade as protection from the drying influence of full sun. During the growing season the longan needs abundant moisture. It is propagated by seeds, layering, and grafting. Seedlings commonly take a long time before they fruit. The longan thrives in any fertile soil.

EUPLOCA. See Heliotropium.

EUPRITCHARDIA. See Pritchardia.

EUPTELEA (Eup-tèlea) This genus, considered by some botanists and here to be a distinct family called the EUPTELEACEAE, of which it is the only member, by others is included in the trochodendron family TROCHODENDRACEAE. It comprises two or three species of deciduous trees or shrubs of eastern Asia. The name, from the Greek *eu*, well or handsome, and *ptelea*, the elm, alludes to the appearance of the fruits.

Eupteleas have alternate, pinnately-veined, hairless, toothed leaves. Their flowers, without sepals or petals, appear in small axillary clusters in spring before or with the foliage. They have many stamens and eight to eighteen stalkless pistils. The flat, wedge-shaped fruits are technically samaras.

Native of Japan, ***Euptelea polyandra*** is sometimes planted for variety, but is mostly restricted to botanical collections. A shrub or slender tree up to 45 feet in height, it has roundish-ovate to triangular-ovate, irregularly-toothed, longish-stalked leaves up to 6 inches long. Their undersides are pale green and have sparsely-hairy veins. When young the foliage is reddish. In fall it becomes yellow and red. The fruits are usually one-seeded. Differing chiefly in the toothing of its leaves, which is more regular than that of the last, and in its fruits frequently having two or three seeds, ***E. pleiosperma*** (syn. *E. franchetii*) is indigenous to central China.

Garden and Landscape Uses and Cultivation. Of graceful appearance and with attractive foliage, eupteleas are worth planting for variety where a small tree will serve. Those mentioned above are hardy in southern New England and grow satisfactorily in ordinary soil. They need no regular pruning or other particular care. They are propagated by seeds and cuttings.

EUPTELEACEAE—Euptelea Family. The characteristics of this family of dicotyledons are those of its only genus, *Euptelea.*

EUROTIA (Eurò-tia)—Winter Fat. Seven species of western North America, Europe, North Africa, and temperate Asia constitute *Eurotia* of the goosefoot family CHENOPODIACEAE. Its name, referring to the whitish appearance of the stems and foliage, comes from the Greek *euros*, mold. Eurotias are low shrubs with abundant stellate (star-shaped) hairs. The slender leaves, alternate or clustered, are not lobed or toothed. The inconspicuous flowers are bisexual or unisexual. The males have a four-parted perianth and four protruding stamens. The females are without perianths, but have two bracts, joined to above their middles, and two protruding styles. The small fruits are four-angled and have two short appendages.

Winter fat, white-sage, sweet-sage, or lamb's tails (***E. lanata***) is native in more or less alkaline soils from California to New Mexico and Saskatchewan. It is a desert and semidesert species, 1 foot to 2½ feet in height, densely-white-hairy or becoming rust colored with age. The linear to lanceolate leaves have their margins rolled under. The main ones are 3⁄4 inch to 2 inches long; much shorter ones are clustered in the axils of the main ones. The flowers, in small axillary clusters, are congregated in spikes at the branch ends. This species is much esteemed as forage.

Garden Uses and Cultivation. Winter fat is occasionally planted in gardens, chiefly in arid and semiarid regions. It needs dryish, porous soil and full sun. It is propagated by seed.

EURYA (Eù-rya). This genus must be distinguished from closely related *Cleyera*. Both belong in the tea family THEACEAE, both consist of evergreen shrubs and small trees. In gardens *Eurya japonica* and *Cleyera japonica* (syn. *Eurya ochnacea*) are sometimes confused. The chief differences are that *Eurya* has unisexual flowers, with the sexes on separate plants, and anthers that are not hairy. It comprises about 100 species, mostly of southern and eastern Asia, but represented in Central America. Its name is of unknown derivation.

Euryas have alternate, usually toothed leaves, and stalkless or short-stalked small flowers, solitary or in clusters in the leaf axils. There are five sepals, and five petals joined at their bases for about one-third of their lengths. The males have usually many, but sometimes only five, stamens, and the females, two to five styles, joined or nearly separate. The fruits are many-seeded berries.

Native to Japan, the Ryukyu Islands, Taiwan, Korea, China, India, and Malaysia, ***E. japonica*** is a shrub or low tree with thinly-hairy or hairless, short-stalked, leathery, elliptic to oblong-lanceolate, toothed leaves up to 3½ inches long by up to 1½ inches wide, each with a very prominent mid-vein and the other veins scarcely noticeable. Their uppersides are glossy and dark green. Their undersides are paler. The greenish-yellow flowers are up to 1⁄4 inch wide, as are the spherical, purplish-black fruits. Variety *E. j. australis*, of Japan, is distinguished by having hairy branchlets and flower stalks and hairs on the backs of its sepals. In *E. j. yakushimensis*, also of Japan, the leaves are not over 1¼ inches in length or ½ inch in width. Leaves handsomely variegated from their margins inward with creamy-white distinguish *E. j. variegata*. As its name suggests, a native of China, ***E. chinensis***, differs in its young shoots, and buds that terminate the branches, being hairy, and in having leaves more finely-toothed.

The Japanese ***E. emarginata*** differs from *E. japonica* in its branchlets being densely-hairy, an its narrowly-obovate leaves having rounded, often indented apexes. In the wild it frequently grows near seashores. Its short-stalked or stalkless, blunt-toothed leaves are 3⁄4 inch to 1½ inches long. Solitary or in twos or threes, the yellow-green flowers, also short-stalked, are not over 1⁄4 inch wide. The purplish-black, spherical fruits are about 1⁄4 inch in diameter.

Garden and Landscape Uses. Euryas are handsome furnishings for borders, beds, foundation plantings and similar accommodations, and for planting as single specimens. Their claims to beauty rest upon good growth habits and handsome foliage. They grow slowly and succeed in ordinary soils that do not become excessively dry, but are satisfactorily drained. They do well in partial shade. None is hardy in the north.

Cultivation. No special care or regular pruning are needed. Occasionally a little cutting may be required to shape or contain the plants to size. This is best done in spring. Propagation is by late summer cuttings 2 to 4 inches long, planted in a cold frame or greenhouse propagating bench or under mist, and by seeds sown in sandy peaty soil.

EURYALE (Eury-àle). The only species of the very distinctive aquatic genus *Euryale* of the water-lily family NYMPHAEACEAE somewhat resembles the giant water lily *Victoria,* but its leaves are considerably smaller, are without upturned rims, and are covered with prickles on their upper surfaces. A less obvious difference is that in *Euryale* all the stamens of the flowers are fertile. The name, pronounced with four syllables, is that of a thorny-locked Gorgon of Greek mythology and alludes to the thorny character of the plant.

Although most often grown as an annual and attaining full size in its first year, ***E. ferox*** is a tender perennial. It has circular, puckered, floating leaves 2 to 4 feet in diameter, purple beneath and olive-green above. The leafstalk is attached at about the center of the blade without any opening from the top of the leafstalk to the leaf margin. The solitary, violet-red flowers, on spiny stalks, have thorny calyxes. About 2 inches in diameter and open during the day, they have twenty to thirty petals, which are shorter than the reddish calyx lobes, and numerous stamens. The fruits are berries containing many seeds that where the plant is native are used as food. The young stalks and roots are also edible.

Garden Uses and Cultivation. Although much less dramatic in appearance than the giant Amazon water-lily *Victoria,* Asian *Euryale* is a quite striking ornament for pools large enough to support its vigorous growth and display to advantage its platter-like leaves. It is much hardier than *Victoria.* As far north as Philadelphia and St. Louis and even at New York City it self-sows and renews itself each year from seeds. It needs full sun and, for its best development, an ample amount of fertile soil. A cubic yard is by no means too much, but quite good specimens can be had in less. The soil surface may be 6 inches to 1 foot under water. This plant needs essentially the same conditions as tropical water-lilies. Propagation is by seeds sown in pots or pans immersed so that the seeds are under an inch or two of water, which should be at a temperature of 70 to 75°F. As soon as they are large enough to handle comfortably the young plants are potted individually and as they increase in size are repotted and gradually lowered in the water to their eventual depth. Finally, when the weather is really mild and settled and the water warm, they are planted, usually from 5- or 6-inch pots, into the soil beds or large containers of soil in which they are to grow. This is usually done about two weeks after it is safe to plant tomatoes outdoors.

Euryale ferox

EURYCHONE (Eury-chò-ne). Scarcely known in cultivation, this tropical African genus of the orchid family ORCHIDACEAE consists of two species closely related to *Angraecum.* The name, alluding to the form of the flowers, comes from the Greek *eurys,* broad, and *chone,* a funnel.

Eurychones are without pseudobulbs. They have short stems and broadly-oblanceolate leaves unequally bilobed at their apexes. Their flowers are borne from the leaf axils. They are spurred, with the spur, which approximates the blade of the lip in length, broadening to a wide mouth that merges into the blunt, obscurely three-lobed lip, which is very much broader than the two petals and three sepals, which are more nearly of equal shape and size.

Native of Uganda, ***Eurychone rothschildianum*** (syn. *Angraecum rothschildianum*) has a stem about 3 inches long by up to slightly over 2 inches wide. Its flowering stalks, about 2½ inches long, carry up to five very fragrant blooms, closely and alternately arranged. They are white with a center band of pale green on the sepals and slightly shorter and broader petals. The sepals and petals are ¾ to 1 inch long. The spur and broadly-white-margined emerald-green, nearly circular lip together are nearly 1½ inches long.

Garden Uses and Cultivation. Little is reported about the cultivation of this rare orchid. It probably responds to conditions and care that suit *Angraecum.* For more information see Orchids.

Eurychone rothschildianum

EURYCLES (Eùry-cles) — Brisbane Lily. Closely related to *Pancratium,* this Old World genus of two or perhaps three species of tropical bulb plants belongs to the amaryllis family AMARYLLIDACEAE. Its name, alluding to the often imperfect staminal cups of the blooms, comes from the Greek *eurys,* broad, and *kleio,* close up.

The leaves of *Eurycles* are stalked and have broad blades with conspicuous veins curving outward and upward from the midribs in the manner of those of some plantain-lilies (*Hosta*). The leafless flowering stalks terminate in umbels of slender-tubed, six-petaled, white blooms. Their white stamens, attached in the throat of the flower, have stalks shorter than the anthers, their lower halves united in a more or less perfect cup. The fruits are capsules.

Commonest in cultivation, ***E. sylvestris*** (syn. *E. amboinensis*) is a handsome native of northern Australia, the Philippine Islands, and Malaya. This has much the aspect of the Amazon-lily (*Eucharis grandiflora*), but in foliage is even more handsome. It has bulbs 1½ to 2 inches in diameter. Its leaves, 6 inches to 1 foot across, are broadly-heart-shaped to nearly round. The flowers, smaller than those of the Amazon-lily and not fragrant, have slender corolla tubes and spreading petals. The staminal cup is deeply six-parted. The Brisbane-lily (***E. cunninghamii***) differs from the last in its flowers having the corona (cuplike center) two-thirds instead of under one-quarter as long as the petals and in the stigma being three-lobed instead of lobeless.

Garden and Landscape Uses and Cultivation. These are as for *Eucharis grandiflora*.

EURYOPS (Eurỳ-ops). Alternate-leaved, evergreen subshrubs and shrubs of the daisy family COMPOSITAE constitute this genus. The name, the greek *euryops*, having large eyes, refers to the showy flower heads of some of its fifty species. As a native this group occurs from South Africa to Arabia and the island of Socotra.

The leaves of *Euryops* are lobeless to pinnately-lobed. The solitary, yellow, daisy-like flower heads are on leafless stalks from the leaf axils or ends of the branches. Each has a center eye of tubular, five-toothed disk florets, encircled by spreading, petal-like, female ray florets. The seedlike fruits are achenes.

The kinds now to be described are natives of South Africa. Its foliage resembling that of southernwood (*Artemisia abrotanum*), 2- to 3-foot-tall ***E. abrotanifolius*** has hairless leaves 1 inch to 2 inches long, pinnately-cut into mostly nine to thirteen, usually threadlike lobes ¼ to ¾ inch long. Atop slender stalks up to 6 inches long, the flower heads are 1½ to 2 inches wide, with fourteen to eighteen ray florets. From the last, ***E. pectinatus*** differs in having stems, leaves, and other parts densely-felted with gray or white hairs. The upper parts of its leaves are divided into sixteen to twenty lobes ½ to ¾ inch long often toothed at their apexes. The flower heads, on stalks up to 6 inches long, have thirteen to twenty ray florets and are 1½ to 2 inches in diameter.

Euryops pectinatus

Frequently misidentified as *E. evansii*, a species not known to be cultivated, ***E. acraeus*** forms a rounded bush 1½ to 2 feet in height with its foliage chiefly clustered at the branch ends. The leathery, silvery-gray, oblong leaves, ½ to ¾ inch long, have turned-back margins, which with age become strongly reflexed. About 1 inch wide, the flower heads have about eleven ray florets. An erect, bushy, essentially hairless shrub, 1½ to 6 feet tall, ***E. chrysanthemoides*** has spreading, flat, shallowly to deeply pinnately-lobed, oblanceolate leaves 1½ to 4 inches long. The flower heads, terminal on the branches and with stalks 2 to 8 inches long, have eleven to twenty yellow ray florets ½ to ¾ inch long.

Euryops chrysanthemoides

Favoring sandy soils, ***E. athanasiae*** is an erect, much-branched, hairless shrub 2 to 5 feet tall. Sometimes parts of it stems and foliage are coated with a thin layer of white wax. The green, leathery leaves, channeled on their upper and lower surfaces, are pinnately-divided into four to seven pairs of almost threadlike, pointed lobes up to 2 to 3 inches long. The flower heads have twenty to thirty-two ray florets, and are 2 to 3½ inches in diameter. From 2 to 3 feet tall, ***E. spathaceus*** has hairless stems and foliage. Only the younger branches are leafy; the older ones are naked and scarred. Lobeless, toothless, and up to 3 inches in length, its leaves are oblong. The flower heads supported on 3-inch-long stalks have six to eight ray florets.

Euryops athanasiae

Garden and Landscape Uses and Cultivation. These are sun-loving plants for ordinary, well-drained soils. They add color to shrub beds and other plantings but are hardy only where little or no frost occurs. They are sometimes grown in greenhouses. Outdoors they prosper in climates of the Mediterranean type, such as that of California. No special care is needed. Any pruning deemed necessary may be done when flowering is through. New plants are easily raised from cuttings and seeds. As greenhouse plants these succeed in any ordinary potting soil in well-drained pots. They need full sun and on all favorable occasions free ventilation. From fall to spring temperatures may be 50°F at night and, depending on the brightness of the weather, five to fifteen degrees higher by day.

EUSCAPHIS (Eús-caphis). Not hardy in the north, what most botanists accept as the only species of this genus of the bladder nut family STAPHYLEACEAE has beautiful foliage and bears decorative seed pods. Native of Japan and China, it is a hairless, deciduous shrub or small tree. Its name derives from the Greek *eu*, good, and *scaphis*, a vessel, and alludes to the appearance of the fruits.

The leaves of ***Euscaphis japonica*** are opposite and pinnate, with seven, nine, or eleven short-stalked, lanceolate-ovate, finely-toothed leaflets 1½ to 3 inches long or a little longer. The yellowish-green or whitish blooms, ⅕ inch in diameter, are in many-flowered, erect, long-stalked, terminal panicles 2 to 4½ inches wide. They have five each sepals, petals, and stamens and two or three styles often joined at their tips. They are succeeded by reddish or purplish fruits consisting of one to three tiny, spreading, leathery, boat-shaped pods that split to reveal dark blue seeds.

Garden and Landscape Uses and Cultivation. For regions of mild winters this is a pleasing change from more commonly planted shrubs. It adapts to ordinary well-drained soils and is propagated by seeds

Eschscholzia californica

Euonymus alata (fall foliage)

Euonymus radicans carrierei, in fruit

Eucharis grandiflora

Euonymus fortunei gracilis

Euphorbia epithymoides

Euphorbia flanaganii

Euphorbia fulgens

Euphorbia milii splendens

Euphorbia palustris

Exacum affine

sown in spring in a cold frame or greenhouse, and from summer cuttings 3 to 4 inches long planted under mist or in a propagating bed in a humid cold frame or greenhouse. Any pruning needed to keep the plant shapely and to size is best done in late winter.

EUSTEPHIA (Eu-stèphia). Closely related to *Phaedranassa* from which it differs in its leaves not being stalked and in the stamens of its flowers having winged stalks, *Eustephia* consists of six species of nonhardy bulb plants native to the Andean region. It belongs in the amaryllis family AMARYLLIDACEAE. The name, alluding to the circle of stamens, comes from the Greek *eu*, well or good, and *stephos*, a crown.

Eustephias have all basal, linear to narrowly-lanceolate leaves and erect, somewhat flattened flowering stalks. The latter terminate in an umbel of more or less pendulous, short-tubed, narrowly-funnel-shaped blooms with six red and green or orange-red petals (more properly tepals), the same number of stamens, and one style. The fruits are capsules.

Native of Argentina, ***E. pamiana*** has about six erect and arching, linear leaves 1 foot long or a little longer forming a tuft encircling the base of the flowering stalk which is 1 foot to 1¼ feet tall. The umbel consists of about six stalked, nodding to pendulous blooms approximately 1¼ inches long, their lower halves flesh-colored, but above green with red apexes. The name ***E. jujuyensis*** is provisionally applied to a species collected in 1936 in the province of Juju in northern Argentina and since offered by commercial dealers in bulbs. This kind has an elongated, black-coated bulb and brilliant orange-red, individually short-stalked flowers in umbels atop stalks 1½ to 2 feet tall.

Garden Uses and Cultivation. These are as for *Phaedranassa*.

EUSTOMA (Eù-stoma)—Bluebells, or Prairie-Gentian, Catchfly-Gentian. This genus of three species is native from North America to South America and the West Indies. Its name is from the Greek *eu*, well or good, and *stoma*, a mouth, in allusion to the wide-mouthed corollas of the blooms.

Eustomas are annuals, biennials, or short-lived perennials of the gentian family GENTIANACEAE. They are tap-rooted and more or less glaucous. From rosettes of basal foliage they send up stems with opposite, undivided, toothless, stemless, and often stem-clasping leaves, from the upper axils of which develop, singly or in panicles, the long-stalked flowers. They have keeled calyxes with long, pointed lobes and blue to white, bell-shaped corollas with five or six lobes (petals). There are five or six stamens and a slender style ending in a conspicuously two-lipped stigma. The fruits are many-seeded capsules.

Texan bluebells or prairie-gentian (***Eustoma grandiflorum*** syns. *E. russellianum, Lisianthus russellianus*) is a handsome annual or biennial up to 3 feet tall. It has ovate to elliptic-oblong or elliptic-lanceolate, distinctly three-veined leaves up to 3½ inches long by 1½ inches wide. The flowers, on stalks up to 2½ inches long, have corollas 2 to 2½ inches long and wide. They are blue-purple, pinkish, white, white tinged with purple, or yellowish. This inhabits moist prairies and fields from Nebraska to Colorado, Texas, and Mexico. From it ***E. exaltatum*** (syn. *E. selenifolium*), a biennial or perhaps perennial of damp, frequently saline and alkaline soils in the southern and southwestern United States, Mexico, Central America, and the West Indies, differs most noticeably in having decidedly smaller blooms.

Garden and Landscape Uses. In climates similar to those of their natural ranges eustomas are pretty garden decoratives and, in their home territories, are welcome additions to native plant gardens. They are also attractive in greenhouses. Cut, their flowers last well in water. Of the kinds discussed *E. grandiflorum*, because of its larger blooms, is superior for most garden purposes.

Cultivation. Eustomas are considered somewhat difficult to grow. One reason for this is their intolerance of root disturbance. Good success can be had by treating them as biennials. The seeds are sown in June and the seedlings transplanted, as soon as they have their second pair of leaves, individually into small pots in porous soil. When these are filled with roots the plants, in regions where they are hardy outdoors, are planted where they are to bloom, in a sunny place and in soil that is fertile and porous, yet does not suffer from lack of moisture during the season of growth. A neutral or slightly alkaline earth is best. Between plants 6 to 9 inches is satisfactory.

In greenhouses, eustomas are transplanted from the small pots to pots 4 inches in diameter, and finally to 5- or 6-inch pots or to benches. The objective is to have, by fall, specimens with strong basal rosettes. Through the winter they are grown in full sun where the night temperature is 50 to 55°F, and by day five to ten degrees higher. Whenever weather is favorable the greenhouse is ventilated freely to prevent the atmosphere from becoming excessively humid, and care is taken not to keep the soil too wet. With the coming of spring discreet applications of dilute liquid fertilizer are helpful. Flowering is in spring or early summer.

EUSTREPHUS (Eu-strèphus)—Wombat Berry, Australian Climbing-Lily. The only species of *Eustrephus*, of the Lily family LILIACEAE, is a native of Australia. Its name, from the Greek *eu*, well, and *strepho*, to climb, alludes to the growth habit.

The wombat berry or Australian climbing-lily (*E. latifolius* syn. *Luzuriaga latifolia*) is a nonaggressive, tuberous-rooted, nontwining, freely-branching vine or when grown in exposed, dryish locations, a shrub. As a vine it attains heights of 10 feet or more. Its attractive foliage consists of nearly stalkless, lustrous, linear, ovate-lanceolate to broadly-elliptic leaves 2 to 4 inches long by 1/10 inch to 2 inches wide, with prominent, parallel, longitudinal, fine veins. The starry, white, pink, or lavender, ¼-inch-long blooms, on slender stalks ¼ to ¾ inch in length, are in clusters of two to ten in the upper leaf axils. They have three fringed inner petals and three thicker, fringeless outer ones. The decorative fruits are bright orange berries about ½ inch in diameter. The tubers developed at the root are sweet and edible.

Garden Uses and Cultivation. Adaptable for outdoor cultivation in southern Florida and similar warm, frost-free or nearly frost-free regions, and in cool greenhouses, the wombat berry is interesting as a rather uncommon foliage and fruiting vine. Its flowers are not showy. It succeeds with little attention in ordinary well-drained, moistish soil in part-day shade. It may be used to clothe pergolas and other supports. In greenhouses a winter night temperature of 50°F is satisfactory. By day at that season a rise of five to fifteen degrees is permissible. At other seasons the greenhouse should be ventilated freely to maintain cool, airy conditions. Watering is done to keep the soil always moderately moist, and somewhat drier in winter than at other times. Propagation is by seed and by division.

EUSTYLIS (Eù-stylis)—Pinewoods-Lily. This small genus of mostly South American plants is by some botanists included in *Nemastylis*; others separate it because of technical differences in the stamens and fruits. It belongs in the iris family IRIDACEAE. From the Greek *eu*, good or well, and *stylis*, a little pillar, its name alludes to the style.

The only kind cultivated, the pinewoods-lily (***Eustylis purpurea*** syn. *Nemastylis purpurea*) is native of sandy soils in grassy places and open woodlands in Texas and Louisiana. It is herbaceous and has brown, ovoid bulbs about ¾ inch in diameter. Its leaves are mostly basal, longitudinally pleated, up to 2 feet long, and about ¾ inch wide. Their bases sheath the lower parts of the stems. The flowers are two to three together from spathes (bracts) at the ends of often zigzag, branching, wiry stems. They are very short-lived, those of each spathe opening in succession. They expand in the morning and close around noon. The blooms have six perianth segments (petals, or more correctly, tepals), the outer three somewhat longer and broader than the others and convex at their bases so that they form a central cup to the bloom as with the flowers of tigridias. The petals are prevailingly purple to rose-purple, but sometimes are whitish, and toward their bases are yellow spotted with reddish-brown. The outer three are obovate and about 1 inch long, the inner ones at their apexes are cupped and crimped. There are three stamens. The style has three recurved branches. About 2 inches long, the seed capsules are ellipsoid.

Garden Uses and Cultivation. In the wild the pinewoods-lily grows in sandy soils in open woods and prairies. Conditions similar to those of its natural habitats should be afforded cultivated specimens. This is a plant for the collector of the rare and unusual rather than for the general gardener. Just how hardy it is has not been determined, but it surely will not survive in the north, and probably not where its bulb is subjected to freezing. Propagation is by seed.

EUTAXIA (Eu-táxia). At least one of the nine species of this Australian genus of the pea family LEGUMINOSAE is grown in California and elsewhere where mild, dry climates prevail. The group consists of opposite-leaved, evergreen shrubs with pea-shaped flowers. The fruits are pods. Its name derives from the Greek *eutaxia,* modesty, and alludes to the delicate appearance of the blooms.

A heathlike, hairless or nearly hairless shrub, 2 to 4 feet in height, much branched, sometimes spiny, and of diffuse habit, ***Eutaxia empetrifolia*** has ovate-oblong to linear leaves up to ¼ inch long, without midribs. From the leaf axils on the upper parts of the branches and branchlets, solitary or in pairs, the very short-stalked, red-and-yellow flowers are borne. They have standard petals up to ⅓ inch in length, and are succeeded by ¼-inch-long seed pods.

Garden and Landscape Uses and Cultivation. The species discussed is suitable for shrub borders, rock gardens, and other places where a low flowering shrub can be used to advantage. It succeeds in well-drained soil in full sun. Propagation is by seeds and by summer cuttings.

EUTERPE (Eutér-pe)—Cabbage Palm, Manac Palm. Fifty feather-leaved species of tropical America and the West Indies belong in *Euterpe* of the palm family PALMAE. The name is that of one of the nine Muses of Greek mythology. Euterpes are often called cabbage palms because their terminal buds or "cabbages," also called hearts of palm, are considered to be among the most delicately flavored and tender of those of all palms. The name cabbage palm is especially commonly used for *E. oleracea,* the buds of which are eaten pickled and as a salad.

Euterpes are somewhat similar in general aspect to royal palms (*Roystonea*), but differ in being mostly lower, sometimes having more than one trunk, and in technical characteristics of their flowers, which develop from below the crownshaft in branched clusters and are succeeded by pea-like fruits. Unlike the flowers of *Roystonea* and *Prestoea,* the latter closely related to *Euterpe,* those of *Euterpe* are in depressions in the stalks that bear them. In their native lands euterpes grow under very varied conditions. In Colombia, for example, from sweltering swampy forests at sea level (several species) to altitudes of more than 10,000 feet (*E. dasystachys*). This wide variation of habitats suggests the desirability of exploring the possibilities the various sorts have as decorative subjects for tropical and subtropical plantings.

The assai or cabbage palm (***E. oleracea***) is one of the commonest palms of the Amazon region of Brazil. In and around Belem it is a very familiar feature of yards and gardens and is planted largely for its crimson, cherry-like fruits, which are produced in abundance. Their pulp is used to make a popular refreshing beverage called assai. This palm has a smooth, slender trunk up to 100 feet in height from the top of which rises a crownshaft looking like a continuation of the trunk and composed of the basal sheathing portions of the leafstalks. The leaves are comparatively few, 4 to 6 feet in length and curved downward toward their ends. They have many narrow, pendulous leaflets up to 2 feet long. Common south of Amazonia, ***E. edulis*** sometimes attains a height of 90 feet and has leaves 6 to 9 feet long with drooping leaflets up to 3 feet long. The plant formerly called *E. globosa* is *Prestoea montana.*

Garden and Landscape Uses and Cultivation. These beautiful palms, little known to gardeners and horticulturists, should be more commonly grown as ornamentals.

Euterpe edulis in Panama

They may be expected to prosper under the same conditions as royal palms, but recorded experience of their outdoor cultivation is not adequate. In greenhouses they thrive without extraordinary care if accorded a minimum temperature of 60 to 65°F in winter, a constantly humid atmosphere, and shade from strong sun. At other seasons the night temperature should be higher and always the day temperature five to ten degrees above that maintained at night. A coarse, fertile, porous, well-drained soil suits. From spring through fall generous amounts of water are needed, somewhat less in winter, but the soil must never be dry. Established specimens benefit from biweekly applications of dilute liquid fertilizer from spring through fall. Propagation is by fresh seeds sown in sandy peaty soil in a temperature of 75 to 90°F. For more information see Palms.

EUTOCA. See Phacelia.

EVAX (È-vax). This temperate-region genus of about two dozen species of annuals and herbaceous perennials is little known horticulturally. It belongs to the daisy famly COMPOSITAE and has a name said to be that of an Arabian chief.

Low, tap-rooted plants that usually branch freely low down, evaxes have alternate, undivided, generally spatula-shaped leaves, ordinarily not over about ½ inch long, and small densely-woolly terminal clusters of little flower heads of all disk florets (the kind that form the central eyes in daisy-type flower heads), with collars of leafy bracts beneath them. They differ from *Gnaphalium* in the flower heads having both male and female florets. The seedlike fruits are achenes.

Sometimes cultivated, ***Evax verna*** (syns. *E. multicaulis, Filago nivea*) is native from Georgia to Texas. From 3 to 9 inches in height, it has branched stems clothed all along with oblong to oblong-lanceolate leaves up to ½ inch long. The numerous heads, with their encircling bracts, are densely-white-woolly and about ½ inch across.

Garden Uses and Cultivation. Sunny, dryish places in rock gardens are likely locations for this plant, which is raised from seeds sown in early spring in gritty well-drained soil. The seedlings are thinned to prevent undue crowding.

EVENING-PRIMROSE. See Oenothera.

EVENING SNOW is *Linanthus dichotomus.*

EVENING STOCK is *Matthiola longipetala.*

EVERGREEN. As an adjective this is correctly applied to all plants that retain their foliage throughout the year. There are evergreen herbaceous perennials, including some bulb plants, as well as evergreen shrubs, trees, and other more or less woody-stemmed plants. By expansion of this definition the term is also sometimes applied to kinds such as brooms (*Cytisus* and related genera) that have minute leaves that soon drop and a plentitude of stems green at all seasons. As a noun evergreen has a more restricted meaning. See Evergreens.

EVERGREEN CANDYTUFT is *Iberis sempervirens.*

EVERGREENS. In its widest context the term evergreens includes all plants that retain foliage throughout the year, all that are not deciduous. In the moist tropics and subtropics the vast majority of kinds do this and in warm climates with long dry seasons gardeners, by supplementing natural rainfall, succeed with numerous evergreen plants that otherwise could not survive. Because of this in most mild climates evergreens form such a substantial part of the natural and planted vegetation that they belong in the landscape without being thought of as a distinctive group.

In temperate regions the reverse is true. Except where a few species, such as pines or spruces, dominate great stretches of forest, deciduous trees and shrubs are highly visible elements in the landscape and one tends to group plants as deciduous or winter-leaf-losing and evergreen. The evergreens in turn are segregated into two groups, narrow-leaved and broad-leaved. The former have no flowers that nonbotanists are likely to recognize as such. Many of the others present glorious displays.

As ornamentals cultivated evergreens are superb. They are chiefly responsible for the well-furnished winter appearance of landscapes in which they thrive in contrast to those where they are scarce or absent. The abundance of broad-leaved evergreens is chiefly responsible for the noticeably different aspect of gardens and other landscapes in mild climates as compared to those of colder ones where chief reliance for winter greenery is of necessity upon narrow-leaved evergreens. This discussion is concerned with evergreen trees, shrubs, vines, and groundcovers and their horticultural employments and cultural needs, with especial emphasis on temperate regions where the ground freezes in winter to a greater or lesser degree.

Narrow-leaved evergreens are in the broad acceptance of that term all conifers, but not all bear easily recognized cones. Junipers, podocarpuses, yews, and some others have soft, berry-like fruits. All kinds are greatly admired for their distinctive forms and foliage and particularly for their great importance in bringing life and warmth to winter landscapes. According to sort they are prized as single specimens, for grouping, as screens, as backgrounds to flowering plants, as hedges and groundcovers, and for providing shelter.

Hardy kinds include arbor-vitaes (*Thuja*), cedars (*Cedrus*), cryptomerias (*Cryptomeria*), Douglas-firs (*Pseudotsuga*), false-cypresses (*Chamaecyparis*), firs (*Abies*), hemlocks (*Tsuga*), incense-cedar (*Calocedrus*), junipers (*Juniperus*), pines (*Pinus*), plum-yews (*Cephalotaxus*), spruces (*Picea*), umbrella-pine (*Sciadopitys*), and yews (*Taxus*). Among these are many that withstand much more cold than all but a few very low broad-leaved sorts. So in climates harsher than that of Washington, D.C. among evergreens, especially evergreen trees, narrow-leaved kinds strongly predominate. Less hardy narrow-leaved evergreens suitable for mild climates include only certain species of some of the genera previously listed and in addition araucarias, cunninghamias, cypresses (*Cupressus*), giant-sequoia (*Sequoiadendron*), Hiba-arborvitae (*Thujopsis*), redwoods (*Sequoia*), and torreyas.

Narrow-leaved evergreens: (a) Cedar (*Cedrus deodara*)

(b) False-cypress (*Chamaecyparis lawsoniana*)

(c) Fir *(Abies nordmanniana)*

(d) Giant-sequoia *(Sequoiadendron giganteum)*

(e) Hemlock *(Tsuga caroliniana)*

(f) Pine *(Pinus nigra austriaca)*

(g) Podocarpus *(Podocarpus totara)*

(h) Spruce *(Picea pungens glauca)*

(i) Umbrella-pine *(Sciadopitys verticillata)*

(j) Yew *(Taxus baccata stricta)*

Broad-leaved evergreens encompass all that are not conifers, all with flowers easily recognizable as such. Unlike narrow-leaved evergreens they belong to many different plant families and exhibit considerably more diversity in appearance. They include plants only inches high, taller shrubs, and small to tall trees. Many are splendid ornamentals esteemed for their foliage, often for their handsome flowers or fruits. The majority are natives of warm-temperate, subtropical, and tropical regions, especially where rainfall is fairly well distributed throughout the year. Others inhabit lands with Mediterranean-type climates where summers are dry. Few occur in regions of severe winters. Not all broad-leaved evergreens, the heaths (*Erica*), for example, have wider leaves than narrow-leaved kinds, and some few of the latter, more especially those suited to mild climates such as araucarias and certain kinds of *Podocarpus*, have distinctly wide, flat leaves.

Broad-leaved evergreens: (a) Andromeda *(Pieris japonica)*

(b) *Aucuba japonica variegata*

(c) Boxwood *(Buxus sempervirens)*

(d) *Ceratonia siliqua*

(e) Holly *(Ilex altaclarensis)*

(f) *Magnolia grandiflora*

(g) Mountain laurel *(Kalmia latifolia)*

(h) Rhododendron

(i) *Skimmia reevesiana*

(j) *Viburnum rhytidophyllum*

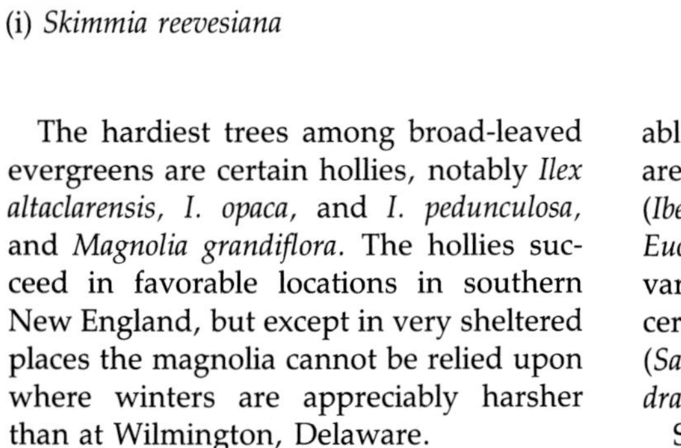

The hardiest trees among broad-leaved evergreens are certain hollies, notably *Ilex altaclarensis, I. opaca,* and *I. pedunculosa,* and *Magnolia grandiflora.* The hollies succeed in favorable locations in southern New England, but except in very sheltered places the magnolia cannot be relied upon where winters are appreciably harsher than at Wilmington, Delaware.

The hardiest broad-leaved evergreen shrubs of medium to tall height are selected sorts of andromedas (*Pieris*), barberries (*Berberis*), boxwoods (*Buxus*), hollies including inkberry (*Ilex*), rhododendrons including azaleas, mountain-laurel (*Kalmia*), *Prunus laurocerasus schipkaensis,* and *Yucca filamentosa.* Among lower kinds usable as groundcovers and some as vines are bearberry (*Arctostaphylos*), candytuft (*Iberis sempervirens*), English ivy (*Hedera*), *Euonymus fortunei coloratus* and some other varieties of *E. fortunei,* heather (*Calluna*), certain heaths (*Erica*), lavender-cotton (*Santolina*), myrtle (*Vinca minor*), *Pachysandra terminalis,* and some sedums.

Slightly less hardy broad-leaved evergreens growable in sheltered, favorable locations at least as far north as New York City are *Aucuba japonica,* some camellias, *Cotoneaster salicifolia floccosa, Elaeagnus pungens,* some kinds of *Leiophyllum, Leucothoe, Mahonia, Pyracantha,* and *Skimmia, Sarcococca humilis hookeriana, Trochodendron aralioides,* and *Viburnum rhytidophyllum.*

Nonhardy broad-leaved evergreens suitable for mild climates only, in addition to species of certain of the genera already listed include kinds of the following genera: *Acacia, Arbutus, Ardisia, Buddleia, Callistemon, Carissa, Ceratonia, Citrus, Cotoneaster, Danae, Daphne, Dombeya, Eriobotrya, Eucalyptus, Eugenia, Fatshedera, Fatsia, Feijoa, Ficus, Fuchsia, Gardenia, Gaultheria, Gaylussacia, Geijera, Gelsemium, Grevillea, Hakea, Hibiscus, Hymenosporum, Hypericum, Jacaranda, Jasminum, Lagunaria, Ligustrum, Lonicera, Loropetalum, Macadamia, Maytenus, Mitchelia, Myrica, Myrtus, Nandina, Nerium, Olea, Osmanthus, Pernettya, Persea, Photinia, Pittosporum, Plumbago, Prunus, Quercus, Rhaphiolepis, Ruscus, Schinus, Tris-*

tania, and *Vaccinium*, and in addition species of many genera of aroids, bamboos, palms, cactuses, and succulents listed in this Encyclopedia under those entries.

In choosing evergreens have in mind kinds likely to prosper locally and in the places and soils you wish to furnish. Keep in mind the purposes you want them to serve and the space needed to allow for their growth over a period of ten, fifteen, or perhaps twenty years or more. Remember that although some kinds can be restrained by appropriate pruning or shearing, others do not respond well to this treatment, and in any case, except with hedges and where formal effects are desired, this makes necessary work that would otherwise be avoidable.

The management of evergreens is dictated to a considerable extent by the fact that they are never without foliage. This means that in winter as well as at other seasons they lose by transpiration considerable amounts of moisture. This must be replaced by the roots if the plant is to prosper. Anything that speeds loss of moisture beyond the roots' capability to replace it, and in winter low soil temperatures severely lowers their power of absorption, or that reduces the extent of the root system poses a threat to the well-being or even survival of evergreens.

Exposure to drying winds is harmful to nearly all. Some pines and a few other kinds the foliage of which is protected by a thick, waxy skin or in other ways from excessive moisture loss, stand such exposure, but most evergreens do much better in humid atmospheres where exposure to wind is limited. Much winter killing of foliage of evergreens in the north results from dehydration, not low temperatures.

Transplantation, by reducing the amounts of roots, limits the ability of a plant to absorb water. With deciduous trees and shrubs it is common practice to transplant when they are leafless and thus loss of moisture by transpiration is minimal and often this is further reduced by more or less severe pruning. But evergreens transpire considerably at all seasons and often it is impracticable or undesirable to prune. An acceptable balance of input with output of water must be assured in other ways.

Horticultural practices especially important in the care of evergreens are those designed to prevent dehydration. So far as possible limit transplanting to times of the year when top growth is firm and mature and the soil is warm enough to encourage the speedy growth of new roots. The most favorable periods are in spring before new growth begins and early fall, beginning in late August, after the season's growth is complete and the shoots have hardened. With all except very small specimens it is important to dig evergreens to be transplanted with their roots contained in intact balls of soil rather than bare-rooted as is satisfactory with many deciduous trees and shrubs. Let the balls, proportionate to the specimens, be of generous sizes.

Spraying the foliage at transplanting time with an antidesiccant (antitranspirant) is helpful in reducing water loss. Sometimes judicious pruning or shearing can be done to achieve the same end without detracting from the appearance or too severely reducing the size of the plant. In other cases, with hollies, for example, it may be advantageous to reduce the transpiration surface by picking off up to about one-third of the leaves.

Watering very thoroughly immediately following transplanting and at regular intervals until root systems have been re-established is tremendously important. And so is supplying water to established evergreens during long spells of dry weather. Except for a few kinds, such as many pines and junipers that are drought-resistant, keeping the soil fairly moist and relatively cool at all times is necessary. Because of this, sorts such as camellias and plants of the heath family including azaleas and other rhododendrons, heathers, and pierises that root near the surface are greatly benefited by the maintenance of permanent mulches.

Do not allow evergreens to go into the winter dry is old-time advice often reiterated. Actually in most regions this rarely or never happens. Normally there is adequate precipitation before the ground freezes. But it is not unusual to experience long rainless periods well in advance of the ground freezing, and it is during dry falls from early to mid-October on, rather than immediately before the freezes, that deep watering of evergreens is most urgent. Before that, in late summer and early fall, a moderate degree of dryness is likely to be helpful by encouraging the season's growth to ripen well in preparation for winter.

In exposed locations excessive loss of moisture by evergreens in winter, with consequent danger of dehydration to the point of winter killing, can be substantially reduced by sheltering them from wind and affording some shade by erecting temporary screens or tentlike or boxlike structures of coarse burlap or of snow fencing or by spraying them in late December and about a month later with one of the antidesiccants sold for the purpose.

EVERLASTING-PEA is *Lathyrus latifolius*.

EVERLASTINGS. This is a group name for plants, some sometimes called immortelles, that have flowers with papery or chaffy parts that retain their forms and colors after being dried. This makes them suitable for use in dried arrangements and in other decorative ways. Drying is done by cutting them with long stems when they are without moisture on them, just before they attain fullest maturity, tying them together by their stalks in small bundles, and suspending them until quite dry upside down in a cool, shaded place where there is good air circulation. The best known everlastings belong in these genera: *Acrolinium, Ammobium, Anaphalis, Antennaria, Armeria, Celosia, Gomphrena, Helichrysum, Helipterum, Humea, Limonium, Waitzia*, and *Xeranthemum*.

In addition to these acknowledged everlastings, the blooms of some other plants not usually included under the term everlasting may be dried in the same way and used similarly. Ornamental grasses and cat tails belong here, and the blooms of hydrangeas.

Kinds not perhaps cultivated in gardens, the dried and often dyed blooms of which are imported and used for making the artificial dwarf trees or bonsai called Ming trees, are *Syngonanthus elegans* and *S. niveus* of the daisy family COMPOSITAE. They are natives of Brazil.

EVE'S NECKLACE is *Sophora affinis*.

EVODIA (E-vòdia). Widely dispersed in the wild from its chief center of distribution in eastern and southeastern Asia, to Madagascar, Polynesia, and Australia, *Evodia*, of the rue family RUTACEAE, is known in cultivation from a few deciduous representatives of its approximately fifty deciduous and evergreen species. Its name is also spelled *Euodia*. It consists of trees and shrubs related to *Phellodendron* and *Zanthoxylum* and having much the same aspect. It is easily distinguished from the first by its axillary buds being clearly visible and not hidden within the leafstalk, and from *Zanthoxylum* by its opposite leaves. The name comes from the Greek *euodia*, fragrance, and alludes to the aromatic foliage of some kinds.

Evodias have opposite, undivided or pinnate leaves, smooth-edged or slightly toothed, and, if pinnate, with an uneven number of leaflets. As is common in the rue family, they are besprinkled with tiny translucent dots visible when held to the light. The small, unisexual blooms are in lateral or terminal clusters or panicles. They have usually four, occasionally five, each sepals and petals and the same number of stamens. The styles are cylindrical. The fruits consist of clusters of four or five small pods, sometimes beaked, each splitting from the tip downward to show small, glossy, brownish-black or black seeds.

One of the best, ***E. daniellii***, a native of Korea and northern China, and hardy in southern New England, is an open-branched tree up to about 25 feet tall. It has pithy shoots, at first pubescent, and leaves 9 inches to 1¼ feet long, with seven to eleven, very short-stalked, ovate, finely-toothed leaflets 2 to 5 inches long that on their green undersides have long hairs on the midribs and in the vein axils.

Evodia daniellii

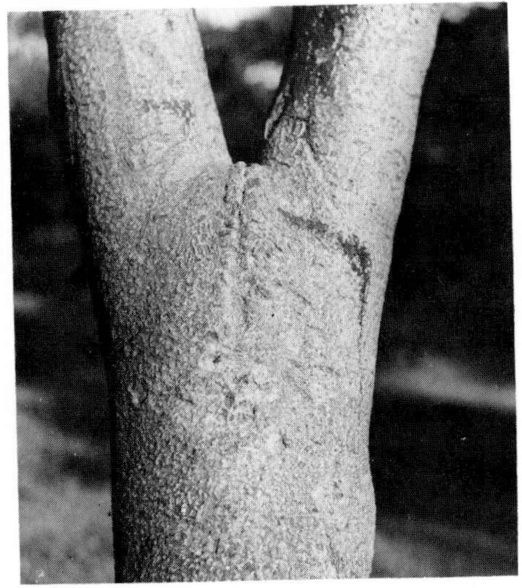

Evodia daniellii (trunk)

Evodia daniellii in fruit

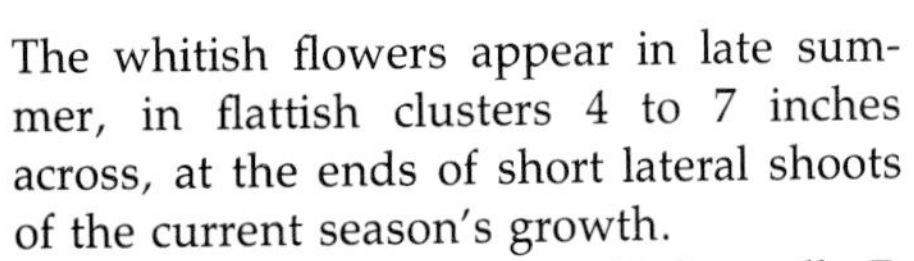

The whitish flowers appear in late summer, in flattish clusters 4 to 7 inches across, at the ends of short lateral shoots of the current season's growth.

In the wild up to about 60 feet tall, ***E. hupehensis*** is about as hardy as the last. Its young shoots are without hairs. Its leaves of seven to nine or rarely fewer, distinctly stalked, ovate to oblongish leaflets 2½ to 5 inches long are pale and somewhat glaucous on their undersides and have either no hairs or very short hairs on the midribs there. The whitish flowers come in late summer in broadly pyramidal clusters or panicles from 4 to 7 inches wide. This species is Chinese.

Evodia hupehensis

Another native of China, ***E. henryi*** attains a height of 20 to 35 feet. Its young shoots are pubescent. Its leaves, 6 inches to 1 foot long, have five to nine oblong-ovate to lanceolate-ovate, toothed, pointed, stalked leaflets, pale and somewhat glaucous beneath, eventually hairless, and up to 4 inches long. Its clusters of pinkish-white flowers are 2 to 2½ inches wide. This is hardy in southern New England.

Garden and Landscape Uses and Cultivation. Evodias are satisfactory landscape trees, useful because they bloom in summer when the flowers of most trees are past, and for their handsome foliage. Neither their flowers nor fruits are as showy as those of many more familiar trees, but they are not without interest. Evodias make no special demands as to soil, being satisfied with one well-drained and reasonably fertile. They prosper in full sun or part-day shade. Generally they are not long-lived, being good perhaps for twenty to forty years. They are propagated by seed.

EVOLVULUS (Evól-vulus). Chiefly natives of warm regions almost exclusively in the Americas, the 100 species of *Evolvulus* belong in the morning glory family CONVOLVULACEAE. They include annuals, herbaceous perennials, undershrubs, and shrubs. From related *Convolvulus* they differ in their stems never twining, hence their name, from the Latin *evolvere,* to unroll.

The leaves of *Evolvulus* are undivided, toothless, and of small to medium size. The small flowers are solitary or in clusters from the leaf axils or are in terminal spikes or heads. They have five sepals and a wheel-, saucer-, or bell-shaped corolla, distinctly or obscurely five-lobed or angled, and usually purple, blue, or white, rarely yellow. There are five slender stamens and two styles. The fruits are capsules.

Possibly occasionally cultivated in warm regions, but not hardy, ***E. tenuis longifolius*** (syn. *E. bocasanus*) of the West Indies and northern South America, is a slender-stemmed, freely-branched herbaceous perennial about 2 feet tall. Its lanceolate, stalkless leaves are up to 2 inches long, its white-centered, bright blue, solitary or clustered flowers ½ inch in diameter. This may be planted in open locations in well-drained soil. It is raised from seed. A newer introduction to cultivation, ***E. glomeratus*** of Brazil is a variable, nonhardy, slender-stemmed herbaceous perennial or subshrub 9 inches to 1½ feet tall or sometimes prostrate. Freely branched, it has short-stalked, ovate-oblong to oblanceolate or spatula-shaped, hairy leaves ½ inch to 1¼ inches long that with age lose most of the hair on their upper surfaces. The usually bright blue flowers, ⅓ inch to ½ inch wide, are in tight or loose, headlike clusters from the leaf axils and ends of the shoots.

Evolvulus glomeratus

EVONYMUS. See Euonymus.

EXACUM (Éx-acum). Some forty species of the Old World tropics constitute *Exacum,* of the gentian family GENTIANACEAE. The name, used by Pliny, is from the Greek *ex,* out, and *ago,* to arrive. It has no obvious application to this genus.

Exacums are annuals, biennials, and herbaceous perennials not hardy in the north. They have opposite, undivided, stalkless or stalked leaves, and tubed, nearly wheel-shaped flowers in forking clusters and from the leaf axils. The calyxes are four- or five-cleft, and there are four or five spreading lilac, purple-blue, or white corolla lobes (petals) and the same number of stamens. The fruits are nearly spherical capsules.

Socotra in the Indian Ocean near the entrance to the Red Sea is home to ***E. affine.*** There it is plentiful near streams below 300 feet elevation both in limestone and nonlimestone soils. It is pretty, erect, branching, and 1 foot or more in height. The ovate-elliptic, stalked, three- or five-veined leaves have blades ½ inch to 1½ inches long and ⅓ to 1 inch wide. The terminal, forking, leafy-branched flower clusters are of somewhat asymmetrical, very short-tubed, usually lavender-blue to violet-blue blooms about ½ inch in diameter. Their stamens are clustered and have yellow, down-curving anthers. The blooms have somewhat the aspect of those of woody nightshade (*Solanum dulcamara*) and are fragrant. Darker lavender flowers are borne by *E. a. atrocaeruleum* and there is an inferior white-flowered variety.

Very much larger blooms are characteristic of ***E. zeylanicum,*** an annual or a biennial native of Sri Lanka (Ceylon) that has four-angled stems, nearly stalkless, slender-pointed, lanceolate leaves up to 3

Exacum affine

Exacum affine (flowers)

inches long with three conspicuous veins, and lovely rich violet-blue flowers 1½ inches across. The sepals have broad wings. The petals are blunt. Closely related and perhaps only a variety of *E. zeylanicum* is ***E. macranthum,*** which differs most obviously in having cylindrical stems and pointed-petaled, rich purple-blue flowers about 2 inches in diameter. This, the most impressive of cultivated exacums, is native of Sri Lanka.

Exacum macranthum

Garden and Landscape Uses and Cultivation. These easily grown plants are attractive for outdoor beds in warm, fairly humid, tropical and subtropical gardens, and for greenhouse display. They are also worth attempting in window gardens. They are propagated by seed. Fertile soil that drains readily is to their liking, and good light with a little shade from the strongest sun is needed. The soil should be kept moderately moist, not constantly wet. Applications of dilute liquid fertilizer made after the final pots are filled with roots encourage good growth. In greenhouses the largest specimens are raised from seed sown in late August or September. The young plants are potted individually or three in a container and are transferred to bigger containers, without disturbing those planted in threes, when their roots fully occupy the available soil. Good single specimens from fall sowings can be finished in 4- or 5-inch pots; final containers for plants grown three together may be 5 or 6 inches in diameter. Smaller specimens are had from seeds started from January to March. Greenhouse winter temperatures at night should be 60°F, and five to fifteen degrees warmer by day. Exacums produce a display of flowers over a long period in summer and fall.

EXBUCKLANDIA (Exbuck-lándia). One or possibly two species of nonhardy trees of the witch-hazel family HAMAMELIDACEAE belong here. Native of warm-temperate, subtropical, and tropical Asia, Java, and Sumatra, *Exbucklandia* has alternate, stalked leaves, sometimes three-lobed, but not toothed, with the chief veins radiating from the bases of the blades. At the bottoms of the leafstalks are a pair of large appendages (stipules) that are folded together to protect the unopened buds. In 1-inch-wide heads of ten to twenty, the inconspicuous, unisexual and bisexual, yellowish flowers have more or less bell-shaped, five-lobed calyxes. The bisexual blooms have a variable number of petals and ten to fourteen stamens, the female flowers four rudimentary petals, but no stamens. There are two recurved styles. The fruits are woody capsules, congregated in globose heads. The name honors Dr. Buckland, a professor of botany at Oxford, England.

An erect, highly ornamental evergreen tree up to 60 feet tall or sometimes taller, ***E. populnea*** (syns. *Bucklandia populnea, Symingtonia populnea*) prospers in southern California and similar mild-climate regions. It has broadly-triangular-ovate leaves, 3 to 6 inches long by nearly as wide, blood-red beneath when young, and red stalked and veined. The conspicuous, early deciduous stipules (appendages at the bases of the leafstalks), are 1 inch to 2 inches in length.

Exbucklandia populnea

Garden and Landscape Uses and Cultivation. As an ornamental evergreen tree the species described is very satisfactory. It thrives in ordinary soils, and may be propagated by seeds and by cuttings. It is known to survive where temperatures drop occasionally to 25°F.

EXOCHORDA (Exo-chòrda)—Pearl Bush. The "ch" in *Exochorda* is pronounced as k. The genus is an Asian one of five species of the rose family ROSACEAE. Its name, alluding to the structure of the fruits, is from the Greek *exo,* external, and *chorde,* a cord.

Exochordas are hardy, deciduous, spirea-like shrubs with much larger blooms than spireas, and alternate, undivided, stalked leaves. The often unisexual flowers are in terminal racemes. They have broadly-top-shaped calyx tubes ending in a wide disk and five each sepals and spreading white petals. There are fifteen to thirty stamens and five pistils, the styles of which meet, but are not united. The star-shaped fruits contain winged seeds.

The most commonly cultivated, but not the finest pearl bush is ***E. racemosa*** (syn. *E. grandiflora*). Up to 10 feet tall, this has

Exochorda racemosa

elliptic to oblong-obovate leaves with blades up to 2½ inches long, sometimes toothed above their middles. Their undersides are whitish. The flowers, 1½ to 1¾ inches wide, in racemes of ten or fewer, have petals that narrow abruptly to a basal claw (shaft). The lower blooms of the racemes have stalks about ⅕ inch long. There are fifteen to twenty-five stamens. Variety *E. r. prostrata* has prostrate branches. The chief distinctions between ***E. giraldii*** and *E. racemosa* are that all the flowers of the former are without stalks or have extremely short ones, and the petals narrow gradually to a basal claw. There are twenty to thirty stamens.

The most handsome pearl bush is a variety of the last. It is *E. giraldii wilsonii.* This attains heights up to 15 feet and from its typical species differs in bearing more profusely flowers 2 inches in diameter that have twenty to twenty-five stamens. Noteworthy because its foliage appears very early in spring, ***E. korolkowii,*** of Turkestan, is narrower and more erect than the other kinds discussed here. From *E. giraldii* it can be distinguished by having at the bases of at least some of its leaves stipule-like appendages. The flowers, in racemes of eight or fewer, have narrowly-obovate petals and twenty-five stamens. As an ornamental this is less attractive than other exochordas discussed above. Its blooms are about 1½ inches in diameter.

Hybrid ***E. macrantha*** has *E. korolkowii* and *E. racemosa* as parents. It generally resembles *E. racemosa,* but is more vigorous and more erect. Sometimes 15 feet tall, it has obovate to oblong-obovate leaves 1½ to 3½ inches long, mostly without teeth. The flowers, 1½ to 1¾ inches wide, and with about twenty stamens, are up to eight in each raceme.

Garden and Landscape Uses and Cultivation. Except for their rather brief period of bloom, pearl bushes are without outstanding ornamental value, but when in flower they are lovely. They are suitable for inclusion in mixed shrub plantings and in spacious landscapes in isolated groups. It is usually advisable to set them in groups of three or more. Solitary specimens are apt to look rather thin and leggy. Ordinary, moderately fertile soil that is well drained suits. Full sun is needed. To encourage the best blooming, weak and crowded branches and shoots must be cut out as soon as blooming is through. Propagation is by seeds and by leafy cuttings made from firm, but not hard young shoots taken from plants forced gently in a greenhouse into early growth. The cuttings are planted in a greenhouse propagating bench. Cuttings taken from plants outdoors often fail to root satisfactorily. Layering is another means of increase.

EXOGONIUM (Exo-gònium). Closely related to *Ipomoea,* and sometimes included there, *Exogonium,* of the morning glory family, CONVOLVULACEAE, is native to tropical and subtropical America including the West Indies and Florida. The name, from the Greek *exo,* outside, and *gony,* a knee joint, presumably alludes to the jointed stems.

Consisting of about twenty-five species of perennial, woody-stemmed, twining vines, this genus has alternate leaves and morning-glory-like flowers from the leaf axils. The blooms have five somewhat unequal sepals, a salver- or funnel-shaped, tubular corolla, and five stamens that, like the united styles, more or less protrude. The fruits are capsules.

Native in Florida and parts of the West Indies, ***E. microdactylum*** (syn. *Ipomoea microdactyla*) has a large, twisted, tuberous root and stems sometimes slightly spiny toward their bases. Its slender-stalked, variously shaped, ovate to lanceolate, fleshy leaves are lobeless or have several palmately (in hand-fashion) arranged lobes. The flowers, few together in short-stalked clusters, make a good display. Scarlet to carmine-red, they have slender corolla tubes 1 inch to 1½ inches long and 1 inch or somewhat more in diameter. This vine attains a height of 8 to 10 feet.

Garden and Landscape Uses and Cultivation. In sunny locations in ordinary soil in the tropics and warm subtropics the species here described is useful for draping walls, tree stumps, trellises, and other supports. It grows robustly without special care. Propagation is by seed and division. A form introduced from the Bahamas by the Fairchild Tropical Garden of Miami, Florida, seems to be more vigorous and showier in bloom than the plants native in Florida.

EXOTIC. An exotic plant is one not native to the country or region in which it is growing, a kind introduced from elsewhere. It is not necessarily, as is sometimes thought, especially striking, colorful, or showy. Cabbages, English ivy, forsythias, and sweet peas are exotics in North America. Exotics that persist without cultivation and maintain and reproduce in competition with natives are said to be naturalized. Among such in North America are many weeds including creeping buttercup and dandelions. Compare with Endemic and Indigenous.

EXPERIMENT STATIONS. See State Agricultural Experiment Stations.

EYE. The central portions of disks of the flower heads of daisies and daisy-type flowers are called eyes (the name daisy is a corruption of day's eye). The term is also used for growth buds on tubers (as the eyes of potatoes) and on stems. Gardeners may refer to pruning a shoot back to one or two or more eyes and to make single-eye cuttings.

F

FABA or FAVA BEAN is *Vicia faba*. See Bean.

FABIANA (Fabi-àna). Of heathlike aspect, the twenty-five shrubs that belong here are natives of temperate parts of South America. None is hardy in the north. They belong in the nightshade family SOLANACEAE and bear a name commemorating Francisco Fabiano y Fuero, a Spanish Archbishop and devotee of botany, who died in 1801.

Fabianas are erect, branching, and sometimes sticky evergreens. Their numerous small leaves are crowded on many short lateral shoots. The usually numerous white flowers, terminal or opposite the leaves, have five-lobed, bell-shaped calyxes, tubular corollas swollen toward their tips, sometimes only on one side, and slightly or not five-lobed. There are five stamens. The fruits are oblong capsules.

Native of Chile, ***Fabiana imbricata*** is from 3 to 8 feet tall, and ultimately about

Fabiana imbricata

as broad as high. As a young specimen its branches are erect, with age they tend to spread. They are heavily clothed with erect leafy twigs up to 2 inches in length that give to them a plumelike appearance. The pointed, scalelike, three-angled, overlapping leaves are under 1/10 inch long. From the ends of the twigs, except the uppermost ones, develop in summer, an up-pointing, narrowly-bell-shaped, pure white flower ½ to ¾ inch long, with five shallow, rounded, reflexed corolla lobes. A similar species, ***F. violacea,*** of Chile, is reported to be slightly hardier than *F. imbricata.* As a young plant its branches are more spreading. Its flowers are mauve or lilac.

Garden and Landscape Uses. In California and other mild-climate regions where summers are not oppressively humid and little winter freezing is experienced, *F. imbricata* is highly satisfactory for sunny locations in well-drained, ordinary soils. Its soft-textured appearance contrasts well with larger-foliaged plants, and it can be used effectively in foundation plantings and other shrub groupings, in rock gardens, and in many other locations.

Cultivation. No special care is required. If the plants become straggly they may be pruned to shape as soon as blooming is

Fabiana imbricata (flowers)

through. Increase is easily had by cuttings inserted in a greenhouse propagating bench, in a cold frame, or under mist in summer, or seeds may be sown in sandy peaty soil.

FACHEIROA (Fach-eiròa). The name of this genus of the cactus family CACTACEAE, which some authorities include in *Espostoa,* is an adaptation of a Brazilian one used for several sorts of cactuses. Its only species is endemic to Brazil.

Erect, much-branched, and columnar, ***Facheiroa ulei*** (syns. *F. publiflora, Espostoa ulei*) is up to 18 feet high and has stems about 3 inches in diameter with fifteen to twenty low ribs and brown-felted areoles (places of origin of spines) spaced less than ½ inch apart. Three or four centrals and ten to twelve shorter, radial, brown spines come from each areole. The centrals are ¾ to 1 inch long. The blooms arise from a development called a pseudocephalum, a strip that extends down one side of the top of the stem for up to 8 inches, which is ¾ inch to 1½ inches broad and densely furnished with reddish-brown, woolly hairs. The short-tubed flowers, about 1¼ inches long and 2½ inches across, are hairless on their insides. They have white perianth segments (petals), nonprotruding stamens, and a style ending in a stigma with about ten lobes. Their ovaries and perianth tubes are densely-scaly on their outsides and have ½-inch-long hairs in the axils of the scales. Green and gelatinous inside, the fruits are spherical and about ¾ inch in diameter. The seeds are black.

Garden and Landscape Uses and Cultivation. These are the same as for other columnar cactuses. See Cactuses.

FAGACEAE—Beech Family. About 900 species are included in this cosmopolitan family of six or according to some authorities eight genera of deciduous and evergreen dicotyledonous trees or less commonly shrubs. Most familiar are beeches, chestnuts, and oaks. Besides supplying many shade trees and other ornamentals the beech family is an important source of lumber, cork, edible nuts, dyes, tanning materials, and pharmaceuticals. Family characteristics include undivided, often lobed or toothed leaves that are usually alternate, rarely whorled (in circles of three or more), and on the same or separate plants. The tiny blooms have usually six-parted perianths. Male flowers have six to numerous stamens, females three or six styles and frequently staminodes (nonfunctional stamens). The fruits are nuts or achenes contained singly or in twos or threes in cups or burs. Cultivated genera are *Castanea, Castanopsis, Fagus, Lithocarpus, Nothofagus,* and *Quercus.*

FAGOPYRUM (Fago-pŷrum)—Buckwheat, India-Wheat. Buckwheat flour is a product of one species of this genus of fifteen, the name of which comes from the Latin *fagus,* the beech, and the Greek *pyros,* wheat, and alludes to the seeds being shaped like those of beech trees. Native to Asia and Europe, the group belongs in the buckwheat family POLYGONACEAE and consists of annuals with alternate, stalked leaves, and small, bisexual white or pink flowers in racemes or clusters. The blooms have a calyx deeply divided into five persistent, petal-like parts, no petals, eight stamens, and three styles. The fruits are three-angled achenes (thin-walled, one-seeded nutlets).

Buckwheat (*Fagopyrum esculentum*) and India-wheat or kangra buckwheat (*F. tataricum*), both naturalized in North America and elsewhere, and both, apparently, originally natives of central Asia, are very similar. They differ in that the fruits of ***F. esculentum,*** the common source of buckwheat flour, are shining and have smooth

Fagopyrum esculentum

angles whereas those of ***F. tataricum*** are dull and are irregularly indented along the angles. Also the greenish-white flowers of *F. esculentum* are often tipped with pink, those of *F. tataricum* are usually all greenish. These plants are 8 inches to 2 feet tall

Fagopyrum tataricum

and have slender stems with lines of hairs in their upper parts, but are without these below. Broadly-triangular-arrow- to heart-shaped, and up to 3 inches long, the leaves are rather widely spaced on the hollow stems. The lower ones are long-stalked, those above are progressively shorter-stalked until the uppermost are nearly stalkless. In long-stalked, crowded, more or less rounded clusters, the fragrant flowers are bell-shaped, and have sepals up to ⅙ inch long. The fruits are often twice as long or longer than the sepals.

Garden Uses and Cultivation. As agricultural crops the buckwheats are cultivated, but their garden uses are confined to growing them as summer cover crops to prevent erosion and to be spaded or plowed under to add organic matter to the soil. The common buckwheat is somewhat hardier than the other. It is a good bee plant. For further information see Cover Crops.

FAGUS (Fà-gus)—Beech. Here belong ten species of Northern Hemisphere deciduous trees of the beech family FAGACEAE. They inhabit the temperate zone and Mexico and often form extensive forests. All are handsome and produce useful lumber. The name is the ancient Latin one for the beech tree. Beeches very much resemble each other so that if one is familiar with one kind others are likely to be recognized. From the Southern Hemisphere genus *Nothofagus,* which is called southern-beech, true beeches may be distinguished by their male flowers being in many-flowered heads and the females usually in twos or sometimes threes surrounded by bracts from out of which peep the red stigmas.

Beeches are tall trees of noble appearance with thin, gray bark and alternate, flat leaves with parallel lateral veins in two ranks. The leaves are toothed or nearly entire according to kind. The winter buds are characteristically long, slender, and sharp-pointed. The flowers are unisexual with both sexes on the same tree. The male flowers have eight to sixteen stamens surrounded by a four- to seven-lobed calyx. Female flowers have three styles. The fruits are triangular nuts borne in pairs enclosed or partly enclosed in a soft-prickly woody husk. A curious feature of the European beech, and perhaps of others, is that it fruits only every five to twelve years depending upon local conditions. Its first flowering and fruiting does not occur until it is fifty years old or older.

The wood of beeches is not durable when exposed to moisture and is not used to any extent for construction or interior trim. It is, however, esteemed for furniture, butchers' blocks, tool handles, clothes pins, and other small articles and it is one of the finest fuel woods. It is believed that Johan Gutenburg, the inventor of movable type, used blocks of beech wood upon which to cut his letters, and the German word for letters, *buchstaben,* is probably derived from *buche,* the German word for beech. Beech nuts are delicious and in Europe are collected, especially by boys, and eaten. They are also great favorites of squirrels, wild turkeys, deer, and bears. The nuts, together with chestnuts and acorns, are known collectively as mast and in Europe and Asia form an important part of the fall diet of domestic hogs. An oil expressed from the nuts of the European beech is used for cooking.

The two most important beeches from horticulturists' points of view are the European beech and the American beech. Strangely, although the former thrives in many parts of North America, the latter grows poorly in Europe. Both are stately trees of magnificent appearance and proportions.

The European beech (***F. sylvatica***) occasionally attains a height of 100 feet or even considerably more, but more usually

Fagus sylvatica

Fagus sylvatica (leaves and fruits)

it matures at about 80 feet. It forms a dense, wide-spreading, domed crown. Its trunk and branches are medium to dark gray. The leaves are ovate or elliptic, up to 4 inches long, with small distantly spaced teeth and eight or nine pairs of veins angled upward and outward from the midrib. They are dark green above, paler beneath, and when young are fringed with fine hairs and are silky-pubescent on their undersides. In fall the foliage assumes a warm shade of bronze-brown. The European beech is native from England and Norway to the Mediterranean region, Iran, and the Crimea. It apparently was not an original native of Ireland or Scotland. There are several very handsome varieties of this tree.

The purple beech (*F. s. atropunicea* syn. *F. s. purpurea*), is one of the most outstanding varieties. It has foliage of an intense purple or copper-purple hue. The exact shade of seedling trees varies considerably and selected forms are sometimes given identifying names. Thus those with coppery-purple foliage are often known as *F. s. cuprea* and are called copper beeches. An especially dark-leaved form is called *F. s. riversii.* The weeping purple beech (*F. s. atropunicea pendula*) has markedly drooping branches. The fern-leaved purple beech (*F. s. atropunicea rohanii*) has foliage like that of *F. s. laciniata,* but purple.

All purple- and copper-leaved beeches in existence are descendants of trees that developed spontaneously in at least three different places. The oldest record is of five discovered by a doctor in Buchs in the canton of Zurich, Switzerland. These are mentioned in a work dated 1680. The armorial shield of the village of Buchs displays a purple beech. The other records are of a single tree that was discovered in the Hantlier forest, near Sonderhausen, Thuringia, Germany, in 1770, and of one that appeared in the Darney forest in the Vosges. The Swiss location is the best known.

The green-leaved weeping European beech (*F. s. pendula*) exists in several forms. The commonest makes a specimen wider than it is tall and with branches forming a wide skirt that touches the ground all around. Other types have more erect trunks with horizontal and drooping branches.

Other very distinct varieties of *F. sylvatica* include the Dawyck beech (*F. s.* 'Dawyck'), sometimes known as *F. s. fastigiata,* a narrow columnar tree that originated at Dawyck in Scotland. It is very distinct and handsome. The fern-leaved

Fagus sylvatica atropunicea at Princeton University, New Jersey

Fagus sylvatica riversii

Fagus sylvatica pendula: (a) Fairly young

(b) Middle-aged

(c) At maturity

Fagus sylvatica fastigiata

beech (*F. s. laciniata*) is especially lovely. Its leaves vary greatly in shape; some are long and slender (up to 4 inches in length by not more than ¼ inch wide), others are much broader, but are lobed almost to their midribs. Many intermediates occur. The trees are shapely. The oak-leaf beech sometimes called *F. s. quercifolia*, seemingly at most is a slight variant of the fern-leaved beech. Much less vigorous than most varieties is the tricolor beech (*F. s. tricolor*). Its leaves are mostly white or pinkish-white spotted or blotched with green and with irregular pink margins. Selected sprays of the young foliage are attractive, but the tree itself is disappointing and is really only for the collector of the unusual. A variety with purple leaves that have an irregular light pink margin is called *F. s. roseo-marginata*. A number of varieties have leaves variegated with white, the commonest being *F. s. albo-variegata*, which has its leaves striped with white. In *F. s. luteo-variegata* the leaves are striped yellow. A very distinct and charming beech is *F. s. rotundifolia*. Its leaves, about 1 inch across, are round with somewhat heart-shaped bases and are set very closely together. This is a slow-growing variety. Another slow grower is *F. s. cris-*

Fagus sylvatica laciniata

Fagus sylvatica rotundifolia at the Arnold Arboretum, Jamaica Plain, Massachusetts

tata, distinct because of its deeply-toothed, curled and clustered leaves. Scarcely deserving of its common name, the golden beech (*F. s. zlatia*) has leaves that are yellow in spring, but this phase soon passes and for most of the season the foliage is scarcely distinguishable from that of the typical species. Very large and broad, scarcely-toothed leaves characterize *F. s. latifolia*.

The American beech (***F. grandifolia***) differs from the European beech in that its leaves are less glossy and more sharply-toothed and have nine to fourteen pairs of veins. Its bark is lighter colored and it frequently produces suckers from its roots that grow up to form a colony around the parent. The crown of the American beech is narrower and less rounded than that of its European relative. The American beech ordinarily grows up to 80 or 90 feet in

Fagus grandifolia in summer

Fagus grandifolia in winter

Fagus grandifolia with abundant suckers around its base

height, more rarely 120 feet. In fall its foliage turns a lovely golden-bronze. This tree is indigenous from New Brunswick and Nova Scotia to Minnesota, Florida, and Texas. Unlike the European beech, it has produced no outstanding horticultural varieties. A natural variety, *F. g. pubescens* has leaves more or less softly-pubescent on their under surfaces, and *F. g. caroliniana* has less prickly fruits and less conspicuously toothed leaves of a firmer texture than those of the type.

Other species are sometimes cultivated. For those now to be described substantial records of hardiness are unavailable, but it is probable that all are hardy in southern New England. ***F. crenata*** (syn. *F. sieboldii*), of Japan, round-topped and up to 90 feet tall, has leaves broadest below their middles. There are leafy appendages at the bottoms of the fruits. ***F. englerana,*** a native of central China, usually has a trunk divided near the base into several branches. It attains a height of 80 feet and has leaves with ten to fourteen pairs of veins that curve before they reach the sinuate margins. The leaves are slightly glaucous beneath and hairless except for silky hairs on the veins. ***F. hayatae*** is endemic to a small area around Taoyuan, Taiwan. It attains a height of about 45 feet and has leaves with seven to ten pairs of veins, glabrous on both surfaces. ***F. longipetiolata,*** a native of central and western China, and usually single-trunked, is up to about 80 feet tall. The undersides of its leaves are downy. ***F. lucida*** attains a height of about 30 feet and has leaves with both surfaces lustrous green that are glabrous except for silky hairs on the midribs on their upper sides and on the leafstalks. The nuts protrude slightly from the husks. ***F. orientalis,*** the Oriental beech, sometimes 120 feet in height, is a native of Asia Minor, Iran, and the Caucasus. Its leaves have seven to ten pairs of veins that curve before they reach the margins. It differs from the European beech and the American beech in having bractlike rather than awl-shaped appendages at the bases of the fruits.

Garden and Landscape Uses. Both the American and European beeches are magnificent large trees for landscape purposes. Because of their size they are not well suited for small gardens nor are they appropriate as street trees, but where space permits their uncrowded display they are unexcelled. At all seasons they are beautiful and serve well whether planted as single specimens or in small groves. They make good screens and the European kind lends itself to shearing to form tall hedges. It may also be pleached. Thought should be given before planting beeches in much-visited public places because their thin, smooth bark is especially inviting to pocket knife vandals who seem unable to resist decorating the trunks with such carved devices as entwined hearts pierced with arrows and legends of such significance as "Bill loves Mary," "John Doe, June 19—," "B.K.D." and the like. Many of the varieties of the European beech appeal only to the collector of the unusual, but others are among the most handsome and useful of trees. Here belong the purple and copper beeches, the weeping beech, the fern-leaved beech and the Dawyck beech, all varieties of distinctive habit of growth or foliage. Beeches are for dryish, but not dry soils. They may be seen at their best on well-drained, fertile loams, and they prosper in soils that overlie limestone. This is especially true of the European beech. They have, however, a quite wide tolerance and excellent specimens occur on neutral and slightly acid soils.

A hedge of the European beech (*Fagus sylvatica*)

Cultivation. Beeches make no special demands of the gardener. The European kind and its varieties transplant with ease even as large trees, but the American beech is much less tolerant of root disturbance. Losses following its transplanting are greater and recovery is slower than with its European relative. Necessary pruning does not prejudice the well-being of these trees, but no regular systematic pruning is needed. When grown as hedges shearing should be done in spring before new leaf growth begins. Because of their thin bark beeches moved from where their trunks have been shaded, as by neighbor trees in nursery rows, are very subject to suffer from sun scald and splitting of the bark following their removal to more exposed locations. To protect them, the trunks should be wrapped in burlap or tree-wrap paper for a year or so following transplanting and finally this should be removed gradually. For the same reason lower branches should not be removed. Their foliage affords some shade for the trunks. Beeches are surface-rooters and cast heavy shade. Once established the number of kinds of plants that can be grown beneath them is strictly limited. The propagation of species of beech is best by seeds, which should be sown in fall or stratified and sown in spring. The seedbed must be protected from mice and other rodents. Varieties are increased by grafting onto seedling understocks in a greenhouse in late winter or early spring. While in the nursery young beeches should be transplanted every two or three years; otherwise they are likely to form taproots and be difficult to transplant later.

Diseases and Pests. The chief diseases of beeches are a bleeding canker and a leaf mottle disease for which there are no known cures. The avoidance of bark wounds, especially near the base of the trunk, is a preventative measure with reference to the former. Adequate fertilization and watering thoroughly during droughts may be helpful in combatting the latter. A bark disease that results from infection with a *Nectria* fungus following infestations of an insect called woolly beech scale can also be troublesome. This is prevented by controlling the scale insect by spraying in August or September, when the young larvae are crawling, with malathion or some other material recommended for the control of scales, and by applying lime-sulfur as a dormant spray in late winter.

FAIR-MAIDS-OF-FRANCE See *Ranunculus aconitifolius* and *Saxifraga granulata.*

FAIR-MAIDS-OF-KENT is *Ranunculus aconitifolius.*

FAIRY. As a component of the common names of plants the word fairy appears in these: fairy bells (*Disporum*), fairy duster (*Calliandra eriophylla*), fairy fans (*Clarkia breweri*), fairy lantern (*Calochortus*), fairy-lily (*Zephranthes*), fairy primrose (*Primula malacoides*), fairy rose (*Rosa chinensis minima*), and fairy wand (*Chamaelirium luteum*).

FAIRY RINGS. This is the name given to circles and partial circles of mushroom-like fungi and to rings of grass of distinctly deeper green than surrounding areas that these fungi cause in lawns and close-cropped meadows. The explanation of fairy rings is more prosaic than the name suggests. Some soil fungi that inhabit grasslands, when the soil is to their liking and growing conditions are favorable, expand by growing outward from the centers. As they do this the mycelia (the threadlike parts of the fungus that live underground) of the older, central parts of the patches die, and the younger, living portions of mycelia, growing ever outward, form a ring around the dead part that gradually becomes bigger as the size of the patch increases. The outer and inner rims of the rings are usually deeper green than the space between them, which commonly consists of brownish, debilitated or dead grass. The reasons for this are believed to be that the grass in the center band of the ring weakens or dies from dryness caused by the mycelia preventing water being absorbed by the soil. The dark green color and vigor of the grass of the outer rim of the ring is thought to result from the actively growing fungus making available to the grass roots additional nitrogen. That of the inner edge of the ring is believed to be stimulated by extra nitrogen that comes from the decay of dead mycelia in that area.

Fairy rings do no great harm, but under some circumstances they are unsightly. Recommended means of eradicating them are to soak the soil when it is moist after rain or irrigation with sulfate of iron dissolved in water at the rate of one pound to one gallon and to repeat ten days later with a solution of one-half a pound to one gallon. Two or three soakings at ten-day intervals with Bordeaux mixture may also be effective.

FALL COLOR. See Autumn Color.

FALL-DAFFODIL is *Sternbergia lutea.* Fall-blooming species of *Narcissus* may be called fall daffodils.

FALL PLANTING. In many sections of North America fall is the ideal season for planting a wide variety of trees, shrubs, and other plants. Then, top growth for the year is completed, but roots remain active as long as the soil is sufficiently warm, which with most hardy kinds is as long as it stays above 35 or 40°F.

Because of this, plants set in fall begin reestablishing their root systems immediately, continue growing until the earth becomes too cold, and resuming as early in spring as ground temperatures permit, which is usually well before it is practicable to do spring planting.

Fall-planted stock generally develops more extensive root systems than spring-set plants before being called upon to withstand the stresses of summer. Of course this is not true if planting is delayed too long. Move deciduous trees and shrubs as soon as most of their leaves have shed, herbaceous perennials after the first killing frost. The best time to transplant a great many evergreen trees and shrubs is early fall. With these it is a mistake to delay too long.

Mulching immediately after planting keeps the soil in a condition conducive to root growth for longer than it would remain otherwise. Staking may be needed to protect newly-set trees from movement by winter storms.

Not always is fall to be preferred to spring for transplanting. With plants on the borderline of hardiness the latter season is generally preferable and a few thick- or fleshy-rooted plants, such as magnolias, transplant better in spring than fall.

FALLOW. The practice of allowing land to lie fallow, that is without cropping it for a fairly extended period, with the purpose of improving its condition, is a very ancient one that has only limited application in gardens. Land allowed to lie fallow, if there is no danger of erosion, benefits from the cumulative effects of exposure to sun, rain, wind, and especially alternate freezing and thawing. When time permits it is a splendid restorative to clayey soil that has been mauled with heavy equipment in the construction of buildings or suchlike facilities.

FALLUGIA (Fal-lùgia)—Apache Plume. One deciduous shrub, native from Utah to Texas, Arizona, and Mexico, and related to *Cowania* and *Potentilla,* is the sole representative of this genus of the rose family ROSACEAE. Its name commemorates Virgilio Fallugi, a seventeenth-century Italian botanist. From *Potentilla* our plant is distinguished by its persistent, feathery styles, and from *Cowania* by having bracts at the base of the calyx, the pistils being numerous, and the foliage glandular.

Of slender, upright habit and branching freely from the base, the apache plume (***Fallugia paradoxa***), up to about 4½ feet tall, has alternate, somewhat clustered, pinnately-three- to seven-lobed leaves, wedge-shaped toward their bases and ⅓ to a little over ½ inch long. Their undersides are finely-white-hairy, their margins are recurved. From the ends of the shoots in summer the flowers arise. They are white, 1 inch to 1¼ inches in diameter, and solitary or in twos or threes. They have a calyx of five ovate sepals with five small bracts alternating with them at their bases. There are five rounded, spreading petals and numerous stamens. The many pistils form cone-shaped centers to the blooms. The fruits with their persistent, purplish, feathery, tail-like styles are in decorative heads about 1½ inches wide.

Garden and Landscape Uses and Cultivation. Both its summer presentation of flowers and early fall display of fruiting heads render *F. paradoxa* worthwhile. It is hardy in southern New England and thrives in dryish, very well-drained soils in full sun. It will not tolerate "wet feet." It may be used with good effect in shrub beds and similar plantings. No special care beyond the removal of faded seed heads is needed. Increase is by seeds and by summer cuttings in a greenhouse or cold frame propagating bed, or under mist.

FALSE. This word forms part of the colloquial names of many plants, among them: false-acacia (*Robinia pseudoacacia*), false-aloe (*Manfreda virginica*), false-aralia (*Dizygotheca*), false-arbor-vitae (*Thujopsis*), false-asphodel (*Tofieldia*), false-baby's-breath (*Galium*), false-bittersweet (*Celastrus scandens*), false-buckthorn (*Bumelia*), false-camomile (*Matricaria recutita* and *Tripleurospermum maritimum*), false-cypress (*Chamaecyparis*), false-day-flower (*Commelinantia anomala*), false-dragonhead (*Physostegia*), false-foxglove (*Aureolaria*), false-garlic (*Nothoscordium*), false-hellebore (*Veratrum*), false-indigo (*Amorpha* and *Baptisia*), false-Jerusalem-cherry (*Solanum capsicastrum*), false-lily-of-the-valley (*Maianthemum*), false-mallow (*Malvastrum*), false-mitrewort (*Tiarella*), false-olive (*Elaeodendron orientale*), false-pineapple (*Pseudananas sagenarius*), false-saffron (*Carthamus tinctorius*), false-Solomon's-seal (*Smilacina*), false-spirea (*Sorbaria*), and false-tamarisk (*Myricaria*).

FAME FLOWER. See Talinum.

FAMILY. A botanical family is a unit of classification that consists sometimes of a single genus, more commonly of two or more genera considered sufficiently distinct from all other genera to warrant family recognition. When more than one, the genera of which a family is composed are more closely related to each other than to other genera. Well-known examples of botanical families are the pea family *Leguminosae,* which includes among its about 600 genera *Arachis* (peanuts), *Lupinus* (lupines), *Pisum* (peas), and *Trifolium* (clovers). The rose family ROSACEAE contains approximately 100 genera, among them *Malus* (apples), *Prunus* (apples, apricots, peaches, plums, etc.), and *Rosa* (roses). The family GINKGOACEAE consists of the single genus *Ginkgo,* which consists of one species, *G. biloba.* For the common and botanical names of families recognized in this Encyclopedia see Plant Families.

FANWORT. See Cabomba.

FAREWELL-TO-SPRING is *Clarkia amoena.*

FARFUGIUM. See Ligularia tussilaginea.

FARKLEBERRY is *Vaccinium arboreum.*

FARSETIA. See Fibigia.

Abnormal fasciation in stem of petunia

FASCIATION. For reasons not fully understood, although virus infections are believed to be responsible in some cases, the stems of plants are sometimes distorted by being much broadened and flattened as a result of abnormal cell growth. Both vegetative and flowering stems may be affected. Fasciated flowering stems usually carry exceptionally large numbers of blooms. Fasciation occurs as an occasional phenomenon in a large number of plants, fairly frequently in lilies, sweet-peas, and a few other commonly grown kinds. It is heredity in cockscomb.

Heredity fasciation in cockscomb (*Celosia argentea cristata*)

FASCICULARIA (Fascicu-lària). The name *Fascicularia,* that of a Chilean genus of five species, comes from the Latin *fasciculus,* a little bundle, and alludes to the arrangement of the foliage. Evergreen, stemless or very short-stemmed, nonhardy herbaceous perennials, fascicularias belong in the pineapple family BROMELIACEAE. Unlike so many of that company they do not grow as epiphytes perched on trees, but like pineapples, are terrestrial.

Fascicularias have rosettes of numerous, narrow, spine-margined leaves, with at their centers, stalkless clusters of blue flowers. The blooms have six perianth parts (petals) and six stamens. The fruits are berries.

Inhabiting steep cliffs near the coast of central Chile, ***F. pitcairniifolia*** (syn. *Rhodostachys pitcairniifolia*) has sharp-spined foliage suggesting that of pineapples. The leaves are up to 3 feet long by ⅝ inch wide. The inner ones have scarlet bases and, encircling the dense clusters of small, light blue flowers, are a colorful decorative feature. The blooms are up to 2 inches long.

Similar ***F. bicolor,*** a native of perpendicular cliffs near the Chilean coast, has sometimes pink-flushed, glaucous-gray-green leaves, with brownish undersides that are up to 2 feet long by a little over ½ inch wide. The 1½-inch-long, pale blue flowers are associated with little ivory-white bracts in heads of twenty to forty. The surrounding leaves are red at their bases.

Fascicularia pitcairniifolia

Garden and Landscape Uses and Cultivation. Of easy cultivation outdoors in California and places with similar dryish, nearly frost-free climates and in greenhouses, fascicularias are adapted for rock gardens, banks, and other well-drained, sunny sites where a low spiny-leaved plant can be used advantageously. Full sun is needed. Poorish soil and dry conditions in summer favor flowering. Also appropriate for sunny, airy greenhouses devoted to succulents and other desert plants, these plants succeed in pots or ground beds in porous soil watered rather sparingly. A minimum winter temperature of 45 to 50°F is satisfactory. Seeds germinate readily, but propagation is more often effected by offsets. To handle them expeditiously it is advisable to wear leather gloves or use long-handled tools (knives and tongs).

FATSHEDERA (Fatshéd-era). The only kinds of this bigeneric hybrid are *Fatshedera lizei* and a variety of it. Their parents are *Fatsia japonica moseri,* of Japan, and a large-leaved variety of English ivy (*Hedera helix hibernica*), of Europe. The name is derived from those of the parent genera, both members of the aralia family ARALIACEAE. Seeds that resulted from the artificial pollination in France in 1910 of the *Fatsia* with pollen from the *Hedera* gave rise to the intermediate, which when exhibited by Lize Frères at Nantes in 1912, was awarded a grand silver-gilt medal. The new plant was formally described and named in 1923. Introduced into the United States in 1926, for years *F. lizei* attracted little attention from gardeners, but after it was found to be a good indoor houseplant, as well as a fine outdoor decorative for mild climates, its popularity increased rapidly and it became common.

Slender and erect, ***F. lizei*** has sparsely-branched stems that are upright while the plants are comparatively small, but later need support. They are not vining like those of English ivy, nor do they send out adventitious roots, but they are less stout and sturdy than those of *Fatsia.* The leaves and their arrangement on the stems are similar to those of *Hedera.* They are evergreen, thick and leathery, three- or five-lobed, and 5 to 10 inches broad. The greenish flowers, in large compound clusters, are imperfectly developed, their stamens being abortive. No fruits are produced. Hardy in sheltered places about as far north as Washington, D.C., *F. lizei* when grown outdoors usually does not exceed 6 feet in height, but under favorable conditions in greenhouses it may become more than twice as tall. Variegated-leaved *F. l. variegata,* has irregular cream-colored bands along its leaf margins. It is a much weaker grower than the green-leaved type.

Garden and Landscape Uses. A tough plant that stands adverse conditions, *F. lizei* does well even in poor light. It stands temporary dry soil conditions and soon recovers from drought; nevertheless it grows best in fairly good, always reasonably moist earth. If hooks, nails, or wires are available to which its stems can be tied it can be used to clothe walls, pillars, and other supports. An excellent foliage plant, *Fatshedera* can be used to ornament places unsuitable for many other indoor plants. It is used as an understock upon which to graft English ivies to grow as standard (tree-form) specimens.

Cultivation. Few plants are easier to grow than *Fatshedera.* Ordinary soil kept moderately moist suits. Branching can be induced by pinching out the tips of the shoots. Pruning to shape may be done without fear of harming the plants. Indoors, this plant has a wide temperature tolerance. It prospers as well on sun porches and other locations where the night temperature in winter may be 40°F or even lower, as in living rooms where the temperature seldom goes below 70°F. Container-grown specimens that have filled their pots with roots benefit from once- or twice-a-month applications of dilute liquid fertilizer. It is desirable to sponge their foliage with soapy water occasionally to remove grime. The variegated-leaved variety is much more finicky than the green-leaved kind. It will not stand exposure to strong sun, nor does it tolerate extremely dry soil conditions. Propagation of both green-leaved and variegated-leaved kinds is commonly by cuttings. These root most surely if the basal cut is made through a node rather than a little distance below as is done with most kinds of plants. The top of the cutting is sliced off about one-half inch above the first node above the base, leaving the leaf at that node attached; thus a cutting consists of two nodes and a single leaf. Early fall is the best time to take the cuttings. Air layering affords an alternative means of propagation.

Fatshedera lizei, in bloom

Fatshedera lizei

FATSIA (Fát-sia). Two or possibly more species of evergreen shrubs of the aralia family ARALIACEAE, natives of eastern Asia, constitute this genus. One and varieties of it are well known as cultivated plants. The one is parent of the remarkable bigeneric hybrid *Fatshedera,* with English ivy (*Hedera helix*) as the other parent. The name is derived from a Japanese one. The plant sometimes called *Fatsia papyrifera* is *Tetrapanax papyrifera.*

Unlike many genera of the aralia family, *Fatsia* is without spines or prickles. Its members have large, stalked, alternate leaves with the blades deeply-palmately (in handlike fashion) veined and lobed. The small, whitish flowers are in umbels displayed in large, stiff-branched panicles. The terminal umbels are of bisexual blooms, the lateral ones of all male flowers. Each longish-stalked bloom has a five- to seven-toothed calyx, five to seven petals, and as many stamens. The usually five, persistent styles of the bisexual flowers are separate, the short, nonfunctional ones of the males are united. Containing flat seeds, the fruits are spherical, fleshy berries.

For long called *Aralia sieboldii,* the species now named ***Fatsia japonica*** (syn. *Aralia japonica*) is native of South Korea, Japan, and the Ryukyu Islands. It is a handsome, bold-foliaged, broad shrub that in the wild favors woodlands near the sea. Young

Fatsia japonica

Fatsia japonica, in bloom

parts of its stems are clothed with rusty hairs, older sections have conspicuous scars from where leaves have fallen. Nearly orbicular in outline with generally markedly heart-shaped bases, deeply-seven- to nine-lobed, and toothed, the glossy green leaf blades, 8 inches to 1¼ feet in diameter, are on long stalks. Terminal on the stems, the panicles of many-flowered umbels of whitish flowers are 8 inches to 1¼ feet in length. The flowers have five slender styles. Suborbicular, the black fruits are almost or quite ¼ inch in diameter. Variety *F. j. moseri* is more compact and has larger, yellowish-veined leaves, pubescent on their undersides. In *F. j. variegata* the leaves are irregularly edged with creamy-white.

Fatsia japonica variegata

Garden and Landscape Uses. In addition to its very considerable usefulness as an indoor foliage plant, *F. japonica* and its varieties are quite handsome outdoor ornamentals for mild climate regions and are suitable for shrub beds, foundation plantings, and similar places as well as for furnishing planters, urns, and other containers. Because of their bold foliage they show to good advantage in architectural surroundings. Fatsias stand considerable shade, but prosper in sun if they have adequate moisture; the variegated-leaved variety needs partial shade for its best comfort. The species is hardy about as far north as Virginia; its varieties are a little more tender.

Cultivation. Although not very fastidious as to soil, a hearty, nutritious loamy one, sufficiently porous to prevent stagnation, and not lacking in organic matter, best suits fatsias. Little care is required by established plants outdoors; the only pruning needed is the removal or shortening of an occasional branch that becomes straggly. As indoor pot plants fatsias need rather cool conditions. Night temperatures, during the time of the year when outdoor ones permit, should not be over 55°F, and lower temperatures, down to 40°F, are better. Day temperatures may be five to ten degrees higher. The atmosphere should be fairly humid, and good light with shade from strong sun is required. Watering to keep the soil always pleasantly moist, but not constantly saturated is routine. Well-rooted specimens appreciate occasional applications of dilute liquid fertilizer. Propagation is by seeds, which germinate readily in a temperature of 60 to 70°F, and by air layering, cuttings, and root cuttings.

Pests. Mealybugs and scale insects are sometimes troublesome. Mites are especially partial to fatsias and may crinkle and distort the young foliage badly.

FAUCARIA (Fauc-ària)—Tiger's Jaw. Among the most popular and easy-to-grow small succulents of South Africa are the sorts of *Faucaria* of the carpetweed family AIZOACEAE. These highly satisfactory plants are *Mesembryanthemum* relatives. They number about thirty-five species. The name, from the Latin *faux*, a jaw, has reference to the appearance of the leaves.

When not in bloom faucarias somewhat suggest baby aloes sitting directly upon the ground, and stemless or nearly so. Each plant body has four, six, or rarely more opposite, spreading leaves, each pair resting against, and at right angles to, the ones below and above. Flowering, which occurs in late summer or fall, quickly dispels any false assumptions that these are *Aloe* relatives. The stalkless or almost stalkless blooms, which open in the afternoons, are of the typical *Mesembryanthemum* kind, daisy-like at first sight, yet differing markedly from daisies in that they are solitary blooms and not heads composed of numerous florets. Yellow, or rarely white, each has five threadlike stigmas and five thick glands. The fruits are capsules that, unlike those of other members of the family, are deeply-cup-shaped.

Very characteristic of the genus and a feature of most species, that earns for one the common name tiger's jaw, are the coarse, sharp teeth of the leaves, which do indeed call to mind the threatening jaws of an animal, possibly a crocodile more than a tiger. In no other genus of the *Mesembryanthemum* complex is toothing of the leaves so prominently developed. In many cases pale horny bands along the leaf margins and keels, and irregular spottings on the leaf surfaces add to the beauty of the plants.

Hybridity is common in *Faucaria* and cultivated plants are frequently of mixed parentage. This adds to the difficulty of identifying them and tends to confuse the botanically minded, who look with greater favor on pure species than upon bastard progeny. Aesthetically there is little to choose between the two groups; all faucarias are worth growing.

Tiger's jaw (***F. tigrina***) is most commonly represented in cultivation by hybrids of the true species. The latter has thick, triangular leaves up to 2 inches long and at their broadest about one-half as wide. Their flat upper sides are fringed with nine or ten recurving short teeth tapered to hairlike points. Rounded, the undersides of the leaves are keeled toward their chinlike ends. Characteristically gray-green with many white dots, when grown in full sun the leaves become reddish. The flowers, solitary or in twos and about 1 inch in diameter, are golden-yellow. Variety *F. t. splendens* has redder leaves. From the tiger's jaw ***F. felina*** differs in having less conspicuously spotted leaves with fewer, shorter, nonbristly teeth, and golden-yellow blooms 2 inches in diameter. Variety *F. f. jamesii* is more distinctly spotted and has leaves more compressed at their ends and flowers with pinkish outsides.

The largest-flowered species, ***F. speciosa*** has blooms that sometimes exceed 3 inches in diameter and are golden-yellow with paler centers and reddish-purple undersides to the petals. From 1 inch to 2 inches in length, the leaves are strongly compressed toward their apexes and have five or six bristle-tipped, broad teeth along each margin.

Other sorts in cultivation include these: ***F. acutipetala*** has golden-yellow flowers about 1½ inches wide. Its pointed, spreading, white-spotted, olive-green leaves barely 1 inch long by approximately one-third as wide and deep are lanceolate to triangular and have three to five white-bristled teeth along each of their white edges. ***F. albidens*** has yellow flowers 1¼ to 2 inches wide. The five or six pairs of leaves, each about one-third as wide as long and up to 1¼ inches long, are glossy-green sprinkled with tiny translucent dots. They have whitish, horny margins from which sprout on each side of the leaf two to four backward-pointing, short teeth. ***F. bosscheana*** has golden-yellow blooms 1¼ to 1½ inches in diameter, and on each shoot six or eight lustrous green, narrow-lanceolate leaves less than 1½ inches long, about ⅓ inch wide, without spots and with along each white, horny margin two or three backward-pointing teeth. Variety

Faucaria acutipetala

Faucaria britteniae

F. b. haagei (syn. *F. haagei*) has flowers 2 inches wide or wider and leaves up to 2¼ inches long, with white horny bands along their edges and keels. ***F. britteniae*** has 2- to 2½-inch-wide yellow flowers, with purplish-pink undersides to the petals. Its ovate-triangular leaves, compressed at their apexes to form a strong chin, have somewhat hollowed upper surfaces. Gray-green sprinkled with tiny dots of darker gray, they have white, horny margins each with three or four recurved teeth. ***F. lupina*** has solitary yellow flowers 1¼ to 1½ inches wide. Its spreading, slightly recurved, tapered, lanceolate leaves, up to 1¾ inches long and ½ inch or slightly more wide, have chinlike apexes and seven to nine hair-tipped teeth along each narrow, horny margin, and others along the keel. Green and with many small dots, they have flat upper surfaces and narrow horny margins. ***F. tuberculosa*** has yellow flowers 1½ inches in diameter. It is distinctive because its dark green triangular-ovate leaves ¾ to 1 inch long by about one-half as wide have their hollowed upper surfaces sprinkled with raised, white-tipped tubercles resembling incipient teeth, and their rounded undersides white-dotted. ***F. duncanii*** somewhat resembles *F. tuberculosa* but is without the prominent tubercles of that species. Its smooth, green, boat-shaped leaves, up to 1 inch long, are lined with red dots toward their tips. They have flat upper surfaces with turned-up, conspicuously-toothed margins, and rounded lower ones sharply keeled toward their apexes. The yellow flowers are about 2 inches wide.

Faucaria tuberculosa

Faucaria duncanii

Garden Uses. Faucarias should be in every collection of succulents, not only because of the ease with which they grow, but also for the charm and beauty of their forms, foliage, and flowers. These responsive plants are not only for specialists; they are highly adaptable and lend themselves well to growing in windows and for use in dish gardens and suchlike arrangements. They are not hardy outdoors where frosts are experienced nor in wet climates. Children of the desert, they flourish in the open only in arid regions. Elsewhere they must be kept where they are shielded from excessive cold and wetness.

Cultivation. Faucarias prosper in porous, well-drained soil watered moderately from spring to fall, less copiously during winter when the plants rest. When repotting is needed, attend to it in spring. Sun-lovers, faucarias succeed remarkably well, although they may not bloom, without direct exposure provided the light is good. Because of this they are useful for dish gardens in rooms with little or no direct sun. When grown in full sun, and this is certainly preferable, specimens in small pots are likely to be most comfortable if they are sunk to the rims of their containers in larger ones filled with sand or gravel. This keeps the roots cooler and reduces the need for frequent watering. Propagation is very easy by seeds and cuttings. For additional information see Succulents.

FAURIA (Faù-ria). Native from Washington to Alaska, the Kurile Islands, and Japan, the only species of this genus is a member of the gentian family GENTIANACEAE or, according to those who recognize the split, to the buckbean family MENYANTHACEAE.

An inhabitant of wet soils, ***Fauria crista-galli*** (syn. *Nephrophyllidium crista-galli*) differs from its relative the buckbean (*Menyanthes trifoliata*) in having undivided leaves. A hardy, hairless, rather fleshy, herbaceous perennial, it has creeping rhizomes from which arise long-stalked, kidney-shaped, blunt-toothed leaves 4 or 5 inches wide and shorter than wide. The stalked white blooms, ½ inch across, are in terminal twice- or thrice-branched clusters on stalks 1 foot to 2 feet tall. They are short-funnel-shaped with a five-lobed calyx, a corolla with five deep lobes (petals), five stamens, and a style with a shield-shaped stigma. The fruits are cylindrical, many-seeded capsules.

Garden and Landscape Uses and Cultivation. This is an interesting species for bog gardens and watersides where the soil is constantly wet and contains abundant organic matter. It is easily propagated by division, preferably in spring, and by seeds sown as soon as they are ripe and before they have had opportunity to dry out. The seed soil must be constantly wet. This can be assured by keeping the seed pods standing in water up to one-half their depth.

FAVA or FABA BEAN is *Vicia faba*. See Bean.

FAWN-LILY. See Erythronium.

FEATHER. The word feather is a component of the common names of these plants: feather fleece (*Stenanthium gramineum*), feather-geranium (*Chenopodium botrys*), feather-hyacinth (*Muscari comosum monstrosum*), golden feather (*Chrysanthemum parthenium aureum*), parrot's feather (*Myriophyllum brasiliense*), and prince's feather (*Amaranthus hybridus hypochondriacus, Polygonum orientale,* and *Saxifraga umbrosa*).

FEATHERFOIL. See Hottonia.

FEATHERLING. The rush featherling is *Pleea tenuifolia*, the white featherling *Tofieldia glabra*.

FEATHERTOP is *Pennisetum villosum*.

FEBRUARY, GARDENING REMINDERS FOR. Outdoor tasks this month make minimal demands on northern gardeners' time. In the main they are those suggested in the "January, Gardening Reminders For" entry in this Encyclopedia. Delay no longer anything that can be done to lighten the spring rush soon to come. Have lawn mowers sharpened, rototillers and other equipment put into good condition. Professional servicers are less busy now than later and so you are more likely to benefit from superior workmanship.

In all but the very coldest parts complete pruning grapes and as weather permits deciduous trees and shrubs, particularly any that have been neglected and are much tangled or overgrown. It is much easier to see what you are doing when branches are leafless than later. If they are infested with scale insects, or for other reasons need a dormant spray, this can be most economically and effectively applied after surplus growth is cut away.

You may now prune free-standing and espaliered hardy deciduous shrubs of kinds that bloom at midsummer or later on shoots of the current season's growth. Such sorts include butterfly bush (*Buddleia davidii* varieties), chaste tree (*Vitex agnus-castus*), other vitexes, peegee hydrangea (*Hydrangea paniculata grandiflora*), and rose-of-Sharon (*Hibiscus syriacus*). Do not, unless they are old, overgrown specimens in need of serious renovation and you are willing to sacrifice at least some of this season's floral display, prune until after blooming is through shrubs that bear their flowers on last year's shoots. Such kinds include *Buddleia alternifolia, Hydrangea macrophylla* varieties (here belong the popular

Prune selected hardy deciduous trees and shrubs as weather permits: (a) Apple

(b) Gooseberry

(c) Peegee hydrangea

(d) Butterfly bush (*Buddleia davidii*)

(e) Chaste tree (*Vitex agnus-castus*)

Wait until after flowering to prune shrubs that bloom on shoots of the previous year: (a) Hortensia hydrangea

(b) *Buddleia alternifolia*

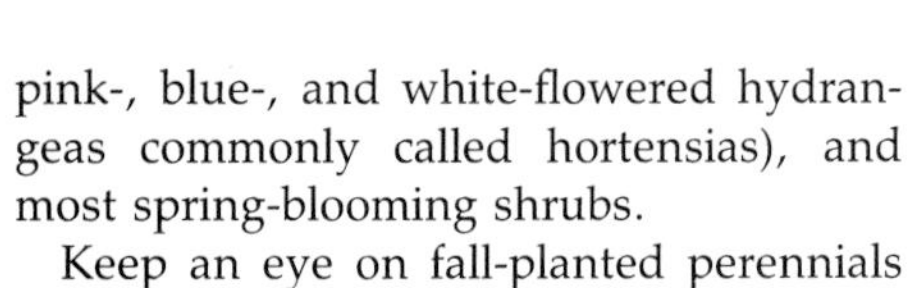
(f) *Spiraea bumalda* 'Anthony Waterer'

(c) Bridal Wreath (*Spiraea prunifolia*)

(d) *Deutzia*

(e) *Philadelphus*

pink-, blue-, and white-flowered hydrangeas commonly called hortensias), and most spring-blooming shrubs.

Keep an eye on fall-planted perennials and biennials. Alternate freezing and thawing may, unless they are well protected by a winter cover, heave them partly out of the ground. If this happens push them back promptly. Make sure any winter covering blown off by wind is raked back into place. Other troubles may come to such plants, both newly planted and more established ones, from puddles of water collecting about their crowns and causing harm by excluding air and freezing into blocks of ice. This trouble can often be averted by constructing miniature trenches with a hoe or spade to drain superfluous water away from the crowns of the plants.

Hardy bulbs, particularly those planted too shallowly or in sheltered places, such as at the foot of a south-facing wall, sometimes produce above-ground growth prematurely and this is liable to damage by severe cold. If this happens cover the sprouts before they begin to expand with 2 or 3 inches of peat moss or peat moss and sand. Pay particular attention to cannas, dahlias, gladioluses, and other bulb plants in storage. Be sure temperatures are neither too high nor too low, atmospheres too moist or too dry. Remove promptly bulbs that rot or slice out affected parts and dust the cut surfaces with sulfur or fermate.

Seeds left over from last year may or may not be viable. Test the percentage of germination of kinds you may think of sowing later indoors or in the garden by sowing twenty-five, fifty, or one hundred seeds in a pot of moist soil or sand. Cover with a sheet of glass or plastic and put in a temperature of 60 to 70°F.

In greenhouses February is a busy month. Sometime between its middle and end light shade will be needed by ferns and many other foliage plants, as well as African-violets, other gesneriads, most begonias, and all flowerng plants that prefer moderate to intense light.

Annuals, geraniums, and other sun-loving plants respond favorably to longer days of brighter light. They will now show much improved growth. Unless it was done last month, pot fall-sown clarkias, schizanthuses, and other annuals into the containers in which they are to bloom in April and later. Selective repotting of long-term or permanent greenhouse inhabitants may be begun. It is best to attend to this when the first signs of renewed growing activity become evident, certainly before new growth is much advanced. First, attend to such tropical plants as anthuriums, crotons (*Codiaeum*), dieffenbachias, dracaenas, palms, and philodendrons. Cool house inhabitants, such as asparagus-ferns, grevilleas, heaths, and blue-flowered plumbago and its white-flowered variety, can be left until somewhat later but potting all permanent greenhouse plants in need of this attention should be completed before mid-March. Those not repotted generally benefit if some of the surface earth is pricked away and replaced with a top dressing of rich new soil.

The potting season provides the opportunity to inspect and overhaul your plants. Some perhaps should be discarded, others perhaps made use of as propagating stock. Certain of those retained are likely to need pruning, others staking. Make sure the drainage of the pots is not clogged, that excess water can escape freely through the holes in the bottoms of the containers. Inspect each plant carefully for aphids, scale insects, mealybugs, and other pests and diseases and take suitable action to free them of these.

Rambler and baby rambler roses pruned moderately and potted in fairly heavy soil in containers just big enough to accommodate the roots comfortably, may be started now for spring bloom. Set them where the night temperature is 45 to 50°F. Spray their tops lightly with water on bright days to encourage the buds to break. After growth is well started higher temperatures are in order. Astilbes, variegated-leaved hostas, and lily-of-the-valleys force easily now and make charming decorative plants. Sturdy clumps of rhubarb stood closely together in the dark in a temperature of 60 to 70°F soon produce welcome pickings. Fresh mint is easy to have from strong roots planted in flats or pots and grown in a sunny place in a 50 to 60°F temperature. Continue to bring into the greenhouse successive batches of hyacinths, narcissuses, tulips, and other hardy bulbs to be forced for early bloom. Start into growth early batches of achimenes, caladiums, gloxinias, and tuberous begonias.

Regular applications of dilute liquid fertilizer benefit plants in active growth that have filled their containers with roots. Here belong annuals, except recently potted ones, as well as such bulbs as freesias, gladioluses, and ixias. Hippeastrums now making active growth and that have been in their pots a year or more also benefit from fertilizing.

Seed sowing begins now to assume importance, but except for a few sorts the first half of February is too early to sow indoors annuals and vegetables for outdoor planting later. You may sow lobelias, verbenas, wax begonias, and others men-

Repot permanent greenhouse plants that start new growth early: (a) Anthuriums

(b) Crotons (*Codiaeum*)

(c) Dieffenbachias

(d) Dracaenas

(e) Palms

(f) Philodendrons

Lily-of-the-valleys are easily forced into bloom: (a) Potting the "pips" closely together

(b) Three weeks to one month later in full bloom

Hardy bulbs well started in plunging beds outdoors or in cold frames may now be brought indoors for forcing

Seeds to sow indoors in February include: (a) Verbenas

(b) Wax begonias

(c) Madagascar-periwinkle (*Catharanthus*)

(d) Silk-oak (*Grevillea*)

(e) Coleuses

(f) Snapdragons

tioned in the Encyclopedia entry "January, Gardening Reminders For" that need a very long period of growth indoors to bring them to a respectable size by planting out time. It is also appropriate to sow seeds of asparagus-ferns, Australian silk-oak (*Grevillea*), blue gum (*Eucalyptus globulus*), coleuses, and *Melianthus major* to produce attractive foliage plants for indoors or for temporary use in summer beds. Toward the end of the month sow hunnemannias, snapdragons, and stocks. Then too make a first sowing of early cabbage, cauliflower, and lettuce to give plants for setting in the garden later.

Propagation by cuttings of a wide variety of plants is an important task in February. The earlier the cuttings of plants for

Cuttings of many kinds root readily now, among them:
(a) Abutilons

(b) Fuchsias

(c) Geraniums

(d) Heliotropes

outdoor use later, such as those of abutilons, fuchsias, geraniums, and heliotropes, are taken the bigger the plants will be when the time arrives to plant them out. These kinds root satisfactorily in a night temperature of 60°F. Those of more tropical plants, such as diffenbachias, dracaenas, and philodendrons, respond better to ten degrees or so higher. Cuttings of fall- and winter-flowering cool-greenhouse plants that are raised anew each year, such as *Buddleia asiatica, B. farquhari,* chrysanthemums, *Salvia leucantha,* and stevia (*Piqueria*), root readily in a temperature of 50 to 55°F. Those started now will make large pot specimens. Smaller ones can be had by delaying taking the cuttings until March or April.

Cold frames need watching carefully from now on. It is as important that temperatures for English daisies, forget-me-nots, pansies, and other plants wintered in a dormant condition not be raised too high by sun heat as it is that they be protected from excessive cold in severe weather and at nights. Ventilate whenever the interior temperature is sufficiently high that frost on the glass begins to melt, but do so in such a way that the plants are not exposed to sweeping winds or drafts.

Houseplant care in February is essentially the same suggested in "January, Gardening Reminders For." Read that and also the remarks above about the February care of greenhouse plants. Remember that as a rule it is well to delay potting and propagating houseplants until somewhat later than it is practicable to begin these operations with the same kinds of plants in greenhouses. Otherwise, much the same procedures are appropriate. It is feasible now, in preparation for potting next month, to prune leggy abutilons, begonias, geraniums, hibiscuses, lantanas, and the like. Be careful not to overwater them afterward. The idea is to reduce the size of plants, shape them, and to encourage new breaks that will be showing as fat buds or incipient shoots when potting time comes.

Continue to bring in successive batches of branches of outdoor shrubs for forcing in water. They force more readily as their natural season for growth nears. Not only flowering branches, but tree branches that produce fresh green foliage can be very pretty. Try some of horse-chestnuts and maples.

In the south, February brings a variety of tasks. Pruning roses, shrubs, and trees that carry blooms on shoots of the current year, as do butterfly bushes, crape-myrtles, oleanders, poincianas, and rose-of-Sharons, should in the deep south be done early in the month, by its end further north. Do not, unless you are undertaking a major renovation of old overgrown specimens and are willing to sacrifice much or all of this year's display, prune before they flower, trees, shrubs, or climbing roses that bloom on shoots developed last year. These are chiefly spring-bloomers. Prune or shear evergreens, including hedges, that need attention before new growth starts.

Complete any spring planting you contemplate, of trees, shrubs, and roses before they begin making new growth. If you must plant after new growth is well started, use container-grown plants. But even they are better set out early. The only possible advantage of late planting is convenience. The same remarks apply to herbaceous perennials, many of which can successfully be transplanted and divided in spring before they begin new growth.

Lawns in areas where grass has been practically dormant through the winter will need making ready for the first mowing of the season. Rake off dead grass and debris. Fork over, fertilize, and patch with sod or sow grass seed on bare spots. Fertilize the entire area.

Fertilizing much of the garden may be done now. This includes, in addition to lawns, most trees, shrubs, vines, evergreens, roses, herbaceous perennials, and bulb plantings. With a few subjects, azaleas and camellias for example, it is better to delay fertilizing until they are through blooming.

Outdoor sowings to be made in the middle and upper south as soon as the soil can be worked into suitable crumbly condition (in the lower south most of these could be made last month) include all annuals except the most tender and tropical ones and such vegetables as beets, broccoli, cabbages, carrots, cauliflowers, Chinese-cabbage, peas, lettuce, mustard, onions, radishes, spinach, and turnips. Plant onion sets as early as possible.

West Coast garden activities in February closely parallel those detailed for the south. As there, it is of first importance to complete as soon as possible all pruning that appropriately may be done in spring and to install before their new growth begins any trees, shrubs, roses, and herbaceous perennials that are to be planted. In southern California February is a good time to set out avocados and citrus fruits.

Lifting and dividing perennials to limit the sizes of the clumps of spreading kinds or to secure increase is in order. If perennial beds are to be planted or replanted this spring lose no time in attending to that. Groundcovers may be planted and in milder sections many sorts of spring-blooming annuals and biennials. Among such are calendulas, cinerarias, English daisies, pansies, primroses, snapdragons, stocks, and wallflowers.

In the Northwest get ground into condition for planting and seed sowing. Sow seeds of annuals and vegetables indoors or in cold frames to give plants to set in the garden later for early crops. As soon as there is no longer chance of snow or freezing, condition lawns by raking off debris, patching where necessary, seeding barish spots, and topdressing or fertilizing. Toward the end of the month sow cabbages, carrots, peas, spinach, turnips, and all hardier sorts of annuals including larkspurs and sweet peas outdoors.

Fertilize perennials and early-flowering bulbs before new growth is more than 2 to 3 inches out of the ground. Scatter the fertilizer evenly and scratch it lightly into the surface. Give lawns attention now. Patch bare spots by sodding or seeding after first improving the soil by forking in compost or other organic matter and fertilizer. Fertilize or top-dress all other lawn areas.

Start indoors if you did not last month, tubers of achimenes, begonias, caladiums, cannas, elephant's ears, gloxinias, and yellow calla-lilies. Prune abutilons, fuchsias, lantanas, and other such shrubby and semishrubby pot plants to shape and repot or top-dress them. In the Southwest plant outdoors cannas, galtonias, gladioluses, montbretias, and tuberoses for garden ornament and cut flowers.

FEDIA (Fèd-ia)—African-Valerian. One of this genus of perhaps three species of the Mediterranean region is sometimes grown in flower gardens and rarely as a salad plant. It belongs in the valerian family VALERIANACEAE. The import of its name is unknown. The African-valerian (***Fedia cornucopiae***) is a hairless leafy annual with branching, purplish stems 1 foot or so tall, and opposite, more or less ovate or spatula-shaped, shining green leaves 2 to 4 inches long. Its attractive, small red or fuchsia-colored flowers are in dense, terminal clusters. They have a calyx with four small teeth, a tubular, slightly two-lipped corolla, and two stamens. The fruits resemble grains of wheat. There is a variety, *F. c. candidissima,* with white flowers, and another, *F. c. floribunda-plena,* with double, reddish-pink blooms.

Garden Uses and Cultivation. This is an unusual and quite delightful addition to flower beds and borders and for use in informal garden areas. It comes readily from seeds sown in early spring, or in mild climates in fall, where the plants are to bloom. If the ground where plants grow is left undisturbed volunteer seedlings often spring up the following year. The plants should be thinned to about 6 inches apart. This species prospers in ordinary well-drained garden soil in sun and needs no special care.

FEIJOA (Fei-jòa)—Pineapple-Guava. South American evergreen shrubs or small trees of the myrtle family MYRTACEAE constitute this genus, named after J. da Silva Feijo, a Director of the Natural History Museum, San Sebastian, Spain. They have opposite, pinnate-veined leaves, green on their upper sides and densely-white-hairy beneath. Their long-stalked, solitary flowers, arising from the leaf axils, have four sepals, four wide-spreading petals, and a conspicuous central cluster of red stamens. The fruits are ellipsoid to nearly round and wear the dried remains of their calyxes at their tips. Embedded in their pulp are many small seeds. Technically the fruits are berries, although they do not look like what that is generally thought to mean. One of the only two species is cultivated.

The pineapple-guava (***Feijoa sellowiana***) inhabits southern Brazil, Paraguay, Uruguay, and Argentina and is widely cultivated in the subtropics and warm-temperate regions including the southern and southwestern United States. It is so named because its fruits combine to some

Feijoa sellowiana

extent the flavors of pineapple and strawberry and in appearance resemble guavas (*Psidium*). Ordinarily a bushy plant up to 18 feet in height, it is often broader than tall. Its oval-oblong leaves are 2 to 3 inches long and its flowers 1 inch to 1½ inches in diameter. The latter are quite beautiful, white-woolly on their outsides and white lightly tinged with purple within. The stamens, longer than the petals, are rich dark red. Dull gray-green sometimes tinged with red, and 1 inch to 3 inches long, the fruits are round, egg-shaped or oblongish. They are dusted with a powdery, whitish bloom and contain a jelly-like pulp in which the seeds are embedded. Jelly and seeds are surrounded by whitish flesh.

Garden and Landscape Uses. In the United States the chief value of this shrub is as a flowering ornamental, although its fruits are edible and palatable and they may be eaten out of the hand or made into jelly. Strangely enough its petals are also edible; they are pleasingly sweet. The pineapple-guava is hardy throughout Florida, along the Gulf Coast, and elsewhere where the temperature does not drop below about 15°F. Dense and slow-growing, this shrub is suitable for beds, borders and single lawn specimens. It shears well and makes effective hedges. When in full bloom it is particularly lovely, at all times it has the appearance of quality. In addition to its suitability for garden planting, this is a good item for growing in pots or tubs for terrace and patio adornment and for greenhouses; it blooms when quite small.

Cultivation. Any ordinary soil and a location in full sun suits the pineapple-guava; it responds favorably to fertilizer and adequate moisture. No regular pruning beyond any necessary to keep it from becoming straggly or growing out of bounds is required, but it stands cutting well and there need be no hesitation about pruning if that seems desirable. Regular shearing should be routine when it is used as a hedge. Propagation is easily achieved by seeds and by cuttings and layers, but the last two methods are slow. A few selected varieties have been given horticultural names such as 'Andre', 'Choice', 'Coolidge', and 'Superba'; these are best increased by grafting onto seedling understocks. For container-grown specimens the soil should be fertile, well-drained, and at all times moist, but in winter somewhat drier than at other seasons. Necessary pruning and repotting is done in late winter and spring. From spring to fall weekly applications of dilute liquid fertilizer should be given well-rooted specimens. A cool greenhouse with a minimum winter temperature of 40 to 50°F provides satisfactory accommodation, but the plant also does well in a warmer environment provided it receives maximum light.

FELICIA (Fel-ícia)—Blue-Marguerite or Blue-Daisy, Kingfisher-Daisy. Most of the sixty species of this genus of nonhardy shrubs, subshrubs, annuals, and perhaps herbaceous perennials of the daisy family COMPOSITAE are endemic to South Africa. A few are natives of tropical Africa. The name commemorates a German official, Herr Felix, who died in 1846. Botanically the group, which is related to *Aster*, is in need of study. Its nomenclature is much confused. Some botanists include in *Aster* some of the species treated here as *Felicia*.

Felicias have alternate or opposite, toothed or toothless leaves and generally long-stalked, prevailingly solitary, daisy-type flower heads consisting of a central eye of usually fertile, bisexual florets and one or two rows of blue, mauve-blue, or less often white, female, commonly fertile ray florets. The involucres (collars of bracts at the backs of the flower heads) are of two to many rows. The fruits are compressed, seedlike achenes. The kinds described here are natives of South Africa.

Blue-marguerite or blue-daisy (***F. amelloides*** syns. *F. capensis, Agathaea coelestis*) is an erect, sparsely-hairy subshrub 1 foot to 2 feet tall. It has opposite, elliptic to roundish-ovate, scarcely toothed or toothless leaves about 1 inch long, with short, winged stalks. It bears masses of long-stalked, yellow-centered, sky-blue to darker blue, 1- to 1¼-inch-wide flower heads. The closely set leaves of subshrubby ***F. echinata*** (syn. *Aster echinatus*) are lanceolate to ovate-oblong, ½ to 1 inch long, spine-tipped and usually with bristly teeth along their margins. The quite freely produced flower heads, ¾ inch to 1½ inches across, have canary-yellow centers and purple-blue to white rays. A perennial about 8 inches tall, best grown as an an-

Felicia amelloides

Felicia amelloides (flowers)

Felicia bergerana

nual, ***F. elongata*** (syn. *Aster elongatus*) is a subshrub about 8 inches high, with spreading branches, slender leaves about 1 inch long, and white, pink, mauve, or blue flower heads 1¼ to 1½ inches across. They have yellow centers and ray florets with cyclamen-purple bases. Another subshrub or shrub, ***F. fruticosa*** (syn. *Aster fruticosus*) is 2 to 3 feet tall. It has woody stems and linear to spatula-shaped leaves up to about 1 inch long with recurved margins. The slender-stalked, solitary flower heads 1 inch wide or wider have yellow centers and violet-purple ray florets. Low, bushy, hairy ***F. petiolata*** (syn. *Aster petiolatus*) has prostrate branches and alternate, short-stalked, obovate to lanceolate, coarsely-toothed leaves ½ to 1 inch long. On slender stems, the flower heads are carried well above the foliage. From ¾ inch to 1½ inches wide, they have yellow centers and mauve-pink ray florets that become bluer with age.

Kingfisher-daisy (***F. bergerana*** syn. *Aster bergeranus*) is one of the most beautiful annuals. Attaining a maximum height of about 8 inches, freely-branched and densely-hairy, this has opposite or alternate, obovate-oblong, toothed leaves 1 inch to 1½ inches long. The flower heads, terminating long, hairy stalks, have sterile, yellow to nearly black disk florets and bright blue rays. Also hairy and annual, of delicate appearance, charming ***F. tenella*** (syn. *F. fragilis*) is 4 to 8 inches tall. It has slender-linear leaves up to 2 inches long and flower heads with yellow centers and light violet-blue rays. Up to 1 foot tall, bushy, and rough-hairy, annual ***F. rotundifolia*** (syn. *Aster capensis rotundifolius*) has opposite, oblong-ovate to roundish-elliptic leaves ¾ inch to 1¼ inches long. Its pretty, yellow-centered flower heads are ¾ inch to 1¼ inches across.

Garden and Landscape Uses. Felicias are delightful for the fronts of borders, rock gardens, as pot plants, and some for growing in hanging baskets. Because the ray florets of the flower heads of many kinds roll up in dull light it is essential to give felicias sunny locations. None of the perennials is hardy in the north. The annuals fail in torrid weather, but make good displays as long as nights are not extremely hot nor days excessively hot and humid. Where summers are marked by such extremes they can be used for early displays and replaced by more heat-tolerant plants after they cease flowering.

Cultivation. Perennial felicias are easily raised from seeds and cuttings. They prosper in well-drained soils including those of limestone derivation kept moderately, not excessively moist. Occasional pinching out of the tips of the shoots during their early stages of growth is needed to encourage bushiness. Where hardy and grown permanently outdoors, it is good practice to shear the plants as soon as flowering is through or just before new growth begins in spring. The annuals, and perennials grown as annuals, are raised from spring-sown seeds. Where summers are cool, sowing may be done outdoors as soon as the weather moderates and it is safe to sow the majority of hardy annuals and vegetables. Where hot summers are the rule, it is better to sow indoors some eight weeks earlier so that the plants have a good start and will bloom before damaging hot weather arrives. A sunny greenhouse or equivalent accommodation where the night temperature is 50°F and that by day five to ten degrees higher is a suitable place to grow young felicias until all danger of frost is passed and they can be planted in the garden. Spacing outdoors may be about 6 inches apart for the annuals and somewhat more for the perennials. As decoratives to display in cool greenhouses felicias are very satisfactory. They bloom for long periods and are easy to satisfy. Pots and hanging baskets are suitable containers. Nourishing, well-drained, sandy soil suits, and exposure to full sun is needed. The winter night temperature should be about 50°F, increased by day to 55 to 60°F.

FELT, FELTED. Descriptive botanical terms for dense coverings of often more or less matted hairs.

FENCES. Like a lie, according to the small boy's classical definition, a fence can be "an abomination in the sight of the Lord and an ever ready aid in times of trouble." The fact is, wisely chosen and appropriately located, fences can be enchantments to properties rather than abominations and can afford marvelous backgrounds for plantings and supports for vines. The beauty of fences well selected and used to advantage must be apparent to anyone who has visited Colonial Williamsburg, Virginia. Fine examples are to be seen in most parts of the United States. As aids in times of trouble fences can serve to break the force of winds or deflect them, and to keep neighbors' dogs, sometimes cats, and—can it be said?—children, off properties or out of gardens.

Traditionally, fencing front yards or gardens has been less done in America than Europe, indeed in some areas it has been quite unacceptable. To this convention many communities owe the distinctively American parklike appearance of streets flanked by lawns and gardens stretching from front doors to sidewalks. These are beautiful prospects where space and density of population permit, but as lots and gardens become smaller and the number of people per acre larger, defining front boundaries becomes more important and fences carefully employed are one means of doing this effectively, and attractively. Certainly no considerate home owner would indulge in the spite fence type of installation, nor will most communities permit them.

The chief uses of fences, other than discouraging trespass, are to achieve privacy and a sense of containment, to screen undesirable views or focus on pleasing ones, and to provide suitable shelter and pleasing backgrounds for plantings. As architectural features they can be attractive, harmonizing elements in garden design.

Fences may be individually designed of locally available materials, thus making possible originality and allowing for artistic expression, or may be of any one of numerous types available commercially. They may be of wood, metal, plastic, or combinations of these, or of other materials. Stone fences are more properly walls. Whatever the material used it should be durable and sturdy enough to serve for many years without sagging, warping, or falling into disrepair. If wood is chosen, especially for posts set in the ground, it should be of a kind highly resistant to decay. Such woods as black locust, white-cedar, swamp-cypress, red-cedar, and redwood are best for fence posts. Before their installation have treated or treat parts that will be in the ground with a good preservative such as copper napthalanate or pentachlorophenol. Creosote is excellent, especially if applied under pressure, but it is toxic to plants. The same woods or others, such as ash, hickory, and red maple, can be used for the rails, boards, or pickets that form the above-ground parts of the fence. No wooden members other than the posts should be in contact with the ground. Some few woods last for many years and weather beautifully without treatment with a preservative, even in humid climates, and more do in dry ones, but the majority need protection against moisture-induced rotting. Treatment with a preservative that scarcely changes the color of the wood or one that stains it brown is one way of achieving this. Or the fence may be painted the color of one's choice. White fences are traditionally American, but other hues can

Fences: (a) Chain link

(b) Cast iron

(c) Vertical boards of different widths without spaces between them

(d) Slightly separated vertical boards of different widths

(e) Plain picket

(f) Alternate pickets of different lengths

(g) Horizontal board

(h) Horizontal woven board

(i) Peeled log

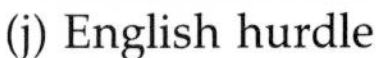
(j) English hurdle

(k) Post and rail

be employed effectively, especially in warm, sunny regions where building in the Spanish or Mediterranean tradition is appropriate.

Fences are of many types and designs. Among the more prosaic are chain link and similar woven-wire kinds yet these are among the most effective in limiting trespass. In gardens such installations are generally acceptable only if they are masked by hedges or other plantings.

Wooden fences can be very satisfactory and decorative if in accord with their surroundings. They may be solid or open, may stop or divert winds or permit them to pass freely through. They can be of pickets, slats, or boards set vertically or be of rail or clapboard design with members other than the supporting posts horizontal. What have been called good neighbor fences are those with both sides faced attractively, say with boards or pickets rather than one side revealing the structural supports. There is much scope for ingenuity and artistic expression in designing fences. For a discussion of gates to be used with fences see Gates.

Solid fences, especially around small areas, sometimes produce too great a feeling of restriction, almost of imprisonment. This can be relieved by planting a few evergreens or other shrubs in front, not to hide the fence completely, but only to interrupt long, unbroken expanses of solid board, woven paling, and similar faces. Vines can be used similarly and along low fences so can narrow flower borders.

Open fences include simple rail types that impede the view not at all and add delightful rural touches, to lattice ones of various designs and degrees of openness and those of vertical boards with apertures perhaps not more than an inch wide between their vertical members. Or they may be of vertical or horizontal louvers, which allow air circulation and ventilation while giving complete privacy.

Space posts for fences 5 to 8 feet apart. If of dressed lumber let them be 4 inches square. The butt ends of natural, rounded posts, such as locust, should be 4 to 6 inches in diameter depending upon the height of the fence. Sink the posts at least 3 feet into the ground and pack them around with crushed stone or gravel or with concrete. Do not use concrete under the posts however. To do so forms little basins that holds water that promotes decay.

FENDLERA (Fend-lèra)—Fendlerbush. Three or four species of deciduous, spring-blooming shrubs of the western United States and Mexico constitute *Fendlera* of the saxifrage family SAXIFRAGACEAE. Only one appears to be cultivated, although the others are worthy of this attention. The genus differs from mock-oranges (*Philadelphus*), to which it is closely related, in its flowers having eight instead of numerous stamens. Its name commemorates August Fendler, a German botanical explorer of New Mexico, who died in 1883.

Fendleras are deciduous or sometimes nearly evergreen shrubs with nearly stalkless, opposite, undivided leaves. Their flowers, one to three together on short branchlets, have four persistent sepals longer than the calyx tube, four clawed (abruptly narrowed toward their bases) approximately ovate petals, slightly toothed at their margins, eight stamens flattened above and having appendages slightly longer than the anthers, and four separate styles. The fruits are capsules.

Hardy in southern New England, ***F. rupicola*** attains a height of 6 feet and is a graceful, slender-branched, ornamental. Its grayish-green, elliptic to oblong-lanceolate leaves are ½ inch to 1¼ inches long and usually are rough-hairy on both surfaces. Mostly solitary, the white flowers are 1¼ to 1½ inches across and have erect stamens one-half as long as the wide-spreading petals. In bud they are squarish and tinged pink on their outsides. The seed capsules are from slightly under to slightly over ½ inch long. Variety *F. r. falcata* is distinguished by its larger, narrowly-elliptic to linear-lanceolate, often sickle-shaped and usually nearly hairless leaves and its larger blooms. The species is indigenous in Texas and New Mexico, the variety from New Mexico to Arizona and Colorado.

Garden and Landscape Uses and Cultivation. The kinds discussed are hardy in southern New England. For satisfactory results they need porous, well-drained soil and a hot, sunny location. They do not succeed in cool, moist places. Propagation is by seeds, and by summer cuttings under mist or in a greenhouse propagating bench. Pruning consists of thinning out old flowering branches and weak and crowded new ones as soon as blooming is through. As horticultural subjects fendleras can be used in much the same way as mock-oranges, as solitary specimens, in groups, or in beds with other shrubs.

FENDLERBUSH. See Fendlera.

FENESTRARIA (Fenes-trària). The remarkable and delightful windowed plants of South Africa are fascinating examples of practical adaptations to a highly specialized environment. To them, as the name, derived from the latin *fenestra*, a window, indicates, the two species of *Fenestraria* belong. They are members of the *Mesembryanthemum* tribe of the carpetweed family AIZOACEAE.

Fenestrarias are nonhardy, stemless, succulent perennials with rosette-like clusters of erect, cylindrical-club-shaped leaves, opaque except for their blunt, translucent, glassy tips. Through these windows light is admitted to the chlorophyll-containing cells below. In the wild fenestrarias are nearly completely buried in the earth.

Exochorda racemosa

Fagus sylvatica atropunicea (*F. s. cuprea* of gardens)

Fagus sylvatica atropunicea (*F. s. riversii* of gardens)

Fatsia japonica

Ficus sagitata variegata

Forsythia intermedia spectabilis

Fothergilla major

Fothergilla (flowers)

Fothergilla major (fall foliage)

Fouquieria splendens

Only the windowed tips of their leaves, which suggest the peering eyes of sand lizards, also given to burying themselves in the desert floor, show. This mode of growth is an efficient protection against browsing animals and, most importantly, against dehydration by the scorching sun, arid atmosphere, and drying winds of the deserts these plants inhabit.

Undoubtedly *Fenestraria* is closely related, as the structure of its flowers show, to *Cephalophyllum*. It resembles in appearance *Frithia*, but from it is easily separated by its opposite, smooth-surfaced leaves and its sepals and petals being free to their bases instead of their lower parts forming tubes. The fruits of *Fenestraria* are capsules.

The more beautiful kind is ***F. aurantiaca,*** which has underground stems that at intervals form tubers and develop clusters of six to eight leaves. The latter are ¾ inch to 1½ inches in length. The golden-yellow blooms, which appear in late summer or early fall on stalks up to 2 inches long, are 1¼ to 2½ inches in diameter and have many wide-spreading, narrow petals. They suggest, although they are structured very differently, the flower heads of daisies. Very similar, but with shorter, fatter leaves, and white blooms up to 1¼ inches across, ***F. rhopalophylla*** is perhaps a little more difficult to grow and less free flowering than the other.

Garden Uses and Cultivation. Fenestrarias are intriguing for inclusion in collections of choice succulents. They require very sandy soil, and the general care needed by *Lithops* and other dwarf, very fleshy plants of the *Mesembryanthemum* relationship. Although in the wild they are mostly subterranean this manner of growth will not do in more humid climates. In cultivation it is important to plant them in more conventional fashion with their leaves entirely above ground. If this is not done the plants are likely to rot and die. Fenestrarias rest in winter and then should be kept dry. At other times they are watered moderately. They resent root disturbance and some growers recommend, instead of normal repotting of specimens in need of more root room, that their pots be placed inside larger pots and be packed around with sandy soil. The roots then find their way through the drainage hole in the pot into the new earth. Propagation is by seed and division. To obtain seeds the plants must be cross-pollinated; they do not set seeds to their own pollen. For additional information see Succulents.

FENNEL. See Foeniculum. Giant-fennel is *Ferula*. Fennel flower is *Nigella*.

FENUGREEK is *Trigonella foenum-graecum*.

FERDINANDA. See Podachaenium.

FERMATE or FERBAM. An organic fungicide (ferric dithiocarbonate) used as a spray or dust, often with sulfur.

FERN. The common names of many true ferns together with their botanical equivalents are listed at the conclusion of the entry on Ferns. The word fern also forms part of the names of a few plants that are not ferns. These include asparagus-fern (*Asparagus*), fern grass (*Catapodium rigidum*), fire-fern (*Oxalis hedysaroides rubra*), and sweet-fern (*Comptonia peregrina*). The so called air- or neptune-fern sold by florists as a novelty is not a plant and is not alive. For more about it see Air-Fern or Neptune-Fern.

FERN ALLIES. See Ferns.

FERN BALLS. These, formerly imported in dormant condition from Japan, are made by tying into compact spheres the rhizomes of certain ferns, most usually *Davallia bullata*. After soaking thoroughly in water the fern balls are suspended like hanging baskets in a warm, humid, shaded greenhouse. There, they soon sprout new foliage and become attractive globes of feathery greenery.

Once started into growth, fern balls must be kept moist. This is most easily done by immersion. Once every two weeks or so a little fertilizer may be added to the water in which they are soaked.

To construct a fern ball allow a *Davallia* that has a large number of rhizomes to become dormant. To do this withhold water in winter. When the foliage has died shake or wash away the soil and with aluminum wire or nylon string tie the rhizomes securely into a tight ball.

FERNS. Ferns constitute the most important group of the pteridophytes. Other groups that belong, often referred to as fern allies, include selaginellas, lycopodiums, equisetums, and psilotums. The term fern allies is something of a misnomer because true ferns are believed to be more closely related to flowering plants, which evolved from ancient "seed ferns," than to the plants we call fern allies.

Ferns and so-called fern allies differ

Fern fiddleheads

Underside of fern frond showing clusters of spore capsules

from the great majority of plants gardeners grow in having no flowers, not even primitive ones, and no seeds. Their sexual reproduction is by spores, but more of that later.

These are plants of ancient lineage. The world was largely populated by them and related vegetation now extinct long before the coming of the first flowering plants. Over 200,000,000 years ago they dominated vast expanses of the surface of the globe and formed great forests of striking aspect in which dinosaurs and other creatures now extinct disported. But long before dinosaurs appeared on earth and millennia before man made his tardy entrance only thousands of years ago, ferns existed.

The ferns and fern allies of those ancient times included many sorts gigantic in comparison to the great majority of kinds extant. The largest living tree ferns reflect inadequately the magnificence and beauty of their prehistoric ancestors.

Slowly evolution proceeded. New kinds of plants and animals appeared. Ancient ones became extinct. Mighty forests were submerged. Their denizens and deep accumulations of organic debris that underlaid them, under the influences of heat and compression, were slowly converted into vast beds of coal, a fossil fuel of tremendous importance to man in his relentless efforts to satisfy his needs for sources of energy. When coal is burned radiant energy received many millions of years ago from the sun and stored by ferns and other plants later transformed into coal is released as heat. Mankind is deeply indebted to ferns and related plants.

Present-day ferns number perhaps 10,000 species widely distributed throughout temperate, subtropical, and tropical regions. Most grow in the ground. Some are tree-perchers (epiphytes). There are even a few aquatic ferns. All are perennials, some evergreen, some deciduous. They vary tremendously in size from miniatures such as curly-grass (*Schizaea pusilla*), a slender-leaved mite usually not over 3 or 4 inches tall, to giant trees such as certain cyatheas and dicksonias, which may attain heights of 50 feet or considerably more.

Certain characteristics possessed by ferns are readily recognizable. Their leaves or fronds in their young stages are coiled and expand by unwinding from base to tip. The coiled young fronds of many resemble and are called fiddleheads. The spores are reproductive organs of dustlike fineness, each of a single cell. They are contained in spore capsules or sacs, which in turn are assembled in clusters called sori (singular sorus).

Unlike seeds, spores are not the result of a sexual mating. They carry only the genes of one parent, the plant that bore them. And unlike seeds, they do not directly produce offspring resembling the plant that bore them. Instead, they give rise to a totally different kind of plant bearing no resemblance to a conventional fern, and called a prothallium (plural prothallia). The prothallium does not produce roots, stems, or leaves. It remains, often for many months, a flat, roundish, heart-shaped, or butterfly-shaped, scale-like body anchored to the ground or other support by rootlike hairs and slowly increasing in size up to a maximum (although this is unusual) of about 1 inch in diameter. Its under surface develops sexual reproductive organs, archegonia (singular archegonium) containing female gametes and antheridia (singular antheridium), containing male ones. Under suitable conditions of temperature and moisture male gametes (sperms) propel themselves through films of water, enter the archegonia, and unite with the females (eggs). The results of these matings are new plants of the type of the original fern of familiar form. Thus with ferns there is a distinct alternation of quite distinct and separate generations. The plants we know as ferns (the sporophyte generation) produce spores that develop into the gametophyte generation individuals of markedly different appearance, which, after a sexual procedure, produce another generation of sporophytes.

Ferns, at least the sporophyte generations we are about to consider, which are the only ones ordinarily familiar, are generally easily recognizable as such. Their usually much-divided leaves give reason for the adjective ferny, meaning lacy or finely-divided. Yet a few ferns have fronds

Ferns come in many forms: (a) Boston fern (*Nephrolepis exaltata*)

(b) Maidenhair fern (*Adiantum*)

(c) Bird's nest fern (*Asplenium nidus*)

(d) Water fern (*Ceratopteris*)

(e) Staghorn fern (*Platycereum*)

that depart from this familiar pattern. There are those with undivided, more or less paddle-shaped fronds, such as the bird's nest fern (*Asplenium nidus*), and some with slender, grasslike ones, such as the curly-grass (*Schizaea pusilla*).

Hardy ferns include deciduous and evergreen sorts with a preponderance of the former. Most of those commonly grown are native North Americans. A few are European or Japanese. All lend themselves to planting in woodland gardens and rock gardens, beneath trees and tall shrubs, at watersides, on the shady sides of buildings, and in other places where there is insufficient light for the majority of flowering plants. Old stone walls such as are found throughout New England afford excellent backgrounds and niches for hardy ferns and so do abandoned cellar excavations. Special gardens in which ferns predominate can be charming, their greens cool and refreshing, the patterns and delicacy of their foliage intriguing.

(f) Tree fern (*Dicksonia*)

The needs of hardy ferns vary considerbly according to kind. For their best satisfaction the great majority require shade from at least strong sun, with deeper gloom for a few. A very few, such as the sensitive fern (*Onoclea sensibilis*) and ostrich fern (*Matteuccia struthiopteris*) will grow in full sun if moisture supplies are ample. Shelter from strong winds, a humid atmosphere, and soil containing an abundance of organic matter and that does not dry excessively are other common environmental needs. Cool conditions are to the liking of these ferns.

Soil preferred by most hardy ferns is damp, but well drained, coarse-textured, loose rather than compact. See that it contains a considerable proportion of leaf mold, peat moss, decayed sawdust, bagasse, or other partially rotted plant debris and keep it surfaced with a mulch of such material. Slightly to strongly acid soil is required for most ferns, but some prefer or must have limestone, among them maidenhair spleenwort (*Asplenium trichomanes*) and walking fern (*Camptosorus rhizophyllus*). This can be provided by adding limestone chips and small pieces of limestone to the soil or by mixing in crushed oyster shells or ground limestone. Some species, such as royal fern (*Osmunda regalis*) and marsh fern (*Thelypteris palustris*), need wetter soils than most.

Select locations for ferns believed to be congenial to individual kinds and give some thought to the known habits of each. Do not set vigorous, spreading kinds such as hay-scented fern (*Dennstaedtia punctilobula*) or the New York fern (*Thelypteris noveboracensis*) beside delicate neighbor plants they are likely to outgrow and overrun. Do select the most promising spots for the choicest kinds. Give special consideration to the placement of evergreen ferns so they can be enjoyed to their fullest in winter as well as other seasons.

Space the plants with consideration for the size and vigor of individual kinds. Fern specialist F. Gordon Foster recommends as a guide to minimal distances what he calls the 1, 2, 3 rule, 1 foot apart for small spleenworts and other miniatures, 2 feet apart between medium-sized plants, such as shield ferns (certain *Dryopteris*), 3 feet between large growers like Goldie's fern (*Dryopteris goldiana*). Heed the admonition to allow ample space. Do not crowd ferns.

When planting, make holes big enough to spread the rhizomes and roots without bunching or crowding. Set the plants at the depth they were previously. Fill among and about the roots suitable soil, mixed if you will with small stones to assure aeration and encourage rooting. Water the newly set plants thoroughly, mulch around them, and if there is danger of harm to choice specimens from wind and exposure protect them temporarily with bushel baskets, boxes, or other ventilated coverings.

Companion plants for hardy ferns include many native woodlanders, such as bleeding hearts, bloodroots, camassias, dutchman's breeches, jack-in-the-pulpits, and trilliums. Non-native species that enjoy woodland or equivalent conditions are appropriate too. Here belong English bluebells, forget-me-nots, plantain-lilies, primulas, and many other lovers of partial shade.

Maintaining happily located ferns calls for little effort. Watering, deep soakings, may be needed in dry weather if natural supplies run low or fail. It is especially important to have the soil moist in fall when new fronds for the following year are forming. Taking guidance from nature's procedure, provide them with a mulch or covering of leaves or other protection from the vagaries of winter. Where leaves are likely to blow away salt hay or straw held in place by a few branches or wire screening or branches of evergreens afford needed protection. Do not cut off the dead foliage of deciduous ferns in fall. It serves as a natural winter covering.

In spring remove any winter covering that cannot readily be crumbled to mulch. Do this by hand. Rakes, forks, and other tools are likely to be serious damage to the awakening crosiers. Decayed leaves and similar organic mulches ordinarily adequately supply the nutrient needs of ferns. If you feel a little additional nourishment is desirable apply in spring a light dressing of rotted manure, dried cow manure, dried sheep manure, or bonemeal. Do not use fertilizers that release nitrogen rapidly as do most inorganic ones.

Housekeeping care consists chiefly of curbing the growth of over-ambitious spreaders and weeding out tree seedlings and other unwelcome volunteers. The first is accomplished by reducing the offender's size by use of the spade or spading fork, the other by hand-pulling. Never use cultivators or hoes near ferns. They are too damaging to surface roots, and too threatening to young fronds. But if the ground is kept covered, as it should be, with a loose mulch weeds will not be many and those that come will pull easily.

Nonhardy, tender, or greenhouse ferns can be grown in shaded places outdoors and in lath houses in Florida, southern California, Hawaii, and other warm-climate regions, elsewhere in greenhouses. Some, more resistant to dry air than most, are satisfactory houseplants. Among the better known of these are bird's nest fern (*Asplenium nidus*), Boston fern (*Nephrolepis exaltata bostoniensis*) and related varieties, holly fern (*Cyrtomium falcatum*), mother spleenwort (*Asplenium bulbiferum*), and button fern (*Pellaea rotundifolia*). Occasionally, where conditions are particularly favorable, tolerably good specimens of maidenhair ferns (*Adiantum*) and *Pteris* succeed in dwellings. A wide selection of small tender ferns is suitable for terrariums, and the button fern and some other small ones last well, if not permanently, in dish gardens. Aquatic ferns that float or root in mud beneath the surface of water include *Azolla, Ceratopteris, Marsilea, Pilularia, Regnellidium,* and *Salvinia*. None is hardy.

Tender ferns come in much greater variety than hardy ones. In addition to a vast

Silver fern (*Pityrogramma*)

number of low and moderate-sized sorts there is a not inconsiderable number of tree ferns, several at maturity in their native lands 50 feet or more tall. Here belongs cibotiums, cyatheas, dicksonias, and others. Then there are tree-perching (epiphytic) ferns, among the most remarkable and imposing, the staghorns. There are floating ferns (*Ceratopteris*) and climbing ferns (*Lygodium*). Nor must the beautiful gold and silver ferns (*Pityrogramma*) be overlooked, the undersides of their fronds conspicuously dusted with powders that give just reason for their vernacular names.

Indoor ferns are grouped as tropical and subtropical or warm greenhouse and cool greenhouse kinds according to their temperature needs. The cool group succeeds with night levels during the cool part of the year of 45 to 50°F. For warm greenhouse ferns 55 to 65°F is needed. For all, day temperatures should exceed night ones by five to ten or fifteen degrees according to the brightness of the weather. In summer considerably warmer conditions night and day are in order.

Fairly high to very high relative humidity of the atmosphere is necessary for practically all ferns. A few are content with more arid environments. An outstanding example, *Pteris vittata* prospers in greenhouses along with cactuses and other succulents. Ferns of the *Nephrolepis* genus, although not as tolerant as the last mentioned, thrive under much drier conditions than most ferns, which does much to explain the once great popularity of Boston ferns as window plants. It is better to maintain the desired high atmospheric humidity by frequently wetting floors, walls, benches between the plants, and other surfaces, than by overhead spraying with water although this is sometimes helpful. A danger of wetting the foliage of some kinds is that it encourages the spread of leaf nematodes.

Shade is another requirement. This is not needed in November, December, or January, but during the remainder of the year the great majority of ferns indoors need relief from direct sun. Yet too heavy shade results in soft foliage, and if much too heavy, weak growth. Allow the maximum light possible without risking scorching the foliage. Acceptable light intensity varies considerably with different sorts. Boston ferns and their close kin for example will stand and indeed for their well-being require more light than maidenhairs. Use a little trial and error and careful observations of the plants' responses as guides.

Watering is very important. Fronds that wilt from lack of moisture, and this soon follows drying of the soil, scarcely ever recover. A wilted frond is a lost one. Except for the relatively few deciduous ferns that must be kept dry through their dormant seasons, keep the earth at all times evenly damp, not so wet however that air is driven from the soil with the result that the roots rot and die. It is generally undesirable to keep potted ferns standing in saucers of water.

As greenhouse plants ferns are grown in pots and pans (shallow pots) and in ground beds. Some lend themselves well for use in hanging baskets and on rafts (usually suspended, openwork platforms of moisture-resistant wood such as cedar, cypress, or teak). Epiphytic sorts do well attached to cork bark posts or similar supports fitted with pockets or crotches filled with a suitable medium in which the fern roots will grow. Potting is normally done in late winter or very early spring just when new growth is about to begin. Young plants of fast-growing kinds may need another shift to larger containers before midsummer. Have the soil as coarse as can reasonably be worked among and around the roots, more lumpy for large specimens than smaller ones. Let it be loose and spongy, not finely sifted or of a texture that under the influence of repeated watering will compact. A good general mixture may consist of one part turfy loam, one part coarse sand or perlite, two parts coarse leaf mold or peat moss, and one-half part dried cow manure with, in addition, bonemeal at the rate of a pint to the bushel and a generous dash of finely broken charcoal. For epiphytic ferns use little or no loam in the mixture. Substitute osmunda fiber, tree fern fiber, or bark chips of the kinds used for orchids. These are general suggestions, many other soil mixes that will be just as suitable and perhaps even better for individual kinds of ferns can be concocted.

Take care not to over-pot. In the main ferns do best when their roots are a little cozy rather than in containers so large that they are packed around with considerable quantities of soil they have not penetrated.

When repotting first examine the plants appraisingly. Look for scale insects. Remove all heavily infested fronds, clean others of the pests. Sometimes, with maidenhairs, for example, it is advisable to cut off all foliage, clean the stubs of leafstalks left and encourage the plants to make a complete new growth of foliage. If there are decaying or dead roots remove these and pick away with a pointed stick any old soil that can be loosened from the ball without doing damage to healthy roots. If there are many dead roots wash the ball free of soil and cut out all dead portions before repotting in the smallest container into which the remaining roots will fit comfortably. Division of many kinds of ferns may be done at potting time, either to reduce the size of the specimens or to effect propagation. For the last purpose select strong, healthy pieces from the circumferences of the plants rather than worn out central parts.

As soon as potting is finished water the new soil very thoroughly with a fine spray and return the plants to a warm, humid, shaded place. Assure houseplants extra humidity by covering them for two weeks to a month with a polyethylene plastic

Hanging basket of maidenhair fern (*Adiantum*)

bag. Do not water the soil oftener than necessary, but do not allow it to become dry. A damp atmosphere is essential, a saturated soil is not.

Hanging baskets and rafts are often more suitable than pots for certain ferns, especially for those with drooping fronds, such as some kinds of sword ferns (*Nephrolepis*) and some maidenhairs. They also are agreeable for davallias and others with thick, creeping rhizomes. Planting in baskets is not greatly different from potting. On rafts, chiefly used for epiphytic ferns, a coarse rooting mixture, usually containing generous proportions of largish pieces of osmunda, tree fern fiber, coarse, half-rotted leaves, bark chips of the kinds in which orchids are planted, and suchlike, is heaped. The fern is planted in this and then wired to the raft.

For outdoor landscaping in agreeable climates nonhardy ferns can be used in the same ways and have similar needs except so far as temperature is concerned as hardy ones in colder regions. In addition, the epiphytes can be displayed on palm trunks, crotches of trees and in other aerial locations.

Propagation of ferns, hardy and tender, is affected by division and spores or more rarely, as with the walking fern (*Camptosorus*) and mother spleenwort (*Asplenium bulbiferum*), by plantlets that develop on the fronds. There are several techniques of growing from spores, none difficult. Collect spores by placing a mature spore-bearing frond or portion of frond in a paper bag. Hang this in a dry place for a few hours to allow the ripe spores to be discharged and fall to the bottom of the bag. The spores are dust fine. Be careful they do not blow away. Usually ferns breed true to type, but if spores of closely related species are sown together hybrids may result.

Raising ferns from spores is mostly limited to kinds that do not lend themselves to propagation by division. It is an interesting, indeed fascinating procedure. Fresh spores give the best results. If they must be stored keep them in a cool, dry place. The environment essential for raising plants from spores is the same as for growing from seeds, except that very high humidity must be maintained. The ideal is a saturated atmosphere in which the prothallia are constantly bathed in moisture. Sterile conditions are very important. Liverworts and other unwanted low forms of plant life revel in conditions favorable to the germination of spores and may crowd out or destroy the prothallia.

Pots or pans are convenient receptacles in which to sow. Make sure the soil consists largely of lifted leaf mold or peat

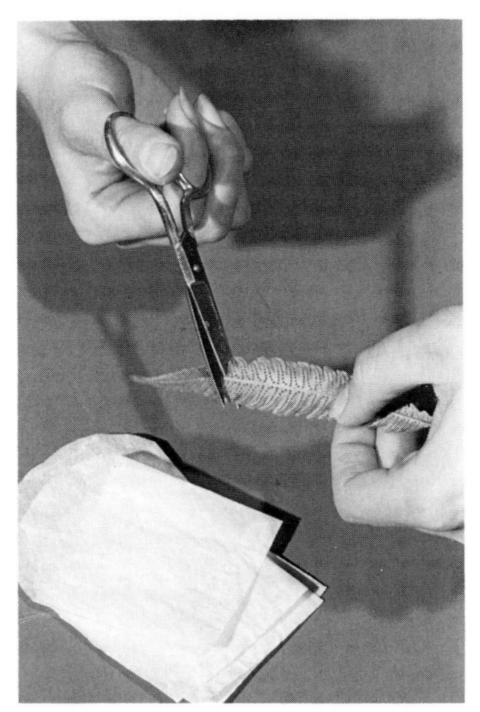

Collecting spores: (a) Snipping fronds into small pieces

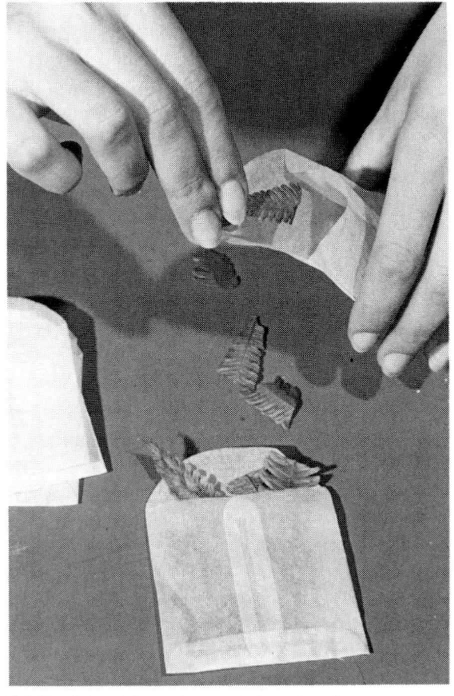

(b) Putting the pieces into glassine envelopes to contain the spores

Raising ferns from spores: (a) Provide ample drainage in bottom of pan before filling in soil

(b) Surface with a sifting of very fine soil

moss and contains a generous amount of coarse sand or perlite as well as a smaller proportion of loamy topsoil. Before sowing, sterilize the containers, soil, and pieces of glass to be used for covering the pots or pans by baking or steaming for one hour in a temperature of 250°F.

The technique employed in filling the containers is the same as is used for sowing seeds in soil. The surface should be made very fine and absolutely level. Because of the minuteness of the spores, sowing should be attempted only in a place where the air is quite still. The spores are scattered evenly and rather thinly over the prepared surface, but are neither pressed in nor covered with soil. After sowing, cover the receptacles with glass or put them in a humid propagating case. Keep them shaded from direct sun, but exposed to light of low to moderate intensity. The spores will not germinate in complete darkness. Stand each pan in a saucer and fill this with water (first boiled and cooled to approximately air temperature) sufficiently often to make sure the surface soil never becomes dry. After the prothallia begin to form keep the containers in a light place out of direct sun. Make sure the atmosphere is humid and the soil does not dry. If the prothallia become crowded or if they begin to die off, transplant them at once.

Fern spores may be germinated quite readily on the moist surface of a flower pot or rough brick. A pot 2½ or 3 inches in diameter is of convenient size, and it should be filled tightly with sphagnum moss and stood bottom side up in a saucer. If brick is used select a soft, porous sample rather than one hard or vitrified. Break each brick in two and stand each half in a saucer or, as an alternative, break the brick into pieces about the size of walnuts and fill a saucer with these. To prevent contamination, pots, bricks, saucers, and sphagnum are sterilized before use by baking, steaming, or boiling, as also are the bell jars used for covering them. Now thoroughly wet the surface of the pot or brick with water (which has first been boiled) and pour sufficient of this into the saucer to cover its bottom. In a still atmosphere scatter the spores over the brick or over the outer surfaces of the flower pot

(c) Press the surface moderately to firm it

(d) Sow the spores evenly over the surface

(e) Cover with a sheet of glass and put in a shaded place

(g) Typical ferns (the sporophyte generation) soon develop

(f) Prothallia (the gametophyte generation) appear

and then place the saucer and its contents in a propagating case beneath a bench or in some other shaded part of the greenhouse and cover with a bell jar or equivalent protection. Subsequent treatment is the same as for fern spores sown on soil.

For garden purposes ferns are somewhat loosely grouped as hardy and as nonhardy or tender. The first survive outdoors in the north, the others either will not withstand low temperatures at all or certainly not more than an occasional light frost.

Pests and Diseases. In general ferns are less subject to pests and diseases than many popular flowering plants. Among the commonest troubles are snails and slugs, which delight in consuming the tender young fronds. With maidenhairs they do this so systematically that the unobservant gardener may worry as to why his plants seemingly fail to put forth any new growth. Scale insects are also likely to infest some ferns, and novices may confuse these with the collections of spore capsules (sori), which are quite normal on fertile fronds. Scale insects will often be found on the stalks as well as the blades of fronds, sori on the blades only. Leaf nematodes mar the beauty of some tender ferns by marking them with more or less triangular patches of dark brown. Refraining from wetting the foliage at any time is the best control for this. Aphids and thrips also infest some ferns. Diseases include leaf spots and disfiguring black molds, the latter usually growing on honeydew secreted by scale insects. The prothallia of ferns raised from spores are subject to damage by various damping-off and rot-causing organisms.

Genera of ferns described in this Encyclopedia include the following: *Acrostichum, Adiantum, Aglaomorpha, Anopteris, Arachniodes, Aspidotis, Asplenium, Athyrium, Azolla, Blechnum, Botrychium, Camptosorus, Campyloneurum, Ceratopteris, Cheilanthes, Cibotium, Coniogramme, Cryptogramma, Cyathea, Cyrtomium, Cystopteris, Davallia, Dennstaedtia, Dicksonia, Diplazium, Doodia, Doryopteris, Drynaria, Dryopteris, Elaphoglossum, Gymnocarpium, Hemionitis, Humata, Hypolepis, Lygodium, Marsilea, Matteuccia, Microlepia, Nephrolepis, Onoclea, Onychium, Ophioglossum, Osmunda, Pellaea, Phyllitis, Pilularia, Pityrogramma, Platycerium, Polypodium, Polystichum, Pteridium, Pteris, Pyrrosia, Regnellidium, Sadleria, Salvinea, Schizaea, Sphenomeris, Stenochlaena, Tectaria, Thelypteris, Todea, Woodsia,* and *Woodwardia.*

Common names of ferns include these: Adder's tongue (*Ophioglossum*), ball fern (*Davallia mariesii*), bamboo fern (*Coniogramme japonica*), bear's foot fern (*Humata tyermannii*), bear's paw fern (*Aglaomorpha meyeniana*), beech fern (*Thelypteris*), bird's nest fern (*Asplenium nidus*), bladder fern (*Cystopteris*), Boston fern (*Nephrolepis exaltata bostoniensis*), bracken or brake fern (*Pteridium aquilinum*), brittle fern (*Cystopteris fragilis*), button fern (*Pellaea rotundifolia* and *Tectaria cicutaria*), cabbage fern (*Platycerium angolense*), chain fern (*Woodwardia*), Christmas or dagger fern (*Polystichum acrostichoides*), cinnamon fern (*Osmund cinnamomea*), claw fern (*Onychium*), cliff-brake fern (*Onychium* and *Pellaea*), climbing fern (*Lygodium*), Clinton's fern (*Dryopteris clintoniana*), cup fern (*Dennstaedtia*), curly-grass fern (*Schizaea pusilla*), deer fern (*Blechnum spicant*), elephant-ear fern (*Ela-*

phoglossum crinitum and *Platycerium angolense*), elkhorn fern (*Platycerium*), fancy or florists' fern (*Dryopteris austriaca*), felt fern (*Pyrrosia lingua*), floating fern (*Ceratopteris*), flowering fern (*Anemia* and *Osmunda*), glade fern (*Athyrium*), grape fern (*Pityrogramma*), Goldie's fern (*Dryopteris goldiana*), grape fern (*Botrychium*), hare's foot fern (*Phlebodium aureum*), Hartford fern (*Lygodium palmatum*), hart's tongue fern (*Phyllitis scolopendrium*), hay-scented fern (*Dennstaedtia punctilobula*), holly fern (*Cyrtomium* and *Polystichum*), interrupted fern (*Osmunda claytoniana*), Japanese painted fern (*Athyrium goeringianum pictum*), lace fern (*Cheilanthes gracillima* and *Sphenomeris chusana*), lady fern (*Athyrium filix-femina*), licorice fern (*Polypodium glycyrrhiza*), lip fern (*Cheilanthes*), maidenhair fern (*Adiantum*), male fern (*Dryopteris felix-mas*), marsh fern (*Thelypteris palustris*), Massachusetts fern (*Thelypteris simulata*), moonwort fern (*Botrychium lunaria*), mosquito fern (*Azolla*), New York fern (*Thelypteris noveboracensis*), oak fern (*Gymnocarpium*), oak leaf fern (*Drynaria quercifolia*), ostrich fern (*Matteuccia struthiopteris*), parsley fern (*Cryptogramma crispa*), pod fern (*Aspidotis densa*), polypody fern (*Polypodium*), rabbit's foot fern (*Davallia fejeensis*), rattlesnake fern (*Botrychium virginianum*), resurrection fern (*Polypodium polypodioides*), royal fern (*Osmunda regalis*), sago fern (*Cyathea medullaris*), saw fern (*Blechnum serrulatum*), sensitive fern (*Onoclea sensibilis*), shield fern (*Dryopteris* and *Polystichum braunii*), Sierra water fern (*Thelypteris nevadensis*), silver fern (*Pityrogramma*), spleenwort fern (*Asplenium* and *Athyrium pycnocarpum*), squirrel's foot fern (*Davallia mariesii*), staghorn fern (*Platycerium*), strap fern (*Campyloneurum phyllitidis*), strawberry fern (*Hemionitis palmata*), sword fern (*Nephrolepis*), tongue fern (*Pyrrosia lingua*), tree fern (*Cibotium, Cyathea,* and *Dicksonia*), walking fern (*Camptosorus*), wall fern (*Polypodium vulgare*), water-clover (*Marsilea*), water fern (*Ceratopteris*), and wood fern (*Dryopteris*).

FEROCACTUS (Fero-cáctus)—Barrel Cactus. Ferocious cactus is a literal translation of *Ferocactus*. In view of the viciously spiny character of most kinds not inapt, the name derives from the Latin *ferox*, ferocious or savage, and the word cactus. Natives of the southwestern United States and Mexico, there are thirty-five species of *Ferocactus*. They belong in the cactus family CACTACEAE. The fruits of some kinds are important foods for small animals and birds.

Often large and handsome, ferocactuses have globular to massive-columnar stems 6 inches to 10 feet tall by 2 inches to 2 feet or sometimes more in diameter according to kind. Except following injury to their tops they do not branch. They have conspicuous, continuous, generally thick ribs. The clusters of straight or hooked spines, the largest up to 6 inches long, originate from usually large areoles (specialized locations on the stems of cactuses from which spines and flowers originate) that when young are felted with hairs. Generally with very short tubes, the bell- to broadly-funnel-shaped blooms 1½ to 3 inches wide arise from young areoles of the current season's development near the centers of the tops of the stems. The ovary, flower tube, and fruits have scales without hairs in their axils. The dryish, thick-skinned, fleshy fruits when ripe open by a slit between their middle and bottom. Botanists have held and hold different opinions about the classification of plants treated here as *Ferocactus* and related species. Most species of *Ferocactus* have synonymous names under *Echinocactus*. Some species treated in this Encyclopedia as *Hamatocactus* are by some authorities included in *Ferocactus*.

Kinds likely to be represented in cactus collections include these: ***F. acanthodes,*** of southern California, has a usually solitary, thick-columnar to less often ovoid stem up to 10 feet tall, averaging 1 foot in diameter. There are thirteen to twenty-seven blunt ribs furnished at intervals of up to ¾ inch with clusters of spreading, curved, but not hooked, white, yellowish, pink, or bright red spines. Each cluster consists of twelve to twenty radials 1½ to 2½ inches long, the inner six to eight resembling the centrals, the others more slender, and one to four thin, flexible centrals 2 to 4½ inches or sometimes more in length. The yellow to orange flowers are 1½ to 2½ inches long. The fruits are yellow with some red, approximately 1½ inches long by about ¾ inch in diameter. ***F. covillei,*** of Arizona and adjacent Mexico, 2 to 8 feet tall and 1 foot to 2 feet thick, has a barrel-shaped or sometimes columnar stem with usually twenty to thirty ribs with spine clusters 1¼ inches or a little more apart. Red or white, but not yellow, in each cluster there are five to nine straight or backward-curved radials up to 3 inches long, and one hooked, flat, or cylindrical central 2 to 4 inches long. The purplish-red to yellow blooms, about 3 inches long, have toothed inner petals. The fruits are yellow, oblong, and 1 inch to 1¾ inches in diameter. ***F. diguetii,*** of Baja California, up to 12 feet tall and 3 feet in diameter and with twenty-five to thirty-nine ribs, inhabits islands in the Gulf of California. Its spine clusters are about ½ inch apart. Each is of six to nine similar curved, yellow spines about 2 inches in length. The funnel-shaped flowers with yellow inside petals and reddish-brown outer ones are about 1½ inches long. ***F. echidne*** is a Mexican with globular stems up to 8 inches in diameter. They have about fourteen ribs with spine clusters ¾ inch to 1¼ inches apart. Each cluster has seven or eight slightly curved, yellowish to grayish radial spines and one longer central up to 1½ inches long. The yellow flowers are 1¼ inches in length. ***F. fordii,*** of Baja California, is spherical with a slightly hollowed top. Up to somewhat over 1 foot in diameter, it has twenty- to twenty-two slightly notched ribs, and spine clusters spaced about ¾ inch apart. The about fifteen white, needle-like radial spines of each cluster are additional to four centrals of which the largest, hooked at its tip and spreading, is up to 1½ inches in length. About as long, the flowers are pink. ***F. glaucescens*** is Mexican. It has a spherical, glaucous-green stem that elongates somewhat with age and is from 8 to 16 inches high. There are eleven to fifteen sharp ribs with spine clusters less than ½ inch apart. There are six radial spines to each cluster, bright yellow and up to 1½ inches long, and one central of similar size and color, or none. Funnel-shaped and 1 inch to 1¼ inches long, the yellow flowers have finely-toothed petals. The fruits are yellow. ***F. gracilis,*** globose when young, becomes columnar and up to 10 feet tall later. Native of Baja California, it has stems that attain 1 foot in diameter, with about twenty-four notched ribs. Of the straight or hooked spines up to 1½ inches long there are in each cluster ten to fourteen needle-like whitish radials and seven to thirteen awl-shaped and flattened, dark red to nearly black centrals. The 1½-inch-long flowers have straw-yellow petals striped with red on their undersides. The fruits are yellow. ***F. histrix*** (syn. *F. melocactiformis*) is Mexican. Its usually solitary stems, somewhat flattened at the top, are spherical to about twice as tall as wide. They are up to two feet high, have as many as twenty-five sharp ribs when adult, fewer as juveniles. From ¾ inch to 1½ inches apart, the spine clusters have each about eight straight or slightly curved radials up to ¾ inch long, pale to deeper yellow with reddish or grayish apexes, and three or four central spines longer and thicker, but otherwise similar to the radials. The funnel-shaped, 1½-inch-long flowers are pale to bright yellow, often reddish on the outside. ***F. latispinus,*** of Mexico, is more or less spherical with a flattened or depressed top. Up to about 1¼ feet tall, it has fifteen to twenty-four notched ribs at maturity, fewer when young. Up to 1½ inches apart, the spine clusters have six to eighteen yellowish or red radials ¾ inch to 1½ inches long and four stouter, brighter, red to reddish, somewhat longer centrals, the lower ones with hooked apexes, and sometimes very broad. The whitish to reddish-purple or violet blooms, 1¼ inches long, have toothed inner petals. ***F. macrodiscus*** is spherical to short-cylindrical with a decidedly hollowed top. Native of Mexico and up to 1½ feet in diameter, it has thirteen to twenty-one sharp, notched ribs. Its spine clusters are from a little over ½ inch to 1¼ inches apart. Each

Ferocactus latispinus

has six to eight curved, yellow to blood-red radials ¾ to 1 inch long and four thicker more strongly curved centrals up to 1½ inches long. The 2-inch-long blooms are funnel-shaped, carmine-red to purple-red. ***F. peninsulae,*** of Baja California, cylindrical to club-shaped, is up to 8 feet tall. Its stems, up to 1 foot thick, have twelve to twenty-one ribs. The spine clusters, 1¼ to 1½ inches apart, consist of about eleven nearly straight radials about 1 inch long and four to six centrals, the longest 4 to 6 inches long. All are red with yellow apexes. The yellow to purple blooms are 2 inches long or a little longer. *F. p. viscainensis* (syn. *F. viscainensis*), of Baja California, is up to 5½ feet tall by 1¼ feet wide. The spine clusters are of one to four grayish-white radials and five to nine unequal, flattened centrals up to 5 inches in length. About 1¾ inches long, the light yellow flowers have a purple stripe down the outside of each petal. ***F. rafaelensis*** (syn. *F. victoriensis*), of Mexico, resembles a small edition of *F. histrix.* At first spherical, later columnar, its flattish-topped stem has fourteen to twenty-two sharp ribs. The spine clusters, spaced up to 1½ inches apart, have each about eight mostly recurved, yellowish, brownish, or grayish radials 1¼ to 1½ inches long and one straight, downward-pointing, stouter, similarly colored central 2 to 2¾ inches long. The flowers are yellow. A plant sold as *F. victoriensis* seems not to be this species, but more closely related to *Echinocactus texensis* (syn. *Homalocephala texensis*). ***F. rectispinus,*** of Baja California, is globular at first, and eventually columnar. It attains a height of 6 feet, a diameter one-third that. Its stems have twelve to twenty-four blunt ribs with spine clusters spaced 1¼ inches or a little more apart. Each cluster consists of seven to twelve straight or curved whitish or reddish radials 1¼ to 2¼ inches in length and one stout, reddish-brown, flat central up to 5½ inches long. From 2¼ to 3½ inches long, the flowers are yellow with the outer petals stained brown. ***F. recurvus*** (syn. *F. nobilis*) is a handsome Mexican. Its solitary, nearly spherical stems, up to 1 foot tall, have twelve to fifteen notched ribs. The spine clusters are 1¼ inches or a little more apart. In each there are eight stiff, reddish, ¾-inch to 1-inch-long radials and one similarly colored stout central 2 to 2¾ inches long, hooked at its apex and with a flat upper surface. The flowers, 1 inch to 2 inches long, have pink petals with darker midribs. ***F. robustus,*** unlike many members of its genus, makes offsets abundantly and eventually forms wide mounds up to 15 feet across. Native of Mexico, this kind has spherical to ovoid, glossy green stems 4 to 8 inches in diameter, with about eight ribs. From ¾ inch to 1½ inches apart, the spine clusters are of ten to fourteen yellow, bristle-like radials and four to six usually flattened, yellow-based, brownish centrals up to 2¼ inches long. The funnel-shaped, yellow blooms are about 1½ inches long. ***F. stainesii*** (syn. *F. pilosus*), a Mexican, at first globular, becomes columnar and up to 5 feet tall and 1 foot in diameter. It develops basal offsets when old. Its stems have thirteen to twenty notched ribs and, 1¼ to 1½ inches apart, clusters of four to six stiff, straight or slightly curved radial spines up to ¾ inch long and interspersed with numerous long, white bristles and four central spines. When young all the spines are bright red. Later they fade to grayish or yellowish. The bell-shaped blooms up to 1½ inches long are orange-red or yellow. *F. s. haematacanthus* (syn. *F. haematacan-*

Ferocactus robustus

thus) is a spherical or short-columnar native of Mexico. Up to about 1½ feet in height and bright green, it has stems up to 1 foot in diameter. They have twelve to twenty broad, notched ribs and yellow-tipped, blood-red spines, six radials and four centrals in each cluster and straight except for the lowest radial, which is down-curved. Longer and stouter than the others, the centrals are up to 2¼ inches in length. The purplish-red flowers are 2¼ inches long. The purple fruits are 1¼ inches long. ***F. townsendianus*** of Baja California, spherical to short-cylindrical, has about sixteen broad, deeply-notched ribs. Spaced about 1¼ inches apart, the clusters of spines consist of fourteen to sixteen brown to grayish radials about 1¼ inches long and five slightly flattened, hooked, brown centrals. The flowers, 2 to 2½ inches long, are greenish-yellow with a pink stripe down the centers of the petals. ***F. viridescens,*** of Baja California, occasionally produces basal offshoots. At first globular with a hollowed top, as it ages it becomes more cylindrical. From 1 foot to 1½ feet tall and up to about 1 foot in diameter, this has thirteen to twenty-one broad, blunt, slightly-notched ribs. Spaced up to 1 inch apart, the spine clusters are of nine to twenty-one about ¾-inch-long radials and four stouter centrals up to 1½ inches long. All are red when young, and later greenish-red. The flowers, about 1½ inches long, have toothed, yellowish-green petals. The 1-inch-long fruits are reddish. ***F. wislizenii,*** of Arizona and adjacent Mexico, is barrel-shaped to massive-columnar, 2 to 10 feet tall by 1 foot to 2 feet in diameter. The stems have twenty to twenty-eight ribs with spine clusters each of twelve to twenty ashy-gray, curved radials, up to 1¾ inches long, and four sometimes hooked centrals not lying flat against the stem, up to 2 inches long. The orange-yellow blooms, 1¾ to 2½ inches long, have petals with toothed edges. The fruits are barrel-shaped, yellow, clothed with nearly circular scales, and up to 1¾ inches long and wide.

Garden and Landscape Uses and Cultivation. These handsome cactuses are much prized. Well-grown, unmarred specimens, especially sizable ones, add distinction to collections. They succeed outdoors in warm, dry climates such as that of southern California and in greenhouses in porous, freely-drained soil. In general they appreciate warmth and sunshine and abhor excessive watering. Propagation is by seeds and with the smaller, clustering kinds, offsets. See Cactuses.

FERONIA (Fer-ònia)—Wood-Apple. The only species of *Feronia,* of the rue family RUTACEAE, is closely related to *Citrus.* Its name honors the Roman nymph Feronia who presided over groves and woodlands.

Native from India to southeast Asia and Java, the wood-apple is in its native range freely cultivated for its edible fruits, which are eaten out of hand or cooked, and are used for making jelly. They are also employed medicinally. The wood of this tree is useful and a gum from it called feronia gum is used as a substitute for gum arabic in the manufacture of water colors and for other purposes.

A moderate-sized, deciduous, thorny tree, the wood-apple (***F. limonia*** syn. *Limonia acidissima*) has small pinnate leaves with one to four pairs of opposite leaflets and a terminal one. The leaflets, obovate to elliptic and often toothed near their ends, are 1 inch to 1½ inches long by ½ to 1 inch wide. The dull red small flowers are many together in terminal and axillary panicles. They have calyxes with five or rarely six lobes and the same number of spreading or down-bent petals. There are eight to twelve, usually ten, of these. Spherical to slightly pear-shaped, the hard-shelled fruits, 2 to 3 inches in diameter, contain pinkish, acid pulp in which are embedded numerous woolly seeds. The flowers and leaves are anise-scented.

Garden and Landscape Uses and Cultivation. The wood-apple is not hardy in the north. For its successful cultivation it needs a mild climate with little or no frost. It is planted somewhat in Florida and California, chiefly as an item of interest. It succeeds under conditions that suit *Citrus* and may be propagated by the same techniques.

FERONIELLA (Feron-iélla). Closely allied to *Limonia* and with a name that is a diminutive of its synonym *Feronia,* the genus *Feroniella* belongs in the rue family RUTACEAE. It comprises three species and is native to southeast Asia and Java.

Feroniellas are spiny trees differing from *Limonia,* the only orange-relative they closely resemble, in not being deciduous, in having about four times as many stamens as petals, and also, in having at the bases of the stamen stalks, hairy and partly separate appendages. The leaves are pinnate, and have an uneven number of small leaflets. The leafstalks and the midribs between the leaflets are winged or not. The very fragrant flowers, bisexual or male, come in branched panicles from the leaf axils. The calyx has four to six lobes. There are four to six sharp-pointed, lanceolate to oblong-lanceolate, white petals and sixteen to twenty stamens. The style is slender. The fruits are large, spherical to ellipsoid, woody-shelled, and contain glutinous sap. Technically they are berries. Those of *F. lucida* are eaten in Java.

Two quite similar species are cultivated as ornamentals in southern Florida and elsewhere in warm climates. In its native Indochina, ***F. oblata*** is sometimes 60 feet tall, but in cultivation is considerably smaller. It has extremely spiny younger branches and bright green leaves with seven or nine elliptic to obovate leaflets ¾ inch to 1¼ inches long. The flowers are ½ to 1 inch in diameter. In clusters of three or four, the flattened-spherical fruits are 2 to nearly 3 inches in diameter. From this, ***F. lucida*** of Java differs in its leaves having five to nine leaflets, its sepals being bigger, and its fruits being spherical instead of flattened at their tops and bottoms.

Garden and Landscape Uses and Cultivation. In frost-free and nearly frost-free climates feroniellas are sometimes planted for interest and ornament. Because of their spines they can be used to form barrier hedges. They succeed under conditions that suit oranges, but may be more susceptible to cold. They are propagated by seed.

FERRARIA (Ferràr-ia). Two African species constitute *Ferraria* of the iris family IRIDACEAE. The name commemorates the Italian Jesuit botanist Giovanni Battista Ferrari, who died in 1655.

Ferrarias are nonhardy, deciduous bulb plants closely related to *Tigridia* and with something of the appearance of *Iris.* Best known, ***F. crispa*** (syn. *F. undulata*), of South Africa, attains a height of 1 foot or somewhat more. It has flat, long-pointed, sword-shaped leaves and erect, branched stems clothed with overlapping leaves that from below up become successively shorter. The flowers, which last but for a day, are 2 inches wide, with triangular, spreading petals and crisped margins joined below in a short corolla tube. They are greenish-brown or olive-green blotched with cream and purple. The stamens are joined to form a cylindrical tube. The style is slender with three small, petal-like branches. This species was illustrated as early as 1646 by the botanist after whom it is named. It has been cultivated in gardens for at least three hundred years.

Garden Uses. In warm climates where a dry season alternates with a wetter growing one, as in California, this is an interesting plant for flower borders and rock gardens. It is also worth growing as something of a curiosity in cool greenhouses, but is scarcely showy enough to warrant planting in large numbers. Its blooms are unusual enough to attract the attention of those who enjoy something out of the ordinary.

Cultivation. No special difficulties attend the cultivation of *F. crispa.* In frost-free or nearly frost-free climates it thrives in warm, well-drained, moderately fertile soil in full sun under approximately the same conditions that suit *Tigridia.* When grown indoors, the rhizome-like corms (bulb-like organs) may be planted in August or September fairly closely together

Ferraria crispa

in well-drained pots or pans containing porous, fertile soil. Immediately after planting they should be watered thoroughly and placed in a cold frame or cool greenhouse. Until roots have taken possession of the new soil care should be taken to give only enough water to keep it just moist, but later more copious watering is in order, and after the containers are well-filled with roots, weekly or semiweekly applications of dilute liquid fertilizer are helpful. A night temperature of 45 to 50°F is adequate, day temperatures may be five to ten degrees higher and a buoyant atmosphere and free circulation of air should be maintained. Humid, stagnant, oppressive atmospheric conditions are to be avoided. Full sun is needed. Under these conditions the plants bloom in late winter or early spring. When after blooming the foliage begins to die down naturally, fertilizing should cease and watering be gradually reduced and finally, after the leaves have died completely, watering should be entirely withheld. During the summer resting period the corms may be stored in the pots of soil in which they grew, in a dry place protected from mice. Propagation is by offsets and by seeds. The latter germinate well and give plants that bloom in their second year. They may be sown indoors in pots of sandy soil in late summer or early fall.

FERTILITY. As applied to soils fertility refers to their abilities to produce crops. Contrary to a common belief this is not entirely or even mainly dependent upon the amounts of nutrients they contain, although this is an important factor. To be fertile a soil must in addition to being nutritious be in good physical condition so that drainage is satisfactory and the roots of plants receive sufficient air, its acidity or alkalinity must be within an acceptable range, and it must contain moisture in adequate amounts for the plants' needs. In addition it must be tolerably free of plant pests and diseases and substances toxic to plants. Finally, it must be of sufficient depth to accommodate roots of plants.

Whereas it is relatively easy to measure the pH of a soil and determine the amounts of the most important nutrients it contains, there is no simple test for overall fertility. Tests such as these and others to determine the mechanical makeup of the soil, supplemented by observation, judgment, and experience, are the bases for reaching reliable conclusions.

FERTILIZATION. Very different meanings attach to the word fertilization as it applies to plants. A rather obvious one is the application of fertilizers (plant nutrients) to stimulate growth. The other meanings pertain to sexual processes. In this context fertilization most correctly alludes to the union of a male reproductive cell (sperm) with a female reproductive cell (ovule or egg) and the subsequent development of a seed or a new plant. Less accurately it is sometimes employed as synonymous with pollination, which is the transfer of pollen from anther to stigma; this may or may not result in fertilization.

FERTILIZERS. As generally understood and here used the word fertilizer embraces all the many materials and combinations of materials added to soil in relatively small amounts for the primary and usually only purpose of supplying plant nutrients. Other additives, such as raw manure, compost, and lime, add greater or lesser amounts of these, but the first two are used in considerable bulk and all three serve other substantial purposes. They are not included in the understanding of the word fertilizer as it is here employed. For more about them see Compost, Lime, and Manures.

Of the about fourteen elements necessary for the growth of plants most, the notable exception carbon obtained from the air, are taken from the soil. Nearly all soils contain adequate amounts of all needed elements except three: nitrogen, phosphorus, and potassium. In a few regions deficiencies of others such as boron and zinc occur, but this is unusual. Cooperative Extension Agents, State Agricultural Experiment Stations, and local purveyors of fertilizers will gladly advise whether or not elements other than the big three are needed. Such, generally known as trace elements because plants need them in minute quantities only, are often included in complete mixed fertilizers sold in the few regions where the soil is deficient in one or more of them.

The ratios of the three chief nutrients they contain must by law be clearly stated on all containers and packages of fertilizer sold, expressed in percentages of total nitrogen, available phosphoric oxide, and soluble potash. Gardeners usually allude to these as percentages of nitrogen, phosphorus, and potassium. We shall follow that convention. The percentages are expressed in a series of three numbers, when written, separated by dashes, for example 5-10-5 or 8-6-4. The first of these indicates how many pounds of nitrogen, the second how many pounds of phosphorus, and the third how many pounds of potassium are in each 100 pounds of the fertilizer.

Complete fertilizers are those that contain significant amounts of all three chief nutrients. Those that supply one or two only are called incomplete. The latter are by no means less important in the economy of gardens than complete fertilizers. They make it possible to supply a needed element when others are not in short supply and adding more would be at best wasteful, at worst harmful. For example, nitrogen-supplying nitrate of soda, sulfate of ammonia, or urea can be used to give a boost to cabbages, lettuce, and other leafy vegetables as well as many other plants at times when they have no need for additional phosphorus or potassium. By contrast, excess nitrogen encourages such root vegetables as beets, carrots, and turnips to develop too much leafage and poorly developed roots. If the ground in which they are to be sown is rich in nitrogen, as it well may be if it has been heavily manured for a previous crop, superphosphate to supply needed phosphorus may be all that is needed or desirable to mix in before sowing.

Other factors than the actual nutrient content must be taken into account in selecting and using fertilizers. Chief of these is the form in which the nitrogen is carried. It may be in natural organic substances, such as fish meal, tankage (slaughterhouse refuse) cottonseed meal, or animal manure, in artificial organic materials, such as urea; or inorganic ingredients, such as sulfate of ammonia and nitrate of soda may supply the nitrogen. The nitrogen of natural organic materials is designated protein nitrogen to distinguish its sources from synthetic organics. Until the coming of urea as a fertilizer all organics were of the protein type. It is important to remember that the actual nitrates absorbed by the roots of plants are precisely the same whether from natural materials or synthetics. Natural ingredients that supply nitrogen are much costlier than synthetic and inorganic ones, but until the development of slow-release synthetics they had the great advantage of having more lasting effects, releasing their nitrogen slowly over long periods instead of rapidly over a brief space of time as do such synthetics as nitrate of soda, sulfate of ammonia, and urea. Because of this, less frequent applications are needed and natural organics are gentle in the sense that there is less danger of excessive applications causing harm to roots or of them "burning" foliage they fall upon.

Another point to consider under some circumstances is whether or not the fertilizer has a decidedly acid or alkaline reaction. In most cases this is of little significance, but where acid-soil plants, such as azaleas, blueberries, and rhododendrons, are concerned it can be important. Sulfate of ammonia, superphosphate, sulfate of potash, and muriate of potash tend to make the soil more acid, nitrate of soda, carbonate of soda, and some phosphates, more alkaline. Clay soils are made sticky and pasty by applications of nitrate of soda. Fertilizers containing nitrogen in some other form should be chosen for them.

Fertilizers are available in powdered, granulated, pelleted, and liquid forms. Per unit of nutrients the first are cheapest, but are less convenient and less pleasant to apply and also often have the disadvantages of caking in storage and of being more likely than other types to adhere to foliage in amounts that may be damaging. Concentrated liquid fertilizers are easy to mix with water and apply with a simple proportioner device through a garden hose. Being highly soluble, their nutrients are immediately available to the roots and small amounts are absorbed by the foliage. But such fertilizers are not long lasting. They are leached rapidly from the soil. Because of the considerable dilution at which concentrated liquid fertilizers are applied there is very little danger of damage to leaves.

Dilute liquid fertilizer is recommended in many places in this Encyclopedia for application to particular plants. One-half a century and more ago it was the practice to prepare what was called liquid manure for such use. This was made by steeping a burlap bag containing a generous amount of fresh cow manure or chicken manure in a covered barrel or other large container of water for a few days. The resulting liquid was diluted before use by adding more water, the amount based on the user's judgment. English gardeners were given to recommending diluting to the "color of weak tea," but with chicken manure, which darkens water less markedly than does cow manure, such a concentration could be harmful. Similar products were made by steeping soft coal soot in the same way. Liquid manures and soot water, as it was called, although excellent stimulants, are little used today.

Modern dilute liquid fertilizers are prepared by adding the amounts of water recommended by the manufacturers to concentrated liquid kinds or by dissolving in water such wholly soluble solid fertilizers as nitrate of soda, sulfate of ammonia, or urea. They are easy to prepare and apply and are without objectionable odor.

Home gardeners rely to a very large extent upon complete fertilizers. These come in various formulations, some for fairly general use, others compounded for particular types of vegetation, such as lawns, acid-soil plants, and tomatoes. The latter are usually more expensive per unit of nutrients than general-purpose fertilizers. Also offered for sale are complete fertilizers containing such additives as trace elements, weed killers, and insecticides. The cost of these is commonly higher than if fertilizers and additives are purchased separately. Furthermore they are often wasteful. A preemergence crab grass herbicide contained in a fertilizer applied in fall is useless, as is a broad-leaf weed killer as an ingredient of a fertilizer spread on a lawn that is free of dandelions and broad-leaved weeds. Other disadvantages of such combinations are that as a result of too frequent application trace elements may build up in the soil to toxic levels (these micronutrients should generally only be used on the advice of a Cooperative Extension Agent or other competent authority) and that trees and shrubs may be damaged or killed as a result of too much or too frequent applications of fertilizers containing weed killers to grass beneath them. On the whole the best and safest procedure is to purchase and use fertilizers, weed killers, insecticides, and trace elements separately.

In comparing the cost of different fertilizers one must take into consideration the amounts of nutrients each pound supplies. Only fertilizers of identical analyses can be directly equated with the prices of unit amounts. Having made such a comparison, one then may be willing to pay a higher price for a fertilizer in which the nitrogen is protein organic, for a pelleted preparation, or other presumed advantage.

What you really want to buy are plant nutrients, not inactive materials, dross or filler that make up most of the remainder of the weight of fertilizers. An 80-pound bag of a 10-20-10 fertilizer supplies the same amount of nutrients and in the same proportions as two such bags of a 5-10-5. It will of course cost more than one 80-pound bag of the latter, but because of differences in transportation, handling, and other charges it may well sell for less than two 80-pound bags of the 5-10-5 of which it is equivalent. To achieve the same results the rate of application of the 10-20-10 is exactly one-half that of the 5-10-5. There are many fertilizer formulations, often prepared to meet the needs of particular plants. Lawn fertilizers for example, generally have a comparatively high percentage of nitrogen, whereas those recommended for vegetables are likely to be proportionately richer in phosphorus. Although it is somewhat wasteful of nutrients to apply a fertilizer richer in any element than the crop or plants need, in small gardens it is usually less expensive to purchase a sizable container of one complete fertilizer, a 5-10-5 say, and use it for all or most fertilizing than it is to buy several smaller amounts of fertilizers of different formulations prepared for the specific needs of different plants. The unit price of nutrients rises rapidly as the package size and weight of its contents decreases.

A general-purpose fertilizer of the type suggested is as satisfactory for lawns, trees, shrubs, and flower gardens as it is for vegetables and fruits. Only perhaps for decidedly acid-soil plants is there much advantage for home gardeners to purchase fertilizers formulated for particular plants. The basic 5-10-5 or equivalent can always be fortified for those plants it is thought need more nitrogen by applying a little nitrate of soda, sulfate of ammonia, urea, or a ureaform fertilizer. If you want to beef up the amount of phosphorus apply a light dressing of superphosphate, if the need seems to be for additional potassium, sulfate of potash, muriate of potash, carbonate of potash, or wood ashes will do the trick. But for most plants the 5-10-5 or its equivalent will suffice.

It may be, of course, that ease of handling, storage problems, or other reasons justify paying the higher rates charged for small as compared with bigger amounts of fertilizer, but for the greatest economy the best plan is to determine the total amount of each kind needed for one season and buy the largest available packages.

Fertilizers are not the answer to all problems. Too often beginners seem to think they are. Whenever a plant becomes sickly they inquire, "What fertilizer should I give it?" Unless other environmental factors, including soil structure and aeration, moisture content, and acidity or alkalinity, are conducive to growth, applying fertilizer is of little avail and under some circumstances may be detrimental. Nor will fertilizers materially affect the prevalence of pests and diseases or alter atmospheric conditions that may be responsible for unhealthy plants.

Here is a selection of natural organic fertilizers. The analyses of the three principle nutrients given are approximations. In natural products these are likely to vary somewhat in different samples.

Bone ash, 0-35-0, nutrient fairly readily available. Reaction alkaline. Apply 3 to 6 ounces per 10 square feet.

Bonemeal (raw), 4-22-0 is slow acting, long lasting, slightly alkaline. Apply 4 to 8 ounces per 10 square feet.

Bonemeal (steamed), 2.5-24-0, nutrients more quickly available, less long lasting than those of raw bonemeal, slightly alkaline. Apply 2 to 4 ounces per 10 square feet.

Castor pomace, 5-2-1, nutrients fairly rapidly available. Acid reaction. Apply 6 to 8 ounces per 10 square feet.

Cattle manure (dried and shredded), 2-1.5-2, moderately slow acting. Apply 1 pound to 2 pounds per 10 square feet.

Cottonseed meal, 7-3-2, nutrients fairly rapidly available, fairly long lasting. Acid reaction. Apply 4 to 8 ounces per 10 square feet.

Dried blood, 12-0-0, nutrients rapidly available. Apply 1 ounce to 2 ounces per 10 square feet.

Fishmeal, 9-7-3, nutrients fairly rapidly available. Apply 2 to 3 ounces per 10 square feet.

Guano, Peruvian, 13-12-2.5. The excreta of sea birds deposited in desert regions. Excellent, quick acting. Apply 2 to 3 ounces per 10 square feet.

Horn and hoof meal (natural), 14-1-0, nutrients released very slowly over an extended period. Apply 2 to 4 ounces per 10 square feet.

Poultry manure (dried), 5-3-1.5, nutrients fairly rapidly available. Apply 1 pound to 1½ pounds per 10 square feet.

Seaweed (dried kelp), approximately 2.5-1.5-15, fairly rapidly available. Apply 1 pound to 2 pounds per 10 square feet.

Sewage sludge (activated), 6-3-0.5, fairly quick acting. Milorganite is one of the best known fertilizers of this type. Apply 3 to 4 ounces per 10 square feet.

Sewage sludge (dried), 2-2-0, nutrients fairly rapidly available. Apply 4 to 8 ounces per 10 square feet.

Sheep manure (dried), 2-1.5-3, nutrients fairly rapidly available. Apply 4 to 8 ounces per 10 square feet.

Soybean meal, 7-1.5-2.5, nutrients fairly rapidly available, fairly long lasting, acid reaction. Apply 6 to 8 ounces per 10 square feet.

Tankage (animal), 2-10-0.5, fairly rapidly available. Has alkaline reaction. Apply 4 to 5 ounces per 10 square feet.

Tankage (garbage), 2.5-3-1, nutrients fairly rapidly available. Apply 3 to 4 ounces per 10 square feet.

Tankage (process), 9-0.5-0, fairly rapidly available. Has acid reaction. Apply 1½ to 2 ounces per 10 square feet.

Wood ashes (dried hardwood), 0-2-5, nutrients rapidly available. Reaction alkaline. Some samples contain higher percentages of potash. Must be kept dry. Potash is highly soluble, easily leached out. Apply ½ to 1 pound per 10 square feet.

Straight goods is the trade term for fertilizers other than those prepared by blending suitable ingredients (often including "fillers" of little or no nutrient value) to meet particular formulations. The latter are called mixed fertilizers. Most straight goods are less popular with American gardeners than Europeans, partly because they are not advertised as compellingly as brand-name mixed fertilizers, partly because some are less pleasant or easy to handle. But based on the unit price of the nutrients they supply, they are often less costly than formulated fertilizers, and because they permit greater latitude in selecting sorts that provide only one or two nutrients under some circumstances they are less wasteful. Many straight goods are employed as ingredients of mixed fertilizers. They can also be added to such fertilizers to increase the proportion of one or more nutrients.

There are natural and synthetic straight goods, kinds that supply protein nitrogen, others nitrogen not of animal or vegetable origin. The rates at which their nutrients become available to plants varies greatly. With bone meal, rock phosphate, and some others this is related to the degree of fineness to which they are ground. With nitrogen fertilizers it depends upon the form in which the nitrogen is present.

Fertilizers not of animal or plant origin are commonly called inorganic (although some contain carbon compounds that are chemically organic). They are also called chemical fertilizers. The nitrogen they contain, except in slow-release products of the ureaform type and those, such as Osmocote, sold as pellets coated with plastic through which the nutrients pass by osmosis, is readily available, but has no long-lasting effect. Most chemical fertilizers contain only one chemical nutrient. Some are compounds containing two or three.

Nitrogen-supplying chemical fertilizers include these:

Cyanamid (calcium cyanamide) contains 21 percent nitrogen and has an acid reaction. Is toxic to some plants, but if applied a few weeks before planting, in moist soil, it soon loses its toxicity. Apply ½ to 1 ounce per 10 square feet.

Nitrate of ammonia (ammonium nitrate) contains 32.5 to 34 percent immediately available nitrogen. Unless packed in moisture-proof bags it cakes badly in storage. Also, in ill-ventilated storage it can spontaneously catch fire or even explode. Apply ½ to ¾ ounce per 10 square feet.

Nitrate of lime (calcium nitrate) contains 15.5 percent immediately available nitrogen. It slightly decreases soil acidity. Apply ½ to 1 ounce per 10 square feet.

Nitrate of soda (sodium nitrate) contains 16 percent immediately available nitrogen. Increases alkalinity slightly so is unsuitable for acid-soil plants. It tends to increase the stickiness and plasticity of clayey soils. Natural nitrate of soda from Chile is called Chile saltpeter or Chilian nitrate. Apply ½ to 1 ounce per 10 square feet or ¼ to ½ ounce per gallon dissolved in water.

Sulfate of ammonia (ammonium sulfate) has about 20 percent nitrogen, which is available to the plants ten days or so after application. It has an acid reaction. Apply ½ to 1 ounce per 10 square feet or ¼ to ½ ounce per gallon dissolved in water.

Urea has 42 percent readily available nitrogen. Its reaction is somewhat acid. Apply ¼ to ½ ounce per 10 square feet.

Ureaform fertilizers are chemical combinations of urea and formaldehyde that have the great virtue of releasing the nitrogen they contain slowly over a period of many weeks or months. They are sold under many brand names. Apply as recommended by the manufacturers.

Chemical fertilizers used as sources of phosphorus include these:

Basic slag (Thomas slag) contains from 2 to 16 percent available phosphoric acid. It also has a content of calcium and small one of magnesium. It reduces acidity and is especially useful on clay soils. Apply 4 to 8 ounces per 10 square feet.

Calcium metaphosphate supplies approximately 53 percent available phosphoric acid. Its reaction is neutral. Apply 1 ounce to 1½ ounces per 10 square feet.

Defluorinated phosphate rock (fused phosphate rock, fused tricalcium phosphate) is heat-treated phosphate rock containing 8 to 24 percent available phosphoric acid. It reduces soil acidity slightly. Apply 1 ounce to 3 ounces per 10 square feet.

Dicalcium phosphate supplies approximately 40 percent available phosphoric acid and has a slightly acid-reducing effect. Apply ½ to ¾ ounce per 10 square feet.

Phosphate rock-magnesium silicate glass (calcium-magnesium phosphate) contains about 17 percent each available phosphoric acid and magnesium oxide. It reduces acidity slightly. Apply 1 ounce to 2 ounces per 10 square feet.

Rock phosphate is a natural rock untreated by heat or other means. Unless exceedingly finely ground it is of little use as a fertilizer and even then the 20 to 30 percent phosphoric acid it contains is very slowly available. It is more readily used by leguminous crops than others. Is not effective in soils with a pH above 6.2. Apply 3 to 4 ounces per 10 square feet.

Superphosphate is prepared by treating rock phosphate with sulfuric acid. Originally it was prepared by treating bones similarly. It contains 14 to 20 percent available phosphoric acid and in excess of 50 percent calcium sulfate. Its reaction is neutral. Apply 1 ounce to 2 ounces per 10 square feet. Double, treble, or triple superphosphate is a high analysis product of similar merit and usefulness that supplies 43 to 49 percent available phosphoric acid and a small amount of calcium sulfate. Apply ½ ounce per 10 square feet.

Chemical fertilizers that are sources of potassium include these:

Kainite contains 12 to 22 percent water soluble potash, about 20 percent magnesium. It is slow acting. On clay soils heavy applications can be harmful. Apply 2 to 3 ounces per 10 square feet.

Manure salts has 25 to 40 percent water soluble potash. It is slow acting. Apply 1 ounce to 1½ ounces per 10 square feet.

Muriate of potash (potassium chloride) contains 48 to 62 percent water soluble potash, about 15 percent sodium chloride. Less satisfactory than sulfate of potash for greenhouse use and for some crops including roses, tomatoes, and potatoes. Apply ½ to 1 ounce per 10 square feet.

Potassium-magnesium sulfate or double manure salts supplies 30 percent water soluble potash and up to 56 percent magnesium. Apply ½ to 1 ounce per 10 square feet.

Sulfate of potash (potassium sulfate) contains 48 to 52 percent water-soluble potash. For acid-soil plants this is better than muriate of potash. Apply ½ to 1 ounce per 10 square feet.

Compound chemical fertilizers that contain more than one of the three basic nutrients include these:

Ammoniated superphosphate has 2 to 4 percent nitrogen, 14 to 49 percent available phosphoric acid. Apply ½ ounce per 10 square feet.

Diamonium phosphate has 21 percent nitrogen, 53 percent available phosphoric acid. Apply ½ ounce per 10 square feet.

Nitrate of potash (potassium nitrate) supplies 13 percent immediately available nitrogen, 44 percent water-soluble potash. Apply 1 ounce to 2 ounces per 10 square feet.

Nitrophoska has approximately 12 percent each nitrogen and available phosphoric acid, 21 percent water-soluble potash. Apply 1 ounce to 2 ounces per 10 square feet.

FERULA (Fér-ula)—Giant-Fennel. Not to be confused with fennel (*Foeniculum*) of the vegetable garden, the giant-fennels number more than 130 species of thick-rooted, tall, stout, herbaceous plants of southern Europe, North Africa, and adjacent Asia. Only one, the common giant-fennel (*Ferula communis*) is much cultivated. The genus belongs in the carrot family UMBELLIFERAE. Its name is an old Latin one of doubtful reference; possibly it is derived from *ferula*, a schoolmaster's rod. It is thought that the stems may in ancient times have been used for punishing pupils. Gum resins of medicinal importance are obtained from species of *Ferula*. The best known is asafedida, a product of *F. asafoetida* and *F. narthex*, of southwestern Asia, native from Iran to Afghanistan. Another, galbanum, is obtained from *F. galbaniflua*. Gum ammoniac is obtained from a variety of *F. communis*.

The giant-fennels are hairless, often somewhat glaucous plants with leaves more than once-pinnately-divided, the ultimate segments being small and thread-like. The tiny flowers, very numerous, are in broad, compound umbels (in effect the flower cluster is an umbel of umbels). The dry fruits are flattened and ovate to round. From the fennel (*Foeniculum*) the giant-fennels (*Ferula*) differ in having the rays of their umbels of almost equal lengths and their fruits conspicuously flattened.

Common giant-fennel (***F. communis***), native to the Mediterranean region, is an imposing species that becomes 8 to 12 feet tall and forms a great clump of pretty, light green, feathery foliage topped by large umbels of greenish-yellow flowers. The terminal or central umbel is almost stalkless, but the secondary ones that arise from below it are long-stalked; they consist of mostly male flowers. The leafstalks have large and conspicuous sheaths.

Other kinds are ***F. asafoetida,*** 6 to 12 feet tall, with greenish-yellow flowers in stalked umbels, ***F. glauca,*** 6 to 8 feet tall, with yellow flowers, a native of southern Europe, ***F. galbaniflua,*** a tall native of Iran that has yellow flowers, ***F. narthex,*** 5 to 8 feet in height, with yellowish flowers, and ***F. tingitana,*** 6 to 8 feet tall, and with yellow

Ferula communis

blooms. The last is indigenous to North Africa.

Garden and Landscape Uses. The great landscape value of these plants is their attractive foliage. Their flower clusters, because of their size and structure rather than any striking display of color, are not without interest, but their display value is secondary to that of the mounds of graceful, light green foliage. This is at its best in spring; after midsummer it tends to fade and lose its freshness. This can to a large extent be prevented by cutting out the flower stalks as soon as they appear. Sacrifice of the blooms is no great loss, and as a result the leaves are preserved in better condition. To be seen to good advantage the giant-fennels should stand alone or be located some little distance in front of much larger and darker-leaved subjects, such as rhododendrons. They can be installed effectively in lawn beds and are elegant by watersides. For good results the soil must be deep and fertile and not lacking in moisture. These are hardy plants of perennial duration.

Cultivation. Like many of the carrot family, giant-fennels do not take kindly to transplanting, especially when large. It can be accomplished, and the plants can even be lifted, divided, and replanted if this is done with great care. But a better and surer procedure is to raise new plants from seeds and set them in their permanent locations while they are yet fairly small. Or the seeds can be sown where the plants are to remain, and the young seedlings thinned so that those that remain are appropriately spaced. Sowing is best done in summer as soon as ripe seeds are available, but may be delayed until spring. Established plantings should be fertilized each spring, and it is of benefit to keep the soil around them mulched with compost or other nourishing cover. In dry weather generous applications of water are beneficial. After the foliage dies in fall it is cut down and carried away. Beyond that little attention is needed. Most giant-fennels do not bloom until the plants have stored within themselves sufficient food to make this effort possible. This may be several years before the first blooming.

FESCUE. See Festuca.

FESTUCA (Fes-tùca)—Fescue. This genus of grasses, widely distributed throughout temperate regions and to some extent on mountains in the tropics, consists of about eighty species, often difficult to separate as to kind. The group belongs to the grass family GRAMINEAE. Its name is based on the Latin one for a grass stem. Several species are grown for forage and hay. Some, chiefly varieties of red fescue (*Festuca rubra*), are useful lawn grasses. For more information about these see Lawns, Their Making and Renovation.

Festucas are chiefly perennials. Annuals formerly included, none of horticultural importance, belong in *Vulpia*. Fescues are tufted or spread by stolons. They have slender to thick stalks and flat or rolled leaf blades. The panicles of blooms, loose or compact, are of stalked, compressed, two- to many-flowered spikelets.

The only kind commonly grown for ornament is blue fescue (***F. ovina glauca*** syn. *F. glauca*), a hardy native of central

Festuca ovina glauca

and southern Europe. This forms compact tufts or clumps of fine, almost hairlike, silvery-blue foliage 4 to 6 inches high, overtopped in season by pale panicles of flowers. It does not have creeping stolons. The leaves are rolled almost into slender tubes.

Garden and Landscape Uses and Cultivation. Blue fescue is useful for rock gardens and permanent edgings and also as temporary planting in summer beds. It thrives in sun in dryish locations. It is very easily increased by division in spring or early fall.

FETTERBUSH. See Leucothoe and Lyonia.

FEVER BUSH. See Garrya.

FEVERFEW is *Chrysanthemum parthenium*.

FEVERWORT. See Triosteum.

FIBIGIA (Fib-ígia). Sometimes included in *Farsetia*, the fourteen species of *Fibigia* belong in the mustard family CRUCIFERAE. The genus ranges in the wild from the Mediterranean region to Afghanistan. Its name commemorates Johann Fibig, a German physician and professor of natural history. He died in 1792.

Erect, herbaceous, and subshrubby, fibigias are perennials with alternate, linear to spatula-shaped leaves. They have flowers with four sepals, four petals that spread in the form of a cross, six stamens of which two are shorter than the others, and a single style. Their fruits are flattened pods with a beaklike tip and two to eight winged seeds in each compartment.

Native of Greece, where it inhabits cliffs, ***F. lunarioides*** is much-branched, up to 1 foot tall or slightly taller, woody at its base, and with many erect, leafy stems. Its leaves, covered with velvety, ashy-white, stellate (star-shaped) hairs, are obovate-lanceolate to narrowly-spatula-shaped and up to 2½ inches long; they have wavy margins. The bright yellow blooms about ½ inch across, in dense clusters, are succeeded by round to elliptic seed pods ½ to 1 inch long and at least one-half as wide as long. Yugoslavian ***F. triquetra*** is also a cliff dweller. It differs from the last in having obovate to oblanceolate leaves and seed pods not over ⅓ inch wide and generally at least twice as long as they are wide. From the Balkans and Italy comes ***F. clypeata,*** a somewhat taller and usually greener plant. Its seed pods are elliptic, ½ to 1 inch in length, and almost one-half as broad as long. The kind named ***F. eriocarpa,*** from Greece, differs from the last only in its seed pods having long, branchless hairs.

Fibigia eriocarpa

Garden Uses and Cultivation. These fibigias may be planted in full sun in hot, dryish places where the soil drains rapidly. They are suitable for rock gardens, dry walls, and similar places, but may not be hardy where winters are severe. In North America they are probably best suited for Pacific Coast gardens. They are easily raised from seeds. The long stalks of seed pods of *F. clypeata* and *F. eriocarpa* are attractive for use in dried flower arrangements.

FICUS (Fì-cus)—Fig. The remarkable and extensive genus *Ficus* includes 800 species of tropical and subtropical plants of very diverse growth habits, almost all containing milky sap. Most familiar is the common edible fig (*F. carica*). For a discussion of it and its cultivation see Fig. Other well-known members of the group are the common rubber plant (*F. elastica*), the giant banyan tree (*F. benghalensis*), and the sacred bo tree of India (*F. religiosa*). In addition to trees and shrubs there are included tiny-leaved creepers and vigorous vines. Many begin life perched on other trees as epiphytes and eventually smother their hosts. The story of how the flowers are pollinated is one of the most fascinating of the plant world. Belonging in the mulberry family MORACEAE, this genus is kin of the osage-orange and nettles. Its name is an ancient Latin one.

Despite their very different habits of growth, all figs are remarkably alike in their floral structures, which, of course, is the basis for plant classification. The tiny, petal-less flowers are attached, few or many, on the insides of hollow, globular or pear-shaped receptacles that are commonly but somewhat mistakenly called fruits (botanically the true fruits are the seeds that develop inside the swollen receptacles but here we shall follow the common and convenient practice of referring to the receptacles themselves as fruits). Each receptacle has a tiny aperture at its end protected by small overlapping bracts. There are three types of flowers, fertile males, fertile females, and sterile females called gall flowers. Some species produce all three kinds in the same receptacle and are self-fertile. The receptacles of others, self-sterile kinds, contain either all female flowers or male and gall flowers only, and indiviual plants produce only one type of receptacle, thus there are separate male and separate female plants. The transference of pollen from male to female flowers is done by insects called gall wasps and, so far as is known, there is a different kind of gall wasp for each species of fig. The procedure can best be made clear by explaining that the gravid female insect enters the receptacle through the aperture at its apex and lays eggs on the ovaries of the flowers. The male insects of the resulting progeny fertilize the females of the same brood without leaving the receptacle. After mating, the males die, but the fertilized females escape through the opening at the end of the receptacle and as they emerge are dusted with pollen from male flowers crowded around the exit. This they carry to the female flowers in the receptacles they select as repositories for their eggs. When the eggs hatch the grubs of the gall wasps feed on the endosperm of the gall flowers. Between insects and plants there is a mutually advantageous relationship, the insects pollinate the flowers, the flowers supply nourishment to the larvae of the insects. Most species of *Ficus* bear their fruits on their younger shoots, but some develop theirs from old branches and from the trunks and one group, called earth figs, produces from its trunks slender stems up to 30 feet in length that bury themselves shallowly in the soil and develop fruits along their lengths.

Most figs are evergreen, a few, including the common edible one, lose their foliage for a part of each year. Most extraordinary are the so-called strangler figs of which there are many kinds in the humid tropics. These normally begin life in a crotch or bark crevice of another tree, or sometimes on the roof or other part of a building that affords parking space for a seed dropped by a fruit-eating bird, bat, or monkey. For some time after the seed germinates the new plant exists as an epiphyte, it lives on the host tree but derives no nourishment directly from it as it would were it a parasite. The young fig accepts lodging, but not board, as it were. At first its nourishment comes from whatever the rains bring and from organic debris that may have accumulated within reach of its roots. But as it increases in size it soon develops one or more roots that descend perpendicularly and it begins to take nourishment from the soil. Lateral roots grow from the primary vertical ones and both kinds gradually envelop the trunk and branches of the host in a strong basket-like complex or, in the case of a building, extend themselves over its surfaces and enter joints and openings and may do serious damage to masonry, woodwork, and metalwork. As the strangler grows its branches and foliage invade the top of its nurse tree and cut off needed light. At the same time the roots of the fig multiply, thicken, exert pressure on the bark of the host, and eventually check the flow of its sap to such an extent that the supporting tree is seriously weakened or killed. Even so, the strangler continues to flourish. Ordinarily, strangler figs develop no substantial trunk from the top of their basketwork of roots, but sometimes the roots coalesce to produce the effect of a massive trunk. Some strangler figs have the banyan habit of sending from their branches, at considerable distances from the central trunk, roots that establish themselves in the ground and develop into stout trunklike supports.

Several figs have religious associations including the common fig (*F. carica*), which presumably provided raiment for Adam and Eve. The sycomore of the Bible is not the sycamore of North America, a plane tree, nor the sycamore of Great Britain, a maple. It is the sycamore fig (*F. sycomorus*), a huge, wide-spreading native of Africa that is common in the Sudan and Egypt and is evergreen or deciduous according to local climate. It has a yellowish trunk, leaves 2 to 3 inches long, and fruits in large clusters that are great favorites of birds. The wood of this species was used by the ancient Egyptians for their sarcophagi. Both the banyan (*F. benghalensis*) and the bo tree or peepul (*F. religiosa*) are held sacred by the peoples of India. Hindus believe that Brahma, the Creator, was transformed into a banyan tree. Both Hindus and Buddhists venerate the bo tree. Under one the Hindu deity Vishnu is believed to have been born, under another Gautama Buddha meditated for six years and received enlightenment. Bo trees are planted in India near temples because of this association with the Buddha and near homes to assure happiness and prosperity. In New Delhi and elsewhere they are used as street trees. Believers will not prune or cut down a banyan or a bo or peepul tree. That work, when necessary, is done by others. There is a saying in India "it is better to die a leper than pluck a leaf of a peepul."

Considering the extent of the genus and excepting the common fig (*F. carica*), few of its members are of much importance other than as shade trees and ornamentals. A few have edible fruits eaten by local populations, and from the bark of some tapa cloth and native cordage is made. So much silica is contained in the leaves of the sandpaper tree (*F. exasperata*) and *F. asperifolia* that in their home continent Africa native carpenters use them as sandpaper to smooth wood and calabashes. The harsh leaves of *F. odorata*, of the Philippines, are used locally for scouring. When dried they are fragrant, hence the epithet *odorata*. Native peoples in some parts of the tropics use the sticky latex of some species to entrap birds. The rubber plant (*F. elastica*), so commonly grown in pots, was the original source of India-rubber, first used by natives of Assam and later exploited by Westerners who employed it chiefly for erasers. In the latter part of the nineteenth century this species was cultivated in Assam and the trees tapped for their latex, but the product is so inferior to that of South American *Hevea brasiliensis* that its commercial exploitation ended when *Hevea* rubber became available. The wood of fig trees is not durable under moist conditions and is little used. Occasionally it is employed for posts and other minor purposes, but the genus provides no commercial lumber. In South America foresters consider figs weed trees even though some attain heights of 150 feet or even more. In Hawaii, southern Florida, southern California, and other warm regions a number of kinds are grown outdoors for ornament. They are also cultivated in greenhouses and as houseplants.

A strangler fig, ***F. aurea***, an endemic ornamental of southern Florida, attains a height of 60 feet. It has attractive, evergreen, oblongish leaves 3 to 4 inches long, and small, nearly stalkless, orange-yellow fruits. Another strangler indigenous to southern Florida and about the same size, ***F. citrifolia*** (syn. *F. brevifolia*) has whitish bark, broadly-ovate, evergreen leaves 1¼ to 4 inches long, and edible, but rather tasteless fruits that hang on distinct stalks. This kind also inhabits the West Indies and

Ficus benghalensis

ranges as far south as Paraguay. It is the commonest native fig of Puerto Rico, where it is called jaguey blanco. It is a good shade tree.

The banyan (***F. benghalensis***), of tropical Asia, is a magnificent evergreen, sometimes 100 feet in height. Like so many figs it begins life as an epiphyte, but once established in the ground its massive limbs spread horizontally and as they extend send down roots that develop into secondary, pillar-like supporting trunks. Over a period of years a single tree may come to occupy a tremendous area. One famous specimen near Poona in India spread so widely that its periphery measured 2,000 feet and it was estimated that it could shelter 20,000 people. The sparsely-veined, ovate leaves of the banyan are up to 10 inches long by a little over one-half as wide. Its globose, red fruits, borne in pairs, are about ¾ inch in diameter.

The bo or peepul tree (***F. religiosa***), a strangler fig of Asia, grows as tall as the banyan and like it develops roots from its branches, but these are usually few in number and the tree is not as aggressive as the banyan in its takeover of adjacent areas. The foliage of the bo tree is among the most beautiful of all figs. Rich, glossy green, and heart-shaped, its leaves, up to 7 inches long, have their tips extended into long tails. Their flattened stalks permit every breeze to set the leaves aflutter. The new foliage of the bo tree is pink and

Ficus religiosa

green. The fruits, about ¼ inch in diameter and purple-black, are beloved by birds. The bo tree is normally deciduous for part of each year, but in constantly humid climates may be evergreen, as it normally is when grown in greenhouses.

The common rubber plant (***F. elastica***) is familiar to almost everybody, at least as a container-grown specimen. In its native India and Malaya it normally starts as an epiphyte and develops into a massive tree 100 feet or more in height. It supports its great branches with prop roots that become secondary trunks. On specimens of tree size the shiny, leathery leaves, which have numerous fine parallel lateral veins

Ficus elastica

at nearly right angles to their midribs, are often up to 6 inches long, but on young specimens they may be 1 foot long or longer and about one-half as broad. A prominent feature of the rubber plant is the large rosy-pink stipules that envelop the leaf buds until the leaves unfold, when they drop. The yellowish-green fruits, egg-shaped and about ½ inch long, grow in pairs on the shorter twigs. In Hawaii, where trees of the rubber plant at least 100 years old exist, the fruits drop early and do not produce seed because the particular gall-wasp needed to pollinate the flowers of this species is not present.

Several attractive varieties of common rubber plant are cultivated. Most familiar is broad-leaved *F. e. decora*, a variant believed to have originated in the 1930s as a seedling in a Belgian nursery. It is possible that more than one variant is grown under this name. Names that have been applied to closely similar horticultural varieties include *F. e. belgica*, the leaves of which have cream to pink midribs. A vigorous variety

Ficus elastica decora

Ficus elastica doescheri

Ficus elastic schrijvereana

with broad leaves variegated with green, gray-green, creamy-yellow, and white, *F. e.* ***schrijvereana*** has red leafstalks. The leaves of *F. e. doescheri* are variegated with green, gray-green, creamy-yellow, and white. They have pink stalks. Distinctly maroon-red young leaves and older leaves with reddish midribs are typical of *F. e. rubra.* The foliage of *F. e. variegata* is like that of typical *F. elastica,* except that its leaves are variegated with gray-green and margined with creamy-yellow.

Chinese banyan or Indian laurel (***F. retusa***) is an attractive, dense-headed, large evergreen tree of moist regions in southeast Asia, Malaysia, China, and the Philippine Islands. It produces some pendulous adventitious roots from its branches and these sometimes thicken and become trunklike. Its short-stalked and characteristically blunt, broad-oval leaves are up to 4 inches in length. The Chinese banyan grows up to 60 feet in height and develops a wide-spreading crown. In axillary pairs, its white to purplish fruits are ⅓ inch in diameter. They are appreciated by birds. This is a strangler that when established in the soil develops prominent and extensive surface roots. Somewhat similar, but with drooping branches and glossy,

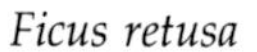

Ficus retusa

Ficus benjamina

Ficus rubiginosa variegata

trunk. Its orange, pink, red, or purple fruits are ½ to ¾ inch in diameter.

The Port Jackson fig (***F. rubiginosa***) is an Australian species of large size. It has a heavy crown and blunt, oval leaves up to about 3 inches long. Its younger shoots are furnished with short brown hairs and so, to some extent, are the undersides of its leaves. The globose, warty fruits are about ⅓ inch in diameter. This species, which develops aerial roots that may become trunklike, is sometimes known as *F. australis.* Variety *F. r. variegata,* has leaves with beautiful cream-colored variegation. Another Australian, the Moreton Bay fig (***F. macrophylla***) also has an extensive root system. Its leathery leaves are oval or oblong and up to 10 inches long. Like those of the rubber plant, when in bud they are encased in rose-pink sheaths that drop as the leaves unfold. The globose fruits are purple and about 1 inch in diameter. Variety *F. m. pubescens* has somewhat downy leaves.

The fiddle-leaved fig (***F. lyrata***), of tropical Africa, is often misnamed *F. pandurata* which name properly belongs to another, uncultivated, species. Handsome and very distinct, in the wild *F. lyrata* begins life as an epiphyte. It attains a height of 50 feet. Its fiddle-shaped leaves up to 1½ feet long by almost 1 foot broad, its fruits about 1½ inches across. One of the most beautiful figs (***F. aspera*** syn. *F. parcellii*) is a thin-leaved shrub of islands of the Pacific. It

Ficus lyrata

pointed-ovate, evergreen leaves up to 5 inches in length, the weeping fig (***F. benjamina***) is indigenous to the more humid parts of India and Burma and of various islands including the Philippines. Like the Chinese banyan it sends roots from its branches and develops an extensive system of surface roots that radiate from its

Ficus lyrata (fruits)

Ficus aspera

Ficus aspera (fruits)

has deciduous or evergreen, pointed-ovate to oblong, asymmetrical leaves up to 8 inches long, conspicuously marbled with creamy-white. Its handsome fruits, globose and about ¾ inch in diameter, are varicolored with green, white, and pink.

The mistletoe fig (***F. deltoidea*** syn. *F. diversifolia*), of Malaya, is interesting. A bushy shrub up to about 6 feet high, as a wildling it usually grows as an epiphyte, but sometimes in the ground. It has curious blunt-ended, more or less triangular leaves rusty-olive-colored beneath and with forked midribs. They are up to 2 inches long. The plant seems always to be heavily fruited. The fruits, about the size of peas, are in pairs in the leaf axils. As they ripen they change from yellow to orange to red.

A small-leaved vine, native from Indochina to Japan, ***F. pumila*** (syn. *F. repens*) creeps over the ground or climbs high by attaching itself by rootlets to wood, masonry, or other supports. In its juvenile stage, which may persist for many years, it lies flat against the support to which it attaches itself, clothed with thin, ovate leaves, heart-shaped at their bases and up to 1 inch long. If it has opportunity to climb high it eventually develops its mature, fruiting branches. These stand out from the support and bear leaves 2 to 4 inches long, elliptic or oblong and without heart-shaped bases. The 2-inch-long, pear-shaped fruits are yellow. Variety *F. p. minima* is a juvenile form that differs from the typical young form of *F. pumila* in having

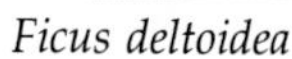

Ficus deltoidea

Ficus deltoidea (leaves and fruits)

Ficus pumila (juvenile foliage)

smaller, more slender leaves. Variety *F. p. variegata* has small leaves with attractive cream-colored variegation. More robust than *F. pumila* is ***F. sagittata*** (syn. *F. radicans*), native from the Himalayas to the Philippine Islands and Caroline Islands. This sort has creeping stems and pointed, oblong-lanceolate, hairless, wrinkled leaves 2 to 4 inches long. The leaves of *F. s. variegata* are irregularly marked with grayish-green and creamy-white.

Ficus pumila (leaves and fruit)

Ficus sagittata variegata

Ficus brandegeei in Baja California

Other species in cultivation include these: ***F. afzelii*** (syn. *F. eriobotryoides*) is a large tropical African tree with stiff, oblanceolate leaves up to 1¼ feet long by about 4 inches broad. Its usually yellow- to red-hairy fruits are 1 inch to 1½ inches in diameter. ***F. altissima*** is a large spreading tree with oval or elliptic leaves up to 8 inches long and clusters of orange-red fruits. It is native to India. ***F. auriculata*** (syn. *F. roxburghii*), of the Himalayas, is a tree or shrub. Its broad-ovate to nearly round leaves may be 1¼ feet long and are pubescent on their undersides. The flat-

Ficus brandegeei, young plant

tened-spherical to pear-shaped, brown to purplish fruits, 2 to 2½ inches in diameter, are in clusters on the trunks and branches. ***F. brandegeei,*** of Baja California, scarcely differs from *F. palmeri* except that its leaves are quite hairless. ***F. cannonii*** (syn. *Artocarpus cannonii*) of the Society Islands is a handsome shrub up to about 9 feet tall. It has thin, glossy, veiny, broad-ovate to somewhat heart-shaped, rarely lobed leaves 4 to 6 inches long and with red stalks, red mid-veins, and red undersides. The upper surfaces are bronzy-crimson tinged with purple. The nearly spherical fruits are less than 1 inch in diameter. ***F. dryepondtiana,*** of west tropical Africa, grows natively as an epiphyte. It has few-branched stems and pointed-oblong leaves 9 inches to 1 foot long with purple under surfaces. In clusters of up to a dozen, the spherical, hairy fruits, about 1½ inches in diameter, are spotted greenish-white. ***F. montana*** (syn. *F. quercifolia*) of southeast Asia and Malaysia is a more or less prostrate shrub with leaves, as its synonymous name indicates, in shape suggesting those of an oak (*Quercus*). Elliptic-ovate to obovate, they are 3 to 6 inches long, have lobed margins, and are rough-hairy. The stalked, spherical fruits are up to ½ inch in diameter. ***F. nekbudu*** (syn. *F. utilis*) is a big African tree with elliptic to obovate leaves up to 1¼ feet long and pubescent fruits ½ inch across. ***F. palmeri,*** of Baja California, is a variable species with a white-barked trunk. Very drought-resistant, it has more or less pointed-heart-shaped leaves up to 7 inches long by 5 inches broad. Its globular fruits, in pairs, are white and ½ inch in diameter. ***F. petiolaris,*** of Mexico, is a small to big tree with round-heart-shaped leaves up to 3 inches across tipped with brief points and with tufts of hair in the chief vein axils. The more or less spherical fruits, densely-hairy when young, are about ½ inch in diameter. ***F. pseudopalma*** is a small tree or shrub of the Philippine Islands with few or no branches and stiffish, linear, coarsely-toothed leaves up to 2 feet long by 4 inches wide clustered near the ends of its branches. The ovoid, 2- to 3-inch-long fruits are solitary or in pairs. ***F. racemosa*** (syn. *F. glomerata*), the cluster fig, has ovate leaves up to 7 inches long, bears big clusters of reddish, 1-inch-wide fruits, and is a native of India. ***F. ulmifolia,*** of the Philippine Islands, is a shrub or small tree with mostly oblong or elliptic, rough leaves up to 7 inches long and toothed or lobed. Its orange-red fruits are about ½ inch long. ***F. virens*** (syn. *F. infectoria*) is a deciduous tree of India with thin, ovate leaves up to 5 inches long. Its whitish fruits, ¼ inch in diameter, are spotted with red.

Garden and Landscape Uses. Where ficuses thrive outdoors their primary uses, with the exception of the common fig (*F. carica*), are as shade and ornamental trees employed as single specimens, in avenues, and as street trees. For these purposes they rank high. Prostrate *F. pumila* is admirable for clothing walls and other surfaces, and both it and *F. radicans* are attractive groundcovers. As container-grown specimens for indoors, and for porches and terraces during warm weather months, several kinds are attractive and decorative. One of the most reliable, and not the least handsome, is the rubber plant (*F. elastica*). This long-time favorite persists with minimum care in a wide variety of environments from dimly lighted apartments and old fashioned curtained bay windows to modern less discreetly draped picture windows. It seems to be a favorite of barbers and may often be seen in company with an orange tree and an avocado grown from seed, a pot containing basil, an aloe, and other assorted flora in their shop windows. Occasionally quite splendid surroundings serve as its foil. Huge, branched and venerable, artistically asymmetrical, rubber plants have graced the professional plant displays that decorate the great glass buildings and stores of New York's most famous avenues. Where light is too dim or other environmental conditions too tough for most indoor plants, the rubber plant is

Ficus cannonii

Ficus cannonii (fruits)

likely to persist. Its broad-leaved variety, *F. e. decora*, is pretty tough also, but will not tolerate as poor conditions as *F. elastica*. The variegated-leaved varieties are even more finicky. The mistletoe fig (*F. deltoidea*) is an attractive and interesting plant for window gardens and other locations where neatness of form and foliage are advantageous. Its little fruits attract attention. For window gardens and for wall brackets low *F. sagittata* is useful, and *F. pumila* can be used to advantage to provide a tracery of greenery over masonry surfaces in sunroom or greenhouse. At Longwood Gardens near Wilmington, Delaware, it is employed with magnificent effect to clothe tall walls and pillars in the conservatory. The desert species *F. brandegeei* and *F. palmeri* are adapted for outdoors in warm, dry regions and are sometimes grown along with cactuses and other succulents in greenhouses.

Ficus pumila at Longwood Gardens, near Wilmington, Delaware

Cultivation. The ornamental figs make no special demands of the gardener. They prosper in full sun if they receive sufficient water, but stand a good deal of shade and are excellent for locations that are without sun for much of the day or receive only dappled sunlight. They thrive in any ordinary soil that is not waterlogged. No systematic pruning is necessary, but any needed may be done with impunity and small-leaved kinds respond well to shearing. In this way they may be trained to shape and kept to convenient size. Pruning may be done at any time, but usually immediately before the beginning of the growing season is preferred because then the cut ends of branches are soon hidden by new growth.

Specimens grown in pots or other containers should have free drainage. The soil should be moderately rich and coarse, not of a type that is likely with the passage of time to become so compact that aeration of the roots is seriously impeded. If the topsoil used in the potting mixture is heavy (clayey), its texture can be improved by mixing crushed brick with it. Repotting is required only at long intervals, several years in the case of large specimens. Whereas excessive dryness is harmful to *Ficus* in containers, over-wetness of the soil can be equally or more disastrous. A middle-of-the-road course should be maintained. Between applications of water the soil should be allowed to become fairly dry. When water is given, the whole body of soil should be saturated. Plants that have filled their containers with healthy roots benefit from an application of dilute liquid fertilizer about once a month from March through September. None is needed during the remainder of the year. Indoors these ornamental figs do best where the minimum temperature is 55 to 60°F. When potted rubber plants and other sorts become leggy they may be cut back to any degree desired. The best time to attend to this is late winter or early spring. Increase of ficuses is effected by air layering, stem cuttings, and by single leaves, with a piece of stem and an axillary bud attached. The creeping kinds can be increased by division. For cultivation of the common edible fig see Fig.

Diseases and Pests. Anthracnose, leaf spots, and crown gall are the chief diseases that affect these plants. Among pests the most likely to be damaging are mealybugs, including root mealybugs, scales, and thrips. Root knot nematodes also sometimes attack *Ficus*.

FIDDLE WOOD, FLORIDA is *Citharexylum fruticosum*.

FIDDLEHEADS. Ferns expand their young leaves (fronds) in a manner unusual among other plants. They gradually uncoil their fronds from their base to their tips. At one stage of this process the uncoiled portion at the top of the leafstalk of some kinds suggests the curved top of a violin. Because of this they are called fiddleheads.

FIELD-SCABIOUS is *Knautia arvensis*.

FIESTA FLOWER is *Pholistoma auritum*.

FIG. Without modification the word fig applies to plants of the genus *Ficus*, especially to *F. carica*. Other plants that have common names incorporating the word fig are these: Hottentot-fig (*Carpobrotus edilus*), Indian-fig (*Opuntia ficus-indica*), and sea-fig (*Carpobrotus chilensis*).

FIG. All members of the genus *Ficus*, of the mulberry family MORACEAE, are sometimes called figs. Here we shall concern ourselves only with horticultural aspects of the common edible fig, *Ficus carica*, and its varieties. For botanical details about this and for information about other kinds see Ficus.

The common fig is a deciduous shrub or tree 15 to 30 feet tall and a native of the

The common fig

The common fig (foliage)

The common fig (fruits)

Mediterranean region. It is chiefly grown for its fruits, but sometimes also as an attractive ornamental in tubs or pots to decorate patios, terraces, and similar locations. Espaliered against walls and fences figs are decorative and useful. Their roughish, deeply-three- to five-lobed leaves form attractive patterns. The structures commonly called fruits (and that is how we shall refer to them here), are not fruits in the botanical sense. Morphologically they are fleshy receptacles formed into hollow vessels with

The common fig espaliered, Williamsburg, Virginia

tiny flowers and later true fruits (the seeds as they are generally called) inside. The only significance of this from the growers' point of view is that some kinds do not develop sizable, edible fruits unless the flowers are cross-fertilized by pollen from a special kind of fig called a caprifig, and this is done only by a tiny gall-wasp that lives for part of its life in the fruits of caprifigs. The process of pollen transfer is called caprification.

Figs are exacting in their climatic needs, to the extent that commercial production is limited to comparatively few regions where winters are mild, summers not excessively hot, wet, or humid, and destructive winds are generally not experienced or shelter from them is provided. In the United States the chief commercial production is in California, Oregon, Washington, and Utah. To a lesser extent figs are raised from Maryland to Florida and Louisiana. In home gardens the range is considerably greater. With winter protection they are raised outdoors as far north as Newport, Rhode Island and are commonly grown in and near New York City. In colder climates figs can be had in tubs wintered in a cool, frost-free cellar or similar accommodation, and summered outdoors.

Soils of most types from sandy to clayey suit figs. They do not prosper, however, in alkali earths, especially those of the kind called black alkali. For their best satisfaction deep, river bottom land that does not dry excessively, yet is well drained, is ideal.

Planting is done in mild climates in winter or spring, in the latter season in colder climates. Small-growing varieties may be spaced 10 to 12 feet, stronger-growing sorts 15 to 20 feet apart. Exceptionally vigorous kinds, such as 'Mission', when grown commercially are allowed 40 to 50 feet between individuals, other strong-growers 30 to 40 feet.

Take care that the roots are not allowed to dry while the trees are out of the ground. Make holes of ample size to accommodate them. Set the trees about 2 inches deeper than they were in the nursery. Spread the roots, then work soil among them and pack it firmly. Prune newly planted trees by heading them back to a height of about 2 feet. Water thoroughly after planting and mulch the ground to keep the soil uniformly moist. During the first summer whenever necessary to keep the soil from becoming excessively dry, soak the ground deeply.

Routine care of established figs involves keeping weeds in check either by frequent shallow cultivations (feeding roots are damaged if the ground is stirred deeply) or by mulching with partly rotted manure, compost, hay, straw or other suitable material. Fertilize with considerable discretion. Excess nitrogen results in over-vigorous shoot and foliage growth and a decrease in fruit production. If the ground is decidedly acid liming to bring it to a slightly acid or neutral condition is beneficial. In dry climates irrigation at regular intervals throughout the summer is needed for good results. In eastern North America and other summer-rainfall regions this is not necessary.

Pruning is done to restrict the trees to a single main trunk and to encourage the development of well-spaced branches not unduly subject to storm damage. Good results are not possible if the trees are allowed to become so crowded with branches that some are denied exposure to light and free air circulation. Remember when pruning that where summers are sufficiently long figs bear two crops, the first on shoots of the previous year's growth (if these are pruned off in spring obviously the fruits they would have borne are lost), the second on shoots of the current season's growth. Also have in mind the need to encourage enough new shoots to assure a second crop. The amount and style of pruning must be geared to the needs of particular varieties, the condition of individual trees, and if the trees are espaliered, to the placement of the branches with that form of training in mind.

Caprification is necessary for Smyrna type varieties and some others. Check this before ordering from the nursery. For home gardens, varieties that fruit without caprification are usually preferred. If you decide on others be sure to have one or more caprifig trees available as a source of pollen. The fruits of caprifigs are usually not palatable. To effect caprification pick the caprifigs just before the wasps are ready to leave them, which is when the pollen on the flowers inside is beginning to shed. Put the fruits in baskets or other containers and suspend them from the branches of the fig trees to be pollinated or hang them from the branches on fine wires or strings. Replace the fruits with fresh ones every three or four days for about three weeks. Depending upon the size of the tree, from two to six or eight caprifigs are needed at one time so that twelve to three or four times that number will suffice for the three-week period.

Winter protection is commonly needed in climates more severe than that of Washington, D.C. A common method is to wrap the trees to insulate them from the severest cold. Do not do this until well after leaf fall and the approach of quite severe weather, usually after the top inch or two of the ground has frozen and winter has definitely begun. Then, prune out any superfluous branches, pull into as tight a bundle as possible without breaking those that remain, and wrap the entire tree in several layers of stout paper (such as roofing paper), burlap, or polyethylene plastic film and tie to hold the branches in position. Next, enclose the bundle with mats, old blankets, straw, or burlap and finally with a well-secured layer of tar paper or polyethylene plastic. Make sure the wrapping is done so that water cannot enter the top. Remove the coverings in early spring,

The common fig wrapped for winter protection, New York City

about the time the buds of other trees begin to sprout.

In tubs or large pots figs thrive in well-drained fertile soil. Containers 1 foot or more in diameter are most suitable and specimens should be repotted every second year in fall at about the time the leaves drop or in early spring before growth begins. In alternate years prick away as much topsoil as possible without seriously harming the roots, and replace it with new. At repotting time young specimens may be transferred to containers a couple of inches more in diameter than those they previously occupied. Bigger specimens that have achieved the largest pots or tubs they are to be allowed are at potting time taken from these and after pricking away as much old soil from the outside of the root ball as conveniently possible, removing the old drainage material, and shortening coarse thick roots, are repotted into containers of the same size that have had fresh drainage material put in their bottoms.

When container-grown figs are about to start into spring growth take them from winter storage (cellar, garage, or other light place where the temperature does not go much below 30°F or above 40 or 45°F) and put them in a cool greenhouse, sunroom, or even a sheltered, sunny place against a wall outdoors where they can be protected with a tentlike covering of polyethylene plastic film with a few holes in it to prevent the temperature inside becoming dangerously high. Spray the branches with water on sunny days and water the soil to keep it evenly moist. Summer the plants in a sunny place outdoors. Give dilute liquid fertilizer weekly or biweekly to specimens that have filled their containers with healthy roots.

Propagation of figs is easy by seeds and cuttings. The former method is usually employed only when breeding new varieties is the objective. Not all seeds produced are fertile. To determine which are, dump them in a container of water; fertile seeds sink, infertile ones float. Much more usual and generally more practicable is increase by hardwood cuttings 8 to 10 inches long taken in fall or winter and made of two- or three-year-old shoots, which are not too thick or pithy. Remove all buds except the topmost from the cuttings. This eliminates or at least reduces the growth of undesirable sucker shoots from the bases of the trees.

Layering, air layering, and grafting are alternative methods of increase. All are easily done. Grafting, cleft-grafting is the preferred method, is only used to change the variety of old trees. Bark grafting, which is similar to budding except that a scion with several buds instead of a single bud is inserted into the understock, is less frequently employed as a means of changing varieties.

Varieties of figs that fruit successfully without caprification, although their fruits are generally bigger if caprified, include 'Adriatic', satisfactory in the east, 'Brown Turkey', one of the hardiest, also satisfactory in the east, 'Brunswick' (syn. 'Magnolia'), satisfactory in the east and in the west (often misidentified as 'Brown Turkey'), 'Celeste', one of the best for the Southeast, 'Dottato' (syn. 'Kadota'), a favorite in the West for canning, 'Mission', a popular black-fruited variety, 'Ronde Noire', best suited for comparatively cool coastal parts of California. Smyrna type figs, such as 'Lob Injur' (syn. 'Calimyrna'), popular in California for drying, fruit satisfactory only if caprified. Named varieties of caprifigs include 'Roeding No. 3', and 'Stanford'.

Pests and Diseases. The most serious pests of figs are scale insects (the most troublesome the Mediterranean fig scale), fig rust mites, and root nematodes. Avoid planting on nematode-infected land and take care to plant only nematode-free stock. Scales and rust mites are controllable by spraying as recommended by Cooperative Extension Agents and State Agricultural Experiment Stations. Consult these authorities too for recommendations for controlling the most prevalent diseases, anthracnose, leaf spots, rust, and internal rot of the fruits.

FIGWORT. See Scrophularia.

FILAGO. See Evax.

FILBERT. Filberts are kinds of hazelnuts. They are varieties and hybrids of the Old World species *Corylus avellana* and *C. maxima*. Some varieties have oblongish nuts shorter than the tubular husks that enclose them, in others the husks are about as long as the angular, roundish nuts. In Europe the latter are distinguished as cob nuts, the others as filberts, but such distinctions are not usually made in America.

Growing filberts for their nuts is pretty much limited in North America to the Pacific Northwest, with lesser adventures having had some success in northern peach-growing regions in the east. Not only climate, especially below zero Fahrenheit temperatures and late frosts, but the prevalence on native hazels of an easily transmissible fungus blight to which filberts are very susceptible limits success in the east.

Filberts grow satisfactorily in a wide variety of well-drained soils managed and fertilized as for apples. In the Pacific Northwest recommended spacing between trees is 25 feet. In the east 18 to 20 feet is enough. The site should be one that will not induce flowering before danger of late frosts has gone. In the east north-facing slopes afford the best likelihood of this. Although filberts carry male and female flowers on the same tree, they are not fertile to their own pollen. For nut production it is necessary that plantings consist of at least two varieties.

Pruning at planting time consists of cutting the trees back to a height of approximately 2 feet. The development of four to six well-placed framework or scaffold branches is encouraged. In subsequent years do a moderate amount of thinning out of branches that are in danger of becoming crowded, but do not head back branches. In cold climates over-severe pruning may be followed by winter killing of shoots and branches. As young specimens, some varieties of filberts tend to produce an abundance of suckers from the bases of the trees. This tendency subsides as the trees age. It is better that all suckers be promptly removed.

Harvesting is done by collecting nuts from the ground. They are then allowed to dry and are stored in an unheated place indoors. The nuts soon spoil if stored where temperatures are high.

Propagation is generally by layering, which may be done by hilling soil to a height of several inches around sucker shoots in spring or by pegging sucker shoots into trenches a few inches deep radiating from the tree, and hilling soil in two or three operations about the bases of the erect lateral shoots that develop from these.

Varieties chiefly favored in the Northwest are 'Barcelona', 'Daviana', and 'Du Chilly'. The first is the chief nut producer. The others are planted mainly to ensure cross-pollination. In the east 'Cosford', 'Italian Red', and 'Medium Long' have proved superior croppers and hardier than those favored in the west. There are also hybrids between European varieties and varieties of the native hazel (*Corylus americana*). Among these 'Bixby' and 'Buchanan' seem most promising. Varieties 'Rush' and 'Winkler' of the native *C. americana* have smaller nuts, but are considerably hardier than the European filberts.

Filipendula vulgaris

FILIPENDULA (Filipén-dula)—Dropwort, Meadowsweet, Queen-of-the-Meadow, Queen-of-the-Prairie. Several of this group of ten or more species of the rose family ROSACEAE, are attractive. In the wild confined to temperate parts of the Northern Hemisphere, filipendulas are hardy herbaceous perennials. At one time they were united with *Spiraea*, a name now correctly restricted to a group of shrubs, but still sometimes applied by gardeners to *Filipendula*. The name comes from the classical one for *F. vulgaris* and is thought to allude to the manner in which the tubers of some kinds are linked.

Filipendulas have alternate, pinnately-divided leaves, commonly with a comparatively large terminal leaflet or lobe, and generally with lateral leaflets staggered at unequal distances along the central axis or midrib, usually small ones alternating with bigger ones. The bisexual flowers, white, cream, pink, or purplish, are small and numerous. They are in plumelike panicles or flattish clusters at the tops of the stems. Each bloom has a five-lobed calyx, generally five, but occasionally four or six petals, and numerous stamens. There are five to fifteen pistils. The fruits are small heads of seedlike achenes.

Dropwort (***F. vulgaris***, syns. *F. hexapetala, Spiraea filipendula*) is popular. Native to dryish grasslands in Europe, and sparsely-hairy, it has tuber-bearing roots and usually branchless stems 1½ to 3 feet tall. Its deeply-pinnately-lobed, fernlike leaves, mostly basal, are 6 inches to 1½ feet long and up to 2 inches wide. They have eight to twenty-five pairs of large leaflets, which are pinnately-lobed, with the lobes often toothed. The flower clusters are up to 4 inches across and wider than long. Usually the creamy-white flowers have six petals ⅕ inch long or longer, and purplish on their undersides. The stamens are about as long as the petals. The follicles of the fruits are not spirally twisted. Double-flowered *F. v. flore-pleno* is especially beautiful. Its larger flowers are longer lasting than those of the single-flowered kind.

Filipendula vulgaris flore-pleno

Queen-of-the-meadow (***F. ulmaria*** syn. *Spiraea ulmaria*) differs from the last in having basal leaves with up to five pairs of large leaflets ¾ inch long or longer, and the usually five petals of its flowers being up to ⅕ inch in length. This native of Europe is naturalized in parts of North America. Variable, and more or less hairy, it has branching or branchless, erect stems 3 to 7 feet tall. The terminal leaflets of its leaves are three- to five-times lobed. The ovate lateral leaflets are double-toothed and sometimes shallowly-lobed. Generally taller than broad, the panicles of blooms are up to 10 inches in length. The flowers have usually five petals, and stamens longer than the petals. Commonly this kind favors dampish soils. The foliage of *F. u. aureo-variegata* is strikingly variegated with golden yellow.

Filipendula ulmaria

Filipendula ulmaria aureo-variegata

Queen-of-the-prairie (***F. rubra*** syns. *Spiraea lobata, S. palmata*) is native of moist soils from New York to Minnesota, North Carolina, and Kentucky. From the last it differs in the terminal leaflet of its leaves being seven- to nine-lobed, and the lateral ones, of which there are two to five pairs, being deeply-three- to five-lobed instead of only coarsely-toothed or at most very shallowly-lobed. Queen-of-the-prairie has erect stems 3 to 7 feet tall, and hairs only on the veins on the undersides of the leaves.

Rare in gardens, ***F. palmata*** (syn. *F. digitata*), of Siberia and Kamchatka, is similar to *F. rubra,* but does not exceed 3 feet in height. Its pale pink flowers fade to white as they age. The plant cultivated as *F. palmata* is usually *F. purpurea.*

Japanese ***F. multijuga*** is nearly hairless and from 1 foot to 3 feet tall. Its leaves have a few pairs of ovate, toothed and sometimes lobed, lateral leaflets, or these may be absent from the upper leaves, and a much larger, nearly round, five- to seven-cleft terminal leaflet 2 to 4 inches in diameter. The usually pink flowers are in terminal clusters.

The most striking ornamental, ***F. purpurea,*** introduced to European and American gardens from Japan, is thought by Japanese botanists to be possibly a hybrid between *F. multijuga* and *F. auriculata,* both natives of their country. Be that as it may, it is very fine. From 1 foot to 4 feet tall, it has leaves with a large five- to seven-lobed, double-toothed, pointed terminal leaflet and few or no lateral leaflets. The rich pink or carmine flowers are in large, red-stalked clusters. In *F. p. purpurascens* the foliage is purplish, *F. p. alba* has paler green foliage and white flowers.

Tallest of filipendulas is ***F. camtschatica*** (syn. *F. camtschatica*) which occasionally attains heights of 10 feet, and commonly 5 to 8 feet. Its leaves have a large three- to five-lobed, toothed, heart-shaped terminal leaflet and are either without lateral leaflets or have a few very small ones. They are hairy or not. The clusters of white or pink-tinged flowers are very large and have hairy stalks. This native of Japan, Manchuria, and Kamchatka, too tall for many garden locations, can be used effectively in naturalistic areas.

Garden and Landscape Uses. Filipendulas are excellent plants for flower borders and informal landscapes and show to advantage at watersides. In addition, their flowers are useful for cutting. They are easy to grow where the soil is fairly moist, but not waterlogged. With the exception of dropwort, they do not flourish in dryish earth. All prosper in full sun and, except dropwort, will stand part-shade. Dropwort thrives in limestone soils.

Cultivation. No special attentions are needed. Annual fertilizing in spring and lifting, dividing, and replanting when the plants become so crowded that flowering deteriorates are routine. Propagation is by division in spring or early fall, and by seed. In cold climates with little snow cover a winter mulch of branches of evergreens or other loose material is helpful.

Filipendula purpurea

FILMY FERNS. The ferns known by this name constitute the filmy fern family HYMENOPHYLLACEAE. Conservatively considered to consist of two genera, *Hymenophyllum* and *Trichomanes,* by most modern authorities the family is divided into more than thirty genera and up to 400 species. One species, *Trichomanes boschianum,* inhabits North America, a few Europe. The vast majority are natives of the humid tropics and New Zealand. Mostly they grow on wet, shaded cliffs, rocks, and trees, always where the atmosphere is very moist.

Filmy ferns generally have slender, branched rhizomes and extremely thin, frequently much-divided fronds (leaves) consisting, except for the veins (not always present), of tissue only one cell thick. Rarely the leaves are undivided and often shredlike, with each consisting of a narrow band of tissue on either side of a vein. The clusters of spore capsules are at the leaf margins. They are raised on little stalks and are enclosed in cup-shaped indusiums (covers).

Rarely cultivated except in botanical collections, filmy ferns require very special environments. Most do best under fairly cool conditions, in a quite heavily shaded greenhouse where the temperature ranges from 40 to 60°F. Inside the greenhouse they should be accommodated in closed terrarium-type cases. Some growers advocate never watering filmy ferns overhead. They rely upon frequently wetting the moss, stones, and other strata on which they grow with water from the spout of a small can. But the very considerable collection at the Royal Botanic Gardens, Kew, England remained in fine condition for many years when misted overhead several times daily with a fine spray.

These ferns succeed best attached to pieces of tree fern trunk or, some kinds, planted in a mixture of leaf mold, peat moss, charcoal chips, and a little loamy soil set on rock ledges or between stones or largish pieces of rock. Often some difficulty is encountered in persuading newly planted specimens to take hold, but once established in a suitable environment they give little trouble to the gardener.

FINGER-LIME, AUSTRALIAN is *Microcitrus australasica.*

FINGERNAIL PLANT is *Neoregelia spectabilis.*

FINOCCHIO is *Foeniculum vulgare dulce.*

FIORIN is *Agrostis gigantea.*

FIR. This is the common name of all members of the genus *Abies.* The name China-fir is applied to *Cunninghamia.* The Douglas-fir is *Pseudotsuga menziesii.* Joint-fir is the colloquial name of *Ephedra,* summer-fir of *Artemisia sacrorum viridis.*

FIRE. This is used as part of the common names of these plants: fire-fern (*Oxalis hedysaroides rubra*), fire-lily (*Cyrtanthus* and *Zephyranthes tubiflora*), fire-on-the-mountain or Mexican fire plant (*Euphorbia heterophylla*), and fire pink (*Silene virginica*).

FIRE BLIGHT. This destructive bacterial disease infects many members of the family ROSACEAE and is especially serious on pears, apples, and quinces. Among ornamentals it attacks are cotoneasters, crab apples, mountain-ashes, and flowering quinces. Caused by *Erwinia amylovora,* fire blight invades the host plants through the flowers, the bacteria being wind-blown or carried by insects from cankers on the branches. First evidence of infection is a blackened blighting of the blooms followed by the destruction of the young fruits and death of foliage on infected branches. The withered leaves hang on giving the effect of whole branches or portions of branches having been scorched by fire. Cankers form on branches, mostly on those over ½ inch in diameter. The most practical, but by no means completely satisfactory control is had by pruning out all cankered branches and excising with a knife or chisel all cankers from the main trunk. Do this in winter, thus removing sources of infection before the flowers appear. Make the pruning cuts at least 4 inches lower on the branch than the canker and cut away apparently sound tissues for at least 1 inch around cankers on the trunks. Cover cut surfaces with tree-wound paint. If blighted branches are removed during the growing season make the cuts 6 inches to 1 foot below the visibly blighted parts and after each cut and before the next is made sterilize the tool with denatured alcohol or formaldehyde. Fire blight is most likely to attack plants that have made fast, succulent growth. Avoid stimulating such development by the excessive use of nitrogen fertilizers. For further information consult your Cooperative Extension Agent.

FIRECRACKER PLANT is *Dichelostemma ida-maia.*

FIRECRACKER VINE is *Manettia inflata.*

FIRETHORN. See Pyracantha.

FIREWEED is *Epilobium angustifolium.*

FIREWHEEL TREE is *Stenocarpus sinuatus.*

FIRMIANA (Firmiàn-a)—Chinese Parasol Tree or Phoenix Tree or Japanese Varnish Tree. Only one of the fifteen East African, Indomalasian, and Australian species of

Fittonia verschaffeltii

Firmiana, of the sterculia family STERCULIACEAE, is commonly known in American horticulture. The name commemorates Karl Josef von Firmian, a Governor of Lombardy, who died in 1782.

Firmianas are nonhardy trees and shrubs. They have alternate, large, usually lobed, palmately-veined leaves. The flowers have sepals, no petals, and ten or more stamens. The fruits are membranous. They separate into leaflike carpels.

Deciduous and with smooth, green bark, the Chinese parasol tree, phoenix tree, or Japanese varnish tree (***F. simplex*** syn. *Sterculia platanifolia*) has palmately-three- to five-lobed leaves that may be 1 foot or more across and look something like those of plane trees (*Platanus*). The lobes are sharp-pointed. The yellowish-green flowers, in large panicles, are of little ornamental value. They are succeeded by wrinkled fruits that separate into four leaflike lobes with two or three pea-sized seeds on their inner edges. Before the fruits mature they are filled with a brownish-black liquid that is released when the pods open. This species is a native of China, Japan, and Taiwan.

Garden and Landscape Uses and Cultivation. The Chinese parasol tree is planted for shade and ornament in the south and is hardy as far north as Washington, D.C. It grows satisfactorily in any fairly good soil and needs no special care. It is easily increased by seed. Self-sown seedlings usually spring up plentifully beneath and near fruiting specimens. Seedlings not needed for transplanting should be removed during their first year. They are deep-rooted, and if left until later may have to be dug out rather than pulled out or hoed. In the Orient *F. simplex* is employed as a street tree and a fiber is obtained from its bark. Its wood is used for furniture.

FISH-GRASS is *Cabomba caroliniana.*

FISHBONE THISTLE is *Cirsium diacantha.*

FISHMEAL. This is a complete organic fertilizer containing from 8 to 10 percent nitrogen, 4.5 to 9 percent phosphoric acid, and 2 to 4 percent potash. It is used in spring and summer at a rate of 2 to 3 ounces to 10 square feet.

FITTONIA (Fit-tònia). Natives of tropical Peru, the two species of *Fittonia* belong in the acanthus family ACANTHACEAE. Their name commemorates Elizabeth and Sarah Mary Fitton, who died early in the nineteenth century. They were the authors of *Conversations on Botany,* published in 1817.

Fittonias are trailing, creeping, or low evergreen herbaceous perennials with

Fittonia verschaffeltii argyoneura

hairy stems and broad, lobeless, toothless leaves decorated with colored veins. Their small, yellowish to reddish flowers, in conspicuously-bracted, erect terminal spikes have little display value. They have deeply-five-lobed calyxes, slender, tubular, two-lipped corollas, two fertile stamens, and a slender style terminated with a two-notched stigma. The fruits are capsules.

The most familiar kinds are ***F. verschaffeltii*** and its variety *F. v. argyoneura*. These are trailers about 6 inches tall, with broadly-ovate leaves up to about 4 inches long. The dullish green leaves of the species are enlivened by a tracing of carmine veins. In *F. v. argyoneura* the leaves are brighter green and the veins white. Another variety, *F. v. pearcei*, is distinguished by its bright green leaves, somewhat glaucous on their undersides, with carmine veins. From the above, ***F. gigantea*** differs in being more erect and up to 1½ feet tall. The elliptic leaves are rather larger and more pointed than those of *F. verschaffeltii*. They are lustrous green veined with carmine.

Garden and Landscape Uses. Only in the humid tropics and in tropical greenhouses and terrariums can these plants be cultivated satisfactorily. They are lovers of high temperatures, high humidity, subdued light, and moist, rich, organic earth. They are good groundcovers, useful for planting under greenhouse benches, in ground beds in conservatories, and in terrariums. They make good cover for the soil of tubs in which palms and other tropical plants are growing. Hanging baskets planted with *F. verschaffeltii* and its varieties are attractive.

Cultivation. In a minimum temperature of 60°F and the environment described above fittonias grow with greatest ease and are very readily propagated by cuttings. The latter are best taken in spring. Pans (shallow pots) afford the best accommodation for fittonias grown in containers. Several plants, spaced 2 to 3 inches apart, should be planted in each. When well established regular applications of dilute liquid fertilizer are beneficial.

Pests. The most common pests are scales, mealybugs, and slugs.

FITZROYA (Fitz-ròya). One conifer rare in cultivation is the only member of *Fitzroya*,

Fittonia verschaffeltii pearcei

Fittonia gigantea

of the cypress family CUPRESSACEAE. A native of the Andes of southern Chile and Argentina, *F. cupressoides* (syn. *F. patagonica*) is named after an officer of the British Navy, Captain R. Fitzroy, who died in 1865. The critical distinguishing feature of the genus is that its cones are of three whorls (tiers or rows) of three scales each, of which the lower whorl is sterile, the middle sterile or fertile, and if fertile with one seed to each scale, and the upper fertile and with two to six seeds to each scale.

As known in cultivation, ***F. cupressoides*** does not ordinarily exceed 50 feet in height and is often no more than a tall shrub, but in its native habitat it is much more impressive, commonly from 80 to 150 feet tall. The maximum recorded size of this tree is 240 feet, with a trunk diameter at breast height of 15 feet, surely one of the giants among conifers. Fitzroyas grow slowly and are estimated to live up to a maximum of 1,000 years. Young trees are broadly-pyramidal and densely-branched, but in their native habitats, with age, their lower limbs die and only the extreme tops of the trees are green. So many of the branches may be dead that when viewed from a distance a forest of old trees appears to consist of mostly defunct trees.

Fitzroyas have reddish, furrowed, peeling bark and slender, drooping branchlets furnished with scalelike leaves about ⅛ inch long that are in alternating whorls of threes. The top sides of the leaves are impressed with two whitish bands and there is a white band on each side of the midrib on the undersides of the leaves. Individual trees may have flowers of one or both sexes. The male catkins are solitary in the upper leaf axils, the females on short leafy shoots. The fruiting cones are ¼ inch or a little more in diameter. They mature the first season.

Often inhabiting swampy soils, *Fitzroya* is the only native Chilean tree that forms extensive forests of one species. It is exploited for its very good lumber, the best produced in Chile. Both the tree and its wood are called *alerce,* the Spanish name for larch, but neither has much similarity to the true larch (*Larix*) of the Northern Hemisphere.

Garden and Landscape Uses and Cultivation. This unusual tree is only likely to be grown in arboretums and other special collections, where mild winters are the rule. Little is known of its behavior in North America, but it is most likely best adapted for parts of the Pacific Coast. So far as is known it has no special soil requirements. It may be increased by seeds sown in sandy peaty soil and by cuttings taken in summer and planted under mist or in a propagating case in a humid greenhouse or cold frame. For general information see Conifers.

FIVE. As parts of colloquial names the word five appears in these: five fingers (*Potentilla*), five leaves (*Isotria verticillata*), and five spot (*Nemophila maculata*).

FL. PL. This is a commonly used abbreviation for the Latin *flore-pleno* and *flore-plena* employed in plant names. It means double-flowered.

FLACOURTIA (Flacoúrt-ia) — Governor's-Plum or Ramontchi or Batoko-Plum. The colloquial name governor's-plum applied to the best-known species of this genus alludes to Etienne de Flacourt, Governor of Madagascar and Director of the French East India Company, who died in 1660, and after whom the genus was named. A member of the flacourtia family FLACOURTIACEAE, the genus consists of fifteen species, natives of the warmer parts of Asia and Africa and associated islands. They are shrubs or small trees with short-stalked, pinnately-veined, usually toothed leaves, and small unisexual or bisexual flowers commonly in axillary clusters or racemes. The flowers are without petals and have four or five scalelike sepals and numerous stamens. Female blooms frequently have non-functional stamens called staminodes. The berry-like fruits contain several seeds each surrounded by a hard shell.

The governor's-plum, ramontchi, or batoko-plum (***Flacourtia indica***), a native of southern Asia and Madagascar, is most familiar in cultivation. In warm countries it is grown for its fruits and for ornament. It thrives in essentially frost-free parts of Florida and in some places has become naturalized. A hairless shrub or bushy tree up to 25 feet in height, this species may be spiny or spineless. It has pointed, ovate to elliptic, crenately-toothed leaves 2 to 3 inches long, and yellowish flowers, usually with only one sex on a plant. The edible, nearly spherical, plum-red to purplish-black fruits contain sweet or acid jelly-like pulp and are ½ to 1 inch in diameter. Each has several flat seeds. The fruits are usually eaten uncooked. They are astringent before they are fully ripe.

Other cultivated species of *Flacourtia* are the paniala (***F. cataphracta***) and the rukam (***F. rukam***). The former is native of India and Malaya, the latter of Malaysia. Both have edible fruits ¾ to 1 inch long and more acid than those of the governor's-plum. This is especially true of the rukam, the fruits of which even when ripe are slightly astringent. Both species are similar in appearance to the governor's-plum, but are of more upright growth, more spiny, and have narrower leaves. The fruits of the rukam are dark purple, those of the paniala dark red to purplish. Both are used for jellies.

Garden and Landscape Uses and Cultivation. The governor's-plum and others discussed above are grown in tropical and near-tropical climates for their fruits and

Flacourtia indica

Flag: (a) African-violet in normally moist soil

(b) African-violet flagging (wilting) from lack of adequate moisture

also as ornamentals. They make effective hedges. When grown for their fruits an occasional male plant should be set among or near the females. Some essentially female plants of the governor's-plum set fruits without a male plant being near, but more abundant crops are assured if one is. No difficulty is usually encountered in growing flacourtias. They succeed in limestone and non-limestone soils that are adequately drained, but not excessively dry, and in full sun. They are propagated by seeds, cuttings of mature shoots, and by grafting selected forms onto seedling understocks. Pruning consists of judicious thinning to prevent the plants becoming too dense or, in the case of hedges, shearing periodically to increase their density. When grown for their fruits irrigation may be needed when the fruits are developing to prevent the soil from drying excessively.

FLACOURTIACEAE—Flacourtia Family. Inhabiting the tropics and subtropics of the Old and the New Worlds, the flacourtia family comprises ninety-three genera of dicotyledons containing approximately 1,000 species, a few of its kinds produce edible fruits, a few useful lumber, and a few oils employed medicinally.

Members of this group are trees, shrubs, or rarely vines, with generally alternate, leathery, evergreen leaves displayed in two ranks. The flowers, bisexual or unisexual, are symmetrical and solitary or are variously clustered or in racemes. They have four to fifteen each sepals and petals or petals may be lacking. There are many stamens and one or as many styles as there are carpels. The fruits are usually berries or capsules. Genera most familiar in cultivation are *Azara, Berberidopsis, Carrierea, Dovyalis, Flacourtia, Idesia, Olmediella, Oncoba, Poliothyrisis,* and *Xylosma.*

FLAG (AS A VERB). When leaves or other soft parts of plants wilt or droop they are said to flag. Flagging results from water being lost faster than it can be replaced. This may be because the soil is too dry or because the roots are unable to absorb or the stems transport enough water to keep the tissues turgid. Damage to roots, injuries to stems, and certain disease organisms that clog the conducting vessels are possible reasons for flagging, as are too-low soil temperatures and exposure to wind or dry air that induces loss of water at rates greater than the capacity of the roots to replace.

FLAG. This is a common name for various irises. It also forms part of the common names of some other plants including these: African spiral flag (*Costus lucanusianus*), crimson flag (*Schizostylis*), and sweet flag (*Acorus*).

FLAMBOYANT is *Delonix regia.* The yellow-flamboyant is *Peltophorum pterocarpum.*

FLAME. This word is incorporated in the common names of these plants: bush-flame-pea or flame-mock-orange (*Chorizema*), flame flower (*Tropaeolum speciosum*), flame-of-the-forest (*Butea monosperma*), flame-of-the-woods (*Ixora*), flame tree (*Brachychiton acerifolium* and *Delonix regia,*) and flame-violet (*Episcia*).

FLAMING-POPPY is *Stylomecon heterophylla.*

FLAMINGO FLOWER. See Anthurium.

FLANNEL BUSH is *Fremontodendron californicum.*

FLANNEL FLOWER is *Actinotus helianthi.*

FLATS. These are shallow boxes or trays in which seeds are sown, cuttings planted, and young and small plants raised. Where more than a few plants are needed they are often more convenient than pots or pans, being simpler to make ready, often requiring less detailed care in watering and other attentions, and making it easier to move numbers of plants from place to place.

Standard flats were for long made only of wood, preferably a decay-resistant one such as cypress or redwood. Such flats are still much used and are excellent. Also available are plastic and metal ones. Sizes vary. Professionals frequently prefer those 1½ feet long, 1 foot wide, and 3 inches deep. Amateurs may find smaller sizes easier to handle and more suitable for their needs. For special purposes, as for wintering chrysanthemum roots in a cold frame, a depth of 4 or 5 inches is advantageous. For seed sowing 2 to 2½ inches is usually adequate. For one-season use flats made

Wooden flats

from wooden boxes and crates used for shipping are adequate. Drainage is provided by holes ½ to ¾ inch in diameter bored in the bottoms of the flats or by having the latter of two or three pieces set lengthwise with narrow slits between them.

FLAX. See Linum. New-Zealand-Flax is *Phormium tenax*, yellow-flax *Reinwardtia indica*, flax-lily *Dianella*.

FLEABANE. See Erigeron. Marsh-fleabane is *Pluchea*.

FLEECE VINE is *Polygonum aubertii*. Feather fleece is *Stenanthium gramineum*, golden fleece *Dyssodia tenuiloba*.

FLEMINGIA. See Maughania.

FLEUR D'AMOUR is *Tabernaemontana divaricata flore-pleno*.

FLEUR-DE-LIS. See Iris.

FLIES. True flies are insects of the order *Diptera* characterized by having only one pair of wings or rarely none and very large eyes. In place of a second pair of wings they have two curious knob-tipped, stalk-like organs called halteres or balancers that vibrate while the creatures are in flight. Belonging are such familiar sorts as crane flies, houseflies, bluebottle flies, and horseflies, as well as fruit flies, no-see-ums, mosquitoes, and midges. But many insects with colloquial names that include the words fly or flies are not true flies, but belong entomologically elsewhere. Such sorts include dragonflies, fireflies, harvest flies, and mayflies.

Adult flies feed by piercing and sucking or lapping. As larvae, called maggots, they may live on or within tissues upon which they feed. They reproduce from eggs and pupate between the larval and adult stages.

A number of sorts are pests of plants, among them the carrot rust fly, cattleya fly, various fruit flies, mushroom flies or fungus gnats, the lesser bulb fly, the narcissus bulb fly, and some sorts of which the larvae are leaf miners.

As with all insect pests preventive and control measures must be attuned to the particular sort, the kind of plant, and often in some measure to geographical location and time of the year. About these matters consult Cooperative Extension Agents, State Agricultural Experiment Stations, or other authoritative sources.

FLOATING FERN. See Ceratopteris.

FLOATING HEART. See Nymphoides.

FLOATING-MOSS is *Salvinia rotundiflora*.

FLOCCULATE. An important objective in the management of clay soils for horticultural and agricultural purposes is to maintain the colloidal fraction of the clay in a flocculated state. Then, the ultramicroscopic particles cling together in larger aggregates, forming crumbs that behave much like the particles of a coarser-textured soil. A flocculated clay soil allows free passage of air and water. Flocculation is promoted by certain cultural practices, such as liming, and by exposure to alternate freezing and thawing. Deflocculation, which causes the clay to become pasty and impervious to water and air, results from walking on, using heavy equipment on, and working clay soils while they are wet.

FLOERKEA. See Limnanthes.

FLOPPERS is *Kalanchoe pinnata*.

FLOR DE MUERTO is *Lisianthus nigrescens*.

FLORA'S PAINT BRUSH. See Emilia.

FLORENCE FENNEL. See Foeniculum.

FLORETS. Tiny flowers that compose the flower heads of plants of the daisy family COMPOSITAE, the spikes or panicles of grasses, and crowded flower heads of some other plants are called florets.

FLORICULTURE. As generally understood this is a specialized branch of horticulture concerned with the study and practice of growing cut flowers and pot plants in greenhouses and outdoors and marketing them. Although in the fullest meaning of the term floriculture embraces all aspects of growing flowers, the management of flower gardens for ornament is usually considered separately as a branch of ornamental horticulture.

FLORIDA. The word Florida forms part of the names of these plants: Florida fiddle wood (*Citharexylum fruticosum*), Florida tree fern (*Ctenitis sloanii*), and Florida velvet-bean (*Mucuna deeringiana*).

FLORISTS' FLOWERS. Flowers popular with florists for growing for cutting or as pot plants, each kind available in a considerable number of horticultural varieties derived from one or few species, are sometimes known as florists' flowers. Most have been bred to a high state of perfection for the needs they are to serve, which commonly includes shipping quality as well as display value. Examples of florists' flowers are African-violets, camellias, carnations, cattleyas, chrysanthemums, cyclamens, dahlias, fuchsias, gladioluses, hyacinths, narcissuses, pelargoniums, roses, and tulips.

FLOSS-SILK TREE is *Chorisia speciosa*.

FLOWER. In the most generally accepted sense of the word, flowers are peculiar to the great group of seed-bearing plants botanists designate as angiosperms. The very primitive equivalents in the seed plants called gymnosperms are scarcely recognizable as flowers and are not considered here. A flower consists of a short shoot, the leaves of which are modified into usually four sets of organs called sepals, petals, stamens, and pistils, respectively, and arranged in that order from the lowermost (or outermost) to the uppermost (or innermost). One or more of these sets may be wanting. Typically green or greenish, but sometimes brightly colored or white and petal-like, the sepals collectively form the calyx. They may be separate or more or less united. Generally brightly colored or white, but sometimes greenish, the petals, which may be separate or more or less united, collectively form the corolla. Calyx and corolla collectively are called the perianth, a term especially favored when, as for example in lilies and tulips, the sepals and petals are very similar. Then each member of the perianth may be called a tepal. The stamens, each of which usually consists of a stalk (filament) terminated by an anther, which produces pollen, although the former may be lacking, are the male parts of the flower. Nonfunctional stamens modified in various ways and called staminodes occur in some flowers. The pistil, the female organ of the flower, consists of one or more carpels. Most usually it comprises an ovary (which contains the ovules or eggs, one or more styles or stalk-like parts, and one or more stigmas that receive the pollen). Sometimes the styles are lacking.

Most flowers have functional male and female parts, but the latter are not always responsive to pollen from the former. If they are not, the plants that bear them are said to be self-sterile. Most flowering plants are self-fertile. Some flowers are unisexual, having only male or female organs functional. Then the sexes may be on different plants [as in the tree-of-heaven (*Ailanthus*) and some hollies (*Ilex*)] or on the same plant (as in begonias), or unisexual and bisexual flowers may be on the same plant. Some structures often referred to as flowers, the inflorescences of almost all members of the daisy family (COMPOSITAE) and of the arum family (ARACEAE), and those of certain members of the spurge family (EUPHORBIACEAE) and of the dogwood family (CORNACEAE) for example, are not single flowers but aggregates or clusters of many small flowers (in some cases called florets), sometimes with showy and colorful attendant bracts, spathes, or other organs.

FLOWER. This word is used as part of the common names of these plants: Barbados flower fence (*Caesalpinia pulcherrima*), cassia flower tree (*Cinnamomum loureiri*), flower-of-an-hour (*Hibiscus trionum*), flower of Jove (*Lychnis flos-jovis*), monkey flower

Fuchsia triphylla

Fuchsia hybrida variety, trained as a standard

Flower beds: Formal beds with tulips, spring

Flower bed: Formal bed, white- and red-flowered wax begonias edged with dwarf feverfew, summer

Flower bed: Island bed with chrysanthemums and dahlias, fall

Flower border: A narrow border of tulips, spring

Flower border: Informal border with light blue phlox, yellow basket-of-gold, blue grape-hyacinths, and red tulips, spring

Flower border: A border in California with calla-lilies and bird-of-paradise, edged with pansies, spring

Flower border: Mixed border with pink phlox and *Echinacea*, yellow rudbeckias and *Heliopsis*, white *Veronica*, and purple-leaved *Perilla*, summer

Flower border: Mixed border with white phlox, yellow, pink, and red zinnias, and red salvias, summer

A selection of flower beds: (a) Daffodils underplanted with grape-hyacinths

(b) Hyacinths, blue and white

(c) Tulips

(d) Pansies edged with *Veronica incana*

(e) Snapdragons

(f) Wax begonias and sweet alyssum

tree (*Phyllocarpus septentrionalis*), and peacock flower fence (*Adenanthera pavonina*).

FLOWER BEDS and BORDERS. Beds or borders of bloom are important features of most gardens. If trees, shrubs, and hedges are regarded as verdant walls and furniture, and lawns and groundcovers carpets, then patches of annuals, biennials, and herbaceous perennials serve as pictures in a room. They add brightness, pinpoint attention. Employ them as you would pictures, with decent restraint. Locate them with care. The "riots of color" that some advocate gardens should be can become tiresome. The eye needs places to rest as well as places that titillate. It is the contrast of the quiet of stretches of lawn and other greenery and of fences and walls of muted or neutral colors that give flowers their finest values in gardens.

In contrast to naturalistic landscaping in which flowering plants are grouped or scattered as they might be in the wild, their assemblage in beds and borders establishes formality or at least semiformality. The influence of the designer is apparent. And so placement, dimensions, forms or shapes, and the relationship of the beds and borders to other features, are of great importance. Give these matters proper thought before proceeding.

(g) A carpet bed of succulents

Beds and borders afford the opportunity to grow and display a much greater variety of flowers than would be meet in a naturalistic garden. Frankly horticultural achievements, such as the most magnificent African marigolds, day-lilies, delphiniums, lilies, peonies, phloxes, salvias, snapdragons, zinnias, and various bulb plants, can be set in juxtaposition with equanimity, without thought as to whether they "look natural," consideration being given only to such matters as relative heights and pleasing color combinations. Assemblage into beds and borders also simplifies cultural procedures.

There are two ways of managing flower beds and borders. The bedding out plan involves complete replanting, usually twice each year, with plants that are at or almost at blooming size when set out. The objective is continuous massed displays of bloom from spring to fall. For more about this see Bedding and Bedding Plants.

The other scheme calls for growing hardy plants, perennials or annuals or both, *in situ* often with the addition of patches of spring or summer bedding plants and tender bulb plants, such as dahlias and gladioluses, to beef up the display. It is this manner of planting that is our concern here. And let it be said that there is no clear difference between a bed and a border except that the latter is generally several times longer than wide and, as the term suggests, usually margins a boundary, path, or other feature.

It is of course quite practicable to have beds or borders of one kind of plant, asters, chrysanthemums, delphiniums, irises, or lilies, for example, but those containing a variety of sorts set in neighborly groups or patches are more common. If all the plants used are kinds that persist from year to year, you have what is correctly termed a perennial border. If they all normally are sown, bloom, and die in one season the result is an annual border. If, as more often happens, annuals, biennials, and hardy and sometimes tender perennials are used together, you have a mixed border. This is the most popular.

Annuals are especially appropriate for rented properties where tenancy is likely to be short, or if you do not wish to invest the considerable effort and probably not small cost of establishing perennials. You may select annuals that can be sown *in situ*, and more respond to this method of cultivation than is generally supposed, or

A flourishing border of annuals

you may set out young plants started indoors, or employ a combination of both methods.

Borders devoted entirely to annuals provide brilliant displays. They are not colorful early in the year, however, and some kinds, such as baby's breath, candytuft, calliopsis, cornflowers, gaillardias, larkspurs, leptosynes, linarias, love-in-a-mist, mignonette, poppies, rudbeckias, salpiglossis, strawflowers, and chrysanthemums, bloom for a comparatively short season. When these cease flowering, pull them out and set in their places plants of balsams, browallias, calendulas, candytuft, globe amaranths, marigolds (both African and French), flowering tobacco, phloxes, statices, sunflowers, and zinnias, raised in the reserve or nursery garden from late sowings. If you do not want to go to this trouble then use only such annuals that have a long blooming season as ageratums, browallias, calendulas, carnations, celosias, cosmos, four o'clocks, globe amaranths, ice plants, lobelias, marigolds, Mexican tulip-poppy, nasturtiums, flowering tobacco, petunias, portulacas, salvias, sanvitalia, snapdragons, snow-on-the-mountain, spider flower, star-of-Texas, sweet-alyssum, sunflowers, tithonia, torenia, verbenas, vincas, wax begonias, and zinnias.

Plant annuals in spring or early summer. Some can be sown directly where the plants are to bloom, others must be, or may be, raised indoors and later transplanted. For more information see Annuals.

Perennial plantings of the elaborate type, once greatly promoted by purist herbaceous or perennial border enthusiasts, are practically of the past. Actually they rarely worked out well in the United States.

The concept of these painstakingly planned and executed creations originated in Great Britain and was developed chiefly in the twentieth century as part of the revolt against the overelaborate temporary bedding, often based largely on subtropical plants, that dominated much late nineteenth-century flower gardening.

As originally conceived, these herbaceous or perennial borders were to free gardeners from dependence upon greenhouses and make possible magnificent floral displays with minimum work. The first they certainly did, for the plants employed are all hardy, and in favorable climates with careful planning truly grand displays can be had. But to make and maintain a successful all-perennial border even in Great Britain calls for much work and the examples as elaborate as those of yesteryear are now rarely seen. A few are kept up, like one of the best known at Hampton Court near London, by considerable "cheating," by planting as replacements during the summer successions of annuals and other plants brought into full bloom in greenhouses.

American gardeners caught from the British something of the fervor for perennial borders, and in the days of the great estates quite elaborate examples were designed and installed by landscape architects and talented gardeners. But even before World War II change came. The labor of maintaining such developments was costly but more importantly beds of nothing but perennials proved less satisfactory where summers are predominantly hot, dry, and sunny than under the cooler, moister, less brilliant skies of the British Isles. Because plants pass much more quickly out of bloom under such circumstances the display at any one time is very much reduced, and it is virtually impossible in most parts of America to have anything approaching a lavish show throughout the season by relying upon perennials alone.

Mixed beds and borders in which hardy perennials form the backbone of the planting and are supplemented with groups of annuals, biennials, and summer bulb plants, such as dahlias, galtonias, gladioluses, montbretias, and tuberoses, and perhaps with some nonhardy perennials, such as fuchsias, geraniums, and lantanas, are most practicable for most American gardens.

They are relatively inexpensive, easy to care for, and are adapted to a wide variety of sites. They fit gardens large or small and provide a variety of bloom from spring to fall.

A well-located border of perennials in an English garden

A selection of flower borders: (a) With tulips, blue *Phlox divaricata,* and pansies

(b) Doronicums and tulips backed by a hedge of yew

(c) Foxgloves and sweet williams behind a low boxwood hedge

(d) An informal border of perennials and annuals

(e) Summer phloxes, delphiniums, and clematises

(f) Chrysanthemums, snapdragons, and summer phloxes, edged with sweet alyssum

(g) Double borders flanking a path

Such plantings are not new. Until greenhouses became practicable and common, plants of the kinds used in them were the only ones available. Flowers in gardens of colonial America and contemporary Europe were chiefly these sorts. Not until well along in the nineteenth century did lavish displays of nonhardy bedding plants typical of the late Victorian age and early twentieth century become popular or possible.

Sites for flower beds and borders should be reasonably level, and sunny for at least one-half of each day. Longer exposure is advantageous. Unless you are content to grow only bog and wetland plants there must be good subsurface drainage. If trees, shrubs, or hedges form backgrounds their roots invading the border may present problems, especially those of such hungry kinds as privets and hornbeams. To circumvent this, if space permits, it is a good plan to have an unplanted strip 3 to 4 feet wide of grass or gravel or other paving between the background and the back of the border. This permits easy access to give needed attention to both shrubs or hedges and the plants at the rear of the border.

A reasonable depth of good soil is necessary. Less than 1 foot of fertile topsoil is limiting. More is desirable. Prepare it thoroughly, particularly for perennials. Once planted, you will have no opportunity to improve the soil beneath them for several years. Annual borders are, of course, made anew each year.

Locate flower beds and borders so that they are important in the garden plan. Fit them pleasingly into the landscape. Place them where they will be seen from an important window, sitting area, or main path, or incorporate them as part of a formal or semiformal garden.

An island bed of mixed perennials

Shapes, sizes, and the relationship of flower beds and borders to each other and other garden features must be accommodated to the site. What are called island beds of perennials have gained considerable popularity in Great Britain since World War II and have sometimes been adopted and adapted by American gardeners. These are free-form beds cut in lawns and viewable from all sides. More traditionally, flower beds, other than those accommodating temporary bedding plants, and nearly always borders, have been located where walls, fences, hedges or shrubs, serve them as agreeable backgrounds.

Where space permits, mixed and perennial borders and beds should be 6 to 12 feet wide. A reasonable relationship between length and breadth is necessary. Usually a short, fairly wide area is preferable to one that is long and excessively narrow. The narrower the border the more difficult it is to have bold groupings of larger and more important plants. Such emphatic points of interest are the very soul of good border arrangement. Annual borders need be not over 4 to 6 feet, in appropriate places not over a foot or two, wide. Long, very narrow examples planted with the same kind of plants throughout are called ribbon borders.

Although a width of 6 feet or more is desirable if large plants are to be accommodated, it is by no means impossible to design narrower borders that are pretty and effective. If you undertake one of these, confine yourself to plants of small or moderate growth. You should then be able to attain commendable results in a strip 3 to 4 feet wide.

Borders may be straight or curved. They may be uniform or variable in width. Logical endings are important. Do not let borders with backgrounds just finish—straggle off into nothingness. Walls, fences, hedges, or groups of shrubs or evergreens can be used effectively to define the extremities. The ends of the borders need to be held in, as it were.

In planting, repeat the same kinds of plants at intervals along the border. Such repetition gives character and strength. The repetition need not be of precisely the same varieties. In July, different sorts of day-lilies will do. Phloxes in groups of various colors are effective in August. Earlier in the season bold clumps of peonies of various hues spaced along the length of the border are good. At the front, clumps of lower kinds, such as evergreen candytuft, basket-of-gold, flax, and nepeta repeated at intervals give a good effect. Use enough of these basic plants to assure color and interest throughout the summer. Fill in with others to give diversity.

Set most plants in groups of three, five, or more of the same kind. Too many single specimens give a spotty effect that lacks restfulness. Let the size of the individual groups vary. Use, however, a few individual plants. Solitary specimens of peonies, perennial baby's breath, and other kinds of rounded outline are effective when used sparingly.

Do not plant all tall-growing kinds at the back of the border and all low ones at the front. A planting carefully graded in this way is uninteresting. Instead, let bays of lower-growing kinds recede to the rear and promontories of taller ones extend forward. Here and there let solitary plants or small groups tower somewhat above their neighbors, but avoid hiding low kinds behind tall groups.

Give attention to the color scheme. This is highly personal. So long as the result pleases you that is all that matters. Some like nothing but pastels, others appreciate brighter hues. Some like contrasts such as orange and deep blue planted together, others respond to subtle harmonies. Avoid such dreadful combinations as magenta and scarlet and strong pink and bright yellow. White flowers and gray foliage, such as that of silver king artemisia and dusty miller, make grand breaks between patches of color that otherwise would clash. Use them freely.

Leave ample spaces in your mixed border between the perennials to accommodate the summer plantings. Each group of annuals, biennials, summer bulbs, or tender perennials will usually need at least a square yard. You must not expect satisfaction if you jam them among crowded perennials. Spade and fertilize the spaces in which the summer plants are set shortly before planting.

Perennials may be planted in early fall or in early spring, hardy bulbs, such as narcissuses, tulips, hyacinths, grape-hyacinths, squills, snowdrops, crocuses, glories-of-the-snow, summer snowflakes, camassias, English and Spanish bluebells, and lilies, in fall.

Biennials are raised in the nursery or reserve garden. They are usually planted in their flowering places in spring. Here belong foxgloves, honesty, pansies, violas, wallflowers, canterbury bells, English daisies, forget-me-nots, mulleins, rose-campion, and, sweet williams.

Once planted, the care of perennials consists of, in colder climates, covering them lightly with salt hay, evergreen branches, dry leaves, or similar loose material after the ground is well frozen, of pushing back into the soil in spring any that have heaved by frost, of removing the winter covering gradually when spring growth begins, of fertilizing and shallowly forking over the surface soil in spring, and of keeping it shallowly cultivated or mulched throughout the summer. Also stake and tie to give support, water thoroughly every few days during dry periods, remove promptly all faded blooms, and spray or dust to control pests and diseases.

Replanting perennial beds and borders needs attention from time to time. Every third or fourth year is satisfactory. When the time arrives, dig up all the plants (except possibly specimens of gas plant and a few others known to resent disturbance) in early fall. Plant them temporarily

Stirring the soil with a short-handled cultivator

Cleaning up a border in fall

closely together in rows in another part of the garden and improve the soil by spading it deeply, by adding compost, manure, or other humus material, and by fertilizing and liming as deemed desirable.

Before the plants are reset divide those that are too large. Replant only strong, vigorous pieces. See that they do not dry before they are installed in the ground. The considerable work of remaking a perennial border may be spread by undertaking one-third or one-fourth of it every year rather than doing the whole every three or four years.

Replanting perennials in mixed beds and borders calls for less rigid adherence to a timed work schedule. Commonly it is done in more piecemeal fashion, the patches reserved for annuals and biennials being spaded and conditioned each year just before planting and the perennials being managed on a more or less individual basis, lifting, dividing, fixing the soil for, and replanting as need arises. Even in mixed borders the perennials may become so overgrown and the soil so depleted that it pays to occasionally dig them all up, condition the earth, and replant as is recommended for borders of all perennials. See also Annuals, Biennials, and Perennials.

FLOWER GARDEN. Areas devoted mainly to the cultivation of annuals, biennials, and perennials for decoration are called flower gardens. If their chief function is the production of blooms for cutting, they are known as cut flower gardens or cutting gardens. More often they are designed to enhance the landscape with colorful and sometimes fragrant blooms.

Flower gardens may be chiefly or entirely of one kind of plant, such as irises, peonies, or roses, or may include many sorts. They may be planted mostly with perennials or with seasonally changed bedding plants. For their best success most types need fairly sunny locations. Generally they should receive sun for one-half of each day; longer is better. The soil must be reasonably deep and fertile. If it is capable of producing good vegetables it will be satisfactory.

Except cutting gardens, which can often be conveniently combined with a vegetable garden, flower gardens are usually best located near the house where they can be enjoyed from advantageously placed windows or outdoor sitting areas. They may be of formal, semiformal, or informal design and may include a pool, fountain, or other water feature or appropriate seats, statuary, or other furnishings. Special type of flower gardens include bog gardens, water gardens, and rock gardens. For more information see Annuals, Bedding and Bedding Plants, Biennials, Cutting Garden, and Flower Beds and Borders. See also Bog Gardens, Planning Gardens and Home Grounds, Rock and Alpine Gardens, and Water Gardens.

FLOWER SHOWS. Flower shows owe their origins to natural and deep-seated desires craftsmen and artists have to pridefully display their skills and achievements and engage in friendly competition with others who practice their arts. Such exhibitions afford these satisfactions to gardeners, landscapers, nurserymen, flower arrangers, and others interested in horticultural pursuits and also bring much pleasure to the many who participate only to the extent of visiting these exhibitions. In addition, the shows serve educationally by conveniently displaying in one place rare, improved, and new kinds of plants, products, and techniques. To give emphasis to this education function it is important that all plants be labeled with their correct names. Often, the bigger shows especially, provide nurserymen, florists, and other professionals opportunity to display and advertise their products and services. They serve most excellently as get-togethers for all interested in growing and using plants.

Just when the first flower shows began is not precisely known. They were probably an outgrowth of the earliest horticultural societies, organizations of keenly interested amateurs and professionals who met to discuss horticultural matters and assemble and disseminate information about their subject. Certainly plants, fruits, vegetables, and flowers were being displayed fairly regularly before such groups both in America and Europe by the latter part of the eighteenth and early nineteenth centuries. Before the twentieth century they were being held regularly in many places including some where one would scarcely expect interest in gardening to run high. For example, in coal-mining regions of Lancashire, England, gooseberry shows that elicited fierce competition among miners and were often staged in local pubs, were popular in the late 1800's. The exhibitors engaged not only in growing and showing gooseberries, but also in breeding them. Some varieties still popular in Great Britain are of that origin. Other groups interested themselves in auriculas and other special plants.

In modern America flower shows are regular events in many communities. They range from small to large ones arranged by individual garden clubs, local horticultural societies, and similar organizations to such elaborate, extensive, expensive, and splendid ones as the Boston and the Philadelphia flower shows, to cite two of the most important in the east. But magnitude is not the only, or necessarily the most important, measure of value. Enthusiasm and interest on the part of exhibitors and the community and desirable competitiveness and good sportsmanship on the part of those who exhibit is often as intense in smaller as larger flower shows. And such events do encourage a certain sporting spirit. Although good losers may not be as plentiful as cheerful winners they constitute the majority of those who fail to win blue, red, or white ribbons or honorable mentions, and even the occasional poor loser is likely to be back the following year to try again.

Gardens displayed at New York flower shows: (a) Tulips

(b) Mixed spring bulbs

(c) Roses

(d) Geraniums and chrysanthemums

Cut chrysanthemums exhibited at a flower show

Pot plants exhibited at a flower show

The operative organization of a flower show, small or large, must be practical. The best success comes only as the result of the sum of numerous wise judgments based on knowledge of local conditions and population and from much good planning and hard work by many people. The organizational work of even the largest shows is done largely, that of the smaller shows entirely, by volunteers. For large shows it is usual and surely wise to engage the services of an experienced flower show manager who, in addition to directing the installation of the show, its operation, and final dismantling, can be of immense aid in counseling the flower show committees, in helping to secure exhibits, and in other ways promoting the event.

Initiating and organizing a flower show is usually a function of one or more garden clubs, horticultural societies, or similar groups. It calls for much work and good judgment. As a preliminary, a survey and appraisal of probable interest in and support for such an event should be made. To be a success, even on a small scale, willing and capable volunteer workers as well as exhibitors are necessary. Most established shows began in a small way and grew as experience was gained and enthusiasm increased.

The first step, if the result of the preliminary survey is favorable, is to determine the general character of the show, its chief purposes, and when and where it is to be held. Unlike British flower shows, which are mostly held in tents, American shows are nearly always staged in a large hall or building that affords suitable space. Questions to be settled include whether it is to be a general show or to be concerned wholly or chiefly with one kind of flower, such as roses or chrysanthemums, and also whether the show is to have a particular theme. Another question that must be settled is whether or not commercial exhibits are to be permitted and encouraged. Once a show has been decided upon, the next move is to appoint an overall committee whose duty it will be to do the general planning and supervise all aspects of the effort from then until the conclusion of the exhibition. The choice of the chairman of this committee is of great importance. If he or she has had experience with flower shows before all to the good. At least leadership experience in chairing committees and working smoothly and effectively with others toward common goals is a must. Committee members should be of diverse backgrounds, but all should have well-founded enthusiasms for the task before them. Needless to say they, too, must be people of good judgment willing and able to work together as a team. Next to that of chairman of the flower show committee, the most important position is that of secretary whose duty it is to keep minutes of meetings and carry out all correspondence and other paper work con-

(b) Chrysanthemums and evergreen foliage

Flower arrangements at a flower show: (a) Anthuriums, carnations, and yew foliage

(c) Witch-hazel flowers and *Leucothoe* foliage

(d) Roses

(e) Hyacinth flowers in bud, begonia leaves, leafless twigs, and rocks

An exhibit of vegetables staged in a tent at Southport Flower Show, England

nected with the show. In a small show the secretary may take charge of receiving and recording entries and making ready entry cards, for larger ones it is wise to have an entries chairman.

Special committees to perform particular functions are next established. These include a planning and staging committee to work out the design or layout of the area and allotments of space for specific types of exhibits. For instance, landscaped features are usually allotted ground space, pot plants, cut flowers, and flower arrangements are most often accommodated on tables. This committee must ever bear in mind the importance of maximum convenience for exhibitors and visitors. Easy access for installation, servicing, and removal of exhibits is imperative for the former, easy traffic flow for the latter. Fire laws and other ordinances must be considered and respected. For the best effect, attention must be given to establishing a certain uniformity in "props" used for staging the show. Tables may be covered with burlap or other suitable material. It is best that vases be of the same color and design, but of different sizes. In large shows the hall itself is often decorated by dressing pillars and posts with branches of evergreens or cork bark.

The schedule committee must get to work early. Its duty is to prepare a schedule of classes for which entries are invited. This should include a list of awards to be made and a statement of the point systems or other methods of judging that will be employed. The rules governing exhibitors and others should be stated as clearly and unequivocally as possible. Every effort must be made to have the schedule ready for distribution several months in advance of the date of the show, to allow exhibitors ample time to plan and prepare their entries. Members of the schedule committee should have a good knowledge of plants and be especially familiar with those in which there is the greatest interest. To be a success the schedule must appeal to local gardeners. It should be practical, varied, and imaginative. In planning it accommodation should be made for various groups of possible exhibitors. Separate classes for commercial and amateurs may be scheduled. Classes for children only perhaps, and possibly for those who have not previously won an award at a flower show, should be considered. The schedule should carry the name of the sponsoring organization or organizations, place, date, and duration of the show, its theme if such is established, and information about submitting entries and the latest dates such can be accepted. The names of at least the chairmen of committees should be included and the name, address, and telephone number of the secretary.

The judges committee has very important functions. Not only is it responsible for selecting and securing the services of the most experienced judges available, it must see that they are advised of the general character of the show and of the rules governing it. It must arrange for clerks to accompany the judges on their rounds, record their decisions, and paste stickers denoting awards on the entry cards. The judges committee also takes care of receiving and entertaining the judges on the day of the show. This last is particularly important if the judges come from afar.

Other special committees and functionaries may be appointed, with the flower show committee and particularly its chairman in close touch with all. There may be need for a committee on prizes, for one on ticket sales and distribution, for a clean-up committee, and so on. A publicity chairman is needed. If large attendance is an objective this is a critical post best filled by someone with experience and a flair for public relations. Attention must be given to sending out timely announcements to newspapers and radio and television stations and to having made and distributed well in advance of opening date display posters and other advertising for the show. Invitations to the news media to send reporters must be sent and reporters met and given facilities to get stories and pictures. Prompt releases of the names and addresses of winners of awards and prizes is important.

FLOWERING. This word is part of the common names of these plants: flowering almond (*Prunus*), flowering apricot (*Prunus*), flowering ash (*Fraxinus ornus*), flowering cherry (*Prunus*), flowering currant (*Ribes*), flowering dogwood (*Cornus*), flowering fern (*Anemia* and *Osmunda*), flowering-maple (*Abutilon*), flowering-moss (*Pyxidanthera*), flowering peach (*Prunus*), flowering plum (*Prunus*), flowering-quince (*Chaenomeles*), flowering raspberry (*Rubus odoratus*), flowering-rush (*Butomus umbellatus*), flowering spurge (*Euphorbia corollata*), flowering tobacco (*Nicotiana*), flowering-willow (*Chilopsis*), flowering-wintergreen (*Polygala paucifiora*), and Japanese flowering ash (*Fraxinus longicuspis*).

FLUELLEN. See Kickxia.

FLUORESCENT LIGHT GARDENING. See Indoor Light Gardening.

FLY. As part of their common names the word fly is associated with these plants: crane fly orchid (*Tipularia discolor*), fly catcher (*Bejaria*), fly honeysuckle (*Lonicera*), fly orchid (*Ophrys muscifera*), fly poison (*Amianthium muscaetoxicum*), shoo-fly plant (*Nicandra physalodes*), and Venus' fly trap (*Dionaea muscipula*).

FOAM FLOWER is *Tiarella cordifolia*. The meadow foam is *Limnanthes douglasii*.

FOCKEA (Fóck-ea). This is a highly specialized African genus of milky-juiced succulents of the milkweed family ASCLEPIADACEAE. It has ten species. Its name commemorates a German physician interested in botany, Dr. G. W. Focke.

Fockeas have large tuberous roots and slender stems. Their flowers are unisexual. They are extremely rare in cultivation and are likely to appeal mainly to keen collectors of succulents. In their native regions their tubers and sometimes other parts are eaten.

For more than a century, from 1799 to 1909, a solitary specimen of *Fockea crispa* (syn. *F. capensis*), was believed to be the only one of its kind in the world, the lone survivor of a presumed extinct species. This grew, and still grows, in the Schoenbrunn garden near Vienna, Austria, where it has lived for over 170 years, surely a record for a greenhouse pot plant. It was originally brought from South Africa by a collector sent by Emperor Joseph II. Schoenbrunn, which belonged to the emperor, then boasted the finest conservatories in the world. Today, this old specimen is very little bigger than when it was first collected, indicating the remarkably slow growth of this species. It proved incapable of being reproduced by cuttings and, although it bloomed regularly, because it is unisexual, no seeds were forthcoming. From 1883 on, Professor R. Marloth, distinguished South African botanist, pursued a quest to rediscover *Fockea crispa* in the wild. In 1905 he succeeded.

Fockea crispa

An amazing plant, ***F. crispa***, in its native habitat forms tubers commonly 1 foot to 2 feet long, 6 to 10 inches in diameter and weighing 50 lb. or more. Reportedly, occasionally specimens attain 9 feet in diameter. When large tubers die bees occupy them as nests. In the wild most of the tuber is buried, usually under shrubs over which the stems of the *Fockea* clamber. The latter are long, slender vines with opposite, narrow-ovate, wavy-edged dark green leaves ¾ inch to 1¼ inches long and ⅓ to ¾ inch wide. Gray-green, brown-spotted flowers, rather starfish-like and with five narrow corolla segments (petals), are borne in fall in the leaf axils, three to five together. The fruits are marbled, hornlike pods containing numerous seeds, which, it is believed, do not retain their ability to germinate for long. So far as is known the largest cultivated specimen is in the Botanic Garden, Stellenbosch, South Africa. Its tuber is 20 inches long

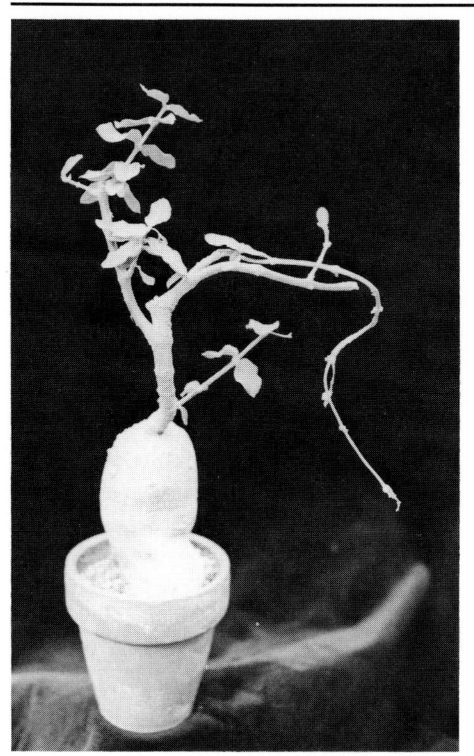

Fockea edulis

and has a circumference of 26 inches. Also sometimes cultivated, ***F. edulis*** as it grows in its native South Africa has a tuber that may weigh well over 50 lb. with, sprouting from its top, only a few delicate, leaf-bearing stems. Other species, with smaller tubers than those of the sorts described here, are *F. angustifolia* and *F. multiflora*.

Garden Uses and Cultivation. Fockeas are intriguing for collections of succulents, outdoors in warm, dry climates, and in greenhouses. They need sharply-drained, porous, loamy soil kept dry throughout their season of rest, when they are without foliage, and at other times moist, but the soil allowed to become nearly dry between thorough soakings. The specimen that has lived so long in Austria is planted with its tuber vertical, most of it exposed above the soil. Even though in their natural habitats the tubers are subterranean best success is had, at least in greenhouses, when the Austrian procedure is followed. Exposure to full sun, both outdoors and in greenhouses, is needed. Indoors, winter night temperatures of 50 to 55°F with an increase of five to ten degrees by day are appropriate. In general indifferent success has been had with specimens dug from the wild. If their tubers are damaged the plants rot. Fresh seeds germinate readily and plants raised in this way are not difficult to grow.

FOENICULUM (Foen-ículum) — Fennel. Florence fennel or finocchio and carosella are horticulturally useful varieties of sweet fennel. The genus comprises five European and Mediterranean region plants of the carrot family UMBELLIFERAE. Its name is an ancient Latin one, a diminutive of *foenum*, hay, it alludes to the odor.

Foeniculums are biennial and perennial herbaceous plants with alternate leaves three- or four-times-pinnately-divided into slender segments. Their small yellow blooms are in umbels of smaller umbels. They are without sepals, but have five petals inflexed at their tips, not lobed or toothed. There are five stamens. The slightly flattened, more or less egg-shaped, seed-like fruits are small and strongly ridged. They are used for flavoring.

Sweet fennel (***Foeniculum vulgare*** syn. *F. officinale*), native of southern and western Europe, often near the sea, is naturalized in North America. It is biennial or perennial, hairless, and more or less glaucous. From 3 to 6 feet tall, it has stalked leaves approximately triangular in outline, divided into numerous almost hairlike segments up to 2 inches long that spread in many directions. Florence fennel (*F. v. dulce*) is distinguished by the bases of its leaves being much enlarged, succulent, and compressed. Carosella (*F. v. piperitum*) is a perennial with the divisions of its leaves fleshy, rigid, and less than ½ inch long. Its young shoots are eaten.

FOG FRUIT. See Phyla.

FOLIAGE PLANTS. In its most common usage this term is employed as a group identification for a considerable variety of plants grown for ornament in greenhouses, homes, offices, and other places, but it can properly be applied to some hardy plants such as eulalias (*Miscanthus*), *Festuca ovina glauca*, and English ivy, esteemed for their foliage rather than their blooms.

As the name suggests, foliage plants are admired chiefly for their leafage, which may be green or otherwise colored or patterned. The majority of sorts, except ferns and other cryptogams, when they attain a certain age and size under favorable circumstances bloom, some quite handsomely. And so there is no sharp line of distinction between foliage and flowering plants as these terms are used by gardeners. Rex begonias and *Aphelandra squarrosa*, for instance, might well be placed in either group. And jacarandas, which are magnificient flowering trees in warm climates, as young specimens in pots are attractive foliage plants.

Among the best-known foliage plants are these: acalyphas, aglaeonemas, alocasias, amaranthuses, anthuriums, asparaguses, aspidistras, aucubas, begonias, calatheas, coleuses, colocasias, cordylines, crotons (*Codiaeum*), dracaenas, English ivy (*Hedera*), fatshederas, ferns, ficuses, grape-ivy (*Cissus*), grevilleas, marantas, Norfolk-Island-pine (*Araucaria*), pandanuses, palms, peperomias, philodendrons, pick-a-back plant (*Tolmiea*), pileas, sanchezias, snake plants (*Sansevieria*), and xanthosomas. In addition there are many fine succulents such as aeoniums, agaves, aloes, cotyledons, crassulas, echeverias, furcraeas, gasterias, haworthias, pachyphytums, sedums, and sempervivums that qualify as foliage plants.

FOLIAR FEEDING. For long scientists were of the opinion that the only nutrient absorbed by plants through their leaves was carbon dioxide from the atmosphere. This, despite the fact that practical gardeners had discovered many decades previously that by spraying the foliage of such crops as cucumbers, melons, and chrysanthemums with dilute liquid manure or with water in which a bag of soft coal soot has been soaked, the size, color, and quality of the foliage was improved as though a nitrogenous fertilizer had been applied to the soil. Nonsense, said the scientists, plants obtain all their nitrogen from the soil. The development of techniques making use of radioactive tracers enabled scientists to probe further. They then found that the empirical knowledge of gardeners was factual, that plants are indeed capable of absorbing nitrogen and other nutrients through their foliage. Once this was established, the practice of foliar feeding became more widely accepted as a cultural routine. It is employed in commercial crop production as well as by amateur gardeners.

Foliar feeding can be helpful. It should not, however, be thought of as a substitute for fertilizing the soil, but only as a supplement to the practice. It is especially useful to supply shot-in-the-arm treatments during critical periods by providing a quickly available major nutrient, such as nitrogen or phosphorus, or a needed trace element, such as iron. A remarkable aspect of the technique is the very small amount of nutrients needed. Too strong solutions can damage foliage, flowers, and fruit, this is especially true of those containing trace elements such as boron, copper, iron, and zinc. Never use these unless a soil test indicates deficiencies exist. Foliar sprays of nitrogen, phosphorus, potassium, and calcium, used in reasonable amounts, are less likely to do harm or to result in a dangerous buildup of the amounts absorbed.

Remember when foliar spraying that the material is absorbed by tiny pores (stomata) chiefly on the undersides of the leaves; therefore direct the spray upward and in no greater quantity than is needed to wet both surfaces of the foliage.

FOLLICLE. This is the technical name for a fruit developed from a single ovary that is dry and podlike and consists of one chamber with a seam down one side that at maturity splits open. The fruits of delphiniums, milkweeds, monkshoods, and peonies are follicles.

FONTANESIA (Fontan-èsia). Two species of deciduous, somewhat privet-like shrubs are the sole representatives of *Fontanesia* of the olive family OLEACEAE. Their name commemorates the French scholar and botanist René Louiche Desfontaines, who died in 1833.

Fontanesias have quadrangular shoots, opposite, undivided, toothless or sometimes minutely-toothed leaves, and leafy panicles of tiny greenish-white flowers that come in late spring and have a minute, four-parted calyx, four small petals, two stamens, and a two-lobed stigma. The fruits are winged, flat nutlets.

Of graceful habit, ***F. fortunei,*** native to China, and there reported to attain heights of 30 to 40 feet, as known in cultivation rarely exceeds 15 feet. It has bright green, long-pointed, hairless, lanceolate leaves 1 inch to 4½ inches long and up to 1 inch broad. Its flowers are in slender panicles 1 inch to 2 inches long. This is hardy in southern New England.

Less tolerant of cold, and probably not hardy north of Virginia, closely related ***F. phillyreoides,*** a twiggy shrub 6 to 10 feet tall, is a native of western Asia. It has ovate-lanceolate to broadly-elliptic leaves 1 inch to 2½ inches in length and under ¾ inch wide. Its panicles of greenish-white blooms, with conspicuously protruding stamens, are up to 1 inch long.

Garden and Landscape Uses and Cultivation. Although not among the most ornamental of shrubs, fontanesias may be usefully employed to give variety to plantings. They have about the same decorative values as privets and like them may be sheared to form hedges. They prosper in sun or part-day shade and are easily raised from seeds and from leafy cuttings in summer and leafless hardwood ones in fall.

FORCING. In its broadest sense, perhaps, the term forcing can be used to cover all procedures that bring flowers, fruits, and vegetables to usable maturity out of season, but in practice and in this Encyclopedia its use is generally more restricted. In its narrower meaning it refers only to the employment of temperatures higher than outdoor ones normal for the season to excite the more or less rapid development of foliage or flowers chiefly at the expense of food materials stored in the plants at the time forcing began. This last is important. Merely bringing plants into bloom out of season is not necessarily forcing. Such kinds as calendulas, carnations, chrysanthemums, and snapdragons grown in greenhouses throughout most of their lives under conditions that approximate those that suit them outdoors develop the food supplies they need to grow and flower during their season of growth. The greenhouse gardeners' concern with them is to grow them, not merely to force from them the stored results of previous growing periods. Manipulation of day length by the use of artificial light and black cloth shading is not considered forcing under the restricted use of the term.

Examples of plants commonly forced are Easter lilies, which normally bloom outdoors after midsummer, being brought into flower in greenhouses in time for the holiday season, and hardy spring-blooming bulbs, such as hyacinths, tulips, and daffodils, being persuaded to present their displays long before their normal outdoor time. In like manner some shrubs, such as azaleas and hydrangeas, commonly are and many others can be flowered ahead of season. Early crops of chicory, dandelion greens, rhubarb, and seakale are easily had by forcing. Cut branches of many sorts of spring-blooming shrubs are readily forced into bloom indoors.

Principles governing successful forcing are few and easily learned. First, plants force more readily and rapidly as their normal season approaches than earlier. Thus it is easier and the forcing period needed is shorter to have daffodils bloom in March than in January. Second, plants well established and well rooted in their containers in general give much better results than those less well furnished with roots. With most hardy bulbs it is imperative that a sufficient time for plentiful root development at comparatively low temperatures be allowed before forcing begins. Shrubs that have been growing in their containers a full year or at least several months before being introduced to forcing temperatures give superior results to those dug and put into a forcing environment right away. Exceptions to this need for a rooting period before forcing are a few plants in which the food reserve is stored in thick roots or rhizomes. Examples are chicory, lily-of-the-valley, and rhubarb. A third principle is that the best results are had by beginning forcing in moderate warmth and gradually raising the temperature after growth is well started.

Breaking (starting into growth) of plants being forced is stimulated by keeping their tops moist. With cut branches this can be done by standing them in containers of water in a cellar or similar place and keeping their tops wrapped in damp burlap or similar material until the buds begin to burst. In greenhouses the same results are had by misting the tops of the plants with water two or three times a day, but not so late that the moisture does not dry before nightfall. The soil or other rooting medium must be kept moist throughout.

Darkness is needed if the finished product, such as rhubarb or whitloof chicory, is to be blanched and may be desirable during the first part of the forcing period to encourage long stems with such plants as hyacinths and lily-of-the-valleys, but usually and certainly where the production of green foliage is important, the plants must be in full light for most of the forcing period.

Temperatures employed for forcing vary according to the kind of plant. With flowering sorts it is usual to lower the temperature a few degrees during the last few days or week before the blooms are at their best. This hardens the flowers so that they last longer on the plant or when cut. A little shade given after the blooms begin to expand often results in more intense color and delays opening of tulips and other flowers that are at their best when the petals are not fully expanded.

Most hardy plants cannot be successfully forced two years in succession, but most, if properly cared for, will recover if planted outdoors and will give satisfaction in future years. To allow the best chance for such recovery give them the best possible care after forcing and after they are through blooming. This involves keeping them watered and growing in a light place where the temperature is favorable to growth, but lower than that in which they were forced, until the outdoor weather is acceptably warm, then placing them in locations appropriate to their kinds, and keeping them well watered afterward.

FORCING CUT BRANCHES. A simple and pleasant way of anticipating spring is to force indoors into early foliage or bloom cut branches of responsive shrubs and trees. Blooming sorts that lend themselves to this are those that develop flower buds in fall that normally open outdoors in late winter or spring. Here belong alders, azaleas, Cornelian-cherry, forsythias, flowering cherries and crabapples, flowering dogwood (*Cornus florida*), hazels, early blooming magnolias, pussy willows, and redbuds. Kinds that have attractive foliage when forced include beeches, birches, hickories, horse chestnuts, maples, oaks, and witch-hazels.

Forcing may begin as soon as the buds become plump or at least start to swell. This is as early as mid-January with some of the more adventurous, such as Cornelian-cherry, witch-hazels, and forsythias, later with such kinds as flowering dogwoods and magnolias. If bloom is the objective, be sure to select branches well furnished with flower buds, which are plumper and rounder than leaf buds. Give thought to the well-being of the bush or tree from which the branches are cut. You may be able to accomplish needed pruning, to improve the shape and appearance of the plant, while gathering material for indoor arrangements.

Cut branches at just above a node or back to a main branch. Make clean, slanting cuts with sharp shears or a knife. If the cut is made through wood not over a year old the flowers are likely to be bigger than if older wood forms the base of the severed branch, this because young tissues absorb water more rapidly than older ones. If possible choose a warm, sunny day for cutting branches for forcing. If they must be taken while frozen submerge them in cold water for three or four hours before proceeding as with branches cut on a warmer day.

Stand newly cut branches that are not

frozen with their lower halves immersed in warm, not hot, water for one-half hour or so. Then transfer them to cold water and put them in a cool cellar or room out of direct sun for about a week. During this period spray them lightly with water a couple of times a day or keep them wrapped in burlap, cotton sheeting, or other porous material that can be kept moist.

Some people recommend mashing the basal 3 or 4 inches of the stems with a mallet before placing them in water, but this does not seem to give any important advantage. Remember that the later in the season the branches are cut, that is the nearer to their normal outdoor flowering time, the more readily they will force and the more quickly come into bloom or leaf.

FORESTIERA (Forest-ièra)—Swamp-Privet. An entirely New World genus, *Forestiera,* comprises fifteen species of the olive family OLEACEAE. It is therefore related to ashes, privets, and lilacs. Its name commemorates a French physician and naturalist, Charles Le Forestier. These plants are sometimes grown under the synonym *Adelia.* Ranging in the wild from Illinois to the West Indies and Brazil, this group of mostly deciduous trees and shrubs is of slight horticultural importance.

Forestieras have opposite, short-stalked, toothed or toothless leaves and unisexual and bisexual flowers in clusters or racemes from the previous year's shoots, appearing before the leaves. They are without petals but usually have four to six, unequal-sized, quickly deciduous sepals. There are two to four stamens and a slender style. The one- or rarely two-seeded, usually black fruits are small and like miniature plums.

A shrub or broad-crowned deciduous tree up to about 30 feet in height, the swamp-privet (***F. acuminata***) inhabits moist stream banks and wet soils from Illinois to Missouri, Georgia, and Texas. Its leaves, ovate to ovate-elliptic, pointed and toothed, are 2 to 5 inches long, hairless, and have ½-inch-long, slender stalks. The flowers are inconspicuous. The wrinkled fruits, dark blue at maturity, are ½ to ¾ inch long. This species is hardy in southern New England. A deciduous shrub, ***F. neo-mexicana,*** occurs from Colorado to Texas and New Mexico. Up to 10 feet in height, it has spreading branches, and pointed or blunt, hairless leaves up to 1½ inches long. Its fruits are ⅙ inch in length. It is about as hardy as the last species. Slightly more tender, ***F. ligustrina*** is a deciduous shrub up to 10 feet tall with blunt, elliptic to oblong-obovate leaves up to approximately 1½ inches long and pubescent beneath. Its fruits are about ⅙ inch long. It is indigenous from Tennessee to Florida.

Garden and Landscape Uses and Cultivation. Of secondary horticultural importance, the swamp-privet and the others described here are sometimes cultivated. They have no special virtues to commend them except that they provide variety and are suited for moist and wet soils. There are many better shrubs and trees. Forestieras are increased by seeds and layers.

FORGET-ME-NOT. See Myosotis. The Chatham-Island-forget-me-not is *Myosotidium hortensia,* the Chinese-forget-me-not *Cynoglossum amabile,* creeping-forget-me-not *Omphalodes verna,* the forget-me-not-anchusa *Brunnera macrophylla.*

FORKS, GARDEN. Garden forks of various types are among the most useful of implements. Spading forks, as their name suggests, are used in the manner of spades for turning over soil and mixing in coarse organic material, such as manure and compost. In this function they, like spades, serve gardeners as plows do farmers and other cultivators of large tracts. Many gardeners prefer forks to spades on heavy, clayey soils. Forks are also employed to do the work harrows are used for on large areas in breaking up soil previously spaded or plowed. Spading forks have usually four nearly straight prongs or tines. Manure forks, used for lifting and moving manure and other coarse, loose material, differ in having slimmer, strongly curved, tines. Hand forks are much smaller. They serve many of the purposes for which trowels are used and have the advantage of not slicing through the roots of plants lifted by them. They are not as useful as trowels, however, for digging holes.

FORMAL and SEMIFORMAL GARDENS. A formal garden is one laid out in a strictly geometrical pattern usually exhibiting bilateral symmetry. Such gardens strongly emphasize man's domination over nature. Their boundaries are clearly defined by walls, hedges, fences, or in other ways and such features as avenues, parterres, topiary, statuary, sundials, geometrically-shaped

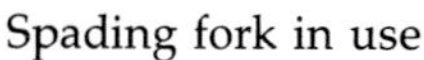
Spading fork in use

Manure fork in use

Formal gardens

Semiformal gardens

pools, fountains, and well heads are commonly introduced as main elements in the design, accents, or focal points.

The gardens of the ancients were all of this type, as were those of medieval Europe. Renaissance gardens, developed after the fashion of earlier enclosed sanctuaries, but on a much grander scale, were formal. These culminated in the vast extravagancies of Italy and France and later England. Noteworthy among these are the gardens of Isola Bella, Villa d'Este, Versailles, and Hampton Court Palace.

In colonial and post-Revolutionary America, garden makers followed European tradition. Their gardens were formal, as those at Williamsburg and Mount Vernon so clearly demonstrate. Much later and much more elaborate examples of formal gardens in America may be seen at Dumbarton Oaks, Washington, D.C. and Longwood Gardens, Kennett Square, Pennsylvania. Those of the latter are integrated into a much larger informal landscape in the fashion that became popular on large estates in Great Britain and America from the eighteenth century on.

Strictly formal gardens are in strong contrast to purely informal or naturalistic gardens (see Naturalistic Gardens), but intermediates that can quite properly be called semiformal are common. These lack the rigid, rectilinear patterns of formal gardens without achieving the apparently natural scenery of naturalistic gardens. The efforts of the gardener manifest themselves as working with rather than subduing and dominating nature. Slopes are not flattened or terraced, but are preserved or pleasingly graded. Paths may curve with the contours or to accommodate trees or shrubs. Free-form pools and other water features replace geometrical ones. Statuary and other furnishings are employed with great restraint, if at all. But the work of the designer and gardener betrays itself in the shaven lawns, cared-for shrubbery, gay flower beds and borders, and neat housekeeping.

Modern gardens are much smaller than those of great European and American estates developed prior to World War II. Most are semiformal. Where largish areas are involved they are almost always naturalistic. They reflect, as did earlier gardens, a life-style of the day. In this new age of gardening purely formal treatments are mostly confined to city gardens, for which they are generally well suited, and in some cases to areas in the immediate vicinity of the dwelling.

FORMALDEHYDE. This is a colorless, suffocating, tear-provoking, toxic gas dissolvable in water and deadly to plant life. It is employed as a soil fumigant and for certain other horticultural purposes. Although commonly referred to as formaldehyde, the 37 to 40 percent aqueous solution generally available and ordinarily used is more correctly named formalin. This is diluted with water at the rate of a pint to 6¼ gallons and soil bare of desir able plants is then drenched with it. Application at the rate of ½ gallon to each square foot, following which the ground is covered for twenty-four hours with polyethylene plastic film or other material to prevent the escape of the fumes, is usual.

FORSTERA (For-stèra). The stylidium family STYLIDIACEAE includes *Forstera,* a genus of five species of hairless, perennial herbaceous plants natives of New Zealand and Tasmania. The name honors Johann Reinhold Forster, German professor of natural history, who died in 1798.

The stems of forsteras are erect or procumbent. Their leaves are undivided. Those of most kinds overlap like shingles on a roof. Their bases are persistent. Usually solitary or paired, but sometimes in groups of up to five, the slender-stalked flowers have tubular calyxes with five or six lobes and slightly asymmetrical, short bell-shaped corollas with five to nine lobes (petals). There are two or rarely three stamens surrounding a style with a two-lobed stigma. The fruits are capsules.

Endemic to moist, mountain places in New Zealand, ***F. tenella*** has slender stems, prostrate and rooting toward their bases, ascending beyond. The extremely short-stalked leaves, not overlapping but spreading and usually separated from each other, are narrowly-oblong to obovate. They are mostly up to ½ inch in length, and have more or less recurved margins. On very slender stalks 2 to 4 inches long are carried one to three white blooms, ⅓ inch across and with five- or six-lobed corollas. Another native of moist alpine and subalpine regions in New Zealand, ***F. sedifolia*** is up to 1 foot tall and sparsely-branched. Its stalkless, stem-clasping, ovate to oblong-obovate leaves up to 3 inches long at first are erect; they later spread horizontally and finally point downward. The flowers, somewhat over ½ inch across and on stalks up to 4 inches long, are white.

Garden Uses and Cultivation. North American experience with forsteras is extremely limited. All that can be said is that they are worth trying by specialist cultivators of alpine plants. They are most likely to succeed in parts of the Pacific Northwest and in cool greenhouses where conditions are satisfactory for the growth of other choice alpines. Seed and division afford means of increase.

FORSTERONIA (Forster-ònia). Most of the fifty species of this genus of the dogbane family APOCYNACEAE are without merit as ornamentals, but the one described here is attractive and since its introduction in 1943 has proved to be a good flowering vine for Flordia and other warm regions. The genus, named after the British botanist Thomas Furley Forster, who died in 1825, is a native of Central America, tropical South America, and the West Indies.

Native of Cuba and Hispaniola, ***Forsteronia corymbosa*** is slender-stemmed and dainty. It has leathery, obovate to broadly-elliptic, privet-like leaves up to 3 inches long and almost or quite hairless. Its numerous carmine-red flowers are in flat, terminal clusters 1 inch to 2 inches wide. Especially appropriate for planting where space is limited, this pretty vine blooms over a long period. It stands very little frost, and needs full sun. Ordinary garden soil suits. Propagation is by cuttings and by seeds.

FORSYTHIA (Forsýth-ia). Golden Bells. The species of *Forsythia* number seven or fewer, all but one natives of eastern Asia. The exception is European. In addition, there are many beautiful varieties and hybrids of special interest to gardeners. The genus, which belongs in the olive family OLEACEAE, is named in honor of William Forsyth, a British horticulturist who died in 1804. The so-called white-forsythia is *Abeliophyllum distichum.*

Forsythias are deciduous or partially evergreen, yellow-flowering shrubs with mostly arching or spreading branches. Some sorts have a hollow central chamber through the center of the branch with pith only at the joints, others have lamellate pith, that is pith consisting of thin plates spaced all along the central chamber. They have opposite leaves, undivided or less often of three leaflets, and flowers with a four-lobed calyx, four spreading, strap-shaped petals, two stamens, and one style. Some individuals of each species have styles longer than the stamens and others have styles shorter than the stamens. Cross pollination between long- and short-styled flowers is usually necessary for seed production. The fruits are capsules. The flowers are one to six together from axillary buds that develop in fall and remain dormant through the winter, ready to expand and reveal their gold at the first evidence of spring. If the fall is mild a few of the flower buds are likely to bloom then, but such precocious conduct does not noticeably diminish the spring display.

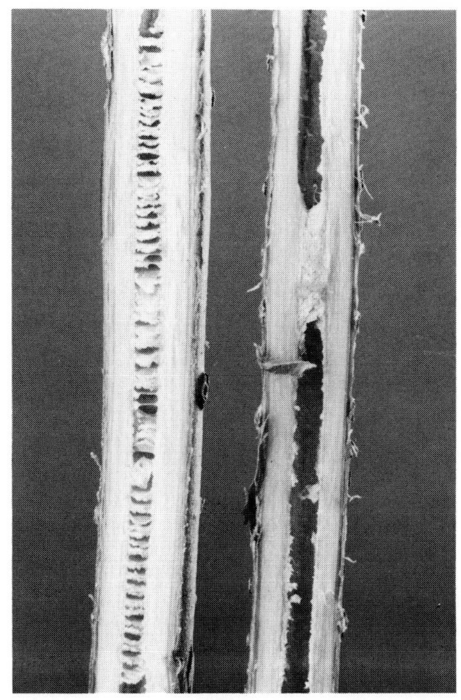
Split stems of *Forsythia* showing pith: Left, *F. intermedia* with lamellate pith; right, *F. suspensa* at joints (nodes) only

So ready to anticipate spring are forsythias that branches cut, brought indoors, and stood in containers of water any time from January to spring are in flower within ten days to three weeks. These forced branches bring a breath of spring ahead of season to many a northern home. Almost all forsythias are winter hardy in New England and adjacent Canada, although the flower buds of many may be damaged if the temperature remains for long below −10°F. Where very cold winters prevail it is not unusual for these shrubs to flower on their lower branches, that have been covered with snow, and to be without blooms on branches exposed above the snow line.

Considering the popularity of forsythias it is rather surprising that they were not known in the United States until after the middle of last century and that European gardeners first became acquainted with them only a very few years earlier. Interestingly, the native European species, which is confined as a wildling to the mountains of Albania, was not the first kind discovered, indeed *Forsythia europaea* was not found until 1897 and was not brought into cultivation until two years later. The first of the golden bells to come to the attention of Westerners was *F. suspensa* which was imported into Holland from a garden in Japan in 1833. The plant is not a native of Japan, but of China. About twenty years later it was introduced into England and not long afterwards into the United States.

Forsythia ovata

Forsythia suspensa

The earliest to bloom, but not the best, is ***F. ovata,*** a native of Korea brought into cultivation in 1918. Besides flowering about ten days earlier than other kinds, its flower buds are less susceptible to damage by severe cold than are those of most forsythias. It has the fault of blooming somewhat erratically, freely some years, much more sparingly in others, without apparent reason. It is an upright shrub, 4 to 5 feet in height, with spreading branches, yellowish with lamellate pith, and cylindrical, hairless branchlets. The broad-ovate, toothed or almost toothless leaves terminate in short points. They are 1½ to 3½ inches long by up to 2¾ inches wide. The flowers, mostly solitary, are pale yellow and nearly ¾ inch wide.

Distinct from other species in cultivation because its branches are hollow except at the nodes, where they are filled with pith, ***F. suspensa*** of China is cultivated in two chief varieties. Up to 10 feet in height, *F. s. fortunei* has erect, arching branches that spread widely. Its ovate to oblong-ovate, toothed leaves, up to 4 inches long, are often deeply-lobed or even divided into three leaflets. The golden-yellow flowers, 1¼ to 1½ inches across, on stalks scarcely longer than the buds, are solitary or in clusters. Their petals spread widely. This variety was introduced from China to England in 1862 and to America by the Arnold Arboretum in 1878. Very long, slender, drooping, vinelike branches that root freely in contact with the ground are characteristic of *F. s. sieboldii,* which has undivided, usually lobeless leaves and slightly smaller flowers with petals that spread less widely than those of *F. s. fortunei.* The flower stalks are about twice as long as the buds. Variants intermediate between *F. suspensa* and *F. s. fortunei* are cultivated. Other varieties are *F. s. atrocaulis,* which has purplish young shoots, and *F. s. variegata,* the leaves of which are variegated with yellow.

Less handsome than *F. suspensa* and somewhat less hardy to winter cold, Chinese ***F. viridissima*** was introduced to England in 1844. A deciduous or partially evergreen, stiffish shrub with erect branches and up to 10 feet in height, this sort has stems with lamellate pith and

Forsythia viridissima koreana

Forsythia viridissima bronxensis

Forsythia intermedia spectabilis

branchlets that are distinctly quadrangular and green. Its elliptic to lanceolate leaves, up to 6 inches long, are toothed only above their middles. Its 1¼ inch-wide flowers, borne a week or two later than those of *F. suspensa* varieties, have a slight greenish tinge. This is the only forsythia with leaves that turn a definitely purple-red before they drop. Hardier than the typical species is *F. v. koreana,* which has slightly larger and more brightly colored flowers. Variety *F. v. bronxensis,* a seedling raised at The Boyce Thompson Arboretum, Yonkers, New York, from seed received from Japan in 1938, grows extremely slowly and does not exceed 1 foot to 1½ feet in height. Its branches spread stiffly and have many side branches. This is the only really compact, dwarf forsythia. Its leaves, set closely together on the shoots, are 1 inch to 1½ inches long. Its pale yellow flowers, on older specimens borne freely but often sparingly produced by young plants, are solitary or in twos or threes.

Except for some recently developed varieties with nonlatinized horticultural names, by far the most beautiful forsythias in bloom are the hybrids between *F. suspensa* and *F. viridissima* that are grouped as ***F. intermedia.*** The original cross appeared in Germany before 1885 and is intermediate between its parents; for instance, its branches are rarely hollow, but usually have solid pith at the nodes and are lamellate or partly so between. Sometimes on vigorous shoots the leaves tend to be three-parted. Variety *F. i. densiflora* has a profusion of crowded, rather pale flowers. Originally discovered in Ireland as a branch sport on *F. i. spectabilis,* variety *F. i.* 'Lynwood' (syn. *F. i.* 'Lynwood Gold') was introduced to America in 1953. A trifle paler in color than its 'parent', its flowers are less bunched, open wider, and are about 1¾ inches across. A compact form that may grow to a height of about 5 feet, *F. i. nana* has greenish-yellow flowers. Variety *F. i. primulina* has beautiful pale yellow flowers. One of the deepest-yellow-flowered and showiest forsythias, *F. i. spectabilis* was introduced from Germany to America in 1906. This magnificent and popular variety is only equaled in intensity of flower color by the much newer 'Beatrix Farrand'. The Spaeth Nursery in Berlin originated *F. i. spectabilis.* Discovered as a branch sport of *F. i. primulina* in Ohio about 1930, variety *F. i.* 'Spring Glory' has a profusion of clear, pale yellow flowers about 2 inches in diameter. Originated by the Spaeth Nursery in Berlin about 1906, *F. i. vitellina* has deep yellow, smallish flowers.

Several newer varieties, the result of experimental breeding, are noteworthy. 'Arnold Dwarf' is from a cross made at the Arnold Arboretum in 1941 between *F. intermedia* and *F. japonica.* It has prostrate branches that root readily into moistish

Forsythia intermedia spectabilis (flowers)

Forsythia intermedia primulina

An unusual employment of *Forsythia intermedia* as a hedge, National Botanic Gardens, Dublin, Ireland

soil. It may eventually grow up to 5 feet in height, but remains lower for a long time and spreads vigorously. Its small, greenish-yellow flowers are produced sparingly. 'Beatrix Farrand', originated by Dr. Karl Sax at the Arnold Arboretum in 1939, is a cross between a colchicine-induced tetraploid of *F. intermedia spectabilis* and the normal *F. i. spectabilis*. Of erect growth, it bears a profusion of rich yellow flowers that are sometimes 2½ inches across. 'Karl Sax' was raised at the Arnold Arboretum and named after its originator, a one-time director of that institution.

Less frequently cultivated species likely to be found in botanical collections include these: ***F. europaea*** is upright, lanky, and up to 10 feet tall. The pith of its branches is lamellate. Its ovate leaves have few teeth or none. This European, the least ornamental of forsythias, is hardy in southern New England. ***F. giraldiana*** is related to the last and is as hardy. Native to northern China, it has elliptic-oblong to lanceolate-oblong leaves and short-stalked blooms that appear very early. ***F. japonica,*** allied to *F. ovata,* is a spreading kind, inferior as an ornamental. It has ovate leaves and solitary flowers and is native to Japan. Variety *F. j. saxatilis,* of Korea, differs but slightly.

Garden and Landscape Uses. Except for grand displays of bloom, ease of cultivation, and general freedom from pests and diseases, forsythias have little to recommend them. They are undistinguished in foliage, without attractive fruits, and have no bright fall colors. Nevertheless, the virtues they have, together with the fact that they can be grown successfully over a large part of North America, are enough to ensure their popularity among gardeners. Forsythia is the official flower of Brooklyn, New York. Like most shrubs that bloom on naked branches, forsythias are displayed to best advantage in front of, but some little distance away from, evergreens. Rich backgrounds of yew, hemlock, and other evergreens make perfect foils for the golden blooms. Forsythias are at their worst when sheared into globes and other formal shapes, a type of mutilation that seems to appeal to those over-inflicted with a mania for horticultural tidiness. Planted immediately behind the top of a high retaining wall, *F. suspensa sieboldii* is especially effective. Its branches hang down for many feet as a curtain of foliage in summer, and in spring a cascade of bloom. Variety 'Arnold Dwarf' is an appropriate groundcover for large banks and can be kept to a height not exceeding 2 feet by occasionally scything or pruning any taller shoots that develop. For rock gardens the slow-growing, dwarf *F. viridissima bronxensis* is an interesting shrublet. Forsythias are easy to bloom in pots or tubs in greenhouses in late winter, and cut branches are easily forced into bloom in water indoors.

Cultivation. Few shrubs are as tolerant of as wide a variety of conditions or are as easy to grow as forsythias. Although they flourish best in a deep, fertile loam they get along well in almost any medium that is not parched in summer or waterlogged. To bloom well they need full sun, but they will grow in light shade. They may be transplanted, even when large, with every expectation that they will quickly reestablish themselves; at moving time the tops should be pruned to reduce their size by one-third to one-half. The need for regular yearly pruning is often over-stressed. The annual thinning after flowering that may be desirable in northern Europe is not necessary in America when summer sunlight is so much more intense. More prolific flowering is had if attention to pruning is given only when the branches are obviously overcrowded and threaten to become unruly tangles; this may be every two or three years. When pruning is done, all very weak shoots as well as old crowded and ill-placed ones should be cut out so that light can reach all the new shoots that develop. The best time to prune is immediately after blooming, but in the case of long neglected and much-tangled specimens, it is more practical to do the work in late winter even at the expense of sacrificing potential flowers, because it is easier then to see what one is doing. If deemed advisable an old shrub may be cut down close to the ground. If this is done, the new shoots should be thinned drastically to prevent crowding. Propagation of forsythias is easily accomplished by leafy cuttings made in summer from firm shoots, and planted under mist or in a greenhouse or cold frame propagating bed, and by hardwood cuttings taken in fall after the foliage has gone. Layering branches that can be bent to the ground, and in some cases division of old plants, are other practical ways of securing increase. For directions for blooming forsythias early in greenhouses see Forcing.

Diseases and Pests. These plants are remarkably little harmed by diseases and pests. Occasionally they develop stem galls, leaf spots, and a die-back disease. The four-lined plant bug and root knot nematodes sometimes infest them.

FORTUNELLA (Fortun-élla) — Kumquat. From four to six, depending upon the interpretations of different botanists, species of small evergreen trees or shrubs closely related to the orange constitute *Fortunella.* Called kumquats, they are probably natives of southern China. They belong in the rue family RUTACEAE and have a botanical name that honors the distinguished English traveler and plant collector Robert Fortune, who first brought kumquats to Europe. He died in 1880.

Kumquats differ technically from oranges (*Citrus*) in having ovaries with six or fewer cells, each with two ovules, and in the stigmas of their flowers being hollow. From all except *Citrus aurantium myrtifolia,* they differ more obviously in their fruits being much smaller, often not over 1 inch in diameter.

Fortunellas may have spines in their leaf axils or not. Their blunt or slightly notched leaves are thickish, undivided, lobeless and toothless, very densely-glandular-dotted on their under surfaces, and have stalks usually with narrow wings. Solitary or in clusters or few from the leaf axils, the white flowers commonly have five, sometimes three, four, or six each sepals and white petals, sixteen to twenty stamens cohering in bundles, and a short style. The aromatic fruits, technically berries, are ellipsoid to globular, and resemble little oranges. They are made into preserves, candied, and eaten raw.

The oval or Nagami kumquat (***F. margarita***) is a thornless or nearly thornless tree 10 to 12 feet tall, with lanceolate leaves 1½ to 3 inches long or occasionally longer. Toward their apexes they are slightly round-toothed. The short-stalked flowers, at maximum, scarcely exceed ½ inch in diameter. The oblongish or ellipsoid, orange-yellow fruits, 1 inch to 1½ inches long, have harsh-flavored, acrid skins and acid pulp. Attached to them are the persistent rudiments of the style.

The round or Marumi kumquat (***F. japonica***) is a small bushy shrub, with or without spines, differing from the oval kumquat in its leaves being proportionately broader, its bright orange fruits, about 1 inch in diameter, being spherical, and not retaining the remains of the style. Their skins are less acrid than those of the oval kumquat.

The Hong Kong kumquat (***F. hindsii***) as its designation indicates is native of Hong Kong, as well as of adjacent mainland China. A small, spiny tree, it has wing-stalked, elliptic leaves, and subspherical bright flame-orange-red, nearly juiceless fruits under 1 inch in diameter. Dwarf *F. h.* 'Chintou' has bigger, thicker, and somewhat narrower leaves, 1½ to 3¼ inches long by up to 1 inch wide, than the typical Hong Kong kumquat, and slightly larger fruits. Both the Hong Kong kumquat and its variety in fruit are very ornamental.

The Meiwa kumquat (***F. crassifolia***) is likely of hybrid origin. A sparsely-spine-bearing or spineless shrub, it has thick leaves with narrowly-winged stalks, and thick-skinned, nearly seedless, broadly-ellipsoid to nearly spherical fruits up to 1½ inches in diameter.

Hybrids between *Fortunella* and *Citrus*, between *Fortunella* and *Poncirus*, and even a trigeneric hybrid involving all three genera, are known. Those between *Fortunella* and *Citrus aurantifolia* are called limequats, those between *Fortunella* and *Citrus reticulata* orangequats (the calamondin probably belongs here), those between *Fortunella* and *Poncirus* citrumquats. Other more complex hybrids have been raised. See Citrofortunella.

Fortunella margarita

Garden and Landscape Uses and Cultivation. For discussions relative here see Kumquat.

FORTUNERIA (Fortun-èria). One deciduous shrub of the witch-hazel family HAMAMELIDACEAE is the only representative of this Chinese genus. It is hardy in southern New England, and by name commemorates a great English botanical explorer of China, Robert Fortune, who died in 1880.

Distinctive ***Fortuneria sinensis*** attains a height of about 20 feet, and has shoots densely clothed with stellate (star-shaped) hairs. Its obovate to oblong, unequally-toothed leaves have short, hairy stalks. They are 3 to 6 inches long and differ from those of its near relative *Corylopsis* in their veins being much branched instead of straight. The inconspicuous, green flowers, about ⅙ inch wide, appear with the leaves, and have very slender petals and short-stalked stamens. There are bisexual and male blooms, the former in racemes up to 2¼ inches long, with usually one to three leaves at their bases, the latter, which develop in fall, but do not open until spring in short, catkin-like spikes without basal leaves. The flowers have five-lobed calyxes, five petals shorter than the calyx lobes, five stamens, and a pistil with two styles; in the male flowers the pistil is rudimentary. The fruits are capsules with two dark brown, shining seeds.

Garden and Landscape Uses and Cultivation. Other than for botanical collections this shrub has little to recommend it. It responds to environments and care appropriate for *Hamamelis.*

FOSTERELLA (Foster-élla). This name commemorates Mulford B. Foster, a twentieth-century American horticulturist and traveler referred to by the author of the genus as "discoverer extraordinary of new species of *Bromeliaceae*." It is to be regretted that a more splendid genus of the pineapple family BROMELIACEAE was not selected to carry the name of such an indefatigable enthusiast. There are thirteen species of *Fosterella*, ranging in the wild from Mexico to Argentina.

Fosterellas are not, like so many of their kin, tree-perchers. They grow in the ground. They are stemless, evergreen perennials with spiny- or smooth-margined leaves in rosettes. The small flowers, in loose, long-stalked, branched panicles from the centers of the rosettes have three each separate sepals and petals, six stamens, and a three-parted style. The anthers are fixed to the stalks of the stamens by their bases, not as in nearly related *Cottendorfia*, at their middles. The fruits are capsules.

Sometimes cultivated by bromeliad fanciers ***F. penduliflora*** (syn. *Lindmania penduliflora*) of Peru has spreading, pointed-strap-shaped leaves with wavy, spineless

Fosterella penduliflora

Fosterella penduliflora (flowers)

margins. Their upper sides are light green, beneath they are silvery. The inconspicuous, bell-shaped, nodding, whitish blooms are in slender, erect panicles.

Garden and Landscape Uses and Cultivation. These plants are unlikely to be grown by other than enthusiastic collectors of bromeliads (plants of the pineapple family). They respond to conditions that suit other terrestrial kinds. For more information see Bromeliads or Bromels.

FOTHERGILLA (Fothergíl-la). The genus *Fothergilla* comprises four or five species of deciduous shrubs natives of the southeastern United States. A member of the witch-hazel family HAMAMELIDACEAE, it has a name honoring John Fothergill, the English eighteenth-century physician and friend of Benjamin Franklin and John Bartram, who championed the cause of the American colonists during the Revolution and introduced many American plants into cultivation in England.

Fothergillas are beautiful hardy shrubs with alternate, stalked, toothed leaves and handsome terminal, feathery, dish-mop heads of fragrant, creamy-white to white, bisexual flowers that come in spring. The petal-less flowers have bell-shaped, four- to seven-lobed calyxes, and fifteen to twenty-four stamens that thicken from base to apex to form the showy parts of the flowers. There are two styles. The fruits are two-seeded capsules. In fall the foliage colors brilliantly, turning to shades and tones of yellow, orange, red, and purplish-red.

The most ornamental, ***F. major,*** of the southern Appalachians, 3 to 6, rarely nearly 20 feet tall, upright or spreading, is partly in leaf when the flowers appear. When fully grown the broad-ovate, obovate, or nearly round leaves, toothed in their upper two-thirds are up to 4 inches long. Their upper sides are green or sometimes glaucous, and essentially hairless, their undersides glaucous and along the veins at least have starlike hairs. The flower spikes are 1 inch to 2 inches long. Closely allied ***F. monticola,*** by some authorities included in *F. major,* differs in having more spreading branches and in being less hairy. It is wild from North Carolina to Alabama.

Native along the margins of swamps from Virginia to Alabama, ***F. gardenii*** (syns. *F. alnifolia, F. carolina*) rarely exceeds 3 feet in height. It has much smaller and narrower, green or less often glaucous leaves than those of *F. major,* toothed only above their middles or not at all. The flower spikes, appearing before the foliage, are from ¾ inch to 1¼ inches long. Up to about 2 feet tall, ***F. parvifolia*** has broad-ovate to nearly round leaves up to 2¼ inches long and gray-hairy on their undersides. Its flowers appear before the leaves. This species is native from North Carolina to Florida.

Garden and Landscape Uses. Just why these really magnificent native American plants are not more widely used in gardens and other plantings is difficult to say. Surely they are among the loveliest of spring-blooming shrubs. In quality and choiceness they compare favorably with the finest other kinds, and they offer no serious problems to the cultivator or propagator. They provide two prime seasons of interest, in spring when they are in bloom and in fall when their foliage colors so attractively. Throughout the summer they are of neat and satisfactory appearance, unobstrusive and blending without conflict with other foliage. Fothergillas can be used with good effect in shrub beds, foundation plantings, and in naturalistic gardens. They associate well with azaleas and other favorite shrubs and are seen at their very best when planted a little distance in front of evergreens. The dark, rich green foliage of the latter provides a perfect foil for the abundant creamy-white, fluffy flower heads of the fothergillas. They flourish in sun or light shade and are best in fairly moist, but not wet, somewhat acid soil.

Cultivation. These shrubs may be transplanted without difficulty. It is well to mix with the soil, before planting, liberal amounts of compost, peat moss, or other decayed organic material. A mulch of the same type, maintained around established specimens is beneficial. During periods of drought regular watering is desirable. Little or no pruning is necessary, but if plants become too large for their allotted space they may be cut back severely in late winter, with sacrifice of the immediate season's bloom, and, then if fertilized and kept watered in dry weather, they soon renew themselves from low down. Propagation is by seeds sown in cold frames in fall, which do not germinate until the second spring after sowing, by layers, by leafy cuttings in summer, preferably under mist, and by root cuttings.

FOUNDATION PLANTING. Grouping shrubs around the foundations of houses is distinctively American. It is not much done in Europe. It is a charming idea that, if carried out correctly, adds immensely to the appearance of the home and to the community, because part of the planting is ordinarily visible from the street.

Unfortunately too many installations are not well done. They suffer from one or more of the following faults, too many plants used, too much variety, poor selection of kinds, poor arrangement.

Fothergilla major

Fothergilla monticola

Fothergilla gardenii

Well-planned foundation plantings add greatly to the beauty of these homes

Foundation planting should tie the house to the ground, partly frame it, emphasize good architectural features, conceal or soften poor ones. It may serve as a background for borders of low flowering plants or lawns. In houses occupied the year around it must look well at all seasons. The character of the planting should depend to a large extent upon the house. A good planting does not compete for attention with the building, it complements and enhances it.

Using too many plants often results from not knowing how fast shrubs grow and their ultimate dimensions. Desire for immediate, filled-in, finished appearance is also responsible. Where the latter is important use a few, large specimens rather than many smaller ones. The cost may not be much greater, the effect will be better, and the shrubs will not crowd and spoil each other quickly. Alternatively, set moderate-sized plants that are to remain permanently at distances appropriate to their mature sizes and fill in with less important shrubs that will be taken out when the permanent specimens begin to crowd. This is satisfactory if you do not fail to remove the fillers before crowding harms the permanent plants, All too often this is not done.

Too many kinds of plants of different appearances produce an uneasy picture. A few that go well together are better. Excellent results may be had by using no more than three kinds at any one side of the house, although good effects may be had with more.

Fouquieria splendens in the Mojave Desert

Poor selection is common. Often young forest giants are set in foundation plantings. Within a few years they grow past the roof, lose lower branches, and need replacing. Ascertain ultimate height and spread of kinds used and, equally important if they are large growers, whether they can be kept to moderate size by pruning. Some, such as yew, firethorn, abelia, and holly, can, others, such as spruces, firs and pines, cannot. Consider shapes. Young plants in nurseries are often sheared to artificial outlines. It is their natural forms that are important because if regularly sheared, shrubs rarely look well in foundation plantings (discreet pruning without formal shaping is often desirable). Plants with very definite forms, such as severely conical spruces, erect, exclamation-point-like red-cedars, and globular arbor-vitaes are unsuitable for most plantings, although conical forms can be used successfully with gabled houses. In most cases less emphatic forms are better; their softening effects and blending qualities are more pronounced.

Avoid highly colored foliage, it lacks the dignity and restraint needed. Flowers are less important than foliage. If you use them be sure the colors go with your house—magenta azaleas against red brick are disturbing! Evergreens should form a substantial part of all foundations. They may be used alone or in combination with leaf-losers. Use plenty of groundcovers to tie the shrubs together.

Group the shrubs, do not string them in a single line. Place the chief groups where strong vertical architectural lines meet the ground, at corners of the house and on either side of the doorway. Unless the house is very low, connect these with lower plantings. If the house is tall extend the corner groups for some way from it, thus creating a wider base line. With low houses use plants of spreading, squatty form that emphasize horizontal lines. Avoid strict symmetry. Balance must be achieved, but need not depend upon planting each side of center identically. Asymmetrical balance is more interesting than the symmetrical kind. It is quite necessary when the house is an off-center type.

Do not feel you must cover all the foundation with plants. Unless it is ugly this is not desirable. Avoid planting massive or tall-growing shrubs under windows. There, a low shrub, a vine trained to the wall, or a groundcover, will suffice. If you plant under overhanging roofs make sure the shrub can be adequately watered. Beware of setting kinds subject to breakage where snow from roofs can slide on them. Snow guards on the roof will check this. If the house is without gutters, take care not to center the shrubs directly under the drip line of the roof. They should be at least 1 foot farther out from the building.

FOUNTAIN PLANT is *Russelia equisetiformis.*

FOUNTAIN TREE. See Spathodea.

FOUQUIERIA (Fou-quièria)—Ocotillo or Candlewood or Coachwhip. Endemic to Mexico and the southwestern United States and comprising nine species, *Fouquieria,* of the fouquieria family FOUQUIERIACEAE, has a name that commemorates Pierre Edouard Fouquier, a nineteenth-century French professor of medicine. By some authorities the boojum tree of Baja California, in this Encyclopedia treated as *Idria,* is included in *Fouquieria.*

Fouquierias are resinous inhabitants of deserts. Some are trees with erect, stout, branched trunks, others are much-branched shrubs, and a third type, the ocotillo group, are shrubs with erect or leaning, mostly branchless, wandlike stems. The trunks and stems contain chlorophyll and synthesize food in the manner of green leaves. The blades of the leaves fall early, leaving the stalks to develop as formidable thorns from the axils of which, later, secondary leaves sprout. In terminal racemes or panicles, the often showy flowers make attractive displays. They have five sepals, five petals joined in their lower parts into a tube, ten stamens or more, and a three-branched style. The blooms of some sorts are pollinated by birds. The fruits are capsules containing flattened seeds with hair-like appendages.

Fouquieria splendens (foliage and flowers)

Only one species is much cultivated. It is the ocotillo, candlewood, or coachwhip (***F. splendens***), a native of dry, usually rocky sites from Texas to California and Mexico. It occurs in the Arizona-Sonora, Mojave, and Colorado deserts and has the widest natural distribution of any species. From the base it produces several to many erect or leaning canes from 6 to 20 feet, but more commonly 8 to 10 feet, long and branchless or branched from their bases. These are gray with darker longitudinal furrows and have stout thorns spreading at about right angles. The fleshy, single-veined leaves are oblong-obovate and ½ to 1 inch long. In slender panicles 4 to 10 inches in length, the numerous flowers are crowded. They have tiny, rounded sepals and narrow, cylindrical, orange-scarlet or scarlet corollas ¾ to 1 inch long, with rolled-back, rounded lobes. There are ten to seventeen protruding stamens and three styles. The white seeds are in capsules almost ¾ inch long.

In the desert the strange, slender wands of ocotillo stand leafless for many rainless months. During that time the roots, although deep in the soil, obtain little or no water. Yet the ocotillo lives. This is possible because of its xerophytic nature (its ability to conserve moisture). Unlike many cactuses and other succulents, the ocotillo does not store considerable amounts of water, but it is extremely parsimonious with the meager supplies it obtains. When leafless, its expenditure of moisture from above-ground parts is very small indeed. with the coming of rain, the seemingly dead stems are transformed, as by magic, into wands of bright green. Along their lengths leaves appear. During their brief lives these work hard to photosynthesize the food for the plants' needs. When the moisture supply is exhausted the leaves drop promptly and the plant relapses into somnambulance, until, many months later perhaps, it is again awakened into activity by the availability of moisture. The name candlewood is applied because the dry stems, if ignited, burn from one end like candles.

Garden and Landscape Uses. In warm desert and semidesert regions the ocotillo is a favorite garden shrub. It is used to add grace and variety to landscapes composed largely of fat and rigid cactuses and other succulents and is esteemed for its slender banners of brilliant blooms. It is much used as an impenetrable hedge plant. In other areas, collectors of succulents appreciate this species and it is not uncommonly included in greenhouses as well as in outdoor collections. It is hardy only in regions of slight or no frost. It is the most cold-resistant species.

Cultivation. The conditions under which these remarkable plants grow naturally suggest their needs when ministered to by gardeners. Perfectly drained soil that is dry for a considerable part of each year, exposure to full sun, and a minimum winter temperature, when grown in greenhouses, of 50 to 55°F are right. Indoors, following a winter of dryness, watering should be initiated in late spring and practiced until early fall, unless the plant drops its foliage before then. If the foliage falls, watering should cease. Propagation is easy by seeds. Cuttings of the ocotillo root readily.

FOUQUIERIACEAE — Fouquieria Family. The only genera of this family of dicotyledons are *Fouquieria* and *Idria.* Its sorts are trees and shrubs, native of dry regions in Mexico and the southwestern United States. They have alternate, undivided leaves, the blades of which soon fall, leaving the stalks to harden into persistent spines. Clusters of secondary leaves grow from the axils of the spines. In terminal racemes or panicles, flowers are small, but numerous and those of some sorts showy. They have five sepals, a five-lobed corolla, and ten to seventeen stamens. The fruits are capsules.

FOUR O'CLOCK is *Mirabilis jalapa.*

FOXGLOVE. See Digitalis. False-foxglove is *Aureolaria,* Mexican-foxglove *Tetranema.*

FOXTAIL. This word is employed as part of the common names of these plants: foxtail grass (*Alopecurus*), foxtail-lily (*Eremurus*), foxtail-millet (*Setaria*), and foxtail orchid (*Rhynchostylis*).

FRAGARIA (Frag-ària)—Strawberry. Strawberries cultivated for their fruits are the best-known members of this genus. There are many varieties, the result of careful selection and hybridizing over a long period of years. They are discussed under Strawberry. Here we are concerned chiefly with the wild species and varieties of them that are grown for interest and ornament.

The genus *Fragaria* belongs in the rose family ROSACEAE. It consists of fifteen species, natives of temperate parts of Europe, Asia, North America, South America, and Hawaii. Its name comes from the Latin *fragum,* fragrant, and refers to the pleasant odor of the fruits.

Strawberries are low perennial herbaceous plants, stemless except for long runners that develop young plantlets. Their leaves characteristically have three leaflets. The flowers somewhat resemble single roses or potentillas. Typically they are white or yellowish, rarely pinkish, and are in raceme-like clusters. The termination of the stem to which the flower parts are attached, which botanists call the receptacle, enlarges and becomes fleshy after the blooms fade and forms what is popularly called the berry or fruit. The true fruits, commonly referred to as seeds, are attached to the outsides of the receptacles. In this discussion the words fruits and seeds are employed in the popular sense. Strawberry flowers normally have five sepals, five spreading petals, and numerous stamens. The fruits are red or, less commonly, white.

Fragaria chiloensis

The chief parent of strawberries cultivated for their fruits is ***F. chiloensis.*** Its horticultural varieties are grouped under the name *F. c. ananassa.* The species is indigenous from Alaska to California, and in Chile. A compact plant, 8 inches or so tall, it has thickish leaves, slightly glossy green above and bluish-white and more or less hairy on the veins beneath. The coarsely-toothed leaflets are ovate-wedge-shaped. The white flowers, about ¾ inch in diameter, are succeeded by red fruits that have their seeds sunken in tiny pits. The fruiting stalks are shorter than the foliage. The runners mostly develop after flowering.

The American ***F. virginiana,*** a frailer plant than the last, develops its runners earlier. Its leaves are of thin texture and

Fragaria virginiana

Fragaria vesca as a groundcover

Fragaria vesca variegata

are green beneath. Borne on upright stalks usually shorter than the leaves, the clusters of ¾-inch-wide blooms are of up to twelve flowers. The fruits are light red and have their seeds sunk in deep pits. Botanists recognize several varieties of this species, which is widely distributed throughout eastern North America to as far north as Hudson Bay.

Native of Europe and North America, ***F. vesca*** differs from the kinds described above in that its seeds are not in pits, but sit on the surfaces of the fruits. Also, its flowers are mostly smaller and in clusters that are usually taller than the foliage. The teeth of the leaf edges are more spreading than those of *F. chiloensis* and *F. virginiana.* The fruits of *F. vesca* are typically red, those of *F. v. alba,* white. There is also a variety with white-margined leaves and white berries, *F. v. variegata,* and one, *F. v. monophylla,* in which the leaves have only one leaflet.

Garden and Landscape Uses and Cultivation. Wild strawberries are obvious candidates for inclusion in native plant gardens within their natural ranges. They may also be used effectively as groundcovers and edgings. For these purposes *F. chiloensis* is particularly well adapted and so is *F. vesca variegata.* In soil that is overly rich, the latter is likely to produce green-leaved crowns. These should be removed as soon as they appear, otherwise, being stronger growers than the variegated crowns, they are likely to become predominant. Fragarias thrive in fertile, moderately moist, well-drained soil in sun. Propagation is by plantlets that develop along the runners and by seeds.

FRAGRANT-BALM is *Monarda didyma.*

FRAGRANT PLANTS and FLOWERS. Most people associate gardens with fragrance, and many popular plants and flowers gratify the nose as well as the eye. One has only to call to mind carnations, gar-

Fragaria chiloensis as a groundcover in California

denias, lavender, lilacs, lily-of-the-valley, orange blossom, roses, sweet peas, and violets to realize that this is true. There are numerous other fragrant plants, some suitable for cut flowers, others not. What appears to be a fairly well-founded belief is that flowers of certain plants that have been bred intensively by hybridists to obtain larger blooms in wider ranges of colors and forms tend to lose their fragrance. Roses and carnations are often cited as examples of this. But among such sorts there are usually available some varieties that have retained their perfume, even though they may not be as fine in other respects as the less highly scented varieties.

Fragrance is, by definition, a sweet or pleasing scent. Not all plant odors are this. Some, those of skunk cabbage, certain stapelias, and species of *Amorphophallus* for example, are offensive sometimes to the point of being nauseating. Such are not our concern. Our discussion is about plants with scents usually accepted as agreeable.

But fragrances affect people in different ways. What is pleasant in mild whiffs may be disagreeable in stronger concentrations. The heady perfumes of gardenias, tuberoses, and some lilies are cases in point. And reactions to some scents are distinctly subjective; what one person finds pleasing, another may not.

The flowers of certain plants, including *Cestrum nocturnum,* lilacs, lilies-of-the-valley, mignonette, roses, and sweet-peas, pervade the air with fragrance by night and day. Others, of which examples are *Akebia quinata, Gladiolus tristis,* nicotianas, and evening stock, do so only at night. The fragrances of yet a third group are intensified at night although perceptible by day. Here belong honeysuckles, pinks, and tuberoses. Other circumstances may influence perfume intensities. Humid weather with little or no breeze makes most readily discernible the fragrances of the musk grape-hyacinth and many other plants, hot sun does this with others, and sometimes exposure to light frost seems to strengthen fragrance. This is true for heliotropes.

Fragrant foliage is characteristic of many plants. With most it is necessary to crush or at least brush against the leaves to detect the scent, but some few including English boxwood and sweet briers, perfume the air to some extent without contact.

Certain fruits when ripe are deliciously fragrant, but rarely are the plants that bear them planted with this advantage in mind.

Among fragrant fruits are osage-orange, trifoliate-orange (*Poncirus trifoliate*), grapes, peaches, and strawberries.

Here are selections of fragrant plants in various categories, some hardy, some not.

Trees: acacias (some kinds), black locust (*Robinia*), *Citrus* (many kinds), crab apples (some kinds), *Eucalyptus citriodora, E. globulus,* fringe tree (*Chionanthus*), honey-locust (*Gleditsia*), *Hymenosporum flavum,* lindens (*Tilia*), and magnolias (some kinds).

Shrubs: angel's trumpet (*Brugmansia*), artemisias (several kinds), azaleas (some kinds), boronias, bouvardias (some kinds), brooms (some kinds), buffalo currant (*Ribes odoratum*), bush honeysuckles (*Lonicera*), button bush (*Cephalanthus*), *Carissa* (several kinds), Carolina-allspice (*Calycanthus*), *Cestrum diurnum, C. nocturnum, C. parqui,* clerodendrums (some kinds), *Daphne cneorum, D. mezereum,* English hawthorn (*Crataegus*), frangipani (*Plumeria*), gardenias, laurestinus (*Viburnum tinus*), lavender (*Lavandula*), lilacs (most kinds), loquat (*Eriobotrya*), Mexican-orange (*Choisya*), michelias, mock-orange (*Philadelphus*), myrtle (*Myrtus*), orange-jessamine (*Murraya*), *Osmanthus fragrans,* pittosporums (some kinds), rosemary (*Rosmarinus*), roses (many kinds), salvias (some kinds), sarcococcas, spice-bush (*Lindera*), sweet pepperbush (*Clethra*), *Viburnum burkwoodii, V. carlesii, V. fragrans,* Virginia-willow (*Itea*), winter sweet (*Chimonanthus*), and witch-hazels (*Hamamelis*) several kinds.

Perennial vines: *Akebia quinata,* Carolina yellow-jessamine (*Gelsemium*), cinnamon vine (*Dioscorea*), *Clematis* (some kinds), honeysuckles (*Lonicera*) many kinds, jasmine (*Jasminum*) several kinds, moonflower (*Ipomoea*), potato-bean (*Apios*), *Stephanotis,* and wisterias.

Herbaceous perennials: auriculas (*Primula auricula* varieties), carnations, *Clematis recta, Dianthus* (some kinds), irises (some kinds), lily-of-the-valley, meadowsweet (*Filipendula*), peonies (some kinds), pinks, and violets (some kinds).

Bulb plants: *Allium tuberosum, Amaryllis belladonna,* Amazon-lily (*Eucharis*), crinums, *Gladiolus tristis,* grape-hyacinths (*Muscari*) some kinds, hyacinths, jonquils (*Narcissus jonquilla*), lilies (*Lilium*) many kinds, *Lycoris squamigera, Narcissus poeticus* varieties, spider-lily (*Hymenocallis*), and tulips (some kinds).

Annuals: candytuft (*Iberis*), daturas, English wallflowers (*Cheiranthus*), evening stock (*Matthiola*), flowering tobacco (*Nicotiana*), heliotrope (*Heliotropium*), mignonette (*Reseda*), night-phlox (*Zaluzianskya*), stocks (*Matthiola*), sweet-alyssum (*Lobularia*), sweet-peas (*Lathyrus*), sweet rocket (*Hesperis*), sweet scabious (*Scabiosa*), sweet sultan (*Centaurea*), sweet william (*Dianthus*), and verbenas.

With fragrant foliage: boxwood (*Buxus*), camomile (*Chamaemelum*), *Coleonema album,* lemon-verbena (*Aloysia*), mints (*Mentha*), myrtle (*Myrtus*), pineapple sage (*Salvia*), rosemary (*Rosmarinus*), scented-leaved geraniums (*Pelargonium*) many sorts, sweet bay (*Laurus*), and thymes (*Thymus*). Additional plants with fragrant foliage are discussed under Herb Gardens.

FRAILEA (Fràil-ea). Widely distributed as natives of South America, chiefly of its southern part, the dozen or so somewhat uncertainly defined species of *Frailea* belong to the cactus family CACTACEAE. Their name honors Manuel Fraile, one-time caretaker of the cactus collection of the United States Department of Agriculture.

Fraileas are small, their plant bodies, spherical, flattened-spherical, or shortly-cylindrical, are often clustered. They have shallow, usually more or less knobby ridges, with clusters of weak spines. Arising from near the tops of the plant bodies, the funnel- to bell-shaped flowers open during the day or not at all (they are often self-fertile and capable of producing seeds without cross pollination). Their calyx tubes have small scales within their axils woolly hairs and bristles. The fruits are small and dry.

Globular ***F. cataphracta*** forms clusters of plant bodies each when fully grown about 1¾ inches wide. They have fifteen poorly defined, warted ribs with areoles (places from which spines develop) with clusters of five to nine all radial, tiny yellow to gray spines. The flowers are up to 1½ inches long. This is native to Paraguay. Also inhabiting Paraguay, flattish-spherical ***F. grahliana*** is brownish-green sometimes tinged with purple. Its plant bodies, about 1½ inches in diameter, have thirteen indistinct, shallowly-warted ribs, their areoles with clusters of usually nine to eleven harmless, gray or brown, all radial, curved spines about ⅛ inch long. The red-centered, pale yellow blooms are 1½ inches long. Another globular species, this a native of Colombia, ***F. columbiana*** has bright green plant bodies about 1½ inches wide and generally with eighteen rather indefinite ribs bearing clusters of yellow to white spines each of fifteen to twenty ¼-inch-long radials and two to five centrals. The 1-inch-long blooms are yellow. Paraguay and Argentina are homelands of ***F. schilinzkyana.*** Its solitary or clustered plant bodies are about 1 inch in diameter and flattened-spherical. They have ten to thirteen indistinct ribs and clusters of twelve to fourteen tiny, flattened radial spines and one longer and thicker central spine. The flowers are small, the fruits yellowish.

Differing from the kinds described above, ***F. gracillima*** of Paraguay has usually branchless, grayish-green stems about 4 inches tall by 1 inch wide. They have about thirteen indistinct, somewhat spiraled, warty ribs and clusters of twenty to nearly thirty spines of which four to eight centrals, up to ⅓ inch long, considerably exceed the others. The yellow flowers are a little over 2 inches in length.

Garden Uses and Cultivation. Among many small cactuses esteemed by fanciers of succulent plants, fraileas rank highly. They grow satisfactorily under conditions that suit mammillarias and plants of similar growth habits. For more information see Cactuses.

FRANCISCEA. See Brunfelsia.

FRANCOA (Fran-còa)—Maiden's Wreath or Bridal Wreath. The four or five species of *Francoa,* of the saxifrage family SAXIFRAGACEAE, are by some authorities accepted as representatives of one variable species. They are natives of Chile. The name commemorates a sixteenth-century physician and patron of botany, Francisco Franco, of Spain. The names maiden's wreath and bridal wreath are also applied to *Spiraea prunifolia* and *S. trichocarpa.*

Francoas are nonhardy, softly-hairy, evergreen herbaceous perennials with thick rootstocks and clustered, short-stalked crowns of chiefly basal and near-basal foliage. The leaves are alternate, oblanceolate to fiddle-shaped, 6 inches to 1 foot long. They have a few deep, blunt lobes, the terminal one much the largest. Like the stems, the foliage is slightly glandular-sticky. White to pink, the flowers are in crowded, terminal, more or less one-sided, spirelike racemes or panicles of racemes that reach above the foliage to heights of 3 to 5 feet or sometimes more. They are graceful and beautiful and when cut last well in water. The blooms have usually four, more rarely five, persistent lanceolate sepals and the same number of petals, or only two, that usually narrow to claws at their bases and generally have a darker spot there. There are four slender-stalked stamens with alternating staminodes (nonfunctional stamens). The fruits are erect, leathery capsules.

Maiden's wreath or bridal wreath (***F. ramosa*** syn. *F. glabrata*) has much-branched flowering stalks bearing a multitude of

Francoa ramosa

white flowers. The blooms of *F. sonchifolia* are pink marked with a darker hue. The leafstalks of this are broadly-winged at their bases. Sometimes spotted with red, the pale pink flowers of *F. appendiculata* are in mostly compact, little-branched arrangements.

Garden and Landscape Uses and Cultivation. Maiden's wreath or bridal wreath is suited for permanent outdoor cultivation only in dryish, nearly frost-free climates, such as that of California. It is excellent for perennial borders and other flower garden beds and gives a fine display over a long summer season. In colder climates this plant may be treated as a biennial, wintering it in a cool greenhouse or frostproof or nearly frostproof cold frame. It can be raised from cuttings, but seed is usually preferred. This is sown in spring or early summer in a temperature of about 55°F. The seedlings, as soon as they are big enough to handle easily, are transplanted to small pots, flats, a cold frame, or an outdoor nursery bed. Later they are transferred from the small pots or flats to larger pots if they are to be grown in greenhouses, or in climates where they are winter hardy to where they are to bloom. When grown throughout in containers, as finals they need pots or tubs 8 to 10 inches in diameter. In permanent locations in garden beds they may be spaced about 2 feet apart. Fertile loamy soil, well drained, and not lacking moisture, but not for long periods wet, suits francoas. They need full sun and free air circulation. In greenhouses winter night temperatures of 40 to 50°F are adequate, with not more than a few degrees increase by day. A dryish rather than a very humid atmosphere is appreciated.

FRANGIPANI. See Plumeria.

FRANKENIA (Fran-kènia) — Sea-Heath. Lovers of seaside environments and saline and gypsum soils, the about twenty-five species of *Frankenia,* of the frankenia family FRANKENIACEAE, are natives of tropical, subtropical, and temperate regions. They are evergreen, more or less heathlike, annuals, herbaceous perennials, and subshrubs. A few kinds are cultivated for ornament. The name honors John Frankenius, a Swedish professor of botany, who died in 1661.

Frankenias have stalkless or nearly stalkless leaves, opposite or in fours, and often on short branchlets. Their flowers, small, white or pink, and stalkless or nearly so, are solitary or clustered. They have five-lobed, tubular, persistent calyxes and five separate petals. There are four or six stamens and a three- or four-cleft style. The fruits are three-angled capsules.

Native to Texas, New Mexico, and Colorado, *F. jamesii* is a much-branched shrub 1 foot tall or somewhat taller. It has

Frankenia hirsuta

pointed-linear leaves up to ⅓ inch long, with strongly rolled-under margins. Its stalkless, white flowers, solitary or clustered, have petals about ⅓ inch in length.

Mat-forming, and with prostrate, woody, often red stems 1 foot long or somewhat longer, from which come many short, erect, leafy branches, *F. hirsuta* (syn. *F. capitata*) ranges in the wild from the Mediterranean region to India and South Africa. Its minutely-stalked, waxy, more or less hairy leaves are linear and up to ⅓ inch long. Most have strongly rolled-under margins. In conspicuous, stalkless clusters at the ends of the main stems and branches, the deep purplish-pink to white flowers are ¼ to nearly ½ inch across. This variable species inhabits sand and shingle strands and saline soils inland.

The sea-heath (*F. laevis*) is a mat-forming perennial. Its procumbent, much branched stems, up to 1 foot long or somewhat longer, are more or less hairy. Unlike those of *F. hirsuta,* the starry blooms, solitary or in small clusters, are scattered throughout the upper parts of the stems instead of being in terminal clusters. They are purplish to white, and ¼ to nearly ½ inch across. This is native from southern England to the Mediterranean region.

Native of deserts in North Africa, pretty *F. thymifolia,* an inch or two high, has creeping stems and crowded, grayish, thick, linear to oblong leaves with turned-under margins. Its flowers, solitary or in groups of few, are pink.

Garden and Landscape Uses and Cultivation. Just how hardy these rarely cultivated plants are is not known, but it is improbable that the first two discussed above will survive in the north, the other may. They are suitable for rock gardens and groundcovers, and because they appreciate salt air, saline soils, and sun, they thrive particularly near the sea. They are propagated by seeds and by cuttings.

FRANKENIACEAE—Frankenia Family. Of minor horticultural significance, this family of dicotyledons consists of four genera totaling ninety species of herbaceous plants or rarely subshrubs of saline soils of tropical, subtropical, and warm-temperate regions. Characteristics of the family are jointed stems and undivided, opposite, often heathlike leaves with inrolled margins, alternate pairs of which are set at right angles to each other. In terminal or axillary clusters or solitary, the small, symmetrical flowers have four to seven sepals, as many petals as sepals, four to seven, most commonly six, stamens, and one style. The fruits are capsules enclosed in persistent calyxes. Only *Frankenia* is likely to be cultivated.

FRANKLIN TREE is *Franklinia alatamaha.*

FRANKLINIA (Frank-lín-ia)—Franklin Tree. Named in honor of Benjamin Franklin, who died in 1790, *Franklinia,* of the tea family THEACEAE, consists of one species. It was discovered by John Bartram and his son William near Fort Barrington close to the mouth of the Altamaha River, Georgia in 1765, and was seen there again by William in 1773. There is no record of it having been found in the wild since 1803, when John Lyon, a nurseryman plant hunter, reported finding a few specimens on not more than one-half an acre of ground near the site of the old Fort Barrington. Fortunately, Bartram collected plants and seeds and established *Franklinia* in his garden at Philadelphia. All known specimens of this interesting tree are descendants of Bartram's original collections.

Franklinia alatamaha

The Franklin tree (*F. alatamaha* syn. *Gordonia alatamaha*) is deciduous, pyramidal, and about 30 feet in height. It has downy twigs and alternate, obovate-oblong, lustrous leaves up to 6 inches long that are pubescent beneath; they turn brilliant orange or red in fall. Its white, shallowly-cup-shaped flowers, about 3 inches in diameter, have a five-lobed, silky-hairy calyx, five petals, many conspicuous, yellow stamens, and a style tipped with a lobed stigma. The blooms resemble single

Franklinia alatamaha (leaves and flower)

camellias and are borne over a long period in late summer and fall. The fruits are woody capsules. This tree is hardy in somewhat sheltered locations in southern New England, but north of New York City is likely to succeed better if pruned to maintain it as a tall shrub rather than as a tree.

In the 1970s hybrids between the Franklin tree and related *Gordonia lasianthus* were raised in New Jersey at Rutgers University. Intermediate between their parents in appearance, these grew vigorously and bloomed freely but unfortunately proved susceptible to a wilt disease that caused them to die when one to three years old.

Garden and Landscape Uses and Cultivation. The Franklin tree is highly ornamental as well as interesting because of its unique history. It is admirable for including in mixed plantings and for displaying as a single specimen. It succeeds in acid to alkaline soil and may be increased by seeds, cuttings, and layers. Propagation is most likely to succeed if an acid rooting medium is used.

FRASERA (Fras-èra) — Green-Gentian, American Columbo. In the wild confined to North America, this genus of fifteen species of coarse, gentian-like, biennials and herbaceous perennials is named after an English nurseryman and collector of American plants, John Fraser, who died in 1811. It belongs in the gentian family GENTIANACEAE and is related to *Swertia.* The chief differences are that the flowers of *Frasera* have their parts in fours, and quite evident styles, whereas those of *Swertia* have extremely short styles or none, and flower parts usually in fives.

Fraseras have deep taproots, erect stems, and undivided, toothless, thickish leaves, opposite or in whorls (circles of more than two). Their blooms, in terminal panicles, have deeply-four-lobed calyxes and short-tubed, wheel-shaped corollas divided nearly to their bases into four lobes (petals), each with one or two more or less fringed glands. There are four stamens and a slender style that ends in a two-lobed stigma. The fruits are many-seeded capsules.

American columbo (***F. caroliniensis***) is 3 to 6 feet tall or sometimes taller. A perennial that occurs as a native from New York to Wisconsin, Georgia, Mississippi, and Arkansas, it inhabits rich woodlands. In whorls (circles) of mostly four, its leaves become progressively shorter upward. The basal ones may exceed 1 foot in length and are oblanceolate. Greenish-yellow spotted with purple, the blooms, 1 inch to 1¼ inches across, are in loose panicles up to 2 feet long.

Western American ***F. speciosa*** (syns. *Swertia radiatum, Tessaranthium radiatum*) is 3 to 6 feet tall. It has short-hairy leaves 4 to 10 inches long varying from oblanceolate to obovate below to lanceolate-oblong above. Those on the stems are in whorls of three to seven. The greenish-white, purple-dotted blooms have two long-fringed glands on each ½- to ¾-inch-long petal. They are in narrow panicles 1 foot to 2 feet in length. This species occurs in dry and moist places at high altitudes from the Rocky Mountains to Washington and California.

White-edged leaves and flowers with one gland only on each petal are characteristic of *F. parryi* (syn. *Swertia parryi*) and *F. albicaulis* (syn. *Swertia albicaulis*), the former an inhabitant of dryish places in southern California and Baja California, the other of dry or moist soils from California to Washington and Montana. From 1 foot to 2 feet tall, ***F. parryi*** is without hairs. Its basal leaves have short, winged stalks and are oblanceolate and 4 to 8 inches long. The stem leaves are opposite, stalkless, smaller, and lanceolate-ovate. The black-dotted, greenish-white blooms, in broad panicles 6 inches to 1 foot long, are 1 inch or a little more in diameter. Up to about 1½ feet tall, ***F. albicaulis*** has stems and leaves with short hairs and narrow panicles of flowers 2 to 6 inches long. Its lower leaves are narrowly-oblanceolate to slender-spoon-shaped, have winged stalks, and are up to 6 inches long. The stem leaves are few, opposite, stalkless, and smaller than the basal ones. Greenish-white to bluish, the blooms ½ to ¾ inch across, are in panicles with short lengths of stalk showing between the circles of flowers. Variety *F. a. nitida* is hairless and has violet-tinged, often violet-dotted flowers.

Garden and Landscape Uses and Cultivation. Chiefly in regions where they are natives fraseras are sometimes planted in gardens, usually after having been transplanted from the wild. They are easy to raise from seeds, however, and the perennials, which all of those discussed here are believed to be, can be increased by careful division. They are suitable for inclusion in collections of native plants and for naturalistic areas. Usually they adapt quite well to garden environments that reasonably duplicate those to which they are accustomed in the wild.

FRAXINELLA. See Dictamnus.

FRAXINUS (Fráxin-us)—Ash. Seventy species of Northern Hemisphere deciduous trees or rarely shrubs comprise *Fraxinus.* Most are natives of North America, eastern Asia, or the Mediterranean region. The southernmost inhabit Mexico and Java. They belong in the olive family OLEACEAE. The generic name is the ancient Latin one.

Ashes mostly have attractive, ornamental foliage, some have showy flowers.

Fraxinus americana

Fraxinus americana (trunk)

Fraxinus americana (young leaves and flowers)

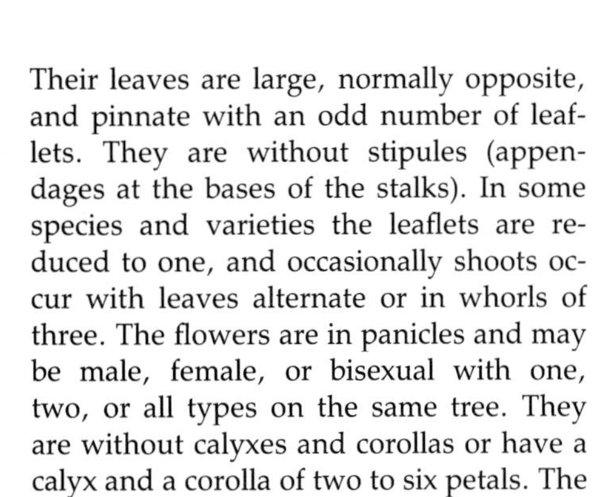

Their leaves are large, normally opposite, and pinnate with an odd number of leaflets. They are without stipules (appendages at the bases of the stalks). In some species and varieties the leaflets are reduced to one, and occasionally shoots occur with leaves alternate or in whorls of three. The flowers are in panicles and may be male, female, or bisexual with one, two, or all types on the same tree. They are without calyxes and corollas or have a calyx and a corolla of two to six petals. The stamens are usually two and there are two stigmas. The fruits, technically samaras, are one-seeded nutlets with a wing attached at the apex.

In addition to some kinds being used as ornamentals, certain ashes are important sources of lumber. The wood of many is strong, elastic, and straight-grained, but that of others, for instance the pumpkin ash (*F. tomentosa*), is soft, light, and brittle. The wood of those exploited as lumber is used for the handles of such tools as spades, shovels, rakes, and hoes. It is inferior to that of hickory where great impact shock must be absorbed, as is the case with hammer and axe handles. It is also employed for oars, baseball bats, agricultural implements, kitchen furniture, and interior trim. A sugary, mildly laxative nutrient called manna is obtained from the manna ash (*F. ornus*) by cutting its bark in summer and collecting the exudate, which is used medicinally. In China *F. chinensis* and *F. mariesii* are used as pasture for wax-prod-

Fraxinus americana (leaf)

Fraxinus americana (fruits)

ucing coccus insects that are bred on *Ligustrum lucidum* and other kinds of privet. In April the insect cocoons are transported by night, often for considerable distances, to ash trees and are fixed to their branches. Some four months later the branches, with the attached insects encrusted in wax, are cut off and boiled to extract the wax, called pe-la, which is used for glossing silk and coating pills and candles.

From a horticultural point of view ashes may be divided into two groups: those that have flowers with showy corollas, which are generally known as "flowering ashes," and those with an insignificant floral display which are grown as shade trees and for other decorative purposes. The latter are more commonly cultivated in America. The blooms of the "flowering ashes" are in terminal and axillary panicles and develop with or after the foliage. Those of the other kinds appear from lateral buds, usually before the leaves.

The white ash (***F. americana***), of eastern North America, is one of the most familiar to gardeners throughout the eastern and central United States. It occurs both wild and cultivated. Of erect habit and with a rounded top, this species attains a height of about 120 feet and grows rapidly. Its leaves usually have seven stalked leaflets, but sometimes five or nine. They are dark green above and glaucous and usually hairless beneath. In fall the foliage turns interesting shades of purple or yellow before it drops. The rachis to which the leaflets are attached is not winged. The flowers of the white ash are bisexual, their calyxes are persistent on the fruits. The wood of this tree, a native from Nova Scotia to Florida and Texas, is of considerable commercial importance. Variety *F. a. ascidiata* is distinguished by having leaves slightly hollow or pitcher-shaped at their bases. The red-purple fruits of *F. a. iodocarpa* are showy in summer. Variety *F. a. pendula* has drooping branches. Silvery-white undersides to its leaves and thick leaflets are characteristic of *F. a. subcoriacea.* A chiefly northern variety, *F. a. juglandifolia* has leaves duller than those of the typical tree. They are less glaucous and more or less pubescent beneath and have more obviously toothed margins.

The red or green ash (***F. pennsylvanica***), is native from Nova Scotia to Manitoba, Georgia, and Mississippi. Up to 60 feet in height, with an irregular crown of stout, upright branches, it has densely-downy, twigs, flower panicles, and leafstalks. The leaves, pubescent beneath, are of five to nine stalked leaflets. The flowers have calyxes that persist with the fruits, which are ¾ inch to 1½ inches long. This variable species merges almost imperceptibly into its variety *F. p. lanceolata.* Most typically the latter is a round-topped tree with slender, spreading branches, hairless twigs, leafstalks, and flower panicles, and leaflets

Fraxinus velutina

that if downy are so only on their midveins beneath. This variety is native from Maine to Saskatchewan, Florida, and Texas. A variety of horticultural origin, *F. p. aucubaefolia,* has yellow-mottled foliage.

The velvet ash (***F. velutina***), of the southwestern United States and Mexico, much less hardy than the white and red ashes, is appropriate for planting in the warmer parts of North America. Especially suited for dry climates, it and its varieties adapt well to saline and alkaline soils. The velvet ash attains a height of 25 to 50 feet and has a slender or rounded crown and leaves of three to five short-stalked or nearly stalkless leaflets, hairy on their undersides. The flower panicles are also hairy. The calyxes are persistent on the fruits. The fruits are about ¾ inch long. The modesto ash (*F. v. glabra*), a variety with glabrous twigs and leaflets native from Arizona to Mexico, in warm dry climates is much favored as a street tree. The Montebello ash (*F. v. coriacea*) has less hairy, more leathery leaves than the typical species.

Another somewhat tender kind, perhaps hardy as far north as Long Island, N.Y., ***F. latifolia*** (syn. *F. oregona*), is native from Washington to southern California. About 80 feet tall, it has unisexual flowers that come before the leaves. Unlike those of somewhat similar *F. platypoda*, the leafstalks are not swollen at their bases. The leaves are of five to nine, but mostly seven, generally stalkless leaflets, pubescent beneath, when young on their upper sides also. The flower panicles are glabrous. The calyxes are persistent on the fruits.

The single-leaf ash (***F. anomala***) is very distinct. Native from Colorado to Utah and southern California, it is a shrub or tree up to 20 feet in height, with four-angled glabrous twigs and leaves that usually consist of only one leaflet, but sometimes of two or three. They are broad-ovate to nearly orbicular, dark green above, paler beneath, and, when young pubescent on their undersides. The flowers, bisexual or unisexual, appear in short panicles together with the leaves. The calyx is persistent on the fruits. As something of a curiosity this ash has interest.

The blue ash (***F. quadrangulata***) has bisexual flowers with minute, deciduous calyxes. Its twigs, distinctly four-angled, are usually slightly-winged. The leaves are of seven to eleven short-stalked, sharply-toothed leaflets of a yellowish-green hue that are smooth except for some hairs along the base of the midrib beneath. This native from Michigan to Tennessee and Arkansas on occasion reaches a height of 120 feet, but is usually lower. It has a slender crown of spreading branches. The blue ash, so called because its inner bark crushed in water produces a blue dye, has scaling bark that gives to very old trunks an appearance not unlike that of the shagbark hickory and is conspicuous even on younger trunks.

The European ash (***F. excelsior***) is a round- or oval-headed tree up to 140 feet in height brought to America in early colonial times. Except for the interest of some of its distinct horticultural varieties, it has little to recommend it above native kinds to American gardeners. In fall its leaves drop without change of color. This has glabrous twigs, large black winter buds, and bisexual and unisexual flowers without calyxes. Its leaves of nine to thirteen leaflets are dark green above, paler beneath, and hairless except for some hair on the midribs below. The European ash is indigenous to Europe and western Asia, and there is one of the most important timber trees. It is hardy far north. Varieties of the European ash include *F. e. asplenifolia,* with leaflets up to ¼ inch wide; *F. e. aurea,* a slow grower with yellow branchlets; *F. e. aurea pendula,* with drooping branches and yellow young shoots; *F. e. crispa,* with leaves with dark green, crumpled, twisted or curled leaflets; *F. e. diversifolia,* with leaves with one to three usually lobed or toothed leaflets; *F. e. erosa,* with leaves with very narrow, jagged-toothed leaflets; *F. e. globosa,* a dwarf, rounded, densely-branched, small-leaved

Fraxinus excelsior pendula

Fraxinus angustifolia

Fraxinus ornus (flowers)

Fraxinus spaethiana

shrub often grown as a standard by grafting high on a taller-growing understock (varieties *F. e. nana* and *F. e. polemoniifolia* are similar if not identical); *F. e. pendula*, with drooping branches; *F. e. spectabilis*, pyramidal in outline; *F. e. verrucosa*, with warty branches; and *F. e. verticillata*, with leaves in threes or sometimes alternate.

European ***F. angustifolia*** is an elegant, somewhat less hardy species than *F. excelsior* from which it is readily distinguished by its brown winter buds, narrower leaflets, and shoots and leaves completely without pubescence. Its leaves have from seven, or more commonly from nine, to thirteen leaflets. The flowers are without calyxes. Distinguished by its leaflets spreading from the rachis widely rather than pointing forward and by their being spaced further apart is *F. a. lentiscifolia*. A variety with drooping branches is *F. a. pendula*. Another native of southern Europe, one that occurs also in western Asia, is ***F. oxycarpa***. It has dark brown winter buds and leaves with five to eleven, but usually seven or nine, stalkless leaflets, glabrous except for long hairs on their midribs beneath. This is hardy in southern New England. Variety *F. o. aureafolia* is offered and described as having golden-yellow bark and green leaves that turn yellow in summer. Another variety, offered as *F. o.* 'Flame', is described as having foliage that turns rich wine color in fall. This may be identical with the older variety *F. o.* 'Raywood'.

The best known of the "flowering ashes," and the one most generally called flowering ash or manna ash, is ***F. ornus***, of southeast Europe and western Asia. Becoming 50 to 60 feet in height, this handsome tree should be planted more often. It blooms in May or June and delights with its fragrance. Its flowers are white and are produced in dense, showy panicles up to 5 inches long. The leaves have usually seven stalked leaflets that are reddish-pubescent on their midribs beneath. The winter buds are grayish or brownish. Variety *F. o. rotundifolia* has broad-elliptic to nearly round leaflets. This species is hardy in southern New England. A hardier "flowering ash" is ***F. bungeana***, a shrubby kind from northern China, not over 15 feet in height, and with leaves with glabrous leaflets, usually five in number. Its winter buds are nearly black. Perhaps the most lovely of all "flowering ashes" is ***F. mariesii***, a native of central China adapted for growing in mild climates only. This 20-foot-tall tree has leaves of three, five, or rarely seven, closely set, very short-stalked, hairless leaflets. Its leafstalks and leaflet stalks are often purple toward their bases. This ash has large panicles of attractive creamy-white flowers and deep purple fruits, ornamental in summer.

Two Japanese "flowering ashes" are *F. sieboldiana* and *F. spaethiana*. The first is often known as *F. longicuspis* although that name appears to belong to another species not of the "flowering ash" group. Attaining a height of 50 feet, ***F. sieboldiana*** has winter buds covered with reddish hairs and basal leaflets very much smaller than those above. In fall its leaves, which are usually of five, but sometimes seven leaflets, turn purple. From the last ***F. spaethiana***, of Japan and Korea, differs in its leafstalks being enormously enlarged at their bases and its leaflets not being stalked. On vigorous shoots the leaves may be as much as 1½ feet long. The four-petaled flowers are in terminal panicles. The entire tree has a somewhat yellowish appearance.

Other kinds of *Fraxinus* cultivated are: ***F. biltmoreana***, a handsome tree up to 50 feet tall, similar to *F. americana*, but with pubescent twigs and leaves, and a native of the southeastern United States. ***F. caroliniana***, the water ash, native to swamps from Virginia to Florida and Texas and not hardy in the north, is about 50 feet in height and has brown winter buds and leaves of five to seven stalked leaflets. ***F. chinensis***, of China, about 50 feet tall, has leaves of five to nine short-stalked or almost stalkless leaflets pubescent on their midribs beneath, and flowers that appear with or after the leaves. *F. c. acuminata* has more sharply-toothed, slenderer leaflets than the type. *F. c. rhynchophylla* has leaves with usually five broad-ovate or obovate leaflets larger than those of the type. ***F. cuspidata***, one of "flowering ashes," is a shrub or slender-branched tree up to 20 feet tall with leaves of seven slender-stalked, glabrous leaflets and stamens with very short filaments. It is native to the south-western United States and Mexico. ***F. dipetala***, rarely exceeding 15 feet in height, is shrubby and has quadrangular twigs and leaves of usually five stalked, glabrous leaflets. The flowers, in axillary panicles with the foliage, have two white petals. A native of California, this is hardy in mild climates only. ***F. elonza***, of unknown origin, but related to *F. angustifolia*, is a small tree with whitish warts on its gray-green or yellowish branchlets and leaves of nine to thirteen leaflets, hairy on their midribs beneath. ***F. floribunda***, of the Himalayas, is up to 120 feet in height and has leaves with slightly-winged rachises and seven to nine leaflets, pubescent on their mid-veins beneath. One of the "flowering ashes," its large panicles of bloom appear with or after the leaves. It is hardy only in the far south. ***F. greggii***, native of Texas and New Mexico, and hardy only in the far south, is about 25 feet in height and has leaves of three small leathery leaflets. ***F. griffithii*** is a nearly evergreen "flowering ash" up to 45 feet tall with large flower panicles and leaves of five to eleven leaflets that are not toothed. It is native from China to Malaya and the Philippine Islands and hardy in mild climates only. ***F. mandshurica***, the Manchurian ash, is a native of northeast Asia much like *F. nigra* but it does not usually thrive in North America. Attaining a height of 100 feet, it has bluntly-four-angled stems and leaves of nine to eleven nearly stalkless leaflets, rusty tomentose at their bases and more or less hairy beneath and sometimes above. ***F. nigra***, the black ash, native of moist lowlands from Newfoundland to West Virginia and Arkansas, attains a height of 80 feet and has leaves with seven to eleven stalkless leaflets with a rusty tomentum below at their bases and along the midribs. This is an important timber tree. ***F. paxiana,*** a handsome native of China and the Himalayas, is up to 60 feet in height and hardy in southern New England. One of the "flowering ashes," it has leaves of seven to nine stalkless leaflets larger (up to 7 inches long) than those of *F. ornus* and flowers as or after its foliage develops. ***F. platypoda,*** of China, about 70 feet in height, is similar to *F. oregona,* but has the bases of its leafstalks conspicuously enlarged. Its leaves of seven to eleven stalkless leaflets are paler beneath and are pilose on the veins. This is hardy in southern New England. ***F. potamophila,*** a native of Turkestan, up to about 30 feet in height, has green branchlets and leaves of seven to thirteen (usually nine to eleven) small, stalked leaflets sometimes hairy on their mid-veins beneath. It is hardy in southern New England. ***F. rotundifolia,*** of southern Europe and western Asia, is a shrub or small tree up to about 15 feet in height with slender, often purplish branches and leaves of seven to thirteen stalkless broad leaflets. It is hardy about as far north as Long Island, N.Y. *F. r. pendula* has pendulous branches. ***F. sogdiana*** is a small tree with leaves of seven to eleven glabrous leaflets. A native of Turkestan, this is hardy in southern New England. ***F. syriaca,*** of central and southwest Asia, is a small tree with sea-green leaves, whorled and rather crowded on the branches, of three, five, or occasionally seven leaflets. ***F. texensis,*** hardy in southern New England, indigenous to Texas, and closely related to *F. americana,* and up to 50 feet tall, has leaves of usually five leaflets. ***F. tomentosa,*** the pumpkin ash, is possibly a hybrid between *F. americana* and *F. pennsylvanica.* Native from New York to Florida and 50 to 100 feet tall, it has leaves up to 1½ feet in length that have seven to nine leaflets nearly hairless above, softly-hairy beneath. ***F. uhdei,*** the Shamel ash, an evergreen or almost evergreen Mexican species is much planted as a street tree in California. In Hawaii it is sometimes known as the Hawaiian ash. Closely related to *F. americana,* it differs in its leafstalks being channeled instead of cylindrical. ***F. xanthoxyloides*** is a small tree native from the Himalayas to Afghanistan. It has spreading branches and leaves with five to nine closely-set, nearly stalkless leaflets, slightly-hairy on their midribs beneath. This is not hardy in the north. *F. x. dimorpha,* a shrubby variety native of North Africa, differs in having shorter and broader leaflets.

Garden and Landscape Uses. Selected and located with discrimination, ashes can be used effectively in landscape plantings. Except for the "flowering ashes," they are generally too big for small properties, but they can be employed to good purpose in parks and as street trees. They are usually fast growers and the foliage of many turns yellow or purple in fall. They are susceptible to attacks of oyster shell scale, the control of which calls for regular spraying as part of maintenance. Some kinds are well adapted for growing on limestone and other alkaline soils, and some survive saline soil conditions better than most trees. A few are notably wetland trees.

Cultivation. Most ashes grow in any reasonably fertile soil that is not excessively dry and certain kinds, notably *F. cuspidata, F. ornus,* and *F. oxycarpa,* prosper even in those drier than average. The black ash (*F. nigra*) is more distinctly moisture loving and the pumpkin ash (*F. tomentosa*) and the water ash (*F. caroliniana*) are definitely wet soil species. Ash trees transplant without difficulty. In their early years care should be taken when pruning to prevent the development of weak crotches that result when two branches of equal vigor develop from a pair of buds at the termination of a cut-back shoot. In such cases one of the new branches should be eliminated early. Older trees call for no special or regular pruning. Propagation is by seeds sown as soon as they ripen in fall or stratified and sown in spring. Usually they germinate promptly, but sometimes not until the second spring. Horticultural varieties are increased by grafting indoors in spring and by budding in summer. They usually succeed best when the understocks are of the species to which the variety belongs.

Diseases and Pests. Ashes are subject to leaf spot diseases, rusts, and cankers. Their chief insect and mite enemies are borers, carpenter worm, sawfly, webworms, leaf miners, flower gall (caused by mites), and scales.

FRECKLEFACE is *Hypoestes phyllostachya.*

FREESIA (Freè-sia). Chiefly familiar to gardeners because of its many splendid, fragrant-flowered hybrids that are grown in greenhouses, and in mild climates outdoors, *Freesia* is a genus, divided by modern botanists into up to twenty species, but previously considered as being one or two variable ones. Endemic to South Africa, and a member of the iris family IRIDACEAE, it has a name that commemorates Friedrich Heinrich Theodor Freese, a German student of South African plants, who died in 1876.

Freesias are small perennials with bulb-like organs called corms and deciduous stems and foliage. The stems are slender, branched, and flexible. The leaves are chiefly in basal fans, with a few smaller ones on the stems. They are linear, up to ½ inch wide, and longitudinally parallel-ribbed. The flowers of freesias face upward in apparently one-sided spikes held more or less horizontally. They have tubular perianths, slender below, becoming broadly-funnel-shaped above, with six (or in double-flowered horticultural varieties more) perianth lobes (petals or more correctly tepals) shorter than the tubes. In the species the blooms are more or less two-lipped, but this is less true of most horticultural varieties. There are three stamens and a three-branched style ending in two-cleft stigmas. This last characteristic distinguishes *Freesia* from nearly related *Tritonia.* The fruits are capsules.

Notable for its intensely sweetly fragrant blooms that are pure white except for yellow toward the base of the corolla, ***F. alba*** (syn. *F. refracta alba*) is less than 1 foot high. It has erect leaves up to 7 inches long, and occasionally up to 1 inch wide but more commonly only a little over ½ inch wide. The flowers, about 2½ inches long, have petals of nearly even size. From 9 inches to 1½ feet tall, ***F. refracta*** has slender, usually branched stems, zigzagged in their upper parts. The linear leaves are up to about 6 inches long. In loose spikes of five to eight, the flowers, very fragrant, two-lipped, and about 1¼ inches long by one-half as wide, are creamy-yellow to greenish-yellow with the lower petals marked with brownish-yellow and having a purple center line, and their outsides stained with dull purple or violet. The slightly bigger blooms of *F. r. leichtlinii* are citron-yellow blotched with ochre-yellow. Similar to *F. refracta* but more vigorous and with orange-throated, white flowers deeply margined and flushed with rose-pink, ***F. armstrongii*** has played an important part as one parent of the handsome *F. hybrida* complex.

Horticultural varieties of freesias, for convenience grouped as ***F. hybrida,*** include numerous beautiful named kinds in a wide range of colors including pink, mauve, lavender, blue, red, cream, yellow, and coppery tones, and some with double flowers. These are described and sometimes illustrated in catalogs of dealers in bulbs. Their development represents a triumph of twentieth-century plant breed-

Freesia hybrida, variety

Freesia hybrida, double-flowered

ing that in a span of thirty years produced from somewhat dowdy parents an array of superior brilliant offspring. Later developments brought taller stems, more numerous and larger flowers. These are the freesias commonly grown today. Their parentage is chiefly referable to *F. refracta* and its varieties and *F. armstrongii*.

Garden and Landscape Uses. In California-type climates and in the deep south freesias are successfully raised outdoors for garden decoration and cut bloom, but by far their most important use is for the production of flowers in winter in greenhouses. Commercial florists grow large quantities for this purpose. Freesias are also delightful as flowering pot plants. It is even possible to grow them with some degree of success in cool sunrooms.

Cultivation. Outdoors or in, soil for freesias needs to be porous, fertile, and well-drained. The bulbs (corms) need no period of low temperature treatment before planting as do those of tulips, daffodils, hyacinths and some other hardy bulbs. Normal planting time is August to October. Outdoors, the bulbs are set in groups, or for cut flowers in rows, at a depth of approximately 3 inches with about the same distance between individuals.

In greenhouses freesias can be grown in beds or benches of soil, in deep flats, or in pots or pans (shallow pots). For late August and September planting, to have flowers in late December and January, use bulbs at least ¾ inch in diameter. For plantings made from October to December to have flowers from the end of January to April, bulbs ⅝ or even ½ inch in diameter are satisfactory. Set the bulbs ½ to 1 inch deep and 2 inches apart in pots or deep pans. After planting, water thoroughly and put containers planted early in a shaded cold frame and cover them with about 2 inches of moss or peat moss to keep them evenly moist and minimize the need for watering. Put later-planted bulbs directly in the greenhouse and, until shoots begin to appear, shade them. A good way of doing this is to stand them under the benches and cover them with a sheet of polyethylene plastic. This will not only reduce the need for watering, but will prevent drip from the benches causing over-wetness. As soon as new growths show at soil level remove the pots to a sunny location. Shade them lightly from strong sun for the first week. Cool conditions with a free circulation of air are necessary to grow good freesias. They resent too much heat. Do not allow the night temperature in winter to exceed 50°F. It is better to be lower than higher than that. Temperatures as low as 40°F will do no harm. Water moderately until good root growth has been made, then generously. Lack of sufficient soil moisture severely retards growth and causes the tips of the leaves to brown. Give weekly or twice-weekly applications of dilute liquid fertilizer from when the flower stalks begin to show until the blooms start to open. Neat staking is needed. This may be done with very slender canes, wires, or twiggy, light brushwood.

Most gardeners buy new bulbs each year. If flowers are cut with foliage this is necessary, and in any case it saves work. If the bulbs are to be kept for a second year do everything possible to build up their strength. As soon as flowering is through cut off the faded blooms but leave the foliage intact. Resume the fertilizing program and keep the soil moist until the leaves begin to turn yellow naturally. When this happens cease fertilizing and dry the bulbs off by gradually increasing the intervals between waterings, and finally ceasing watering. Then, store the bulbs in an airy, shaded place, either in the pots of soil in which they grew or shaken free of soil and placed in bags or flats. In August grade them according to size, repot them in fresh soil, and start the growth cycle again. Freesias are chiefly increased by offsets, but the species may be raised from seeds, and this method is also used in the production of new varieties.

FREMONTIA. See Fremontodendron.

FREMONTODENDRON (Fremonto-déndron) — Flannel Bush. The sterculia family STERCULIACEAE, to which this attractive genus belongs, is mostly tropical, but *Fremontodendron* as a native is confined to southern California, Arizona, and Baja California. Previously named *Fremontia,* it consists of two species of tall shrubs or small trees that are evergreen, but in arid summers may lose a considerable portion of their foliage. The name commemorates the distinguished early explorer of western North America, Major-General John Charles Fremont, who died in 1890.

Fremontodendrons branch freely. They have stellate (star-shaped) hairs. Their leaves are alternate, undivided, and three- to seven-lobed. Usually they are dull green, but plants with brighter green foliage occur. The showy, solitary blooms, mostly bisexual, are opposite the leaves on young shoots or are on short, spurlike twigs. They have a calyx-like collar of three bractlets and a large deeply-five-lobed calyx that is bright yellow and looks like a handsome five-lobed corolla. There are no petals. The five stamens are joined from their middles downward and encircle the slender style, which projects beyond them. The fruits are persistent, bristly capsules.

The flannel bush (***F. californicum***) is a spreading shrub 4½ to 12 feet tall or rarely taller, with densely-hairy shoots. Its thickish, roundish to elliptic-ovate, more or less three-lobed leaves, 1 inch to 3 inches long, have heart-shaped bases from which radiate one to three veins. They are usually

Fremontodendron californicum

Fremontodendron mexicanum

Fremontodendron 'California Glory'

Fremontodendron californicum napensis

Fremontodendron 'California Glory' (flowers)

dull green and sparingly-hairy above and densely covered with tawny hairs on their undersides. The flat, clear yellow flowers are 1½ to 2¼ inches across. From the small pits or glands at the bottoms of the calyx lobes long hairs usually sprout. This species, native to southern California and Arizona, makes great massed displays of flowers that open more or less at one time. Varieties of it, by some botanists regarded as distinct species, are *F. c. napensis, F. c. obispoensis, F. c. crassifolium,* and *F. c. decumbens.* The first is distinguished by its slender twigs, its leaves being under 1½ inches wide. The leaves of *F. c. obispoensis* are usually not lobed or not much lobed, and its seed capsules are twice as long as wide. Its flowers are from 1½ to 2 inches broad. Thickish leaves with three veins conspicuous on their upper surfaces and flowers 2 to 2¼ inches in diameter are characteristic of *F. c. crassifolium.* A shrub up to 4 feet tall and of wider spread, *F. c. decumbens* has light orange to brownish flowers up to 1⅜ inches wide.

Differing from *F. californicum* and its varieties in its leaves having five to seven veins radiating from their bases and the pits at the bottoms of the calyx lobes being without long hairs, ***F. mexicanum*** of southern California and Baja California is 6 to 18 feet tall. It has densely-tomentose shoots and thickish, rounded, shallowly-lobed leaves 1 inch to 3 inches across, dark green and somewhat pubescent above, and densely-tawny hairy on their undersides. Its saucer-shaped blooms, 2¼ to 4½ inches across, are orange-yellow tinged reddish at their bases on the outsides. Because they are more hidden among the foliage than those of *F. californicum* the blooms are not so well displayed, but a succession is produced over a long season.

A splendid hybrid with large, richly colored yellow blooms that make a grand, massive show, is *F.* 'California Glory'. Its parents are *F. californicum* and *F. mexicanum.*

Garden and Landscape Uses. Fremontodendrons are not hardy in the north, but in climates where little or no freezing is experienced and where summers are not exceedingly humid they are among the most beautiful and satisfactory shrubs. They need well-drained, dryish soil and

Fremontodendron 'California Glory', espaliered

full sun, and are displayed to fine advantage as single specimens, in groups, or espaliered.

Cultivation. These shrubs do not transplant well and only container-grown plants should be set out. They soon become established and bloom freely even when young. No regular pruning other than any needed to keep the plants shapely or to remove unwanted crowded shoots is necessary. Seeds afford a ready means of perpetuating the species and botanical varieties. The hybrid 'California Glory' and other especially desirable plants can be propagated by cuttings.

FRENCH. The word French appears as part of the common names of these plants: French-honeysuckle (*Hedysarum coronarium*), French marigold (*Tagetes patula*), French-mulberry (*Callicarpa americana*), French-oak (*Catalpa longissima*), and French sorrel (*Rumex scutatus*).

FREREA (Frèr-ea). Something of a botanical curiosity, *Frerea* is the only member of the *Stapelia* relationship of the milkweed family ASCLEPIADACEAE, which some botanists segregate as the stapelia family STAPELIACEAE, to have well-developed, normal, flat leaves. This is the only characteristic that sets this genus apart from *Caralluma.* There is only one species. Its name, given in 1865, commemorates Sir Henry Bartle Frere who, it is stated, was "always the enlightened encourager and promoter of scientific researches in India." He died in 1884.

As its name indicates, a native of India, ***F. indica*** is a sparsely-branched herbaceous perennial 4 to 5 inches tall. Its cylindrical, succulent, smooth, whitish stems, somewhat under ½ inch thick, carry on their upper parts short-stalked, fleshy, elliptic leaves ¾ to 1 inch in length. The wheel-shaped flowers, from the branch ends, are solitary and nearly ½ inch across or somewhat larger. They have flat, five-sided, shallowly-five-lobed corollas, finely-hairy, and shallowly-depressed at their centers. They look much like small, purplish-red stapelias. The central coronas or crowns are dark purple and ruby-red. The fruits are pod-like follicles.

Garden Uses and Cultivation. This rare species is suitable for inclusion in collections of choice succulents. It needs the same conditions and care as *Stapelia,* with perhaps slightly more moisture. For more information see Succulents.

FREYCINETIA (Freycin-ètia). Except for two species of *Sararanga* not known to be cultivated, *Freycinetia* and more familiar *Pandanus* are the only genera of the screw-pine family PANDANACEAE. There are about 100 species of the former, natives from Ceylon to Polynesia and New Zealand. Their name honors the French navigator Charles Louis de Freycinet, who died in 1842.

Freycinetias differ technically from pandanuses in botanical characteristics of their fruits, more obviously in being slender vines, sometimes climbing to the tops of tall trees, and in having the bracts of their flower clusters often brightly colored and showy. Male and female blooms are on separate plants. Neither has sepals or petals. The males consist of several short-stalked stamens, the latter of many ovaries. What are commonly called fruits are conglomerates of many fleshy or hard drupes (the true fruits).

Common in the forests of the Hawaiian Islands, and endemic there, the ié-ie (***F. arborea***) frequently climbs to the tops of tall trees by clinging firmly to their trunks and branches with slender, aerial roots. At other times it flops along the ground to form impenetrable jungles. Its stems are slim and yellow. They branch freely, each branch ending in a cluster of glossy, tapering leaves up to 2½ feet long, and at their bases, 1½ inches wide. Their margins, and on their undersides, their midribs are spiny. Both male and female cone-like flower clusters are in threes at the centers of a collar of brilliant rosy-red, fleshy bracts. An unusual feature of freycinetias is that their flowers, at least those of some kinds, are pollinated by mammals, often bats but with the ié-ie of Hawaii by rats, which are fond of the sweet bracts that accompany the blooms.

Native of New Zealand and the only species occurring there, ***F. banksii*** is a lofty climber with recurved leaves 2 to 3 feet long and about 1 inch wide. The fleshy white or pale lilac bracts that accompany the flowers are 3 to 5 inches in length. These were much appreciated as food by the aboriginals, and are still used to make a pleasant strawberry-flavored jelly. The bright yellow male flower heads are 3 to 4 inches long and about ⅓ inch wide. The females are shorter. The composite oblong-cylindrical fruits are about 5 inches long by 2 inches wide.

A Philippine Island native, ***F. cumingiana*** (syn. *F. luzonensis*) has very slender, climbing stems and narrowly-ovate leaves 4 to 6 inches long, usually softly spiny on their margins and on the midribs of their undersides. Male plants bear quite ornamental flowers with three papery, orange bracts encircling clusters of club-shaped stamens. The red fruits of female plants are fairly attractive. Also an endemic of the Philippine Islands, high-climbing ***F. multiflora*** has narrowly-lanceolate leaves up to 1 foot long. The fruits, which have associated with them partly-colored bracts, are three to five together, erect, almost cylindrical, and about 3 inches long.

Garden and Landscape Uses and Cultivation. In the humid tropics and warm subtropics these evergreen vines are cultivated for ornament. To be effective they need considerable room and tree trunks or other supports up which they can climb or rocks over which they can scramble. They are also occasionally grown in large tropical conservatories. Any good soil is satisfactory and they respond to conditions that suit pandanuses. They are easily increased by seeds, cuttings, and air layering.

FRIABLE SOIL. Friable simply means crumbly, and a friable soil is one that works easily without sticking unpleasantly to spades, forks, rakes, or other tools and implements. Unlike tenacious, clayey soil, friable earth does not form difficult-to-pulverize clods. This crumbly condition is a ideal state for all garden soils.

The friability of tenacious, clayey soils can be greatly improved by proper management. Of first importance is to avoid walking upon, trucking over, spading, forking, raking, or otherwise working such soils when they are so wet that they stick to shoes and tools. Of next importance is to mix in liberal amounts of semi-decayed organic matter, such as strawy manure or compost, or green manures, such as winter rye and buckwheat. The addition of coarse sand sometimes helps, but not always. Under some conditions it forms a concrete-like mixture with the clay. Gritty coal cinders without admixture of fine ash are far better. Other gritty non-humus-forming materials, such as crushed brick and crushed oyster shells, can also be used. The application of lime or of gypsum is also very helpful and so is spading or plowing the soil in fall and leaving it in large lumps over winter to be ameliorated by the crumbling effects of alternate freezing and thawing.

FRIENDSHIP PLANT. See Billbergia.

Formal garden: A center pool, flowering dogwoods, and azaleas

Informal garden: An apple tree, azaleas, and tulips

Franklinia alatamaha

Freesia hybrida variety

Fremontodendron 'California Glory' (flowers)

Fritillaria pudica

Fritillaria recurva

Fritillaria imperialis

Fuchsia hybrida variety

FRINGE or FRINGED. These words appear as part of the common names of these plants: fringe bell (*Shortia soldanelloides*), fringe cups (*Tellima grandiflora*), fringe-myrtle (*Calythrix*), fringe tree (*Chionanthus*), fringed-lily (*Thysanotus*), and fringed orchid (*Habenaria*).

FRITHIA (Fríth-ia). Closely resembling *Fenestraria,* the only species of *Frithia* is not as nearly related to that genus as at first sight it seems, although both belong in the vast *Mesembryanthemum* association of the carpetweed family AIZOACEAE. An important difference is that the leaves of *Frithia* are alternate rather than opposite. Also, they have finely-warted instead of smooth surfaces. Both their sepals and petals are united at their bases into short tubes which is not true of *Fenestraria.* The name honors Mr. Fred Frith of South Africa, a twentieth century collector of succulents.

A desert perennial, ***F. pulchra*** is one of the remarkable windowed plants of South Africa. It grows with most of its plant body

Frithia pulchra

buried in the ground. Only the translucent tips of its leaves are exposed. These windows admit light to the chlorophyll-containing cells below in which photosynthesis takes place. The advantage of this chiefly subterranean life are protection from dehydration by brilliant sun, arid atmosphere, and scorching winds, and from browsing animals. Stemless *F. pulchra* consists of a rosette-like cluster of six to about nine erect, club-shaped, fleshy, gray-green, flat-topped leaves about ¾ inch long. Appearing in late summer and early fall, the blooms, about 1½ inches in diameter, are carmine with the bases of the many petals forming a yellowish central zone. In appearance they have the apparent daisy-like form characteristic of *Mesembryanthemum,* but are, of course, structurally very different from daisies, each flower being a solitary bloom rather than a head composed of many florets. The fruits are capsules.

Garden Uses and Cultivation. This rare species is a treasure for collectors of non-hardy succulents. It needs the same general treatment as *Fenestraria, Lithops* and other South African windowed and pebble plants, but is more exacting in its needs than the above-mentioned two. Although in the wild it grows buried in the ground, in climates more humid than its desert home it should be planted in more conventional fashion with its leaves fully exposed. The resting period of *Frithia* is summer. At that time it should be kept quite dry. Careful watering is needed at other seasons. Propagation by seeds and by cuttings is readily accomplished. For further information see Succulents.

FRITILLARIA (Fritil-lària)—Fritillary. Crown Imperial, Guinea Hen Flower or Checkered-Lily or Snake's Head. Native of north-temperate regions, chiefly in mountains, *Fritillaria,* of the lily family LILIACEAE, contains among its about eighty-five species many very good garden plants, some of easy cultivation, others decidedly difficult. Most are hardy. The name, alluding to the markings of the flowers of some kinds, derives from the Latin *fritillus,* a checkerboard or dice box.

Fritillaries are allied to lilies (*Lilium*), tulips, and *Nomocharis.* They have bulbs covered with smooth tunics or composed of thick, fleshy, overlapping scales. Native American species belong to the latter group. The stems of fritillaries are erect and without branches. Alternate, whorled, or less often opposite, the leaves are narrow and parallel-veined. The nodding, bell- to funnel-shaped, dull or brightly colored flowers are solitary and terminal or are in umbels or racemes. They are accompanied by leafy bracts. Each bloom has six perianth segments or tepals (commonly called petals) with large nectaries at their bases inside, six slender-stalked stamens, and a slender, three-branched or grooved style. The fruits are ovoid to oblongish capsules. For the plant previously named *F. thomsoniana* see Notholirion.

The crown imperial (***F. imperialis***), long known in cultivation, is native from Iran to India and Afghanistan. Majestic in aspect and 2 to 4 feet tall, this has bulbs about 6 inches in diameter. Its many erect, lanceolate leaves are alternate and often over 1 inch wide. The strongly, somewhat unpleasantly musk-scented flowers hang on curved stalks, many together in large umbels or clusters from just below a conspicuous terminal crown or tuft of short leaves. The blooms, brick-red, yellow-red, yellow, or purplish, have large drops of nectar inside. They are 1½ to 2 inches long and have three-parted styles. Horticultural varieties of crown imperial include *F. i.* 'Aurora', with red-orange blooms; *F. i. lutea,* with yellow flowers; *F. i. maxima,* robust and with orange-red blooms; and *F. i. rubra,* with very big, red flowers. Similar to the crown imperial, earlier-flowering ***F.***

Fritillaria, unidentified species

Fritillaria, unidentified species

Fritillaria imperialis

Fritillaria meleagris

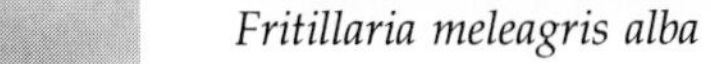

Fritillaria meleagris alba

raddeana has narrower, glaucous leaves. Its pale greenish-yellow flowers are 1 inch to 1¼ inches long. This is a native of western Asia.

The guinea hen flower, checkered-lily, or snake's head (***F. meleagris***) ranges in the wild from Great Britain to southwestern Asia. From 9 inches to 1½ feet tall, it has stems with a few alternate, linear to obovate, clasping leaves and one or sometimes up to three broadly-squarish-bell-shaped blooms up to 2 inches long and 2 to 3 inches wide. Typically, they are strongly veined and checkered with maroon-red or purple against a lighter ground. Those of *F. m. alba* are white, those of *F. m.* 'Artemis' grayish-purple and checkered, those of *F. m.* 'Pomona' checkered with violet on a white ground, those of *F. m. purpurea* purplish.

Spain and Portugal are the home countries of ***F. lusitanica,*** a usually solitary-bloomed kind 8 inches to a little over 1 foot tall. Its alternate leaves, near the tops of the stems, are linear and up to 3½ inches in length. Outside, the 1- to 1½-inch-long blooms are reddish-brown. Inside, they are yellow checkered with maroon-red. This kind grows naturally in pine woods. As its name indicates, native of the Pyrenees, where at high elevations it flourishes on dry slopes, ***F. pyrenaica*** is 6 inches to 1 foot tall, or in a giant form sometimes 2 feet. Its chiefly linear leaves are somewhat glaucous, its one to three flowers have recurved petals, mahogany-red on their outsides, polished yellow within. They are 1½ inches long by two-thirds as wide. A variant with pure yellow blooms is known. Favoring limestone soils, ***F. tubaeformis*** (syn. *F. delphinensis*), of southwestern Europe, attains a height of about 6 inches and has lanceolate leaves, or the upper ones linear. The solitary flowers, 1½ inches long and wide and red-purple checkered with a deeper hue, have three-branched stigmas. Variety *F. t. moggridgei* has yellow blooms speckled with brown.

Central and southern Europe and adjacent Russia are the home territories of ***F. nigra,*** a variable species often found on limestone soils. About 1 foot tall, this has linear to linear-lanceolate leaves, and purple or yellow-spotted-purple flowers, solitary or paired, and about 1 inch long. Very similar, differing chiefly in its upper leaves ending in tendrils, is ***F. ruthenica,*** of southeastern Europe and adjoining Asia.

Native of Syria, Asia Minor, and Cyprus, where it grows in limestone soils in fields and woodlands, ***F. acmopetala*** is 1 foot to 1½ feet tall. Its alternate leaves are linear and about 3 inches long. Usually one, occasionally two, on a stem, the flowers have recurved petals, their outsides green streaked with purple, their inner surfaces yellowish and glossy. They are about 1½ inches long by 1 inch wide. From Israel, Syria, and Iran comes ***F. persica,*** which attains a height of 2 to 3 feet and has alternate, lanceolate leaves up to 6 inches long. The fifteen to twenty 1-inch-wide, shallowly-bell-shaped flowers are violet-blue to plum-purple. A vigorous variant with green instead of purple stems is offered as *F. p.* 'Adiyaman'. Inhabiting deep, moist soils in the mountains of Iraq, ***F. assyriaca*** is 9 inches or so tall. It has solitary flowers 1 inch long by one-half as wide, maroon-red on their outsides, golden-bronze marked with faint black lines inside. Native of Asia Minor, rare ***F. dasyphylla*** is 4 to 6 inches tall. Its lower leaves are narrowly-elliptic to spatula-shaped and alternate, and the uppermost are linear and often opposite. Each stem carries one or two more or less nodding, narrowly-bell-shaped flowers about ¾ inch long, as wide across the mouth, and with a branchless, scarcely lobed style. Their outsides, except for yellow tips to the petals, are dull wine-purple; within they are yellow.

Fritillaria dasyphylla

Himalayan ***F. roylei*** grows in nonlimestone soils at high altitudes and is said to favor, in the wild, places grazed by sheep. In its home region it is used medicinally.

Up to 2 feet tall, this kind has linear-lanceolate leaves, in whorls of up to six, and greenish-yellow flowers, faintly purple-checkered inside, and 2 inches long by 1½ inches wide. Ranging from the Himalayas to western China, ***F. cirrhosa*** is a high altitude species 6 inches to 3 feet tall; it inhabits nonlimestone soils. The uppermost of its linear-lanceolate leaves terminate in tendrils. The one, or rarely two or three flowers, 1½ to 2 inches long, are usually cream to pale green, sometimes purple on their outsides, and checkered within. They have a three-branched stigma. Very satisfactory in cultivation, ***F. pallidiflora,*** of Siberia, is 1 foot to 1½ feet tall. Its glaucous leaves, alternate, opposite, or in whorls, are lanceolate, those near the base of the stem broader and up to 2 inches wide. There may be as many as twelve blooms, 1½ to 2 inches long by two-thirds as wide as they are long, and delicate yellow-green lightly flecked with brown. From China, Japan, and other parts of temperate Asia comes ***F. verticillata,*** which is 1½ to 2 feet tall and has lanceolate leaves usually in whorls, the upper ones ending in tendrils. The three to six green-flecked, cream blooms are 1 inch long by slightly over one-half as wide. They have three-branched styles. *F. v. thunbergii* differs in technical floral details.

Western North America is home to nearly twenty species of *Fritillaria,* among them those now to be considered. Widespread from Alaska to Washington, as well as, as its name implies, a native of eastern Asia, ***F. camschatcensis*** (syn. *Sarana camschatcensis*) has bronze-purple blooms so dark that they are nearly black. From 1 foot to 1¼ feet tall, the stems carry whorls of lanceolate leaves and one to several flowers 1¼ inches long and almost as broad. This kind favors part-shade. It is the latest blooming species. Its bulbs, sometimes called Eskimo-potatoes, are edible by animals and humans. The unlovely, yet appropriately named stink bells (***F. agrestis***), of California, favors sunny places at low elevations. It has a stem about 1 foot tall, abundantly foliaged in its lower portion with alternate, lanceolate to linear leaves. Its one to eight green-veined, greenish-yellow blooms are about 1¼ inches long. Another offensively scented kind fond of open, sunny locations, the mission bells or black fritillary (***F. biflora***), of California, grows at low and medium altitudes. This, 6 inches to 1½ feet in height, has leaves in clusters low on the otherwise bare stems. They are 2 to 4 inches long and about 1 inch wide. The up to seven 1-inch-long, half-nodding, dark purple-brown flowers are veined with green and red. They have three-forked stigmas. The yellow fritillary (***F. pudica***) occurs from British Columbia to California and New Mexico in a wide variety of soils and locations, and at various altitudes. Blooming early, it is 3 inches to 1 foot in height. Its few alternate leaves are oblanceolate to lanceolate. The deep yellow to orange, purple-tinged flowers, solitary or occasionally up to three on a stem, are 1 inch long and wide. The pink fritillary or adobe-lily (***F. pluriflora***), endemic to California, has very deep-seated bulbs, and stems, with foliage mostly low down, 8 inches to 1 foot tall. The oblong-lanceolate leaves are about 1½ inches long by somewhat less than 1 inch wide. Distinctive in color, the blooms of this species are purplish-pink, 1¼ inches long, and up to twelve on each stem. The pink fritillary favors clay soils, wet during the rainy season, arid in summer. It occurs at low to medium elevations. Most brilliant of the genus, the scarlet fritillary (***F. recurva***), in the wild native at fairly high altitudes from Oregon to California, often grows in clay soils. It may become 2 feet or more high, but often is much lower. Its linear to lanceolate leaves are opposite or in circles of more than two. In racemes of up to six, the up to 1½-inch-long, tubular, scarlet blooms are checkered with yellow or orange. The bulbs of the scarlet fritillary produce many rice-grain-sized bulblets. The white fritillary (***F. liliacea***) is confined in the wild to California, where it occurs at low altitudes in open places near the coast. One of the earliest of the genus to bloom, it is up to 1 foot tall and has linear to oblanceolate, 1-inch-wide leaves congregated in basal rosettes. The up to three half-pendent, 1-inch-long and 1-inch-wide blooms are dull creamy-white marked faintly with green streaks. They have three-forked stigmas.

Other western North Americans include high alpine ***F. atropurpurea,*** in the wild the most widely distributed American fritillary. Native from North Dakota to Oregon and California, usually in dry, gravely soil, it is 1 foot to 2 feet tall and has narrowly-linear leaves, alternate or in whorls, and one to four or rarely more brown-purple flowers spotted with greenish-yellow and white. They have three-forked stigmas. The Siskiyou-lily (***F. glauca***) inhabits the Siskiyou Mountains at medium to high altitudes. About 6 inches tall, it has a few glaucous, alternate or whorled, oblanceolate leaves and usually one, occasionally up to three, semipendent purple blooms flecked to a greater or lesser extent with green and about 1 inch long by 1½ inches wide. Bulbs that produce numerous tiny bulblets give reason for the vernacular name of the rice-grain fritillary (***F. lanceolata***), a native of moist, open woodlands at low to medium elevations from British Columbia to California. Of variable height and aspect, this kind is from 2 feet or lower to as much as 4 feet tall. Its leaves are ovate-lanceolate, its 1½-inch-long flowers, up to four on each stem, dark purple mottled with greenish-yellow. Charming ***F. multiflora*** (syn. *F. parviflora*) has up to twelve bell-shaped blooms ½ inch long by 1 inch wide on stems 1 foot to 3 feet high. Maroon-purple to greenish, flecked with green and yellow, they have three-forked stigmas. The bulbs of this kind produce many rice-grain bulblets. Its alternate or whorled leaves are linear to linear-lanceolate. This species inhabits moist woodlands at low to medium elevations in California. One of the prettiest American species, Californian ***F. purdyi,*** is a variable kind that grows wild in poor, shallow soils at medium to high altitudes. From 6 inches to 1 foot tall, it has few linear leaves above the basal cluster of ovate ones. The 1-inch-long flowers, from one to six on each stem, are pale mauve flecked with chocolate-brown and green. They have three-branched stigmas.

Garden and Landscape Uses. In the main, fritillaries are for rock gardeners and native plant enthusiasts to toy with and test their skills, although a few, inappropriate for collections of native plants because of their exotic origins, are manifestly too big for any but the largest rock gardens. Here belongs the dramatic crown imperial. Such giants can be given places in flower borders, at the fringes of shrub plantings, and in open places in woodlands. Their needs are those of many plants that enjoy dappled, but certainly not heavy shade. Deep, moist, but not wet soil well furnished with organic matter in the form of leaf mold, old compost, or peat moss is needed. Manure, so long as it is sufficiently decayed so that its true character is almost disguised comes not amiss. Among easier fritillaries to grow in eastern American gardens are the quaint guinea hen flower and its lovelier white-flowered variety, brilliant red-flowered *F. recurva* and less glowing *F. camschatcensis* and *F. pudica.* These, and undoubtedly some others, are likely to settle down and make themselves at home in deep, well-drained sandy or gritty soil that has a high organic content and is obligingly damp. With yet others, gardeners must be prepared for trial and error, taking their cues so far as available from known preferences of the specific kinds in the wild. Certainly some western American fritillaries, as well as many that hail from Mediterranean lands and eastward thereof, know agonizingly dry summers following mild moist winters and springs. Except in alpine greenhouses and cold frames these conditions cannot be approximated in many parts of North America.

Cultivation. Under favorable conditions the bulbs of fritillaries multiply encouragingly. Because of this, many kinds, unless they are transplanted about every third or fourth year, are likely to become overcrowded and produce fewer flowers of poorer quality. An exception is the crown imperial, which often thrives for years without disturbance and is best left well alone so long as it is doing well. Surface cultivation around these plants, except that occasioned by hand pulling weeds, should not be practiced. Instead, a mulch, organic in the case of kinds that like semiwoodland conditions, or of grit or small stones with others, is appreciated.

Transplanting should be done and new plantings made shortly after the tops of the plants have died down or as early as the bulbs can be obtained from dealers. This means July in the case of crown imperials and some others, slightly earlier or later with certain other kinds. An important point to remember is that the bulbs must never be allowed to dry. If they must be out of the ground they should be packed in slightly damp peat moss, vermiculite, sand, or similar material. The bulbs of many fritillaries grow naturally fairly deep in the ground. This is to be remembered at planting time. Those of the crown imperial are best set with their tips 5 inches beneath the surface. It is important, of course, that a deep bed of agreeable earth underlies the bulbs.

Increase of fritillaries is had by natural multiplication of the bulbs, by seeds, and in some cases by scales detached from the parent bulbs and handled, like the bulb scales of lilies, by planting them vertically, with their tips just showing above the soil surface, in sandy peaty soil in flats or in a cold frame. From seed sowing, four to six years are likely to pass before first flowering. The seeds are usually sown in pots or pans (shallow pots) in a greenhouse or cold frame as soon as possible after they are ripe. The seedlings are not disturbed until the second year, when they are transplanted to a deep bed of porous soil in a cold frame, there to remain until after their first blooming.

FRITILLARY. See Fritillaria.

FROG FRUIT. See Phyla.

FROG'S BIT. This is the common name of *Hydrocharis morsus-ranae* and *Limnobium.*

FROND. Leaves of ferns are called fronds, so sometimes are those of palms.

FROSTS and FREEZES. As used in warm-temperate and subtropical parts of the United States, regions including Florida and southern California, where many trees, shrubs, and other plants highly sensitive to cold are cultivated, a distinction is made between a frost and a freeze. The first, often quite local, alludes to a drop in air temperature to 32°F or below that results in ice crystals developing on the ground and other surfaces. This, called a white or hoarfrost, commonly occurs on clear, still nights when heat from the ground is lost rapidly by radiation.

A freeze, generally preceded by rain, with high winds, but usually without ice crystals forming, is a cold wave that affects a large area and is commonly accompanied by a marked drop in barometric pressure.

Both frosts and freezes can do damage, which varies with the kind of plant and other circumstances, from slight to catastrophic. Several methods are employed by growers of citrus fruits and other commercial crops to minimize damage and some can be adapted by home gardeners to protect sensitive ornamentals.

Banking clean, weed-free earth around the trunks in fall before the first slight frost and removing it in early spring often prevents complete killing even though severe damage may be done to the exposed upper parts of trees. Alternatively, wrapping trunks and lower branches with several sheets of newspaper enclosed in an outer layer of waterproof paper or polyethylene plastic film may be effective. Be sure the wrappings reach completely to the ground. Wrapped trees may also have the bases of their trunks banked with earth.

Citrus growers frequently employ fire heat to combat frost. Contrary to earlier beliefs the creation of great volumes of heavy smoke by smudge pots is much less effective than actual heat from wood, coke, or oil fires. The successful employment of these costly devices calls for judgment and experience. Information about the practicability of using them in any particular locality and about the number of heaters needed for a given area is available in fruit-growing regions from Cooperative Extension Agents and other authorities. In such areas warnings of impending frosts and freezes are issued and broadcast by the United States Weather Bureau. For information about protecting plants in colder areas against damage by low temperatures see Protection for Plants.

FROSTWEED. See Helianthemum.

FRUIT. Botanically, a fruit is a mature ovary of a flower. It contains the seed or seeds. It may or may not be edible, juicy, fleshy, or quite dry. Fruits come in many forms; some of the most familiar are achenes, berries, capsules, nuts, and pods. Some plant parts commonly called fruits, such as blackberries, raspberries, strawberries, and pineapples, are really conglomerates of numerous little fruits each developed from a separate ovary. In this Encyclopedia the term fruit is used for these as well as for what are more correctly single fruits.

FRUIT GROWING. As used here the word fruit has a more restrictive meaning than its strictly botanical one. It refers only to edible kinds eaten raw or cooked and that in common parlance are called fruits. The last limitation eliminates such sorts as cucumbers, eggplants, squash, and tomatoes, botanically just as much fruits as apples, peaches, and pears, but which by growers and users are commonly classed as vegetables or salads. Also disregarded are the botanical fruits identified as nuts and discussed under that heading elsewhere in this Encyclopedia.

Fruits form a highly important part of normal diets and are grown and marketed commercially in vast quantities. Many sorts are popular for home gardens. With the notable exception of strawberries and melons they are products of trees, shrubs, or woody vines, some of native American origin, most originally brought from distant lands. A few are hybrids between domestic and foreign species.

Home gardeners interested in growing fruits should first make a careful survey of the possibilities afforded by the sites, soils, and other conditions. Not all fruits do well in all areas and not all varieties grown commercially are well adapted to home gardens. One or two cherry trees in a populated area are much more likely to be devastated by birds than is a cherry orchard in rural surroundings. To produce clean fruit tall apple and pear trees need spraying several times a season with equipment more powerful than is usually readily available to amateur gardeners. In making a preplanting survey note carefully the fruits that are being grown successfully in your locality and then check with your Cooperative Extension Agent or State Agricultural Experiment Station who can usually supply excellent bulletins giving highly helpful information about how to proceed.

Because, except for melons and strawberries, fruits occupy the land for much longer periods than other food crops it is of the utmost importance to match sites and sorts with care and to bring the ground into as nearly ideal condition as possible for the sort to be planted.

Choice of sites is for home gardeners usually quite limited. It is well to remember that many sorts if planted in hollows that are frost pockets or at the bottoms of slopes where cold air collects are subject to frost damage that may reduce or destroy crops.

It is equally as important to choose varieties best suited to the locality and your special needs and in some cases, with apples for instance, to obtain trees grafted on dwarfing understocks so that they will never become too big to manage easily. Pears, apples, and some other sorts may be espaliered on trellis or walls and grapes may be grown on arbors or pergolas to serve decoratively as well as productively.

When you look into the matter of varieties be sure to ascertain whether they are self-fertile (self fruitful) or whether they are partially or completely self-sterile and need another variety located nearby to produce satisfactory crops. If you plant a self-sterile variety be sure the companion you set near it is of a variety that will effectively pollinate it.

To produce satisfactory crops all fruits call for regular attention to such matters as fertilizing, pruning, disease and pest control, and of course harvesting. To be effective some tasks, particularly those associated with disease and pest control, must receive attention at just the right times. Others, fertilizing and pruning, for example, call for considerable judgment and certain skills, which can be acquired.

If approached sensibly growing fruit at home can for those who enjoy the challenge of producing useful crops be immensely rewarding and furthermore has the potential for adding delightfully and substantially to the family food supply. Ill-conceived and poorly carried out, amateur fruit growing is likely to fall far short of this ideal. For more information consult the many entries in this Encyclopedia that appear under the names of individual fruits such as Apple, Blueberry, Grape, Orange, Peach, and Strawberry. Also see Orchards.

FUCHSIA. The genus *Fuchsia* is considered in the next entry. Some other plants have the word incorporated in their common names as for instance Australian-fuchsia (*Correa*), California-fuchsia (*Zauschneria californica*), and Cape-fuchsia (*Phygelius capensis*).

FUCHSIA (Fùchs-ia). The pronunciation few-sha used by most English-speaking people as the name of the beautiful plants that belong here disguises its intent of honoring the distinguished German botanist and professor of medicine Leonhard Fuchs, who died in 1566. Obviously fooks-ia would be a truer rendering, and one helpful to those who find spelling the word troublesome. The genus *Fuchsia* belongs in the evening-primrose family ONAGRACEAE and is unique in that group in having berries as fruits. That the berries are edible is probably not well known to most growers of these delightful ornamentals.

Although the first fuchsia was described botanically in 1703 it was 1788 before plants were brought into cultivation. In that year *F. coccinea* was being grown at the botanic garden at Kew in England, and in the nearby greenhouses of a nurseryman named Lee. The Kew plants had been brought from South America by a Captain Frith. According to a pleasant story, Lee's stock came from a plant he discovered as he passed along the street, growing in the window of a widow's home on the other side of London. He said that after considerable trouble he overcame the lady's reluctance to sell and purchased the plant for eight guineas. It had, according to Lee, been brought from South America by the widow's sailor son. This account may be true, but on the other hand, it is not impossible that Lee obtained his stock less ethically from the botanic gardens at Kew and that the widow story served as something of a smokescreen explanation as well as a sentimental touch to justify Lee's not immoderate prices for the new plant. Be that as it may, Lee made a good thing out of it. By 1793 he was selling fuchsia plants at ten to twenty guineas (then fifty to one hundred dollars) each. And money was, of course, worth much more in those days than now.

Fuchsias are evergreen and deciduous shrubs, small trees, or rarely climbers or trailers. There are 100 species, natives of the Americas from Mexico southwards and in the West Indies, and in New Zealand and Tahiti. They have undivided leaves, alternate, opposite, or in whorls (circles of more than two). Most often their flowers, bisexual or unisexual, are showy. They come from the leaf axils, or are sometimes in terminal panicles or racemes. In most kinds they are pendulous and are red, purplish, whitish, or parti-colored. The calyxes have cylindrical to bell-shaped tubular portions and four spreading, or reflexed petal-like lobes (sepals). There are usually four petals, but in some kinds these are represented by small, scalelike structures, and in others are absent. The stamens number eight, and like the style often protrude considerably from the throats of the flowers.

The most commonly cultivated fuchsias, of which there are many hundreds of magnificent named varieties in a wide selection of sizes, growth habits, and forms and colors of bloom, are hybrids chiefly of *F. magellanica* and *F. fulgens*. They include both single- and double-flowered kinds and are grouped under the name ***F. hybrida.*** From *F. magellanica* they differ most obviously in their thicker, sturdier stems and branches, larger, broader, longer-stalked leaves, and bigger blooms, with longer calyx tubes, broader sepals, and petals shorter than the calyx lobes. From *F. fulgens* the varieties of *F. hybrida* differ in the calyx tubes usually not being longer than the sepals, and in the stamens and style being long-protruding. Despite their great variability, varieties of *F. hybrida* agree in many respects. They have erect or pendulous stems that become woody with age, and when young are usually reddish. The leaves, pointed, strongly-toothed, and broadly-ovate to ovate-oblong, have stalks from slightly under ½ to a little under 1 inch long, and blades 1½ to 3 inches long. The blooms, hanging on stalks nearly or quite 2 inches in length, have usually crimson calyxes with tubes up to 1 inch long and twice or more times the length of the ovary, but not exceeding the sepals. The latter are 1 inch long or longer by about ⅓ inch wide or wider. Shorter than the calyx lobes, the purple, rose-pink, or white petals are ½ to 1 inch long. Commonly, the stamens and style protrude conspicuously.

Fuchsia hybrida, double-flowered

Fuchsia hybrida, single-flowered

The hardiest fuchsia is a horticultural variety of *F. magellanica* not easily differ-

Fuchsia magellanica riccartonii

entiated from the species by its appearance, but capable of surviving considerably lower temperatures. Called *F. m. riccartonii*, it survives outdoors in sheltered locations in the vicinity of New York City. True, in such cold climates it often is killed back severely, but if the roots live they send up new growths that bloom the first year. Parent of numerous hybrid fuchsias, and native of Chile and Argentina, ***F. magellanica*** is a hairless or nearly hairless, upright or sprawling, slender-stemmed shrub up to about 3 feet in height. Its usually toothed leaves, in pairs, threes, or fours, have stalks scarcely over ½ inch long and often much shorter, and lanceolate to ovate blades rarely exceeding 1 inch in length. The dainty blooms, solitary or in pairs from the upper leaf axils, are on slender stalks ¾ inch to 1¼ inches long. They have crimson calyxes with tubes up to ⅜ inch in length, and sepals about ¾ inch long by under ¼ inch wide. The petals are purplish. Variety *F. m. macrostema*, which intergrades with the typical species, has leaves 1 inch to 2 inches long, and flower stalks 1½ to 2 inches in length. The leaves are up to ¾ inch wide. Pink calyxes, stamens, and styles, and lilac-pink petals are distinguishing characteristics of *F. m. molinae*.

Mexican ***F. fulgens*** is rather sparsely branched and commonly 3 to 4 feet tall. In the wild it sometimes grows as an epiphyte (a plant that perches on trees without extracting nourishment from its host), more commonly on rocky ledges or in the ground. Its shoots and leaves are reddish and pubescent. The latter are opposite, broadly-ovate to ovate-oblong, and shallowly-toothed. They have stalks ¾ inch to 4 inches long, and blades 2 to 7 inches long. In several-flowered racemes 2 to 8 inches long, the flowers hang from the branch ends on stalks ½ to 1 inch or slightly more in length. They have calyxes with very narrowly-trumpet-shaped, dull-red tubes 2 to 2½ inches long, and yellowish to greenish sepals, sometimes tinged red at their bases, about ½ inch in length. The bright red petals are ⅓ inch long. The stamens do not protrude, the style only slightly. The nearly hairless, green sepals of *F. fulgens* are a ready means of distinguishing it from related ***F. boliviana,*** which has wider-spreading, dark red sepals, hairy on their outsides. This last shrub or small tree, which is often wrongly tagged with the name of quite different and seemingly not cultivated *F. corymbiflora,* is up to 12 or occasionally 20 feet tall. It and its varieties range widely through much of South America. The typical species has densely-hairy shoots and softly-hairy to nearly hairless, oblong-lanceolate or oblong-ovate leaves, usually opposite but sometimes alternate or in threes. They have blades 2 to 6 inches long on stalks ¾ inch to 1½ inches in length. Variety *F. b. luxurians* is softly-hairy. It has blooms with calyx tubes 2 inches long or longer, and sepals ¾ inch long or longer, all of which dimensions considerably exceed those of the flowers of the typical species.

Less well known than the sorts described above, ***F. denticulata*** (syn. *F. serratifolia*) of Bolivia and Peru is a shrub that in the wild may scramble among trees to a height of 30 feet or more. As known in cultivation it is a shrub with red stems and with, mostly in whorls (circles) of three or four or sometimes opposite, stalked, elliptic to oblong-lanceolate, toothed leaves 2½ to 4 inches long. The flowers hang on slender stalks from the leaf axils and are 1½ to 2 inches long. They have a long, cylindrical, red calyx tube, green-tipped red sepals, and carmine, brick-red, or tangerine petals. The stalks of the stamens and the anthers are cream. The oblong fruits are purplish. A white-flowered variant, perhaps a hybrid, is known.

Calyx tubes very much longer than the sepals are also characteristic of *F. triphylla* and *F. splendens.* Endemic to Haiti and Santo Domingo, ***F. triphylla*** is densely-pubescent and has few-branched stems about 1½ feet tall. In whorls (circles) of three, or sometimes in pairs or in fours, the short-stalked, slightly-toothed, lanceolate to lanceolate-ovate leaves, often red-dish-veined, have blades 1½ to rarely 4 inches long. The blooms, several together in crowded, terminal racemes, have slender stalks rarely over ¾ inch long. The pubescent, red calyx tubes, slender below and broadening markedly above, are 1 inch to 1½ inches long. The sepals, erect rather than spreading, are under ½ inch long. The stamens do not protrude.

Completely different in habit, ***F. splendens,*** of Mexico and Central America, is sometimes a small tree, more frequently a shrub 2 to 8 feet high, often with drooping branches. The leaves are opposite, pointed, ovate-heart-shaped, and toothed. They have stalks ¾ inch to 2 inches in length and blades 1½ to 4 inches long. Drooping from the leaf axils or clustered on short side branches, the flowers, on stalks up to 2 inches in length, have bright pink or red calyxes with tubes ¾ inch long or sometimes longer, and pubescent sepals up to

Fuchsia fulgens

Fuchsia denticulata

Fuchsia triphylla

Fuchsia arborescens

scarcely over ½ inch long that are green with sometimes reddish bases. The greenish petals are up to ⅓ inch in length. The stamens are conspicuously exerted.

Flowers with calyx tubes not over ¼ inch long distinguish ***F. coccinea*** (syns. *F. montana, F. pubescens*) from all species dealt with above. Even *F. magellanica* has somewhat longer calyx tubes, and it differs, too, in its almost hairless shoots. Those of *F. coccinea* are decidedly short-hairy. Native of Brazil, *F. coccinea* is a sturdy, erect, sprawling, or semiclimbing shrub 3 to 10 feet tall. Its usually toothed, extremely short-stalked leaves, opposite or in threes, are pubescent. They are ovate, and ¾ inch to 3 inches long by under 1 inch wide. Solitary from the upper leaf axils, on slender stalks 1 inch to 1½ inches long, the blooms have red sepals ½ to ¾ inch long, and violet-purple petals up to ⅓ inch long. The stamens notably protrude.

A tree up to 25 feet in height, ranging down to a small shrub, and as known in cultivation of shrub dimensions, ***F. arborescens*** (syn. *F. syringaeflora*), of Mexico, in the wild is sometimes epiphytic (perching on trees without extracting nourishment from them). Variable also in other ways, this sort is hairless or nearly so. Its pointed leaves, opposite or in threes, are elliptic to oblong-oblanceolate, toothless or nearly so, paler beneath than on their upper sides, and have blades up to 8 inches long. The numerous little, fragrant flowers are in crowded, pyramidal, terminal, lilac-like panicles 2 to 10 inches long and broad. The flower stalks are erect, the calyx tube and narrow, spreading to reflexed sepals, reddish to wine-purple. The petals, longer than the stamens, are lilac to lavender and shorter than the sepals, which are up to about ¼ inch long. About ¼ inch in diameter, the subspherical berries are purplish with a bluish, waxy bloom.

Several of a group of small-flowered Mexican species are cultivated. They are erect shrubs with opposite leaves, and solitary blooms from the leaf axils. The stamens do not protrude, but the style does. The fruits are pea-sized and black. One of this group is ***F. thymifolia,*** a slender-stemmed, 3-foot-tall shrub with mostly opposite, elliptic to broad-ovate, minutely-hairy leaves, paler on their undersides than above, with slender stalks, and blades from ⅓ to a little over 1 inch long. The blooms, in the bud stage and when they first open, are white. As they age they gradually change to red. They have calyx tubes about ⅙ inch long that are not more than twice as long as the sepals. The latter, at first spreading, later become reflexed. Very similar, but easily distinguished by its minute blooms, the calyx tubes of which are not more than 1/25 inch long, is 3- to 10-foot tall ***F. minimiflora.*** Frequently misidentified as *F. thymifolia,* but differing in its cerise-red flowers, with petals that turn ruby-red with age, not being white in the bud stage or when they first open, and in having calyx tubes twice or more as long as the sepals, ***F. microphylla*** ranges in height from 1 foot to 6 feet. A dainty species that for long traveled under a wrongly applied name, that of *F. reflexa,* is ***F. cinnabarina.*** Characteristics that most readily distinguish it from the other small-flowered species discussed here are its cinnabar-red to vermilion blooms, about ½ inch long, and its short-stalked, ½-inch-long leaves having very strongly turned back (revolute) apexes and margins. With a mode of branching quite distinct from that of other small-flowered fuchsias here considered, ***F. minutiflora*** has branchlets that spread in one plane to form flat sprays. Its shoots are conspicuously hairy, its leaves generally lanceolate, sometimes ovate. They are under ½ inch long. Flower color in this kind ranges from light pink to cerise-pink.

A curious group of four more or less intergrading species is endemic to New Zealand. An allied species inhabits Tahiti. These form a botanically recognized distinct section of the genus, characterized by the flowers being without petals or having only tiny scalelike structures in place thereof, and having sepals separate to their bases and strongly reflexed. The stamens have anthers with blue pollen. A group, not known to be cultivated, of South American species are similarly petal-less or essentially so, but their sepals are spreading or erect, not turned backward, and are more or less joined at their bases. The New Zealand species include both the tallest and dwarfest fuchsias. The blue pollen of the species next discussed was used by Maori women to decorate their faces.

A wide-spreading deciduous tree up to 40 feet tall, ***F. excorticata*** has trunks and branches with light brown, papery bark that peels in thin sheets. Its slender-stalked leaves are alternate, oblong-ovate to lanceolate-ovate, and sometimes obscurely toothed. They have blades 2 to 4 inches long, with sometimes red-flushed, whitish or silvery undersides. The pendulous flowers, solitary from the leaf axils, and at first greenish, change to purplish-red. Their scalelike, dark purple petals are erect. Under ½ inch long, the fruits are al-

Fuchsia procumbens

Double-flowered fuchsia as a 5-inch pot plant

most black-purple. Scarcely differing from the last except that it is a shrub, branching freely from the base instead of developing a single, massive trunk, and in having leaves with blades rarely more than 1¾ inches long, is ***F. colensoi.*** A few-branched, woody climber, ***F. persicandens*** is a New Zealander that inhabits woodlands. Its blooms are similar to those of the above-mentioned kinds. Its leaves are alternate, broad-ovate to nearly round, and have blades from somewhat less to rather more than 1 inch long.

Prostrate or trailing ***F. procumbens*** in its native New Zealand inhabits sites bordering the ocean barely beyond reach of ordinary tides and washed by exceptionally high ones. It has thin, woody stems and alternate, slender-stalked leaves with broad-ovate to nearly round blades from a little under ½ inch to ¾ inch long, and paler beneath than on their upper sides. The solitary, axillary blooms are without petals. They have ¼-inch-long, sharply reflexed, purplish-tipped green sepals. The ¾-inch-long berries are bright red to magenta.

Garden and Landscape Uses. As flowering shrubs for permanent outdoor display in mild climates fuchsias are superb. They prosper in semishade or part-day shade and are useful for growing as isolated specimens and in groups and for training against walls. When grown in this last manner some kinds attain heights of 20 feet or more. They make very satisfactory hedges, especially free-branching kinds of the *F. magellanica* complex. Fuchsias are also excellent in planters and other containers for porch, patio, and terrace decoration, and those with more or less pendulous stems for hanging baskets. Erect-growing sorts are easily trained as standards (specimens with a naked trunk topped with a mop head of branches, foliage, and flowers).

Where not winter-hardy they are among the finest plants for temporary summer displays. They bloom continuously and in

Single-flowered fuchsia in a hanging basket

A standard-trained fuchsia

usefulness, willingness, and general merit, compare with geraniums, although their blooms are somewhat less flamboyant. Their cultural needs too are much the same as those of geraniums, with the notable exception that they are more successful and bloom more freely where they get some shade than in full sun. The reverse is true of geraniums. As with geraniums, 1-year-old or older plants may be carried over winter in greenhouses, sunrooms, and in similar environments, or cuttings can be rooted in late summer, fall, or winter, and the young plants grown on for the following season's display. The last method is most favored to produce plants for summer beds, window and porch boxes, and suchlike, but if large specimens trained as pyramids, standards, or other forms are desired the wintering indoors of older specimens is often done. Another excellent use for fuchsias, more popular years ago when dwellings were less uniformly heated to high temperatures than is common today, is as window plants.

Cultivation. Fuchsias can be raised from seeds, but because cuttings root so readily and ensure uniform propagations, for practical purposes that method is nearly always used. Cuttings made from the ends of vigorous shoots serve best. They may be 3 inches or so long, and root readily in sand, vermiculite, or perlite, in a greenhouse propagating bench, under a glass jar, or wherever a fairly humid atmosphere is maintained. Cuttings may be taken at any time, but late summer and early spring are favorite seasons.

To have plants in 4-inch pots for setting out in late spring for display in summer beds, window and porch boxes, and similar places, take cuttings in January or February from plants stored since fall in a cool greenhouse or other frostproof shelter. To have such cuttings available move the stock plants a month or five weeks earlier from cooler conditions to a greenhouse with a night temperature of 55 to 60°F and daytime temperatures five to ten degrees higher. Spray their tops lightly with water whenever the weather is bright or sunny, but keep the soil only moderately moist. As soon as the buds begin to swell prune the plants to shape and cut out all thin, spindling shoots. At the same time shake out as much old soil as possible from the roots, and repot the plants into the same sized containers, in sandy, fertile soil. Continue the spraying and moderate watering. Under this treatment new shoots soon develop. As soon as these are big enough make cuttings and plant them in a propagating bed in a temperature of about 60°F. Pot the rooted cuttings in 3-inch pots. Pinch out the tips of the stems when they have three or four sets of leaves and repeat this once or twice later to ensure well-branched plants. Grow the young plants throughout in a night temperature of 55°F with a daytime increase of up to fifteen degrees. Shade them lightly from strong sun. Transfer to 4-inch pots as soon as the smaller ones are filled with roots. Plants in 5- or 6-inch pots can be had by May from cuttings rooted in September and the resulting plants kept growing without check throughout the winter, and transferred to successively larger pots as growth makes necessary.

Fine large specimen plants in pyramid or tree forms can be had the following summer from cuttings of strong-growing varieties taken in August. As soon as their roots are an inch or two long pot them individually in sandy soil in 3-inch pots. Keep the plants in a lightly shaded greenhouse where the night temperature is 60°F and that by day five or ten degrees higher. Before the roots mat tightly transfer the plants to pots 4 inches in diameter, and to larger sizes as growth makes necessary, and always before the roots are tightly packed. For pottings subsequent to the first use coarse, loamy, fertile soil that permits free drainage, but is not excessively sandy nor composed of more than one-quarter to one-third leaf mold, peat moss, or other organic matter. If you wish the plants to be pyramidal pinch out their tips when the stems have four or five sets of leaves, but do not do this if you want standard (tree-form) specimens. The pinched plant will develop several shoots. Pinch out, just beyond the first leaves, the tips of all except the stronger of the upper two. Tie the unpinched shoot to a stake to become a leader (part of the main stem), and when it has five or six pairs of leaves pinch its tip out. Once again, select the stronger of the upper two shoots that develop from just below the point of pinching, tie it in as leader, and pinch the tips out of the other shoots beyond their first leaves. After the new leader has five or six sets of leaves pinch out its tip, and repeat what was done before. Continue this procedure until the specimen attains the height desired, and systematically pinch the tips out of all branches and side shoots to develop a symmetrical, conical specimen. Cease pinching about six weeks before you want the plants in bloom.

Standard-trained (tree-form) fuchsias are developed in exactly the same way except that only one stem is allowed to grow (pinch out all side shoots while they are small). Keep the stem neatly tied to a stake, and do not remove its tip until the stem is almost the height you wish the trunk of the finished specimen to be. From 2 to 3 feet is usually considered about right, but you can allow it to grow taller if you wish. Pinch out the tips of all laterals that develop following the pinching of the main stem as soon as they have four to six sets of leaves, and repeat this process until about six weeks before you wish the plants to bloom.

Wintering older plants, including those trained in pyramid form and tree form and stock plants brought in from outdoors to serve as sources of cuttings to be rooted in January or February, is usually done in a cool greenhouse, frostproof cold frame, or similar place. Plants for stock are dug from garden beds or outdoor boxes just before frost, pruned lightly, and potted in sandy soil in containers only just large enough to accept the root balls. After potting, water them thoroughly and stand them in a shaded place where the atmosphere is fairly humid. Spray lightly with water on sunny days, but be careful not to drench the soil. This treatment encourages new roots to form. After two or three weeks subject them to more airy conditions and keep them in a temperature of about 40°F, with moderate light, until it is time to start them into growth to give the cuttings you want. Large specimen plants can be satisfactorily wintered in a greenhouse or similar place where the temperature is 40 to 45°F and moderate light is available. In late winter or spring prune them to shape, repot them if needed or prick away some of the surface soil and renew it with fresh, move them to where the night temperature is 50 to 55°F, and resume normal watering and regular fertilizing.

Diseases and Pests. Fuchsias are not unduly troubled by pests and diseases, but like most popular groups of plants they have their share of possible afflictions. Among pests the most common are aphids, mealybugs, mites, scale insects, thrips, white-fly, beetles, and nematodes. Root rot, wilt diseases, and gray mold blight may appear.

FUJIWARARA. This is the name of trigeneric orchids the parents of which include *Brassavola*, *Cattleya*, and *Laeliopsis*.

FULL SUN. When gardeners speak of a plant needing full sun, they mean it should be exposed to maximum natural light without shade.

FULLER'S TEASEL is *Dipsacus sativus*.

FUMANA (Fu-màna). So closely related to sunroses (*Helianthemum*) are the about nine species of *Fumana* of the rock-rose family CISTACEAE that some botanists include them there. The chief differences are that in *Fumana* the outer stamens are sterile and have filaments (stalks) resembling strings of tiny beads. The group is native to Europe and western Asia. Its name comes from the Latin *fumus*, smoke, and alludes to the gray coloring of the shoots and stems.

Fumanas are low shrubs with usually alternate, narrow, ovate-lanceolate to linear or needle-like leaves. The flowers, generally clustered, sometimes solitary, are yellow. They have five sepals, the two outer larger than the others, five petals, numerous stamens, and a threadlike style. The fruits are capsules.

Native to the Mediterranean region and favoring limestone soils, ***F. ericoides*** is erect or more or less straggling-erect and up to about 10 inches tall. It has linear leaves up to ½ inch long and scattered among the foliage in twos to fives or located near the shoot ends, flowers about ⅓ inch wide. A native of rocky and sandy soils in western and southern Europe, ***F. procumbens*** (syns. *F. nudifolia*, *Helianthemum procumbens*) is 4 to 8 inches tall, has much-branched, finely-hairy, procumbent stems and linear leaves ½ to ¾ inch long. Its flowers, solitary from the leaf axils, are mostly in groups of two to four. They are about ⅓ inch wide.

Garden Uses and Cultivation. These are as for *Helianthemum*. The species described are hardy in southern New England.

FUMARIA (Fu-mària)—Fumitory. The name *Fumaria* comes from the Latin *fumus*, smoke, but is of uncertain application. The genus it identifies consists of forty or more species of annuals, biennials, and herbaceous perennials belonging to the fumaria family FUMARIACEAE. Natives of temperate parts of the Old World, they have much the appearance of *Corydalis*. They are not sufficiently attractive to serve as garden ornamentals, but one is sometimes cultivated because of its medical and related virtues. These were recognized by the ancient Greeks, and Dioscorides commented on the supposed effectiveness of common fumitory "being smeared on with gum it is of force also not to suffer ye hairs pulled from off ye eyebrows to grow again." Later herbalists put great store upon fumitory as a purifier of the blood, and used it for other purposes as well.

The genus *Fumaria* differs from *Corydalis* in that its seed pods are subspherical rather than elongated and do not open spontaneously to release the seeds. Fumarias have much-dissected, ferny foliage, and racemes of slender, asymmetrical blooms spurred at their bases. The flowers have two small sepals, four long, forward-pointing petals, the two upper ones of which are expanded near their ends, and the two inner joined at their tips over the stigma, and six stamens.

Common fumitory (***F. officinalis***), a native of Europe, is naturalized in North America. It is a variable, hardy, rather pretty, delicate-looking, weedy plant with usually erect, slender, branching stems up to 2 feet tall and leaves divided two to four times into small lance-shaped to linear leaflets. Its many little flowers are pink with the tips of the petals and entire wing petals blackish-red. They are up to ⅓ inch in length, and in dense, stalked racemes ¾ inch to 1½ inches long. The fruits are approximately kidney-shaped.

Garden Uses and Cultivation. Fumitory is deserving of a place in medicinal gardens and in those devoted to plants of the herbalists. Otherwise it is without garden merit. It prospers in ordinary soil in sun, and is easily raised from seeds sown in spring where the plants are to remain. The seedlings are thinned out sufficiently not to crowd to the extent that they harm each other.

FUMARIACEAE—Fumitory Family. This family is so closely related to the poppy family PAPAVERACEAE that by some botanists it is included in that group. The chief differences are the asymmetrical flowers and watery sap of plants of the fumaria relationship as opposed to the symmetrical blooms and milky juice of the poppy clan. Chiefly native of the Old World, the fumitory family, which includes such familiar garden plants as bleeding hearts and Dutchman's breeches, consists of 450 dicotyledonous species classified in sixteen genera. Belonging are annuals, biennials, and herbaceous perennials, some vining, some with tubers, and all with watery (not milky) sap. The leaves are alternate, usually thin and variously dissected, much more rarely undivided. Generally in racemes, the flowers are markedly asymmetrical. They have two sepals, four petals, the inner pair of which are crested and one or both of the outer pair with basal spurs. There are six stamens and a slender style tipped with a headlike stigma. The fruits, usually with glossy, black seeds, range from small one-seeded nutlets to slender or bladder-like capsules. Genera in cultivation are *Adlumia*, *Corydalis*, *Dicentra*, *Fumaria*, and *Rupicapnos*.

FUMIGATION. Saturating the air in a greenhouse or other confined space or in the ground with a gas or a vapor is called fumigation. It is an effective method of controlling certain pests, diseases, and to a lesser extent, weeds. Because most fumigants carelessly used present serious hazards to health or even life they must always be employed with proper caution and in strict accord with the manufacturer's directions.

Formerly deadly hydrocyanic gas generated from sodium cyanide and sulfuric acid, or more slowly and consequently more safely from calcium cyanide dusted over moist surfaces, was used. Other fumigants employed especially in greenhouses included nicotine smokes and vapors from burning or heated tobacco products, vaporized sulfur, and evaporated naphthalene and paradichlorobenzene crystals. Formaldehyde was sometimes used in greenhouses emptied of plants.

Modern gardeners rely chiefly on aersol dispersal of approved fumigants. Small "bombs" are available for the use of home greenhouses, larger canisters with adjustable measuring valves for larger operations. Be sure the kind you use is safe for plants. Not all made for controlling household pests are.

Soil fumigation to sterilize soil if properly carried out is extremely effective. It is much used outdoors and in greenhouses to control nematodes and other soil pests and disease organisms and incidentally to eliminate weeds. The condition of the soil, especially its moisture content and temperature, has an important effect on the success of the operation. Optimum conditions vary with different fumigants. Among those commonly used, often sold under trade names, are chloropicrin, formaldehyde, and ethylene dibromide. The fumes are confined in the soil for a period by covering with plastic sheets, tarpaulin, or a water seal.

FUMITORY. See Fumaria. Climbing-fumitory is *Adlumia fungosa*.

FUNGI or FUNGUSES. One of the major divisions of the plant world, the fungi (singular fungus) consists of more than 40,000 known species and probably as many or more not yet identified and described. Only one, the common mushroom, is cultivated by gardeners, but many others, according to kind, aid or plague him. Of the latter, mildews, molds, rots, rusts, smuts, and others responsible for plant diseases are among the best known and most important. Also familiar are mushrooms, toadstools, and bracket fungi. Toadstools often grow in lawns, woodlands, and other places, the often large, shelflike bracket fungi on trees and stumps.

Fungi (or funguses) employed usefully by man include the epicurean delicacies collected from the wild called truffles and morels. Certain others, such as puffballs, are edible, but it is decidedly unwise for any other than knowledgeable experts to eat them because poisonous kinds, some deadly, are plentiful. Among the most useful fungi are certain penicilliums and other kinds that produce antibiotic drugs, the yeasts so important in bread making, wine making, and brewing, and kinds essential to the manufacture of certain cheeses.

Others extremely important are the many that live in the soil and aid in the transformation of dead organic matter into end products nourishing to plants and that in other ways promote fertility. These are the sorts that, with bacteria and other lowly organisms, convert vegetable and animal wastes into compost. Yet other fungi maintain a sympathetic relationship called symbiosis with higher plants. Each benefits the other. Here belong the mycorrhiza, which live in intimate association with the roots of many plants, notably members of the heath family ERICACEAE, pines, and orchids, and are highly important or essential to their well-being. Mycorrhiza benefit their hosts by absorbing for them water and nutrients from the soil. In return they obtain from the hosts the foods they need. Lichens are other examples of fungi living in symbiotic relationships with other plants, in this case algae.

Fungi are nonflowering plants without chlorophyll. They are incapable of elaborating food from simple elements. They depend, as do animals, upon other organisms for the supplies they need. Some are parasitic on living plants or animals. Some, called saprophytes, live on dead organic matter. Yet others do both. Certain fungi, such as cedar-apple rust and wheat rust, need to complete their life cycles on two different kinds of plants as hosts and alternate between them.

A few fungi consist of single cells. The remarkable slime molds are without well-defined cells, at certain phases of their lives are mobile, and in some ways appear to be intermediates between plants and animals. Most typically the vegetative parts of fungi are the slender, multicelled filaments. The reproductive bodies or spores are often in special developments called fructifications of which mushrooms, toadstools, puffballs, and bracket fungi are familiar types.

FUNGICIDES. Strictly speaking fungicides are substances, usually chemicals, employed to inhibit the growth of or kill fungi. Gardeners often include with them materials used to kill bacteria. The latter are more properly bactericides. Hundreds of commercial formulations of fungicides and fewer bactericides are available under trade names. In addition to these, gardeners not infrequently make use of nonproprietary substances such as sulfur.

Fungicides are usually applied as dusts or as sprays or dips. The sprays or dips may be emulsions or suspensions in water of wettable powders. Whichever kind are used, it is important to follow strictly the manufacturer's directions or the recommendations of Cooperative Extension Agents, State Agricultural Experiment Stations or other reliable sources of information. This is of particular importance with treatments applied to crops to be used for food.

Timing applications is of the utmost importance. With very many plant diseases, notably those caused by rust fungi, and black spot of roses, success depends upon preventing the fungi from penetrating the plant tissues rather than controlling it afterward. In the instances cited this last is impossible. Applications are timed and spaced therefore to provide a surface coating of fungicide over the foliage and other plant parts susceptible to infection at all times when the fungus spores are being distributed. Some fungicides, those containing sulfur, for example, are damaging if applied in hot, sunny weather.

The most commonly used fungicides include Bordeaux mixture and various other copper compounds, sulfur, various compounds of mercury, formaldehyde (for soil sterilization or disinfection), and a number of newer products such as captan, ferbam, folpet (Phaltan), manate, thiram, zineb, and ziram. Not all are equally effective against all diseases. Most have no effect against some. Read the label or check with authorities before applying.

Never mix two or more fungicides together or a fungicide with an insecticide or other such substance without first ascertaining that they are compatible. It is not unusual for substances harmless when used alone to be noneffective or harmful when mixed. And of course never mix a fungicide in a receptacle or use it in a spray or other container in which weed-killer has been used.

FUNKIA. See Hosta.

FUNTUMIA (Fun-tùmia). Inhabitants of tropical Africa, the large trees that compose the genus *Funtumia* number three species. They belong in the dogbane family APOCYNACEAE. The name is derived from an African vernacular one. One species, *F. elastica*, has been exploited as a source of rubber.

In West Africa up to 100 feet in height, ***F. elastica*** is an upright tree with short, slender branches and numerous branchlets. Its opposite, oblongish, short-pointed, toothless leaves are 4 to 8 inches long, dark green and glossy, and have well-marked, pinnately-disposed veins. The small white to yellowish flowers, in dense clusters, are succeeded by woody capsules 4 inches long or longer that contain seeds with long hairy beaks.

Garden and Landscape Uses and Cultivation. A good ornamental for warm, humid climates, *F. elastica* grows rapidly and adapts to most soils. It is propagated by seeds and by cuttings.

FURCRAEA (Fur-cràea). The name of this genus, which commemorates the distinguished French chemist Antoine Francois de Fourcroy, who died in 1809, is sometimes spelled *Fourcroya*. It applies to a group of twenty species endemic to warm parts of the Americas. The genus belongs in the amaryllis family AMARYLLIDACEAE and is closely related to *Agave*, differing in its flowers being white or whitish rather than greenish and in having their six perianth segments (petals) spreading widely instead of forming funnel-shaped blooms. Also, the six stamens have cushion-like swellings at their bases. These are lacking in *Agave*.

Furcraeas are desert and semidesert plants with very short to tall trunks, great tufts or rosettes of evergreen, succulent foliage, and flowers in immense, loose panicles. Their leaves have spiny or finely-toothed margins and stout, keel-like midribs. Mostly, the species are monocarpic, they bloom and fruit once only and then die. The age and size at which blooming occurs is unpredictable. It is not impossible, under some circumstances, that individual plants live for a century or more before they flower, but most surely the usual life-span is considerably shorter. Furcraeas do not develop suckers from their bases, but great numbers of plantlets are produced on their flower stalks and these provide ready means of multiplication. Less frequently fruits, capsules, are borne.

Mauritius-hemp and Cuban-hemp, commercial fibers of importance, are products of *Furcraea*, the former from *F. foetida*, the other from *F. hexapetala*. These fibers are of great length and softness and are used alone or with other kinds, for twine, sacks, hammocks, and other purposes.

Among furcraeas with sizable trunks ***F. longaeva*** is outstanding because, exceptionally, its branchless main stem may be 40 to 50 feet high. It has dense rosettes of lanceolate leaves and flower panicles 15 feet long. The leaves are 4 to 5 feet long and 3 to 6 inches wide. Their upper surfaces are concave, their margins roughened with fine teeth. This is native to Mexico and Guatemala. Mexican ***F. roezlii*** has a trunk 8 to 12 feet tall skirted with dead leaves. Its spreading, rather soft, lanceolate, bluish-green leaves with finely-toothed, upturned margins are 3 to 4 feet long by about 3 inches broad. At first erect, they become pendulous with age. The flower panicles are 15 to 20 feet long and have branches with drooping ends. The blooms are mostly in threes and are about 1¾ inches long. With a trunk not over 3 feet tall and rather stiff, spreading, flattish, pointed-lanceolate leaves 1½ to 2 feet long by 2 to 3 inches broad, very finely-toothed at their margins, ***F. bedinghausii*** is a native of Mexico. It has rather narrow panicles of bloom 10 to 15 feet long with pendulous branches. The flowers are 1½ inches in length. Differing from kinds considered above, Mexican ***F. macdougalii*** has leaves with spiny margins. They are erect or spreading at first, become pendulous with age, and finally die and remain as heavy skirts about the trunks. In dense rosettes, they are 7½ feet long by under 3 inches wide, linear and fleshy. The panicle of blooms is terminal, has erect branches and is up to 25 feet long by about 4½ inches wide. The flowers are whitish with the insides of the petals greenish. Native to Colombia, ***F. selloa*** has a trunk 3 feet long and rough, narrow-lanceolate leaves slendered at their bases and 3 feet long or longer by 3 to 3½ inches wide. They have distantly-spaced, big, hooked, brown spines on their margins. The loosely-branched flower panicle is up to 20 feet in length. In *F. s. marginata* the leaves are margined with yellow, those of *F. s. edentata* are similar but are without spines and tinged with pink.

Some species are trunkless or essentially so. Here belongs ***F. pubescens,*** native of tropical America. Its numerous leaves are lanceolate, and conspicuously thickened at their bases on both surfaces. They are up to 4½ feet long by under 3 inches wide, and bright green. Their margins are armed with triangular spines that bend toward the leaf apexes. Up to 20 feet long, the flower panicles branch freely and have fragrant, greenish-yellow blooms about 1¾ inches long. Their branches are minutely-pubescent. Native of Cuba and Haiti, ***F. tuberosa*** is almost or quite trunkless and has nearly flat, lanceolate leaves 8 to 10 inches wide and 5 to 6 feet long. Toward their apexes their margins bend upward. The margins are furnished with irregularly-spaced spines but these are sometimes missing toward the tops and bottoms of the leaves. Probably Mexican, ***F. undulata*** is a trunkless kind with dark green, wavy-edged, lanceolate leaves about 1½ feet long by 2 inches broad, ending in a brown spine and with brown, triangular, marginal spines. From 9 to 12 feet tall, the flower panicles have shortly-pubescent branches.

The Mauritius-hemp or green-aloe (*F. foetida* syn. *F. gigantea*), native of southern Brazil, is an important commercial crop in Mauritius, Madagascar, South Africa, parts of South America, and other places in the tropics. It is grown as an ornamental elsewhere in the tropics and subtropics. Even more popular for decoration is its handsome variegated-leaved variety *F. f. mediopicta* (syns. *F. variegata, F. watsoniana*), which has a broad, creamy-yellow band down the center of each leaf. The Mauritius-hemp (***F. foetida***) has a trunk up to 3 or 4 feet tall. Its sword-shaped leaves, flat or nearly so, have slightly-wavy edges and roughish undersides. From 4 to 7 feet long and 4 to 8 inches wide at their middles, they may or may not have triangular, hooked, marginal teeth toward their bases. The flower stems are up to 40 feet long and have rather narrow, strongly-scented, greenish-white flowers over 1½ inches in length. This kind blooms annually. It is naturalized in Hawaii, India, Mauritius, and Madagascar.

The Cuban-hemp (***F. hexapetala*** syn. *F. cubensis*) has little or no trunk and glossy green, lanceolate leaves, with rounded, roughish undersides with keeled midribs. The teeth on the leaf margins are nearly straight and 1 inch or nearly 1 inch apart. Up to 15 feet in height, the flower stalks carry panicles of 2-inch-long blooms with the inner sides of the petals milky-white. This species is native to Cuba.

Garden and Landscape Uses. As landscape subjects furcraeas serve the same purposes as agaves, but although they withstand considerable drought, are more adaptable than agaves to humid conditions. They have a bold, insistent appearance that contrasts strongly with plants of less formal habit. Because of this, they lend themselves for planting as single specimens and in groups and when well placed produce striking and handsome effects. They are of particular interest in bloom, but as this occurs rarely and sporadically and as the plants die afterward, flower production is not a primary objective in planting furcraeas. In large greenhouses and conservatories, particularly those devoted to succulents, these plants are occasionally grown. The variegated-leaved kinds are sometimes cultivated in large tubs to decorate terraces, patios, and similar places. Furcraeas are hardy only in frost-free or almost frost-free places. As a group they require more warmth than agaves and yuccas.

Cultivation. Excellent drainage, fertile soil, and full sun best suit these plants. The soil should be dryish, but not excessively so. Although furcraeas stand considerable drought, they need moderate supplies of water to give of their best. In greenhouses a minimum winter temperature of 55°F is most satisfactory. Propagation is by seeds and by plantlets that develop among the flowers.

FURROW. The narrow trench cut into the ground by a plow is called a furrow and so sometimes are shallower grooves made with a hoe or other tool for the purpose of sowing seeds or planting. These last are also known as drills.

FURZE is *Ulex europaeus.*

FUSTIC TREE is *Chlorophora tinctoria.*